Semantics of Parallelism

Non-Interleaving Representation of Behaviour

Springer
London
Berlin
Heidelberg
New York
Barcelona
Budapest
Hong Kong
Milan
Paris
Santa Clara
Singapore
Tokyo

Michael W. Shields

Semantics of Parallelism

Non-Interleaving Representation of Behaviour

Springer

Michael W. Shields, BA, PhD
Department of Mathematical and Computing Science,
University of Surrey, Guildford, Surrey GU2 5HX

ISBN-13:978-3-540-76059-7 Springer-Verlag Berlin Heidelberg New York

Springer-Verlag Berlin Heidelberg New York
British Library Cataloguing in Publication Data
Shields, M. W. (Mike W.)
Semantics of parallelism : non-interleaving representation of behaviour
1.Parallel processing (Electronic computers)
I.Title
004.3'5
ISBN-13:978-3-540-76059-7

Library of Congress Cataloging-in-Publication Data
Shields, M. W. ((Michael William), 1950-
Semantics of parallelism : non-interleaving representation of behaviour /
Michael W. Shields.
p. cm.
Includes index.
ISBN-13:978-3-540-76059-7 e-ISBN-13:978-1-4471-0933-4
DOI: 10.1007/978-1-4471-0933-4

1. Parallel processing (Electronic computers) 2. Programming languages (Electronic computers) - - Semantics. I. Title.
QA76.58.S537 1997
004'.35'015118 - - dc21
DNLM/DLC
for Library of Congress 96-51077

Softcover reprint of the hardcover 1st edition 1997

Typesetting: Camera-ready by author

34/3830-543210 Printed on acid-free paper

To my dear wife Myra, and to Ben and Rebecca

Preface

> No, so holp me Petault, it is not a miseffectual whyacinthinous riot of blots and blurs and bars and balls and hoops and wriggles and juxtaposed jottings linked by spurts of speed: it only looks as like it as damn it; ...
>
> JAMES JOYCE
> Finnegans Wake.

Since the pioneering work of researchers such as C. A. Petri in the early 1960's, numerous theories of concurrent systems have been proposed. A theory, in this sense consists of a notation for specification or description, the syntax of the theory, and a means of interpreting specifications or descriptions so as to infer properties of the behaviour of a system so specified or described, the semantics of the theory.

There is a wide variation among proposed theories of concurrent systems, not only in their syntax but, more disturbingly perhaps, in their semantics. This is in contrast to, for example, the theory of electronic circuits, in which there is a standard syntax, in this case, graphical, and a standard semantics. For instance, the semantics of the symbol for an ideal resistor is given by Ohm's 'Law'.

In this book, we describe a means for comparing these theories at the semantic level, based on an automata theory for concurrent systems. The automata in question, rather than generating sets of sequences, effectively generate systems of labelled pre-ordered sets which model possible behaviours of the system, the preorder being interpreted as 'no later than'.

Comparison at the semantic level consists of providing each theory with a semantics based on the automata, that is, a mapping $\sigma_S : S \rightarrow A$ from the class of all possible specifications to the class of automata in such a way that this semantics is consistent with that of the theory in the following sense. If $\sigma_T : S \rightarrow T$ is the semantics of the theory, then there is a mapping $j_S : S \times A \rightarrow T$ such that for each specification $s \in S$, $j_S(s, \sigma_S(s)) = \sigma_T(s)$. In some cases, j_S is determined by a function $\tilde{j}_S : A \rightarrow T$, in the sense that $j_S(s, a) = \tilde{j}_S(a)$, all $s \in S$.

Characteristically, these automata and the objects which represent their possible behaviours are designed to model both concurrency and/or simultaneity explicitly. This is in contrast the interleaving approach, favoured, for example, by Process Algebraists. Our rationale for this is the not entirely indefensible one that if one wishes to model concurrent systems then it is reasonable to wish to model concurrency. Only then is one in a position to decide whether or not abstracting away from concurrency and/or simultaneity is legitimate or useful in any particular case. For example, in the structure theory of determinacy treated in chapter 11, the interleaving approach would seem quite inappropriate, as it would in any examination of phenomenon arising from the interaction of concurrency and non-determinacy, such as the glitch phenomenon in electronics.

This book is directed at post-graduate students and researchers both in academia and industry, working in the area of Theoretical Computer Science, particularly those with an interest in concurrent systems.

The reader should be well acquainted with the languages of Set Theory and Formal Logic. An appendix outlines the basic mathematics required to understand the text. The mathematics is not particularly deep, but there is quite a lot of it.

Guildford Mike Shields
February 1997

Acknowledgements

The author would like to express his gratitude to Professor Peter Lauer, who introduced him to the study of concurrent systems, to Dr. C. A. Petri, who drew his attention to non-interleaving models and to Professor Robin Milner, who first exposed him to the elegant world of Process Algebra. His debt to the work of Professor G. Winskel and A. Mazurkiewicz should be evident from the text.

The author would also like to thank Professor Eike Best, Professor Bernard Cohen, Mr. Tim Denvir, Professor Matthew Hennessy, Dr. Marta Kwiatkowska, Dr. David Pitt, Dr. Judith Segal, Professor Steve Schuman and Professor Dan Simpson, for all their help and encouragement over the years. But don't blame any of them for mistakes which remain in the text; these are the responsibility of the wretched author.

Whatever its merit, this book would not have seen the light of day had it not been for the inspiration of the lady who is now my wife. I dedicate it to her and to her children.

Contents

1. Whys and Wherefores	1
1.1. Why Models of Parallelism ?	1
1.2. Specification in a Variety of Styles.	5
1.3. An Overview of the Basic Model.	7
1.4. The Contents of the Book.	13
2. Behavioural Presentations	16
2.1. Introduction.	16
2.2. Examples.	18
2.3 Basic Properties of Behavioural Presentations.	20
Notes.	25
3. Discrete Behavioural Presentations	26
3.1. Introduction: Zeno's Paradoxes.	26
3.2. Discreteness.	28
3.3. A Characterisation of Discrete Behavioural Presentations.	30
Notes.	37
4. Order Theoretic Properties of Left-Closed Behavioural Presentations	38
4.1. Introduction.	38
4.2. Consistent Completeness.	39
4.3. Prime Algebraicity.	40
Notes:	43
5. Event Structures and Closures of Behavioural Presentations	44
5.1 Introduction.	44
5.2. Event Structures and Behavioural Presentations.	45
5.3. Closure of Behavioural Presentations.	49
Notes.	54

6. Another Event Structure Model 56

6.1 Introduction. 56
6.2. The Model. 57
6.3. Behavioural Presentations Which Represent Event Structures. 62
Notes. 65

7. Properties of Behavioural Presentations 66

7.1. Introduction. 66
7.2. Asynchrony. 66
7.3. Left-Closure and Discreteness. 67
7.4. Determinacy. 68
7.5. Concurrency. 72
7.6. Closure. 74

8. Karp and Miller Computation Graphs 75

8.1. Introduction. 75
8.2. The Model and its Node-Set Sequence Semantics. 76
8.3. Behavioural Presentation Semantics. 78
8.4. Comparing the Semantics. 86
Notes. 91

9. Trace and Vector Languages 93

9.1. Introduction. 93
9.2. Basic Definitions and Properties. 94
9.3. Equivalence of the Models. 102
Notes. 104

10. Order Theoretic Properties of Vector Languages 105

10.1. Introduction. 105
10.2. Consistent Completeness. 105
10.3. Prime Algebraicity. 113
10.4. Left-Closed Vector Languages. 117
10.5. Parallel Composition. 118
Notes. 120

11. Determinacy 122

11.1. Introduction. 122
11.2. Strongly Forward Stable Vector Languages. 122
11.3. Simple Vector Languages. 125
Notes. 133

12. Linguistic Behavioural Presentations 134

12.1. Introduction. 134

12.2. The Structure of $B_{L,A}$. 136
12.3. Strong Forward Stability Revisited. 149
Notes. 151

13. Unambiguous Transition Systems 152
13.1. Introduction. 152
13.2. Transition Systems. 154
13.3. Properties of Unambiguous Transition Systems. 161
13.4. Bisimulation. 164
13.5. Computation Graphs Revisited. 167
Notes. 170

14. General Transition Systems 173
14.1. Introduction. 173
14.2. Behaviour of Transition Systems. 174
14.3. Strong Equivalence. 177
14.4. Strict Equivalence. 178

15. Asynchronous Transition Systems 183
15.1. Introduction. 183
15.2. The Automata. 184
15.3. Equivalences. 187
Notes. 189

16. Condition/Event Systems: Process and Occurrence Net Semantics 190
16.1. Introduction. 190
16.2. Net Theory, C/E Systems and Process Semantics. 191
16.3. Occurrence Net Semantics. 197
Notes. 199

17. Condition/Event Systems: Asynchronous Transition System Semantics 200
17.1. Introduction. 200
17.2. Asynchronous Transition System Semantics. 201
17.3. Comparison of the Semantics: Preliminaries. 203
17.4. From Behavioural Presentations to Processes … 206
17.5. …And Back Again. 213

18. Loosely Coupled Systems 224
18.1. Introduction. 224
18.2. Pure Loosely Coupled Systems. 225
18.3. Loosely Coupled Systems with Transitions. 229

19. Categorical Connections 231

19.1. Introduction. 231
19.2. Asynchronous Transition Systems. 232
19.3. Trace Languages. 235
19.4. Discrete Behavioural Presentations. 240

20. Categorical Applications 251

20.1. Introduction. 251
20.2. Reduction Morphisms. 251
20.3. Products. 258

21. An Application: Realisation by Nets 266

21.1. Introduction. 266
21.2. Conditions for Realisation. 267
21.3. Representation of Loosely Coupled Systems. 273
Notes. 280

22. Hybrid Transition Systems 281

22.1. Introduction. 281
22.2. The Model. 282
22.3. Classes of Hybrid Transition Systems. 299
Notes. 303

23. The Keller Parallel Computation Model 304

23.1. Introduction. 304
23.2. The Model and its Transition System Semantics. 305
23.3. A Hybrid Transition System Semantics for General Nets. 308
23.4. Hybrid Transition System Semantics for Parallel Programs. 314
23.5. Comparison of the Semantics. 318
Notes. 319

24. From Path Expressions to TCSP 320

24.1. Introduction. 320
24.2. Syntax and Marked, Labelled Net Semantics. 322
24.3. Hybrid Transition System Semantics. 328
24.4. Comparison of the Semantics. 332
24.5. Embedding COSY Into TCSP. 337
Notes. 340

25. The Nivat/Arnold Process Model 342

25.1. Introduction. 342
25.2. The Model and its Vector Semantics. 343
25.3. Process Structure. 348

25.4. Hybrid Transition System Semantics. 351
25.5. Comparison with the Path/Process Notation. 358
25.6. Process Structure and Occurrence Nets. 363
Notes. 367

26. Constructions on Hybrid Transition Systems 368
26.1. Introduction. 368
26.2. Recursion. 368
26.3 Non-Recursive Constructions. 375
Notes. 382

27. Process Calculus (CCS) 383
27.1. Introduction. 383
27.2. Syntax and Transition System Semantics. 384
27.3. Hybrid Transition System Semantics. 388
27.4. Properties of Expressions. 394
27.5. Comparison of the Semantics. 399
Notes. 413

28. Theoretical CSP 415
28.1. Introduction. 415
28.2. TCSP and its Failure/Divergence Semantics. 416
28.3. Hybrid Transition System Semantics. 429
28.4. Comparison of the Semantics. 432
28.5. Extending the Hybrid Transition System Semantics. 446
Notes. 450

29. Conclusions: Theories of Parallelism 452

Appendix. Review of Mathematical Prerequisites 457

References 463

Index 471

Chapter 1
Whys and Wherefores

> `Where are you from, Mr. Thurl?' she asked me one day. I told her Ohio, and she said, 'Ooooh, to be sure!' as if I had given her a clue to my crazy definitions, my insensitivity to the ordinary household nouns, and my ignorance of the commoner migratory birds. 'Semantics, Ohio,' I said. 'Why, there's one of them in Massachusetts, too,' said Della. 'The one I mean,' I told her, 'is bigger and more confusing.' 'I'll bet it is,' said Della.
>
> JAMES THURBER
> What Do You Mean It Was Brillig?

1.1. Why Models of Parallelism ?

By the term *parallelism* we shall understand that property of an entity by virtue of which it is capable of several distinct actions at the same time. This may be the case if the entity is composed of spatially separated sub-entities which co-ordinate their activities in some way, but are capable of autonomous action. In such cases, it is natural to speak of such entities as *distributed.* Of course, distributed entities usually possess parallelism, although one does not always conceive of an entity possessing parallelism as being distributed. I may be able to walk and chew gum at the same time, but I do not consider myself a distributed system (although an anatomist might). Other terms used for such entities are *asynchronous* and *concurrent*; we offer formal interpretations of 'asynchronous' and 'concurrent' in chapter 2.

This involves clarifying what we mean by 'at the same time'. As we shall see, there are two fundamental relations on occurrences of events, one of which, denoted by $\rightarrow$, is to be interpreted as 'occurred not later than'. If it is the case that $o_1 \rightarrow o_2$ and $o_2 \rightarrow o_1$ then we shall say that o_1 and o_2 are *simultaneous.* If neither of these relationships hold and the two occurrences belong to the same behaviour, then we shall say that they are *concurrent.* Both concurrency and simultaneity are relations of 'at the same time as'.

Parallelism, then, refers to the behaviour of an entity; we shall use the term *system* to describe an entity that may be considered to possess behaviour, to perform actions. This book is concerned with the representation of systems possessing parallelism.

The practical need for such a theory is a reflection of the changing rôle of the computer, from being a mere programmable calculating machine to a device capable of exercising control within physical systems; for example, the telephone network.

Of course, automatic, or at least semi-automatic, control existed before the development of the electronic computer in the 1940's. The electro-mechanical Strowger exchange is a fine example of a distributed system as is the nineteenth century railway signalling system. Such systems are distributed by necessity.

As pointed out by A. Pnueli in [1] modern computer systems are not in general merely *transformational*, that is, establishing a relation between an initial (input) state and final (output) state with all intermediate activity going on in isolation inside a 'black box', but *reactive*, that is requiring an interaction between the system and its environment. Operating systems are reactive. So are computerised gambling machines, which require continuous input (of money) and persistently 'reward' the user with outbursts of appalling 'music' (and occasionally with money).

A computer acting within an embedded application necessarily acts in a reactive fashion; it must interact with that which it is controlling. Consequently, an understanding of the computer's function in such a situation cannot deal with the machine in isolation; the machine must be considered together with its environment. In such circumstances there is always a potential for parallelism. In some cases, the environment itself has parallelism in it; any system with at least two 'users' must be capable of dealing with parallel input/output.

Of course, this is not the only reason why parallelism is of interest to computer scientists; there is also the issue of *parallel computation*, in which a computation is composed of a number of tasks, some of which may be executed at the same time. The value of this becomes apparent in cases in which a vast amount of numerical calculation needs to be performed, which would take a prohibitive amount of time if the calculations were carried out serially. This has long been recognised by commercial organisations such as banks, which require recurrent updating of their financial positions, a distributed computation involving more than one clerk.

So parallel systems occupy an important position within computer science. The increased use of computers in embedded applications leads to an increasing requirement for the production of such systems. The challenge to the computer science community is to build parallel systems that work properly. We are immediately faced with two problems. The first is that parallel systems or parallel programs are difficult to design and understand, much more so than conventional sequential systems or programs, which are bad enough. The programmer's 'intuition', no matter how well developed in the case of conventional programs, does not equip him or her to deal with pitfalls arising from the completely new set of phenomena that can occur in parallel systems.

The second problem concerns the issue of reliability. D. L. Parnas has argued in a sequence of short papers [4] that the proposed US Strategic Defence Initiative (SDI) is unfeasible 'because of the extreme demands on the system and our inability to test it, we will never be able to believe, with any confidence, that we have succeeded'. In one of the papers, entitled `Why Software is Unreliable', Parnas compares computer science with established engineering disciplines.[1]

> 'When most engineering products have been completed, tested and sold, it is reasonable to expect that the product design is correct and that it will work reliably. With software products, it is usual to find that the software has major "bugs" and does not work reliably for some users. These problems may persist for several versions and sometimes worsen as the software is "improved". While most products come with an express or implied warranty, software products often carry a specific disclaimer of warranty.'

Indeed, the software I am using to produce this text is presumably written in a sequential language, but David Pitt in the office next to mine can always tell when I'm using it from the volume of profanity penetrating the wall.

The problem of reliability - which is exacerbated when one moves from conventional sequential programming to parallel programming - has occupied the attention of the theoretical computer science community from the beginning of its existence. For example, work by Turing [6] on program correctness dates back to 1950. The comparison with engineering practice has inspired the notion of attempting to reproduce the essentials of such practice in the computing milieu.

What can we identify in engineering practice that might be usefully imitated by computer science?

All mature engineering disciplines use formal and abstract descriptions of the objects which their practitioners study, design, build, maintain and repair. For example, those electronic engineers who are concerned with electrical circuits or logic circuits use circuit diagrams to represent them. Civil engineers draw technical plans of the bridges or jib-cranes or whatever. Such representations are *formal* in the sense that they are subject to *syntactic constraints* - not every possible combination of lines and symbols makes sense as an electronic circuit diagram - and *abstract* in that they deal only with certain aspects of the object - the colour of a box-girder bridge is not the

---

[1]E. W. Dijkstra cleverly satirises the practices of software designers by imagining what mathematics would be like if carried on in the same way. In a number of short articles, the hacks of Mathematics Inc. [5] are described delivering bug-ridden 'proofs' of the Four Colour Conjecture, assuring the customers that the next release will have some of the bugs removed and that the customer service department has people standing by.

professional concern of a civil engineer; the stresses and strains in its component parts are. These representations are what we described in the introduction as the contents of the *syntax* of a theory.

These representations have several functions. Firstly, they allow a design to be expressed in a precise manner. A representation may act as the design specification of a product. The specification, perhaps augmented (with details such as colour), forms the basis for implementation. In many cases, such a representation is essential. One rarely finds the architect of a building laying bricks or shovelling dolomite; that is done by others acting under the guidance of someone who can interpret the specification.

It is vital here that the interpretation of such a specification be absolutely unambiguous, which means that some form of precise *semantics* is essential. This applies particularly if the system in question is distributed and its individual components are to be implemented by more than one group of persons. Inconsistent implementations of the components could lead to a system which could not work.

The second function of a representation is to allow *analysis* of a design prior to its implementation. It is important here that the formal language in which the design is written should have some kind of formal semantics, which allows the designer to use the design notation to generate a mathematical model of the object of the design, the mathematics in question being that in which the semantics is couched. For example, the 'meaning' of a circuit diagram is a collection of functions describing current flow and voltage changes at various points in the circuit. These may be derived from an interpretation of the component symbols of the circuit (for example, Ohm's Law provides an interpretation of the symbol for a resistor), and from rules which say what happens when these elements are combined together (for example, Kirchoff's Current Law explains what happens when several branches are joined together at a node). Likewise, a Civil Engineer knows how to apply the appropriate branches of Newtonian mechanics to deduce the forces present in some structure and accordingly whether it will fall down or not.

The practice of using formal representations possessing a semantics for specification and analysis is my understanding of the term, much used but seldom defined, *formal methods*. We may sum up the foregoing by saying that conventional engineering is seen to owe its success to the use of formal methods which is why computer scientists have attempted to introduce formal methods into their own discipline. One ingredient of a formal method, as we have explained, is a *formal description language* and a *formal semantics*.

Hence, our interest in theories of parallelism.

Accordingly, the computer science community has developed numerous theories of parallel programs and parallel systems. This effort dates back at least to the beginning of the 1960's with important work by C. A. Petri [7, 8], R. M. Karp, R. E. Miller [9] and A. Holt [10]. Many alternative models have been proposed since then, notably R. Milner's Calculus of Communicating Systems [3, 11] which has inspired numerous imitations.

The commercial community also has gradually come to accept the need for formal description of complex systems. A good example of this is provided by the attempt of the International Standards Organisation (ISO) to establish an international standard computer network protocol, the Open Systems Interconnection (OSI) [12]. Realising that the precise specification of something as complex as the OSI in a natural language, such as English, could be hazardous, the ISO has set itself the task of specifying the OSI formally and to that end have designed their own formal specification languages LOTOS [13], which is derived from CCS, and ESTELLE [14], based on state machines. Likewise, the CCITT has developed the SDL Specification and Description Language.

To sum up, there is a perceived need to develop formal methods for the design and verification of parallel systems, and this requires the development of specification languages or notations with well defined, sound, formal semantics.

As we have said, much work has been done in this area. Indeed, it may well be argued that there are enough theories of parallelism around without adding to their number. However, the fact is that, while there are a number of well established theories of parallelism, they take quite different views of the subject., as we shall see[2] .

It is one of the aims of this book to examine relationships between these views within the context of a simple general framework in which they may all be embedded, so to speak.

1.2. Specification in a Variety of Styles.

We illustrate this observation with a simple example. We consider a system consisting of two entities *E1* and *E2*. *E1* repeatedly performs the actions a_1 and b cyclical forever. *E2* repeatedly performs the actions a_2 and b cyclical forever. b is a shared action (for example, a handshake communication); it must be performed by both entities acting together. There are no other constraints on the system.

The system can be described formally in a number of ways. Figure 1.1 shows a *Petri Net* representation of the system.

We may also describe it using the Karp and Miller *Computation Graph* pictured in figure 1.2 and by the pair of *path expressions* in (1.1).

---

[2]The situation is reminiscent of that obtaining in the field of programming languages. As is well known, the multiplicity of languages used in their installations prompted a certain Defence Department to seek to replace them all by a single high-level language. The resulting language is named after the daughter of certain English poet; this poet, incidentally, was once described as 'mad, bad and dangerous to know'.

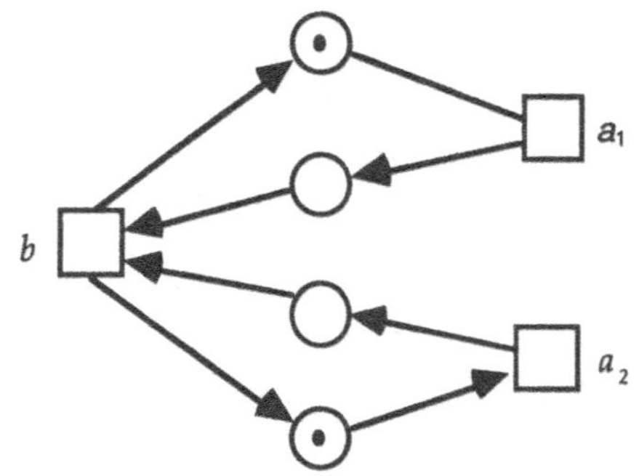

Figure 1.1.

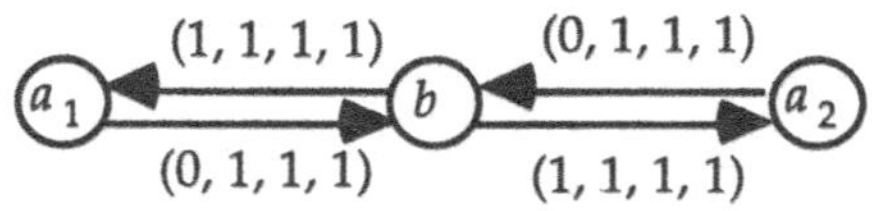

Figure 1.2.

path $a_1;b$ **end**
path $a_2;b$ **end** (1.1)

If we allow b to be an unobservable (τ) action resulting from a handshake communication between two processes, then the example can be described in CCS in (1.2).

$$(fix_1(X_1 = a_1.c.X_1) \mid (fix_2(X_2 = a_2.\bar{c}.X_2))\backslash c \quad (1.2)$$

It may also be described in TCSP in the program of (1.3).

$$\mu X{:}\{a_1,b\}.(a_1 \to (b \to X)) \,\|\, \mu Y{:}\{a_2,b\}(a_2 \to (b \to Y)) \quad (1.3)$$

or in OCCAM in the fragment of (1.4) below

```
while true
  seq
    par
      a1
      a2
  b
```
(1.4)

Now we would like to claim that the net, path expressions, computation graph and CCS equations and TCSP and OCCAM programs define 'the same system'. How may such a claim be justified? Informally, if we were to translate each of the descriptions into words, the results would be more or less the same as the description of the system with which we began. It is also true that the net of figure 1.1 may be shown to correspond to the paths of (1.1) under the net semantics of P. E. Lauer and R. H. Campbell [15], so that they define the same system by fiat as it were. But what about the others?

One problem is that nets, computation graphs, CCS and TCSP have different *operational semantics*.[3] For example, Net Theory emphasises the notion of *concurrency* and models it explicitly at the level of *condition/event systems*. The operational semantics of computation graphs, on the other hand, is given by Karp and Miller in terms of sequences of mutually simultaneous event occurrences, while that of CCS is in terms of sequences of event occurrences and that of TCSP is in terms of sequences, each of which is associated with a so-called refusal set and some of which are distinguished as divergent.

We shall attempt to resolve this problem by proposing a general operational semantics for parallel systems. Apart from anything else, we shall then be in a position interpret each of the six descriptions in a uniform manner and be in a position to compare their 'meanings' directly and formally.

1.3. An Overview of the Basic Model.

The key object in our approach is a kind of generalised automaton called an *asynchronous transition system*.[4] It transpires that in our general semantics, all six descriptions give rise to the same asynchronous transition system, pictured in figure 1.3 below. It is in this sense that we can assert that the six descriptions are of the same system.

To be more precise, each description maps to an object (*hybrid transition system*) having the same 'meaning' as the automaton in figure 1.3, in the sense that these hybrid transition systems give rise to mutually isomorphic *behavioural presentations*.

We shall now explain what asynchronous transition systems are and their dynamics in terms of trace languages and behavioural presentations.

Figure 1.3 resembles an ordinary transition system;[5] it is composed of nodes, representing global states - and we have an initial node, which we have shaded in the diagram - connected by directed arcs labelled with action

---

[3]An operational semantics associates a formal specification or computer program with an object whose structure is intended to reflect the behaviour of the entity specified or the program.

[4]Actually, we need the more general model of *hybrid transition systems* to handle the full class of systems we shall be considering, but it will suffice to consider the asynchronous model to begin with. See chapters 15 and 22.

[5]See chapters 13 and 14.

names. The one difference is that we have shaded in the interior of the lozenge shape. This is to indicate that the actions a_1 and a_2 are *concurrent*. We shall now see how this is represented in the dynamics of asynchronous transition systems.

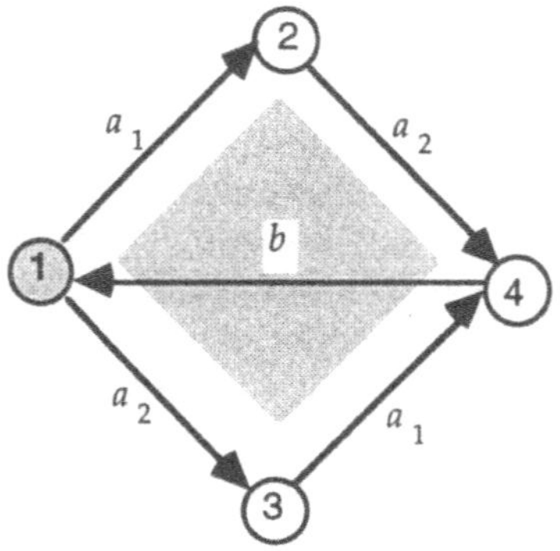

Figure 1.3.

The dynamics of a normal transition system is given in terms of *derivation sequences*, that is, sequences of states and transitions describing a trajectory through the graph of the transition system from some initial state.

For example, the transition system underlying the structure of figure 1.3 has a derivation sequence

$$1 - a_1 \rightarrow 2 - a_2 \rightarrow 4 - b \rightarrow 1 - a_2 \rightarrow 3 \tag{1.5}$$

The transition system of figure 1.3 has a property, characteristic of asynchronous transition systems, which we shall call *unambiguity*; from each state, there is at most one arc labelled with any given action name. As a consequence, for any two derivation sequences

$$\begin{array}{l} q_1 - x_1 \rightarrow q_2 - x_2 \rightarrow \cdots - x_{n-1} \rightarrow q_n \\ r_1 - x_1 \rightarrow r_2 - x_2 \rightarrow \cdots - x_{n-1} \rightarrow r_n \end{array} \tag{1.6}$$

if $q_1 = r_1$ then $q_i = r_i$, all i. It follows that the derivation sequence (1.6) is uniquely determined by q_1 and the *action sequence* $x_1 x_2 \cdots x_{n-1}$. In particular, given that all pertinent derivations in the transition system of figure 1.3 must begin in state 1, the derivation sequence (1.5) is uniquely determined by the derivation sequence $a_1 a_2 b a_2$.

The shading inside the lozenge which indicates that a_1 and a_2 are concurrent tells us that in the sequence $a_1 a_2 b a_2$, the first two occurrences of a_1 and a_2 do not actually occur in that, or any, order. If we go round the lozenge in the other direction we get another sequence; $a_2 a_1 b a_2$. The import of the shading is that this sequence describes the same asynchronous behaviour as $a_1a_2ba_2$. We may express this as

$$a_1a_2ba_2 = a_2a_1ba_2 \tag{1.7}$$

Of course, this is not true for sequences unless $a_1 = a_2$. Accordingly, we do not regard these terms as sequences, but elements in a *semigroup*, generated by the elements $\{a_1, a_2, b\}$ and satisfying the *relation*

$$a_1.a_2 = a_2.a_1 \tag{1.8}$$

The dynamics of the system of figure 1.3 are thus described by a set of execution 'sequences' in which a_1 and a_2 commute. Technically, the set of behaviours of the system of figure 1.3 is a language in a *semi-commutative monoid*.

A. Mazurkiewicz, who pioneered this approach in [16], observes that each element of the monoid, in his terminology, a *traces*,[6] determines a labelled partially ordered set, that is, a set, whose elements we may think of as occurrences, each of which is labelled by the name of the event of which it is the occurrence, equipped with a partial order relation indicating how they are ordered in time.

Let us take (1.7) as an illustration of this. Corresponding to the two sequences, we have the two labelled total orders[7] pictured in figure 1.4.

$$o_1\,(a_1) \rightarrow o_2\,(a_2) \rightarrow o_3\,(b) \rightarrow o_4\,(a_2)$$
$$o_2\,(a_2) \rightarrow o_1\,(a_1) \rightarrow o_3\,(b) \rightarrow o_4\,(a_2)$$

Figure 1.4.

Note that we have chosen the elements of the posets in a consistent way. Thus, o_2 is associated in both cases with the first occurrence of a_1 and a_2 with its second.

Denoting the two underlying total orders by $(O, \leq_1)$ and $(O, \leq_2)$, we may then form a partial order $(O, \leq_1 \cap \leq_2)$. Since

$$<_1 = \{(o_1,o_2),(o_1,o_3),(o_1,o_4),(o_2,o_3),(o_2,o_4),(o_3,o_4)\}$$
$$<_2 = \{(o_2,o_1),(o_2,o_3),(o_2,o_4),(o_1,o_3),(o_1,o_4),(o_3,o_4)\}$$

it follows that

---

[6]Technically, a trace is an element of the factor monoid of A^* determined by a set of relations of the form $a.b = b.a$. See chapter 9 and appendix A1.

[7]See appendix A2.

$$<_1 \cap <_2 = \{(o_2,o_3),(o_2,o_4),(o_1,o_3),(o_1,o_4),(o_3,o_4)\}$$

which is pictured in figure 1.5.

$$\begin{array}{ccc} o_1\,(a_1) \rightarrow & o_3\,(b) & \leftarrow o_2\,(a_2) \\ & \downarrow & \\ & o_4(a_2) & \end{array}$$

Figure 1.5.

Interpreting the ordering as the order, in time, in which the actions have occurred, we see that the occurrence of a_1 and the first occurrence of a_2 both precede b but that neither of them precedes the other. They are indeed concurrent.

This construction illustrates that traces determine labelled partial orders, which themselves model behaviour in a direct way; the ordering reflects the order in which events occurred. It is slightly unsatisfactory in the sense that is does not give a canonical way of constructing the elements of the set of occurrences. This is not so much of a problem with individual traces, but does matter when we begin to consider traces in relation to each other. For example, the statement '**S** is deadlock-free' translates as 'every behaviour of **S** may be extended to a strictly larger one'; which means that we need an ordering on behaviours. For traces, this is easy. We define:

$$u \leq v \Leftrightarrow \exists w : u.w = v \tag{1.9}$$

so that, for example $a_2 \leq a_1a_2$, since $a_1a_2 = a_2a_1$, by (1.8).

Thus, the trace a_1a_2b describes an earlier part of the behaviour given by (1.7). It would give rise to a labelled partial order such as that of figure 1.6.

$$o'_1\,(a_1) \rightarrow o'_3\,(b) \leftarrow o'_2\,(a_2)$$

Figure 1.6.

However, unless we choose the elements of the second partial order so that $o'_1 = o_1$, $o'_2 = o_2$ and $o'_3 = o_3$, there will be no immediate connection between the two. We wish to be able to choose the elements of the ground sets not only in a manner consistent with that for the other total orders associated with the same trace, but all the total orders associated with all the other traces of the system.

We may provide such a 'naming of occurrences' by noting that each occurrence is the unique last occurrence of a unique behaviour. For example, o_4 is the unique last occurrence (of a_2) in the behaviour $a_1a_2ba_2$, and o_2 is the unique last occurrence (of a_2) in a_2 (not in a_1a_2, which doesn't have a unique last occurrence).

We shall say that a trace $a_1 \ldots a_n$ is *prime* if and only if whenever $a_1 \ldots a_n = b_1 \ldots b_m$, then $a_n = b_m$ - there is a unique last element in any factorisation of it.[8] If u is prime, then we denote its unique last element by $\lambda(u)$. Thus, $\lambda(a_1a_2b) = b$.

To each trace x we may associate the set of primes beneath it

$$Pr(x) = \{u : u \leq x \wedge u \text{ is prime}\} \tag{1.10}$$

so that (1.9) and (1.10) associates a poset, labelled by λ with each trace x. For example

$$Pr(a_1a_2ba_2) = \{a_1, a_2, a_1a_2b, a_1a_2ba_2\}$$

and the corresponding labelled poset is represented diagrammatically in figure 1.7. It is clearly isomorphic to the labelled poset of figure 1.5.

$$\begin{array}{c} a_1\ (a_1) \rightarrow a_1a_2b\ (b) \leftarrow a_2\ (a_2) \\ \downarrow \\ a_1a_2ba_2(a_2) \end{array}$$

Figure 1.7.

If we now make the same construction for the trace a_1a_2b, then we obtain the labelled poset of figure 1.8. Not only is this isomorphic to that of figure 1.6, but there *is* now a relationship between that of figure 1.8 and that of figure 1.7. This is a relationship between points in a behavioural presentation which we define in chapter 3 and denote by $\lhd$.[9]

$$a_1\ (a_1) \rightarrow a_1a_2b\ (b) \leftarrow a_2\ (a_2)$$

Figure 1.8.

To sum up, an asynchronous transition system C with initial state q determines a *trace language*, that is, a set of traces, $L(C,q)$ and the trace language determines a set of labelled partial orders.

---

[8] It turns out that such elements are prime in the partial order of traces in the technical sense that if such an element lies below the least upper bound of a set of traces, then it must lie below some element of that set. See chapter 10.

[9] See 3.3.7.

The next observation to make is that we may dispense with the order relation. Define O (for occurrences) to be the set of all primes of $L(C,q)$ and let P (for space/time *points* or *possible worlds*) denote the set of all $Pr(x)$ for $x \in L(C,q)$. We may show that

$$u \leq v \Leftrightarrow u \rightarrow v$$[10]

where

$$u \rightarrow v \Leftrightarrow (\forall p \in P: v \in p \Rightarrow u \in p) \tag{1.11}$$

Hence, we may describe the behaviour of C from q by a quadruple

$$B = (O,P,E,\lambda) \tag{1.12}$$

where E denotes the range of λ. In our example, $E = \{a_1, a_2, b\}$.

The connection between $L(C,q)$ and B is a very intimate one. Indeed, as we shall show, the map that sends a trace x to $Pr(x)$ is an isomorphism of the posets $(L(C,q),\leq)$ and $(P,\subseteq)$.

B is an example of what we call a *behavioural presentation*. The behavioural presentation is the central behavioural model of our approach. In general, behavioural presentations are quadruples of the form (1.12) satisfying the axiom

$$\bigcup_{p \in P} p = O$$

The behavioural presentation model is very general. We may use it to describe continuous as well as discrete systems, as we shall see in chapter 2. However, we shall be almost entirely concerned with a subclass of behavioural presentations, those we defined to be *discrete*. It should also be pointed out that, in general, the relation $\rightarrow$ defined in (1.11) determines not a partial order but a *pre-order*[11]. This allows us to give a uniform treatment of concurrency and simultaneity. We write $o_1 \approx o_2$ to denote that o_1 and o_2 are *simultaneous*, where

$$o_1 \approx o_2 \Leftrightarrow o_1 \rightarrow o_2 \wedge o_2 \rightarrow o_1$$

so that $\approx$ is the equivalence relation determined by $\rightarrow$. We write $o_1\, co\, o_2$ to denote that o_1 and o_2 are *concurrent*, where

---

[10]See (12.1)

[11]That is to say, a reflexive, transitive relation.

$$o_1 \; co \; o_2 \Leftrightarrow (\exists p \in P: o_1, o_2 \in p) \wedge \neg(o_1 \rightarrow o_2) \wedge \neg(o_2 \rightarrow o_1)$$

The system of figure 1.3 is *determinate* [12] in the sense that it has no arbitrary choice in it. (The system doesn't choose to go one way or another around the 'lozenge'; it goes 'through' it). Equivalently, its corresponding behavioural presentation is a lattice. Equivalently, it has a unique maximal infinite behaviour of which the finite behaviours are partial approximations. In general, behavioural presentations may exhibit *non-determinacy*. We write $o_1 \# o_2$ to denote that o_1 and o_2 are *mutually exclusive*, where

$$o_1 \# o_2 \Leftrightarrow (\forall p \in P: o_1 \in p \Rightarrow o_2 \notin p)$$

This gives us a full range of temporal relationships.

To sum up, we have a class of objects, the behavioural presentations, which allow us to model the possible non-sequential behaviour of systems. In particular, these provide a behavioural semantics for asynchronous and hybrid transition systems.

Thus, to provide an operational semantics for a specification language, all we need to do is to define a mapping which associates each object of the language with a pair (H, q), where H is a hybrid transition system and q is an initial state. Its underlying asynchronous transition system determines a behavioural presentation in the manner which we have sketched.

1.4. The Contents of the Book.

We begin in chapter 2 with a definition of behavioural presentations and the basic temporal relations which they determine. Chapter 3 describes the subclass which will uniquely occupy us for the rest of the book; the *discrete behavioural presentations,* and presents a characterisation of these, which we chiefly shall be using in the sequel.[13] In chapter 4, we elucidate the principal order theoretic properties of a class of behavioural presentations which includes those that are discrete. Specifically, we prove that behavioural presentations which are left-closed with respect to a relation $\triangleleft$ are prime algebraic and consistently complete.[14] These properties play a crucial part in relating behavioural presentations to trace and vector languages, thereby building a bridge between order-theoretic and algebraic representation of behaviour.

---

[12]Not to be confused with 'deterministic'; the property that a given state and action determines a unique next state. It is to avoid such a confusion that we have adopted the term 'unambiguous'.

[13]See 3.3.14.

[14]See 4.3.2.

Chapters 5 and 6 examine the relationship between the behavioural presentation model and two kinds of *event structure*. It is shown that event structures correspond to behavioural presentations which are *closed*; equivalently, which are prime algebraic and coherent.[15]

In chapter 7, we give characterisations of important *synchronic*[16] properties, which we use in chapter 8, in which we define a semantics based on behavioural presentations for Karp and Miller computation graphs, to show that the behavioural presentations which are images of computation graphs under this semantics have such properties.

In chapter 9, we turn to algebraic ways of representing behaviour, via trace and vector languages. This chapter is mostly concerned with showing the equivalence of the two representations, in the course of which some useful operations and relations are defined. Chapter 10 is concerned with the order theoretic properties of certain kinds of these languages, those which are left-closed with respect to the ordering on traces or vectors. They turn out to have precisely the same order theoretic properties as discrete behavioural presentations. In fact, every such language determines a discrete behavioural presentation. The details are worked out in chapter 12, in which we use the results of chapter 10 to construct behavioural presentations from left-closed trace languages, as sketched in section 3 of the present chapter. We then characterise, up to isomorphism, the behavioural presentations which may be obtained in this way; the so-called ι-linguistic behavioural presentations.[17] Chapter 11, meanwhile, uses the theory developed in chapter 10 to characterise the structure of vector languages having certain determinacy and finiteness properties.

In chapter 13, we turn our attention to the automata theoretic aspects of the theory, by showing how to associate a (necessarily sequential) behavioural presentation with an unambiguous transition system with initial state. We show that this semantics establishes a bijection between a certain class of equivalence classes of unambiguous transition systems and a certain class of equivalence classes of behavioural presentations.[18] The equivalence relations concerned are *isomorphism*, in the case of behavioural presentations and a form of *strong equivalence* as defined by R. Milner for CCS. This is the first of a sequence of *semantics theorems*, other examples of which we shall meet in later chapters. Chapter 14 extends the theory to general transition systems.

In chapter 15, we begin our consideration of non-sequential automata with the introduction of asynchronous transition systems. These are shown to determine left-closed trace languages and hence ι-linguistic behavioural presentations. We define a relation between asynchronous transition

---

[15]See 5.3.9.
[16]That is, properties referring to the relationship between occurrences in time.
[17]See 15.2.14.
[18]See 13.3.12.

systems with initial states and prove that two such automata are in this relation if and only if they determine isomorphic behavioural presentations.

In chapters 16, we describe one of the earliest models of parallel systems, the condition/event model of Petri, together with a non-interleaving semantics based on process nets. Chapter 17 gives an alternative semantics based on asynchronous transition systems; we show that the two semantics are equivalent.

Chapter 18 describes the Loosely-Coupled Systems model, and we provide the model with a semantics based on asynchronous transition systems.

In chapter 19, we examine notions of morphisms between trace languages, asynchronous transition systems and discrete behavioural presentations. All three classes of object become categories, which are then seen to be strongly structurally related. Chapter 20 looks at constructions on asynchronous transition systems related to certain ideas in category theory and chapter 21 uses ideas from chapter 19 to characterise those asynchronous transition systems which arise from the semantics of chapter 17. As a consequence, we are able to show that every loosely coupled system may be represented by a net, and as a consequence, derive a construction which builds a net from a loosely coupled system in such a way that the two have the same meaning within the asynchronous transition system semantics.

In chapter 22 , we finally meet the general automata model, that of hybrid transition systems. We show that any discrete behavioural presentation may arise as the behavioural semantics of such automata and that in fact such automata have precisely discrete behavioural presentations as their behavioural semantics. Chapter 23 sees an application of hybrid transition systems to provide a non-interleaving semantics for a Net Based model of parallel computation due to R. Keller. In chapters 24 and 25, we do the same for the path/process notation that lies at the centre of the COSY notation and to the process model of Nivat and Arnold.

In chapter 26, we describe constructions on automata and we use these in chapters 27 and 28 to provide a non-interleaving semantics for CCS and a subset of TCSP. In each case, we show that the semantics associated with these notations may be recovered from our semantics, effectively by removing concurrency.

In an appendix, we outline some of the basic mathematics used throughout the book.

Chapter 2
Behavioural Presentations

Something is happening here but you don't know what it is, do you, Mr. Jones

BOB DYLAN
Highway 61 Revisited

2.1. Introduction.

In this chapter, we set out a model of system behaviour, the *behavioural presentation*. We shall use behavioural presentations to model the potential behaviour of a given system. In this way we may equip description formalisms with an operational semantics by defining a function from the set of all legal descriptions to the class of behavioural presentations

Systems do things; they perform actions. A behavioural presentation is thus associated with a set E of events. For example:

1• The 'system' is a person tossing coins. The events of this system might be '*heads*' and '*tails*'.
2• The system is a black box generating two analogue signals f and g. The events of this system might be of the form 'f instantaneously attains a voltage of x'.
3• The system is a microprocessor executing a machine code program. The events of this system might be the set of all machine code instructions together with all possible valid actual parameters.
4• The system is a railway network. The events of this system might include 'train number x enters section s of the track'.

Events occur. We assume a set O of occurrences as well as a function

$$\lambda: O \rightarrow E$$

which attributes occurrences to events. The equation $\lambda(o) = e$ may be read: 'o is an occurrence of e'.

Associated with any system is a collection of '*possible states of affairs*' or '*possible worlds*. A possible world corresponds to the experience of an observer of the system during some possible execution of it. We assume that at least such an observer will have noted that certain events have occurred. We shall identify a possible world with the set of occurrences of events during the period of execution in question. For brevity, we shall refer to possible worlds as *points*.

Associated with every behavioural presentation is thus a set $P \subseteq \wp(O)$ of points.[1]

There is no point in concerning ourselves with things that could never happen. We shall therefore assume that every occurrence belongs to some point, that is

$$\bigcup_{p \in P} p = O$$

These four entities make up our basic behavioural model.

2.1.1. DEFINITION. A *behavioural presentation* is a quadruple $B = (O, P, E, \lambda)$ where

1• O is a set of *occurrences;*

2• $P \subseteq \wp(O)$ is a non-empty set of *points* satisfying $\bigcup_{p \in P} p = O$;

3• E is a set of *events;*

4• $\lambda : O \to E$ is the *occurrence function.*

2.1.2. CONVENTION. Behavioural presentations will mostly be written as 'B' possibly subscripted or 'dashed'. In order not to have to write out the elements of the quadruple every time, we make the convention that all elements of the corresponding quintuple inherit the subscript or dash. Thus, for example, we would have $B_1 = (O_1, P_1, E_1, \lambda_1)$, $\tilde{B} = (\tilde{O}, \tilde{P}, \tilde{E}, \tilde{\lambda})$ and $B_H^{fin} = (O_H^{fin}, P_H^{fin}, E_H^{fin}, \lambda_H^{fin})$.

We extend this convention to all entities which are tuples.

It is always useful to be able to draw pictures of mathematical objects. Figure 2.1 describes a behavioural presentation B with

$O = \{o1, o2, o3, o4\}$
$P = \{\emptyset, \{o1\}, \{o2, o3\}, \{o1, o2, o3\}, \{o1, o4\}\}$
$E = \{a, b, c, d\}$
$\lambda(o1) = a \quad \lambda(o2) = b \quad \lambda(o3) = c \quad \lambda(o4) = d$

---

[1] $\wp(O)$ denotes the powerset of O.

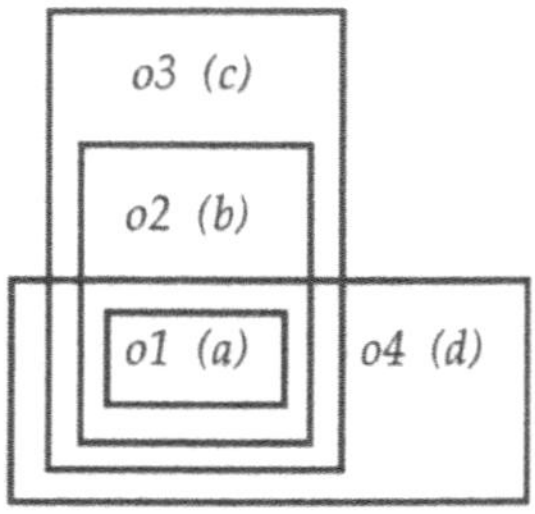

Figure 2.1.

In the figure, *o1 (a)* represents an occurrence *o1* with $\lambda(o1) = a$. Non-empty points are represented by boxes. We omit the empty point.

2.2. Examples.

Now let us look at some examples. These will serve to illustrate the expressive range of the model as well as giving examples of its use.

2.2.1 EXAMPLE. Our first example is a description of a system consisting of a coin being tossed and coming down either heads or tails. Let $E = \{H, T\}$.

We suppose that two occurrences of the same action are the same if and only if they have been preceded by the same sequences of events. The third 'tail' in the sequence *HTHHTT* is not the same occurrence as the third 'tail' in the sequence *HHTHHTT*; they take place in different 'possible worlds'

Pictorially, occurrences are nodes in the tree, part of which is shown in figure 2.2. Each node is uniquely distinguished by the sequence which describes the route through the tree from its root leading to that node.

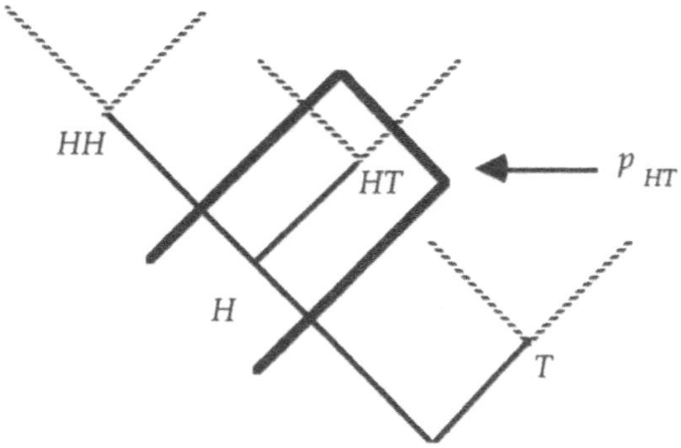

Figure 2.2

We may thus refer to events by giving the sequence of which they are the last occurrence. Therefore, $O=\{o_x | x\in E^+\}$ and $P=\{p_x | x\in E^*\}$,[2] with $p_x=\{o_y\in O\,|\,y\leq x\}$. If $x=u.a$, with $e\in E$, then define $\lambda(o_x)=a$. For example

- $p_\Omega=\varnothing \qquad p_{HTHH}=\{o_H, o_{HT}, o_{HTH}, o_{HTHH}\}$;
- $\lambda(o_H)=\lambda(o_{HTH})=\lambda(o_{HTHH})=H$ and $\lambda(o_{HT})=T$.

2.2.2. EXAMPLE. 2.2.1 dealt with a discrete, sequential system. Behavioural presentations may also be used to describe continuous and non-sequential systems. Consider an electronic black box with two output lines. The function of the box is to generate a signal f at one of its output lines and a signal g at its other output line. We consider the behaviour of the system between times t_1 and t_2. The example is illustrated in figure 2.3.

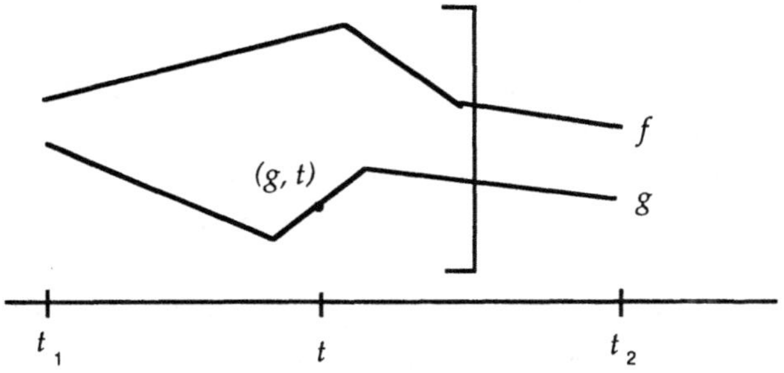

Figure 2.3.

We shall let the occurrences of this system be the attaining of a given voltage at a given time by a given signal, so that

$$O=[t_1,t_2)\times\{f,g\}$$

Points correspond to instants in the time interval, so that

$$P=\{p_t | t\in[t_1,t_2]\}$$

and the set of all things that have happened prior to a time t is therefore

$$p_t=[t_1,t)\times\{f,g\}$$

Finally, $\lambda(t,f)=f(t)$ and $\lambda(t,g)=g(t)$.

2.2.3. EXAMPLE. This example is based on the famous thought-experiment of Einstein in [17]. Two trains are travelling at a constant speed in opposite

---

[2] See appendix A3.

directions along a pair of straight parallel tracks. Observers O1 and O2 are sitting at the middle of the two trains. At a given instant, the two observers are on a line at right angles to the side of the train with a third observer, O3, sitting on the embankment, and at that instant two forks of lightning strike the ends of the first train in such a way that O3 sees them strike simultaneously. Observer O2, travelling towards the light coming from the first strike, sees that before he sees the light coming from the second. Observer O1 travelling towards the light coming from the second strike, sees that before he sees the light coming from the first.[3]

Let o_1 denote the occurrence of the first bolt of lightning striking and let o_2 denote the occurrence of the second bolt of lightning striking, so that

$$O = \{o_1, o_2\}$$

Now, from the point of view of observer O3, there are two distinct time points; $p_0 = \varnothing$ when nothing has happened yet and $p_{both} = \{o_1, o_2\}$, when both have. O3 never sees one without the other. O3 sees the lightning bolts strike simultaneously. O2 sees the first bolt strike before the second, so that from his point of view, there are three points; p_0 and p_{both} and a third point $p_1 = \{o_1\}$, when o_1 has occurred but not o_2. Likewise, O1 has three points; p_0, p_{both} and $p_2 = \{o_2\}$.

Thus, $P = \{p_0, p_1, p_2, p_{both}\}$, $E = \{flash\}$ and $\lambda(o_1) = \lambda(o_2) = flash$.

The resulting behavioural presentation is pictured in figure 2.4.

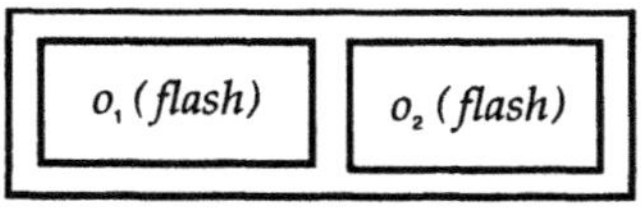

Figure 2.4.

2.3 Basic Properties of Behavioural Presentations.

We shall now consider possible relationships between occurrences. We introduce two fundamental relations.

2.3.1. DEFINITION. Let B be a behavioural presentation and suppose $o_1, o_2 \in O$. Define.

---

[3]If O1 were to see the two strikes as simultaneous, it would appear to him that the light from the first strike were travelling faster than that from the second, contradicting the postulate that the speed of light in vacuo in an inertial frame is a constant.

1• $o_1 \# o_2 \Leftrightarrow (\forall p \in P : o_2 \in p \Rightarrow o_1 \notin p)$;

2• $o_1 \to o_2 \Leftrightarrow (\forall p \in P : o_2 \in p \Rightarrow o_1 \in p)$.

We apply the subscripting convention, 2.1.2, to relations associated with a behavioural presentation where necessary. Thus, if we wish to distinguish between the # relations of B_1 and B_2, we would refer to them as $\#_1$ and $\#_2$ respectively.

These two definitions introduce concepts of *mutual exclusion* and *time ordering*. If $o_1 \# o_2$, then an occurrence o_2 excludes the future occurrence o_1 - and vice versa. It is this relation that allows us to introduce notions of *non-determinacy* into the theory. If $o_1 \to o_2$, on the other hand, then if o_2 has occurred, then so must o_1. We may read $o_1 \to o_2$ as meaning, 'o_1 occurred no later than o_2'.

The following remark gives the basic properties of these two relations. An *independence relation* is an irreflexive, symmetric relation, so that R is an independence relation on X if and only if

$$\forall x, y \in X : (\neg x \, R \, x) \wedge (x \, R y \Rightarrow y \, R \, x)$$

2.3.2. PROPOSITION. If B is a behavioural presentation then.

1• # is an independence relationon O;

2• $\forall o_1, o_2, o_1', o_2' \in O : o_1 \# o_2 \wedge o_1 \to o_1' \wedge o_2 \to o_2' \Rightarrow o_1' \# o_2'$;

3• $\to$ is a pre-order on O.

PROOF. (1) By the tautologies $(o_2 \in p \Rightarrow o_1 \notin p) \Rightarrow (o_1 \in p \Rightarrow o_2 \notin p)$ and $\neg(o \in p \Rightarrow o \notin p)$.

(2) If $o_1 \to o_1'$, $o_2 \to o_2'$ and $\neg o_1' \# o_2'$. then $o_1', o_2' \in p$, some $p \in P$ by 2.3.1 (1), so $o_1, o_2 \in p$ by 2.3.1 (2), that is $\neg o_1 \# o_2$, by 2.3.1 (1).

(3) By the tautologies $o \in p \Rightarrow o \in p$ and

$$(o_3 \in p \Rightarrow o_2 \in p) \wedge (o_2 \in p \Rightarrow o_1 \in p) \Rightarrow (o_3 \in p \Rightarrow o_1 \in p)$$

■

Let us see how these relations manifest themselves in our examples.

1• In 2.2.1, we may check that $o_x, o_y \in p_z \Leftrightarrow x, y \leq z$ and this can be the case only if $x \leq y$ or $y \leq x$. Thus, $o_x \# o_y \Leftrightarrow x \not\leq y \wedge y \not\leq x$. For example $o_{HTHHTT} \# o_{HHTHHTT}$. Here we have an example of non-determinism. The first event follows one choice - after the first H, a T occurred, while the second event followed an alternative choice - after the first H, an H occurred. We may check that $o_x \to o_y \Leftrightarrow x \leq y$.

2• In 2.2.2, $O = p_{t_2}$, from which it follows that $o_1, o_2 \in p_{t_2}$, for all $o_1, o_2 \in O$. Thus, $\# = \varnothing$ in this example. On the other hand, if $x, y \in \{f, g\}$ and $t, t' \in [t_1, t_2)$, then $(t, x) \rightarrow (t', y) \Leftrightarrow t \leq t'$. Again, this is consistent with our view of the system as having no non-determinacy and in which the occurrences are ordered according to the (global) time at which they occurred.

3• In 2.2.3, $O = p_{both}$, so that $\# = \varnothing$ in this example. However, $o_1 \in p_1$and $o_2 \notin p_1$ while $o_2 \in p_2$ and $o_1 \notin p_2$. Thus, neither $o_1 \rightarrow o_2$ nor $o_2 \rightarrow o_1$. As we shall see, this indicates that o_1 and o_2 are concurrent.

Next, we consider the relationships between arbitrary pairs of occurrences.

2.3.3. LEMMA. Exactly one of the following relations holds:

1• $o_1 \# o_2$

2• $o_1 \rightarrow o_2$ and $o_2 \rightarrow o_1$

3• $o_1 \rightarrow o_2$ and $o_2 \not\rightarrow o_1$

4• $o_1 \not\rightarrow o_2$ and $o_2 \rightarrow o_1$

5• $\neg o_2 \# o_1$, $o_2 \not\rightarrow o_1$ and $o_2 \not\rightarrow o_1$.

PROOF. We first note that

$$o_1 \rightarrow o_2 \Rightarrow \neg o_1 \# o_2 \tag{2.1}$$

Indeed, $o_2 \in p$ some $p \in P$ by 2.1.1 (2) so if $o_1 \rightarrow o_2$, then $o_1, o_2 \in p$ by 2.3.1 (2) and so $\neg o_1 \# o_2$. by 2.3.1 (1). Hence, (2), (3), (4) and (5) is the complete set of cases in which $\neg o_1 \# o_2$. The only other possibility is (1).

■

We have already discussed 2.3.3 (1). Let us examine the others.

2.3.4. DEFINITION. $o_1 \approx o_2 \Leftrightarrow (o_1 \rightarrow o_2) \wedge (o_2 \rightarrow o_1)$

$\approx$ is the equivalence relation generated by the pre-order $\rightarrow$, so the following result is standard.

2.3.5. PROPOSITION. $\approx$ is an equivalence relation and if $o, o_1, o_2 \in O$ and $o_1 \approx o_2$, then

1• $o_1 \rightarrow o \Leftrightarrow o_2 \rightarrow o$;

2• $o \rightarrow o_1 \Leftrightarrow o \rightarrow o_2$;

3• $o \# o_1 \Leftrightarrow o \# o_2$.

■

We see that two occurrences in the relation $\approx$ stand in exactly the same relationship to other occurrences. Suppose that the system is clocked, that is, there are occurrences whose attribution is the instantaneous time shown by some clock. Also suppose that we have some means of deriving from the system the 'time' at which o_1 occurred. This would involve the use of our temporal relations $\rightarrow$ and #. From 2.3.5, we see that if o_1 is in a relationship with a clock occurrence - say half-past four on a Friday afternoon - then o_2 must also be in that relation, and vice versa. Thus, if o_1 occurred at half-past four on a Friday afternoon, then o_2 must also have occurred at half-past four on a Friday afternoon. Our interpretation is that o_1 and o_2 are *simultaneous.*

If we now go through our examples, we see that in 2.2.2, $(t,x)\rightarrow(t',y)\Leftrightarrow t\leq t'$ and so $(t,x)\approx(t',y)\Leftrightarrow t=t'$. Thus, $(t,f)\approx(t,g)$. Thus, there is non-trivial simultaneity in this example.

2.2.3 is based on a thought-experiment whose point was to demonstrate the non-objectivity of simultaneity in relativistic mechanics. C.A. Petri in [18] refers to this principle when adopting a different form of contemporaneity for his Net Theory, namely concurrency. In Net Theory, the only type of simultaneity is coincidence, that is to say, two occurrences are simultaneous if and only if they are identical.

Concurrency is also a concept that we can cope with. Indeed, we have seen it in 2.2.3. The two occurrences o_1 and o_2 are clearly contemporaneous - certainly neither precedes the other since we have neither $o_1\rightarrow o_2$ nor $o_2\rightarrow o_1$ and they are not mutually exclusive, since $o_1,o_2\in p_{both}$. Two such events will be said to be concurrent.

2.3.6. DEFINITION. $o_1\,co\,o_2\Leftrightarrow\neg(o_1\#o_2)\wedge(o_1\not\rightarrow o_2)\wedge(o_2\not\rightarrow o_1)$

This deals with 2.3.3 (5). The remaining cases, 2.3.3 (3) and 2.3.3 (4) are merely mirror images of each other.

2.3.7. DEFINITION. $o_1<o_2\Leftrightarrow(o_1\rightarrow o_2)\wedge\neg(o_2\rightarrow o_1)$

then $o_1<o_2$ may be read 'o_1 strictly precedes o_2'. It is easy enough to show that $<$ is a strict pre-order - that is to say, a transitive, irreflexive relation - and *co* is an independence relation. Having dressed up the case analysis of 2.3.3 in (hopefully) more intuitive terms, we may now state:

2.3.8. PROPOSITION. In any behavioural presentation, two occurrences are either

1• strictly ordered in time;
2• simultaneous
3• concurrent;
4• mutually exclusive

but only one of these relations holds.

PROOF. By 2.3.1, 2.3.3 , 2.3.6 and 2.3.7.

■

The systems described in section 2.2 are of different types. In examples 2.2.1, there can be only one occurrence at a time. In examples 2.2.2 and 2.2.3 this is not the case, but in the former case contemporaneous occurrences are simultaneous whereas in the latter they are concurrent. In examples 2.2.2 and 2.2.3, the behaviours of the systems are pre-determined, whereas in example 2.2.1 they are not.

Behavioural presentations may be classified according to the nature of their co, $\approx$ and # relations. First, if X is a set, then let Id_X denote its identity relation.

2.3.9. DEFINITION. If B is a behavioural presentation, then

1• B is *sequential* $\Leftrightarrow co = \varnothing \wedge \approx = Id_O$;
2• B is *synchronous* $\Leftrightarrow \approx \neq Id_O$;
3• B is *asynchronous* $\Leftrightarrow co \neq \varnothing$;
4• B is *hybrid* $\Leftrightarrow co \neq \varnothing \wedge \approx \neq Id_O$;

2.3.10. DEFINITION. If B is a behavioural presentation, then

1• B is determinate $\Leftrightarrow \# = \varnothing$.
2• B is non-determinate $\Leftrightarrow \# \neq \varnothing$.

Our examples classify as follows. 2.2.1 is of a non-determinate, sequential system. 2.2.2 is of a determinate, non-sequentially synchronous system. 2.2.3 is of a determinate, non-sequentially asynchronous system. These give us three out of a possible eight kinds, which are listed in the following result.

2.3.11. PROPOSITION. If B is a behavioural presentation, then precisely one of the following holds.

1• B is sequential and determinate;
2• B is sequential and non-determinate.
3• B is non-sequential, synchronous and determinate.
4• B is non-sequential, synchronous and non-determinate.
5• B is non-sequential, asynchronous and determinate.
6• B is non-sequential, asynchronous and non-determinate.
7• B is non-sequential, hybrid and determinate.
8• B is non-sequential, hybrid and non-determinate.

PROOF. By the law of the excluded middle.

■

There is one further distinction, that between 2.2.2 and the other examples, namely that 2.2.2 deals with a non-discrete system whereas the other examples deal with discrete systems (at least, on the level of abstraction on which they are described). We shall be exclusively concerned with discrete systems from now on. We begin our examination of them in the next chapter.

Notes.

The model presented here owes a great deal to the work of Winskel [19] and the earlier paper of Neilsen, Plotkin and Winskel. [20].

The main difference between behavioural presentations and other models of systems behaviour known to the author is the greater generality of the former. By refraining from corseting the model with restrictions, we obtain something which is capable of describing both the continuous and the discrete, the synchronous and the asynchronous.

Existing models of systems tend to work within one of the above classes of behavioural presentations. For instance, the essentially interleaving models used, for example, by Keller [21], Hoare [2] or Milner[11] are sequential. Models such as that using sets of partial orders to describe behaviours as in [10, 22], process nets for the semantics of Condition/Event systems, [23], Winkowski's Algebras of Arrays [24] or vector firing sequences [25] for the semantics of path expressions are in general asynchronous. The Nivat-Arnold process model [26] or synchronous CCS [27] would appear to be synchronous. The computation graph model of Karp and Miller [9] has a behavioural semantics that suggests a synchronous intention.

In most cases, one may identify a subclass having something like the property of being determinate. Marked graphs in Net Theory [28], GR_0 paths, in the theory of path expressions [25] confluent and determinate CCS systems [11] may all be viewed in this manner. Their chief attraction is that they are easy to analyse; they have nice properties; we look at structural properties of determinate systems in chapter 11.

Chapter 3
Discrete Behavioural Presentations

SOLDIER: You cheat. You dirty rotten cheat!
THE DEVIL: Please be a little more discrete.

C.F. RAMUS, I STRAVINSKI
The Soldier s Tale

3.1. Introduction: Zeno's Paradoxes.

The philosopher Zeno is notorious for his paradoxes which seek to demonstrate the impossibility of motion. In his arrow paradox, for example, he argues that an arrow can never reach its target, since before it can get there it must first travel half the distance, but before that it must travel the first quarter and before that it must travel the first eighth and so on. The paradox involves the construction of a series of points regressing in time towards, but never reaching, the hypothetical point at which the arrow was fired.

A similar paradox concerns Achilles and a tortoise. The two agree to have a race and in acknowledgement of Achilles' greater speed, the tortoise is given a head start. The argument goes that Achilles will never overtake the tortoise, for by the time that he reaches a place where the tortoise was at some previous time, the latter will have moved on. In this paradox, Zeno constructs a series of points extending into the future towards, but never attaining, a hypothetical point at which Achilles and the tortoise are dead level.

We may describe both these situations using behavioural presentations. First of all, we notice that in each case we are given a set of points - points at which assertions are made about the current state of affairs - for example, that the arrow has travelled one-eighth of the distance towards the target.

In the case of the arrow paradox we have a set of events:

$$E_{arrow} = \{e_n \mid n \in \omega\}$$

where ω denotes the set of finite ordinals and e_n is shorthand for 'The arrow attains 2^{-n} of the distance' (so that e_0 represents the arrow arriving at the target). For our set of set of occurrences, we have

$$O_{arrow} = \{o_n \mid n \in \omega\}$$

with associated occurrence function given by

$$\lambda_{arrow}(o_n) = e_n, \text{ all } n \in \omega.$$

Finally, we define $P_{arrow} = \{p_n \mid n \in \omega\}$ where for each $n \in \omega$,

$$p_n = \{o_i \in O_{arrow} \mid i \geq n\}.$$

Applying 2.3.1 (2) to the behavioural presentation B_{arrow}, we see that

$$\cdots \rightarrow_{arrow} o_{n+1} \rightarrow_{arrow} o_n \rightarrow_{arrow} \cdots \rightarrow_{arrow} o_2 \rightarrow_{arrow} o_1.$$

from which it would appear that such a system could never start to execute. There is not even an initial point (which would have to be $\emptyset$) at which nothing has happened yet. Even if we add into P_{arrow} an initial time point $\emptyset$, no event could occur, because there is always another one which must precede it.

In the case of Achilles and the tortoise, we may exhibit a behavioural presentation $B_{Achilles}$ with occurrences $O_{Achilles} = O_{arrow}$, where o_1 denotes the occurrence of Achilles reaching the starting position of the tortoise and o_n represent him reaching the position that the tortoise had reached at occurrence o_{n-1}. The points are $p_n = \{o_i \mid i \leq n\}$ together with a futuremost point, p_∞, which we can think of as the point when Achilles catches up with the tortoise, so that $p_\infty = O_{Achilles}$. Again, applying 2.3.1 (2), we see that

$$o_1 \rightarrow_{Achilles} o_2 \rightarrow_{Achilles} \cdots \rightarrow_{Achilles} o_n \rightarrow_{Achilles} \cdots.$$

This system can get started, but the futuremost point is never reached; Achilles never catches the tortoise.

Zeno's argument may be applied to any system which can be described by a piecewise continuous function of time. It may be used, for example, to 'prove' that a logic inverter never works (take a sequence of time points in the period of its propagation delay which converge to the end of that period). However, this way of giving a discrete description of an (intrinsically) analogue system is surely somewhat perverse.

Practical experiments and common sense tells us that arrows do reach their targets - or at least reach something - and that unless he is hit by a stray arrow, Achilles will catch up with and pass the tortoise. With a little elementary physics and calculus, and given the relevant facts, we may *calculate* the time at which the arrow hits the target and Achilles overtakes the tortoise. So how do we reconcile this with the paradoxes?

The description of the flying arrow given by Zeno and that given by classical mechanics do not contradict each other. Both are valid descriptions of the same system. However, Zeno wants us to draw a particular conclusion from his description and we may be disinclined to do this. We may phrase our observations concerning the first paradox thus: *if* the motion of an arrow were to consist of a sequence of discrete steps one after the other, *then* such motion is impossible. In B_{arrow}, there is no first occurrence; any given occurrence o_n is strictly preceded by another o_{n+1}. Likewise, *if* the motions of Achilles and the tortoise consists of a sequence of discrete steps, *then* Achilles never overtakes the tortoise. In $B_{Achilles}$, no finite sequence of occurrences leaves us at the point where overtaking occurs.

Both these observations have the form $a \Rightarrow b$ and Zeno would like us to accept b. We, however, may decline to do so. Classical logic then obliges us to accept $\neg a$. That is to say, a conclusion that we may draw from Zeno's arguments is not that such motion is *impossible*, but that it is not *discrete*.

3.2. Discreteness.

Computer Science is largely concerned with discrete systems, systems which are conceived of as proceeding in discrete steps. We would like to use our behavioural model to describe such systems and for that purpose would like to discover a subclass of behavioural presentations that we may deem to represent discrete behaviour.

Our discussion leads us to exclude infinite, strictly ascending and descending chains of points. We accordingly introduce the following definitions.

3.2.1 DEFINITION. If B is a behavioural presentation, then B will be said to satisfy the *descending chain condition* (DCC) $\Leftrightarrow$ for every indexed family $(p_n)_{n\in\omega}$ of elements of P:

$$\cdots p_n \subseteq \cdots \subseteq p_2 \subseteq p_1 \Rightarrow \exists N \in \omega\ \forall n \geq N : p_n = p_N$$

3.2.2. DEFINITION. If B is a behavioural presentation, then B will be said to satisfy the *ascending chain condition* (ACC) $\Leftrightarrow$ for every indexed family $(p_n)_{n\in\omega}$ of elements of P, such that $p_n \subseteq p$, all $n \in \omega$, some $p \in P$.

$$p_1 \subseteq p_2 \subseteq \cdots \subseteq p_n \subseteq \cdots \Rightarrow \exists N \in \omega\ \forall n \geq N : p_n = p_N$$

However, behavioural presentations, even without such infinite chains, still lack an essential element of discrete behaviour. We illustrate with an example.

Consider a behavioural presentation with $O=\{c,h,t\}$, $P=\{p_0,p_1,p_2\}$ and $p_0=\varnothing$, $p_1=\{c,h\}$ and $p_2=\{c,t\}$ (ignore λ and E). We can think of it as representing a simple non-deterministic system is which a coin is tossed (occurrence c) and then either lands with its head uppermost (occurrence h) or its tail uppermost (occurrence t). We may check that $c<h$, $c<t$ and $h\#t$.

However, there is something missing. The coin has been tossed; it glitters as it spins through the air and for a heartbeat or two the two opponents wait apprehensively for the outcome - who will be forced to give the *FORTRAN for Social Scientists* course? This is certainly a point in the space-time of the system, which would be represented by a set $\{c\}$ - but this point is not present in P.

As well as the chain conditions, we shall insist that a discrete presentation contains enough time points to separate events which are strictly ordered or non-simultaneous. This is the repletion property, defined below.

3.2.3. DEFINITION. If B is a behavioural presentation, then B will be said to be *replete* $\Leftrightarrow$ whenever $p_1,p_2\in P$ such that $p_1\subseteq p_2$ and $o_1,o_2\in p_2-p_1$, then

$$o_2 \not\to o_1 \Rightarrow \exists p_3\in P:(p_1\subseteq p_3\subseteq p_2)\wedge(o_1\in p_3)\wedge(o_2\notin p_3). \tag{3.1}$$

(3.1) is a bit of a mouthful, so let us explain it in words. Given points p_1 and p_2 such that p_1 is before p_2, and given that o_1 and o_2 occurred between p_1 and p_2, then if o_2 occurred later than or concurrently with o_1, then there is a point in time, p_3, after p_1 and before p_2, at which it is legitimate to assert that o_1 has happened but that o_2 hasn't.

In our example, $p_0\subseteq p_1$, $c,h\in p_1-p_0$ and $h\not\to c$. In order for the behavioural presentation to be replete, we would need a point p_3 such that $p_0\subseteq p_3\subseteq p_1$, $c\in p_3$ and $h\notin p_3$. Thus, $p_3=\{c\}$ which, however does not belong to P, as we have observed. Hence the example is not replete.

There is an additional issue. Consider the example of the tossed coin again, but with the 'initial state', p_0, removed. < and # are unaffected by this modification, but something is again missing. Not only do we not have a point at which the coin has not yet been tossed, but in this new behavioural presentation, the repletion property is invoked in vain; there are no points p, p' such that $p\subset p'$. One further condition will do the trick.

3.2.4. DEFINITION. If B is a behavioural presentation, then B will be said to have a *bottom element* $\Leftrightarrow$ $\varnothing\in P$.

We choose the properties described in 3.2.1 to 3.2.4 as characterising the behaviours of discrete systems.

3.2.5. DEFINITION. If B is a behavioural presentation, then B will be said to be *discrete* $\Leftrightarrow$ B satisfies the DCC and the ACC, is replete and has a bottom element.

3.3. A Characterisation of Discrete Behavioural Presentations.

The rest of this chapter is devoted to elucidating the properties of discrete behavioural presentations. We are chiefly concerned with two aspects.

Firstly, we must reassure ourselves that discrete systems do proceed in an orderly way in discrete steps without giving rise to embarrassing anomalies. In doing this, we need to explain what a step actually *is* in terms of our model. This will be of great importance later when we come to examine the relationships between behavioural presentations and automata.

Secondly, we shall give an alternative characterisation of the discreteness property. This will be in terms of a certain finiteness condition (to cope with the chain conditions) and a closure property. This closure property is related to an ordering on subsets of O.

A step from p to p' may be thought of in the following terms. We are at p, an occurrence o of some event takes place, and we are now at p'. Hence, we obtain p' by adding in o. We note though that if o' is some other occurrence such that $o \approx o'$, then o' must also be in p'. Indeed, it follows easily from 2.3.1 (2) and 2.3.4 that

$$\forall o_1, o_2 \in O: o_1 \approx o_2 \Leftrightarrow (\forall p \in P: o_1 \in p \Leftrightarrow o_2 \in p) \tag{3.2}$$

Thus, in going from p to p', we must add in an entire simultaneity class $X = o/\approx$ and hence $p' = p \cup X$.

An occurrence cannot take place twice; events may re-occur, but each occurrence is distinct. Thus, if we are at p and an occurrence o takes place, then o cannot belong to p. By (3.2), if $X = o/\approx$ then $p \cap X = \emptyset$.

3.3.1. DEFINITION. If B is a behavioural presentation, then a *derivation* or *step* in B is a triple $(p_1, X, p_2) \in P \times \wp(O) \times P$ such that

1• $X \in O/\approx$;

2• $p_1 \subseteq p_2$;

3• $p_2 - p_1 = X$.

We shall write $p_1 \succ^X p_2$ to indicate that (p_1, X, p_2) is a step and we shall refer to it as a step from p_1 to p_2 via (the occurrences in) X. By abuse of terminology we shall refer to the expression $p_1 \succ^X p_2$ as a step. If $X = \{o\}$, for $o \in O$, then we write $p_1 \succ^o p_2$

Looking back at the Zeno examples, we see that in B_{arrow}, $p_{n+1} - p_n \in O/\approx$ for each $n \in \omega$, but that $\varnothing \succ^o p_n$ for no p_n or o - there is no first step. In the athletics example, $\varnothing \succ^{o_1} p_1 \succ^{o_2} p_2 \succ \cdots$, but for no p_n or o- do we have $p \succ^o p_\infty$ - there is no final step.

The arrow paradox is excluded by the banning of infinite descending chains, so let us consider a behavioural presentation B with the DCC. We shall also suppose that B is replete. Let $p \in P$. If p is maximal, that is to say, $p \subset p'$ for no $p' \in P$, then we cannot expect anything to happen after p; p is a 'halt state'. Consider, therefore, the case in which $p \subset p'$ for some $p' \in P$. We aim to show that something may happen at point p, in the sense that there is a step from p.

Of course, if $p' - p \in O/\approx$, then $p \succ^{p'-p} p'$. What if this is not the case?

We would then have $o_1, o_2 \in p' - p$ such that $o_1 \not\approx o_2$, so that by 2.3.4, either $o_1 \not\twoheadrightarrow o_2$ or $o_2 \not\twoheadrightarrow o_1$. In either case, the repletion property entails that $p \subset p_1 \subset p'$ for some $p_1 \in P$. We may repeat the argument with p_1 in place of p' and conclude that either $p \succ^{p_1 - p} p_1$ or there exists $p_2 \in P$ such that $p \subset p_2 \subset p_1 \subset p'$.

Thus, if $p \not\succ^X p''$ for all $p'' \in P$ and $X \in O/\approx$, then we may argue by induction that there exists an infinite descending chain $p \subset \cdots \subset p_2 \subset p_1 \subset p'$, contradicting the assumption that B satisfies the DCC. Thus.

3.3.2. LEMMA If B is a replete behavioural presentation satisfying the DCC, then

$$\forall p, p' \in P : p \subset p' \Rightarrow \exists p'' \in P\, \exists X \in O/\approx : p \succ^X p'' \wedge p'' \subseteq p'.$$

■

3.3.2 assures us that if we take any non-maximal point, then the system may be 'clocked forward' from it, one step at a time. Let us again consider $p, p' \in P$ such that $p \subset p'$. We may 'clock' our system onwards from p towards p', but in general it need not get there. The behavioural presentation describing the paradox of Achilles and the tortoise may be shown to be replete and satisfy the DCC but although $p_n \subset p_\infty$ for each $n \in \omega$, one may never get from p_n to p_∞ no matter how vigorously one 'clocks'. Of course, what is missing from $B_{Achilles}$ is the ACC. In a replete behavioural presentation satisfying both chain conditions, on the other hand, all future points may be attained, otherwise, we could use 3.3.2 repeatedly to construct an infinite ascending chain $p \succ^{X_1} p_1 \succ^{X_2} p_2 \cdots$ with $p_n \subset p'$ for each $n \in \omega$, contradicting the ACC hypothesis. Thus

3.3.3. LEMMA. If B is a replete behavioural presentation satisfying the DCC and the ACC, then for all $p, p' \in P$

$$p \subset p' \Rightarrow \exists p_1, \cdots, p_{n-1} \in P \ \exists X_1, \cdots, X_n \in O/\approx : p \succ^{X_1} p_1 \cdots p_{n-1} \succ^{X_n} p'$$

■

Applying lemma 3.3.3 with $p = \varnothing$, we see that if B is discrete - that is, B satisfies the hypotheses of the lemma together with the hypothesis that $\varnothing \in P$, and if $p' \neq \varnothing$, then $\varnothing \succ^{X_1} p_1 \succ^{X_2} \cdots \succ^{X_{n-1}} p_{n-1} \succ^{X_n} p'$ and hence $p' = X_1 \cup \cdots \cup X_n$. We note that each X_i is a $\approx$-class of O; this inspires the following definition and corollary.

3.3.4. DEFINITION. If B is a behavioural presentation and $X \subseteq O$, then we define $X/\approx = \{o/\approx \mid o \in X\}$. Note that if $p \in P$, then $p/\approx = \{X \in O/\approx \mid X \subseteq p\}$, by the foregoing argument.

B will be said to be *$\approx$-finitary* $\Leftrightarrow \forall p \in P: |p/\approx| < \infty$.

3.3.5. COROLLARY. If B is a discrete behavioural presentation, then B is $\approx$-finitary.

■

There is a partial converse to this corollary. Suppose B is $\approx$-finitary and suppose that $p, p' \in P$ with $|p'/\approx| = N$. If $p \subset p'' \subset p'$ and $p'' \in P$, then p'' is a union of elements of $p'/\approx$, by (3.2). Thus, there can be at most 2^N such p'', which immediately entails the following.

3.3.6. LEMMA. If B is $\approx$-finitary behavioural presentation, then B satisfies the DCC and the ACC.

■

The $\approx$-finitary property corresponds to the chain conditions. We now investigate another property which turns out to be equivalent to repletion. Essentially, a non-replete behavioural presentation has certain points missing; a behavioural presentation with the property we shall investigate is guaranteed to have no essential points missing. We consider sets $X \subseteq O$ which contain all occurrences which are precede or are simultaneous to its elements, that is to say

$$\forall o_1 \in X \ \forall o_2 \in O: o_2 \to o_1 \Rightarrow o_2 \in X \tag{3.3}$$

and $o_1 \# o_2$ for no $o_1, o_2 \in X$. X certainly describes a portion of the behaviour of the system. Could it not be the kind of missing point that might cause B to fail to be replete?

For instance, in our coin-tossing example, we noted that the set $\{c\}$ was a conceptually legitimate point, corresponding to the coin having been tossed but not yet having landed. And since $c < h$ and $c < t$, it follows that $o \to c \Rightarrow o = c$ and hence that $o \to c \Rightarrow o \in \{c\}$. Thus, $\{c\}$ satisfies (3.3). Also $\neg(c \# c)$. We also note that it was the absence of the set $\{c\}$ which caused the example behavioural presentation to fail to be replete.

If a set X satisfying (3.3) and $o_1 \# o_2$ for no $o_1, o_2 \in X$ were added to P as a new point, then $\to$ and # would remain unchanged and B would be replete. So we might be tempted to choose as the desired property that P contains all such sets.

However, other properties of B could change. For example consider the behavioural presentation, B, obtained from $B_{Achilles}$ by removing p_∞. B is certainly $\approx$-finitary, and hence satisfies the DCC and the ACC. It also contains a bottom element and an easy argument shows it to be replete. It is thus discrete. We may also note that p_∞ trivially satisfies (3.3) and $o_1 \# o_2$ for no $o_1, o_2 \in p_\infty$. However, adjoining p_∞ to B gives rise to a behavioural presentation which does not satisfy the ACC and is hence not discrete. As we shall see, the missing points of a non-replete behavioural presentation must lie under existing points. We formalise:

3.3.7. DEFINITION. Let $X, Y \subseteq O$, then X is *left-closed* in Y and we write $X \lhd Y \Leftrightarrow$

1• $X \subseteq Y$;
2• $\forall o_1 \in X \; \forall o_2 \in Y: o_2 \to o_1 \Rightarrow o_2 \in X$.

For instance, in the coin-tossing example, $\{c\} \lhd \{c,h\}$ and $\{c\} \lhd \{c,t\}$.

It is easy to verify that $\lhd$ is a partial order on $\wp(O)$ and that for $p_1, p_2 \in P$, $p_1 \subseteq p_2 \Leftrightarrow p_1 \lhd p_2$. We also observe that if $X \subseteq O$ and $\mathbf{U} \subseteq \wp(O)$, then

$$(\forall Y \in \mathbf{U}: Y \lhd X) \Rightarrow \bigcup_{Y \in \mathbf{U}} Y \lhd X \tag{3.4}$$

For certainly, $\bigcup_{Y \in \mathbf{U}} Y \subseteq X$, giving 3.3.7 (1), while if $o_1 \in \bigcup_{Y \in \mathbf{U}} Y$ and $o_2 \in X$ with $o_2 \to o_1$, then $o_1 \in Y$, some $Y \in \mathbf{U}$ in which case, as $Y \lhd X$, $o_2 \in Y$ and so $o_2 \in \bigcup_{Y \in \mathbf{U}} Y$, giving 3.3.7 (2).

Suppose $X \subseteq O$ and let *max*(X) denote the set of *maximal elements* of X,

$$max(X) = \{o \in X \mid \forall o' \in X: o \to o' \Rightarrow o' \to o\} \tag{3.5}$$

Suppose that $p \in P$, $o_1 \in max(p)$ and $o_1 \approx o_2$. If $o_2 \to o$, $o \in X$, then $o_1 \to o$, by 2.3.4 and 2.3.2 (3) and so $o \to o_1$, by (3.5), and so $o \to o_2$, by 2.3.4 and 2.3.2 (3). Thus,

$$\forall p \in P \,\forall o_1 \in max(p) \,\forall o_2 \in O: o_1 \approx o_2 \Rightarrow o_2 \in max(p) \tag{3.6}$$

Next, suppose that $p \in P$ and $o \in max(p)$; we shall show that $p - o/\approx \lhd p$ Certainly $p - o/\approx \subseteq p$, so 3.3.7 (1) holds. If $o_1 \in p - o/\approx$ and $o_2 \in p$ with $o_2 \to o_1$, then we cannot have $o_2 \in o/\approx$, since that would entail $o_2 \in max(p)$, by (3.6) and consequently, $o_1 \to o_2$, by (3.5) and so $o_1 \approx o_2$ by 2.3.4, giving the contradiction $o_1 \in o/\approx$. Hence, $o_2 \in p - o/\approx$ and 3.3.7 (2) holds. Thus,

$$\forall p \in P \,\forall o \in max(p): p - o/\approx \lhd p \tag{3.7}$$

Our requirement on the set X in (3.3), that it contains all predecessors of its elements, may be formally stated as $X \lhd O$. To avoid negating the chain conditions, it suffices that $X \subseteq p$, some $p \in P$. However, it is easy enough to show that $X \lhd O \wedge X \subseteq p \Rightarrow X \lhd p$. We therefore consider as potential points those sets such that $X \lhd p$ for some $p \in P$ and we shall define a behavioural presentation to be left-closed if it contains all such points.

3.3.8. DEFINITION. A behavioural presentation will be said to be *left closed* $\Leftrightarrow$

$$\forall X \subseteq O: (\exists p \in P : X \lhd p) \Rightarrow X \in P$$

A left-closed behavioural presentation contains all 'reasonable' points. We now show that such a behavioural presentation is replete. Consider $p_1, p_2 \in P$ with $p_1 \subseteq p_2$ and suppose $o_1, o_2 \in p_2 - p_1 \in P$ such that $o_2 \not\to o_1$. We wish to find $p_3 \in P$ with $p_1 \subseteq p_3 \subseteq p_2$ such that $o_1 \in p_3$ and $o_2 \notin p_3$. As a candidate for p_3, we look for the 'smallest' point containing both p_1 and o_1. Since $o_2 \not\to o_1$, this point needs to have $p_1 \cup \{o_1\}$ as a subset and contain all its predecessors. Such a set may be obtained using the *left-closure operator* which we now define.

3.3.9. DEFINITION. Let $X \subseteq O$, we define its *left-closure*, written $\downarrow_B X$ to be the set $\{o \in O \mid \exists o' \in X: o \to o'\}$.

If $X = \{o\}$, $o \in O$, then we write $\downarrow_B o$ for $\downarrow_B X$.

We shall drop the subscript when there is no danger of confusion.

We list some consequences of the definition.

3.3.10. LEMMA. Let $X, Y \subseteq O$, $\mathbf{U} \subseteq \wp(O)$ and $p \in P$, then

1• $X \subseteq \downarrow X$;

2• $X \subseteq Y \Rightarrow \downarrow X \lhd \downarrow Y$ and in particular, $\downarrow X \lhd O$;

3• $\downarrow \bigcup_{X \in \mathbf{U}} X = \bigcup_{X \in \mathbf{U}} \downarrow X$;

4• $p = \downarrow p$;

5• $X \subseteq p \Rightarrow \downarrow X \lhd p$.

PROOF. (1) As $o \rightarrow o$, all $o \in O$.
(2) Suppose $o \in \downarrow X$, then $o \rightarrow o'$, some $o' \in X \subseteq Y$, so $o' \in \downarrow Y$ by (1). Thus, $\downarrow X \subseteq \downarrow Y$. If $o_1 \in \downarrow X$ and $o_2 \in \downarrow Y$ such that $o_2 \rightarrow o_1$, then for some $o \in X$, $o_1 \rightarrow o$, and hence $o_2 \rightarrow o$, that is $o_2 \in \downarrow X$. Thus both conditions of 3.2.7 hold and $\downarrow X \lhd \downarrow Y$.
(3) $o \in \downarrow \bigcup_{X \in \mathbf{U}} X$ if and only if $o \rightarrow o'$, some $o' \in \bigcup_{X \in \mathbf{U}} X$ if and only if $o \rightarrow o'$, some $o' \in X$ and $X \in \mathbf{U}$ if and only if $o \in \downarrow X$, some $X \in \mathbf{U}$, if and only if $o \in \bigcup_{X \in \mathbf{U}} \downarrow X$.
(4) $p \subseteq \downarrow p$, by (1). On the other hand, if $o \in \downarrow p$, then $o \rightarrow o'$, some $o' \in p$, and so $o \in p$, by 2.3.1 (2). Thus, $\downarrow p \subseteq p$.
(5) By (2) and (4), $X \subseteq p \Rightarrow \downarrow X \lhd \downarrow p \Rightarrow \downarrow X \lhd p$.

■

Returning to our investigation of repletion. We are looking for the smallest left-closed set containing $p_1 \cup \{o_1\}$. Define $p_3 = \downarrow(p_1 \cup \{o_1\})$. As $p_1 \cup \{o_1\} \subseteq p_2$, $p_3 \lhd p_2$, by 3.3.10 (5), so that if B is left-closed, then $p_3 \in P$. $p_1 \subseteq p_1 \cup \{o_1\} \subseteq \downarrow(p_1 \cup \{o_1\})$, by 3.3.10 (1), so that $p_1 \subseteq p_3$. Likewise $o_1 \in p_3$ Finally, $o_2 \notin p_1$ and $o_2 \notin \downarrow o_1$, since $o_2 \not\rightarrow o_1$. Thus, $o_2 \notin p_1 \cup \downarrow o_1$. But, $p_1 \cup \downarrow o_1 = \downarrow p_1 \cup \downarrow o_1$, by 3.3.10 (4), and $\downarrow p_1 \cup \downarrow o_1 = \downarrow(p_1 \cup \{o_1\})$, by 3.3.10 (3), and so $o_2 \notin p_3$. We have proved:-

3.3.11. LEMMA. If B is a left-closed behavioural presentation, then B is replete.

■

Since $\varnothing \lhd p$ for every $p \in P$, it follows that left-closed behavioural presentations also have bottom elements. We may thus invoke 3.3.10 and 3.3.11 to conclude the following.

3.3.12. COROLLARY. If B is a $\approx$-finitary, left-closed behavioural presentation, then B is discrete.

■

We shall now show that the conditions $\approx$-finitary and left-closed are not only sufficient to guarantee discreteness, but also necessary. We already know, from 3.3.5, that discrete behavioural presentations are $\approx$-finitary, so it remains to be shown that they are left-closed. First, we need a technical lemma, which says that if Y lies between two points and is contained in the larger of the two, p', then $Y-p'$ contains a minimal element.

3.3.13. LEMMA. Let B be a discrete behavioural presentation and suppose that $p,p' \in P$ and $p' \subset Y \lhd p$ then there exists $o \in Y-p'$ such that $\forall o' \in O: o' < o \Rightarrow o' \in p'$.

PROOF. If not, then there exists a function $f: Y-p' \to O-p'$ such that if $o \in Y-p'$, then $f(o) < o$. But if $o \in Y-p'$, then $o \in p$ and so $f(o) \in p$, by 2.3.1 (2) and 2.3.7. Furthermore, $o \in Y$ and $Y \lhd p$, so $f(o) \in Y$. Therefore, $f(o) \in Y \cap (O-p') \subseteq Y-p'$. $p' \subset Y$ so we may select $o_1 \in Y-p'$ and construct an infinite descending chain $\cdots o_3 < o_2 < o_1$, with $o_n \in Y-p'$, each $n \in \omega$, by iteratively defining $o_{n+1} = f(o_n)$. But now we may apply the repletion property, 3.2.3, to assert the existence of an infinite descending chain $\cdots p_3 \subseteq p_2 \subseteq p_1 \subseteq p$ such that for each $n \in \omega$, $p' \subseteq p_n$ and $o_n \notin p_n$ and $o_{n+1} \in p_n$. Thus, we have an infinite descending chain $\cdots p_3 \subset p_2 \subset p_1 \subseteq p$ This contradicts the DCC and completes the proof.

■

We come to the main result of this chapter.

3.3.14. THEOREM. Let B be a behavioural presentation, then B is discrete $\Leftrightarrow$ B is $\approx$-finitary and left-closed.

PROOF. In view of 3.3.5 and 3.3.12, it suffices to show that discrete behavioural presentations are left-closed. Suppose, therefore, that $Y \subseteq O$ with $Y \lhd p$ for some $p \in P$. If $Y = \varnothing$, then $Y \in P$, as B has a bottom element, by 3.2.5, so suppose that $Y \neq \varnothing$. We shall show that

$$(p',p \in P \wedge p' \subset Y \lhd p) \Rightarrow (\exists X \subseteq O \exists p_1 \in P: p' \succ^X p_1 \subseteq Y) \tag{3.8}$$

By induction, there is a sequence $\varnothing \succ^{X_1} p_1 \succ^{X_2} p_2 \succ \cdots$, such that $p_n \subseteq Y$ and $p_n \in P$, all $n \in \omega$ and since B is discrete, we must have $Y = p_n$, for some $n \in \omega$. Thus $Y \in P$ in this case also. We prove (3.8).

Suppose that, $p,p' \in P$ and $p' \subset Y \lhd p$. By 3.3.13, there exists $o \in Y-p'$ such that $o' < o \Rightarrow o' \in p'$. We claim that

$$p' \succ^{o/\approx} p' \cup o/\approx \subseteq Y \tag{3.9}$$

which means that (3.8) holds, with $X = o/\approx$ and $p_1 = p' \cup X$. We first show that $X \subseteq Y$, from which it follows that $p_1 \subseteq Y$. If $o' \in X$, then $o' \to o$, since

$X = o/\approx$ and so $o' \in p$, since $o \in p$, by 2.3.1 (2). As $o \in Y$, $Y \lhd p$ and $o' \rightarrow o$, $o' \in Y$, by 3.3.7 (2).

Next, we show that $p_1 \lhd p$ which entails $p_1 \in P$, by left-closure. We know that $p_1 \subseteq Y \subseteq p$, giving 3.3.7 (1), so let $o_1 \in p$ and $o_2 \in p_1$ such that $o_1 \rightarrow o_2$. We must show that $o_1 \in p_1$. This is certainly true if $o_2 \in p'$, so suppose that $o_2 \in X$. As $o_1 \rightarrow o_2$, there are then two cases to consider, namely $o_1 \approx o_2$ and $o_1 < o_2$. If $o_1 \approx o_2$, then $o_1 \in X \subseteq p_1$, since $X = o/\approx$. If $o_1 < o_2$, then $o_1 < o$, by 2.3.7, and hence $o_1 \in p' \subseteq p_1$, as o was assumed to satisfy $o' < o \Rightarrow o' \in p'$. In all three cases, $o_1 \in p_1$, as required.

Finally, we prove that $p' \succ^X p_1$. That $X \in O/\approx$ and $p' \subseteq p_1$ follow immediately from the definition, giving 3.3.1 (1) and (2). If $o' \in p' \cap X$, then $o' \approx o$ and so $o \rightarrow o'$, by 2.3.4, and since $o' \in p'$, $o \in p'$, by 2.3.1 (2), contradicting $o \in Y - p'$, so $p' \cap X = \emptyset$. Hence,

$$p_1 - p' = (p' \cup X) - p' = X - p' = X.$$

concluding the proof of (3.9), and hence of the theorem.

■

Notes.

C. A. Petri in [18] introduces the notion of K-density as a property possessed by 'real' processes. To anticipate the chapter on Petri Nets, a process net (also known as an occurrence net) is a partial ordering of event occurrences and holdings of local conditions which represents the evolution in time of a concurrent system. K-density is the requirement that every sequential sub-process (subset maximal with respect to being a total order) and every case (subset maximal with respect to being trivially ordered) has non-empty intersection - in every global state, one may say what local state a sequential sub-process is in. As Petri points out, (the matter is discussed at greater length by E. Best in [29]), the K-density property excludes phenomena such as the dichotomy paradox (Achilles and the tortoise). Best also shows that such a property excludes systems without 'first causes' (such as arise from the arrow paradox).

Chapter 4

Order Theoretic Properties of Left-Closed Behavioural Presentations

Left in the rain, the deputation,
Closed ranks outside the railway station.
Consistently, they missed their trains.
Complete confusion fogged their brains.
Prime suspect was a noted Satanist, an
Algebraic coding theoretician.

F. X. Reid [40]

4.1. Introduction.

We wish to use discrete behavioural presentations to describe and reason about the behaviour of systems, so it would be helpful to know about their properties.

In this regard, 3.3.14 is useful, not only because it gives a more tractable characterisation of discreteness than that given in 3.2.5, but also because it emphasises the importance of the left-closure property. The principal order theoretic properties of discrete behavioural presentations that we shall now establish depend almost without exception on left-closure.

The two characteristic properties of left-closed behavioural presentations are those of being consistently complete and prime algebraic. We shall deal with them in that order.

Our application of these results may be summed up as follows. In chapter 15, we define a class of automata called asynchronous transition systems. To each such automata, together with an initial state, we may associate a subset of a semi-commutative monoid - it's *trace language*. This turns out to be a prime algebraic and consistently complete poset. We may therefore turn it into a behavioural presentation - the occurrence function λ is determined by a property of the prime elements that we saw in chapter 1 - thereby establishing a behavioural semantics for the asynchronous transition systems.

In later chapters, we give semantics for a variety of notations by mapping objects in these notations to asynchronous transition systems with initial states. Each asynchronous transition systems with initial state, in turn, determines a discrete behavioural presentation. In this way, objects in the notation are associated with discrete behavioural presentations.

4.2. Consistent Completeness.

4.2.1. DEFINITION. A poset $(X,\leq)$ is *consistently complete* $\Leftrightarrow$

$$(\exists x \in X: Y \leq x) \Rightarrow \bigsqcup Y \text{ exists}$$

for every $Y \subseteq X$.

Note that if $Y = \varnothing$, then $\forall y \in Y: y \leq x$ is true vacuously, and if $\bigsqcup Y$ exists, it must therefore be a bottom element - we shall usually denote bottom elements by $\bot$.

There is an alternative and equivalent definition of consistent completeness, namely:

1• $(X,\leq)$ has a bottom element;

2• $\bigsqcap Y$ exists, all $Y \subseteq X$ such that $Y \neq \varnothing$.

(1) and (2) are justified as follows. We have already seen that if $(X,\leq)$ is consistently complete then $(X,\leq)$ has a bottom element. If $Y \subseteq X$ such that $Y \neq \varnothing$, and we define $Z = \{x \in X | x \leq Y\}$, then $Z \leq y$ for some (indeed, every) $y \in Y$ and hence $\bigsqcup Z$, exists. And it is easy to see that $\bigsqcup Z = \bigsqcap Y$.

Conversely, if (1) and (2) hold and $Y \subseteq X$ with $Y \leq x$, some $x \in X$, then $\bigsqcup Y$ exists when $Y = \varnothing$, as $\bot \in X$, whereas if $Y \neq \varnothing$, then the set $Z = \{x \in X \mid Y \leq x\}$ is non-empty and so $\bigsqcap Z$, exists, by (2). And it is easy to see that $\bigsqcap Z = \bigsqcup Y$.

4.2.2. LEMMA. If B is a left-closed behavioural presentation, then $(P,\subseteq)$ is consistently complete. Furthermore

1• $\bigsqcap \mathbf{U} = \bigcap_{p \in \mathbf{U}} p$, if $\varnothing \neq \mathbf{U} \subseteq P$;

2• $\bigsqcup \mathbf{U} = \bigcup_{p \in \mathbf{U}} p$, if the former exists.

PROOF. Suppose that $\varnothing \neq \mathbf{U} \subseteq P$, then we may select $p \in \mathbf{U}$. Let $p_{\mathrm{U}} = \bigcap_{p' \in \mathrm{U}} p'$; we prove that $p_{\mathrm{U}} \lhd p$. It is clear that $p_{\mathrm{U}} \subseteq p$. Let $o \in p$ and $o' \in p_{\mathrm{U}}$ such that

$o \to o'$, then for all $p' \in \mathbf{U}$, $o' \in p'$ and so $o \in p'$, by 2.3.1(2). Thus, $o \in p_{\mathbf{U}}$. Hence $p_{\mathbf{U}} \lhd p$, by 3.3.7, and consequently $p_{\mathbf{U}} \in P$, since B is a left-closed. It is clear that $(\forall p' \in \mathbf{U}: p \subseteq p') \Leftrightarrow p \subseteq p_{\mathbf{U}}$. Hence $p_{\mathbf{U}} = \sqcap \mathbf{U}$.

Suppose $\mathbf{U} \subseteq P$ such that $p = \sqcup \mathbf{U}$ exists, then $p' \subseteq p$, each $p' \in \mathbf{U}$. Let $p^{\mathbf{U}} = \bigcup_{p' \in \mathbf{U}} p'$. By 3.3.10 (4) and (5), $p' \lhd p$, each $p' \in \mathbf{U}$ and so $p^{\mathbf{U}} \lhd p$ by (3.4), and so $p^{\mathbf{U}} \in P$, by left-closure. And it is clear that $(\forall p' \in \mathbf{U}: p' \subseteq p'') \Leftrightarrow p^{\mathbf{U}} \subseteq p''$, so $p^{\mathbf{U}} = \sqcup \mathbf{U}$, showing consistent completeness and (2) both hold.

■

4.3. Prime Algebraicity.

Let us now turn to prime algebraicity.

4.3.1. DEFINITION. Let $(D, \leq)$ be a poset. $x \in D$ is a *complete prime* of $D \Leftrightarrow$

$$\forall X \subseteq D: x \leq \sqcup X \Rightarrow \exists y \in X: x \leq y \tag{4.1}$$

where X ranges over the set of all subsets of D such that $\sqcup X$ exists. Define $Pr(D)$ to be the set of all complete primes of D. $(D, \leq)$ is *prime algebraic* $\Leftrightarrow$

$$\forall x \in D: \sqcup pr_D(x) \text{ exists and equals } x \tag{4.2}$$

where $pr_D(x) = \{y \in Pr(D) \mid y \leq x\}$.

We can illustrate this definition with the behavioural presentation of 2.2.1. We shall show that every point of the form p_x, where $x > \Omega$, is a prime in $(P, \subseteq)$. First, $p_x \subseteq p_y \Leftrightarrow x \leq y$. If $\mathbf{U} \subseteq P$ and $\sqcup \mathbf{U}$ exists, then $\mathbf{U} \lhd p_u$, some u and hence $V = \{x \mid p_x \in \mathbf{U}\}$ satisfies $V \leq u$. It is not hard to see that V must be totally ordered and finite and hence contain its own least upper bound, v, say. Thus, if $p_x \subseteq \sqcup \mathbf{U}$, then $p_x \subseteq p_v \in \mathbf{U}$, so p_x is prime. Also, p_Ω cannot be prime, since $p_\Omega \subseteq \varnothing = \sqcup \varnothing$ but p_Ω cannot be less than any element of $\varnothing$.

In this example, $(P, \subseteq)$ is also prime algebraic. Indeed, if $x > \Omega$, then $p_x = \bigcup_{p \in Pr_P(p_x)} p$.

This example generalises to arbitrary left-closed behavioural presentations. The complete primes are the elements $\downarrow o$, $o \in O$; the idea being that if $\downarrow o \subseteq \bigcup_{p' \in U} p'$, then $o \in p$, some $p \in \mathbf{U}$, and consequently $\downarrow o \subseteq p$.

4.3.2 THEOREM. If B is a left-closed behavioural presentation, then $(P,\subseteq)$ is prime algebraic and consistently complete. The complete primes of $(P,\subseteq)$ are the elements $\downarrow o$, $o \in O$.

PROOF. We have proved consistent completeness in 4.2.2.

Let $Pr(B)$ denote the set $\{\downarrow o \mid o \in O\}$. We first show that $Pr(B) \subseteq Pr(P)$.

If $o \in O$, then $o \in p$, for some $p \in P$, by 2.1.1 (2). By 3.3.10 (5), $\downarrow o \lhd p$. This shows that $\downarrow o \in P$, since B is left-closed. Next suppose that $\mathbf{U} \subseteq P$ such that $\downarrow o \subseteq \sqcup \mathbf{U}$, then $o \in \downarrow o \subseteq \bigcup_{p \in \mathbf{U}} p$, and hence $o \in p$, for some $p \in \mathbf{U}$. By 3.3.10, (5) $\downarrow o \lhd p$ and therefore $\downarrow o \subseteq p$. Hence, $\downarrow o$ is a complete prime.

For each $p \in P$, define $pr(p) = \{p' \in Pr(B) \mid p' \subseteq p\}$. By 4.2.2, $\sqcup pr(p)$ exists, since $p' \subseteq p$, all $p' \in pr(p)$, by definition of $pr(p)$. Again, by 4.2.2, $\sqcup pr(p) = \bigcup_{o \in p} \downarrow o$ so that

$$o' \in \sqcup pr(p) \Leftrightarrow \exists o \in p : o' \to o \Leftrightarrow o' \in p.$$

Thus, $\sqcup pr(p) = p$. But now, if $p \in pr(P)$, then $p = \sqcup pr(p)$ so $p \leq \sqcup pr(p)$. By (4.1), $p \subseteq \downarrow o$, some $o \in p$. But $\downarrow o \subseteq p$, by definition, so $p = \downarrow o$, that is $p \in Pr(B)$. Hence, $Pr(P) \subseteq Pr(B)$. We have proved that $Pr(P) = Pr(B)$.

It follows that for all $p \in P$, $pr_P(p) = pr(p)$ and so $\sqcup pr_P(p)$ exists and equals p, that is, $(P,\subseteq)$ is prime algebraic.

■

The correspondence between primes and occurrences suggests that given an abstract prime algebraic poset, we may construct a behavioural presentation from it. Consider a prime algebraic poset $(D,\leq)$. D determines a set of sets $P_D = \{pr_D(d) \mid d \in D\}$. Furthermore, the mapping $pr_D : D \to P_D$ which sends d to $pr_D(d)$ is an isomorphism of the posets $(D,\leq)$ and $(P_D,\subseteq)$. For it is clear that $d \leq d' \Rightarrow pr_D(d) \subseteq pr_D(d')$, while

$$pr(d) \subseteq pr(d') \Rightarrow \sqcup pr(d) \leq \sqcup pr(d') \Rightarrow d \leq d'$$

We can easily construct a behavioural presentation from $Pr(D)$ and P_D. Pick a non-empty set, E and a function $\lambda : Pr(D) \to E$. Let $B_{D,E,\lambda} = (Pr(D), P_D, E, \lambda)$. We have shown that any prime algebraic poset is isomorphic to the set of points of some behavioural presentation.

This sheds new light on 2.2.1. It may be shown that for any non empty set A, A^* is prime algebraic with A^+ as the set of complete primes. Thus, if we take $A = \{H,T\}$, then by our observation, the set $Pr(A^*)$ may be taken as a set of occurrences with sets $pr(x)$ as points. But this is precisely what we have in 2.2.1, because $pr(x) = p_x$. Furthermore, there is a 'natural' way of associating these occurrences with events and an occurrence function. Each element x of A^+ ends in a unique element $a \in A$. (The reader may now care to recall the example given in chapter 1). We can define $\lambda(x) = a$ - and this is indeed how we defined E and λ in the example.

We now examine the relationship between the ordering $\leq$ that comes with D and the ordering $\rightarrow_{D,E,\lambda}$ on $Pr(D)$, considered as the set of occurrences of $B_{D,E,\lambda}$. Let $o_1, o_2 \in Pr(D)$. If $o_1 \rightarrow_{D,E,\lambda} o_2$, then since $o_2 \in pr_{D,E,\lambda}(o_2)$, $o_1 \in pr_{D,E,\lambda}(o_2)$ and hence $o_1 \leq o_2$, by 2.3.1 (2).

Conversely, let $o_1 \leq o_2$, then if $d \in D$, then $o_2 \leq d \Rightarrow o_1 \leq d$, that is $o_2 \in pr_{D,E,\lambda}(d) \Rightarrow o_1 \in pr_{D,E,\lambda}(d)$. Thus, $o_1 \rightarrow_{D,E,\lambda} o_2$. We have shown:

$$o_1 \leq o_2 \Leftrightarrow o_1 \rightarrow_{D,E,\lambda} o_2 \tag{4.3}$$

We are interested here in left-closed behavioural presentations. If $B_{D,E,\lambda}$ is left-closed then $(P_{D,E,\lambda}, \subseteq)$ is consistently complete, and hence $(D, \leq)$ is, as the two posets are isomorphic. Thus, if $(D, \leq)$ isn't consistently complete then $B_{D,E,\lambda}$ isn't left-closed. So, assume that $(D, \leq)$ is consistently complete. We show that $B_{D,E,\lambda}$ is left-closed.

We need to show that if $X \subseteq Pr(P_{D,E,\lambda})$ with $X \lhd pr_D(d)$, some $d \in D$, then $X \in P_{D,E,\lambda}$, that is $X = pr_D(d')$, for some $d' \in D$. Since $X \lhd pr_D(d)$, $X \subseteq pr_D(d)$, and therefore $X \leq d$, whence $\sqcup X$ exists, by consistent completeness. Define $d' = \sqcup X$; we shall show that $X = pr_D(d')$.

By definition $X \leq d'$ and $X \subseteq Pr(P_{D,E,\lambda})$, so $X \subseteq pr_D(d')$. We show the converse inclusion. Let $o_1 \in pr_D(d')$, then $o_1 \leq d' = \sqcup X$ and so by prime algebraicity there exists such $o_2 \in X$ that $o_1 \leq o_2$, and so $o_1 \rightarrow_{D,E,\lambda} o_2$ by (4.3). Also, since $o_1 \in pr_D(d')$, $o_2 \leq d' = \sqcup X \leq \sqcup pr_D(d) = d$, by prime algebraicity, and so it follows that $o_1 \leq d$ and so $o_1 \in pr_D(d)$. But we also have $X \lhd pr_D(d)$, $o_2 \in X$ and $o_1 \rightarrow_{D,E,\lambda} o_2$, from which we deduce $o_1 \in X$, by 3.3.7. Thus, $pr_D(d') \subseteq X$ and hence $X = pr_D(d')$, as required.

To sum up.

4.3.3. THEOREM. Suppose $(D, \leq)$ is prime algebraic and consistently complete and $\lambda : Pr(D) \rightarrow E$, then $B_{D,E,\lambda}$ is a left-closed behavioural presentation and $(P_{D,E,\lambda}, \subseteq)$ is isomorphic to $(D, \leq)$.

■

Notes.

Whereas some of the authors mentioned in the notes at the end of chapter 3 have been concerned with individual behaviours of systems, our concern in this chapter has been with sets of such behaviours, and in particular the structure of such sets. In 'traditional' theories of systems - such as Automata Theory - behaviours have been described by sequences and the mathematical objects used to describe them have stressed binary operations. Formal languages are subsets of free semigroups, after all. Such a preoccupation with an abstract-algebraic approach has characterised work such as that of Winkowski [24] and Starke [22]. To the suggestion that the significant structure is order-theoretic we are indebted to the domain theory based approach, particularly [19, 20, 80], which we shall examine in detail in the next chapter.

Readers familiar with model theory will recognise 4.3.3 as belonging to a family of results, the most famous of which is Stone's representation theorem for Boolean Algebras, in which abstract algebraic objects are shown to be isomorphic to an algebra of sets. For an introduction to this subject the reader is referred to [31]. 4.3.3 is a mild generalisation of 5.3.7 below, which appeared first in [20].

Chapter 5
Event Structures and Closures of Behavioural Presentations

> The world for them is not a concourse of objects in space; it is a heterogeneous series of independent acts. It is successive and temporal not spatial
>
> J. L. BORGES
> Tlon, Uqbar, Orbis Tertius

5.1 Introduction.

In chapter 2, we saw that from a behavioural presentation, we could derive two relations, denoted by $\rightarrow$ and #, which are defined on the set O of occurrences. In chapter 3, when we were discussing what made a set X a 'reasonable' point, we concluded that it should be a set which satisfied at least the two properties $X \lhd O$ and $\forall o_1, o_2 \in X: \neg o_1 \# o_2$. In the event, we also decided that, in order that our behavioural presentation remained discrete, it should also satisfy the condition $\exists p \in P: X \subseteq p$, which, of course, implies the latter of the two properties.

Now, many of the properties of system behaviour that we have discussed have depended upon these two relations $\rightarrow$ and #. For example, the definition of $\lhd$ depends on $\rightarrow$. Indeed, we could quite easily have set up our model from this starting point, that is with a set of occurrences with appropriate relations on it and a rule for extracting time points from the set. This is the approach taken by [20]. The objects in question are known as *event structures*.

In this chapter, we shall examine event structures and investigate the connection between them and behavioural presentations. Our reasons for doing this are twofold.

Firstly, event structures are an important model in their own right and the behavioural presentation model is little more than an extension of the event structure model. Secondly, the relationship between the two involves a form of closure - in a classical sense, as explained in the notes - and such 'closed' behavioural presentations are useful for the definition of liveness properties of systems such as starvation and fairness.

We shall not be concerned with E or λ in this chapter; our results concern only the order structure of P. Let us therefore define an *unlabelled behavioural presentation* to be a pair $B=(O,P)$ where O is a set of occurrences and $P \subseteq \wp(O)$ satisfies $\bigcup_{p \in P} p = O$. All results obtained heretofore apply to unlabelled behavioural presentations and we shall use them without further comment. We shall also apply the results of this chapter to labelled behavioural presentations without comment.

5.2. Event Structures and Behavioural Presentations.

We present a slightly extended notion of event structure. In [20] the ordering relation is a partial order rather than a pre-order.

5.2.1. DEFINITION. An event structure is a triple $S=(O,\rightarrow,\#)$, where

1• O is a set of occurrences;

2• $\rightarrow$ is a preorder on O;

3• # is an irreflexive, symmetric relation (the *conflict relation*) on O satisfying:

$$\forall o_1, o_2, o_1', o_2' \in O : o_1 \rightarrow o_1' \wedge o_2 \rightarrow o_2' \wedge o_1 \# o_2 \Rightarrow o_1' \# o_2'$$

$\rightarrow$ is a time ordering relation; $o_1 \rightarrow o_2$ is to be read 'o_1 occurred no later than o_2'.

is a relation of mutual exclusion; $o_1 \# o_2$ is to be read 'If o_1 has occurred then o_2 never can'.

Looking back to 2.3.2, we see immediately that if B is a behavioural presentation, then $\Sigma(B)$, defined by

$$\Sigma(B)=(O,\rightarrow,\#) \tag{5.1}$$

is an event structure.

For instance, the event structure derived in this way from the example given in figure 2.1 could be pictured as in figure 5.1. The symbolism should be self explanatory.

Conversely, given an event structure, and hence relations $\rightarrow$ and #, we may construct a set of points, as we indicated at the beginning of this chapter. $\triangleleft$ is defined for subsets of event structures exactly as in 3.3.7.

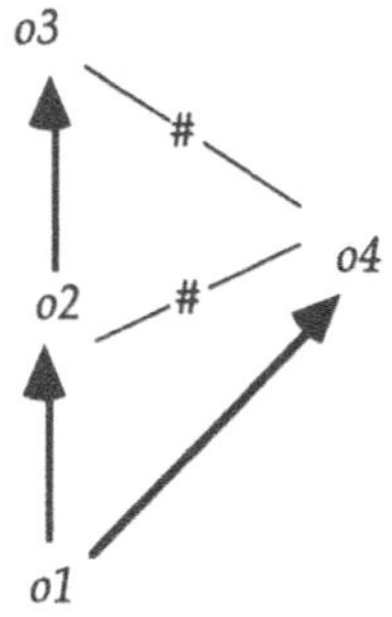

Figure 5.1

5.2.2. DEFINITION. If $S=(O,\rightarrow,\#)$ is an event structure then $X\subseteq O$ is *conflict-free* in $S \Leftrightarrow \forall o_1,o_2 \in X: \neg o_1 \# o_2$.

$X\subseteq O$ will be said to be a *process* of $S \Leftrightarrow X$ is conflict-free and $X \lhd O$. Denote the set of all processes of S by $\Pi(S)$.

If $S=(O,\rightarrow,\#)$ is an event structure and $X\subseteq O$, we define

$$\downarrow_S X = \{o \in O \mid \exists o' \in X: o \rightarrow o'\}.$$

If $X=\{o\}$, then we write $\downarrow_S o$ instead of $\downarrow_S X$.

Finally, define $\Lambda(S)=(O,\Pi(S))$.

We shall show that $\Lambda(S)$ is an unlabelled behavioural presentation. The only thing that needs proving is 2.1.1 (2), that every occurrence lies in some process. The following lemma is central to this proof.

5.2.3. LEMMA. If $S=(O,\rightarrow,\#)$ is an event structure and $X\subseteq O$ is conflict-free in S, then $\downarrow_S X \in \Pi(S)$.

PROOF. Let $o_1,o_2 \in \downarrow_S X$, then by 5.2.2 there exists $o_1',o_2' \in X$ such that $o_1 \rightarrow o_1'$ and $o_2 \rightarrow o_2'$. We cannot have $o_1 \# o_2$, since this would entail $o_1' \# o_2'$, by 5.2.1 (3), contradicting X conflict-free in S. Thus, $\downarrow_S X$ is conflict-free in S.

Next, suppose that $o_1 \in \downarrow_S X$ and $o_2 \in O$ such that $o_2 \rightarrow o_1$, then by 5.2.2 there exists $o_1' \in X$ such that $o_1 \rightarrow o_1'$. But then, $o_2 \rightarrow o_1'$, by transitivity of $\rightarrow$, and so $o_2 \in \downarrow_S X$. We have shown that $\downarrow_S X \lhd O$.

■

We may now exhibit a correspondence between the class of event structures and that of behavioural presentations.

5.2.4. PROPOSITION. If B is an unlabelled behavioural presentation, then $\Sigma(B)$ is an event structure and if S is an event structure then $\Sigma(B)$ is an unlabelled behavioural presentation.

PROOF. If B is a behavioural presentation, then $\Sigma(B)$ is an event structure by 2.3.2. If S is an event structure and $o \in O$, then $\{o\}$ is conflict free, because # is irreflexive, by 5.2.1 (3), and so $\downarrow_S o \in \Pi(S)$, by 5.2.3 Thus, $O \subseteq \bigcup_{p \in \Pi(S)} p$, and consequently $\Sigma(B)$ is an unlabelled behavioural presentation.

■

Let us now look at this relationship in more detail.

Suppose, first of all, that $B = (O, P)$ is an unlabelled behavioural presentation, then B determines a pre-order, which we shall call $\rightarrow_B$, and a mutual exclusion relation $\#_B$ and $\Sigma(B) = (O, \rightarrow_B, \#_B)$. Now, if $p \in P$, then by 3.3.10 (2) and (5), $p = \downarrow p \lhd O$ in B, so $p \lhd O$ in $\Sigma(B)$. Since p is conflict free in B it must also be conflict free in $\Sigma(B)$. Thus $p \in \Pi(S)$. It follows that $P \subseteq \Pi(S)$. We shall find it convenient to make the following definition.

5.2.5. DEFINITION. Let B_1, B_2 be unlabelled behavioural presentations. Define

$$B_1 \subseteq B_2 \Leftrightarrow P_1 \subseteq P_2.$$

Of course, $\subseteq$ is reflexive, antisymmetrical and transitive. We also note that $B_1 \subseteq B_2 \Rightarrow O_1 \subseteq O_2$, since $O_i = \bigcup_{p \in P_i} p$, each i.

We may now express our first observation in the form

$$B \subseteq \Lambda(\Sigma(B))$$

Now consider an event structure $S = (O, \rightarrow_S, \#_S)$. We may construct an unlabelled behavioural presentation, $B = \Lambda(S) = (O, \Pi(S))$ and from that an event structure $\Lambda(\Sigma(B)) = (O, \rightarrow_B, \#_B)$. If $o_1, o_2 \in O$ such that $o_1 \rightarrow_S o_2$, then for every $p \in \Pi(S)$, $o_2 \in p \Rightarrow o_1 \in p$, since $p \lhd O$. Thus, $o_1 \rightarrow_B o_2$. Conversely, if $o_1 \rightarrow_B o_2$ then for all $p \in \Pi(S)$, $o_2 \in p \Rightarrow o_1 \in p$. As we have observed, $\downarrow_S o_2 \in \Pi(S)$, so $o_1 \in \downarrow_S o_2$ and hence $o_1 \rightarrow_S o_2$. It follows that $\rightarrow_S = \rightarrow_B$.

Next, we turn to the mutual exclusion relation. We first observe that if $o_1, o_2 \in O$ such that $o_1 \#_S o_2$, then $\{o_1, o_2\} \not\subseteq p$, for all $p \in \Pi(S)$, by 5.2.2. Conversely, if $\neg o_1 \#_S o_2$, then $\{o_1, o_2\}$ is conflict-free in S and hence $\downarrow_S \{o_1, o_2\} \in \Pi(S)$, by 5.2.3. Thus, if $\{o_1, o_2\} \not\subseteq p$, all $p \in \Pi(S)$, then $o_1 \#_S o_2$.

Hence, $o_1 \#_S o_2 \Leftrightarrow \forall p \in \Pi(S): \{o_1, o_2\} \not\subseteq p$. But $\{o_1, o_2\} \not\subseteq p$, all $p \in \Pi(S)$ is equivalent to $o_2 \in p \Rightarrow o_1 \notin p$ and so:

$$o_1 \#_S o_2 \Leftrightarrow \forall p \in \Pi(S): \{o_1, o_2\} \not\subseteq p \Leftrightarrow o_1 \#_B o_2$$

using 2.3.1 (1), so that $\#_S = \#_B$. To sum up:

5.2.6. PROPOSITION. If B is an unlabelled behavioural presentation, then $B \subseteq \Lambda(\Sigma(B))$ and if S is an event structure then $S = \Sigma(\Lambda(S))$.

■

Let **ES** denote the class of all event structures. Let **BP** denote the class of all unlabelled behavioural presentations and let **CBP** denote the class of all unlabelled behavioural presentations of the form $\Lambda(S)$ for $S \in \mathbf{ES}$. We may first observe that Λ is a bijection between **ES** and **CBP**. For, it is certainly onto, by definition, while if $\Lambda(S_1) = \Lambda(S_2)$ then by 5.2.6,

$$S_1 = \Sigma(\Lambda(S_1)) = \Sigma(\Lambda(S_2)) = S_2$$

Furthermore, Σ restricted to **CBP** is the inverse to Λ. That $S = \Sigma(\Lambda(S))$ is given in 5.2.6. If $B \in \mathbf{CBP}$, then $B = \Lambda(S)$, some $S \in \mathbf{ES}$, and so, again by 5.2.6

$$\Lambda(\Sigma(B)) = \Lambda(\Sigma(\Lambda(S))) = \Lambda(S) = B$$

Therefore, hybrid event structures correspond precisely to behavioural presentations belonging to the class **CBP**. Standard event structures (in which the $\rightarrow$ relation is a partial order) correspond precisely to the *asynchronous* behavioural presentations belonging to the class **CBP**.

Thus, the theory of behavioural presentations is at least not less expressive than that of event structures. In fact, there are behavioural presentations that do not correspond directly to any event structure. To see this, it suffices to note that $\emptyset \in \Pi(S)$ for any event structure S, and hence $(O,P) \in \mathbf{CBP} \Rightarrow \emptyset \in P$. Thus, if $\emptyset \notin P$, then $(O,P) \notin \mathbf{CPB}$. There are certainly behavioural presentations without bottom elements, for instance that described in 5.3.3 below.

Does it really matter that the class of behavioural presentations is 'bigger' than the class of event structures? In one sense it doesn't because, as we have seen, every element of **BP** may be embedded into a particular element of **CBP**. However, for our purposes it does matter for the following reason.

In chapter 12 we describe a construction that will associate certain kinds of behavioural presentation with subsets of semi-commutative monoids. We have seen the converse construction in the analysis of the example system of chapter 1. The aim is to form a link between behavioural presentations and automata-like objects in order to develop a class of 'acceptors' for behavioural presentations. Now, the construction requires that the behavioural presentations involved are discrete - but not all discrete behavioural presentations belong to **CBP** and not all elements of **CBP** are discrete. For example, the unlabelled behavioural presentation associated with the behavioural presentation $B_{Achilles}$ of chapter 2 belongs to **CBP**, as we shall see in 5.3.5, but is not discrete.

5.3. Closure of Behavioural Presentations.

There still remain a number of questions that it seems worth our while to answer. Notably, what are the elements of **CBP** like and how are they related to those of **BP**? We note that $\Lambda \circ \Sigma$ determines a map from **BP** to **CBP**; what is the relationship between $B \in$ **BP** and $\Lambda(\Sigma(B))$? Let us first introduce some notation.

5.3.1. DEFINITION. Suppose $B \in$ **BP**. Define the *closure* of B, denoted by, $\overline{B}$ to be $\Lambda(\Sigma(B))$. If $\overline{B} = (O', P')$, then we observe that $O' = O$ and we shall define $\overline{P} = P'$.

If B is a labelled behavioural presentation, then we define $\overline{B} = (O, \overline{P}, E, \lambda)$.

In each case, we shall call $\overline{B}$ the *closure* of B. If $B = \overline{B}$, then we shall say that B is *closed.*

5.3.2. LEMMA. If $B \in$ **BP** then

1• $B \in \mathbf{CBP} \Leftrightarrow B = \overline{B}$;

2• $\overline{B} = \overline{\overline{B}}$.

PROOF. (1) If $B = \overline{B}$, then $B = \Lambda(S)$, for $S \in$ **ES**, where $S = \Sigma(B)$, by 5.2.4 (2) and 5.3.1. Thus $B \in$ **CBP**. On the other hand, if $B \in$ **CBP**, then $B = \Lambda(S)$, for $S \in$ **ES** and so $\overline{B} = \Lambda(\Sigma(B)) = \Lambda(\Sigma(\Lambda(S))) = \Lambda(S) = B$.

For (2), we observe that $\overline{\overline{B}} = \Lambda(\Sigma(\Lambda(S)))$, where $S = \Sigma(B)$. By 5.2.6, $\Lambda(\Sigma(\Lambda(S))) = \Lambda(S)$ and so $\overline{\overline{B}} = \Lambda(\Sigma(B)) = \overline{B}$

■

By 5.3.2 (1), B is closed if and only if $B \in$ **CBP**, hence the notation. **CBP** stands for 'closed behavioural presentations'.

We may be in a better position to answer the remaining questions concerning the relationship between **BP** and **CBP** after we have looked at a few examples.

5.3.3. EXAMPLE. Define $B = (O, P)$, where

$O = \{o_1, o_2, o_3\}$
$P = \{\{o_1, o_2\}, \{o_2, o_3\}, \{o_1, o_3\}\}$

Observe that $\# = \emptyset$ and $\rightarrow$ is the identity relation. It follows easily enough that $X \lhd Y \Leftrightarrow X \subseteq Y$ and hence, $\Pi(\Lambda(B)) = \wp(O)$. Note that in going from B to $\overline{B}$ we have added in sets such as $\{o_1\}$ and $\emptyset$, so that $\overline{B}$ becomes left closed. But we have also added in the set O itself. What is the significance of this? Since $\# = \emptyset$ and $\rightarrow$ is the identity it follows that no two distinct occurrences of B may be ordered in time, mutually exclusive or simultaneous. By 2.3.8, every two such events must be concurrent. Intuitively, the occurrence of any two of them should not prevent the occurrence of the third, and yet P contains no point when all three have occurred. In passing from B to $\overline{B}$, this extra point is added in. Thus, this construction involves a form of 'closure with respect to concurrency'. It produces a behavioural presentation with a structural property called forward stability.[1]

We mentioned in 5.3.3 that points had been added into B to cause it to become left-closed. Suppose $B \in \mathbf{CBP}$, so that $B = (O, \Pi(S))$ for some event structure $S = (O, \rightarrow, \#)$. If $X \subseteq O$ with $X \lhd p$, some $p \in \Pi(S)$, then $X \lhd O$ and X is conflict free (it is a subset of the conflict free set p). Thus $X \in \Pi(S)$. We thus have:

5.3.4. PROPOSITION. Every $B \in \mathbf{CBP}$ is left-closed.

■

We may thus deduce from 3.3.14 that if S is an event structure, then $(\Pi(S), \subseteq)$ is prime algebraic, with elements $\downarrow_{\Lambda(S)} o$ as the complete primes, and consistently complete.

Closure with respect to concurrency is not the only kind that occurs, as the following example shows.

5.3.5. EXAMPLE. Define $B = (O, P)$, where

$O = \{o_n \mid n \in \omega\}$
$P = \{p_n \mid n \in \omega\}$ where $p_i = \{o_i \mid i \leq n\}$, each i

so that $B = (O_{Achilles}, P_{Achilles} - \{O_{Achilles}\})$.

Observe that $\# = \emptyset$ and $o_i \rightarrow o_j \Leftrightarrow i \leq j$. It may also be verified that $\Pi(\Lambda(B)) = P \cup \{O\}$. Thus $\overline{B} = B_A$, where B_A is the unlabelled behavioural presentation corresponding to $B_{Achilles}$ of chapter 3. In particular, $B_A \in \mathbf{CBP}$.

---

[1]See 7.4.2.

Here, we do not have closure of B with respect to concurrency - B is in fact a discrete, sequential, deterministic behavioural presentation - but we have added on the extra point, O. What may we say about O? It's most immediate property is that $O = \bigcup_{p \in P} p$, that is, $O = \bigsqcup P$. In passing from B to $\overline{B}$, we are effecting a form of 'closure with respect to limits'.

From these examples, it might be guessed that the elements of **CBP** are closed in both these ways.

We have remarked that the 'C' in **CBP** stands for 'closed'. In fact, 'C' could also stand for 'coherent', an order theoretic property that, as we shall see, is possessed by event structures and closed behavioural presentations. In fact, it will turn out that coherence is one of the characteristic properties of closed behavioural presentations.

5.3.6. DEFINITION. Let $(D, \leq)$ be a poset.

$x, y \in D$ are *compatible* and we write $x \uparrow y$, $\Leftrightarrow$ the set $\{x, y\}$ possesses an upper bound.

$X \subseteq D$ is *pairwise compatible* $\Leftrightarrow \forall x, y \in X : x \uparrow y$.

$(D, \leq)$ is *coherent* $\Leftrightarrow$ every pairwise compatible subset of D has a least upper bound.

5.3.7. THEOREM ([20]). Suppose S is an event structure, then $(\Pi(S), \subseteq)$ is prime algebraic and coherent. Its complete primes are the elements $\downarrow o$, each $o \in O$.

PROOF. Suppose that $X \subseteq \Pi(S)$ is pairwise compatible. We wish to show that $\bigsqcup X$ exists. Define $p_X = \bigcup_{p \in X} p$, then it is clear that if $p_X \in \Pi(S)$, then $p_X = \bigsqcup X$. We must therefore show that $p_X \lhd O$ and that p_X is conflict-free.

Suppose that $o \in O$ and $o' \in p_X$ such that $o \rightarrow o'$. We must show that $o \in p_X$. But, by definition of p_X, $o' \in p$, some $p \in X$, and since $p \in \Pi(S)$, $p \lhd O$, by 5.2.2. Thus, $o \in O$ and $o' \in p$ with $o \rightarrow o'$ and $p \lhd O$, and so $o \in p \subseteq p_X$, as required.

Next suppose that $o_1, o_2 \in p_X$, then there exists $p_1, p_2 \in X$ such that $o_i \in p_i$, each i. Since X is pairwise compatible, the set $\{p_1, p_2\}$ has an upper bound p, from which it follows that $o_1, o_2 \in p$. Thus p_X is conflict free.

We have already observed (following 5.3.4) that $(\Pi(S), \subseteq)$ is prime algebraic.

■

We are now in a position to give an order theoretic characterisation of the elements of **CBP**. First, we need a simple technical lemma.

5.3.8. LEMMA. Suppose that B is an unlabelled behavioural presentation and $o_1, o_2 \in O$, then

1• $o_1 \rightarrow o_2$ in $B \Leftrightarrow o_1 \rightarrow o_2$ in $\overline{B}$;
2• $o_1 \# o_2$ in $B \Leftrightarrow o_1 \# o_2$ in $\overline{B}$.

PROOF. Let $S = \Sigma(B)$, then $S = \Sigma(\Lambda(S)) = \Sigma(\Lambda(\Sigma(B))) = \Sigma(\overline{B})$, by 5.2.6 and 5.3.1 and so $\Sigma(B) = \Sigma(\overline{B})$. Using (5.1), we have $o_1 \rightarrow o_2$ in $B \Leftrightarrow o_1 \rightarrow o_2$ in $\Sigma(B) \Leftrightarrow o_1 \rightarrow o_2$ in $\Sigma(\overline{B}) \Leftrightarrow o_1 \rightarrow o_2$ in $\overline{B}$ and $o_1 \# o_2$ in $B \Leftrightarrow o_1 \# o_2$ in $\Sigma(B) \Leftrightarrow o_1 \# o_2$ in $\Sigma(\overline{B}) \Leftrightarrow o_1 \# o_2$ in $\overline{B}$

■

We now have the promised characterisation.

5.3.9. THEOREM. If B is an unlabelled behavioural presentation, then $B \in \mathbf{CBP} \Leftrightarrow (P, \subseteq)$ is left closed and coherent.

PROOF. First, note that if $B \in \mathbf{CBP}$, then $B = \Lambda(S)$ for some $S \in \mathbf{ES}$, by 5.3.1, and that consequently $B = (O, \Pi(S))$, by 5.2.2. But $(\Pi(S), \subseteq)$ is coherent, by 5.3.7. We already know from 5.3.4 that B is left-closed.

Now let us suppose that $(P, \subseteq)$ is left closed and coherent. We show that $B = \overline{B}$ and appeal to 5.3.2 (1). Since we already know that $B \subseteq \overline{B}$, from 5.2.6 and 5.3.1, it suffices to show that $\overline{B} \subseteq B$.

By 5.3.8 (1), $\downarrow o$ denotes the same set, whether we are thinking of o as belonging to B or $\overline{B}$. So, if $p \in \overline{P}$, then $pr(p) \subseteq P$ and if $\bigsqcup pr(p)$ exists in P, then $p = \bigsqcup pr(p) \in P$. We shall use the coherence property to ensure that $\bigsqcup pr(p)$ does exist in P.

Let $\downarrow o_1, \downarrow o_2 \in pr(p)$. As $o_1, o_2 \in p$, it follows that $\neg\, o_1 \# o_2$ in $\overline{B}$ and so $\neg\, o_1 \# o_2$ in B, by 5.3.8 (2) Thus by 2.3.8 there are four cases:

1• $o_1 < o_2$: so $\downarrow o_1 \subseteq \downarrow o_2$ and thus $\downarrow o_1 \sqcup \downarrow o_2 = \downarrow o_2 \in P$;
2• $o_2 < o_1$: analogously to 1, $\downarrow o_1 \sqcup \downarrow o_2 = \downarrow o_1 \in P$;
3• $o_1 \approx o_2$: so $\downarrow o_1 = \downarrow o_2$ and thus $\downarrow o_1 \sqcup \downarrow o_2 = \downarrow o_1 \in P$;
4• $o_1\ co\ o_2$: so $o_1, o_2 \in \tilde{p}$ for some $\tilde{p} \in P$ and hence $\downarrow o_1, \downarrow o_2 \subseteq \tilde{p}$, by 3.3.10 (1). Since B is left closed, $(P, \subseteq)$ is consistently complete, by 4.2.2, and therefore $\downarrow o_1 \sqcup \downarrow o_2 \in P$.

Thus in all four cases $\downarrow o_1 \sqcup \downarrow o_2 \in P$, which shows that $pr(p)$ is pairwise consistent. We now appeal to the coherence property.

■

We have identified event structures as corresponding precisely to those behavioural presentations which are left-closed and coherent. Let us conclude our discussion of the relationship between the two models by describing the relationship between behavioural presentations B and $\overline{B}$.

5.3.10. THEOREM. If B is a behavioural presentation, then $(\overline{P},\subseteq)$ is the smallest left-closed, coherent poset containing P. That is, if $\tilde{B}$ is a behavioural presentation such that $(\tilde{P},\subseteq)$ is left closed and coherent, then $P\subseteq\tilde{P}\Rightarrow\overline{P}\subseteq\tilde{P}$.

PROOF. We know that $P\subseteq\overline{P}$, by 5.2.6. $(\overline{P},\subseteq)$ is left-closed and coherent by 5.3.9.

Now we show that $(\overline{P},\subseteq)$ is the smallest left-closed, coherent behavioural presentation containing P. Suppose $(\tilde{P},\subseteq)$ is left-closed and coherent with $P\subseteq\tilde{P}$. We show that $\overline{P}\subseteq\tilde{P}$. This will follow if we can show that

$$B_1\subseteq B_2\Rightarrow\overline{B_1}\subseteq\overline{B_2} \tag{5.2}$$

for then $P\subseteq\tilde{P}\Rightarrow\overline{P}\subseteq\overline{\tilde{P}}$, and by 5.3.9, $\tilde{B}\in\mathbf{CPB}$, and by 5.3.2 (1), $\overline{P}=\overline{\tilde{P}}$, Thus $\overline{P}\subseteq\tilde{P}$.

To complete the proof, we demonstrate (5.2).

Suppose $B_1=(O_1,P_1)$ and $B_2=(O_2,P_2)$ with $P_1\subseteq P_2$, and let $\overline{B_1}=(O_1,\overline{P_1})$ and $\overline{B_2}=(O_2,\overline{P_2})$. We show that $\overline{P_1}\subseteq\overline{P_2}$.

Let $\tilde{p}\in\overline{P_1}$. As in the proof of 5.3.9, $pr(\tilde{p})\subseteq P_1\subseteq P_2$ and if $\sqcup\, pr(\tilde{p})$ exists in $\overline{P_2}$, then $\tilde{p}=\sqcup\, pr(\tilde{p})\in\overline{P_2}$. It follows that we need to show that $pr(\tilde{p})$ is pairwise consistent in $\overline{P_2}$.

Let $\downarrow o_1,\downarrow o_2\in pr(\tilde{p})$. Since $o_1,o_2\in\tilde{p}$, it follows that $\neg o_1\#\,o_2$ in B_1, since $\tilde{p}$ is conflict free, and hence $\neg o_1\#\,o_2$ in B_2, since $P_1\subseteq P_2$, and hence $\neg o_1\#\,o_2$ in $\overline{B_2}$, since $B_2\subseteq\overline{B_2}$. Therefore as in the proof of 5.3.9, we have four cases: $o_1<o_2$, $o_2<o_1$, $o_1\approx o_2$ and o_1 *co* o_2, and as in that proof, in all four cases it may be shown that $\downarrow o_1\sqcup\downarrow o_2\in P_2$, which shows that $pr(\tilde{p})$ is pairwise consistent in $\overline{B_2}$. We now appeal to the coherence property.

■

The relationship between a behavioural presentation and its closure strongly resembles that between a string language $L\subseteq A^*$ and its closure $\overline{L}\subseteq A^\infty$. Indeed, it is possible to regard a string language L as a behavioural presentation, B_L - as we saw in the coin-tossing example (2.2.1) and as we shall examine further in chapter 13. It may be shown that

$$B_{\bar{L}} = \overline{B_L}$$

This observation allows us to explain more fully the discussion following 5.2.6, as to why event structures are not quite suitable as a basis for the theory that we wish to develop in this book. In general a string language cannot be regarded as a representation of the set of processes of an event structure, although the elements of a language can be considered as describing pieces of sequential behaviour. Any such representation would establish a poset isomorphism between $\Lambda(S)$ and L, where L has the prefix ordering. In general, however, languages are not coherent, so such a morphism cannot exist in all cases.

For example, consider the string language $L = a^*$, where a is a single alphabetical character. L may be considered to represent a system described by the behavioural presentation B_L, where:

$$O_L = \{o_n \mid n \in \omega\}$$
$$P_L = \{\{o_m \mid m \leq n\} \mid n \in \omega\}$$
$$E_L = \{a\}$$
$$\forall n \in \omega : \lambda_L(o_n) = a$$

but B_L is not closed, since $\overline{P_L} \neq P_L$. In fact, $\overline{P_L} = P_L \cup \{O_L\}$.

The problem arises when we try to represent non-sequential event structures in a linguistic form because of complications arising from attempts to 'close' trace or vector languages.

Notes.

In [20], the authors sought to establish a link between General Net Theory, a dominant theory of asynchronous parallelism, and the theory of Scott domains. To do this, they showed that the set of processes of a Net - a set of objects modelling the behaviour of an asynchronous system - could be endowed with an order structure characteristic of partial orders used in denotational semantics.

Roughly speaking, to each marked Condition/Event Net (N, M), they associated a second net N', which they called an *occurrence net*, which exhibited the total potential behaviour of $\overline{B}$ from M. This construction can be thought of as a generalisation of the tree structure that may be imposed on the set of execution sequences of a partial state machine.

Such a net now determines two things; first, a partial order $\leq$ on events and secondly a relation # of mutual exclusion and so an event structure. We return to this subject in chapter 16.

The function that maps B to B has the three characteristic properties of a closure operator [31], namely that if $B, B_1, B_2 \in \mathbf{BP}$, then:

1• $B \subseteq \overline{B}$ by 5.2.6 (1) and 5.3.1

2• $B_1 \subseteq B_2 \Rightarrow \overline{B_1} \subseteq \overline{B_2}$ by (5.2)

3• $\overline{B} = \overline{\overline{B}}$ by 5.3.2 (2).

Chapter 6
Another Event Structure Model

A loutish collector of rents,
Attempted to structure events.
But it didn't work out,
For this hectoring lout
Starved to death, locked up in the Gents.

F. X. Reid [40]

6.1 Introduction.

Winskel has studied a variety of even structure models. The version presented in this chapter was introduced in [19] within the context of non-interleaving semantics for algebraic specification languages.

An event structure in this model comprise a set E and a set $F \subseteq \wp(E)$ of processes, so that if E were the union of the elements of F, then such an event structure would actually turn out to be an unlabelled behavioural presentation, albeit an unlabelled behavioural presentation which is required to satisfy certain order theoretic constraints.

The relationship between the two models, which we shall elucidate, is more subtle than may appear at first sight. The difference lies in the way in which time ordering is defined. In the event structure, time ordering is defined locally; for each $x \in F$, there is a partial order $\leq_x$ defined locally on its elements in the same way as the relation $\rightarrow$ is defined globally on O in a behavioural presentation. It is possible, for example to have $e \leq_x e'$ in a process x and $e \not\leq_y e'$ in a process y. There is no global time ordering as in the model presented in chapter 5.

This means that in certain cases, we may distinguish between the e sitting in x and the e sitting in y. We may now see that the elements of E are events, in our terminology, and that 'e-in-x' names some occurrence of e.

This suggests a formal relationship between event structures of this kind and behavioural presentations. Occurrence are equivalence classes of 'e-in-

x's, a time point is made from $x \in F$ by replacing each e by the equivalence class of 'e-in-x' and the equivalence class of 'e-in-x' is labelled e.

The resulting behavioural presentation is related to the parent event structure in a particular way. There is an isomorphism from F to the set of time points such that each process and its corresponding point are isomorphic as labelled partially ordered sets. In such circumstances, we speak of the behavioural presentation representing the event structure.

We shall show that every event structure may be represented by a behavioural presentation and we also characterise the behavioural presentation which do so. Apart from establishing a formal relationship between the two models, this also illustrates the use of the order theory that has occupied us in chapters 4 and 5. Such applications will recur.

6.2. The Model.

First, we need to following notions.

Let E be a set and let $F \subseteq \wp(E)$. If $X \subseteq F$, then

- X is *compatible* in F - and we write $X \uparrow_F \Leftrightarrow \exists y \in F \forall x \in X: x \subseteq y$
- X is *pairwise compatible* if $x \uparrow_F y$ for all $x, y \in X$.

where, if $x, y \in X$, then we write $x \uparrow_F y$ for $\{x, y\} \uparrow_F$.

6.2.1. DEFINITION. An *event structure* in the sense of [19] (hereafter just 'event structure' providing there is no possibility of confusion) is a pair $S = (E, F)$ where E is a non-empty set of *events* and $F \subseteq \wp(E)$ is a set of *configurations*, satisfying.

1• F is *coherent*: if $X \subseteq F$ is pairwise compatible, then $\bigcup_{x \in X} x \in F$;

2• F is *stable*: $\forall X \subseteq F: X \neq \emptyset \Rightarrow \bigcap_{x \in X} x \in F$;

3• F is *coincidence free*: $\forall x \in F \forall e_1, e_2 \in x: e_1 \neq e_2 \Rightarrow$
$\exists y \subseteq x: (e_1 \in y \wedge e_2 \notin y) \vee (e_2 \in y \wedge e_1 \notin y)$;

4• F is *finitary*: $\forall x \in F \forall e \in x \exists y \subseteq x: e \in y \wedge |y| < \infty$.

Furthermore, if $E = \bigcup_{x \in F} x$, then S is *full*.

Thus, an event structure is rather like an unlabelled behavioural presentation. Indeed, full event structures *are* unlabelled behavioural presentations (though not all unlabelled behavioural presentations satisfy all the conditions of 6.2.1). However, there is a great deal of difference between the ways in which behavioural presentations and event structures are to be understood in modelling system behaviour.

Technically, this difference manifests itself in the way time-ordering of events is defined.

6.2.2. DEFINITION. Let $S=(E,F)$ be an event structure and let $x \in F$. Define

$$e_1 \leq_x e_2 \Leftrightarrow \forall y \in F: y \subseteq x \wedge e_2 \in y \Rightarrow e_1 \in y.$$

$\leq_x$ is clearly reflexive and transitive, while condition 6.2.1 (3) ensures that it is antisymmetrical. Thus $(x, \leq_x)$ is a partial order for each $x \in F$.

$\leq_x$ is defined *locally* on the configurations of S, as opposed to our relation $\rightarrow$ which is defined globally for a behavioural presentation. It is possible that $e_1 \leq_x e_2$ for some $x \in F$, while $e_1 \not\leq_y e_2$ for some $y \in F$.

We illustrate this with an example (which is taken from [19]). Let $S=(E,F)$, where

$$\begin{aligned} &E=\{0,1,2\} \\ &F=\{\emptyset,\{1\},\{2\},\{1,0\},\{2,0\}\} \end{aligned} \tag{6.1}$$

S is an unlabelled behavioural presentation in which $1\, co\, 0$, $2\, co\, 0$ and $1 \# 2$.

Regarded as an event structure, however, we have $1 \leq_{\{1,0\}} 0$ and $2 \leq_{\{2,0\}} 0$. Furthermore, 1 and 2 are in conflict - they never appear in the same configuration. So the event structure describes a system in which either 1 and 2 occur, after which an 0 occurs. However, the two 0s are distinct. In fact, in the terminology of chapter 2, the elements of E stand revealed as *events* rather than *occurrences*.

S considered as a behavioural presentation does not represent the same system as S regarded as an event structure. It is relatively easy to see how we could represent S by a behavioural presentation. Let B by given by

$$\begin{aligned} &O=\{o_0,o_0',o_1,o_2\}, \\ &P=\{\emptyset,\{o_1\},\{o_2\},\{o_1,o_0\},\{o_2,o_0'\}\}, \\ &E=\{0,1,2\} \\ &\lambda(o_1)=1,\ \lambda(o_2)=2,\ \lambda(o_0)=\lambda(o_0')=0. \end{aligned} \tag{6.2}$$

As we shall see, any event structure may be represented by a behavioural presentation, in the sense of the following definition.

6.2.3 DEFINITION. Suppose $S=(E,F)$ is an event structure and $\tilde{B}$ is a behavioural presentation, then $\tilde{B}$ will be said to represent $S \Leftrightarrow$ there exists a bijection $\chi: F \to \tilde{P}$ such that:

1• $\forall x,y \in F: x \subseteq y \Leftrightarrow \chi(x) \subseteq \chi(y)$;

2• $\tilde{E} = E$;

3• For each $x \in F$, $\lambda(\chi(x)) = x$ and the restriction of λ to $\chi(x)$ is a surjection satisfying $\forall o_1, o_2 \in \chi(x): o_1 \to o_2 \Leftrightarrow \lambda(o_1) \leq_x \lambda(o_2)$.

We observe that 6.2.3 (1) entails that χ is injective, so χ will actually be a bijection.

Thus, $\tilde{B}$ represents S providing there is a correspondence between the points of $\tilde{B}$ and the configurations of S such that corresponding points and configurations describe the same labelled partial orders. In the case of (6.1) and (6.2), χ is given by:

$$\chi(\varnothing) = \varnothing \quad \chi(\{1\}) = \{o_1\} \quad \chi(\{2\}) = \{o_2\}$$
$$\chi(\{0,1\}) = \{o_0, o_1\} \quad \chi(\{0,2\}) = \{o_0', o_2\}$$

We shall show that every event structure may be represented by a behavioural presentation, so that in some sense, the class of behavioural presentations contains that of event structures. To show that this is indeed the case, we shall give a construction that turns event structures into behavioural presentations.

The discussion leading to the statement of 4.3.2 suggests that if S is prime algebraic, then we may construct the points of a behavioural presentation from the complete primes of S. S is indeed prime algebraic, as we shall show. First, we prove a technical lemma.

6.2.4. LEMMA. If $S=(E,F)$ is an event structure, then

1• $\forall x \in F\ \forall e \in x: \downarrow_x e \in F$;

2• $\forall x,z \in F\ \forall e \in E: \downarrow_x e \subseteq z \Rightarrow \downarrow_x e = \downarrow_z e$,

where

$$\downarrow_x e = \{e' \in x \mid e' \leq_x e\} \tag{6.3}$$

PROOF. (1) Define $X = \{y \subseteq x \mid e \in y\}$. $x \in X$ so $X \neq \varnothing$ and hence $\bigcap_{y \in X} y \in F$, by 6.2.1 (2). Now

$$e' \in \downarrow_x e \Leftrightarrow \forall y \subseteq x : (e \in y \Rightarrow e' \in y) \Leftrightarrow (\forall y \in X : e' \in y) \Leftrightarrow e' \in \bigcap_{y \in X} y$$

Thus $\downarrow_x e = \bigcap_{y \in X} y \in F$.

(2) If $e' \leq_z e$, then since $\downarrow_x e \subseteq z$ and $z, \downarrow_x e \in F$ and $e \in \downarrow_x e$, $e' \in \downarrow_x e$, by 6.2.2. Hence, $e' \leq_x e$. Conversely, suppose that $e' \leq_x e$, and let $y \subseteq z$ such that $e \in y$, then $x \cap y \in F$, by 6.2.1 (2), $x \cap y \subseteq x$ and $e \in x \cap y$, so $e' \in x \cap y \subseteq y$. Thus for all $y \subseteq z$, $e \in y \Rightarrow e' \in y$, that is $e' \leq_z e$.

■

6.2.5. THEOREM ([19]). If $S = (E, F)$ is an event structure, then $(F, \subseteq)$ is prime algebraic and coherent. The complete primes of $(F, \subseteq)$ are the elements $\downarrow_x e$, $x \in F$, $e \in x$

Write $pr(S) = \{\downarrow_x e \mid x \in F \wedge e \in x\}$.

PROOF. By 6.2.4 (1) $\downarrow_x e \in F$ for each $x \in F$ and $e \in x$.

$(F, \subseteq)$ is coherent by 6.2.1 (1), so we need only prove the prime algebraicity part.

Suppose that $X \subseteq F$ such that $z = \sqcup X \in F$ and that $\downarrow_x e \subseteq z$. Since $z = \bigcup_{y \in X} y$, it follows that $e \in y$, some $y \in X$. But now $\downarrow_x e, \downarrow_y e \subseteq z$, so $\downarrow_x e = \downarrow_z e = \downarrow_y e$, by 6.2.4,. Therefore, there exists $y \in X$ such that $\downarrow_x e \subseteq y$.

Let $p_x^{\cdot} = \{\downarrow_x e \mid e \in x\}$. Clearly

$$\bigcup \downarrow p_x = x \tag{6.4}$$

while if x is prime, then $x \subseteq \bigcup \downarrow p_x$, so that $x \subseteq \downarrow_x e$, some $e \in x$. But, $e \in x$ implies $\downarrow_x e \subseteq x$, so $x = \downarrow_x e$. Thus, $pr(S)$ is the set of complete primes of $(F, \subseteq)$, while (6.4) ensures prime algebraicity.

■

Since coherence entails consistent completeness - any set which is bounded above is easily seen to be pairwise consistent - we may use 6.2.5 and 4.3.3 to construct a behavioural presentation from S. Recall that in the construction of 4.3.3, occurrences are constructed from the primes and points are the primes below given members of - in this case F.

For instance, in example of (6.1), the primes are

$$\downarrow_{\{1\}} 1 = \{1\} \quad \downarrow_{\{2\}} 2 = \{2\} \quad \downarrow_{\{0,1\}} 0 = \{0,1\} \quad \downarrow_{\{0,2\}} 0 = \{0,2\}$$

Note that each prime in the example is associated with a particular event name, the last event to occur in the behaviour it describes. This allows us to define a labelling function on primes. Define $\lambda(\downarrow_x e) = e$. The behavioural presentation thus constructed has the form:

$$\begin{aligned} &O = \{\downarrow_{\{1\}}1, \downarrow_{\{2\}}2, \downarrow_{\{0,1\}}0, \downarrow_{\{0,2\}}0\} \\ &P = \{\varnothing, \{\downarrow_{\{1\}}1\}, \{\downarrow_{\{2\}}2\}, \{\downarrow_{\{1\}}1, \downarrow_{\{0,1\}}0\}, \{\downarrow_{\{2\}}2, \downarrow_{\{0,2\}}0\}\} \\ &E = \{0,1,2\} \\ &\lambda(\downarrow_{\{1\}}1) = 1 \quad \lambda(\downarrow_{\{2\}}2) \quad \lambda(\downarrow_{\{0,1\}}0) = 0 \quad \lambda(\downarrow_{\{0,2\}}0) = 0 \end{aligned} \tag{6.5}$$

Compare this behavioural presentation with that of (6.2). They are clearly isomorphic, with an isomorphism given by

$$\begin{aligned} &f_O(\downarrow_{\{1\}}1) = o_1 \quad f_O(\downarrow_{\{2\}}2) = o_2 \quad f_O(\downarrow_{\{0,1\}}0) = o_0 \quad f_O(\downarrow_{\{0,2\}}0\} = o_0' \\ &f_P(\{\downarrow_{\{1\}}1\}) = \{o_1\} \quad f_P(\{\downarrow_{\{1\}}1, \downarrow_{\{0,1\}}0\}) = \{o_1, o_0\} \\ &f_P(\{\downarrow_{\{2\}}2\}) = \{o_2\} \; f_P(\{\downarrow_{\{2\}}2, \downarrow_{\{0,2\}}0\}) = \{o_2, o_0'\} \; f_P(\varnothing) = \varnothing \end{aligned}$$

Since the behavioural presentation of (6.2) represents the event structure of (6.1) via a bijection χ, it follows that the behavioural presentation of (6.5) represents the event structure of (6.1) via a bijection $\pi \circ \chi$. This is typical, as the next theorem shows. First, we need a definition.

6.2.6 DEFINITION. If $S = (E,F)$ is an event structure, then define $B_S = (O_S, P_S, E_S, \lambda_S)$, where

1• $O_S = Pr(S)$;
2• $P_S = \{p_x \mid x \in F\}$, where $p_x = \{\downarrow_x e \mid e \in x\}$;
3• $E_S = E$;
4• $\lambda_S : O_S \to E_S$ is defined by $\lambda_S(\downarrow_x e) = e$.

6.2.7. THEOREM. Suppose $S = (E,F)$ is an event structure, then B_S is a behavioural presentation and B_S represents S via the map $\chi_S : F \to P_S$ defined by $\chi_S(x) = p_x$.

PROOF. Let $o = \downarrow_x e$. Since $o \subseteq x$ it follows that $o \in p_x$ and so $\bigcup_{p_x \in P_S} p_x = O_S$. Thus B_S is a behavioural presentation. We check that it represents S via χ_S. χ_S is onto by definition. If $\chi_S(x) \subseteq \chi_S(y)$, then $x = \bigcup_{e \in x} \downarrow_x e \subseteq \bigcup_{e \in y} \downarrow_y e = y$, by 6.2.5. Conversely, suppose that $x \subseteq y$. If $\downarrow_x e \in p_x$ then $\downarrow_x e \subseteq x \subseteq y$ and hence $\downarrow_x e = \downarrow_y e \subseteq y$ by 6.2.4. Thus $\chi_S(x) = p_x \subseteq p_y = \chi_S(y)$ and 6.2.3 (1) holds.

$E_S = E$ by definition, giving 6.2.3 (2).

Let $x \in F$, then $e \in \lambda_S(\chi_S(x)) \Leftrightarrow \downarrow_x e \in \chi_S(x) \Leftrightarrow e \in x$, so $\lambda_S(\chi_S(x)) = x$. By 6.2.6 (4), $o \in p_x \Leftrightarrow \lambda_S(o) \in x$, so λ_S restricted to $\chi_S(x)$ is a bijection and 6.2.3 (3) holds.

Finally, we show that if $o_1, o_2 \in \chi_S(x)$, then $o_1 \to o_2 \Leftrightarrow \lambda_S(o_1) \leq_x \lambda_S(o_2)$.

First, suppose that $o_1 \to o_2$ and let $y \subseteq x$ such that $\lambda_S(o_2) \in y$. Then $\downarrow_y \lambda_S(o_2) \in p_y$, and so $o_2 = \downarrow_x \lambda_S(o_2) = \downarrow_y \lambda_S(o_2) \in p_y$, by 6.2.4. Thus, $o_1 \in p_y$ as $o_1 \to o_2$, and so $\lambda(o_1) \in y$. We have shown that if $y \subseteq x$ and $\lambda_S(o_2) \in y$, then $\lambda(o_1) \in y$, that is $\lambda_S(o_1) \leq_x \lambda_S(o_2)$.

Next, suppose that $\lambda_S(o_1) \leq_x \lambda_S(o_2)$ and let $p \in P$ such that $o_2 \in p$. We have $p = p_y$, for some $y \in F$. Let $\tilde{p} = p_x \cap p_y$. Since $(P_S, \subseteq)$ is isomorphic to $(F, \subseteq)$ and $(F, \subseteq)$ is coherent, and hence consistently complete, it follows that $\tilde{p} \in P_S$ and hence $\tilde{p} = p_z$, some $z \in F$. Since $\tilde{p} \subseteq p$, we must have $z = \chi_S(\tilde{p}) \subseteq \chi_S(p_x) = x$. Now $\lambda(o_2) \in z$ and $\lambda_S(o_1) \leq_x \lambda_S(o_2)$, so $\lambda(o_1) \in z$. Since λ is injective on $\tilde{p}$, it follows that $o_1 \in \tilde{p} \subseteq p$. Hence $o_2 \in p \Rightarrow o_1 \in p$ that is $o_1 \to o_2$.

■

6.3. Behavioural Presentations Which Represent Event Structures.

Not all behavioural presentations represent event structures. Let us suppose that B represents $S = (E, F)$ via a function χ. What can we conclude about B?

Firstly, χ establishes a poset isomorphism between $(P, \subseteq)$ and $(F, \subseteq)$. Hence, by 6.2.5, $(P, \subseteq)$ is prime algebraic and coherent. Applying 5.3.8:

6.3.1. LEMMA. If B represents $S = (E, F)$, then B is closed.

■

Secondly, we note that 6.2.1 (3) entails that S contains no non-trivial simultaneity.

This is true of B also. If $x \in F$ and $o_1, o_2 \in \chi(x)$, then by 6.2.3 (3),

$$o_1 \approx o_2 \Leftrightarrow \lambda(o_1) \leq_x \lambda(o_2) \wedge \lambda(o_2) \leq_x \lambda(o_1) \Leftrightarrow \lambda(o_1) = \lambda(o_2).$$

But, λ is injective on $\chi(x)$. Thus $o_1 \approx o_2 \Leftrightarrow o_1 = o_2$, and so

6.3.2. LEMMA. If B represents $S = (E, F)$, then B is asynchronous.

■

Next, we note that according to the finitary condition, 6.2.1 (4), each event of S belongs to a finite set. From this it follows that $\downarrow_x e$ must be finite for each $x \in F$ and $e \in x$. Let us say that B is *prime-finitary* $\Leftrightarrow$

$$\forall o \in O: |\downarrow o| < \infty.$$

If $o \in O$, then $o \in p$, some $p \in P$ and so $o \in \chi(x)$, some $x \in F$, as χ is bijective and $e = \lambda(o) \in x$. By 6.2.1 (4), there exists $y \in F$ such that $|y| < \infty$. and $e \in y \subseteq x$, so if $e' \leq_x e$, then $e' \in y$ and so $\downarrow_x e \subseteq y$ and hence $|\downarrow_x e| < \infty$. Since $\chi(\downarrow_x e) = \downarrow o$ and $\lambda(\downarrow o) = \downarrow_x e$ bijectively, we also have $|\downarrow o| < \infty$. We have proved the following result.

6.3.3. LEMMA. If B represents $S = (E, F)$, then B is prime-finitary.

■

The final peculiarity of behavioural presentations which represent event structures is to do with labelling. We note that χ is a bijection and that $\lambda(\chi(x)) = x$, so that the sets $\lambda(p)$ are all distinct. We also note that λ must be injective on each p. There is one final property, which is not as easily motivated, that is:

$$\forall p_1, p_2 \in P\ \forall X \subseteq p_1: \lambda(X) = \lambda(p_2) \Rightarrow X = p_2 \tag{6.6}$$

A behavioural presentation whose occurrence function λ has these three properties - namely that the sets $\lambda(p)$ are all distinct, that λ is injective on each p and that (6.6) holds - will be called *W-labelled.*

Let us show that (6.6) holds. Suppose that $p_1, p_2 \in P$ and $X \subseteq p_1$. If $\lambda(X) = \lambda(p_2)$, then $\lambda(p_2) \subseteq \lambda(p_1)$, so $p_2 = \chi(\lambda(p_2)) \subseteq \chi(\lambda(p_1)) = p_1$. We now have $X, p_2 \subseteq p_1$ with $\lambda(X) = \lambda(p_2)$ and since λ is injective on p_1, it follows that $X = p_2$. Thus:

6.3.4. LEMMA. If B represents $S = (E, F)$, then B is W-labelled.

■

6.3.5. DEFINITION. We shall say that B is *Winskellian* $\Leftrightarrow$ B is closed, asynchronous, prime-finitary and W-labelled.

6.3.6. COROLLARY. Suppose that B represents $S = (E, F)$, then B is Winskellian.

■

We now show that the converse of 6.3.6 holds.

6.3.7. THEOREM. B is Winskellian $\Leftrightarrow$ there exists an event structure S_B such that B represents S_B. Furthermore, S_B is full $\Leftrightarrow$ λ is surjective.

PROOF. We have shown necessity. Let us show sufficiency. Suppose B is Winskellian and define $S_B = (E_B, F_B)$ where

1• $E_B = E$;
2• $F_B = \{\lambda(p) \mid p \in P\}$.

We shall prove that S_B is an event structure and that B represents it via a map $\chi_B : F_B \to P$ defined so that $\chi_B(\lambda(p)) = p$. Indeed, by the W-labelling property, the map that sends p to $\lambda(p)$ is injective, and it is surjective, by definition of F_B, so χ_B is well-defined and a bijection.

χ_B also satisfies $\lambda(\chi_B(x)) = x$, for $x \in F_B$ only if $x = \lambda(p)$, some $p \in P$ and since $\chi_B(\lambda(p)) = p$, $\lambda(\chi_B(x)) = \lambda(\chi_B(\lambda(p))) = \lambda(p) = x$.

We show that χ_B is a poset isomorphism. Let $x_1, x_2 \in F_B$, so that $x_1 = \lambda(p_1)$ and $x_2 = \lambda(p_2)$ for $p_1, p_2 \in P$. If $x_1 \subseteq x_2$, then $\lambda(p_1) \subseteq \lambda(p_2)$ and so there exists $X \subseteq p_2$ such that $\lambda(p_1) = \lambda(X)$. But, by the W-labelling property, $X = p_1$ and so $\chi_B(x_1) = \chi_B(\lambda(p_1)) = p_1 = X \subseteq p_2 = \chi_B(\lambda(p_2)) = \chi_B(x_2)$. On the other hand, if $\chi_B(x_1) \subseteq \chi_B(x_2)$, then $x_1 = \lambda(\chi_B(x_1)) \subseteq \lambda(\chi_B(x_2)) = x_2$, by 6.2.3 (3).

6.2.3 (2) follows directly from the definition of E_B. For 6.2.3 (3), $\lambda(\chi_B(x)) = x$, as we have seen, and λ restricted to $\chi_B(x)$ is bijective as a map from $\chi_B(x)$ to x by the W-labelling hypothesis. It remains to show that it is also a poset isomorphism from $\chi_B(x)$ to x.

Suppose that $x \in F_B$ - so that $x = \lambda(p)$, some $p \in P$, and hence $\chi_B(x) = p$ - and suppose $o_1, o_2 \in p$.

First, let $o_1 \to o_2$ and let $y \in F_B$ such that $y \subseteq x$ and $\lambda(o_2) \in y$. If $\tilde{p} = \chi_B(y)$, then $\tilde{p} \subseteq p$, because χ_B is a poset isomorphism, and $o_2 \in \tilde{p}$. Since $o_1 \to o_2$, it follows that $o_1 \in \tilde{p}$ and hence that $\lambda(o_1) \in y$. Thus, $\lambda(o_1) \leq_x \lambda(o_2)$, by 6.2.2.

Next, suppose that $\lambda(o_1) \leq_x \lambda(o_2)$ and let $\tilde{p} \in P$ such that $o_2 \in \tilde{p}$. Let $\hat{p} = p \cap \tilde{p}$, then $\hat{p} \in P$, since B is consistently complete by 6.3.5 and 5.3.9, and $\hat{p} \subseteq p$ and $o_2 \in \hat{p}$. Thus $\lambda(\hat{p}) \subseteq \lambda(p) = x$ and $\lambda(\hat{p}) \in F_B$ and $\lambda(o_2) \in \lambda(\hat{p})$. Since $\lambda(o_1) \leq_x \lambda(o_2)$, it follows that $\lambda(o_1) \in \lambda(\hat{p})$ and hence that there exists $o_1' \in \hat{p}$ such that $\lambda(o_1') = \lambda(o_1)$. But $o_1 \in p$ and $o_1' \in \hat{p} \subseteq p$. Since λ is injective on p it follows that $o_1 = o_1' \in \hat{p} \subseteq \tilde{p}$. Thus $o_1 \to o_2$.

Let us show that S_B is an event structure. 6.2.1 (1) follows immediately from the fact that $(P, \subseteq)$ and $(F_B, \subseteq)$ are isomorphic and $(P, \subseteq)$ is coherent, by 5.3.9 and 6.3.5.

Next, suppose that $\emptyset \subset X \subseteq F_B$. λ is injective on $\chi_B(x)$, all $x \in X$, and so λ is injective on $\bigcap_{x \in X} \chi_B(x)$. Thus,

$$\lambda(\bigcap_{x \in X} \chi_B(x)) = \bigcap_{x \in X} \lambda(\chi_B(x)) = \bigcap_{x \in X} x$$

But, $\bigcap_{x \in X} \chi_B(x) \in P$, since B is consistently complete. Thus $\bigcap_{x \in X} x \in F_B$ and so 6.2.1 (2) holds.

Next we show that 6.2.1 (3) holds. Suppose $x \in F_B$ and $e, e' \in x$ with $e \neq e'$. By construction, there exists $o, o' \in \chi_B(x)$ such that $\lambda(o) = e$ and $\lambda(o') = e'$. Since λ is injective on $\chi_B(x)$, we have $o \neq o'$. Since B is asynchronous, by 6.3.5, we have $\downarrow o \neq \downarrow o'$, so without loss of generality we may suppose that $o \notin \downarrow o'$. If $y = \lambda(\downarrow o')$, then (again using injectivity of λ), $e = \lambda(o) \notin \lambda(\downarrow o') = y$, but $e' = \lambda(o') \in \lambda(\downarrow o') = y$.

To complete the proof that S_B is an event structure, we show that 6.2.1 (4) holds.

Suppose $x \in F_B$ and $e \in x$. Let $y = \downarrow_x e$. Clearly $y \subseteq x$ and $e \in y$. Let $p = \chi_B(x)$ and suppose $o \in p$ such that $\lambda(o) = e$, then $o' \to o \Leftrightarrow \lambda(o') \leq_x e$ and so $\chi_B(y) = \downarrow o$. Since $|\downarrow o| < \infty$, by the prime finitary property, and $|\downarrow o| = |\downarrow_x e|$ via the bijection λ, it follows that $|y| < \infty$.

It follows immediately from the definition of S_B that S_B is full $\Leftrightarrow$ λ is surjective.

■

Notes.

We have selected two models treated by Winskel in his studies of semantics. This is by no means the whole story, but we have not the space to go into any more detail here.

Chapter 7
Properties of Behavioural Presentations

Asynchrony! Concurrency! Determinacy!

F. X. REID
Forgotten Slogans of the French Revolution

7.1. Introduction.

When we come to the business of equipping descriptive notations with our own brand of operational semantics we will sometimes be interested in the type of behavioural presentations which arise in the case in question. For example, in the following chapter, we provide a semantics for Karp and Miller computation graphs and we wish to show that the corresponding behavioural presentations are discrete, asynchronous and determinate. The results in this chapter will facilitate this proof.

The properties that concern us will be *determinacy*, and *asynchrony*, and left-*closure*, and hence *discreteness*. We shall also demonstrate a simple but fundamental property relating to *concurrency* which will loom large when we begin our study of asynchronous transition systems in chapter 15. Finally, we discuss properties which are preserved under closure.

7.2. Asynchrony.

First, we prove an elementary lemma.

7.2.1. LEMMA. If B is a behavioural presentation and $o_1, o_2 \in O$ such that $o_1 \approx o_2$, then for all $p_1, p_2 \in P$, $o_1 \in p_2 - p_1 \Rightarrow o_2 \in p_2 - p_1$.

PROOF. As for all $p \in P$, $o_1 \approx o_2 \Rightarrow (o_1 \in p \Leftrightarrow o_2 \in p)$ by 2.3.4, and 2.3.1 (2).

■

Our next result characterises asynchrony in terms of the following.

$$\forall o \in O: \exists p_1, p_2 \in P: p_1 \succ^{o} p_2 \tag{7.1}$$

7.2.2. PROPOSITION. If B is a behavioural presentation then

1• If B satisfies (7.1), then B is asynchronous;
2• If B is left-closed and asynchronous, then B satisfies (7.1).

PROOF. Suppose that there exists $p_1, p_2 \in P$ such that $p_1 \succ^{o} p_2$, then $p_2 - p_1 = \{o\}$, by 3.3.1 and hence $o \in p_2 - p_1$ so that if $o \approx o'$ then $o' \in p_2 - p_1$, by 7.2.1, and so $o = o'$. Thus B is asynchronous.

Next, suppose that B is left-closed and asynchronous. Let $p_2 = \downarrow o$, then $p_2 \in P$, by 4.3.2. By (3.7), $p_1 = p_2 - o/\!\approx\, \lhd\, p_2$, since clearly $o \in max(\downarrow o)$. Hence $p_1 \in P$, by left closure and $p_1 \succ^{o/\approx} p_2$. But $o/\!\approx\, = \{o\}$, since B is asynchronous.

■

7.3. Left-Closure and Discreteness.

Suppose that B is left closed and suppose that $p \in P$ and that $o \in max(p)$, then $p - o/\!\approx\, \lhd\, p$, by (3.8), and so we conclude that if B is left-closed, then

$$\forall p \in P\, \forall o \in max(p): p - o/\!\approx\, \in P \tag{7.2}$$

We shall show that if B is $\approx$-finitary, then (7.2) implies left-closure. Suppose that B is $\approx$-finitary and that (7.2) holds and let $X \lhd p$, $p \in P$. We argue by induction on $|p/\!\approx|$ that $X \in P$. The case $|p/\!\approx| = 0$ is easy, since then $X \subseteq p = \varnothing$ and so $X = p \in P$. Suppose, then, that $|p/\!\approx| > 0$. If $X = p$, then we are done, so let $X \subset p$. We argue that

$$\exists o \in max(p): X \lhd p - o/\!\approx \tag{7.3}$$

for then, as $p - o/\!\approx\, \in P$, by (7.2) and $|p/\!\approx - o/\!\approx| < |p/\!\approx|$, we may apply the induction hypothesis and conclude that $X \in P$. In order to prove (7.3), it suffices to show that $X \subseteq p - o/\!\approx$, some $o \in max(p)$, for in general it is not hard to show that

$$\forall X \subseteq O\, \forall p_1, p_2 \in P: X \subseteq p_1 \subseteq p_2 \wedge X \lhd p_2 \Rightarrow X \lhd p_1$$

Indeed, 3.3.7 (1) holds by hypothesis and if $o \to o'$ with $o' \in X$ and $o \in p_1$, then $o \to o'$ with $o' \in X$ and $o \in p_2$, so $o \in X$, as $X \lhd p_2$, giving 3.3.7 (2).

Hence, if (7.3) does not hold, then $X \cap o/\!\approx\, \neq \emptyset$, all $o \in max(p)$. In such a case, if $o \in max(p)$ and $o' \in X \cap o/\!\approx$, then $o \in p$, $o' \in X$, $X \lhd p$ and $o \to o'$, so that $o \in X$, by 3.3.7. Thus, if (7.3) does not hold, then $max(p) \subseteq X$. We shall argue in this case that $X = p$, which contradicts our assumption $X \subset p$. Let $o \in p$. If we can always find $o' \in max(p)$ such that $o \to o'$, then we are done, for then we would have $o \in p$, $o' \in X$, $X \lhd p$ and $o \to o'$, so that $o \in X$, by 3.3.7, and consequently $p \subseteq X$. Our proof is then completed by the following:

7.3.1. LEMMA. If B is a $\approx$-finitary behavioural presentation and $p \in P$ and $o \in p$, then there exists $o' \in max(p)$ such that $o \to o'$.

PROOF. Suppose not then the set $X = \{o \in p \mid \forall o' \in max(p) : o \not\to o'\}$ is non-empty, and since $\to$ is reflexive, $X \cap max(p) = \emptyset$, so we may construct a function $f : X \to p$, such that if $o \in X$ then $o < f(o)$. Now if $f(o) \to o' \in max(p)$ then $o \to o' \in max(p)$ contradicting $o \in X$, so $f(o) \in X$, all $o \in X$. Thus we may iteratively construct an ascending chain $o_1 < \cdots < o_n \cdots$ of elements of p. where o_1 is some element of X and for each $n \in \omega$, $o_{n+1} = f(o_n)$. But, $o < o' \Rightarrow o \not\approx o'$, by 2.3.8. Hence the sets $o_n/\!\approx\, \subseteq p$ are distinct, contradicting B, $\approx$-finitary.

■

7.3.2. PROPOSITION. If B is a behavioural presentation, then

1• If B is left-closed, then B satisfies (7.2);
2• If B is $\approx$-finitary and satisfies (7.2), then B is left-closed.

■

7.3.3. COROLLARY. If B is a behavioural presentation, then B is discrete$\Leftrightarrow$ B is $\approx$-finitary and satisfies (7.2).

PROOF. By 3.3.14 and 7.3.2

■

7.4. Determinacy.

In this section, we establish some local conditions for determinacy. These characterisations all depend on the order theoretic structure of $(P, \subseteq)$. Our first result gives a sufficient condition $(P, \subseteq)$. It fails to be necessary precisely when B lacks a property called forward stability, which is essentially a failure to be closed under concurrency, as exemplified by example 5.3.3.

7.4.1. PROPOSITION. If B is a left-closed behavioural presentation such that $(P,\subseteq)$ is a lattice, then B is determinate.

PROOF. Suppose that $(P,\subseteq)$ is a lattice and let $o_1, o_2 \in O$. We must show that $\neg o_1 \# o_2$. As B is left-closed, $\downarrow o_1, \downarrow o_2 \in P$, by 4.3.2, and as $(P,\subseteq)$ is a lattice, $\downarrow o_1 \cup \downarrow o_2 \in P$ by 4.2.2. As $o_1, o_2 \in \downarrow o_1 \cup \downarrow o_2$, $\neg o_1 \# o_2$.

■

It is possible for a behavioural presentation to be determinate without being a lattice, as the following example shows.

$$B = (\{o_1, o_2, o_3\}, \{\emptyset, \{o_1\}, \{o_2\}, \{o_3\}, \{o_1, o_2\}, \{o_1, o_3\}, \{o_2, o_3\}\}, \{a\}, \lambda)$$
$$\lambda(o_1) = \lambda(o_2) = \lambda(o_3) = a$$

where $\{o_1, o_2\} \cup \{o_1, o_3\} \notin P$. What is missing is the following property of being 'closed' with respect to concurrency which we have already met in chapter 5. We formally define the concept, that of forward stability, in 7.4.2.

As will become clear in the sequel, the property of forward stability is enjoyed by the vast majority of behavioural presentations deriving from the semantics of specification notations, so the following results have a general application.

7.4.2. DEFINITION. If B is a behavioural presentation, then B is *forward stable* $\Leftrightarrow$ for all $p, p_1, p_2 \in P$ and $X_1, X_2 \subseteq O$:

$$p \succ^{X_1} p_1 \wedge p \succ^{X_2} p_2 \wedge X_1 \, co \, X_2 \Rightarrow p_1 \cup p_2 \in P$$

where

$$X_1 \, co \, X_2 \Leftrightarrow (\forall o_1 \in X_1 \, \forall o_2 \in X_2 : o_1 \, co \, o_2).$$

We shall demonstrate that in the presence of forward stability, determinism is equivalent to strong forward stability, as defined below.

7.4.3. DEFINITION. If B is a left-closed behavioural presentation, then B is *strongly forward stable* $\Leftrightarrow$ for all $p, p_1, p_2 \in P$ and $X_1, X_2 \subseteq O$

$$p \succ^{X_1} p_1 \wedge p \succ^{X_2} p_2 \Rightarrow p_1 \cup p_2 \in P.$$

7.4.4. PROPOSITION. If B is forward stable and determinate, then B is strongly forward stable.

PROOF. Let $p, p_1, p_2 \in P$ and $X_1, X_2 \subseteq O$ such that $p \succ^{X_1} p_1$ and $p \succ^{X_2} p_2$. As $\approx$ is an equivalence relation, either $X_1 = X_2$, in which case $p_1 \cup p_2 = p_1 \in P$, or

$X_1 \cap X_2 = \emptyset$. Now, if $o_1 \in X_1$ and $o_2 \in X_2$, then we cannot have $o_1 \# o_2$, since this would contradict determinacy, and we cannot have $o_1 \rightarrow o_2$ since then $o_1 \in p_2 = p \cup X_2$, by 2.3.1 (2) and 3.3.1. which entails $o_1 \in p$, as $X_1 \cap X_2 = \emptyset$, so that $p \cap X_1 \neq \emptyset$, contradicting $p \succ^{X_1}$. Likewise $o_2 \not\rightarrow o_1$. Hence $o_1 \, co \, o_2$, by 2.3.6 and so $X_1 \, co \, X_2$, and so $p_1 \cup p_2 \in P$, by forward stability.

■

The following is the key technical result which says that in a strongly forward stable behavioural presentation if there is a chain of steps from a point, then the class of events belonging to any step from that point either is included in one of the steps of chain or may be inserted at the head of it. Pictorially, in the second case, we have:

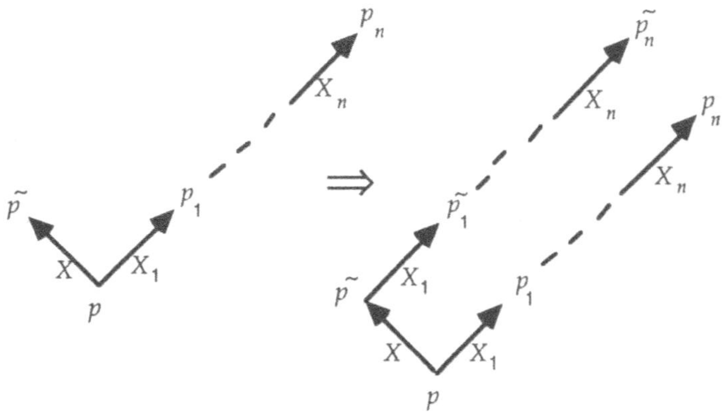

Figure 7.1.

7.4.5. LEMMA. If B is strongly forward stable and $p, \tilde{p}, p_1, \cdots, p_n \in P$ and $X, X_1, \cdots, X_n \subseteq O$ such that $p \succ^X \tilde{p}$ and $p \succ^{X_1} p_1 \succ^{X_2} \cdots \succ^{X_n} p_n$, and $X = X_i$ for no i, then $p \succ^X \tilde{p} \succ^{X_1} \tilde{p}_1 \succ^{X_2} \cdots \succ^{X_n} \tilde{p}_n$, where for each i, $\tilde{p}_i = p_i \cup X \in P$.

PROOF. We argue by induction on n.

If $n = 1$, then $\tilde{p}_1 = p_1 \cup \tilde{p} \in P$ by strong forward stability, $p \succ^X \tilde{p}$, by hypothesis, $\tilde{p} \subseteq \tilde{p}_1$ and, by 3.3.1, $\tilde{p}_1 - \tilde{p} = (p \cup X_1) - (p \cup X) = X_1 - X$. But X and X_1 are $\approx$-equivalence classes such that $X \neq X_1$ and so $X_1 - X = X_1$. Hence, $\tilde{p}_1 - \tilde{p} = X_1$. By 3.3.1, $\tilde{p} \succ^{X_1} \tilde{p}_1$.

If $n > 1$, then by induction, $p \succ^X \tilde{p} \succ^{X_1} \tilde{p}_1 \succ^{X_2} \cdots \succ^{X_{n-1}} \tilde{p}_{n-1}$ and for each $i < n$, $\tilde{p}_i = p_i \cup X \in P$. By definition $p_{n-1} \subseteq \tilde{p}_{n-1}$ and

$$\tilde{p}_{n-1} - p_{n-1} = (p \cup X \cup \bigcup_{i<n} X_i) - (p \cup \bigcup_{i<n} X_i) = (X \cup \bigcup_{i<n} X_i) - \bigcup_{i<n} X_i = X$$

where the latter equality follows from the fact that X and the X_i are $\approx$-equivalence classes such that $X \neq X_i$, each i. Hence $p_{n-1} \succ^{X} \tilde{p}_{n-1}$. But, also, $p_{n-1} \succ^{X_n} p_n$ and so by strong forward stability, $\tilde{p}_n = p_n \cup \tilde{p}_{n-1} \in P$. Again, using that fact that $\tilde{p}_{n-1} \subseteq \tilde{p}_n$ and that X and the X_i are $\approx$-equivalence classes such that $X \neq X_i$, each i, and that consequently $\tilde{p}_n - \tilde{p}_{n-1} = X_n$, we deduce that $\tilde{p}_{n-1} \succ^{X_n} \tilde{p}_n$.

■

7.4.6. PROPOSITION. If B is a left-closed, strongly forward stable behavioural presentation, then $(P,\subseteq)$ is a lattice.

PROOF. Let $p_1, p_2 \in P$. By 4.2.2, $p_1 \sqcap p_2 \in P$.

If $p_1 \subseteq p_2$ or $p_2 \subseteq p_1$, then certainly, $p_1 \sqcup p_2 = p_1 \cup p_2 \in P$, so assume that neither $p_1 \subseteq p_2$ nor $p_2 \subseteq p_1$.

Let $p = p_1 \cap p_2$, then $p \in P$ so by 3.3.3 there exists $p_1^1, \cdots, p_m^1, p_1^2, \cdots, p_n^2, \in P$ and $X_1, \cdots, X_m, Y_1, \cdots, Y_n \subseteq O$ such that $p \succ^{X_1} p_1^1 \succ^{X_2} \cdots \succ^{X_m} p_m^1 = p_1$ and $p \succ^{Y_1} p_1^2 \succ^{Y_2} \cdots \succ^{Y_n} p_n^2 = p_2$.

If $X_i = Y_j$, then $X_i \subseteq p_1 \cap p_2 = p$. But since $p \succ^{X_1} p_1^1 \succ^{X_2} \cdots \succ^{X_m} p_m^1 = p_1$, $X_i \cap p \subseteq X_i \cap p_{i-1}^1 = \varnothing$. Thus $X_i \neq Y_j$, all i, j.

We argue by induction on m that $\hat{p}_m^1 \cup \hat{p}_n^2 \in P$.

If $m = 1$, then apply 7.4.5.

If $m > 1$, then by 7.4.5, $\hat{p} \succ^{\hat{X}_1} \hat{p}_1^1 \succ^{\hat{X}_2} \cdots \succ^{\hat{X}_{m-1}} \hat{p}_{m-1}^1$ and $\hat{p} \succ^{Y_1} \hat{p}_1^2 \succ^{Y_2} \cdots \succ^{Y_n} \hat{p}_n^2$, where $\hat{p} = p \cup X_1$, and for each i, $\hat{p}_i^1 = p_{i+1}^1$, $\tilde{X}_i = X_{i+1}$ and $\hat{p}_i^2 = p_i^2 \cup X_1$. By induction, $\hat{p}_{m-1}^1 \cup \hat{p}_n^2 \in P$, But since $X_1 \subseteq p_m^1$, $\hat{p}_{m-1}^1 \cup \hat{p}_n^2 = p_m^1 \cup p_n^2 \cup X_1$ and $p_m^1 \cup p_n^2 \cup X_1 = p_m^1 \cup p_n^2$.

■

7.4.7. PROPOSITION. If B is a left-closed behavioural presentation, then the following statements are equivalent.

1• $(P,\subseteq)$ is a lattice;
2• B is determinate and forward stable;
3• B is strongly forward stable.

PROOF. (1) $\Rightarrow$ (2): Suppose that $(P,\subseteq)$ is a lattice, then B is determinate, by 7.4.1, and if $p, p_1, p_2 \in P$ and $X_1, X_2 \subseteq O$ such that $p \succ^{X_1} p_1$ and $p \succ^{X_2} p_2$, then $p_1 \cup p_2 \in P$ whether X_1 *co* X_2 or not, so B is forward stable.

(2) $\Rightarrow$ (3): By 7.4.4.

(3) $\Rightarrow$ (1): By 7.4. 6.

■

In the case of closed behavioural presentations, we have a similar result.

7.4.8. PROPOSITION. If B is closed and left-closed, then B is determinate $\Leftrightarrow$ $(P,\subseteq)$ is a complete lattice.

PROOF. The $\Leftarrow$ implication is established in 7.4.1, since a complete lattice is certainly a lattice.

Conversely, suppose that B is closed and determinate. By 2.3.10 (1), $\# = \emptyset$ and hence $\Sigma(B) = (O, \rightarrow, \emptyset)$. Thus, O is conflict-free in $\Sigma(B)$ and since obviously $O \lhd O$, $O \in \Pi(\Sigma(B))$, by 5.2.2, and hence, $O \in \overline{P}$, by 5.3.1. But, B is closed, and so $\overline{P} = P$. Hence, $O \in P$.

Finally, let $\mathbf{X} \subseteq P$. If $\mathbf{X} \neq \emptyset$, then since B is left-closed, $\sqcap \mathbf{X} \in P$, by 4.2.2, (1) and since $\mathbf{X} \lhd O$ and $O \in P$, $\sqcup \mathbf{X} \in P$, by 4.2.2 (2). If $\mathbf{X} = \emptyset$, then $\sqcap X = O \in P$ and $\sqcup X = \emptyset \in P$.

■

We resume our study of determinate behaviour in chapter 11, by which time we will have a representation of behaviour by Mazurkiewicz trace languages. Under certain finiteness conditions, these languages have a characteristically simple structure, which, incidentally, is lost when one takes an interleaving approach to the description of non-sequential behaviour.

7.5. Concurrency.

We begin with a result which, although simple in itself, plays a central rôle in the automata-theoretic part of the theory. Suppose B is a discrete asynchronous behavioural presentation and that $p_1, p_2, p_3 \in P$ and $o_1, o_2 \in O$ such that $p_1 \succ^{o_1} p_2 \succ^{o_2} p_3$ and $o_1 \; co \; o_2$. The derivations seem to impose an order on the occurrences o_1 and o_2 which, on reflection, is quite artificial. We should be able to sequence them in the opposite order, and in that case there would exist $\tilde{p}_2 \in P$ such that $p_1 \succ^{o_2} \tilde{p}_2 \succ^{o_1} p_3$, and by 3.3.1, we would therefore have to have $\tilde{p}_2 = p_3 - \{o_1\}$ or equivalently, $\tilde{p}_2 = p_1 \cup \{o_2\}$. And then $p_1 \subset \tilde{p}_2 \subset p_3$, $\tilde{p}_2 = p_3 - \{o_1\}$ and $p_1 = \tilde{p}_2 - \{o_2\}$, so to complete the proof that $p_1 \succ^{o_2} \tilde{p}_2 \succ^{o_1} p_3$, we need only show that $\tilde{p}_2 \in P$,

For this, we need B to be left-closed, for we may show that $\tilde{p}_2 \lhd p_3$ and then invoke the left-closure hypothesis.

Certainly, $\tilde{p}_2 \subseteq p_3$, giving 3.3.7 (1). To prove 3.3.7 (2), suppose that $o \in p_3$ and $o' \in \tilde{p}_2$ with $o \rightarrow o'$; we must show that $o \in \tilde{p}_2$. Note that since $\tilde{p}_2 = p_1 \cup \{o_2\}$ either $o' \in p_1$ or $o' = o_2$. Since $o_1 \; co \; o_2$, $o_1 \not\rightarrow o_2$ and so if $o' = o_2$, then $o_1 \neq o \neq o_2$ and so $o \in p_3 - \{o_1, o_2\} = p_1$. If $o' \in p_1$, then since $o \to o'$, $o \in p_1$. Thus in both cases, $o \in p_1 \subseteq \tilde{p}_2$, and we are done. To sum up:

7.5.1. PROPOSITION. Suppose B is a left-closed, asynchronous behavioural presentation and that $p_1, p_2, p_3 \in P$ and $o_1, o_2 \in O$ such that $p_1 \succ^{o_1} p_2 \succ^{o_2} p_3$ and $o_1 \; co \; o_2$, then there exists $\tilde{p}_2 \in P$ such that $p_1 \succ^{o_2} \tilde{p}_2 \succ^{o_1} p_3$.

■

The statement of proposition 7.5.1 is illustrated in figure 7.2 , which may explain why we refer to the associated property as the *lozenge property,*

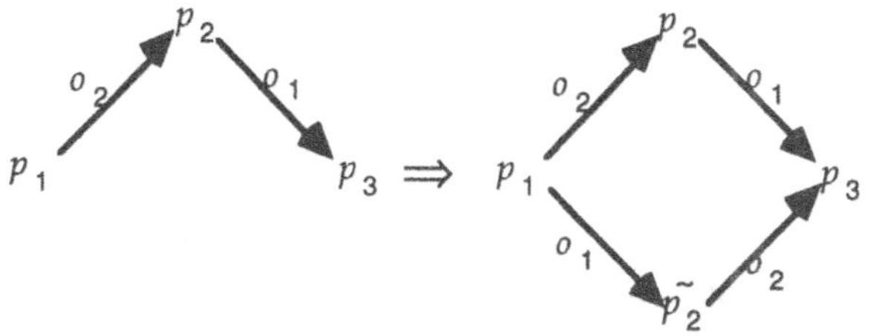

Figure 7.2.

7.5.1 has a sort of converse. First, need a technical lemma.

7.5.2. LEMMA. If B is a left-closed behavioural presentation and $o_1, o_2 \in O$ such that $o_1 \; co \; o_2$ then $\downarrow o_1 \cup \downarrow o_2 \in P$.

PROOF. As B is left-closed $\downarrow o_1, \downarrow o_2 \in P$, by 4.3.2. As $o_1 \; co \; o_2$, then by 2.3.6, $\neg o_1 \# o_2$, and so by 2.3.1 (1), there exists $\bar{p} \in P$ such that $o_1, o_2 \in \bar{p}$. By 3.3.10 (5), $\downarrow o_1, \downarrow o_2 \lhd \bar{p}$ and so $\downarrow o_1, \downarrow o_2 \subseteq \bar{p}$ and so $\downarrow o_1 \cup \downarrow o_2 \in P$, by 4.2.2.

■

7.5.3. PROPOSITION. If B is a left-closed, asynchronous behavioural presentation and $o_1, o_2 \in O$, then

$$o_1 \; co \; o_2 \Leftrightarrow \exists p_1, p_2, \tilde{p}_2, p_3 \in P : p_1 \succ^{o_1} p_2 \succ^{o_2} p_3 \wedge p_1 \succ^{o_2} \tilde{p}_2 \succ^{o_1} p_3.$$

PROOF. ⇒: Suppose that $o_1 \; co \; o_2$ and let $p_3 = \downarrow o_1 \cup \downarrow o_2$, then $p_3 \in P$, by 7.5.2. We argue that $o_2 \in max(p_3)$,

Let $o \in p_3$ such that $o_2 \to o$. By (3.5), we must show that $o \to o_2$. As $p_3 = \downarrow o_1 \cup \downarrow o_2$, either $o \to o_1$ or $o \to o_2$. In the former case, transitivity of $\to$ gives $o_2 \to o_1$, contradicting $o_1 \; co \; o_2$, so $o \to o_2$.

Thus, if $p_2 = p_3 - \{o_2\}$, then by asynchrony of B and (3.7), $p_2 \succ^{o_2} p_3$.

Next, we argue that $o_1 \in max(p_2)$.

Let $o \in p_2$ such that $o_1 \rightarrow o$; we must show that $o \rightarrow o_1$. Since $p_3 = (\downarrow o_1 \cup \downarrow o_2) - \{o_2\}$, either $o \rightarrow o_1$ or $o < o_2$. In the latter case, 2.3.7 and transitivity of $\rightarrow$ gives $o_1 \rightarrow o_2$, contradicting $o_1\ co\ o_2$, so $o \rightarrow o_1$.

Thus, if $p_1 = p_3 - \{o_1, o_2\}$, then by asynchrony of B and (3.8), $p_1 \succ^{o_1} p_2$.

By symmetry, $p_1 \succ^{o_2} \tilde{p}_2 \succ^{o_1} p_3$, where $\tilde{p}_2 = p_3 - \{o_1\}$.

$\Leftarrow$: Suppose that $p_1 \succ^{o_1} p_2 \succ^{o_2} p_3$ and $p_1 \succ^{o_2} \tilde{p}_2 \succ^{o_1} p_3$, then by 3.3.1, we have $o_1 \in p_2$ and $o_2 \notin p_2$ and so $o_1 \not\rightarrow o_2$. Likewise, $o_1 \not\rightarrow o_2$. Finally, by 3.3.1, $o_1, o_2 \in p_3$, and so $\neg o_1 \# o_2$, by 2.3.1 (1). By 2.3.8, $o_1\ co\ o_2$.

■

7.6. Closure.

The only point of this section is to note the following immediate consequence of 5.3.8.

7.6.1. PROPOSITION. Suppose B is a behavioural presentation, then:

1• B is asynchronous $\Leftrightarrow$ $\bar{B}$ is asynchronous;

2• B is determinate $\Leftrightarrow$ $\bar{B}$ is determinate.

■

Chapter 8
Karp and Miller Computation Graphs

> In contrast with conventional digital computer operation, in which only one computation step is initiated at any moment, parallel computation procedures are based on the simultaneous execution of several computation steps.
>
> R. M. KARP and R. E. MILLER [9]

8.1. Introduction.

We illustrate the use of behavioural presentations as a semantic domain with an example. We give a non-interleaving semantics to computation graphs, introduced by R. Karp and R. Miller in 1966 [9] and one of the earliest theories of parallel computation.

The idea behind it is roughly as follows. A computation is to be carried out in parallel by a set of co-operating agents v_i, each of which is capable of performing some specific operation op_i, consuming data at various input ports and placing results on various output ports. Synchronisation is achieved by connecting the agents by FIFO queues[1], linking output ports to input ports.

We therefore assume a finite set V of agents (hereafter, nodes) and a finite set D of FIFO queues (hereafter, directed arcs). We also have a function $inc: D \to V \times V$ with the interpretation that if $inc(d) = (v_i, v_j)$, then the queue d goes from an output port of v_i to an input port of v_j.

The model abstracts away from the details of computation; it only concerns itself with the number of data items resident on any given directed arc at any given time. One component of a computation graph is a function $A: D \to \omega$ with the interpretation that $A(d)$ is the number of items initially on the queue.

---

[1]In a FIFO ('first-in-first-out') queue, items are inserted one at a time and removed in the order in which they are inserted.

A node will only operate if each of its input arcs contains a threshold number of items, given by a function $T: D \to \omega$.

In operating, a node will consume a certain number of items from each input queue and deposit a certain number of items on each output queue. These numbers are given respectively by functions $W: D \to \omega$ ('withdraw') and $U: D \to \omega$ ('update').

There are certain constraints which it is natural to impose. First, given that a node should be capable of execution if it has $T(d)$ items on each input arc d, then the node should not try to consume more than $T(d)$ items from arc d. Hence, we require that $W(d) \leq T(d)$.

Secondly, each node should consume some item from each input queue and deposit some item on each output queue, otherwise the queue is superfluous. Hence, we also require that $U(d) > 0$ and $W(d) > 0$.

This is the static structure; its dynamics, described in terms of sequences of sets of executions of operations, is presented in section 8.2. In section 8.3, we derive a behavioural presentation semantics from the set-sequence semantics and establish that from the latter point of view, computation graphs describe asynchronous, determinate systems. The formal relationship between the two semantics is further investigated in section 8.4. Essentially, we show that the Karp and Miller semantics may be recovered from ours.

8.2. The Model and its Node-Set Sequence Semantics.

8.2.1. DEFINITION. A *Computation Graph* is a septuple $G = (V, D, inc, A, U, W, T)$, where

1• $V = \{v_1, \cdots, v_l\}$ is a finite set of *nodes*;

2• $D = \{d_1, \cdots, d_m\}$ is a finite set of *directed arcs*;

3• $inc: D \to V \times V$ is the *incidence function*;

4• $A, U, W, T: D \to \omega$ satisfy

$$\forall d \in D: W(d) \leq T(d) \wedge U(d) > 0 \wedge W(d) > 0.$$

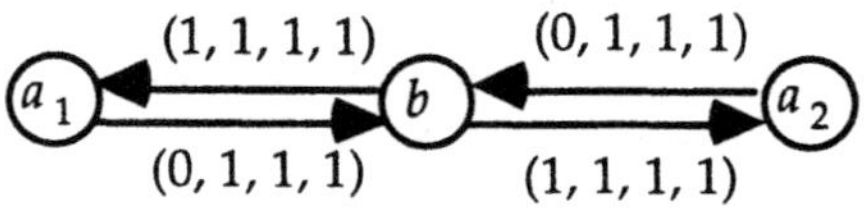

Figure 8.1.

Figure 8.1 is a pictorial representation of a computation graph. Nodes are represented by circles labelled with the name of the node; the quadruple beside arc d is the vector $(A(d), U(d), W(d), T(d))$.

Thus, in figure 8.1, each operation requires and will take exactly one item from its input queue and on execution will place exactly one item on each of its output queues. Initially, there is exactly one item on the input queues to the nodes with operations a_1 and a_2.

The model abstracts away from manipulation of data objects. Instead, an 'instantaneous' state of the system is represented by an allocation of numbers to each arc, giving the numbers of items currently residing in that FIFO queue.

Karp and Miller describe the dynamics of computation graphs in terms of sequences of subsets of the set $\{1,\cdots,l\}$, where $V=\{v_1,\cdots,v_l\}$. The significance of such execution sequences is made clear by the following quotation from [9].

> 'The sequence $S_1, S_2, \cdots, S_N, \cdots$ is interpreted as a possible temporal sequence of initiations of operations throughout the performance of the parallel computation specified by G; the occurrence of S_N denotes the simultaneous initiation of op_j for all $j \in S_N$.'

Karp and Miller model parallel execution explicitly and the use of the word 'simultaneous' and the use of sequences suggests a synchronous interpretation. We shall actually give an asynchronous interpretation.

Before we present the definition of execution sequences, let us tidy up the notation a little. From now on, we identify operations with their associated nodes and take subsets of V rather than of $\{1,\cdots,l\}$ as the elements of our sequences.

Let $\Sigma_G = \wp(V)-\{\emptyset\}$, so that executions will be elements of Σ_G^{∞}, the set of all finite and infinite sequences of elements of Σ_G. If $\mathbf{S}\in\Sigma_G^{\infty}$, then as usual, we denote the length of $\mathbf{S}$ by $|\mathbf{S}|$. If $1\le i\le|\mathbf{S}|$, then we denote the ith element in the sequence $\mathbf{S}$ by $\mathbf{S}(i)$.

If $v\in V$ and $n\in\omega$, then we define

$$\#(\mathbf{S},v,n)=\left|\{i\in\{1,\cdots,n\}\mid v\in\mathbf{S}(i)\}\right|$$

Thus, $\#(\mathbf{S},v,n)$ is the number of sets in the sequence $\mathbf{S}(1)\cdots\mathbf{S}(n)$ which contain v. By convention, $\#(\mathbf{S},v,0)=0$.

We shall also find it convenient to adopt the following notation from Net Theory . If $v\in V$, then define

$$\bullet v=\{d\in D\mid \exists v'\in V: inc(d)=(v',v)\}$$

and

$$v\bullet=\{d\in D\mid \exists v'\in V: inc(d)=(v,v')\}$$

and for $d \in D$, if $inc(d) = (v, v')$, then we define $\bullet d = v$ and $d\bullet = v'$.

We also extend this dot notation to sets so that, for example, if $X \subseteq V$ then $X\bullet = \{d \in D \mid \exists v \in X: d \in v\bullet\}$.

8.2.2. DEFINITION. Suppose that $\mathbf{S} \in \Sigma_G^{\infty}$, $d \in D$ and $i \in \omega$, then

$$\chi(\mathbf{S}, d, i) = A(d) + U(d).\#(\mathbf{S}, \bullet d, i) + W(d).\#(\mathbf{S}, d\bullet, i)$$

1• **S** is a *partial execution* of the computation graph $G \Leftrightarrow$

$$\forall i < |\mathbf{S}|\ \forall d \in \bullet\mathbf{S}(i+1): \chi(\mathbf{S}, d, i) \geq T(d).$$

2• **S** is an execution of $G \Leftrightarrow$ **S** is a partial execution of G and

$$|\mathbf{S}| < \infty \Rightarrow \forall v \in V\ \exists d \in \bullet v: \chi(\mathbf{S}, d, |\mathbf{S}|) < T(d)$$

We shall denote the set of executions of G by $Ex(G)$ and the set of partial executions of G by $Px(G)$.

The reader may check that in the example of figure 8.1,

$$\{a_1, a_2\}\{b\}\{a_2\}\{a_1\}\{b\}\{a_1\} \in Px(G).$$

In explanation of these definitions, $\chi(\mathbf{S}, d, i)$ counts the number of items residing on queue d following the partial execution $\mathbf{S}(1) \cdots \mathbf{S}(i)$, as we shall see, and so (1) merely states that v may be in $\mathbf{S}(i+1)$ if each of its input queues has at least the threshold number of items on it. (2) may now be read as saying that an execution may only be finite if it leaves an insufficient number of items on at least one of the input queues of each node. Formally, we have the following immediate consequence of 8.2.2.

8.2.3. LEMMA. Let G be a computation graph and let $X \in \Sigma_G$ and $\mathbf{S} \in Px(G) \cap \Sigma_G^*$ then

$$\mathbf{S}.X \in Px(G) \Leftrightarrow \forall d \in \bullet X: \chi(\mathbf{S}, d, |\mathbf{S}|) \geq T(d).$$

■

8.3. Behavioural Presentation Semantics.

Let G be a computation graph. We shall define two behavioural presentations, B_G^{fin} and B_G, which is defined to be $\overline{B_G^{fin}}$. B_G is the 'meaning' of G in our semantics.

O_G will be a certain subset of $\tilde{O}_G = V \times \omega$ and $E_G = V$ with λ_G defined by $\lambda_G(v,i) = v$. Intuitively, (v,i) is the ith occurrence of v. It remains to define P_G^{fin}, for then, in order that $B_G^{fin} = (O_G, P_G^{fin}, V, \lambda_G)$ be a behavioural presentation, we are *obliged* to define $O_G = \bigcup_{p \in P_G^{fin}} p$.

P_G^{fin} is defined using a trinary relation on $\wp(\tilde{O}_G)$ denoted by $\triangleright_G$. If (p, X, p') is in this relation, then we shall write $p \triangleright_G^X p'$. The similarity with the notation for a derivation, given in 3.3.1, is not coincidental, for as we shall show, if $o \in O_G$, then $p \succ_G^o p' \Leftrightarrow p \triangleright_G^o p'$.[2] P_G^{fin} is the smallest subset $\wp(\tilde{O}_G)$ containing $\varnothing$ and closed under $\triangleright_G$.

It therefore remains to define the relation $\triangleright_G$. We first need some notation.

8.3.1. DEFINITION. If $q \subseteq \tilde{O}_G$, $d \in D$ and $v \in V$, then we define

1• $\#(q,v) = |\{i \in \omega \mid (v,i) \in q\}|$;

2• $\chi(q,d) = A(d) + U(d).\#(q,\bullet d) - W(d).\#(q,d\bullet)$.

Compare these definitions with 8.2.2. $\#(q,v)$ counts the number of occurrences of v in a behaviour q, while $\chi(q,d)$ counts the number of items left on queue d by a behaviour q, We may use this function to determine which operations may occur given that q has happened so far. These are the elements of the set *next*(q), which we now define.

8.3.2. DEFINITION. If $q \subseteq \tilde{O}_G$, then we define *next*(q) to be the set of all $(v,i) \in \tilde{O}_G$ such that:

1• $\forall d \in \bullet v : \chi(q,d) \geq T(d)$;

2• $i = \#(q,v) + 1$.

We may now define our derivation relation.

8.3.3. DEFINITION. If $q, q', X \subseteq \tilde{O}_G$, then we define $q \triangleright_G^X q' \Leftrightarrow$

1• $\varnothing \subset X \subseteq next(q)$;

2• $q' = q \cup X$.

---

[2]See (8.5) and (8.7).

If $o \in \tilde{O}_G$, then we write $p \triangleright_G^{o} p'$ instead of $p \triangleright_G^{\{o\}} p'$.

Let $\Theta_G = \wp(\tilde{O}_G) - \{\varnothing\}$. If $\mathbf{X} \in \Theta_G^*$, then we shall write $\mathbf{X}(i)$ for the ith element in the sequence $\mathbf{X}$. If $\mathbf{X} \neq \Omega$, then we define

$$frt(\mathbf{X}) = \mathbf{X}(1) \cdots \mathbf{X}(|\mathbf{X}| - 1)$$
$$end(\mathbf{X}) = \mathbf{X}(|\mathbf{X}|)$$

If $\mathbf{X} \in \Theta_G^*$, then we define

$$p_{\mathbf{X}} = \bigcup_{i=1}^{|\mathbf{X}|} \mathbf{X}(i) \tag{8.1}$$

8.3.4. DEFINITION. If $\mathbf{X} \in \Theta_G^*$, then we shall call $\mathbf{X}$ an *establishing sequence,* if and only if either $\mathbf{X} = \Omega$ or $frt(\mathbf{X})$ is an establishing sequence and $end(\mathbf{X}) \subseteq next(p_{frt(\mathbf{X})})$. We denote the set of all establishing sequences by $Est(G)$.

If $\mathbf{X} \in Est(G)$ and $p \subseteq \tilde{O}_G$, then we shall say that $\mathbf{X}$ *establishes* p, and write $\mathbf{X}\ est\ p$, if and only if $p = p_{\mathbf{X}}$.

Let us gather together some properties of establishing sequences. As an immediate consequence of 8.3.4, we have

$$\mathbf{X} \in Est(G) - \{\Omega\} \Rightarrow frt(\mathbf{X}) \in Est(G) \wedge p_{frt(\mathbf{X})} \triangleright_G^{end(\mathbf{X})} p_{\mathbf{X}} \tag{8.2}$$

In particular, $\mathbf{X} \in Est(G) \Rightarrow \mathbf{X} = \Omega \vee frt(\mathbf{X}) \in Est(G)$ and by an easy induction argument, we obtain

$$\mathbf{X} \in Est(G) \wedge \mathbf{Y} \leq \mathbf{X} \Rightarrow \mathbf{Y} \in Est(G) \tag{8.3}$$

Suppose that $\mathbf{X} \in Est(G)$ and let $\mathbf{X}' = frt(\mathbf{X})$, $X = end(\mathbf{X})$ and $\varnothing \subset Y \subset X$. We shall show that $\mathbf{X}'.(X-Y).Y \in Est(G)$. First, $\varnothing \subset X - Y \subseteq X \subseteq next(p_{\mathbf{X}'})$ and so $\mathbf{X}'.(X-Y) \in Est(G)$.

Suppose next that $(v,k) \in Y$. We claim that $(v,k) \in next(p_{\mathbf{X}'.(X-Y)})$. We know that $(v,k) \in next(p_{\mathbf{X}'})$ so, by 8.3.2, $k = \#(p_{\mathbf{X}'}, v) + 1$ and if $d \in \bullet v$, then $\chi(p_{\mathbf{X}'}, d) \geq T(d)$. We must show that $k = \#(p_{\mathbf{X}'.(Y-X)}, v) + 1$ and if $d \in \bullet v$, then $\chi(p_{\mathbf{X}'.(Y-X)}, d) \geq T(d)$

But $\#(p_{\mathbf{X}'.(Y-X)}, v) = \#(p_{\mathbf{X}'}, v) + \#(p_{(Y-X)}, v)$, from 8.3.1 (1), and if $d \in \bullet v$, then

$$\chi(p_{\mathbf{X}'.(Y-X)}, d) = \chi(p_{\mathbf{X}'}, d) + U(d).\#(p_{(Y-X)}, \bullet d) - W(d).\#(p_{(Y-X)}, v)$$

by 8.3.1 (2), since $d\bullet = v$. Our claim will then follow if we can show that $\#(X-Y, v) = 0$. But if $(v,l) \in X - Y$, then also, $(v,l) \in next(p_{\mathbf{X}'})$, and so

$l = \#(q,v)+1 = k$,8.3.2 (2) but then we have $(v,k) \in Y \cap (end(\mathbf{X}) - Y)$, which is impossible. Thus,

$$\mathbf{X} \in Est(G) \wedge \varnothing \subset Y \subset end(\mathbf{X}) \Rightarrow frt(\mathbf{X}).(end(\mathbf{X}) - Y).Y \in Est(G) \qquad (8.4)$$

We proceed to the formal definitions of B_G^{fin} and B_G.

8.3.5. DEFINITION. If G is a computation graph, then we define $B_G^{fin} = (O_G, P_G^{fin}, V, \lambda_G)$, where:

1• $P_G^{fin} = \{p_{\mathbf{X}} \mid \mathbf{X} \in Est(G)\}$;

2• $O_G = \bigcup_{p \in P_G^{fin}} p$;

3• For all $(v,n) \in O_G$, $\lambda_G(v,n) = v$.

We also define $B_G = \overline{B_G^{fin}}$.

8.3.5 (2) guarantees that B_G^{fin} and hence B_G is a behavioural presentation. Let us see what kind of behavioural presentation they are.

We obtain one property easily.

8.3.6. PROPOSITION. If G is a computation graph, then for all $p \in P_G^{fin}$, p is finite, and in particular, B_G^{fin} is ≈--finitary.

PROOF. If $p \in P_G^{fin}$, then by 8.3.5, $p = p_{\mathbf{X}}$, some $\mathbf{X} \in Est(G)$. Now $\mathbf{X}(1)\cdots\mathbf{X}(i) \in Est(G)$, by (8.3), so $\mathbf{X}(i) \subseteq next(\mathbf{X}(1) \cup \cdots \cup \mathbf{X}(i-1))$, by 8.3.4. Thus, if $(v,m),(v,n) \in \mathbf{X}(i)$, then by 8.3.2, $m = \#(\mathbf{X}(1) \cup \cdots \cup \mathbf{X}(i-1)) + 1 = n$. Therefore, $|\mathbf{X}(i)| \leq |V|$, each i, and since V is finite, so is p.

■

Our next result characterises the elements of a typical point and has as a corollary, the assertion that the sets in an establishing sequence must be pairwise disjoint.

8.3.7. LEMMA. If G is a computation graph and $p \in P_G^{fin}$ then $(v,i) \in p \Leftrightarrow 1 \leq i \leq \#(p,v)$.

PROOF. First, let $(v,i) \in p$ and let $\mathbf{X}\ est\ p$. We argue by induction on $n = |\mathbf{X}|$ that $1 \leq i \leq \#(p,v)$. If $n = 1$, then $p \subseteq next(\varnothing)$ so $i = \#(\varnothing, v) + 1 = 1 \leq \#(p,v)$, by 8.3.2, establishing the base case. Suppose next that $n > 1$, then either $(v,i) \in end(\mathbf{X})$ or $(v,i) \in p_{frt(\mathbf{X})}$. If $(v,i) \in end(\mathbf{X})$, then $(v,i) \in next(p_{frt(\mathbf{X})})$ so $i = \#(p_{frt(\mathbf{X})}, v) + 1 = \#(p_{frt(\mathbf{X})}, v) + \#(end(\mathbf{X}), v) = \#(p,v)$, by 8.3.2. If $(v,i) \in p_{frt(\mathbf{X})}$, then by induction, $i \leq \#(p_{frt(\mathbf{X})}, v) \leq \#(p,v)$ and so $1 \leq i \leq \#(p,v)$ in both cases.

Conversely, suppose that $1 \le i \le \#(p,v)$. We argue by induction on $n = |\mathbf{X}|$ that $(v,i) \in p$. If $n = 1$, then $p \subseteq next(\emptyset)$ and so $i = \#(p,v) = 1$ and $(v,1) \in p$. Suppose next that $n > 1$. If $1 \le i \le \#(p_{frt(\mathbf{X})}, v)$, then by induction, $(v,i) \in p_{frt(\mathbf{X})} \subseteq p$. Otherwise, $i = \#(p,v) > \#(p_{frt(\mathbf{X})}, v)$, so for some j, $(v,j) \in end(\mathbf{X})$. Since, $\#(end(\mathbf{X}),v) = 1$, $\#(p,v) = \#(p_{frt(\mathbf{X})}, v) + 1$, and so by 8.3.2 and 8.3.3, $i = \#(p,v) = j$.

■

8.3.8. COROLLARY. If G is a computation graph and then if $\mathbf{X} \in Est(G)$ then the $\mathbf{X}(i)$ are disjoint. Consequently, $p \cap next(p) = \emptyset$, all $p \in P_G^{fin}$.

PROOF. Suppose, that $(v,m) \in \mathbf{X}(i) \cap X(j)$, where $i < j$. By 8.3.7, $m \le \#(\mathbf{X}(1) \cup \cdots \cup \mathbf{X}(i), v) \le \#(\mathbf{X}(1) \cup \cdots \cup \mathbf{X}(j-1), v)$. By 8.3.2 (2), $m = \#(\mathbf{X}(1) \cup \cdots \cup \mathbf{X}(j-1), v) + 1$, a contradiction.

Let $p \in P_G^{fin}$ and let $\mathbf{X}\ est\ p$. If $next(p) = \emptyset$, then $p \cap next(p) = \emptyset$. Otherwise, by 8.3.4, $\mathbf{X}.next(p) \in Est(G)$ and so $\mathbf{X}(i) \cap next(p) = \emptyset$, all i, Thus, $p \cap next(p) = \emptyset$.

■

8.3.9. COROLLARY. If G is a computation graph and $p \in P_G^{fin}$, then if $\mathbf{X}\ est\ p$ and $\emptyset \subset Y \subseteq end(\mathbf{X})$, then $p - Y \in P_G^{fin}$ and $p - Y \triangleright_G^Y p$.

PROOF. By (8.4), there exists $\mathbf{X}' \in Est(G)$ such that $\mathbf{X}'\ est\ p$ and $Y = end(\mathbf{X}')$. By (8.3), $p_{frt(\mathbf{X}')} \triangleright_G^Y p$. By 8.3.8, $p_{frt(\mathbf{X}')} = p - Y$.

■

We now set about proving that the behavioural presentations B_G^{fin} are discrete, asynchronous and strongly forward stable.

First, we look at asynchrony. By 7.2.1, it suffices to show that if $o \in O_G$, then there exists $p_1, p_2 \in P_G^{fin}$ such that $p_1 \subseteq p_2$ and $p_2 - p_1 = \{o\}$. This is an immediate consequence of the following.

8.3.10 LEMMA. If G is a computation graph, then for all $o \in O_G$, there exists $p_1, p_2 \in P_G^{fin}$ such that $p_1 \triangleright_G^o p_2$.

PROOF. By 8.3.3 (2), there exists $p \in P_G^{fin}$ such that $o \in p$. Let $\mathbf{X}\ est\ p$ so that $o \in \mathbf{X}(i)$, some i. by (8.1) and 8.3.4. By (8.3) $\mathbf{Y} = \mathbf{X}(1) \cdots \mathbf{X}(i) \in Est(G)$ and so $p_2 = p_{\mathbf{Y}} \in P_G^{fin}$. By 8.3.9, $p_1 = p_2 - \{o\} \in P_G^{fin}$ and $p_1 \triangleright_G^o p_2$.

■

In view of the remarks preceding 8.3.10, we have:

8.3.11. COROLLARY. If G is a computation graph, then G is asynchronous. ■

We note that if $p_1 \rhd_G^o p_2$, then by 8.3.3, (3), $p_1 \subseteq p_2$ - so that 3.3.1 (2) holds - and $p_2 - \{o\} = (p_1 \cup \{o\}) - \{o\} = p_1 - \{o\}$, so that in order to establish 3.3.1 (3), namely that $p_2 - \{o\} = p_1$, it suffices to show that if $p_1 \rhd_G^o p_2$ then $o \notin p_1$; equivalently, that $p_1 \cap next(p_1) = \varnothing$. But, this is the contents of 8.3.8. Thus

$$\forall o \in O_G\ \forall p_1, p_2 \in P_G^{fin} : p_1 \rhd_G^o p_2 \Rightarrow p_1 \succ_G^o p_2 \tag{8.5}$$

Turning to left-closure, as B_G^{fin} is $\approx$-finitary, by 8.3.6, 7.3.4 applies. In view of this and 8.3.9, in order to guarantee left-closure, it suffices to show that,

$$\forall p \in P_G^{fin}\ \exists \mathbf{X} \in Est(G) : \mathbf{X}\ est\ p \wedge max(p) \subseteq end(\mathbf{X}) \tag{8.6}$$

for by 8.3.9 and 7.3.4 , we then have

$$o \in max(p) \Rightarrow o/\!\approx\, \subseteq max(p) \Rightarrow o/\!\approx\, \subseteq end(\mathbf{X}) \Rightarrow p - o/\!\approx\, \in P_G^{fin}.$$

We shall prove (8.6) in due course. As we shall now argue, (8.6) will also establish determinacy. We shall in fact establish conditions for strong forward stability and appeal to 7.4.7. So assume that $o_1, o_2 \in O_G$ and $p, p_1, p_2 \in P_G^{fin}$ such that $p \succ_G^{o_1} p_1$ and $p \succ_G^{o_2} p_2$. We must show that $p_1 \cup p_2 \in P_G^{fin}$ and in order to do this, it suffices to show that $p \rhd_G^{\{o_1, o_2\}} p_1 \cup p_2$, by 8.3.3. But this latter follows immediately from $p \rhd_G^{o_1} p_1$ and $p \rhd_G^{o_2} p_2$. Thus, in order to guarantee strong forward stability of B_G^{fin}, it suffices to demonstrate the following condition

$$\forall o \in O_G\ \forall p_1, p_2 \in P_G^{fin} : p_1 \succ_G^o p_2 \Rightarrow p_1 \rhd_G^o p_2 \tag{8.7}$$

(8.7) is actually a consequence of (8.6).

8.3.12. LEMMA. If G is a computation graph, then if B_G^{fin} satisfies (8.6) then B_G^{fin} satisfies (8.7).

PROOF. Suppose that B_G^{fin} satisfies (8.6) and suppose that $o \in O_G$ and $p_1, p_2 \in P_G^{fin}$ such that $p_1 \succ_G^o p_2$. Then $o \in max(p_2)$ and by (8.6), there exists $\mathbf{X} \in Est(G)$ such that $\mathbf{X}\ est\ p_2$ and $max(p) \subseteq end(\mathbf{X})$. By 8.3.9, $p_2 - \{o\} \rhd_G^o p_2$. But, by 3.3.1 (2), $p_2 - \{o\} = p_1$. Hence, $p_1 \rhd_G^o p_2$. ■

It remains to show that (8.6) holds for every computation graph G. In order to do this, we need to associate the order structure of P_G^{fin} to its derivation structure. We shall prove the following:

$$\mathbf{X}\ est\ p \wedge o \in max(p) \Rightarrow \exists \mathbf{X}' \in Est(G): \mathbf{X}'\ est\ p \wedge end(\mathbf{X}') = end(\mathbf{X}) \cup \{o\} \quad (8.8)$$

(8.7) follows from (8.8) by induction on $|max(p)|$, which we know to be finite, by 8.3.6. The proof of (8.8) is based on the following construction, which slides o up a member of the sequence $\mathbf{X}$, as pictured in figure 8.2 and defined in 8.3.13. Effectively, the construction involves removing o from $\mathbf{X}(i)$ and placing it in $\mathbf{X}(i+1)$. There are two cases to consider depending on whether o is the only element of $\mathbf{X}(i)$ or not.

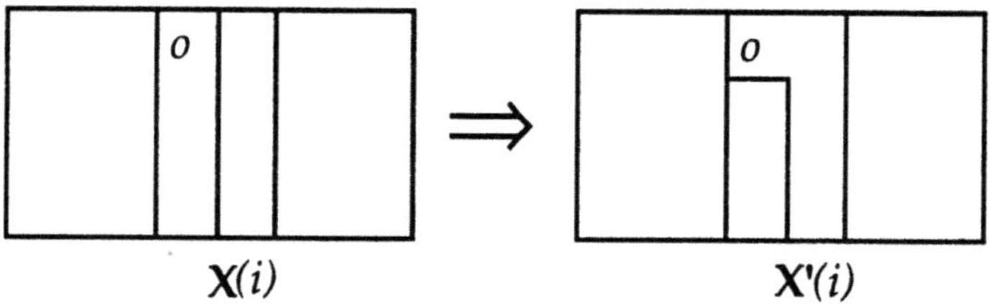

Figure 8.2

8.3.13. DEFINITION. If $\mathbf{X} \in Est(G)$ and $o \in p_{\mathbf{X}}$, then we define $\mathbf{X}[o \rightarrow] = \mathbf{X}'$, where if $n = |\mathbf{X}|$ and $o \in \mathbf{X}(i)$, then

$$\mathbf{X}' = \begin{cases} \mathbf{X} \text{ if } i = n \\ \mathbf{X}(1) \cdots \mathbf{X}(i-1).(\mathbf{X}(i+1) \cup \{o\}) \cdots \mathbf{X}(n) \text{ if } i < n \wedge \mathbf{X}(i) = \{o\} \\ (\mathbf{X}(1) \cdots (\mathbf{X}(i) - \{o\}).(\mathbf{X}(i+1) \cup \{o\}) \cdots \mathbf{X}(n) \text{ otherwise} \end{cases}$$

(8.8) now follows by an easy induction on $|\mathbf{X}|$ from the following

$$\mathbf{X} \in Est(G) \wedge o \in max(p_{\mathbf{X}}) \Rightarrow \mathbf{X}[o \rightarrow] \in Est(G) \quad (8.9)$$

Before we embark on a proof of (8.9), we need to establish some technical results. We begin with an easy consequence of 8.3.7

8.3.14. LEMMA. Suppose that G is a computation graph and $v \in V$ and $(v,l),(v,m) \in O_G$, then $l \leq m \Rightarrow (v,l) \rightarrow (v,m)$.

PROOF. Let $p \in P_G^{fin}$ such that $(v,m) \in p$, then by 8.3.7, $m \leq \#(p,v)$ so $l \leq \#(p,v)$. By 8.3.7, $(v,l) \in p$. By 2.3.1 (2), $(v,l) \rightarrow (v,m)$.

■

8.3.15. LEMMA. Let G be a computation graph and suppose that $p, \hat{p}, \tilde{p} \in P_G^{fin}$ and $o, \hat{o} \in O_G$ such that $p \triangleright_G^{o} \hat{p} \triangleright_G^{\hat{o}} \tilde{p}$ but $\hat{o} \notin next(p)$, then $o \to \hat{o}$.

PROOF. Let $o=(v,k)$ and $\hat{o}=(w,l)$. $p \triangleright_G^{o} \hat{p} \triangleright_G^{\hat{o}} \tilde{p}$, so by 8.3.2 and 8.3.3, $k=\#(p,v)+1$ and $l=\#(\hat{p},w)+1$. If $v=w$, then $l=k+1$, and so $o \to \hat{o}$, by 8.3.14.

So suppose that $v \neq w$. In this case, $\#(\hat{p},w)=\#(p,w)$. If $\chi(p,d) \geq T(d)$, all $d \in v\bullet \cap \bullet w$, then $\hat{o} \in next(p)$, a contradiction, so there exists $d_0 \in v\bullet \cap \bullet w$ such that $\chi(p,d_0) < T(d_0)$. We therefore have:

$$\begin{aligned} A(d_0)&+U(d_0).(k-1)-W(d_0).(l-1)\\ &= A(d_0)+U(d_0).\#(p,v)-W(d_0).\#(p,w)\\ &= A(d_0)+U(d_0).\#(p,\bullet d_0)-W(d_0).\#(p,d_0\bullet)\\ &= \chi(p,d_0) < T(d_0) \end{aligned} \tag{8.10}$$

Now, let $\overline{p} \in P_G^{fin}$ such that $\hat{o} \in \overline{p}$, and suppose that $\mathbf{X}\ est\ \overline{p}$, then $\hat{o} \in \mathbf{X}(i)$, some i, Hence, $\hat{o} \in next(\breve{p})$, where $\breve{p} = \mathbf{X}(1) \cup \cdots \cup \mathbf{X}(i-1)$ and

$$\begin{aligned} A(d_0)&+U(d_0).\#(\breve{p},v)-W(d_0).(l-1)\\ &= A(d_0)+U(d_0).\#(\breve{p},\bullet d_0)-W(d_0).\#(\breve{p},d_0\bullet)\\ &= \chi(\breve{p},d_0) \geq T(d_0) \end{aligned} \tag{8.11}$$

Comparing (8.10) and (8.11), we see that $\#(\breve{p},v) \geq k$, and so $o=(v,k) \in \breve{p} \subseteq \overline{p}$ by 8.3.7. We have proved that $\hat{o} \in \overline{p} \Rightarrow o \in \overline{p}$, that is $o \to \hat{o}$.

■

We prove (8.9). It certainly holds if $o \in end(\mathbf{X})$, so suppose that $o \in \mathbf{X}(i)$, $i<n$. To complete the proof, it suffices to show that $\mathbf{X}(i+1) \cup \{o\} \subseteq next(p_i - \{o\})$, where $p_j = \mathbf{X}(1) \cup \cdots \cup \mathbf{X}(j)$, each $j \leq n$. As $o \in \mathbf{X}(i)$, $o \in next(p_{i-1})$, and as $\chi(p_{i-1},d) \leq \chi(p_i - \{o\},d)$, each $d \in \bullet\lambda_G(o)$, $o \in next(p_i - \{o\})$. As $\mathbf{X}(1)\cdots\mathbf{X}(i)\ est\ p_i$, by (8.3), $p_i - \{o\} \in P_G^{fin}$ and $p_i - \{o\} \triangleright_G^{o} p_i$, by 8.3.9.

Also $\mathbf{X}(i+1) \subseteq next(p_i)$, by 8.3.5, and thus if $\mathbf{X}(i+1) \cup \{o\} \not\subseteq next(p_i - \{o\})$, then there exists $\hat{o} \in \mathbf{X}(i+1)$ such that $\hat{o} \notin next(p_i - \{o\})$. But then by 8.3.3, $p_i \triangleright_G^{\hat{o}} p_i \cup \{\hat{o}\}$, as $\hat{o} \in \mathbf{X}(i+1) \subseteq next(p_i)$. By 8.3.5, $p_i \cup \{(w,l)\} \in P_G^{fin}$ Thus, we now have $p_i - \{o\} \triangleright_G^{o} p_i \triangleright_G^{\hat{o}} p_i \cup \{\hat{o}\}$ with $\hat{o} \notin next(p_i - \{o\})$ and so by 8.3.15, $o \to \hat{o}$. Since $o \in max(p)$, $o \approx \hat{o}$, and as B_G^{fin} is asynchronous, by 8.3.11, $o = \hat{o}$. But by hypothesis, $o \in \mathbf{X}(i)$ and $\hat{o} \in \mathbf{X}(i+1)$, and we now have a contradiction to 8.3.8.

We finally reach the main result of this section.

8.3.16. THEOREM. If G is a computation graph, then

1• B_G^{fin} and B_G are asynchronous;
2• B_G^{fin} is discrete;
3• B_G^{fin} and B_G are strongly forward stable and hence determinate.

PROOF. B_G^{fin} is asynchronous, by 8.3.11. Thus B_G is asynchronous, by 7.6.1 (1).

We have shown that B_G^{fin} satisfies (8.9) from which (8.8) and accordingly (8.6) follow and we have observed that this is enough to guarantee that B_G^{fin} is discrete.

We have just seen that B_G^{fin} also satisfies (8.6) and so by 8.3.12, B_G^{fin} satisfies (8.7), and we have observed that this is enough to guarantee that B_G^{fin} is strongly forward stable. By 7.4.7, B_G^{fin} is determinate. By 7.6.1, (2) B_G is determinate.

■

8.4. Comparing the Semantics.

We now have two ways of ascribing a meaning to a computation graph G, that which associates G with the set of sequences $Px(G)$ and that which associates G with the closed behavioural presentation B_G. We maintain that the behavioural presentation semantics is equivalent to the $Px(G)$ semantics in the sense they distinguish between computation graphs in exactly the same way, that is, if $V_1 = V_2$, then

$$Px(G_1) = Px(G_2) \Leftrightarrow B_{G_1} = B_{G_2}$$

Our proof will involve the construction, given in 8.4.3, of a set of sequences from a given behavioural presentation. The behavioural presentations in question will be required to possess the following properties, which, as we shall show, are possessed by the behavioural presentations B_G^{fin}.

8.4.1. DEFINITION. If B is a behavioural presentation, then B is *plain* $\Leftrightarrow$

1• B is asynchronous and discrete;
2• $\forall o_1, o_2 \in O: o_1 \ co\ o_2 \Rightarrow \lambda(o_1) \neq \lambda(o_2)$.

8.4.2. LEMMA. If G is a computation graph, then B_G^{fin} is plain.

PROOF. By 8.3.16, B_G^{fin} is asynchronous and discrete.

Suppose $(v,m),(w,n) \in O_G$ such that $\lambda_G(v,m) = \lambda_G(w,n)$, then $v = w$, by 8.3.5 (3), and so by 8.3.14 either $(v,m) \rightarrow (w,n)$ or $(w,n) \rightarrow (v,m)$ and so we cannot have $(v,m)\ co\ (w,n)$, by 2.3.6.

■

8.4.3. DEFINITION. Suppose that B is a plain behavioural presentation and $X \subseteq O$. We define $\mathbf{S}_B(X) \subseteq \wp(E)^*$ as follows:

1• If $X = \varnothing$ then $\mathbf{S}_B(X) = \Omega$

2• If $X \neq \varnothing$, then $\mathbf{S}_B(X) = \bigcup_{\varnothing \subset U \subseteq max(X)} \mathbf{S}_B(X-U).\lambda(U)$.

We now embark on a proof that if G is a computation graph and $B = B_G^{fin}$, then $\mathbf{S}_B(P) \subseteq Px(G)$. The iterative nature of 8.4.3 suggests a proof by induction. We establish some preliminary lemmas. The first allows us to strip subsets of $max(p)$ from $p \in P$ giving another element of P. This means that if the set X of 8.4.3 belongs to P, then each of the sets $X-U$ also belongs to P.

8.4.4. LEMMA. Suppose that B is a left-closed, asynchronous behavioural presentation, $p \in P$ and $U \subseteq max(p)$, then $p - U \in P$.

PROOF. We argue by induction on $|U|$. The case $|U| = 0$ is trivial, so suppose that $|U| > 0$ and let $o \in U$. Since B is left-closed and asynchronous, $o/\approx = \{o\}$ so by 7.3.4, $p - \{o\} \in P$. It is immediate that $V = U - \{o\} \subseteq max(p) - \{o\}$. By induction, $p - U = (p - \{o\}) - V \in P$.

■

8.4.5. LEMMA. Let G be a computation graph and let $B = B_G^{fin}$, $p \in P$, $\mathbf{S} \in \mathbf{S}_B(p)$ and $v \in V$, then $\#(p,v) = \#(\mathbf{S},v,|\mathbf{S}|)$.

PROOF. We argue by induction on $|\mathbf{S}|$. If $|\mathbf{S}| = 0$, then $\mathbf{S} = \Omega$ and so $p = \varnothing$ and so $\#(p,v) = 0 = \#(\mathbf{S},v,|\mathbf{S}|)$.

Next, suppose that $|\mathbf{S}| > 0$, then $\mathbf{S} = \tilde{\mathbf{S}}.\lambda_G(X)$, where $\varnothing \subset X \subseteq max(p)$ and $\tilde{\mathbf{S}} \in \mathbf{S}_B(p - X)$, by 8.4.3. Since B is plain, by 8.4.2, $p - X \in P$, by 8.4.4, and hence $\#(p - X, v) = \#(\tilde{\mathbf{S}},v,|\tilde{\mathbf{S}}|)$, by induction. And $v \in \lambda_G(X) \Leftrightarrow \exists n{:}(v,n) \in X$ so if $v \notin \lambda_G(X)$ then

$$\#(p,v) = \#(p-X,v) = \#(\tilde{\mathbf{S}},v,|\tilde{\mathbf{S}}|) = \#(\mathbf{S},v,|\mathbf{S}|)$$

and if $v \in \lambda_G(X)$, then

$$\#(p,v) = \#(p-X,v)+1 = \#(\tilde{\mathbf{S}},v,|\tilde{\mathbf{S}}|)+1 = \#(\mathbf{S},v,|\mathbf{S}|)$$

completing the induction step.

■

8.4.6. COROLLARY. Let G be a computation graph and let $B = B_G^{fin}$, $p \in P$, $\mathbf{S} \in \mathbf{S}_B(p)$ and $d \in D$, then $\chi(p,d) = \chi(\mathbf{S},d,|\mathbf{S}|)$.

■

The following lemma is the basis for the induction step of the proof that if $B = B_G^{fin}$, then $\mathbf{S}_B(P) = Px(G) \cap \Sigma_G{}^*$.

8.4.7. LEMMA. Let G be a computation graph and let $B = B_G^{fin}$, $p \in P$, $\varnothing \subset X \subseteq V$ and $\mathbf{S} \in \mathbf{S}_B(p)$ such that $\mathbf{S} \in Px(G) \cap \Sigma_G{}^*$, then

$$\mathbf{S}.X \in Px(G) \cap \Sigma_G{}^* \Leftrightarrow \exists Y \subseteq O_G\ \exists \tilde{p} \in P : p \triangleright_G^Y \tilde{p} \wedge \lambda_G(Y) = X$$

PROOF. Suppose that $\mathbf{S}.X \in Px(G)$. If $d \in \bullet X$, then $\chi(\mathbf{S},d,|\mathbf{S}|) \geq T(d)$, by 8.2.3, and so $\chi(p,d) \geq T(d)$, by 8.4.6. Thus, for each $v \in X$, $(v,\#(p,v)+1) \in next(p)$, by 8.3.2, and so by 8.3.3, $p \triangleright_G^Y \tilde{p}$, where $Y = \{(v,\#(p,v)+1) \mid v \in X\}$ and $\tilde{p} = p \cup Y$. By 8.3.5, $\lambda_G(Y) = X$.

Conversely suppose that $p \triangleright_G^Y \tilde{p}$ and $\lambda_G(Y) = X$, then $\varnothing \subset Y \subseteq next(p)$, by 8.3.3. If $d \in \bullet X$, then $\chi(p,d) \geq T(d)$, by 8.3.2, so by 8.4.6 $\chi(\mathbf{S},d,|\mathbf{S}|) \geq T(d)$. Thus, $\mathbf{S}.X \in Px(G)$, by 8.2.3.

■

8.4.8. PROPOSITION. If G is a computation graph and $B = B_G^{fin}$, then

$$\mathbf{S}_B(P) = Px(G) \cap \Sigma_G{}^*.$$

PROOF. First, let $p \in P$. Since $|p| < \infty$, by 8.3.6, $\mathbf{S}_B(p) \subseteq \Sigma_G{}^*$. We prove by induction on $|p|$ that $\mathbf{S}_B(p) \subseteq Px(G)$. If $|p| = 0$, then $p = \varnothing$ and so by 8.4.3 (1), $\mathbf{S} = \Omega \in \mathbf{S}_B(p)$. Next let $|p| > 0$ and let $\mathbf{S} \in \mathbf{S}_B(p)$. By 8.4.3 (2), $\mathbf{S} = \tilde{\mathbf{S}}.\lambda_G(X)$, where $\varnothing \subset X \subseteq max(p)$ and $\tilde{\mathbf{S}} \in \mathbf{S}_B(p-X)$. By 8.4.2 and 8.4.4, $p - X \in P$. By induction, $\tilde{\mathbf{S}} \in Px(G)$. By 8.3.11, $p - X \triangleright_G^X p$. By 8.4.7, $\mathbf{S} = \tilde{\mathbf{S}}.\lambda_G(X) \in Px(G)$.

Conversely, suppose that $\mathbf{S} \in Px(G)$. We prove by induction on $|\mathbf{S}|$, that there exists $p \in P$ such that $\mathbf{S} \in \mathbf{S}_B(p)$. If $|\mathbf{S}| = 0$, then $\mathbf{S} = \Omega \in \mathbf{S}_B(\varnothing)$. As B is discrete, by 8.3.16, $\varnothing \in P$. Otherwise, $\mathbf{S} = \tilde{\mathbf{S}}.X$, where $\tilde{\mathbf{S}} \in Px(G)$ and $\varnothing \subset X \subseteq V$. By induction, there exists $\tilde{p} \in P$ such that $\tilde{\mathbf{S}} \in \mathbf{S}_B(\tilde{p})$. By 8.4.7, there exists $Y \subseteq V$ such that $\tilde{p} \triangleright_G^Y p$ and $\lambda_G(Y) = X$.

Since $\tilde{p} \triangleright_G^Y p$, $p = \tilde{p} \cup Y$ and $Y \subseteq next(\tilde{p})$, by 8.3.3, and so $\tilde{p} \cap Y = \varnothing$, by 8.3.8 giving $\tilde{p} = p - Y$. Hence, if $Y \subseteq max(p)$, then as $\tilde{\mathbf{S}} \in \mathbf{S}_B(\tilde{p})$ and $Y \neq \varnothing$, $\mathbf{S} = \tilde{\mathbf{S}}.\lambda_G(Y) \in \mathbf{S}_B(p)$. As $\tilde{p} \in P$ and $\tilde{p} \triangleright_G^Y p$, $p \in P$, by 8.3.5 and so it remains to show that $Y \subseteq max(p)$.

Let $o \in Y$ and $\hat{o} \in p$ such that $o \rightarrow \hat{o}$. If $\hat{o} \in \tilde{p}$, then from $o \rightarrow \hat{o}$ it would follow that $o \in \tilde{p}$, contradicting $\tilde{p} \cap Y = \varnothing$, and so $o' \in Y$. But now, $\hat{o} \in next(\tilde{p})$ and so $\tilde{p} \triangleright_G^{\hat{o}} \tilde{p} \cup \{\hat{o}\}$, by 8.3.3, and so $\tilde{p} \cup \{\hat{o}\} \in P$, by 8.3.5. Since $o \rightarrow \hat{o}$, $o \in \tilde{p} \cup \{\hat{o}\}$. But $o \notin \tilde{p}$, since $o \in Y$ and $\tilde{p} \cap Y = \varnothing$. Thus, $o = \hat{o}$. We have shown that if $o \in Y$ and $\hat{o} \in p$ such that $o \rightarrow \hat{o}$, then $o = \hat{o}$, and so $Y \subseteq max(p)$.

■

We conclude this section by proving the equivalence of the two semantics. First, we need a couple of lemmas which in turn rely on the following notation.

If $(A, \leq)$ is a partially ordered set and $X \subseteq A$, then we define the *prefix* of X, denoted by $\downarrow_A X$, as follows:

$$\downarrow_A X = \{a \in A \mid \exists x \in X : a \leq x\}.$$

It is easy to prove that if X is finite, then

$$X \subseteq \downarrow_A max(X) \tag{8.12}$$

We define the *suffix* of X, denoted by $\uparrow_A X$, as follows:

$$\uparrow_A X = \{a \in A \mid \exists x \in X : a \geq x\}.$$

We drop the subscript A when the partially ordered set in question is clear from the context.

If G is a computation graph, then it is an immediate consequence of 8.2.2 that within the partially ordered set $(Px(G), \leq)$:

$$\downarrow Px(G) = Px(G) \tag{8.13}$$
$$\uparrow Px(G) = Px(G) \tag{8.14}$$

We also observe that

$$Ex(G) = max(Px(G)) \tag{8.15}$$

8.4.9. LEMMA. If G is a computation graph, then:

1• $Px(G) = \downarrow Ex(G)$;

2• $Px(G) = \uparrow(Ex(G) \cap \Sigma_G^*)$.

PROOF. As $Ex(G) \subseteq Px(G)$, $\downarrow Ex(G) \subseteq \downarrow Px(G)$, so $\downarrow Ex(G) \subseteq Px(G)$, by (8.13), and since $Ex(G) = max(Px(G))$, by (8.15), $\downarrow Ex(G) = \downarrow max(Px(G))$, so $\downarrow Ex(G) \subseteq Px(G)$, by (8.12).

(2) Since $Px(G) \cap \Sigma_G^* \subseteq Px(G)$, $\uparrow(Px(G) \cap \Sigma_G^*) \subseteq \uparrow Px(G)$ and so by (8.14), $\uparrow(Px(G) \cap \Sigma_G^*) \subseteq Px(G)$. Conversely, if $\mathbf{S} \in Px(G)$, then as $\Omega \le \mathbf{S}$ and $\Omega \in Px(G) \cap \Sigma_G^*$, $\mathbf{S} \in \uparrow(Px(G) \cap \Sigma_G^*)$. Thus, $Px(G) \subseteq \uparrow(Px(G) \cap \Sigma_G^*)$.

■

8.4.10. LEMMA. If G is a computation graph, then $P_G^{fin} = P_G^F$, where

$$P_G^F = \{p \in P_G \mid |p| < \infty\}.$$

PROOF. As $P_G^{fin} \subseteq P_G$ and $|p| < \infty$, all $p \in P_G^{fin}$, by 8.3.6, it follows that $P_G^{fin} \subseteq P_G^F$.

Conversely, suppose that $p \in P_G^F$. If $o \in p$, then $o \in \tilde{p}$, some $\tilde{p} \in P_G^{fin}$, by 8.3.5 and so $\downarrow o \lhd \tilde{p}$, by 3.3.10. Since B_G^{fin} is left-closed, by 8.3.16 (2) and 3.3.14, $\downarrow o \in P_G^{fin}$. $p = \bigcup_{o \in p} \downarrow o$, by 4.3.2, and $|p| < \infty$, so p is the least upper bound of a finite set of elements of P_G^{fin}. But B_G^{fin} is discrete by 8.3.16 (2) and strongly forward stable, by 8.3.16 (3), and hence P_G^{fin} is a lattice, by 7.4.7. Hence, $p \in P_G^{fin}$.

■

8.4.11. THEOREM. If G_1 and G_2 are computation graphs, then the following statements are equivalent:

1• $B_{G_1}^{fin} = B_{G_2}^{fin}$;

2• $B_{G_1} = B_{G_2}$;

3• $Px(G_1) \cap \Sigma_{G_1}^* = Px(G_2) \cap \Sigma_{G_2}^*$ and $V_1 = V_2$;

4• $Px(G_1) = Px(G_2)$ and $V_1 = V_2$;

5• $Ex(G_1) = Ex(G_2)$ and $V_1 = V_2$.

PROOF. (1) ⇒ (2): If $B_{G_1}^{fin} = B_{G_2}^{fin}$, then by 8.3.5, $B_{G_1} = \overline{B_{G_1}^{fin}} = \overline{B_{G_2}^{fin}} = B_{G_2}$.

(2) $\Rightarrow$ (1): If $B_{G_1} = B_{G_2}$, then $B_{G_1}^{fin} = B_{G_1}^{F} = B_{G_2}^{F} = B_{G_2}^{fin}$, by 8.4.10.
(1) $\Rightarrow$ (3): If $B_{G_1}^{fin} = B_{G_2}^{fin}$, then by 8.4.8:

$$Px(G_1) \cap \Sigma_{G_1}^* = \mathbf{S}_{B_{G_1}^{fin}}(P_{G_1}^{fin}) = \mathbf{S}_{B_{G_2}^{fin}}(P_{G_2}^{fin}) = Px(G_2) \cap \Sigma_{G_2}^*$$

and $V_1 = E_{G_1}^{fin} = E_{G_2}^{fin} = V_2$.

(3) $\Rightarrow$ (1): Suppose that $Px(G_1) \cap \Sigma_{G_1}^* = Px(G_2) \cap \Sigma_{G_2}^*$ and $V_1 = V_2$ and let $p_1 \in P_{G_1}^{fin}$ and $\mathbf{S} \in \mathbf{S}_{B_{G_1}^{fin}}(p_1)$, then $\mathbf{S} \in Px(G_1) \cap \Sigma_{G_1}^*$, by 8.4.8, and so $\mathbf{S} \in Px(G_2) \cap \Sigma_{G_2}^*$. Therefore, by 8.4.8, there exists $p_2 \in P_{G_2}^{fin}$ such that $\mathbf{S} \in \mathbf{S}_{B_{G_2}^{fin}}(p_2)$. By 8.3.7 and 8.4.5:

$$\begin{aligned}(v,i) \in p_1 &\Leftrightarrow 1 \le i \le \#(p_1, v) \Leftrightarrow 1 \le i \le \#(\mathbf{S}, v, |\mathbf{S}|) \\ &\Leftrightarrow 1 \le i \le \#(p_2, v) \Leftrightarrow (v,i) \in p_2\end{aligned}$$

and so $p_1 = p_2$. It follows that $P_{G_1}^{fin} \subseteq P_{G_2}^{fin}$. By symmetry, $P_{G_2}^{fin} \subseteq P_{G_1}^{fin}$ and we have proved that $P_{G_1}^{fin} = P_{G_2}^{fin}$. By 2.1.1 (3), $O_{G_2}^{fin} = O_{G_1}^{fin}$. By hypothesis $E_{G_1}^{fin} = V_1 = V_2 = E_{G_2}^{fin}$. By 8.3.5 (3), $\lambda_{G_1}^{fin} = \lambda_{G_2}^{fin}$.
(3) $\Rightarrow$ (5): Suppose that $Px(G_1) \cap \Sigma_{G_1}^* = Px(G_2) \cap \Sigma_{G_2}^*$ and $V_1 = V_2$, then $\uparrow(Px(G_1) \cap \Sigma_{G_1}^*) = \uparrow(Px(G_2) \cap \Sigma_{G_2}^*)$ and $V_1 = V_2$ so $Ex(G_1) = Ex(G_2)$ and $V_1 = V_2$, by 8.4.9 (2).
(5) $\Rightarrow$ (4): If $Ex(G_1) = Ex(G_2)$ and $V_1 = V_2$, then $Px(G_1) = Px(G_2)$ and $V_1 = V_2$, by 8.4.9. (1)
(4) $\Rightarrow$ (3): If $Px(G_1) = Px(G_2)$ and $V_1 = V_2$ then $Px(G_1) = Px(G_2)$ and $\Sigma_{G_1} = \Sigma_{G_2}$, and so $Px(G_1) \cap \Sigma_{G_1}^* = Px(G_2) \cap \Sigma_{G_2}^*$ and $V_1 = V_2$.

■

Notes.

In [9], Karp and Miller examine various properties of computation graphs, including termination, to which we return when we give a transition system semantics for computation graphs in chapter 13.

Determinacy, in the abstract sense of chapter 2, may be given a more concrete interpretation if we extend the model to interpret nodes as functions. Each queue d now becomes associated with a set $\mathcal{V}_d$ of values and each node v is associated with a function

$$\phi_v : \mathop{\times}_{d \in \bullet v} \mathcal{V}_d^{\,W(d)} \to \mathop{\times}_{d \in v \bullet} \mathcal{V}_d^{\,U(d)}.$$

Let us define an interpreted computation graph to be a tuple $\Gamma = (V, D, inc, U, W, T, \mathcal{V}, \Phi)$, where V, D, inc, U, W and T are as in 8.2.1, $\mathcal{V}$ is the family of sets $\mathcal{V}_d$ and Φ is the family of functions ϕ_v.

The instantaneous state of the system may be modelled by a function $\underline{\theta}$ which maps each queue d to a sequence $\mathcal{V}_d{}^*$. Let the set of all such functions be denoted by $\mathcal{V}_\Gamma$.

Interpreted computation graphs are related to standard computation graphs as follows. Suppose that $\Gamma = (V, D, inc, U, W, T, \mathcal{V}, \Phi)$ and $\underline{\alpha} \in \mathcal{V}_\Gamma$, then we define $G(\Gamma, \underline{\alpha}) = (V, D, inc, A, U, W, T)$, where $A(d) = |\underline{\alpha}(d)|$, each $.d$

We describe how operations of nodes modify the states. Suppose that $d_1, \cdots, d_m$ are the elements of $\bullet v$ and $d'_1, \cdots, d'_n$ are the elements of $v \bullet$. The operation v may be applied in state $\underline{\theta}_1$ transforming it to state $\underline{\theta}_2$ providing there are at least $T(d_i)$ elements in $\underline{\theta}_1(d_i)$, each i. and that $\underline{\theta}_2$ is obtained from $\underline{\theta}_1$ as follows. The sequence $\underline{x}_i$ of length $W(d_i)$ of items from the front of $\underline{\theta}_1(d_i)$ is deleted giving a new function $\underline{\theta}'_1$, and then a sequence $\underline{y}_j$, which will have length $U(d'_j)$, is added to the rear of $\underline{\theta}'_1(d'_j)$, where $\phi_v(\underline{x}_1, \cdots, \underline{x}_m) = (\underline{y}_1, \cdots, \underline{y}_n)$. Write $\underline{\theta}_1 \to_\Gamma^{\,v} \underline{\theta}_2$ to indicate that v effects this transformation of $\underline{\theta}_1$ into $\underline{\theta}_2$.

Determinacy entails that if two nodes operate concurrently, then they do not *both* 'touch' any given data item, so that their combined effect on the global state is well defined. Formally, we have the following

THEOREM. If Γ is an interpreted computation graph and $\underline{\alpha} \in \mathcal{V}_\Gamma$ then there exists a unique function $\vartheta_{\Gamma,\underline{\alpha}} : P_{G(\Gamma,\underline{\alpha})}{}^{fin} \to \mathcal{V}_\Gamma$ such that

1• $\vartheta_{\Gamma,\underline{\alpha}}(\varnothing) = \underline{\alpha}$;

2• $\forall p_1, p_2 \in P_{G(\Gamma,\underline{\alpha})}{}^{fin}\ v \in V : (\vartheta_{\Gamma,\underline{\alpha}}(p_1) \to_\Gamma^{\,v} \vartheta_{\Gamma,\underline{\alpha}}(p_2) \Leftrightarrow \exists n \in \omega : p_1 \succ^{(v,n)} p_2)$.

The theorem allows one to view Γ as defining a partial function from $\mathcal{V}_\Gamma$ to itself. For if $O_{G(\Gamma,\underline{\alpha})}$ is finite, then $P_{G(\Gamma,\underline{\alpha})}$ is a finite lattice, by 7.4.7, and as $O_{G(\Gamma,\underline{\alpha})}$ is the union of the elements of $P_{G(\Gamma,\underline{\alpha})}$, it must therefore belong to $P_{G(\Gamma,\underline{\alpha})}$ and is therefore a finite, terminating compitation. We may thus define a function $\Theta_\Gamma : \mathcal{V}_\Gamma \to \mathcal{V}_\Gamma$ by

$$\Theta_\Gamma(\underline{\theta}) = \begin{cases} \vartheta_{\Gamma,\underline{\theta}}(O_{G(\Gamma,\underline{\theta})}) \text{ if } |O_{G(\Gamma,\underline{\theta})}| < \infty \\ \text{undefined, otherwise} \end{cases}$$

Chapter 9
Trace and Vector Languages

> NAPOLEON: You feel yourself my superior, I know: nay you are my superior: have I not bowed my knee to you by instinct. Yet I challenge you to a test of our respective powers. Can you calculate what the mathematicians call a vector without putting a single algebraic symbol on paper?
>
> G. B. SHAW
> Back to Methuselah

9.1. Introduction.

We resume our study of the representation of behaviour with models having a language theoretic flavour. So far, our models have been order-theoretic in nature. For example, in the behavioural presentation approach, an individual behaviour is represented by a labelled pre-ordered set, $(p,\rightarrow,\lambda)$, where $p \in P$. While, being intuitively satisfying, they are rather cumbersome to manipulate. Indeed, one of the attractions of the interleaving approach to the representation of behaviour is that it is amenable to operations familiar to the student of classical formal language theory.[1]

In this chapter, we look at two extensions of language theory. In this first, the *trace languages* of Mazurkiewicz, languages are subsets not of the freely generated monoid A^* but of a semi-commutative monoid in which the commuting of generators models the potential concurrency of their occurrences. In the second extension, the languages are subsets of a monoid whose generators are tuples of sequences; we refer to these as *vector languages*.

The two models turn out to be equivalent, as we prove in section, 9.3. In a sense, vector languages are to trace languages what matrices are to linear transformations; they afford a more concrete representation which has advantages when it comes to computation or manipulation. We shall exploit

---

[1] Alas.

this in the following chapter where we establish important order theoretic properties of trace and vector languages. These will enable us to relate these language theoretic formulations with behavioural presentations in chapter 12. They will also allow us to build an automata theory for non-sequential systems, which we begin in chapter 15.

9.2. Basic Definitions and Properties.

We begin with trace languages.

9.2.1. DEFINITION. Suppose that ι is an independence relation[2] on a set A, then we define a relation $\equiv_\iota^{(1)} \subseteq A^* \times A^*$ as follows:

$$x \equiv_\iota^{(1)} y \Leftrightarrow (\exists u, v \in A^* \exists a, b \in A : a \,\iota\, b \wedge x = u.a.b.v \wedge y = u.b.a.v)$$

and let $\equiv_\iota$ be the reflexive, transitive closure of $\equiv_\iota^{(1)}$. $\equiv_\iota$ is clearly an equivalence relation. If $x \in A^*$ then we write x_ι for the $\equiv_\iota$-class of x, and write A_ι^* - for the set of $\equiv_\iota$-classes of A^*.

The elements of A_ι^* are the *traces* of A over ι. A *trace language* is a subset of A_ι^*.

The idea behind this construction is as follows. $a \,\iota\, b$ means that whenever a and b are both capable of occurring, then they do so concurrently.[3] Hence, if $x = u.a.b.v$ is a linearisation of a behaviour, then $y = u.b.a.v$ is a linearisation of the same behaviour; symbolically, $x \equiv_\iota^{(1)} y$. Two sequences x and y will now be understood to represent the same behaviour if and only if $x = y$ or there exists sequences $w_1, \cdots, w_k$ such that $x = w_1 \equiv_\iota^{(1)} \cdots \equiv_\iota^{(1)} w_k = y$; symbolically, $x \equiv_\iota y$.

9.2.2. PROPOSITION. Suppose ι is an independence relation on a set A, then

1• A_ι^* is a monoid[4] with respect to a composition defined by $x_\iota . y_\iota = (x.y)_\iota$. The identity of the monoid is Ω_ι.
2• A_ι^* is a partial order with respect to the relation $\leq_\iota$ defined by $x_\iota \leq_\iota y_\iota \Leftrightarrow \exists z \in A^* : x.z \equiv_\iota y$. It has a bottom element Ω_ι.

---

[2] That is, an irreflexive, symmetric relation.
[3] Note that we are talking about a and b in terms of events rather than of occurrences.
[4]See appendix A.1 for basic definitions involving monoids and semigroups.

PROOF. The following properties of $\equiv_\iota^{(1)}$ are almost immediate.

$$x \equiv_\iota^{(1)} y \Rightarrow y \equiv_\iota^{(1)} x \tag{9.1}$$
$$x \equiv_\iota^{(1)} y \Rightarrow x.z \equiv_\iota^{(1)} y.z \wedge z.x \equiv_\iota^{(1)} z.y \tag{9.2}$$

Since $\equiv_\iota$ is the reflexive, transitive closure of $\equiv_\iota^{(1)}$, and $\equiv_\iota^{(1)}$ is symmetric by (9.1), it follows that $\equiv_\iota$ is also symmetric and hence an equivalence relation. From (9.2) and transitivity we get

$$x \equiv_\iota y \Rightarrow x.z \equiv_\iota y.z \wedge z.x \equiv_\iota z.y \tag{9.3}$$

Thus $\equiv_\iota$ is a semigroup congruence. Since $x.\Omega \equiv_\iota x$ and $\Omega.x \equiv_\iota x$, $x_\iota.\Omega_\iota = x_\iota$ and $\Omega_\iota.x_\iota = x_\iota$, that is Ω_ι is the identity.

$\leq_\iota$ is clearly reflexive. If $x_\iota \leq_\iota y_\iota$ and $y_\iota \leq_\iota z_\iota$, then we can find $w_1, w_2 \in A^*$ such that $x.w_1 \equiv_\iota y$ and $y.w_2 \equiv_\iota z$, so by (9.3), $x.w_1.w_2 \equiv_\iota y.w_2 \equiv_\iota z$, and thus, $x_\iota \leq_\iota z_\iota$, giving transitivity of $\leq_\iota$.

It is also immediate from the definition that $x \equiv_\iota^{(1)} y \Rightarrow |x| = |y|$ and so $x \equiv_\iota y \Rightarrow |x| = |y|$. Thus,

$$x \equiv_\iota x.y \Rightarrow y = \Omega \tag{9.4}$$

We may use (9.4) to show that $\leq_\iota$ is antisymmetric. Suppose that $x_\iota \leq_\iota y_\iota$ and $y_\iota \leq_\iota x_\iota$, then we can find $w_1, w_2 \in A^*$ such that $x.w_1 \equiv_\iota y$ and $y.w_2 \equiv_\iota x$. By (9.3), $x.w_1.w_2 \equiv_\iota y.w_2 \equiv_\iota x$. By (9.4), $w_1.w_2 = \Omega$. Hence, $w_1 = \Omega$ and so $x = x.w_1 \equiv_\iota y$, that is $x_\iota = y_\iota$.

■

We shall now describe a representation of trace languages using vectors of strings. These first appeared in [25] as an asynchronous semantics for path expressions. However, they provide a natural semantics for other kinds of notation and this seems a reasonable way of introducing them.

Let us suppose that we have a system which is subject to a set of sequential constraints. That is to say that we have a set $\{R_1, \cdots, R_n\}$ of rules, each of which dictates the orders in which the events of some set A_i may be allowed to occur, and each of which requires that no two of the elements of any A_i may occur concurrently. [5]

How can we check that the system is obeying the constraints? Suppose that for each rule R_i, we employ an observer, whose job it is to record the

---

[5] This is the principle underlying Path Expressions for example, or the Nivat Arnold model. See chapters 24 and 25.

occurrences of events from A_i as the system runs. As soon as rule R_i is infringed, the observer i blows a little whistle and a representative from the manufacturer of the system is summoned.

Imagine that we have our observers, furiously scribbling away as the system churns on, and that the system is suddenly halted. Each observer will have accumulated a sequence of action names - recording the sequence in which these actions have been performed. It follows that the behaviour of the system insofar as it concerns R_i may be described by a string $x_i \in A_i^*$. We may describe the behaviour of the system as described by all n observers using a tuple of strings,

$$(x_1, \cdots, x_n) \in A_1^* \times \cdots \times A_n^*$$

Now $(x_1, \cdots, x_n)$ certainly tells us something about the behaviour of the system before it stopped. We may go further, however, and assert that if the R_i were the *only* constraints satisfied by the system, then $(x_1, \cdots, x_n)$ *must* be a complete description of that behaviour.

Not every member of $A_1^* \times \cdots \times A_n^*$ could represent the behaviour of a system. For example, if an event a belongs to both A_i and A_j then x_i and x_j must contain the same number of a's. Thus, if only the event a had occurred before shut-down, the corresponding *event vector* would satisfy:

$$x_i = \begin{cases} a \text{ if } a \in A_i \\ \Omega \text{ if } a \notin A_i \end{cases} \tag{9.5}$$

Suppose next that two events, a and b had been allowed to occur before the system was switched off. Let us suppose in addition that whenever $a, b \in A_i$, then observer i would see a happen before b. The corresponding behaviour vector $\underline{y}$ has therefore co-ordinates given by the following equation.

$$y_i = \begin{cases} ab \text{ if } a, b \in A_i \\ a \text{ if } a \in A_i \wedge b \notin A_i \\ b \text{ if } b \in A_i \wedge a \notin A_i \\ \Omega \text{ if } a, b \notin A_i \end{cases}$$

Now, $\underline{y}$ may be seen to be the *co-ordinatewise concatenation* [6] of two *event-vectors* [7] $\underline{a}$ and $\underline{b}$, where $\underline{a}$ defined in (9.5) and $\underline{b}$, is defined analogously for

---

[6] See 9.2.6.
[7] See 9.2.5.

b. Inductively, any vector history of the system is the concatenation of such event vectors. We may also see that every co-ordinate entry of $\underline{a}$ is the string prefix of the corresponding co-ordinate entry of $\underline{y}$. In other words, $\underline{a}$ is the *co-ordinatewise prefix* of $\underline{y}$.

It follows that for a system such as we have described, the vectors representing all possible histories of the system may be characterised as products of the event vectors and that this set of 'histories' has both an additive and an order structure. We may say even more. Provisionally, let V denote the set of all vectors corresponding the family of sets A_i. Let $L_i \subseteq A_i^*$ be the set of all sequences allowed by constraint A_i, then the set of all behaviours of the system is completely described by the *vector language* $V \cap (L_1 \times \cdots \times L_n)$.

This vector description appears to apply only to such situations in which a system is composed of a set of sequential subsystems which co-operate in a very specific way, and hence of limited application. In fact, as we have said, vectors may serve as representations of traces, and there are advantages to such a representation. As we shall see in due course, manipulations of vectors is often more straightforward than manipulation of traces. Let us set up the machinery. All we need to begin with is a family of sets of actions; these are the indexed covers.

9.2.3. DEFINITION. An *indexed cover* for a set A, is a map $\alpha : I \to \wp(A)$ such that

$$\bigcup_{i \in I} \alpha(i) = A$$

9.2.4. LEMMA ([22]). Every indexed cover $\alpha : I \to \wp(A)$ determines an independence relation ι_α by

$$a \,\iota_\alpha\, b \Leftrightarrow \forall i \in I : \{a,b\} \not\subseteq \alpha(i)$$

Conversely, if ι is an independence relation, then there exists an indexed cover a such that $\iota_\alpha = \iota$.

PROOF. ι_α is clearly symmetric. If $a \,\iota_\alpha\, a$, then $\{a\} \subseteq \alpha(i)$ for no $i \in I$, which contradicts the fact that α is a cover. Hence, $\neg a \,\iota_\alpha\, a$.

Now suppose that ι is an independence relation and let $I = \{\{a,b\} \subseteq A \mid \neg a \,\iota\, b\}$. Define $\alpha : I \to \wp(A)$ by $\alpha(X) = X$, each $X \in I$. Since $a \not\iota\, a$, it follows that $\{a\} \in I$, each $a \in A$ and hence α is a cover.

Finally, $a \,\iota\, b \Leftrightarrow \{a,b\} \notin I \Leftrightarrow (\forall i \in I : \{a,b\} \not\subseteq \alpha(i)) \Leftrightarrow a \,\iota_\alpha\, b$.

■

From 9.2.4, we see that an indexed cover α of A determines an independence relation ι_α and hence a trace monoid $A_{\iota_\alpha}{}^*$. α also determines a set of string-vectors and we have noted that these may be concatenated together. It transpires that, under this concatenation, vectors form a monoid, which we shall call $A_\alpha{}^*$. 9.3.3 explains the connection between the two; they are structurally identical. First, we need some definitions.

9.2.5. DEFINITION If $\alpha: I \to \wp(A)$ is an indexed cover, then an α-*vector* is a mapping $\underline{v}: I \to A^*$ such that

$$\forall i \in I: \underline{v}(i) \in \alpha(i)^*$$

$\underline{\Omega}_\alpha$ is defined to be the vector $\underline{v}$ satisfying $\forall i \in I: \underline{v}(i) = \Omega$.

If $I = \{1, \cdots, n\}$, then we shall sometimes write an α-vector $\underline{x}$ in the form $(\underline{x}(1), \cdots, \underline{x}(n))$.

For example, if $A = \{a,b,c,d,e\}$, $I = \{1,2,3\}$, $\alpha(1) = \{a,b\}$, $\alpha(2) = \{b,c\}$ and $\alpha(3) = \{a,d,e\}$, then (ab, bc, ae) is an α-vector.

We shall denote the set of all α-vectors by M_α. We may define a componentwise concatenation and ordering on M_α as follows.

9.2.6. DEFINITION. If $\alpha: I \to \wp(A)$ is an indexed cover and $\underline{x}, \underline{y} \in M_\alpha$, then:

1• $\underline{x}.\underline{y}$ is defined to be the α-vector $\underline{z}$ such that $\forall i \in I: \underline{z}(i) = \underline{x}(i).\underline{y}(i)$;
2• $\underline{x} \leq \underline{y} \Leftrightarrow \forall i \in I: \underline{x}(i) \leq \underline{y}(i)$.

For example, if $A = \{a,b,c,d,e\}$, $I = \{1,2,3\}$, $\alpha(1) = \{a,b\}$, $\alpha(2) = \{b,c\}$ and $\alpha(3) = \{a,d,e\}$, then $(ab,c,a).(\Omega,b,e) = (ab,cb,ae)$ and $(ab,c,a) \leq (ab,cb,ae)$.

The assertions in the following remark may be easily deduced using a coordinatewise argument. For example, $\underline{x}.(\underline{y}.\underline{z}) = (\underline{x}.\underline{y}).\underline{z}$ because $\underline{x}(i).(\underline{y}(i).\underline{z}(i)) = (\underline{x}(i).\underline{y}(i)).\underline{z}(i)$ for all $i \in I$.

9.2.7. PROPOSITION. If α is an indexed cover, then:

1• M_α is a monoid with identity $\underline{\Omega}_\alpha$;
2• M_α is a poset with respect to $\leq$, with bottom $\underline{\Omega}_\alpha$.

■

A particular submonoid of M_α is of interest to us, namely that generated by the event vectors of (9.5).

9.2.8. DEFINITION. If $\alpha: I \to \wp(A)$, is an indexed cover then for each $a \in A$, the α-*event vector* of a, denoted by $\underline{a}_\alpha$, is the α-vector such that

$$\underline{a}_\alpha(i) = \begin{cases} a \text{ if } a \in \alpha(i) \\ \Omega \text{ otherwise} \end{cases}$$

Let $A_\alpha = \{\underline{a}_\alpha \mid a \in A\}$ and let $A_\alpha{}^*$ denote the submonoid of M_α generated by A_α (that is, the set of all products of elements of $A_\alpha{}^*$ together with $\underline{\Omega}_\alpha$).

For example, if $A = \{a,b,c,d,e\}$, $I = \{1,2,3\}$, $\alpha(1) = \{a,b\}$, $\alpha(2) = \{b,c\}$ and $\alpha(3) = \{a,d,e\}$, and so

$$A_\alpha = \{(a,\Omega,a),(b,b,\Omega),(\Omega,c,\Omega),(\Omega,\Omega,d),(\Omega,\Omega,e)\}.$$

The most important elementary fact about $A_\alpha{}^*$ is that it has the same commutativity as $A_{\iota_\alpha}{}^*$. For instance, in the example above, $\{a,c\} \not\subseteq \alpha(i)$ for each $i \in I$, and so $a \, \iota_\alpha \, c$. Hence $ac \equiv_{\iota_\alpha}^{(1)} ca$, and so $ac \equiv_{\iota_\alpha} ca$, which gives

$$a_{\iota_\alpha} . c_{\iota_\alpha} = (ac)_{\iota_\alpha} = (ca)_{\iota_\alpha} = c_{\iota_\alpha} . a_{\iota_\alpha}$$

But, calculating:

$$\underline{a}_\alpha \cdot \underline{c}_\alpha = (a,\Omega).(\Omega,c) = (a,c) = (\Omega,c).(a,\Omega) = \underline{c}_\alpha \cdot \underline{a}_\alpha$$

The general situation is explained in the following lemma.

9.2.9. LEMMA. If $\alpha: I \to \wp(A)$ is an indexed cover and $a,b \in A$, then

$$\underline{a}_\alpha \cdot \underline{b}_\alpha = \underline{b}_\alpha \cdot \underline{a}_\alpha \Leftrightarrow a \, \iota_\alpha \, b \vee a = b.$$

PROOF. Suppose first that $a \neq b$ and $a \not\iota_\alpha b$ then $\{a,b\} \subseteq \alpha(i)$ for some $i \in I$, and thus $(\underline{a}_\alpha \cdot \underline{b}_\alpha)(i) = ab \neq ba = (\underline{b}_\alpha \cdot \underline{a}_\alpha)(i)$, so that $\underline{a}_\alpha \cdot \underline{b}_\alpha \neq \underline{b}_\alpha \cdot \underline{a}_\alpha$, Therefore, $\underline{a}_\alpha \cdot \underline{b}_\alpha = \underline{b}_\alpha \cdot \underline{a}_\alpha \Rightarrow a \, \iota_\alpha \, b \vee a = b$.

If $a = b$ then certainly $\underline{a}_\alpha \cdot \underline{b}_\alpha = \underline{b}_\alpha \cdot \underline{a}_\alpha$. If $a \, \iota_\alpha \, b$ then $\{a,b\} \not\subseteq \alpha(i)$, all $i \in I$, and hence only the following three cases are possible

- $a \in \alpha(i)$ and $b \notin \alpha(i)$ so $(\underline{a}_\alpha \cdot \underline{b}_\alpha)(i) = a = (\underline{b}_\alpha \cdot \underline{a}_\alpha)(i)$
- $a \notin \alpha(i)$ and $b \in \alpha(i)$ so $(\underline{a}_\alpha \cdot \underline{b}_\alpha)(i) = b = (\underline{b}_\alpha \cdot \underline{a}_\alpha)(i)$
- $a \notin \alpha(i)$ and $b \notin \alpha(i)$ so $(\underline{a}_\alpha \cdot \underline{b}_\alpha)(i) = \Omega = (\underline{b}_\alpha \cdot \underline{a}_\alpha)(i)$

Hence, $(\underline{a}_\alpha \cdot \underline{b}_\alpha)(i) = (\underline{b}_\alpha \cdot \underline{a}_\alpha)(i)$, all $i \in I$, that is, $\underline{a}_\alpha \cdot \underline{b}_\alpha = \underline{b}_\alpha \cdot \underline{a}_\alpha$.

∎

9.2.9 is a key step in the proof that $A_{\iota_\alpha}{}^*$ is isomorphic to $A_\alpha{}^*$. Intuitively, the two monoids have 'the same' sets of generators and relations and therefore the map from A_{ι_α} to A_α given by $a_{\iota_\alpha} \mapsto \underline{a}_\alpha$ extends to an isomorphism. We need to do a little work first. Mainly, this is because we don't know yet that the only relations on the elements of A_α are those described in 9.2.9. Let us therefore set up some more machinery.

We shall need to argue by induction on the size of a vector. Intuitively, each vector has a 'length', namely the number of elements of A_α that go to make it up. Let us make this precise. First, if $a \in A$ and $x \in A^*$, then define $\#_a x$ to be the number of a's occurring in x. Formally,

$$\#_a \Omega = 0$$

and if $a, b \in A$ and $x \in A^*$, then

$$\#_a x.b = \begin{cases} \#_a x + 1 \text{ if } a = b \\ \#_a x \text{ otherwise} \end{cases} \tag{9.6}$$

An easy induction argument shows that if $a \in A$ and $x, y \in A^*$, then

$$\#_a x.y = \#_a x + \#_a y \tag{9.7}$$

9.2.10. LEMMA. If $a \in A$ and $\underline{x} \in A_\alpha{}^*$, then

$$\forall i, j \in I: a \in \alpha(i) \cap \alpha(j) \Rightarrow \#_a \underline{x}(i) = \#_a \underline{x}(j)$$

PROOF. We argue by induction on n that if $a^1, \cdots, a^n \in A$ and $\underline{x} = \underline{a}^1{}_\alpha \cdots \underline{a}^n{}_\alpha$, then for all $i, j \in I$, $a \in \alpha(i) \cap \alpha(j) \Rightarrow \#_a \underline{x}(i) = \#_a \underline{x}(j)$. This certainly holds if $n = 0$, for then $\underline{x} = \underline{\Omega}_\alpha$. Next, let $n > 0$ and let $\underline{y} = \underline{a}^1{}_\alpha \cdots \underline{a}^{n-1}{}_\alpha$. and $i, j \in I$ such that $a \in \alpha(i) \cap \alpha(j)$. By induction, $\#_a \underline{y}(i) = \#_a \underline{y}(j)$. Therefore, if $a_n \neq a$, then by 9.2.6 (1) and 9.2.8, $\# \underline{x}(i) = \# \underline{y}(i)$ and $\# \underline{x}(j) = \# \underline{y}(j)$ and hence $\#_a \underline{x}(i) = \#_a \underline{y}(i) = \#_a \underline{y}(j) = \#_a \underline{x}(j)$. Otherwise $\underline{x}(i) = \underline{y}(i).a$ and $\underline{x}(j) = \underline{y}(j).a$ and so $\#_a \underline{x}(i) = \#_a \underline{y}(i) + 1 = \#_a \underline{y}(j) + 1 = \#_a \underline{x}(j)$, by (9.6).

■

By 9.2.10, we may unambiguously define $\#_a \underline{x}$ to be $\#_a \underline{x}(i)$, for any $i \in I$ such that $a \in \alpha(i)$. Since α is a cover, $a \in \alpha(i)$ for some $i \in I$, and hence $\#_a \underline{x}$ is always defined. Finally, define

$$|\underline{x}| = \sum_{a \in A} \#_a \underline{x} \tag{9.8}$$

9.2.11. LEMMA. If $a \in A$ and $\underline{x}, \underline{y} \in A_\alpha{}^*$, then

1• $|\underline{x}| = 0 \Leftrightarrow \underline{x} = \underline{\Omega}_\alpha$;

2• $|\underline{a}_\alpha| = 1$;

3• $|\underline{x}.\underline{y}| = |\underline{x}| + |\underline{y}|$.

PROOF. (1) By (9.8), $|\underline{x}| = 0 \Leftrightarrow (\forall a \in A: \#_a \underline{x} = 0) \Leftrightarrow \underline{x} = \underline{\Omega}_\alpha$.
(2) follows directly from 9.2.8.
(3) If $a \in A$ and $i \in I$ such that $a \in \alpha(i)$, then

$$\#_a(\underline{x}.\underline{y}) = \#_a(\underline{x}.\underline{y})(i) = \#_a \underline{x}(i) + \#_a \underline{y}(i) = \#_a \underline{x} + \#_a \underline{y}$$

by (9.7). Hence

$$|\underline{x}.\underline{y}| = \sum_{a \in A} \#_a(\underline{x}.\underline{y}) = \sum_{a \in A} \#_a \underline{x} + \sum_{a \in A} \#_a \underline{y} = |\underline{x}| + |\underline{y}|$$

■

We shall find it convenient to extend the notion of independence from individual actions to vectors. This notion proves useful in connection with least upper bounds of sets of vectors, for example.

9.2.12. DEFINITION. If $\underline{x}, \underline{y} \in M_\alpha$, then $\underline{x}$ and $\underline{y}$ are *independent*, and we write $\underline{x}$ *ind* $\underline{y}$ $\Leftrightarrow$

$$\forall i \in I: \underline{x}(i) > \Omega \Rightarrow \underline{y}(i) = \Omega$$

The following are easy consequences of the definition.

9.2.13. LEMMA. Suppose $a, b \in A$, $\underline{x}, \underline{y}, \underline{z} \in M_\alpha$, then

1• $\underline{x}\ ind\ \underline{y} \Leftrightarrow \underline{y}\ ind\ \underline{x}$;

2• $\underline{a}_\alpha\ ind\ \underline{b}_\alpha \Leftrightarrow a\ \iota_\alpha\ b$;

3• $\underline{x}\ ind\ \underline{y} \Rightarrow \underline{x}.\underline{y} = \underline{y}.\underline{x}$;

4• $\underline{x}.\underline{y}\ ind\ \underline{z} \Leftrightarrow \underline{x}\ ind\ \underline{z} \wedge \underline{y}\ ind\ \underline{z}$.

PROOF. (1) By the tautology $(p \Rightarrow q) \Leftrightarrow (\neg q \Rightarrow \neg p)$.
(2) $a\ \iota_\alpha\ b \Leftrightarrow (\forall i \in I: \{a, b\} \not\subseteq \alpha(i))$, by 9.2.4 (2)
$\Leftrightarrow (\forall i \in I: a \in \alpha(i) \Rightarrow b \notin \alpha(i))$
$\Leftrightarrow (\forall i \in I: \underline{a}_\alpha(i) > 0 \Rightarrow \underline{b}_\alpha(i) = 0)$ by 9.2.8.
(3) If $i \in I$, then we have exactly three cases:

- $\underline{x}(i) > \Omega$ and $\underline{y}(i) = \Omega$, so, $(\underline{x}.\underline{y})(i) = \underline{x}(i) = (\underline{y}.\underline{x})(i)$.
- $\underline{x}(i) = \Omega$ and $\underline{y}(i) > \Omega$, so, $(\underline{x}.\underline{y})(i) = \underline{y}(i) = (\underline{y}.\underline{x})(i)$.
- $\underline{x}(i) = \Omega$ and $\underline{y}(i) = \Omega$, so, $(\underline{x}.\underline{y})(i) = \Omega = (\underline{y}.\underline{x})(i)$.

So $(\underline{x}.\underline{y})(i) = (\underline{y}.\underline{x})(i)$, all $i \in I$.

(4) Let $\underline{x}.\underline{y}\ ind\ \underline{z}$. If $\underline{x}(i) > \Omega$, then $(\underline{x}.\underline{y})(i) > \Omega$ so $\underline{z}(i) = \Omega$. Thus, $\underline{x}\ ind\ \underline{z}$. Similarly, $\underline{y}\ ind\ \underline{z}$. Hence, $\underline{x}.\underline{y}\ ind\ \underline{z} \Rightarrow \underline{x}\ ind\ \underline{z} \wedge \underline{y}\ ind\ \underline{z}$.

Conversely, let $\underline{x}\ ind\ \underline{z}$ and $\underline{y}\ ind\ \underline{z}$. If $(\underline{x}.\underline{y})(i) > \Omega$, then either $\underline{x}(i) > \Omega$ or $\underline{y}(i) > \Omega$ and so $\underline{z}(i) = \Omega$ in either case. Hence, $\underline{x}\ ind\ \underline{z} \wedge \underline{y}\ ind\ \underline{z} \Rightarrow \underline{x}.\underline{y}\ ind\ \underline{z}$.

■

9.3. Equivalence of the Models.

The last piece of machinery that we require for the establishment of the representation of traces by vectors is a useful 'left-cancellation' operator. First, we prove a convenient lemma.

9.3.1. LEMMA. If $a \in A$ and $\underline{x} \in A_\alpha{}^*$ then

$$\underline{a}_\alpha \leq \underline{x} \Rightarrow \exists \underline{x}' \in A_\alpha{}^*: \underline{x} = \underline{a}_\alpha \cdot \underline{x}'$$

PROOF. Since $\underline{a}_\alpha \leq \underline{x}$, we may write $\underline{x} = \underline{x}_1 \cdot \underline{a}_\alpha \cdot \underline{x}_2$ where $\#_a \underline{x}_1 = 0$. If $a \in \alpha(i)$, then $a \leq \underline{x}_1(i).a.\underline{x}_2(i)$ and so $\underline{x}_1(i) = \Omega$. Therefore $\underline{a}_\alpha\ ind\ \underline{x}_1$ and by 9.2.13 (3) $\underline{x} = \underline{x}_1 \cdot \underline{a}_\alpha \cdot \underline{x}_2 = \underline{a}_\alpha \cdot \underline{x}_1 \cdot \underline{x}_2$. Thus the lemma holds with $\underline{x}' = \underline{x}_1 \cdot \underline{x}_2$.

■

9.3.2. PROPOSITION. If $\underline{x}, \underline{y} \in A_\alpha{}^*$, then

$$\underline{x} \leq \underline{y} \Leftrightarrow \exists \underline{z} \in A_\alpha{}^*: \underline{y} = \underline{x}.\underline{z}$$

PROOF. The $\Leftarrow$ implication is trivial. For the $\Rightarrow$ implication, we argue by induction on $|\underline{x}|$. If $|\underline{x}| = 0$, then $\underline{x} = \underline{\Omega}_\alpha$, and the proposition holds with $\underline{z} = \underline{y}$. If, $|\underline{x}| > 0$ then there exists $a \in A$ such that $\underline{a}_\alpha \leq \underline{x}$ and hence $\underline{a}_\alpha \leq \underline{y}$. By 9.3.1, there exists $\underline{x}', \underline{y}' \in A_\alpha{}^*$ such that $\underline{x} = \underline{a}_\alpha \cdot \underline{x}'$. and $\underline{y} = \underline{a}_\alpha \cdot \underline{y}'$. But now, since $\underline{a}_\alpha(i).\underline{x}'(i) \leq \underline{a}_\alpha(i).\underline{y}'(i)$, all $i \in I$, it follows that $\underline{x}'(i) \leq \underline{y}'(i)$, all $i \in I$, that is, $\underline{x}' \leq \underline{y}'$. By induction, there exists $\underline{z} \in A_\alpha{}^*$ such that $\underline{y}' = \underline{x}'.\underline{z}$. Consequently, $\underline{y} = \underline{a}_\alpha \cdot \underline{y}' = \underline{a}_\alpha \cdot \underline{x}'.\underline{z} = \underline{x}.\underline{z}$.

■

If $\underline{x}.\underline{z}_1 = \underline{x}.\underline{z}_2$, then $\underline{x}(i).\underline{z}_1(i) = \underline{x}(i).\underline{z}_2(i)$ each i. and so $\underline{z}_1(i) = \underline{z}_2(i)$ each i and so $\underline{z}_1 = \underline{z}_2$. Hence, the vector $\underline{z}$ of 9.3.2 is unique; we denote it by $\underline{y}/\underline{x}$. We shall use the same notation for sequences; if $x \le y$, then y/x is defined to be the unique string such that $x.(y/x) = y$.

We are now in a position to prove that vectors and traces are structurally identical.

Let $\alpha: I \to \wp(A)$ be an indexed cover and suppose that $\iota = \iota_\alpha$. Define $f_\alpha: A \to A_\alpha{}^*$ by $f_\alpha(a^1 \cdots a^n) = \underline{a}^1{}_\alpha \cdots \underline{a}^n{}_\alpha$. f_α is monoid epimorphism. Suppose that $x, y \in A^*$ with $x \equiv^{(1)}{}_\iota\, y$, so that there exists $u, v \in A^*$ and $a, b \in A$ with $a\,\iota\, b$, $x = u.a.b.v$ and $y = u.b.a.v$. Then, by 9.2.13 (2) and (3)

$$f_\alpha(x) = f_\alpha(u).\underline{a}_\alpha.\underline{b}_\alpha.f_\alpha(v) = f_\alpha(u).\underline{b}_\alpha.\underline{a}_\alpha.f_\alpha(v) = f_\alpha(y)$$

from which we deduce

$$\forall x, y \in A^*: x \equiv_\iota y \Rightarrow f_\alpha(x) = f_\alpha(y)$$

so that there exists a monoid epimorphism $\varphi_\alpha: A_\iota{}^* \to A_\alpha{}^*$ given by $\varphi_\alpha(x_\iota) = f_\alpha(x)$. In fact:

9.3.3. THEOREM. The function $\varphi_\alpha: A_\iota{}^* \to A_\alpha{}^*$ satisfying $\varphi_\alpha(x_\iota) = f_\alpha(x)$, all $x \in A^*$, is both a monoid and poset isomorphism.

PROOF. We first show that φ_α is injective. Since we know that φ_α is a monoid epimorphism, this shows that φ_α is a monoid isomorphism.

Suppose that $X, Y \in A_\iota{}^*$ such that $\varphi_\alpha(X) = \varphi_\alpha(Y)$ and let $x \in X$ and $y \in Y$ be such that if $x' \in X$ and $y' \in Y$, then $|x \wedge y| \ge |x' \wedge y'|$. Here $x \wedge y$ denotes the longest common prefix of x and y. We prove that $x = y$, from which it follows that $X = Y$.

Suppose $x \ne y$, then since $\varphi_\alpha(X) = \varphi_\alpha(Y)$, we may write $x = u.a.v$ and $y = u.b^1 \cdots b^r.a.w$ such that $a \ne b^n$, $n = 1, \cdots, r$. Now, $\varphi_\alpha(X) = \varphi_\alpha(Y)$ means that

$$f_\alpha(u).\underline{a}_\alpha.f_\alpha(v) = f_\alpha(u).\underline{b}^1{}_\alpha \cdots \underline{b}^r{}_\alpha.\underline{a}_\alpha.f_\alpha(w)$$

and so if $\underline{a}_\alpha(i) > \Omega$, then $a \le \underline{b}^1{}_\alpha(i) \cdots \underline{b}^r{}_\alpha(i).a.f_\alpha(w)$, and since $a \ne b^n$, $n = 1, \cdots, r$, we must have $\underline{b}^n{}_\alpha(i) = \Omega$. Thus, for all $n = 1, \cdots, r$, $\underline{a}_\alpha\ ind_\alpha\ \underline{b}^n{}_\alpha$. and hence $a\,\iota_\alpha\, b^n$, by 9.2.9, so if we define $y' = u.a.b^1 \cdots b^r.w$, then $y' \equiv_{\iota_\alpha} y$, and hence $y' \in Y$. But then, $|x \wedge y| < |x \wedge y'|$, the desired contradiction.

Finally, suppose that $x, y \in A$, then

$$x_\iota \le y_\iota \Leftrightarrow \exists z_\iota \in A_\iota^*: x_\iota . z_\iota = y_\iota, \text{ by } 9.2.1$$
$$\Leftrightarrow \exists z_\iota \in A_\iota^*: \varphi_\alpha(x_\iota).\varphi_\alpha(z_\iota) = \varphi_\alpha(y_\iota)$$
$$\Leftrightarrow \exists \underline{z} \in A_\alpha^*: \varphi_\alpha(x_\iota).\underline{z} = \varphi_\alpha(y_\iota)$$
$$\Leftrightarrow \varphi_\alpha(x_\iota) \le \varphi_\alpha(y_\iota), \text{ by } 9.3.2.$$

■

Notes.

Vector languages were originally introduced by the author as a semantic model for unambiguous path expressions [15, 33] in [25] and have subsequently been much studied in that connection, particularly by R. Janicki and P. E. Lauer. [34, 35, 36]. M. Nivat's vector model [37] appeared at much the same time. We discuss the relationship between these two models in chapter 25.

Chapter 10
Order Theoretic Properties of Vector Languages

Round up the usual suspects.

from *Casablanca*

10.1. Introduction.

We have introduced trace and vector languages as models of behaviour. These have an algebraic or language theoretic flavour, in contrast to the order theoretic flavour of behavioural presentations and event structures. The two approaches may be formally related, however, and we establish this in chapter 12.

The question is: given a trace or vector language, how may one construct a behavioural presentation from it? The clue is given by 4.3.3, which says that if $(D,\leq)$ is prime algebraic and consistently complete and $\lambda: Pr(D) \rightarrow E$ then $B_{D,E,\lambda}$ is a behavioural presentation.

In this chapter, we shall show that left-closed trace languages (or, what comes to the same thing, as a consequence of theorem 9.3.3, left-closed vector languages) are indeed prime algebraic and consistently complete and that there is a natural way of constructing a function λ.

The construction of the corresponding behavioural presentations will be deferred to chapter 12, where we shall also give a characterisation, up to isomorphism, of behavioural presentations which arise in this way.

10.2. Consistent Completeness.

Our purpose in setting up the vector machinery was to simplify the work involved in examining the order theoretic structure of trace languages. We have already seen that the partial order on vectors is defined co-ordinate-wise. It turns out that least upper bounds, for example, may be calculated co-ordinate-wise.

For the rest of this chapter, therefore, we will work almost exclusively with vectors. 9.3.3 allows us to transfer results that we obtain about A_α^* into results about $A_{\iota_\alpha}^*$. We also know from 9.2.4 that every monoid A_ι^* equals $A_{\iota_\alpha}^*$ for some indexed cover α.

Let us start by considering greatest lower bounds. First, if $X \subseteq A_\alpha^*$ is non empty and $\underline{y} \leq X$ (that is $\underline{y} \leq \underline{x}$, all $\underline{x} \in X$), then define

$$X/\underline{y} = \{\underline{x}/\underline{y} \mid \underline{x} \in X\}.$$

We prove that if $\emptyset \subset X \subseteq A_\alpha^*$, then $\sqcap X$ exists. The proof is by induction on $|X| = min(\{|\underline{x}| \mid \underline{x} \in X\})$. If this number is zero, then by 9.2.11 (1), $\underline{\Omega}_\alpha \in X$ and so $\sqcap X = \underline{\Omega}_\alpha$. Likewise, if $\underline{a}_\alpha \not\leq X$, for every $a \in A$, then $\sqcap X = \underline{\Omega}_\alpha$. But, if $\underline{a}_\alpha \leq X$, for some $a \in A$, then $X/\underline{a}_\alpha$ is defined and non-empty, and furthermore $|X/\underline{a}_\alpha| < |X|$, by 9.2.11 (3). By induction $\sqcap (X/\underline{a}_\alpha)$ exists, and it now merely remains to show that if $\sqcap(X/\underline{a}_\alpha)$ exists, then

$$\forall a \in A: \underline{a}_\alpha \leq X \Rightarrow \sqcap X = \underline{a}_\alpha.(\sqcap(X/\underline{a}_\alpha)) \tag{10.1}$$

The last step in this argument is not, perhaps, as obvious as it might seem. Let us try to prove it. We wish to show that

$$\forall \underline{y} \in A_\alpha^*: \underline{y} \leq X \Leftrightarrow \underline{y} \leq \underline{a}_\alpha.(\sqcap(X/\underline{a}_\alpha)) \tag{10.2}$$

In the case of string languages, this is straightforward. Indeed, if $a \in A$, $y \in A^*$ and $X \subseteq A^*$ such that $y > \Omega$ and $a, y \leq X$, then $a \leq y$, since in general, if $x \in A^*$, then the set $\downarrow x = \{u \in A^* \mid u \leq x\}$ is totally ordered and since $y > \Omega$, we cannot have $y < a$. And now:

$$y \leq X \Leftrightarrow y/a \leq X/a \Leftrightarrow y/a \leq \sqcap(X/a) \Leftrightarrow y \leq a.(\sqcap(X/a)).$$

However, we have a richer structure in the case of vector languages. It is not necessarily the case that if $\underline{a}_\alpha, \underline{y} \leq X$ and $\underline{y} > \underline{\Omega}_\alpha$, then $\underline{a}_\alpha \leq \underline{y}$. For example, if $A = \{a, b\}$ and $I = \{1, 2\}$, with $\alpha(1) = \{a\}$ and $\alpha(2) = \{b\}$, then $\underline{a}_\alpha, \underline{b}_\alpha \leq \{(a, b)\}$ and $\underline{b}_\alpha > \underline{\Omega}_\alpha$, but $\underline{a}_\alpha \not\leq \underline{b}_\alpha$.

The actual state of affairs is given in the following lemma.

10.2.1. LEMMA. If $\underline{x}, \underline{y} \in M_\alpha$ and $a \in A$, then

$$\underline{a}_\alpha, \underline{x} \leq \underline{y} \wedge \underline{a}_\alpha \not\leq \underline{x} \Rightarrow \underline{a}_\alpha \text{ ind } \underline{x}$$

PROOF. Suppose that $\underline{a}_\alpha, \underline{x} \le \underline{y}$. If it is not the case that, $\underline{a}_\alpha$ *ind* $\underline{x}$ then for some $i \in I$, $\underline{a}_\alpha(i) > \Omega$ and $\underline{x}(i) > \Omega$, so that $a, \underline{x}(i) \le \underline{y}(i)$ and so $a \le \underline{x}(i)$, which means that $\#_a \underline{x} > 0$. Hence, for all $i \in I$, if $\underline{a}_\alpha(i) > \Omega$ then $\underline{x}(i) > \Omega$ and so $a \le \underline{x}(i)$. But then $\underline{a}_\alpha \le \underline{x}$.

■

Returning to our discussion, if $\underline{a}_\alpha, \underline{y} \le X$ and $X \neq \emptyset$, then by 10.2.1, either $\underline{a}_\alpha \le \underline{y}$ or $\underline{a}_\alpha$ *ind* $\underline{y}$. In the former case, $\underline{y}/\underline{a}_\alpha \le X/\underline{a}_\alpha$ - a consequence of the following, which may easily be proved by a coordinatewise argument.

$$\forall \underline{x}, \underline{y}, \underline{z} \in A_\alpha{}^*: \underline{x} \le \underline{y}, \underline{z} \Rightarrow (\underline{y} \le \underline{z} \Leftrightarrow \underline{y}/\underline{x} \le \underline{z}/\underline{x}) \tag{10.3}$$

Thus, $\underline{y}/\underline{a}_\alpha \le \sqcap(X/\underline{a}_\alpha)$ and so $\underline{y} = \underline{a}_\alpha.(\underline{y}/\underline{a}_\alpha) \le \underline{a}_\alpha.(\sqcap(X/\underline{a}_\alpha))$. This makes use of another result that may also be proved by a coordinatewise argument.

$$\forall \underline{x}, \underline{y}, \underline{z} \in A_\alpha{}^*: \underline{x} \le \underline{y} \Leftrightarrow \underline{z}.\underline{x} \le \underline{z}.\underline{y} \tag{10.4}$$

In the second case, that in which $\underline{a}_\alpha$ *ind* $\underline{y}$, we cannot form $\underline{y}/\underline{a}_\alpha$. However, we do have the following result.

10.2.2. LEMMA. If $\underline{y}, \underline{z} \in A_\alpha{}^*$ and $a \in A$, such that $\underline{a}_\alpha \le \underline{z}$ and $\underline{a}_\alpha$ *ind* $\underline{y}$ then $\underline{y} \le \underline{z} \Leftrightarrow \underline{y} \le \underline{z}/\underline{a}_\alpha$.

PROOF. We prove that $\underline{y}(i) \le \underline{z}(i) \Leftrightarrow \underline{y}(i) \le (\underline{z}/\underline{a}_\alpha)(i)$, all $i \in I$. This is certainly the case if $\underline{y}(i) = \Omega$. If, $\underline{y}(i) > \Omega$, then $\underline{a}_\alpha(i) = \Omega$, since, $\underline{a}_\alpha$ *ind* $\underline{y}$ and so,

$$\underline{y}(i) \le \underline{z}(i) \Leftrightarrow \underline{y}(i) \le \underline{z}(i)/\Omega \Leftrightarrow \underline{y}(i) \le \underline{z}(i)/\underline{a}_\alpha(i) \Leftrightarrow \underline{y}(i) \le (\underline{z}/\underline{a}_\alpha)(i).$$

■

Thus, in the second case, $\underline{y} \le X/\underline{a}_\alpha$, and so $\underline{y} \le \sqcap(X/\underline{a}_\alpha)$. We wish to be able to conclude that $\underline{y}/\underline{a}_\alpha \le \sqcap(X/\underline{a}_\alpha))$. But, this follows from 10.2.2.

So far, we have shown that

$$\forall \underline{y} \in A_\alpha{}^*: \underline{y} \le X \Rightarrow \underline{y} \le \underline{a}_\alpha.(\sqcap(X/\underline{a}_\alpha)) \tag{10.5}$$

Now, suppose that $\underline{y} \le \underline{a}_\alpha.(\sqcap(X/\underline{a}_\alpha))$ and let $\underline{x} \in X$. We show that $\underline{y} \le \underline{x}$. If $\underline{a}_\alpha \le \underline{y}$ then, $\underline{y}/\underline{a}_\alpha \le (\underline{a}_\alpha.(\sqcap(X/\underline{a}_\alpha))/\underline{a}_\alpha = \sqcap(X/\underline{a}_\alpha)$, using (10.3). But $\underline{x}/\underline{a}_\alpha \le X/\underline{a}_\alpha$ and thus, $\underline{y}/\underline{a}_\alpha \le \underline{x}/\underline{a}_\alpha$, so by (10.4), $\underline{y} \le \underline{x}$. On the other hand, if

$\underline{a}_\alpha \not\leq \underline{y}$, then as $\underline{a}_\alpha, \underline{y} \leq \underline{a}_\alpha.(\sqcap(X/\underline{a}_\alpha))$, $\underline{a}_\alpha$ *ind* $\underline{y}$, by 10.2.1 and so, by 10.2.2 $\underline{y} \leq (\underline{a}_\alpha.(\sqcap(X/\underline{a}_\alpha))/\underline{a}_\alpha = \sqcap(X/\underline{a}_\alpha))$. Thus, $\underline{y} \leq X/\underline{a}_\alpha$. and so $\underline{y} \leq X$, by 10.2.2, since $\underline{a}_\alpha$ *ind* $\underline{y}$. We have proved the following

$$\forall \underline{y} \in A_\alpha{}^*: \underline{y} \leq \underline{a}_\alpha.(\sqcap(X/\underline{a}_\alpha)) \Rightarrow \underline{y} \leq X \qquad (10.6)$$

and putting (10.5) and (10.6) together, we obtain (10.2). This was the major step in the proof of the following.

10.2.3. PROPOSITION. If $\varnothing \subset X \subseteq A_\alpha{}^*$, then $\sqcap X$ exists.

■

$A_\alpha{}^*$ contains a bottom element, namely $\underline{\Omega}_\alpha$. Thus, by the argument following 4.2.1:

10.2.4. COROLLARY. $A_\alpha{}^*$ is consistently complete.

■

We now look at ways to compute least upper bounds.

10.2.5. LEMMA. Let $\underline{u}, \underline{x}, \underline{y} \in A_\alpha{}^*$ such that $\underline{u} \leq \underline{x}, \underline{y}$ and $\underline{x}/\underline{u}$ *ind* $\underline{y}/\underline{u}$, then

1• $\underline{x}.(\underline{y}/\underline{u}) = \underline{y}.(\underline{x}/\underline{u})$;

2• $\forall i \in I: (\underline{x}.(\underline{y}/\underline{u}))(i) = max(\underline{x}(i), \underline{y}(i))$;

3• $\underline{x} \sqcup \underline{y} = \underline{x}.(\underline{y}/\underline{u})$.

PROOF. Let $\underline{x}' = \underline{x}/\underline{u}$ and $\underline{y}' = \underline{y}/\underline{u}$.

By 9.3.2 and 9.2.13 (3), $\underline{x}.(\underline{y}/\underline{u}) = \underline{u}.\underline{x}'.\underline{y}' = \underline{u}.\underline{y}'.\underline{x}' = \underline{y}.(\underline{x}/\underline{u})$, giving (1).

Let $i \in I$. If $\underline{x}'(i) > \Omega$, then $\underline{y}'(i) = \Omega$, by 9.2.12, so

$$\underline{y}(i) = \underline{u}(i) < \underline{u}(i).\underline{x}'(i) = \underline{x}(i) = \underline{x}(i).\underline{y}'(i) = (\underline{x}.\underline{y}')(i).$$

Similarly, if $\underline{y}'(i) > \Omega$, then $\underline{x}'(i) = \Omega$ and

$$\underline{x}(i) = \underline{u}(i) < \underline{u}(i).\underline{y}'(i) = \underline{y}(i) = \underline{y}(i).\underline{x}'(i) = (\underline{y}.\underline{x}')(i).$$

The only other possibility is $\underline{x}'(i) = \underline{y}'(i) = \Omega$ and here, $\underline{x}(i) = \underline{y}(i) = (\underline{y}.\underline{x}')(i)$. We have proved (2). Using (2), we have

$$\begin{aligned}\underline{x},\underline{y} \leq \underline{z} &\Leftrightarrow \forall i \in I: \underline{x}(i), \underline{y}(i) \leq \underline{z}(i)\\ &\Leftrightarrow \forall i \in I: max(\underline{x}(i), \underline{y}(i)) \leq \underline{z}(i)\\ &\Leftrightarrow \forall i \in I: (\underline{x}.\underline{y}')(i) \leq \underline{z}(i)\end{aligned}$$

■

10.2.6. LEMMA. If $\underline{u}, \underline{x}, \underline{y} \in A_\alpha{}^*$ such that $\underline{u} \leq \underline{x}, \underline{y}$ then

1• If $\underline{x} \sqcup \underline{y}$ is defined, then $(\underline{x} \sqcup \underline{y})/\underline{u} = \underline{x}/\underline{u} \sqcup \underline{y}/\underline{u}$;

2• $(\underline{x} \sqcap \underline{y})/\underline{u} = \underline{x}/\underline{u} \sqcap \underline{y}/\underline{u}$

PROOF. Using (10.3), for every $\underline{v} \in A_\alpha{}^*$

$$(\underline{x} \sqcup \underline{y})/\underline{u} \leq v \Leftrightarrow (\underline{x} \sqcup \underline{y}) \leq \underline{u}.\underline{v} \Leftrightarrow \underline{x}, \underline{y} \leq \underline{u}.\underline{v} \Leftrightarrow \underline{x}/\underline{u}, \underline{y}/\underline{u} \leq \underline{v}.$$

Likewise, for any $\underline{v} \in A_\alpha{}^*$,

$$\underline{v} \leq (\underline{x} \sqcap \underline{y})/\underline{u} \Leftrightarrow \underline{u}.\underline{v} \leq x \sqcap \underline{y} \Leftrightarrow \underline{u}.\underline{v} \leq \underline{x}, \underline{y} \Leftrightarrow \underline{v} \leq \underline{x}/\underline{u}, \underline{y}/\underline{u}$$

■

10.2.7. LEMMA. If $\underline{x}, \underline{y}, \underline{z} \in A_\alpha{}^*$ with $\underline{x}, \underline{y} \leq \underline{z}$, then

$$\underline{x} \; ind \; \underline{y} \quad \Leftrightarrow \quad \underline{x} \sqcap \underline{y} = \underline{\Omega}_\alpha.$$

PROOF. If $\underline{x} \sqcap \underline{y} > \underline{\Omega}_\alpha$, then, for some $i \in I$, $(\underline{x} \sqcap \underline{y})(i) > \Omega$ and so $\underline{x}(i) > \Omega$ and $\underline{y}(i) > \Omega$ and it is not the case that $\underline{x} \; ind \; \underline{y}$.

Conversely, suppose that $\underline{x} \sqcap \underline{y} = \underline{\Omega}_\alpha$, so that $\underline{a}_\alpha \leq \underline{x}, \underline{y}$ for no $a \in A$. We show by induction on $|\underline{x}|$ that $\underline{x} \; ind \; \underline{y}$.

If $|\underline{x}| = 0$, then $\underline{x} = \underline{\Omega}_\alpha$, by 9.2.11 (1) and certainly $\underline{\Omega}_\alpha \; ind \; \underline{y}$.

If $|\underline{x}| > 0$, then $\underline{b}_\alpha \leq \underline{x}$ some $b \in A$. By hypothesis, $\underline{b}_\alpha \not\leq \underline{y}$ and so $\underline{b}_\alpha \; ind \; \underline{y}$, by 10.2.1, and so by 10.2.2, $\underline{y} \leq \underline{z}/\underline{b}_\alpha$. Also, by (10.3), $\underline{x}' \leq \underline{z}/\underline{b}_\alpha$, where $\underline{x}' = \underline{x}/\underline{b}_\alpha$.

Next, we show that $\underline{x}' \sqcap \underline{y} = \underline{\Omega}_\alpha$, that is, $\underline{a}_\alpha \leq \underline{x}', \underline{y}$ for no $a \in A$. Indeed, if $a \in A$ such that $\underline{a}_\alpha \leq \underline{x}', \underline{y}$ then as $\underline{b}_\alpha \; ind \; \underline{y}$, $\underline{b}_\alpha \; ind \; \underline{a}_\alpha$. But then, $\underline{x} = \underline{b}_\alpha . \underline{a}_\alpha . (\underline{x}'/\underline{a}_\alpha) = \underline{a}_\alpha . \underline{b}_\alpha . (\underline{x}'/\underline{a}_\alpha)$, so $\underline{a}_\alpha \leq \underline{x}, \underline{y}$, contrary to hypothesis. Hence, $\underline{x}' \sqcap \underline{y} = \underline{\Omega}_\alpha$. Therefore, $\underline{x}', \underline{y} \leq \underline{z}/\underline{b}_\alpha$ and $\underline{x}' \sqcap \underline{y} = \underline{\Omega}_\alpha$ with $|\underline{x}'| < |\underline{x}|$, so $\underline{x}' \; ind \; \underline{y}$, by induction, and as $\underline{b}_\alpha \; ind \; \underline{y}$, $\underline{x} \; ind \; \underline{y}$ by 9.2.13 (4).

■

10.2.8. PROPOSITION. Suppose $\underline{x}, \underline{y} \in A_\alpha{}^*$, then $\underline{x} \sqcup \underline{y}$ exists $\Leftrightarrow$ $\underline{x}'$ *ind* $\underline{y}'$, where $\underline{x}' = \underline{x}/(\underline{x} \sqcap \underline{y})$ and $\underline{y}' = \underline{y}/(\underline{x} \sqcap \underline{y})$ and then

$$\underline{x} \sqcup \underline{y} = (\underline{x} \sqcap \underline{y}).\underline{x}'.\underline{y}' = \underline{x}.\underline{y}' = \underline{y}.\underline{x}'.$$

PROOF. If $\underline{x}'$ *ind* $\underline{y}'$, then $\underline{x} \sqcup \underline{y} = \underline{x}.\underline{y}'$, by 10.2.5(3).

Conversely, suppose that $\underline{x} \sqcup \underline{y}$ exists, then by 10.2.6(1), $\underline{x}' \sqcup \underline{y}'$ exists. By 10.2.6 (2)

$$\underline{x}' \sqcap \underline{y}' = \underline{x}/(\underline{x} \sqcap \underline{y}) \sqcap \underline{y}/(\underline{x} \sqcap \underline{y}) = (\underline{x} \sqcap \underline{y})/(\underline{x} \sqcap \underline{y}) = \underline{\Omega}_\alpha$$

and since $\underline{x}', \underline{y}' \le \underline{x}' \sqcup \underline{y}'$, $\underline{x}'$ *ind* $\underline{y}'$, by 10.2.7.

Finally, $(\underline{x} \sqcap \underline{y}).\underline{x}'.\underline{y}' = \underline{x}.\underline{y}'$, since $\underline{x}' = \underline{x}/(\underline{x} \sqcap \underline{y})$, and since $\underline{x}'$ *ind* $\underline{y}'$, $\underline{x}.\underline{y}' = \underline{y}.\underline{x}'$, by 10.2.5 (3).

■

For instance, if $\underline{x} = (abc, a)$ and $\underline{y} = (a, ade)$, then $\underline{u} = \underline{x} \sqcap \underline{y} = (a, a)$, $\underline{x}/\underline{u} = (bc, \Omega)$ and $\underline{y}/\underline{u} = (\Omega, de)$. And by 9.2.12, we have (bc, Ω) *ind* (Ω, de) so $\underline{x} \sqcup \underline{y}$ exists and equals (abc, ade).

The next result shows one advantage of vectors over traces: least upper bounds may be calculated in a straightforward manner.

10.2.9. COROLLARY. Let $X \subseteq A_\alpha{}^*$, then $\sqcup X$ exists and equals $\underline{x} \in A_\alpha{}^*$ if and only if:

$$\forall i \in I: \underline{x}(i) = \sqcup\{\underline{y}(i) \mid \underline{y} \in X\} \tag{10.7}$$

PROOF. Certainly, if $\underline{x}$ satisfies (10.7), then $X \le \underline{u} \Leftrightarrow \underline{x} \le \underline{u}$.

Conversely, if $\sqcup X$ exists, then, X must be bounded above, by $\underline{z}$ say. Since there are only a finite number of elements $\underline{u} \le \underline{z}$, it follows that X is finite. The corollary now follows from induction on $|X|$ and 10.2.5.

■

The analogous statement for greatest lower bounds is not true in general. For example, the coordinatewise greatest lower bound of (ab, bd) and (ab, cbd) is (ab, Ω), but $(ab, bd) \sqcap (ab, cbd) = (a, \Omega)$. However, the analogous statement is true when the least upper bound of the two vectors exists, as we see from the next result. First, if $\underline{x}, \underline{y} \in A_\alpha{}^*$, then we define:

$\underline{x} \leftrightarrow \underline{y} \Leftrightarrow \forall i \in I\colon \underline{x}(i) \leq \underline{y}(i) \vee \underline{y}(i) \leq \underline{x}(i)$.

10.2.10. PROPOSITION. Suppose that $\underline{x}, \underline{y} \in A_\alpha{}^*$ with $\underline{x} \leftrightarrow \underline{y}$, then

1• $\forall i \in I\colon (\underline{x} \sqcap \underline{y})(i) = min(\underline{x}(i), \underline{y}(i))$;

2• $\underline{x} \sqcup \underline{y}$ exists.[1]

PROOF. (1) We argue by induction on $|\underline{x}|$. If $|\underline{x}| = 0$, then $\underline{x} = \underline{\Omega}_\alpha$ and hence $\underline{x} \sqcap \underline{y} = \underline{\Omega}_\alpha$ and for all $i \in I$, $min(\underline{x}(i), \underline{y}(i)) = \Omega$.

Next, suppose that $|\underline{x}| > 0$ and let $a \in A$ such that $\underline{a}_\alpha \leq \underline{x}$. We have two cases:

Case 1: $\underline{a}_\alpha$ *ind* $\underline{y}$. If $a \in \alpha(i)$, then $\underline{y}(i) = \Omega \leq (\underline{x}/\underline{a}_\alpha)(i)$, by 9.2.12. If $a \notin \alpha(i)$, then $(\underline{x}/\underline{a}_\alpha)(i) = \underline{x}(i)$ so $(\underline{x}/\underline{a}_\alpha)(i) \leq \underline{y}(i) \vee \underline{y}(i) \leq (\underline{x}/\underline{a}_\alpha)(i)$. Thus, $\underline{x}/\underline{a}_\alpha \leftrightarrow \underline{y}$. By induction, $((\underline{x}/\underline{a}_\alpha) \sqcap \underline{y})(i) = min((\underline{x}/\underline{a}_\alpha)(i), \underline{y}(i))$, all $i \in I$, We now argue that $\underline{x} \sqcap \underline{y} = (\underline{x}/\underline{a}_\alpha) \sqcap \underline{y}$.

Suppose that $\underline{z} \leq \underline{x} \sqcap \underline{y}$, If $\underline{a}_\alpha \leq \underline{z}$, then $\underline{a}_\alpha \leq \underline{y}$, contradicting $\underline{a}_\alpha$ *ind* $\underline{y}$ so $\underline{a}_\alpha \not\leq \underline{z}$. But $\underline{a}_\alpha, \underline{z} \leq \underline{x}$, so $\underline{a}_\alpha$ *ind* $\underline{z}$, by 10.2.1, and hence $\underline{z} \leq \underline{x}/\underline{a}_\alpha$ by 10.2.2. Since $\underline{z} \leq \underline{y}$, $\underline{z} \leq (\underline{x}/\underline{a}_\alpha) \sqcap \underline{y}$.

Conversely, let $\underline{z} \leq (\underline{x}/\underline{a}_\alpha) \sqcap \underline{y}$, then as $\underline{z} \leq \underline{y}$ and $\underline{a}_\alpha$ *ind* $\underline{y}$, it follows that $\underline{a}_\alpha$ *ind* $\underline{z}$ and since $\underline{a}_\alpha, \underline{z}/\underline{a}_\alpha \leq \underline{x}$, $\underline{z} \leq \underline{x}$ by 10.2.2. As $\underline{z} \leq \underline{y}, \underline{z} \leq \underline{x} \sqcap \underline{y}$.

Thus, $\underline{z} \leq \underline{x} \sqcap \underline{y} \Leftrightarrow \underline{z} \leq (\underline{x}/\underline{a}_\alpha) \sqcap \underline{y}$, that is $\underline{x} \sqcap \underline{y} = (\underline{x}/\underline{a}_\alpha) \sqcap \underline{y}$.

If $a \in \alpha(i)$, then $\underline{y}(i) = \Omega$, as $\underline{a}_\alpha$ *ind* $\underline{y}$ so

$$min((\underline{x}/\underline{a}_\alpha)(i), \underline{y}(i)) = \Omega = min(\underline{x}(i), \underline{y}(i)).$$

On the other hand, if $a \notin \alpha(i)$, then $(\underline{x}/\underline{a}_\alpha)(i) = \underline{x}(i)$ and so in this case too we have $min((\underline{x}/\underline{a}_\alpha)(i), \underline{y}(i)) = min(\underline{x}(i), \underline{y}(i))$. Thus for all $i \in I$

$$(\underline{x} \sqcap \underline{y})(i) = ((\underline{x}/\underline{a}_\alpha) \sqcap \underline{y})(i) = min((\underline{x}/\underline{a}_\alpha)(i), \underline{y}(i)) = min(\underline{x}(i), \underline{y}(i)).$$

Case 2: $\neg \underline{a}_\alpha$ *ind* $\underline{y}$. Let $i \in I$ such that $a \in \alpha(i)$ and $\underline{y}(i) > \Omega$, then as $a \leq \underline{x}(i)$ and $\underline{x} \leftrightarrow \underline{y}$, $a \leq \underline{y}(i)$. Thus, $\#_a \underline{y} > 0$ so for every $i \in I$ such that $a \in \alpha(i)$, $\underline{y}(i) > \Omega$, and so $a \leq \underline{y}(i)$. Thus, $\underline{a}_\alpha \leq \underline{y}$. Clearly, $\underline{x}/\underline{a}_\alpha \leftrightarrow \underline{y}/\underline{a}_\alpha$ and hence

[1]We also note that if $\underline{x} \sqcup \underline{y}$ exists then $\underline{x} \leftrightarrow \underline{y}$, as a consequence of 10.2.9.

$((\underline{x}/\underline{a}_\alpha)\sqcap(\underline{y}/\underline{a}_\alpha))(i) = min((\underline{x}/\underline{a}_\alpha)(i),(\underline{y}/\underline{a}_\alpha)(i))$, for all $i \in I$, by induction. By 10.2.6 (2), $(\underline{x}\sqcap\underline{y})/\underline{a}_\alpha = (\underline{x}/\underline{a}_\alpha)\sqcap(\underline{y}/\underline{a}_\alpha)$. Therefore

$$(\underline{x}\sqcap\underline{y})(i) = \underline{a}_\alpha(i).min((\underline{x}/\underline{a}_\alpha)(i),(\underline{y}/\underline{a}_\alpha)(i)) = min(\underline{x}(i),\underline{y}(i))$$

(2) If $\underline{x}/(\underline{x}\sqcap\underline{y}))(i) > \Omega$, then by (1), $(\underline{x}\sqcap\underline{y})(i) = \Omega$ so $(\underline{y}/(\underline{x}\sqcap\underline{y}))(i) = \Omega$. Thus, $\underline{x}/(\underline{x}\sqcap\underline{y})\ ind_\alpha\ \underline{y}/(\underline{x}\sqcap\underline{y})$. By 10.2.8, $\underline{x}\sqcup\underline{y}$ exists.

■

Before we embark on our next task, the examination of prime algebraicity, we present the following useful result, to which we shall refer as the *factorisation lemma.*

10.2.11. LEMMA. Suppose that $\underline{x},\underline{y},\underline{z} \in A_\alpha{}^*$, then

$$\underline{x} \leq \underline{y}.\underline{z} \Rightarrow \exists \underline{u},\underline{v} \in A_\alpha{}^*: \underline{u}.\underline{v} = \underline{x} \wedge \underline{u} \leq \underline{y} \wedge \underline{v} \leq \underline{z} \wedge \underline{v}\ ind\ \underline{y}/\underline{u}$$

Consequently, $\underline{y}.\underline{z} = \underline{x}.(\underline{y}/\underline{u}).(\underline{z}/\underline{v})$.

PROOF. The reader may find the diagram of figure 10.1 useful in picturing the statement of the lemma.

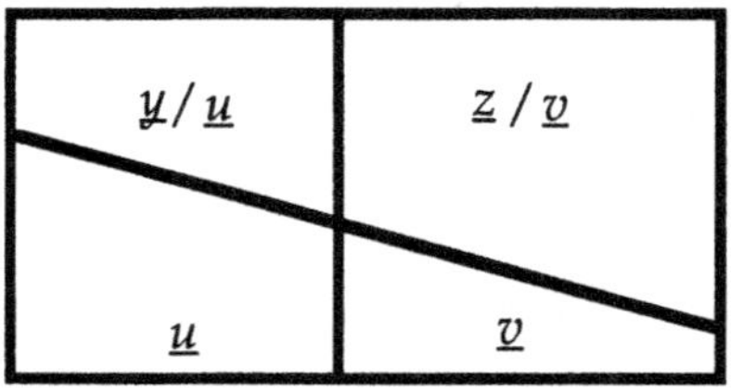

Figure 10.1.

Let $\underline{u} = \underline{x} \sqcap \underline{y}$, then $\underline{u} \leq \underline{x}$ and so we may define $\underline{v} = \underline{x}/\underline{u}$. It immediately follows that $\underline{x} = \underline{u}.\underline{v}$ and that $\underline{u} \leq \underline{y}$.

We next show that $\underline{v}\ ind\ (\underline{y}/\underline{u})$. First, we note that $\underline{v} \leq (\underline{y}/\underline{u}).\underline{z}$ and $\underline{y}/\underline{u} \leq (\underline{y}/\underline{u}).\underline{z}$. Also

$$\underline{v} \sqcap (\underline{y}/\underline{u}) = (\underline{x}/\underline{u}) \sqcap (\underline{y}/\underline{u}) = (\underline{x}\sqcap\underline{y})/\underline{u} = (\underline{x}\sqcap\underline{y})/(\underline{x}\sqcap\underline{y}) = \underline{\Omega}_\alpha$$

by 10.2.6, (2), so that $\underline{v}\ ind\ (\underline{y}/\underline{u})$, by 10.2.7. Finally, from $\underline{u} \leq \underline{y}$ and a coordinatewise argument, we may conclude that $(\underline{y}.\underline{z})/\underline{u} = (\underline{y}/\underline{u}).\underline{z}$, so

$\underline{v} = \underline{x}/\underline{u} \leq (\underline{y}.\underline{z})/\underline{u} = (\underline{y}/\underline{u}).\underline{z}$, using (10.3). This, together with $\underline{v}\ ind\ (\underline{y}/\underline{u})$, entails that $\underline{v} \leq \underline{z}$ by 10.2.2 and induction on $|\underline{z}|$.

■

10.3. Prime Algebraicity.

We now wish to show that left-closed vector languages are prime algebraic. Obviously, it would be a great help if we could say what the complete primes were. We can get a clue about this by considering what the complete primes of behavioural presentations and string languages are. The former, you will remember, are points $\downarrow o$, where $o \in O$, that is, points which contain a unique final simultaneity set $o/\approx$. Now, we don't have any non-trivial simultaneity in vector or trace languages, so that it is a question of behaviours with a unique final occurrence. This prompts the following definition.

10.3.1. DEFINITION. Define $Pr(A_\alpha{}^*)$ as the set of all vectors $\underline{x}$ such that

$$\forall a, b \in A\ \forall \underline{u}, \underline{v} \in A_\alpha{}^*: \underline{u}.\underline{a} = \underline{x} \wedge \underline{v}.\underline{b} = \underline{x} \Rightarrow a = b$$

If $\underline{x} \in A_\alpha{}^*$ then define $Pr(\underline{x}) = \{\underline{u} \in Pr(A_\alpha{}^*) \mid \underline{u} \leq \underline{x}\}$. If $X \subseteq A_\alpha{}^*$, then define $Pr(X) = Pr(A_\alpha{}^*) \cap X$.

For example,

$$Pr((abc, aebd)) = \{(a,a), (a,ae), (ab,aeb), (abc,aeb), (ab,aebd)\}.$$

We shall prove that the elements of $Pr(A_\alpha{}^*)$ are the complete primes of $A_\alpha{}^*$. First, we give a useful construction, which, among other things, allows the elements of $Pr(\underline{x})$ to be computed. The idea is that if $\underline{x} \in A_\alpha{}^*$ and $a \in A$, then we can find a unique vector $\underline{u} \leq \underline{x}$ whose length is minimal with respect to the property $\underline{x}/\underline{u}\ ind\ \underline{a}_\alpha$. It turns out that $\underline{u}.\underline{a}_\alpha$ is prime.

First, if $\underline{x} \in A_\alpha{}^*$ and $a \in A$, then define

$$X_{\underline{x},a} = \{\underline{w} \leq \underline{x} \mid \underline{a}_\alpha\ ind\ \underline{x}/\underline{w}\} \tag{10.8}$$

10.3.2. LEMMA. If $\underline{x} \in A_\alpha{}^*$ and $a \in A$ then $\sqcap X_{\underline{x},a}$ exists and belongs to $X_{\underline{x},a}$

PROOF. $\sqcap X_{\underline{x},a}$ exists, by 10.2.3, since $\underline{x} \in X_{\underline{x},a}$. Define $\underline{u}_{\underline{x},a} = \sqcap X_{\underline{x},a}$, then $\underline{u}_{\underline{x},a} \leq \underline{x}$. By 10.2.10, for each $i \in I$, there exists $\underline{w} \in X_{\underline{x},a}$ such that $\underline{u}_{\underline{x},a}(i) = \underline{w}(i)$.

Thus, if $\underline{a}_\alpha(i) > \Omega$ then $(\underline{x}/\underline{u}_{\underline{x},a})(i) = (\underline{x}/\underline{w})(i) = \Omega$, by (10.8), so $\underline{a}_\alpha$ *ind* $\underline{x}/\underline{u}_{\underline{x},a}$. Hence, $\underline{u}_{\underline{x},a} \in X_{\underline{x},a}$.

■

If $\underline{x} \in A_\alpha{}^*$ and $a \in A$, then we define $pr(\underline{x}, a) = \underline{u}_{\underline{x},a}.\underline{a}_\alpha$.

10.3.3. LEMMA. If $\underline{x} \in A_\alpha{}^*$ and $a \in A$, then $\underline{u}_{\underline{x},a}$ is the unique vector $\underline{u} \leq \underline{x}$ such that $\underline{a}_\alpha$ *ind* $\underline{x}/\underline{u}$ and $\underline{u}.\underline{a}_\alpha \in Pr(A_\alpha{}^*)$.

PROOF. By 10.3.2, $\underline{u}_{\underline{x},a} \in X_{\underline{x},a}$, so by (10.8), $\underline{u}_{\underline{x},a} \leq \underline{x}$ and $\underline{a}_\alpha$ *ind* $\underline{x}/\underline{u}_{\underline{x},a}$. Let us show that $\underline{u}_{\underline{x},a}.\underline{a}_\alpha \in Pr(A_\alpha{}^*)$. Suppose $\underline{u}' \in A_\alpha{}^*$ and $b \in A$ such that $\underline{u}_{\underline{x},a}.\underline{a}_\alpha = \underline{u}'.\underline{b}_\alpha$. We must show that $a = b$. Note that if $a \neq b$, then $\underline{a}_\alpha$ *ind* $\underline{b}_\alpha$, for otherwise, we would have $\underline{u}_{\underline{x},a}(i).a = \underline{u}'(i).b$, some $i \in I$. Hence, whether or not $a = b$, $\underline{a}_\alpha.\underline{b}_\alpha = \underline{b}_\alpha.\underline{a}_\alpha$.

Since $\underline{u}_{\underline{x},a} \leq \underline{u}'.\underline{b}_\alpha$, that there exists $\underline{z}_1, \underline{z}_2 \in A_\alpha{}^*$ such that $\underline{u}_{\underline{x},a} = \underline{z}_1.\underline{z}_2$, $\underline{z}_1 \leq \underline{u}'$, $\underline{z}_2 \leq \underline{b}_\alpha$ and $\underline{z}_2$ *ind* $\underline{u}'/\underline{z}_1$, by the factorisation lemma (10.2.11).

There are two cases to consider, $\underline{z}_2 = \underline{b}_\alpha$ and $\underline{z}_2 = \underline{\Omega}_\alpha$.

If $\underline{z}_2 = \underline{\Omega}_\alpha$, then $\underline{u}_{\underline{x},a} = \underline{z}_1$ so $\underline{u}_{\underline{x},a} \leq \underline{u}'$ and $|\underline{u}_{\underline{x},a}| = |\underline{u}_{\underline{x},a}.\underline{a}_\alpha| - 1 = |\underline{u}'.\underline{b}_\alpha| - 1 = |\underline{u}'|$, so $\underline{u}_{\underline{x},a} = \underline{u}'$. Thus, $\underline{a}_\alpha = (\underline{u}_{\underline{x},a}.\underline{a}_\alpha)/\underline{u}_{\underline{x},a} = (\underline{u}'.\underline{b}_\alpha)/\underline{u}' = \underline{b}_\alpha$

If, on the other hand, $\underline{z}_2 = \underline{b}_\alpha$, then $\underline{u}_{\underline{x},a} = \underline{z}_1.\underline{b}_\alpha$ with $\underline{b}_\alpha$ *ind* $\underline{u}'/\underline{z}_1$. But, $\underline{u}'.\underline{b}_\alpha = \underline{u}_{\underline{x},a}.\underline{a}_\alpha = \underline{z}_1.\underline{b}_\alpha.\underline{a}_\alpha = \underline{z}_1.\underline{a}_\alpha.\underline{b}_\alpha$, so $\underline{u}'/\underline{z}_1 = \underline{a}_\alpha$ and hence $\underline{a}_\alpha$ *ind* $\underline{b}_\alpha$. Now $\underline{x}/\underline{z}_1 = \underline{b}_\alpha.(\underline{x}/(\underline{z}_1.\underline{b}_\alpha)) = \underline{b}_\alpha.(\underline{x}/\underline{u}_{\underline{x},a})$ and since $\underline{a}_\alpha$ *ind* $\underline{x}/\underline{u}_{\underline{x},a}$ and $\underline{a}_\alpha$ *ind* $\underline{b}_\alpha$, $\underline{a}_\alpha$ *ind* $\underline{x}/\underline{z}_1$. Thus $\underline{z}_1 \in X_{\underline{x},a}$ and $\underline{z}_1 < \underline{u}_{\underline{x},a}$, a contradiction to $\underline{u}_{\underline{x},a} = \sqcap X_{\underline{x},a}$.

Finally, we show the uniqueness part. Suppose that $\underline{u}'$ also satisfies $\underline{u}' \leq \underline{x}$ and $\underline{a}_\alpha$ *ind* $\underline{x}/\underline{u}'$. By (10.8), $\underline{u}' \in X_{\underline{x},a}$ and so $\underline{u}_{\underline{x},a} \leq \underline{u}'$, since $\underline{u}_{\underline{x},a} = \sqcap X_{\underline{x},a}$. If $\underline{u}_{\underline{x},a} \neq \underline{u}'$, then there exists $\underline{w} \in A_\alpha{}^*$ and $b \in A$ such that $\underline{u}' = \underline{u}_{\underline{x},a}.\underline{w}.\underline{b}_\alpha$. Now, $\underline{w}.\underline{b}_\alpha = \underline{u}'/\underline{u}_{\underline{x},a} \leq \underline{x}/\underline{u}_{\underline{x},a}$ *ind* $\underline{a}_\alpha$, and therefore $\underline{a}_\alpha$ *ind* $\underline{b}_\alpha$, by 9.2.12 (4), and $\underline{u}'.\underline{a}_\alpha = \underline{u}_{\underline{x},a}.\underline{w}.\underline{b}_\alpha.\underline{a}_\alpha = \underline{u}_{\underline{x},a}.\underline{w}.\underline{a}_\alpha.\underline{b}_\alpha$ with $a \neq b$. Thus, $\underline{u}'.\underline{a}_\alpha \notin Pr(A_\alpha{}^*)$, by 10.3.1.

■

This construction allows one to generate all the primes below a given vector, as the next lemma shows.

10.3.4. LEMMA. If $\underline{x} \in A_\alpha{}^*$, then

$$Pr(\underline{x}) = \{pr(\underline{v}, a) \mid \underline{v} \in A_\alpha{}^* \wedge a \in A \wedge \underline{v}.\underline{a}_\alpha \leq \underline{x}\}.$$

PROOF. Let $P(\underline{x}) = \{pr(\underline{v}, a) \mid \underline{v} \in A_\alpha{}^* \wedge a \in A \wedge \underline{v}.\underline{a}_\alpha \leq \underline{x}\}$. We shall show that $P(\underline{x}) = Pr(\underline{x})$.

Indeed, if $\underline{u}.\underline{a}_\alpha \in Pr(A_\alpha{}^*)$, then $\underline{u}.\underline{a}_\alpha = pr(\underline{u}, a)$, for otherwise, $\underline{u} = \underline{v}.\underline{v}'.\underline{b}_\alpha$ with $\underline{a}_\alpha$ *ind* $\underline{v}'.\underline{b}_\alpha$ and hence $\underline{u}.\underline{a}_\alpha = \underline{v}.\underline{a}_\alpha.\underline{v}'.\underline{b}_\alpha$, Since $\underline{a}_\alpha$ *ind* $\underline{v}'.\underline{b}_\alpha$, $a \neq b$, giving a contradiction to $\underline{u}.\underline{a}_\alpha \in Pr(A_\alpha{}^*)$. Thus $Pr(\underline{x}) \subseteq P(\underline{x})$.

On the other hand, if $\underline{u}.\underline{a}_\alpha \in P(\underline{x})$, then there exists $\underline{v} \in A_\alpha{}^*$ such that $\underline{u}.\underline{v}.\underline{a}_\alpha = \underline{x}$ and $\underline{a}_\alpha$ *ind* $\underline{v}$. But by 9.2.13 (3), $\underline{u}.\underline{a}_\alpha \leq \underline{u}.\underline{a}_\alpha.\underline{v} = \underline{u}.\underline{v}.\underline{a}_\alpha \leq \underline{x}$ so $\underline{u}.\underline{a}_\alpha \in Pr(A_\alpha{}^*)$ and $P(\underline{x}) \subseteq Pr(\underline{x})$.

■

For example, let $\underline{x} = (abc, aebd)$, then $\underline{v}.\underline{c}_\alpha \leq \underline{x}$, where $\underline{v} = (ab, aebd)$ and $\underline{c}_\alpha = (c, \Omega)$, then $X_{\underline{x},c} = \{(ab, aebd), (ab, aeb)\}$ and so $pr(\underline{v}, c) = (abc, aeb)$, which we have seen, in the example following 10.3.1, to be a member of $Pr(\underline{x})$.

We would like to prove that $Pr(A_\alpha{}^*)$ is the set of complete primes of $A_\alpha{}^*$. Let us see how this goes. Suppose that $X \subseteq A_\alpha{}^*$ and that $\underline{u} \in Pr(A_\alpha{}^*)$ with $\underline{u} \leq \sqcup X$. We assume by induction that if $|X| < n$ then $\underline{u} \leq \underline{z}$, some $\underline{z} \in X$ (recall that if X has a least upper bound, then X must be finite). Suppose $|X| = n$.

If $\underline{u} \leq X$, then certainly $\underline{u} \leq \underline{z}$, some $\underline{z} \in X$. Let us suppose, then, that $\underline{u} \nleq X$ and let $\underline{z}_1 \in X$ with $\underline{u} \nleq \underline{z}_1$. Consider $X' = X - \{\underline{z}_1\}$. Since $X' \leq \sqcup X$, X' must itself have a least upper bound, call it $\underline{z}_2$. We will have succeeded in our aim, to show that $\underline{u}$ is prime, if we can show that $\underline{u} \leq \underline{z}_2$, since we would then have $\underline{u} \leq \sqcup X'$ with $|X'| = n-1$ and hence, by induction, $\underline{u} \leq \underline{z}$, some $\underline{z} \in X' \subseteq X$. The necessary inference is provided by the next lemma.

10.3.5. LEMMA. If $\underline{u} \in Pr(A_\alpha{}^*)$ and $\underline{z}_1, \underline{z}_2 \in A_\alpha{}^*$, then

$$\underline{u} \leq \underline{z}_1 \sqcup \underline{z}_2 \Rightarrow \underline{u} \leq \underline{z}_1 \vee \underline{u} \leq \underline{z}_2.$$

PROOF. Indeed, by 10.2.8, we have $\underline{z}_1 \sqcup \underline{z}_2 = \underline{v}.\underline{v}_1.\underline{v}_2$ where $\underline{v} = \underline{z}_1 \sqcap \underline{z}_2$, $\underline{v}_1 = \underline{z}_1/\underline{v}$, $\underline{v}_2 = \underline{z}_2/\underline{v}$ and $\underline{v}_1$ *ind* $\underline{v}_2$. Also, $\underline{u} = \underline{y}.\underline{a}_\alpha$, for some $\underline{y} \in A_\alpha{}^*$ and $a \in A$ and so by the factorisation lemma (10.2.11), we may write $\underline{y}.\underline{a}_\alpha = \underline{w}.\underline{w}_1.\underline{w}_2$, where $\underline{w} \leq \underline{v}$, $\underline{w}_1 \leq \underline{v}_1$ and $\underline{w}_2 \leq \underline{v}_2$, so that as $\underline{v}_1$ *ind* $\underline{v}_2$, $\underline{w}_1$ *ind* $\underline{w}_2$.

Now, we claim that if $\underline{w}_1 \neq \underline{\Omega}_\alpha$, then $\underline{w}_1$ is of the form $\underline{y}_1.\underline{a}_\alpha$ For otherwise, $\underline{w}_1 = \underline{y}_1.\underline{b}_\alpha$, with $b \in A$ and $a \neq b$, and accordingly:

$$\underline{u} = \underline{y}.\underline{a}_\alpha = \underline{w}.\underline{w}_1.\underline{w}_2 = \underline{w}.\underline{w}_2.\underline{w}_1 = \underline{w}.\underline{w}_2.\underline{y}_1.\underline{b}_\alpha$$

contradicting 10.3.1.

Since the $\underline{w}_i$ commute, exactly the same argument may be made for $\underline{w}_2$.

But if neither of the $\underline{w}_i$ equals $\underline{\Omega}_\alpha$, then they must both be of the form $\underline{y}_1 \cdot \underline{a}_\alpha$, which contradicts $\underline{w}_1$ *ind* $\underline{w}_2$. Thus one of them (say $\underline{w}_2$) equals $\underline{\Omega}_\alpha$. And now,

$$\underline{u} = \underline{y}.\underline{a}_\alpha = \underline{v}.\underline{w}_1.\underline{w}_2 \leq \underline{v}.\underline{w}_1 = \underline{z}_1.$$

■

10.3.6. COROLLARY. Each element of $Pr(A_\alpha^*)$ is prime in $(A_\alpha^*, \leq)$.

■

We next wish to show that the elements of $Pr(A_\alpha^*)$ are the only complete primes of A_α^*. The key to this is

$$\forall \underline{x} \in A_\alpha^*: \underline{x} = \bigsqcup Pr(\underline{x}) \tag{10.9}$$

for we may then argue that if $\underline{x}$ is prime, then from $\underline{x} \leq \bigsqcup Pr(\underline{x})$ we may infer that $\underline{x} \leq \underline{u}$ for some $\underline{u} \in Pr(\underline{x})$. But $\underline{u} \leq \underline{x}$ and so $\underline{x} = \underline{u}$. Thus, $\underline{x} \in Pr(\underline{x})$. (10.9) now allows us to conclude that $(A_\alpha^*, \leq)$ is prime algebraic.

10.3.7. THEOREM. $(A_\alpha^*, \leq)$ is prime algebraic and consistently complete with $Pr(A_\alpha^*)$ as its complete primes.

PROOF. We have shown consistent completeness in 10.2.4. By the preceding remarks, it suffices to establish (10.9). The proof is based on the following fact:

$$\forall i \in I \; \exists \underline{y} \in Pr(\underline{x}): \underline{x}(i) = \underline{y}(i) \tag{10.10}$$

(10.9) follows from (10.10) by virtue of 10.2.9.

Suppose $\underline{x}(i) = u.a$ with $a \in A$. First, we shall show that there exists $\underline{z} \leq \underline{x}$ such that $\underline{z}.\underline{a}_\alpha \leq \underline{x}$ and $\underline{z}(i) = u$.

Indeed, let $Y = \{\underline{z}' \leq \underline{x} \mid \underline{z}'(i) \leq u\}$ and let $\underline{z} = \bigsqcup Y$, which exists, by corollary 10.2.4, since Y is bounded above by $\underline{x}$. If $\underline{z}(i) \neq u$, then $\underline{z} \neq \underline{x}$ and hence $\underline{z}.\underline{b}_\alpha \in Y$ for any $\underline{b}_\alpha \leq \underline{x}/\underline{z}$, a contradiction to $\underline{z} = \bigsqcup Y$. A similar contradiction arises if $\underline{a}_\alpha \not\leq \underline{x}/\underline{z}$. Thus, $\underline{z}.\underline{a}_\alpha \leq \underline{x}$ and $\underline{z}(i) = u$.

Now let $\underline{y} = pr(\underline{z}, a)$. By 10.3.3, $\underline{y} \in Pr(\underline{x})$. Suppose $\underline{z} = \underline{y}.\underline{z}'$ with $\underline{a}_\alpha$ *ind* $\underline{z}'$, then since $\underline{a}_\alpha(i) > \Omega$, we must have $\underline{z}'(i) = \Omega$ and consequently $(\underline{y}.\underline{a}_\alpha)(i) = (\underline{y}.\underline{z}'.\underline{a}_\alpha)(i) = (\underline{z}.\underline{a}_\alpha)(i) = u.a$.

■

10.4. Left-Closed Vector Languages.

Sections 10.2 and 10.3 establish the principal order theoretic properties of sets $A_\alpha{}^*$, as part of our attempt to associate vector languages with behavioural presentations. This is part of a larger scheme, the aim of which is to associate a class of automata called asynchronous transition systems, with behavioural presentations. It turns out that asynchronous transition systems, give rise to subsets of $A_\alpha{}^*$ rather than the whole monoid. However, the subsets in question have a characteristic property.

10.4.1. DEFINITION. Suppose that $X, Y \subseteq A_\iota{}^*$, then we shall say that X is *left-closed;* in Y, and write $X \lhd Y \Leftrightarrow$

1• $X \subseteq Y$;
2• $\forall \underline{x} \in X\ \forall \underline{y} \in Y: \underline{y} \leq_\iota \underline{x} \Rightarrow \underline{y} \in X$.

If $X \lhd A_\alpha{}^*$ then we simply say that X is left-closed.
Left-closure of sets of vectors is defined analogously.

10.4.2. COROLLARY. If $X, Y \subseteq A_\iota{}^*$ and $\iota_\alpha = \iota$, then

$$X \lhd Y \Leftrightarrow \phi_\alpha(X) \lhd \phi_\alpha(Y).$$

PROOF. By 9.3.3.

■

The corollary shows that in order to study the properties of left-closed trace languages, it suffices to study those of left-closed vector languages.
We shall now show that left-closed vector languages are prime algebraic and consistently complete. First, some notation.

10.4.3. DEFINITION. If $X \subseteq L \subseteq A_\alpha{}^*$, then we write $\sqcup_L X$ for the least upper bound of X in L (if it exists).
If $\underline{x} \in L$, define $Pr_L(\underline{x}) = Pr(\underline{x}) \cap L$.

10.4.4. LEMMA. If $X \subseteq L \subseteq A_\alpha{}^*$, then $L \lhd A_\alpha{}^* \Rightarrow \sqcup_L X = \sqcup X$.

PROOF. First of all, let $\underline{x} = \sqcup_L X$. Since $X \leq \underline{x}$, $\underline{y} = \sqcup X$ exists, by consistent completeness. Since $L \subseteq A_\alpha{}^*$, we must have $\underline{y} \leq \underline{x}$. As L is left-closed, $\underline{y} \in L$. and as $X \leq \underline{y}$, $\underline{x} = \sqcup X \leq \underline{y} \leq \underline{x}$ and so $\underline{x} = \underline{y}$.

■

We have established the principal order theoretic properties of left-closed vector languages.

10.4.5. THEOREM. If L is left-closed in $A_\alpha{}^*$, then $(L,\leq)$ is prime algebraic and consistently complete.

PROOF. Consistent completeness of L follows easily from l10.4.3 for if $X \leq \underline{x}$ for $\underline{x} \in L$, then $\sqcup X$ exists and since $\sqcup X \leq \underline{x}$ and $\underline{x} \in L$, it follows by left-closure that $\sqcup X \in L$.

From 10.4.3 we may also see that if $\underline{u} \in Pr(L)$ and $\underline{u} \leq \sqcup_L X$, then $\underline{u} \in Pr(A_\alpha{}^*)$ and $\underline{u} \leq \sqcup X$ and so $\underline{u} \leq \underline{x}$, some $\underline{x} \in X$. It is also easy to see that if $\underline{x} \in L$, then $\underline{x} = \sqcup Pr(L) = \sqcup_L Pr_L(\underline{x})$. As in the proof of 10.3.7, we may conclude that $(L,\leq)$ is prime algebraic.

■

10.5. Parallel Composition.

In our introduction to vector languages in section 9.2, we saw how they may arise from families of string languages each of which represents some aspect of the behaviour of a complete system. In this section, we formalise the construction and illustrate its use by proving a simple structure theorem for vector languages.

10.5.1. DEFINITION. If $\mathcal{A} = (A_i)_{i \in I}$ is a family of sets, then we define

1• $A_{\mathcal{A}} = \bigcup_{i \in I} A_i$;

2• $\alpha : I \to \wp(A_{\mathcal{A}})$, given by $\alpha_{\mathcal{A}}(i) = A_i$.

Observe that the function $\alpha_{\mathcal{A}}$, is an indexed cover of $A_{\mathcal{A}}$.

We define a *sorted string language* to be a pair (A,L) with $L \subseteq A^*$.

10.5.2. DEFINITION. If $\mathcal{L} = (A_i, L_i)_{i \in I}$ is a family of sorted string languages then we define

$$\| \mathcal{L} = \{ \underline{x} \in A_\alpha \mid \forall i \in I : \underline{x}(i) \in L_i \}$$

where $(A,\alpha) = (A_{\mathcal{A}}, \alpha_{\mathcal{A}})$. We call $\| \mathcal{L}$ the *parallel composition* of $\mathcal{L}$.

We now demonstrate some elementary properties of parallel composites. First, if $V \subseteq A_\alpha{}^*$, then we define

$$V(i) = \{\underline{v}(i) \mid \underline{v} \in V\} \tag{10.11}$$

10.5.3. PROPOSITION. If $\mathcal{L} = (A_i, L_i)_{i \in I}$ is a family of sorted string languages then

1• If each L_i is left-closed then $\|\mathcal{L}$ is left-closed;
2• If $X \subseteq \|\mathcal{L}$ and $\sqcup X$ exists, then $\sqcup X \in \|\mathcal{L}$.

PROOF. Let $\underline{x} \in \|\mathcal{L}$ and $\underline{y} \in A_\alpha^*$ such that $\underline{y} \leq \underline{x}$, then for each $i \in I$, $\underline{x}(i) \in L_i$ and $\underline{y}(i) \leq \underline{x}(i)$ and as L_i is left-closed, $\underline{y}(i) \in L_i$. Hence, $\underline{y} \in \|\mathcal{L}$.

By 10.2.9, $(\sqcup X)(i) = \sqcup X(i)$. By 10.5.2, $X(i) \subseteq L_i$, each $i \in I$. Since $\sqcup X$ exists, $|X| < \infty$, so $|X(i)| < \infty$, each $i \in I$. Thus for all $i \in I$, $max(X(i)) \in X(i)$ and therefore $(\sqcup X)(i) \in L_i$. By 10.5.2, $\sqcup X \in \|\mathcal{L}$.

■

We shall now consider a class of families of sorted string languages; the conditions imposed on the languages ensure that their behaviours have a characteristically simple structure. First, we need some notation. The *prefix closure* of V is defined to be

$$\downarrow V = \{\underline{x} \in A_\alpha^* \mid \exists \underline{v} \in V : \underline{x} \leq \underline{v}\} \tag{10.12}$$

We use the same notation for prefix closures of string languages.

10.5.4. PROPOSITION. Suppose that $\mathcal{L} = (A_i, L_i)_{i \in I}$ is a family of sorted string languages, $\mathcal{A} = (A_i)_{i \in I}$, $(A, \alpha) = (A_\mathcal{A}, \alpha_\mathcal{A})$ and $V \subseteq A_\alpha^*$, such that

$$\forall i \in I\ \forall v \in V(i)\ \forall x \in A_i^* : v.x \in L_i \Rightarrow x \in L_i \tag{10.13}$$

then $\|\mathcal{L} = \downarrow(V^*) \Leftrightarrow$

1• $\downarrow(V(i)^*) \subseteq L_i$, all i;
2• $\forall \underline{x} \in \|\mathcal{L} - \{\Omega\}\ \exists \underline{v} \in V : \underline{x} \leftrightarrow \underline{v} \wedge \neg \underline{x}\ ind\ \underline{v}$.

PROOF. $\Leftarrow$: If $\underline{x} \in \downarrow(V^*)$, then $\underline{x} \in A_\alpha^*$ and $\underline{x}(i) \in \downarrow(V(i)^*) \subseteq L_i$ for each i, by (1), so $\underline{x} \in \|\mathcal{L}$. Conversely, suppose that $\underline{x} \in \|\mathcal{L}$. We prove by induction on $|\underline{x}|$ that $\underline{x} \in \downarrow(V^*)$. The base case, $|\underline{x}| = 0$ is clear.

For the induction step, suppose that $|\underline{x}| > 0$. By (2), there exists $\underline{v} \in V$ such that $\underline{x} \leftrightarrow \underline{v}$ and $\neg \underline{x}\ ind\ \underline{v}$. By 10.3.10 (2), $\hat{\underline{x}} = \underline{x} \sqcup \underline{v}$ exists and by 10.3.9, for each i, either $\hat{\underline{x}}(i) = \underline{x}(i) \in L_i$ or $\hat{\underline{x}}(i) = \underline{v}(i) \in V(i) \subseteq L_i$, so in either case,

$\hat{\underline{x}}(i) \in L_i$. Thus, $\hat{\underline{x}} \in \| \mathcal{L}$ and by 10.2.8 and 10.3.5 (3) $\hat{\underline{x}}/\underline{v} = \underline{x}/(\underline{v} \sqcap \underline{x})$. But, by (10.13), $\hat{\underline{x}}/\underline{v} \in \| \mathcal{L}$, so $\underline{y} = \underline{x}/(\underline{v} \sqcap \underline{x}) \in \| \mathcal{L}$. Since $\underline{v}, \underline{x} \leq \hat{\underline{x}}$ and $\neg \underline{x}\ ind\ \underline{v}$, $\underline{v} \sqcap \underline{x} \neq \underline{\Omega}_\alpha$, by 10.2.7, so $|\underline{y}| < |\underline{x}|$. By induction, $\underline{y} \in \downarrow(V^*)$, that is $\hat{\underline{x}}/\underline{v} \in \downarrow(V^*)$, and hence $\hat{\underline{x}} \in \downarrow(V^*)$. But $\underline{x} \leq \hat{\underline{x}}$, so $\underline{x} \in \downarrow(V^*)$. We have shown that $\| \mathcal{L} = \downarrow(V^*)$.

$\Rightarrow$: Suppose that $\| \mathcal{L} = \downarrow(V^*)$. If $\underline{x} \in \| \mathcal{L} - \{\underline{\Omega}\}$, then $\underline{x} \in \downarrow(V^*)$ and so there exists $\underline{v} \in V$ and $\underline{w} \in V^*$ such that $\underline{x} \leq \underline{v}.\underline{w}$. Since $\underline{x} \leq \underline{w}$ if $\underline{x}\ ind\ \underline{v}$, we may assume that $\neg \underline{x}\ ind\ \underline{v}$, so that for each i, $\underline{x}(i) \leq \underline{v}(i).\underline{w}(i)$. Since $\downarrow \underline{v}(i).\underline{w}(i)$ is totally ordered, either $\underline{x}(i) \leq \underline{v}(i)$ or $\underline{v}(i) \leq \underline{x}(i)$ for each i, that is $\underline{x} \leftrightarrow \underline{v}$.

Finally, $\downarrow(V(i)^*) = (\downarrow V^*)(i) = (\| \mathcal{L})(i) \subseteq L_i$.

■

Notes.

The relationship between individual traces and labelled partial orders was pointed out by Mazurkiewicz (in a footnote) in [16]. Relationships between sets of partial orders and trace languages were established in [38] for so called ι-linguistic behavioural presentations in the guise of sets of labelled partial orders. Most of the order theoretic properties of traces presented in this chapter appeared in that report. The relationship was used in [38, 39] to derive a class of generalised Mealy automata which describe asynchronous input/output devices. These properties and more are also described in [43] although Bednarcztk deals with unlabelled event structures rather than behavioural presentations, so our own results are stronger.

Trace languages from the point of view of Language Theory are discussed in [44] and [45]. The authors observe that there is a relationship between individual traces and labelled, acyclic graphs (so-called dependency graphs), but again, do not deal with the relationships of languages of traces and languages of graphs. What we have called the Factorisation Lemma, appears in [44, 45] in a different form as a generalisation of *Levi's Lemma* for strings. We shall mostly be using the Factorisation Lemma, but we state and prove the asynchronous generalisation of Levi's lemma here for completeness.

LEVI'S LEMMA. Suppose $\underline{w}, \underline{x}, \underline{y}, \underline{z} \in A_\alpha^*$, then $\underline{x}.\underline{y} = \underline{w}.\underline{z} \Leftrightarrow$ there exists $\underline{u}, \underline{v}, \underline{r}, \underline{s} \in A_\alpha^*$ such that $\underline{x} = \underline{u}.\underline{v}$, $\underline{y} = \underline{r}.\underline{s}$, $\underline{z} = \underline{u}.\underline{r}$, $\underline{w} = \underline{v}.\underline{s}$ and $\underline{r}\ ind\ \underline{v}$.

PROOF. If $\underline{u}, \underline{v}, \underline{r}, \underline{s} \in A_\alpha^*$ such that $\underline{x} = \underline{u}.\underline{v}$, $\underline{y} = \underline{r}.\underline{s}$, $\underline{z} = \underline{u}.\underline{r}$, $\underline{w} = \underline{v}.\underline{s}$ and $\underline{r}\ ind\ \underline{v}$, then $\underline{x}.\underline{y} = \underline{u}.\underline{v}.\underline{r}.\underline{s} = \underline{u}.\underline{r}.\underline{v}.\underline{s} = \underline{z}.\underline{w}$.

Conversely, suppose that $\underline{x}.\underline{y} = \underline{w}.\underline{z}$. Since $\underline{x} \leq \underline{z}.\underline{w}$, by the factorisation lemma, we may find $\underline{u}, \underline{v} \in A_\alpha{}^*$ such that $\underline{x} = \underline{u}.\underline{v}$, $\underline{u} \leq \underline{z}$, $\underline{v} \leq \underline{w}$ and $\underline{x}/\underline{u}$ *ind* $\underline{v}$. Define $\underline{r} = \underline{z}/\underline{u}$ and $\underline{s} = \underline{w}/\underline{v}$, then $\underline{z} = \underline{u}.\underline{r}$ and $\underline{w} = \underline{v}.\underline{s}$ and $\underline{r}$ *ind* $\underline{v}$. Finally,

$$\underline{x}.\underline{y} = \underline{z}.\underline{w} = \underline{u}.\underline{r}.\underline{v}.\underline{s} = \underline{u}.\underline{v}.\underline{r}.\underline{s} = \underline{x}.\underline{r}.\underline{s}$$

and so $\underline{y} = (\underline{x}.\underline{y})/\underline{x} = (\underline{x}.\underline{r}.\underline{s})/\underline{x} = \underline{r}.\underline{s}$.

■

In [46], Mazurkiewicz introduced a generalisation of trace languages, his *multitraces*. Essentially, a multitrace is a set of multisets satisfying a number of properties. A multitrace arises from a trace as follows. For each prefix w of some element of the trace, construct a multiset which has $\#_a w$ copies of the element a. In [47], the author extended the results of this chapter to a subclass of the class of multitrace languages, which he called *hypertraces*.

Chapter 11

Determinacy

riverrun, past Eve and Adam's, from swerve of shore to bend of bay, brings us by a commodius vicus of recirculation back to Howth Castle and Environs.

JAMES JOYCE
Finnegans Wake

11.1. Introduction.

In this chapter, we examine the structure of the behaviours of certain kinds of determinate systems. We begin by recasting some of the properties examined in chapter 7 in terms of vector languages.
In section 11.2, we characterise left-closed, strongly forward stable vector languages in terms of string languages of which they are the parallel composites.
In section 11.3, we derive structure theorems for left-closed, strongly forward stable vector languages which are in some sense finite state.

11.2. Strongly Forward Stable Vector Languages.

11.2.1. DEFINITION $V \subseteq A_\alpha{}^*$ is *strongly forward stable* if and only if

$$\underline{x}.\underline{a}_\alpha, \underline{x}.\underline{b}_\alpha \in V \wedge a \neq b \Rightarrow \underline{a}_\alpha \; ind \; \underline{b}_\alpha \wedge \underline{x}.\underline{a}_\alpha.\underline{b}_\alpha \in V \qquad (11.1)$$

for all $\underline{x} \in A_\alpha{}^*$ and $a, b \in A$.

For left-closed vector languages, we have an order theoretic characterisation of strong forward stability.

11.2.2. PROPOSITION. If $V \subseteq A_\alpha{}^*$ is left-closed, then V is strongly forward stable $\Leftrightarrow$ $(V,\leq)$ is a lattice.

PROOF. $\Rightarrow$: Since V is left-closed, if $\underline{x},\underline{y} \in V$, then $\underline{x} \sqcap \underline{y} \in V$, by 10.4.5. For least upper bounds, we first show that if $\underline{x},\underline{u} \in A_\alpha{}^*$ and $a \in A$, then

$$\underline{x}.\underline{a}_\alpha, \underline{x}.\underline{u} \in V \Rightarrow \underline{a}_\alpha \leq \underline{u} \vee (\underline{a}_\alpha \; ind \; \underline{u} \wedge \underline{x}.\underline{a}_\alpha.\underline{u} \in V) \tag{11.2}$$

We argue by induction on $|\underline{u}|$.

The case $|\underline{u}|=0$ is trivial, so suppose that $\underline{u}=\underline{u}'.\underline{b}_\alpha$, $b \in A$. Then by induction, either $\underline{a}_\alpha \leq \underline{u}'$, so that $\underline{a}_\alpha \leq \underline{u}$, or $\underline{a}_\alpha \; ind \; \underline{u}'$ and $\underline{x}.\underline{a}_\alpha.\underline{u}' \in V$. In the latter case, $\underline{a}_\alpha.\underline{u}' = \underline{u}'.\underline{a}_\alpha$ and so if $a=b$, then $\underline{a}_\alpha \leq \underline{u}$ and if $a \neq b$, then since $\underline{x}.\underline{u}'.\underline{a}_\alpha, \underline{x}.\underline{u}'.\underline{b}_\alpha \in V$, $\underline{a}_\alpha \; ind \; \underline{b}_\alpha$ and $\underline{x}.\underline{u}'.\underline{a}_\alpha.\underline{b}_\alpha \in V$, by (11.1). Hence, in this case, $\underline{a}_\alpha \; ind \; \underline{u}$ and $\underline{x}.\underline{u} \in V$. We have proved (11.2).

Suppose, next, that $\underline{x},\underline{y} \in V$ and let $\underline{w} = \underline{x} \sqcap \underline{y}$, $\underline{u} = \underline{x}/\underline{w}$ and $\underline{v} = \underline{y}/\underline{w}$. We argue by induction on $|\underline{u}|$, that $\underline{u} \; ind \; \underline{v}$ and $\underline{x}.\underline{u}.\underline{v} \in V$.

The case $|\underline{u}|=0$ is trivial, so suppose that $\underline{u}=\underline{u}'.\underline{a}_\alpha$, $a \in A$, then by induction, $\underline{u}' \; ind \; \underline{v}$. and $\underline{w}.\underline{u}'.\underline{v}, \underline{w}.\underline{u}'.\underline{a}_\alpha \in V$, so by (11.2), either $\underline{a}_\alpha \leq \underline{v}$ or $\underline{a}_\alpha \; ind \; \underline{v}$ and $\underline{w}.\underline{u}'.\underline{a}_\alpha.\underline{v} \in V$. But, if $\underline{a}_\alpha \leq \underline{v}$, then $\underline{w}.\underline{a}_\alpha \leq \underline{x},\underline{y}$, contradicting $\underline{w} = \underline{x} \sqcap \underline{y}$, and so $\underline{a}_\alpha \; ind \; \underline{v}$. As $\underline{u}' \; ind \; \underline{v}$, we must have $\underline{u} \; ind \; \underline{v}$. And by 10.2.8 $\underline{x} \sqcup \underline{y} = \underline{w}.\underline{u}.\underline{v} = \underline{w}.\underline{u}'.\underline{a}_\alpha.\underline{v} \in V$.

$\Leftarrow$: Suppose that $\underline{x} \in A_\alpha{}^*$ and $a,b \in A$ such that $\underline{x}.\underline{a}_\alpha, \underline{x}.\underline{b}_\alpha \in V$ and $a \neq b$, then since $(V,\leq)$ is a lattice, $\underline{z} = \underline{x}.\underline{a}_\alpha \sqcup \underline{x}.\underline{b}_\alpha \in V$. Since , $a \neq b$, $\underline{x}.\underline{a}_\alpha \sqcap \underline{x}.\underline{b}_\alpha = \underline{x}$, so by 10.2.8 $\underline{a}_\alpha \; ind \; \underline{b}_\alpha$ and $\underline{x}.\underline{a}_\alpha.\underline{b}_\alpha \in V$.

■

We shall now characterise strongly forward stable, left-closed vector languages in terms of the string languages determined by their component string languages $V(i)$, defined in (10.11). In the following definition and for the rest of the chapter, I will denote the domain of α.

11.2.3. DEFINITION. If $V \subseteq A_\alpha{}^*$, then define $\overline{\overline{V}} = \|_{i \in I}(\alpha(i), V(i))$.

Note that if $\underline{x} \in V$, then $\underline{x}(i) \in V(i)$, all $i \in I$, and so $\underline{x} \in \|_{i \in I}(\alpha(i), V(i))$, by 10.5.2, that is $\underline{x} \in \overline{\overline{V}}$. Hence, $V \subseteq \overline{\overline{V}}$.

The component string languages of strongly forward stable, left-closed vector languages will be seen to have a characteristic structure; they are linear.

11.2.4. DEFINITION. A language $L \subseteq A^*$ will be said to be *linear* $\Leftrightarrow$ L is left-closed in A^* and totally ordered.

We shall prove that if a vector language is left-closed and strongly forward stable, then its components are all linear. First, we need a simple lemma.

11.2.5 LEMMA If $\underline{x} \in A_\alpha{}^*$, $i \in I$ and $u, v \in \alpha(i)^*$ such that $\underline{x}(i) = u.v$, then there exists $\underline{u}, \underline{v} \in A_\alpha{}^*$ such that $\underline{x} = \underline{u}.\underline{v}$, $\underline{u}(i) = u$ and $\underline{v}(i) = v$.

PROOF. We argue by induction on $|u|$. If $|u| = 0$, then the lemma holds with $\underline{u} = \underline{\Omega}_\alpha$ and $\underline{v} = \underline{x}$.

Next, suppose that $u > \Omega$. Let $a \in A$ such that $a \leq u$ and define $u' = u/a$. As $\#_a \underline{x} > 0$, there exists $\underline{y}, \underline{y}' \in A_\alpha{}^*$ such that $\underline{x} = \underline{y}.\underline{a}_\alpha.\underline{y}'$, and $\#_a \underline{y} = 0$. Thus, $a \leq a.u'.v = \underline{x}(i) = \underline{y}(i).a.\underline{y}'(i)$. If $\underline{y}(i) > \Omega$, then $a \leq \underline{y}(i)$, contradicting $\#_a \underline{y} = 0$. Hence $\underline{y}(i) = \Omega$ and so $u'.v = \underline{y}'(i)$. By induction there exists $\underline{u}', \underline{v} \in A_\alpha{}^*$ such that $\underline{y}' = \underline{u}'.\underline{v}$, $\underline{u}'(i) = u'$ and $\underline{v}(i) = v$. Thus, if we define $\underline{u} = \underline{y}'.\underline{a}_\alpha.\underline{u}'$, then $\underline{x} = \underline{y}.\underline{a}_\alpha.\underline{y}' = \underline{y}.\underline{a}_\alpha.\underline{u}'.\underline{v} = \underline{u}.\underline{v}$, $\underline{u}(i) = \underline{y}(i).a.\underline{u}'(i) = a.u' = u$ and $\underline{v}(i) = v$.

■

11.2.6. PROPOSITION. If $V \subseteq A_\alpha{}^*$ and $V(i)$ is linear, all $i \in I$, then $\overline{\overline{V}}$ is left-closed and strongly forward stable.

PROOF. Let $\underline{x} \in \overline{\overline{V}}$ and $\underline{u} \in A_\alpha{}^*$ such that $\underline{u} \leq \underline{x}$, then for all $i \in I$, $\underline{u}(i) \leq \underline{x}(i)$ and so $\underline{u}(i) \in V(i)$, as $V(i)$ is left-closed. Thus, $\underline{u} \in \|_{i \in I}(\alpha(i), V(i)) = \overline{\overline{V}}$, by 10.5.2. Hence, $\overline{\overline{V}}$ is left-closed.

Thus, if $\underline{x}, \underline{y} \in \overline{\overline{V}}$, then $\underline{w} = \underline{x} \sqcap \underline{y} \in \overline{\overline{V}}$ exists, by 10.4.5. Since $V(i)$ is linear, $\underline{x} \leftrightarrow \underline{y}$, so by (2) of 10.2.10, $\underline{z} = \underline{x} \sqcup \underline{y}$ exists and $\underline{z}(i) = max(\underline{x}(i), \underline{y}(i))$, each $i \in I$. Thus, $\underline{z}(i) \in V(i)$, all $i \in I$, and so $\underline{z} \in \overline{\overline{V}}$. We have shown that $\overline{\overline{V}}$ is a lattice. Since $\overline{\overline{V}}$ is left-closed, $\overline{\overline{V}}$ is strongly forward stable, by 11.2.2.

■

11.2.7. PROPOSITION. If $V \subseteq A_\alpha{}^*$ and $|I| < \infty$, then V is left-closed and strongly forward stable $\Leftrightarrow$ $\overline{\overline{V}} = V$ and $V(i)$ is linear, all $i \in I$.

PROOF. Suppose first that V is left-closed and strongly forward stable.

We know that $V \subseteq \overline{\overline{V}}$. Suppose that $\underline{x} \in \overline{\overline{V}}$, then for each $i \in I$, there exists $\underline{x}_i \in V$ such that $\underline{x}(i) = \underline{x}_i(i)$. Since V is a lattice, by 11.2.2, and I is finite, $\underline{z} = \sqcup\{\underline{x}_i \mid i \in I\}$ exists and belong to V. By 10.2.9, $\underline{z}(i) = \sqcup\{\underline{x}_j(i) \mid j \in I\}$, and so for each $i \in I$, $\underline{x}(i) = \underline{x}_i(i) \leq \underline{z}(i)$, that is $\underline{x} \leq \underline{z}$ and since $\underline{z} \in V$ and V is left-closed, $\underline{x} \in V$. We have proved that $\overline{\overline{V}} = V$.

Let $i \in I$. If $x \in V(i)$ and $u \leq x$, then by since $\overline{\overline{V}} = V$, there exists $\underline{x} \in V$ such that $\underline{x}(i) = x$ and there exists $v \in \alpha(i)^*$ such that $u.v = \underline{x}(i)$. Thus, by 11.2.5, there exists $\underline{u}, \underline{v} \in A_\alpha{}^*$ such that $\underline{x} = \underline{u}.\underline{v}$ and $\underline{u}(i) = u$. Since V is left-closed and $\underline{u} \leq \underline{x}$, $\underline{u} \in V$, and so $u \in V(i)$. We have shown that $V(i)$ is left-closed.

Next, suppose that $x, y \in V(i)$, then there exists $\underline{x}, \underline{y} \in V$ such that $\underline{x}(i) = x$ and $\underline{y}(i) = y$. By 11.2.2, V is a lattice, so $\underline{w} = \underline{x} \sqcup \underline{y}$ is defined and belongs to V. By 10.2.9, $\underline{w}(i) = max(\underline{x}(i), \underline{y}(i)) = max(x, y)$. Hence either $x \leq y$ or $y \leq x$. Thus, $V(i)$ is totally ordered. By 11.2.4, $V(i)$ is linear.

Conversely, suppose that $V(i)$ is linear, all $i \in I$ and $\overline{\overline{V}} = V$, then $\overline{\overline{V}}$ is left-closed and strongly forward stable, by 11.2.6, and so V is left-closed and strongly forward stable.

■

11.3. Simple Vector Languages.

We now turn our attention to finite state, strongly forward stable systems. This involves a relation on vectors. Two vectors are equivalent if they have exactly the same continuations; in such a situation they may be considered to determine the same system state.

11.3.1. DEFINITION. Suppose that $V \subseteq A_\alpha{}^*$ and $\underline{x}, \underline{y} \in V$, then define

$$\underline{x} \approx_V \underline{y} \Leftrightarrow (\forall \underline{u} \in A_\alpha{}^*: \underline{x}.\underline{u} \in V \Leftrightarrow \underline{y}.\underline{u} \in V) \tag{11.3}$$

If $L \subseteq A^*$ and $x, y \in L$, then define:

$$x \approx_L y \Leftrightarrow (\forall u \in A^*: x.u \in L \Leftrightarrow y.u \in L) \tag{11.4}$$

and

$$\underline{x} \approxeq_V \underline{y} \Leftrightarrow (\forall i \in I: \underline{x}(i) \approx_{V(i)} \underline{y}(i)) \tag{11.5}$$

We list the basic properties of these relations.

11.3.2. LEMMA If $V \subseteq A_\alpha{}^*$, then

1• $\approx_V$ and $\approx_{V(i)}$ are equivalence relations;
2• If $\underline{x}, \underline{y} \in V$, $\underline{u} \in A_\alpha{}^*$ and $\underline{x}.\underline{u}, \underline{y}.\underline{u} \in V$, then $\underline{x} \approx_V \underline{y} \Rightarrow \underline{x}.\underline{u} \approx_V \underline{y}.\underline{u}$;
3• If $\overline{\overline{V}} = V$, then $\underline{x} \stackrel{\approx}{=}_V \underline{y} \Rightarrow \underline{x} \approx_V \underline{y}$.

PROOF. (1) and (2) are trivial. For (3), suppose that $\overline{\overline{V}} = V$ and that $\underline{x} \stackrel{\approx}{=}_V \underline{y}$. Let $\underline{u} \in A_\alpha{}^*$ such that $\underline{x}.\underline{u} \in V$, then for each $i \in I$, $\underline{x}(i).\underline{u}(i) \in V(i)$, and hence $\underline{y}(i).\underline{u}(i) \in V(i)$, as $\underline{x} \stackrel{\approx}{=}_V \underline{y}$. Therefore, $\underline{y}.\underline{u} \in \overline{\overline{V}} = V$. We have shown that $\underline{x}.\underline{u} \in V \Rightarrow \underline{y}.\underline{u} \in V$. By symmetry, $\underline{y}.\underline{u} \Rightarrow \underline{x}.\underline{u}$. Thus, $\underline{x} \approx_V \underline{y}$.

■

Write $\underline{x}/\approx_V$ for the $\approx_V$-class of $\underline{x} \in V$ and $V/\approx_V$ for the set of all such classes.

11.3.3. DEFINITION If $V \subseteq A_\alpha{}^*$, then

1• V is *finite state* if and only if $|V/\approx_V| < \infty$.
2• V is *simple* if and only if V is left-closed, finite-state and strongly forward stable.

11.3.4. PROPOSITION. Suppose that $V \subseteq A_\alpha{}^*$ such that $V(i)$ is linear, all $i \in I$ and $\overline{\overline{V}} = V$, then for all $\underline{x}, \underline{y} \in V$, $\underline{x} \stackrel{\approx}{=}_V y \Leftrightarrow \underline{x} \approx_V y$.

PROOF. $\underline{x} \stackrel{\approx}{=}_V y \Rightarrow \underline{x} \approx_V y$, by 11.3.2 (3). Conversely, suppose that $\underline{x} \approx_V y$ and let $u \in \alpha(i)^*$ such that $\underline{x}(i).u \in V(i)$, then there exists $\underline{z} \in V$ such that $\underline{z}(i) = \underline{x}(i).u$, by 11.2.3. By 11.2.6, $V = \overline{\overline{V}}$ is left-closed and strongly forward stable, and hence a lattice, by 11.2.2. Therefore, $\underline{w} = \underline{x} \sqcup \underline{z}$ exists and belongs to V. Thus, $\underline{w}(i) = max(\underline{z}(i), \underline{x}(i)) = \underline{x}(i).u$, by 10.2.9, and so if we define $\underline{u} = \underline{w}/\underline{x}$, then $\underline{x}.\underline{u} = \underline{w} \in V$ and $\underline{u}(i) = u$. Since $\underline{x} \approx_V y$, $\underline{y}.\underline{u} \in V$, so $\underline{y}(i).u = \underline{y}(i).\underline{u}(i) \in V(i)$. We have proved that $\underline{x}(i).u \in V(i) \Rightarrow \underline{y}(i).u \in V(i)$. By symmetry, $\underline{y}(i).u \in V(i) \Rightarrow \underline{x}(i).u \in V(i)$. Hence, $\underline{x} \stackrel{\approx}{=}_V \underline{y}$.

■

11.3.5. PROPOSITION. Suppose that $V \subseteq A_\alpha{}^*$ such that $|I| < \infty$, then V is simple $\Leftrightarrow$ $V = \|_{i \in I}(\alpha(i), L_i)$, where for each $i \in I$, L_i is linear and finite state. Furthermore, the L_i may be chosen so that $L_i = V(i)$ each $i \in I$.

PROOF. If V is simple, then by 11.2.7 $V = \|_{i\in I}(\alpha(i), L_i)$, where for each $i \in I$, $L_i = V(i)$ is linear. By 11.3.4, for each $i \in I$ there are functions $f_i : V/\approx_V \to V(i)/\approx_{V(i)}$ satisfying $f_i : \underline{x}/\approx_V = \underline{x}(i)/\approx_{V(i)}$, which are clearly onto. Hence $|V(i)/\approx_{V(i)}| \le |V/\approx_V| < \infty$, that is $V(i)$ is finite state, each $i \in I$.

Conversely, suppose that $V = \|_{i\in I}(\alpha(i), L_i)$, where for each $i \in I$, L_i is linear and finite state. By 11.2.4 and 10.5.3 (1), V is left-closed. By 10.5.2 and (10.11), $V(i) \subseteq L_i$, and since L_i is totally ordered, then so is $V(i)$. Suppose that $x \in V(i)$ and $y \in \alpha(i)^*$ such that $y \le x$, then $x = \underline{x}(i)$, some $\underline{x} \in V$, and by 11.2.5, there exists $\underline{y} \in A_\alpha{}^*$ such that $\underline{y} \le \underline{x}$ and $\underline{y}(i) = y$. Since V is left-closed, $\underline{y} \in V$ and so $y = \underline{y}(i) \in V(i)$. Thus, $V(i)$ is also left-closed and so linear. By 11.2.7, $\overline{\overline{V}}$ is left-closed and strongly forward stable. But, if $\underline{x} \in \overline{\overline{V}}$, then $\underline{x}(i) \in V(i) \subseteq L_i$, each $i \in I$, and so $\underline{x} \in V$. Thus $\overline{\overline{V}} = V$ and so V is left-closed and strongly forward stable. Finally, as $V(i)$ is also left-closed then either $V(i)$ is finite or $V(i) = L_i$, each $i \in I$, and it follows that each $V(i)$ is finite state.

Let S be the set of all functions

$$f : I \to \bigcup_{i\in I} V(i)/\approx_{V(i)}$$

such that $f(i) \in V(i)/\approx_{V(i)}$, each $i \in I$. As $|I| < \infty$ and each $V(i)$ is finite state, S is finite. As $V(i)$ is linear, and L_i is totally ordered each $i \in I$, and $\overline{\overline{V}} = V$, by 11.3.4 there is a bijection $\varphi : V/\approx \to S$ given by $(\varphi(\underline{x}/\approx))(i) = \underline{x}(i)/\approx_{V(i)}$. Hence, V is finite-state.

■

11.3.6. PROPOSITION. $L \subseteq A^*$ is linear and finite state $\Leftrightarrow$

$$\exists x, y \in A^* : L = \downarrow(x.y^*).$$

PROOF. Suppose first that $L \subseteq A^*$ is linear and finite state.

If L is finite, then let $x = max(L)$ and $y = \Omega$. Then, $L \subseteq \downarrow(x.y^*)$, while $\downarrow(x.y^*) \subseteq L$, as L is left-closed in A^*, by 11.2.4. Thus, in this case $L = \downarrow(x.y^*)$.

Otherwise, as L is finite state and totally ordered, there exists $x, w \in L$ such that $x < w$ and $x \approx_L w$. Let $y = w/x$. Thus, $x.y \in L$ and $x \approx_L x.y$ and if $x.y^n \in L$ and $x \approx_L x.y^n$, then by (11.4), $x.y^{n+1} \in L$ and $x.y \approx_L x.y^{n+1}$ and so $x \approx_L x.y^{n+1}$ by transitivity of $\approx_L$. By induction, $x.y^* \subseteq L$, and so $\downarrow(x.y^*) \subseteq \downarrow L = L$, since L is left-closed in A^*. If $u \in L$, then as $y > \Omega$, there exists $n \in \omega$ such that $|x| + n.|y| > |u|$. Since L is totally ordered, $u < x.y^n$, that is to say, $u \in \downarrow(x.y^n)$ and we have $L \subseteq \downarrow(x.y^*)$. Thus, $L = \downarrow(x.y^*)$ in this case also.

Conversely, suppose that $L = \downarrow(x.y^*)$, then L is certainly totally ordered and left-closed in A^*, that is L is linear. We claim that $L/{\approx_L} = \{u/{\approx_L} \mid u < x.y\}$. Certainly, $\{u/{\approx_L} \mid u < x.y\} \subseteq L/{\approx_L}$ while, if $v \in L$, then either $v < x$, and so $v/{\approx_L} \in \{u/{\approx_L} \mid u < x.y\}$, or there exists $n \in \omega$ and $w < y$ such that $v = x.y^n.w$. We now have:

$$x.y^n.w.z \in L \Leftrightarrow w.z \in \downarrow y^* \Leftrightarrow x.w.z \in \downarrow(x.y^*) \Leftrightarrow x.w.z \in L$$

so $v/{\approx_L} = x.w/{\approx_L} \in \{u/{\approx_L} \mid u < x.y\}$. As $\{u/{\approx_L} \mid u < x.y\}$ is finite, L is finite state.

■

11.3.7. PROPOSITION. If $V \subseteq A_\alpha{}^*$ such that $|I| < \infty$, then V is simple $\Leftrightarrow$ there exists $\underline{x} \in A_\alpha{}^*$ and $w_i \in \alpha(i)^*$, each $i \in I$, such that $V = \downarrow(\underline{x}.W)$, where $W = \|_{i \in I}(\alpha(i), \downarrow w_i{}^*)$ and $W(i) = \downarrow w_i{}^*$, each $i \in I$.

PROOF. Suppose first, that V is simple, then by 11.3.5, $V = \|_{i \in I}(\alpha(i), L_i)$, where for each $i \in I$, L_i is linear and finite state and $L_i = V(i)$, each $i \in I$, so for each $i \in I$, $L_i = Pref(x_i.y_i{}^*)$, some $x_i, y_i \in \alpha(i)^*$, by 11.3.6. By 11.3.3 (2), $V = \overline{\overline{V}}$, and so for each $i \in I$ there exists $\underline{x}_i \in V$ such that $\underline{x}_i(i) = x_i$. Since V is a lattice, by 11.2.2, and $|I| < \infty$, the least upper bound $\underline{x}$ of the $\underline{x}_i$ exists and belongs to V. For each $i \in I$, $x_i = \underline{x}_i(i) \leq \underline{x}(i)$, by 10.2.9, and so there exists $n_i \in \omega$ and $v_i < y_i$ such that $\underline{x}(i) = x_i.y_i{}^{n_i}.v_i$. If $w_i = (y_i/v_i).v_i$, then for all $n \in \omega$:

$$\underline{x}(i).w_i{}^n = x_i.y_i{}^{n_i}.v_i((y_i/v_i).v_i)^n = x_i.y_i{}^{n_i+n}.v_i \leq x_i.y_i{}^{n_i+n+1}$$

from which we deduce that $\underline{x}(i).w_i{}^* \subseteq L_i$ and hence, $\downarrow(\underline{x}(i).w_i{}^*) \subseteq L_i$, by left-closure of L_i. If, on the other hand, $u \in L_i$, then $u \leq x_i.y_i{}^n$, some $n \in \omega$ so $u \leq x_i.y_i{}^{n_i+n}.v_i = x_i.w_i{}^n$. We have shown that for each $i \in I$, $L_i = \downarrow(\underline{x}(i).w_i{}^*)$.

If $\underline{w} \in W = \|_{i \in I}(\alpha(i), \downarrow w_i{}^*)$, then for each $i \in I$, $\underline{w}(i) \in \downarrow w_i{}^*$ and hence $(\underline{x}.\underline{w})(i) = \underline{x}(i).\underline{w}(i) \in \downarrow(\underline{x}(i).w_i{}^*) = L_i$. Thus, $\underline{x}.\underline{w} \in \overline{\overline{V}} = V$. We have shown that $\underline{x}.W \subseteq V$, from which it follows that $\downarrow(\underline{x}.W) \subseteq V$, by left-closure of V. Conversely, if $\underline{v} \in V$, then as V is a lattice, by 11.2.2, $\underline{z} = \underline{x} \sqcup \underline{v} \in V$. For each $i \in I$, $\underline{z}(i) \in L_i$ and $\underline{x}(i) \leq \underline{z}(i)$, so $(\underline{z}/\underline{x})(i) = \underline{z}(i)/\underline{x}(i) \in \downarrow w_i{}^*$. Thus, $\underline{z}/\underline{x} \in W$

and consequently $\underline{z} \in \underline{x}.W$. But $\underline{v} \leq \underline{z}$, so $\underline{v} \in \downarrow(\underline{x}.W)$. We have shown that $V \subseteq \downarrow(\underline{x}.W)$.

Conversely, let $V = \downarrow(\underline{x}.W)$, where $\underline{x} \in A_\alpha{}^*$, and $W = \|_{i \in I}(\alpha(i), \downarrow w_i{}^*)$, with $w_i \in \alpha(i)^*$, each $i \in I$. Let, $L_i = \downarrow(\underline{x}(i).w_i{}^*)$. By 11.3.6, each $L_i = \downarrow(\underline{x}(i).w_i{}^*)$ is linear and finite state. Now $\underline{u} \in V$ if and only if $\underline{u} \in A_\alpha{}^*$ and $\underline{u} \leq \underline{x}.\underline{w}$, some $\underline{w} \in W$ if and only if $\underline{u} \in A_\alpha{}^*$ and $\underline{u}(i) \leq \underline{x}(i).\underline{w}(i)$, each $i \in I$, for some $\underline{w} \in W$ if and only if $\underline{u} \in A_\alpha{}^*$ and $\underline{u}(i) \in \downarrow(\underline{x}(i).w_i{}^*) = L_i$. Thus $V = \|_{i \in I}(\alpha(i), L_i)$, By 11.3.5, V is simple.

Finally, we show that $W(i) = \downarrow w_i{}^*$, each $i \in I$. Clearly $W(i) \subseteq \downarrow w_i{}^*$ each $i \in I$. Conversely, suppose that $w \in \downarrow w_i{}^*$, then $\underline{x}(i).w \in L_i = V(i)$, so there exists $\underline{y} \in V$ such that $\underline{y}(i) = \underline{x}(i).w$. As V is simple, V is left-closed and forward stable, by 11.3.3, and hence a lattice, by 11.2.2, and so $\underline{z} = \underline{x} \sqcup \underline{y}$ exists and belongs to V. By 10.2.9, $\underline{z}(i) = \underline{x}(i).w$. Now, $\underline{z} \in V = \downarrow(\underline{x}.W)$ and so $\underline{z} \leq \underline{x}.\underline{w}$, some $\underline{w} \in W$. But $\underline{x} \leq \underline{z}$, so $\underline{z}/\underline{x} \leq (\underline{x}.\underline{w})/\underline{x} = \underline{w}$. Thus, $w = (\underline{z}/\underline{x})(i) \leq \underline{w}(i)$. By 11.2.5, there exists $\underline{w}' \leq \underline{w}$ such that $w = \underline{w}'(i)$. Since $W = \|_{i \in I}(\alpha(i), \downarrow(w_i{}^*))$, where $\downarrow w_i{}^*$ is left-closed, W is left-closed, by 10.5.3 (1), so $\underline{w}' \in W$, and consequently $w \in W(i)$.

■

We now characterise vector languages of the form $W = \|_{i \in I}(A_i, \downarrow w_i{}^*)$. satisfying $W(i) = \downarrow w_i{}^*$, each $i \in I$. First, we need some information on the state structure of languages of the form $\downarrow w^*$.

11.3.8. LEMMA. Suppose that $w \in A^*$ and let $L = \downarrow w^*$ such that $w \approx_L x \Rightarrow x \in w^*$ for all $x \in L$, then:

$$\forall u_1, u_2 < w \ \forall n_1, n_2 \in \omega: w^{n_1}.u_1 \approx_L w^{n_2}.u_2 \Rightarrow u_1 = u_2$$

PROOF. The result is trivial if $w = \Omega$, so suppose that $w \neq \Omega$.

As $\Omega \approx_L w^n$, all $n \in \omega$, $u_1 \approx_L w^{n_1}.u_1 \approx_L w^{n_2}.u_2 \approx_L u_2$. Since L is linear, by 11.3.6, we may assume without loss of generality that $u_1 \leq u_2$. Define $v_i = w/u_i$, each i, then $u_2.v_2 = w \in L$ and since $u_1 \approx_L u_2$, $u_1.v_2 \in L$, by (11.4), and $u_1.v_2 \approx_L u_2.v_2 = w$. By hypothesis, $w \approx_L x \Rightarrow x \in w^*$ and so $u_1.v_2 = w^n$, some $n \in \omega$. Since $u_2 < w$, by hypothesis, $v_2 > \Omega$ and so $u_1.v_2 > \Omega$. But, $u_1 \leq u_2$, and so $0 < n.|w| = |w^n| = |u_1.v_2| \leq |u_2.v_2| = |w|$

Thus, $n = 1$ and so $u_1.v_2 = w = u_2.v_2$. Hence, $u_1 = u_2$.

■

11.3.9. LEMMA. Suppose that $\hat{w} \in A^*$ and $L = \downarrow\hat{w}^*$, then for all $x, y \in L - \{\Omega\}$
$x \approx_L y \Rightarrow \exists x', y' \in L\ \exists a \in A: x = x'.a \wedge y = y'.a \wedge x' \approx_L y'$.

PROOF. The lemma is trivial if $\hat{w} = \Omega$, so suppose that $\hat{w} \neq \Omega$.

Let $w \in L$ be minimal with respect to the property $L = \downarrow w^*$. Suppose that $x \in L$ with $w \approx_L x$. We show that $x \in w^*$.

We have $x = w^n.u$, some $n \in \omega$, where $u < w$. Since $\Omega \approx_L w^n$, $u \approx_L w^n.u = x \approx_L w \approx_L \Omega$. By induction, for each $m \in \omega$ $u^m \in L$ and $u^m \approx_L \Omega$. Hence, $u^* \subseteq L$ and so $\downarrow u^* \subseteq L$. In addition, if $u > \Omega$, then as L is totally ordered for all $y \in L$, $y \leq u^k$, where $|y| \leq k.|u|$. But, then $L \subseteq \downarrow u^*$ and consequently $L = \downarrow u^*$ contradicting the minimality of w. Hence, $u = \Omega$ and so $x \in w^*$

We have shown that the hypothesis of 11.3.8 applies.

Now, suppose that $x, y \in L - \{\Omega\}$ such that $x \approx_L y$, then there exists $u_1, u_2 < w$ and $n_1, n_2 \in \omega$ such that $x = w^{n_1}.u_1$ and $y = w^{n_2}.u_2$. By 11.3.8, $u_1 = u_2$. Hence, if $u_1 > \Omega$ there exists $u < w$ and $a \in A$ such that $u_1 = u.a = u_2$ and so if we define $x' = w^{n_1}.u$ and $y' = w^{n_2}.u$, then $x = x'.a$, $y = y'.a$. Otherwise, $w = u.a$, some $a \in A$ and so if we define $x' = w^{n_1-1}.u$ and $y' = w^{n_2-1}.u$, then $x = x'.a$, $y = y'.a$.

Finally, suppose that $x'.u \in L$. If $u = \Omega$, then $y'.u \in L$. Otherwise $u = a.v$, some $v \in A^*$, that is, $x.v \in L$. Since $x \approx_L y$, $y'.u = y.v \in L$. We have shown that $x'.u \in L \Rightarrow y'.u \in L$. The converse implication is similar.

■

11.3.10. LEMMA. Suppose that $\alpha: I \to \wp(A)$ is an indexed cover and let $a \in A$ and $\underline{\gamma} \in M_\alpha$ such that $\underline{\gamma}.\underline{a}_\alpha \in A_\alpha{}^*$, then $\underline{\gamma} \in A_\alpha{}^*$

PROOF. If $\underline{\gamma}$ *ind* $\underline{a}_\alpha$, then $\underline{a}_\alpha.\underline{\gamma} = \underline{\gamma}.\underline{a}_\alpha \in A_\alpha{}^*$ and so $\underline{a}_\alpha \leq \underline{\gamma}.\underline{a}_\alpha$ and $(\underline{\gamma}.\underline{a}_\alpha)/\underline{a}_\alpha = \underline{\gamma}$ and thus, $\underline{\gamma} \in A_\alpha{}^*$, by 9.3.2. Suppose then that $\neg\underline{\gamma}$ *ind* $\underline{a}_\alpha$, so that $\underline{\gamma} \neq \underline{\Omega}_\alpha$.

Since $\underline{\gamma}.\underline{a}_\alpha \in A_\alpha{}^*$, $\underline{\gamma}.\underline{a}_\alpha = \underline{b}_\alpha.\underline{x}$, some $b \in A$ and $\underline{x} \in A_\alpha{}^*$, so $\underline{b}_\alpha, \underline{\gamma} \leq \underline{b}_\alpha.\underline{x}$ and hence by 10.2.1, either $\underline{b}_\alpha \leq \underline{\gamma}$ or $\underline{b}_\alpha$ *ind* $\underline{\gamma}$. But, if $\underline{b}_\alpha$ *ind* $\underline{\gamma}$, then whenever $b \in \alpha(i)$, $a = (\underline{a}_\alpha.\underline{\gamma})(i) = (\underline{x}.\underline{b}_\alpha)(i) = \underline{x}(i).b$, so $a = b$, contradicting $\neg\underline{\gamma}$ *ind* $\underline{a}_\alpha$. Thus $\underline{b}_\alpha \leq \underline{\gamma}$, so $\underline{\gamma}'.\underline{a}_\alpha \in A_\alpha{}^*$, where $\underline{\gamma}'(i) = \underline{\gamma}(i)/\underline{b}_\alpha(i)$, each i. By induction, on $|\underline{\gamma}.\underline{a}_\alpha|$, $\underline{\gamma}' \in A_\alpha{}^*$ and so $\underline{\gamma} = \underline{b}_\alpha\underline{\gamma}' \in A_\alpha{}^*$.

■

We present a characterisation of languages such as W in proposition 11.3.7. They have a particularly simple structure being asynchronously cyclic.[1] In general, this structure is lost when one takes an interleaved view of behaviour.

11.3.11. THEOREM. Let $\mathcal{A} = (A_i)_{i\in I}$ be a family of sets with $|I| < \infty$ finite, and let $w_i \in A_i^*$, each $i \in I$ and let $A = A_{\mathcal{A}}$, $\alpha = \alpha_{\mathcal{A}}$ and $W = \|_{i\in I}(A_i, \downarrow w_i^*)$. Then if $W(i) = \downarrow w_i^*$, each $i \in I$, there exists $\underline{w} \in W$ such that $W = \downarrow \underline{w}^*$.

PROOF. By hypothesis, $|I| < \infty$ is finite, and by 11.3.6, each $\downarrow w_i^*$ is linear and finite state. Hence, W is simple. By 11.2.2, W is a lattice.

Suppose that $\underline{x} \in W$ then for each $i \in I$ such that $w_i \neq \Omega$, there exists $n_i \in \omega$ such that $|\underline{x}(i)| < n_i.|w_i|$, and hence $\underline{x}(i) < w_i^{n_i} \in W(i)$, since $W(i)$ is totally ordered. Since $W(i) = \downarrow w_i^*$, for each $i \in I$, there exists $\underline{u}_i \in W$ such that $\underline{u}_i(i) = w_i^{n_i}$. Since W is a lattice and $|I| < \infty$, the $\underline{u}_i$ have a least upper bound $\underline{y} \in W$. By 10.2.9, $\underline{x}(i) \leq \underline{y}(i)$, each $i \in I$. We have established that if $\underline{x} \in W$ then there exists $\underline{y} \in W$ such that $\underline{x} \leq \underline{y}$ and for each $i \in I$ such that $w_i \neq \Omega$, $\underline{x}(i) < \underline{y}(i)$.

Thus, we may construct a sequence $\underline{x}_0 \leq \underline{x}_1 \leq \cdots$ in W such that $\underline{x}_0 = \underline{\Omega}_\alpha$ and for each $n \in \omega$, if $i \in I$ such that $w_i \neq \Omega$, then $\underline{x}_n(i) < \underline{x}_{n+1}(i)$. As W is finite state, we may conclude that there exists $\underline{v}, \underline{w} \in W$ such that $\underline{v} \leq \underline{w}$, $\underline{v} \approx_W \underline{w}$ and for all $i \in I$ such that $w_i \neq \Omega$, $\underline{v}(i) < \underline{w}(i)$. We suppose that $\underline{v}, \underline{w} \in W$ are such that $\underline{v}$ is of minimal length and demonstrate that $\underline{v} = \underline{\Omega}_\alpha$.

We argue by contradiction. Suppose that $\underline{v} \neq \underline{\Omega}_\alpha$, then there exists $\underline{v}' \in W$ and $a \in A$ such that $\underline{v} = \underline{v}'.\underline{a}_\alpha$. Suppose that $i \in I$ such that $a \in \alpha(i)$, then $\underline{v}(i) = \underline{v}'(i).a$. Since $\underline{v} \approx_W \underline{w}$, $\underline{v}(i) \approx_{W(i)} \underline{w}(i)$, by 11.3.1, 11.3.4 and 11.3.6, and since $\underline{v} \leq \underline{w}$, $\underline{w}(i) \neq \Omega$. Hence, by 11.3.9, there exists $w_i' \in W(i)$ such that $\underline{w}(i) = w_i'.a$ and $\underline{v}'(i) \approx_{W(i)} w_i'$. Thus, if we define

$$\underline{w}' = \begin{cases} w_i' \text{ if } a \in \alpha(i) \\ \underline{w}(i) \text{ otherwise} \end{cases}$$

then $\underline{w}'.\underline{a}_\alpha = \underline{w} \in A_\alpha^*$ and so $\underline{w}' \in A_\alpha^*$, by 11.3.10. As in addition, $\underline{w}'(i) \in W(i)$, each $i \in I$, we must have $\underline{w}' \in W$. Since $\underline{v}'(i) \approx_{W(i)} \underline{w}'(i)$, each $i \in I$ we have $\underline{v}' \approx_W \underline{w}'$, by 11.3.2. We show that $v_i' < w_i'$ for each $i \in I$ such that $a \in \alpha(i)$. This will ensure that $\underline{v}' \leq \underline{w}'$. and that if $w_i \neq \Omega$, then $\underline{v}'(i) < \underline{w}'(i)$. We will thus have a contradiction to the minimality of $\underline{v}$.

---

[1]The quotation at the head of this chapter is an allusion to Giambattista Vico's cyclical theory of history.

By an argument similar to that of 11.3.9, we may assume that the hypothesis of 11.3.8 holds. So, since $\underline{v}(i) \approx_{W(i)} \underline{w}(i)$, it follows that there exists $m, n \in \omega$ with $n < m$, and $u < w_i$ such that $\underline{v}(i) = w_i^{\,n}.u$ and $\underline{w}(i) = w_i^{\,m}.u$. Thus, $\underline{w}(i) = \underline{v}(i).((w_i / u).u)^{m-n}$ and either $u > \Omega$, in which case $u = u'.a$, some $u' \in \alpha(i)^*$, and $\underline{w}'(i) = \underline{v}'(i).((w_i / u').u')^{m-n}$, or $u = \Omega$, in which case $w_i = w'.a$, some $w' \in \alpha(i)^*$, and $\underline{w}'(i) = \underline{v}'(i).(a.w')^{m-n}$.

We have shown that there exists $\underline{w} \in W$ such that $\underline{\Omega}_\alpha \approx_W \underline{w}$ and that if $w_i \neq \Omega$, then $\underline{w}(i) > \Omega$. As $\underline{\Omega}_\alpha \approx_W \underline{w}$, it follows by induction that for each $n \in \omega$, that $\underline{w}^n \in W$ and $\underline{\Omega}_\alpha \approx_W \underline{w}^n$. Thus, $\underline{w}^* \subseteq W$ and so $\downarrow\underline{w}^* \subseteq W$, by left-closure of W. Conversely, if $\underline{x} \in W$, then for each $i \in I$ such that $w_i \neq \Omega$, $\underline{x}(i) < \underline{w}(i)^{n_i}$, where $|x(i)| < n_i.|\underline{w}(i)|$, since $\underline{w}(i) > \Omega$ and $W(i)$ is totally ordered. Hence, $\underline{x} \leq \underline{w}^n$, where n is the maximum of the n_i and therefore, $W \subseteq \downarrow\underline{w}^*$.

■

11.3.12. THEOREM. If $V \subseteq A_\alpha^*$ such that I is finite, then V is simple $\Leftrightarrow$ there exists $\underline{x}, \underline{w} \in A_\alpha^*$ such that $V = \downarrow(\underline{x}.\underline{w}^*)$.

PROOF. Suppose first that V is simple, then by 11.3.7, there exists $\underline{x} \in A_\alpha^*$ and $w_i \in \alpha(i)^*$, each $i \in I$ such that $V = \downarrow(\underline{x}.W)$, where $W = \|_{i \in I}(\alpha(i), \downarrow w_i^*)$ and $W(i) = \downarrow w_i^*$, each $i \in I$. By 11.3.11, there exists $\underline{w} \in A_\alpha^*$ such that $W = \downarrow\underline{w}^*$. We thus have

$$V = \downarrow(\underline{x}.\downarrow\underline{w}^*) = \downarrow(\underline{x}.\underline{w}^*)$$

Conversely, suppose that there exists $\underline{x}, \underline{w} \in A_\alpha^*$ such that $V = \downarrow(\underline{x}.\underline{w}^*)$, We shall prove that $V = \|_{i \in I}(\alpha(i), L_i)$, where $L_i = \downarrow(\underline{x}(i).\underline{y}(i)^*)$, each $i \in I$. Since each L_i is linear and finite state, by 11.3.6, it then follows that V is simple, by 11.3.5.

If $\underline{v} \in V$, then $\underline{v} \leq \underline{x}.\underline{w}^n$, some $n \in \omega$, and so for each $i \in I$, $\underline{v}(i) \leq \underline{x}(i).\underline{w}(i)^n$, and hence $\underline{v}(i) \in L_i$, each $i \in I$ Thus, $\underline{v} \in \|_{i \in I}(\alpha(i), L_i)$, by 10.5.2. Conversely, suppose that $\underline{v} \in \|_{i \in I}(\alpha(i), L_i)$, For each $i \in I$, $\underline{v}(i) \in L_i$, so there exists $n_i \in \omega$ such that $\underline{v}(i) \leq \underline{x}(i).\underline{w}(i)^{n_i}$, so that if n is the maximum of the n_i, then $\underline{v}(i) \leq \underline{x}(i).\underline{w}(i)^n$, each $i \in I$ that is $\underline{v} \leq \underline{x}.\underline{w}^n$. Hence, $\underline{v} \in V$.

■

Notes.

The author's earliest contribution to Concurrency Theory was a theorem giving necessary and sufficient conditions for adequacy in GR_0-paths. [25]. Effectively, a GR_0-path defines a vector language of the form $W = \|_{i \in I}(A_i, \downarrow w_i^*)$ and adequate means, in this context, that if $\underline{x} \in W$ and $a \in A$, then there exists $\underline{u} \in A_\alpha^*$ such that $\underline{x}.\underline{u}.\underline{a}_\alpha \in W$. So, if $W(i) \neq \downarrow w_i^*$, some i, then each elements of A_i may only appear a finite number of times, and the path cannot be adequate. Hence adequacy implies that there exists $\underline{w} \in A_\alpha^*$ such that $W = \downarrow \underline{w}^*$. and for each $i \in I$, $\underline{w}(i) \neq \Omega$.

Subsequently, E. Best produced an alternative proof; 11.3.9 is essentially due to him.

It is perhaps worth pointing out that this investigation preceded the invention of vectors. However, a certain relation played an important role in it. This relation was subsequently seen to be a Mazurkiewicz trace congruence.

It would appear that even when one is adopting an interleaving approach to the study of parallelism, there are times when one cannot help using the machinery of a non-interleaving approach.

This result generalises one by Lauer and Campbell in [15] for a class of path expressions called GE_0-paths; these are a special case of GR_0-paths kin which the sequences w_i have no repeated elements. In this case, we may actually choose $\underline{w} \in A_\alpha^*$ such that $\underline{w}(i) = w_i$.

Lauer and Campbell derived their result from a theorem on absence of deadlock in a type of Petri Net called a marked graph. The derivation works in the opposite direction; the apparently static criterion for liveness is in fact a restatement of a dynamic criterion.

Chapter 12
Linguistic Behavioural Presentations

What kind of behaviour do you call this!

MRS. D. E. SHIELDS
Private Communication

12.1. Introduction.

10.4.5 shows that we may construct behavioural presentations from left-closed trace languages. We have indeed seen (in 4.3.3) that any prime algebraic and consistently complete poset is isomorphic to some behavioural presentation, where the occurrences of the presentation are in bijection with the complete primes of the poset. Furthermore, since each prime in a left-closed trace language has a unique last element, we have a means of associating each prime with an element of the A.

It is therefore fairly natural to make the following construction, which, as we shall see shortly, gives rise to asynchronous discrete behavioural presentations.

12.1.1. DEFINITION. Suppose that L is left-closed in $A_\alpha{}^*$. Define $B_{L,A} = (O_L, P_L, A, \lambda_L)$ where:

1• $O = Pr(L)$;
2• $P_L = \{Pr(\underline{x}) | \underline{x} \in L\}$;
3• If $a \in A$, then $\lambda_L(\underline{u}) = a$ if and only if $\underline{u} = \underline{x}.\underline{a}_\alpha$, some $\underline{x} \in A_\alpha{}^*$.

If $\underline{u} \in O_L$, then $\underline{u} \in Pr(\underline{u})$. It follows that $O_L \subseteq \bigcup_{\underline{x} \in L} Pr(\underline{x})$ and so $B_{L,A}$ is a behavioural presentation, by 2.1.1 (2). Let us first look at a small example and then find out what kind of behavioural presentation we get from this construction.

Let $L=\{(\Omega,\Omega),(a,a),(a,ab),(ac,a),(ad,a),(ac,ad)\}$.

The occurrences of $B_{L,A}$ are as follows.

$O_L=\{(a,a),(a,ab),(ac,a),(ad,a)\}$

For the points, we have

$Pr(\Omega,\Omega)=\varnothing \qquad Pr(a,a)=\{(a,a)\}$

$Pr(a,ab)=\{(a,a),(a,ab)\} \qquad Pr(ac,a)=\{(a,a),(ac,a)\}$

$Pr(ad,a)=\{(a,a),(ad,a)\} \qquad Pr(ac,ab)=\{(a,a),(ac,a),(a,ab)\}$

and for the label function, we have

$\lambda_L(a,a)=a \qquad \lambda_L(a,ab)=b \qquad \lambda_L(ac,a)=c \qquad \lambda_L(ad,a)=d.$

$B_{L,A}$ is pictured in figure 12.1 below

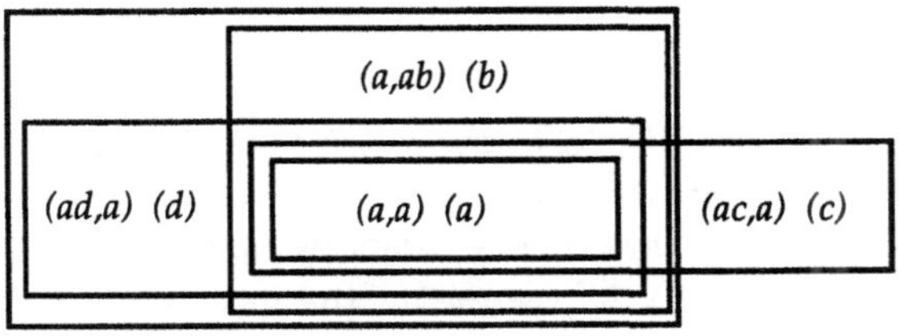

Figure 12.1

If $B_{L,A}=(O,P,A,\lambda)$, then O has two partial orders on it, the order of (2) of 2.3.1 which we shall call $\rightarrow_L$, and that which it inherits from $A_\alpha{}^*$. However, if $\underline{u}\rightarrow_L\underline{v}$, then since $\underline{v}\in Pr(\underline{v})$, $\underline{u}\in Pr(\underline{v})$ and hence $\underline{u}\le\underline{v}$. On the other hand, if $\underline{u}\le\underline{v}$ and $\underline{v}\in Pr(\underline{x})$, then $\underline{v}\le\underline{x}$ and so $\underline{u}\le\underline{x}$, whence $\underline{u}\in Pr(\underline{x})$. Thus $\underline{v}\in Pr(\underline{x})\Rightarrow\underline{u}\in Pr(\underline{x})$ and so $\underline{u}\rightarrow_L\underline{v}$. We have shown

$$\forall\underline{u},\underline{v}\in O_L:\underline{u}\rightarrow_L\underline{v}\Leftrightarrow\underline{u}\le\underline{v} \tag{12.1}$$

Hence for all $\underline{u},\underline{v}\in O_L$

$$\underline{u}\approx\underline{v}\Leftrightarrow\underline{u}\rightarrow_L\underline{v}\wedge\underline{v}\rightarrow_L\underline{u}\Leftrightarrow\underline{u}\le\underline{v}\wedge\underline{v}\le\underline{u}\Leftrightarrow\underline{u}=\underline{v}$$

and so $B_{L,A}$ is asynchronous.

Next, we show that $B_{L,A}$ is discrete. Indeed, if $\underline{x}\in L$, then $|Pr(\underline{x})|\le 2^{|\underline{x}|}$, so $B_{L,A}$ is $\approx$-finitary. For left-closure of $B_{L,A}$, suppose $p\in P_L$ and $X\subseteq O_L$ with

$X \lhd p$. By 12.1.1, $p = Pr(\underline{x})$, some $\underline{x} \in L$. As $X \subseteq p$, $X \leq \sqcup p = \underline{x}$, using prime algebraicity of L, and hence $\sqcup X$ exists and $\sqcup X \leq \underline{x}$. By left-closure of L, $\sqcup X \in L$. Call it $\underline{y}$. We show that $X = Pr(\underline{y})$, from which it will follow that $X \in P_L$.

Clearly $X \subseteq Pr(\underline{y})$. Conversely, let $\underline{u} \in Pr(\underline{y})$, then $\underline{u} \leq \sqcup X$ and so by primality of $\underline{u}$, $\underline{u} \leq \underline{u}'$, some $\underline{u}' \in X$. Now $\underline{u} \leq \underline{y} \leq \underline{x}$ and so $\underline{u} \in Pr(\underline{x})$. Thus, $\underline{u} \leq \underline{u}'$, $\underline{u} \in Pr(\underline{x})$, $\underline{u}' \in X$ and $X \lhd Pr(\underline{x})$. By 3.3.7, $\underline{u} \in X$. Thus $X = Pr(\underline{y})$ and so $B_{L,A}$ is left-closed.

Summarising:-

12.1.2. PROPOSITION. Suppose that L is left-closed in $A_\alpha{}^*$, then $B_{L,A}$ is a discrete, asynchronous behavioural presentation.

■

12.2. The Structure of $B_{L,A}$.

Of course, we would like to know what sort of discrete, asynchronous behavioural presentations derive from left-closed trace or vector languages. For it is not the case that every discrete asynchronous behavioural presentation isomorphic to $B_{L,A}$, some L; the independence relation ι_α has some part in the construction of $B_{L,A}$, and one would expect that this would manifest itself in the structure. In particular one would imagine that ι_α determines which occurrences may be concurrent. Indeed, the results of the following investigation may be considered as merely formalising the intuition that lies behind the use of traces/vectors as representing asynchronous behaviours.

We shall therefore take a closer look at the order structure of $B_{L,A}$. Before we do so, we will find it useful to have the following lemma at our disposal.

12.2.1. LEMMA. Suppose $\underline{u} \in A_\alpha{}^*$ and $a \in A$ such that $\underline{u}.\underline{a}_\alpha \in Pr(A_\alpha{}^*)$ then

$$\forall \underline{v} \in A_\alpha{}^*: \underline{v} < \underline{u}.\underline{a}_\alpha \Rightarrow \underline{v} \leq \underline{u}.$$

PROOF. By the factorisation lemma, 10.2.11, there exists $\underline{v}_1, \underline{v}_2 \in A_\alpha{}^*$ such that $\underline{v} = \underline{v}_1.\underline{v}_2$, $\underline{v}_1 \leq \underline{u}$, $\underline{v}_2 \leq \underline{a}_\alpha$ and $\underline{v}_2$ *ind* $\underline{u}/\underline{v}_1$.

Since $\underline{v}_2 \leq \underline{a}_\alpha$, either $\underline{v}_2 = \underline{\Omega}_\alpha$ or $\underline{v}_2 = \underline{a}_\alpha$. In the former case, $\underline{v}_1 \leq \underline{u}$ and we are done. Let us assume $\underline{v}_2 = \underline{a}_\alpha$ and derive a contradiction.

First, $\underline{u}/\underline{v}_1 > \underline{\Omega}_\alpha$, for otherwise, $\underline{u} = \underline{v}_1$ and hence $\underline{v} = \underline{v}_1.\underline{v}_2 = \underline{u}.\underline{a}_\alpha$, contrary to hypothesis.

Therefore, $\underline{u}/\underline{v}_1 = \underline{w}.\underline{b}_\alpha$, some $\underline{w} \in A_\alpha{}^*$ and $b \in A$. Since $\underline{a}_\alpha = \underline{v}_2$ *ind* $\underline{u}/\underline{v}_1$, it follows that $\underline{a}_\alpha$ *ind* $\underline{b}_\alpha$. In particular, $a \neq b$. But now

$$\underline{u}.\underline{a}_\alpha = \underline{v}_1.(\underline{u}/\underline{v}_1).\underline{a}_\alpha = \underline{v}_1.\underline{w}.\underline{b}_\alpha\underline{a}_\alpha = \underline{v}_1.\underline{w}.\underline{a}_\alpha.\underline{b}_\alpha$$

with $a \neq b$, contradicting primality of $\underline{u}.\underline{a}_\alpha$.

■

Let us consider two occurrences $\underline{u}_1, \underline{u}_2 \in O_L$ such that $\underline{u}_1\ co_L\ \underline{u}_2$. As, $\underline{u}_1, \underline{u}_2 \in Pr(L)$, $\underline{u}_i = \underline{v}_i.\underline{a}^i{}_\alpha$, where $\underline{v}_i \in A_\alpha{}^*$ and $a^i = \lambda_L(\underline{u}_i)$, each i.

Since $\underline{u}_1\ co_L\ \underline{u}_2$, there exists $p \in P_L$ such that $\underline{u}_1, \underline{u}_2 \in p$. Since $p = Pr(\underline{x})$, some $\underline{x} \in L$, it follows that $\underline{u}_1, \underline{u}_2 \leq \underline{x}$ and hence , by 10.2.4, $\underline{u}_1 \sqcup \underline{u}_2$ exists. By 10.2.8, $\underline{u}_1/\underline{w}$ *ind* $\underline{u}_2/\underline{w}$, where $\underline{w} = \underline{u}_1 \sqcap \underline{u}_2$.

Now, if $\underline{w} = \underline{u}_1$, then $\underline{u}_1 \leq \underline{u}_2$ from which we may deduce, using (12.1), that $\underline{u}_1 \rightarrow_L \underline{u}_2$, contradicting $\underline{u}_1\ co_L\ \underline{u}_2$. Thus, $\underline{w} < \underline{u}_1$ and likewise $\underline{w} < \underline{u}_2$

We may apply 12.2.1 to deduce that $\underline{w} \leq \underline{v}_1$ and $\underline{w} \leq \underline{v}_2$. Hence, $\underline{u}_1/\underline{w} = (\underline{v}_1/\underline{w}).\underline{a}^1{}_\alpha$ and $\underline{u}_2/\underline{w} = (\underline{v}_2/\underline{w}).\underline{a}^2{}_\alpha$. But $\underline{u}_1/\underline{w}$ *ind* $\underline{u}_2/\underline{w}$. By 9.2.13 (4), $\underline{a}^1{}_\alpha$ *ind* $\underline{a}^2{}_\alpha$. By 9.2.13 (2), $a^1\ \iota_\alpha\ a^2$. We have proved the following

12.2.2. LEMMA. If L is left-closed in $A_\alpha{}^*$, then

$$\forall \underline{u}_1, \underline{u}_2 \in O_L : \underline{u}_1\ co_L\ \underline{u}_2 \Rightarrow \lambda_L(\underline{u}_1)\ \iota_\alpha\ \lambda_L(\underline{u}_2).$$

■

ι_α does have some relationship to the structure of $B_{L,A}$, as we suspected; two occurrences may be concurrent only if their labels are independent. The converse is not necessarily true, as it is possible, for example, that $\lambda_L(\underline{u}_1)\ \iota_\alpha\ \lambda_L(\underline{u}_2)$, but that some third party gets in the way in the sense that there exists $\underline{u}' \in O_L$ such that either $\underline{u}_1 < \underline{u}' < \underline{u}_2$ or $\underline{u}_2 < \underline{u}' < \underline{u}_1$. Let us pause to identify situations in which such $\underline{u}'$ are not present.

12.2.3. DEFINITION. Let B be a behavioural presentation and let $o_1, o_2 \in O$. We shall say that they are *unseparated*, in B and write o_1 *unsep* $o_2 \Leftrightarrow$

1• $\neg o_1 \# o_2$;

2• $\forall o \in O: o_1 < o \Rightarrow o \not< o_2$;

3• $\forall o \in O: o_2 < o \Rightarrow o \not< o_1$;

We now show that the converse of 12.2.2 holds, providing $\underline{u}_1$ and $\underline{u}_2$ are unseparated in $B_{L,A}$. In other words, $B_{L,A}$ will allow as much concurrency between occurrences with ι_α-independent labels as it possibly can.

12.2.4. LEMMA. If L is left-closed in A_α^*, then

$$\forall \underline{u}_1, \underline{u}_2 \in O_L : \underline{u}_1 \; unsep_L \; \underline{u}_2 \Rightarrow (\lambda_L(\underline{u}_1) \, \iota_\alpha \, \lambda_L(\underline{u}_2) \Rightarrow \underline{u}_1 \; co_L \; \underline{u}_2).$$

PROOF. If $\underline{u}_1, \underline{u}_2 \in O_L$ with $\underline{u}_1 \; unsep_L \; \underline{u}_2$ and $\lambda_L(\underline{u}_1) \, \iota_\alpha \, \lambda_L(\underline{u}_2)$, then $\underline{u}_1 \approx_L \underline{u}_2$ is impossible, by 12.1.2, and $\underline{u}_1 \#_L \underline{u}_2$ is impossible, by 12.2.3, so the only other possibilities are $\underline{u}_1 \to_L \underline{u}_2$, $\underline{u}_2 \to_L \underline{u}_1$ and $\underline{u}_1 \; co_L \; \underline{u}_2$.

Suppose that $\underline{u}_1 \to_L \underline{u}_2$, then by (12.1), $\underline{u}_1 \leq \underline{u}_2$. $\underline{u}_1 \neq \underline{u}_2$, for otherwise, we would have $\lambda_L(\underline{u}_1) \, \iota_\alpha \, \lambda_L(\underline{u}_2)$ (ι_α is irreflexive) and so $\underline{u}_1 < \underline{u}_2$.

Now, $\underline{u}_1 = \underline{v}_1 . \underline{a}_\alpha$ and $\underline{u}_2 = \underline{v}_2 . \underline{b}_\alpha$ with $\underline{v}_1, \underline{v}_2 \in A_\alpha^*$ and $a, b \in A$, so that $a = \lambda_L(\underline{u}_1) \, \iota_\alpha \, \lambda_L(\underline{u}_2) = b$ and hence $\underline{a}_\alpha . \underline{b}_\alpha = \underline{b}_\alpha . \underline{a}_\alpha$, by 9.2.13 (2). Since $\underline{u}_1 < \underline{u}_2$, it follows from 12.2.1 that $\underline{u}_1 \leq \underline{v}_2$, so there exists $\underline{w} \in A_\alpha^*$ such that $\underline{v}_2 = \underline{v}_1 . \underline{a}_\alpha . \underline{w}$. Now we cannot have $\underline{a}_\alpha \; ind \; \underline{w}$, for then it would follow that $\underline{u}_2 = \underline{v}_1 . \underline{a}_\alpha . \underline{w} . \underline{b}_\alpha = \underline{v}_1 . \underline{w} . \underline{a}_\alpha . \underline{b}_\alpha = \underline{v}_1 . \underline{w} . \underline{b}_\alpha . \underline{a}_\alpha$, by 9.2.13 (3), contradicting $\underline{u}_2$ prime, so we may write $\underline{w} = \underline{w}_1 . \underline{c}_\alpha . \underline{w}_2$ where $\underline{a}_\alpha \; ind \; \underline{w}_1$ and $\neg \underline{a}_\alpha \; ind \; \underline{c}_\alpha$.

Let $\underline{x} = \underline{u}_1 . \underline{w}_1$ and let $\underline{u} = pr(\underline{x}, c)$, then since $\underline{x} < \underline{u}_2$, $\underline{u} < \underline{u}_2$. Since $\underline{u}_1 \; unsep_L \; \underline{u}_2$, $\underline{u}_1 \not< \underline{u}$. Since $\underline{u}_1, \underline{u} \in Pr(\underline{u}_2)$, $\neg \underline{u}_1 \# \underline{u}$. Since $B_{L,A}$ is asynchronous, by 12.2.1, $\underline{u}_1 \neq \underline{u}$. Since $\neg a \, \iota_L \, c$, $\neg \lambda_L(\underline{u}_1) \, \iota_L \, \lambda_L(\underline{u})$ and so $\neg \underline{u}_1 \; co_L \; \underline{u}$, by 12.2.2. The only other possibility is $\underline{u} \leq \underline{u}_1$ and since $\lambda_L(\underline{u}) \neq \lambda_L(\underline{u}_1)$, $\underline{u} < \underline{u}_1$ and so $\underline{u} \leq \underline{v}_1$, by 12.2.1.

By 10.3.3, $\underline{u} = \underline{v} . \underline{c}_\alpha$ with $\underline{v} \leq \underline{x}$ and $\underline{x}/\underline{v} \; ind \; \underline{c}_\alpha$. But $\underline{v} < \underline{u} \leq \underline{v}_1 < \underline{v}_1 . \underline{a}_\alpha \leq \underline{x}$ and so $(\underline{v}_1/\underline{v}) . \underline{a}_\alpha = \underline{u}_1/\underline{v} \leq \underline{x}/\underline{v}$, giving $\underline{a}_\alpha \; ind \; \underline{c}_\alpha$, a contradiction.

Hence, $\underline{u}_1 \not\to_L \underline{u}_2$. By symmetry, $\underline{u}_2 \not\to_L \underline{u}_1$ and the only remaining possibility is $\underline{u}_1 \; co_L \; \underline{u}_2$, as required.

■

12.2.2 and 12.2.4 characterise behavioural presentations which derive from trace or vector languages as far as constraints on labelling are concerned. There is one further characteristic property we need to consider.

12.2.5. DEFINITION. Let B be a behavioural presentation and suppose $p_1, p_2 \in P$, then p_1 and p_2 are *isomorphic*, and we write $p_1 \equiv p_2$, if there exists a bijection $\varphi : p_1 \to p_2$ such that

1• $\forall o_1, o_2 \in p_1 : o_1 \to_1 o_2 \Leftrightarrow \phi(o_1) \to_2 \phi(o_2)$;

2• $\forall o \in p_1 : \lambda_1(o) = \lambda_2(\phi(o))$.

B will be said to be *λ-reduced* $\Leftrightarrow \forall p_1, p_2 \in P : p_1 \equiv p_2 \Rightarrow p_1 = p_2$.

We shall show in 12.2.7 that if L is left-closed in A_α^* then $B_{L,A}$ is λ-reduced. The proof is based on the observation that if $\underline{x}_1, \underline{x}_2 \in A_\alpha^*$, then

$Pr(\underline{x}_1) \equiv Pr(\underline{x}_2) \Rightarrow \underline{x}_1 = \underline{x}_2$ and we establish this by induction on $|\underline{x}_1|$. The key idea in this proof is that if $a \in A$ and $\underline{a}_\alpha \leq \underline{x}_1$ then $\underline{a}_\alpha \leq \underline{x}_2$ and $Pr(\underline{x}_1) \equiv Pr(\underline{x}_2) \Rightarrow Pr(\underline{x}_1/\underline{a}_\alpha) \equiv Pr(\underline{x}_2/\underline{a}_\alpha)$. To establish this, we need a closer understanding of the relationship between $Pr(\underline{x})$ and $Pr(\underline{x}/\underline{a}_\alpha)$.

12.2.6. LEMMA. Suppose $\underline{x} \in A_\alpha{}^*$ and suppose $a \in A$ with $\underline{a}_\alpha \leq \underline{x}$. Define

$$X_a(\underline{x}) = \{\underline{u} \in Pr(\underline{x}) \mid \underline{a}_\alpha < \underline{u}\};$$
$$Y_a(\underline{x}) = \{\underline{u} \in Pr(\underline{x}) \mid \underline{a}_\alpha \not\leq \underline{u}\}$$

then

1• $Pr(\underline{x}) = X_a(\underline{x}) \cup Y_a(\underline{x}) \cup \{\underline{a}_\alpha\}$,
2• $Pr(\underline{x}/\underline{a}_\alpha) = X_a(\underline{x})/\underline{a}_\alpha \cup Y_a(\underline{x})$,

where in each case, the unions are pairwise disjoint

3• $|Pr(\underline{x})| = |\underline{x}|$.

PROOF. (1) is clear. Let us show (2).

It is not hard to see that if $\underline{a}_\alpha < \underline{u} \in Pr(\underline{x})$, then $\underline{u}/\underline{a}_\alpha \in Pr(A_\alpha{}^*)$ and $\underline{u}/\underline{a}_\alpha \leq \underline{x}/\underline{a}_\alpha$. Thus $X_a(\underline{x})/\underline{a}_\alpha \subseteq Pr(\underline{x}/\underline{a}_\alpha)$. If $\underline{u} \in Y_a(\underline{x})$, then $\underline{a}_\alpha, \underline{u} \in \underline{x}$ and $\underline{a}_\alpha \not\leq \underline{u}$. By 10.2.1, $\underline{a}_\alpha$ *ind* $\underline{u}$. By 10.2.2, $\underline{u} \leq \underline{x}/\underline{a}_\alpha$. But $Y_a(\underline{x}) \subseteq Pr(A_\alpha{}^*)$, and so $Y_a(\underline{x}) \subseteq Pr(\underline{x}/\underline{a}_\alpha)$, and we have shown that $X_a(\underline{x})/\underline{a}_\alpha \cup Y_a(\underline{x}) \subseteq Pr(\underline{x}/\underline{a}_\alpha)$.

Next, suppose that $\underline{u} \in Pr(\underline{x}/\underline{a}_\alpha)$, then $\underline{u} \in Pr(A_\alpha{}^*)$ and $\underline{u} \leq \underline{x}/\underline{a}_\alpha$.

Since $\underline{u} \in Pr(A_\alpha{}^*)$, there exists $\underline{v} \in A_\alpha{}^*$ and $b \in A$ such that $\underline{u} = \underline{v}.\underline{b}_\alpha$. If $\underline{a}_\alpha.\underline{u}$ is not prime, then there exists $\underline{w} \in A_\alpha{}^*$ and $c \in A$ such that $\underline{a}_\alpha \underline{v}.\underline{b}_\alpha = \underline{a}_\alpha \underline{u} = \underline{w}.\underline{c}_\alpha$ and $b \neq c$. We cannot have $\underline{a}_\alpha \leq \underline{w}$, for then $\underline{u} = (\underline{w}/\underline{a}_\alpha).\underline{c}_\alpha$ and $\underline{u} = \underline{v}.\underline{b}_\alpha$, with $b \neq c$, contradicting $\underline{u} \in Pr(A_\alpha{}^*)$. But otherwise, as $\underline{a}_\alpha, \underline{w} \leq \underline{w}.\underline{c}_\alpha$, $\underline{a}_\alpha$ *ind* $\underline{w}$, by 10.2.1, and so, by 10.2.2 and induction on $|\underline{w}|$, $\underline{a}_\alpha \leq (\underline{w}.\underline{c}_\alpha)/\underline{w} = \underline{c}_\alpha$, whence $a = c$ and $\underline{u} = \underline{w}$. Hence $\underline{a}_\alpha$ *ind* $\underline{u}$. Thus, if $\neg \underline{a}_\alpha$ *ind* $\underline{u}$ then $\underline{a}_\alpha.\underline{u} \in Pr(A_\alpha{}^*)$ and as $\underline{u} \leq \underline{x}/\underline{a}_\alpha$, $\underline{a}_\alpha.\underline{u} \leq \underline{x}$ whence $\underline{a}_\alpha.\underline{u} \in X_a(\underline{x})$ and consequently $\underline{u} \in X_a(\underline{x})/\underline{a}_\alpha$.

We conclude that if $\neg \underline{a}_\alpha$ *ind* $\underline{u}$ then $\underline{u} \in X_a(\underline{x})/\underline{a}_\alpha$.

On the other hand, if $\underline{a}_\alpha$ *ind* $\underline{u}$, then $\underline{a}_\alpha \not\leq \underline{u}$ and since $\underline{u} \leq \underline{x}/\underline{a}_\alpha$, $\underline{u} \leq \underline{x}$, by 10.2.2. Thus, $\underline{u} \in Y_a(\underline{x})$.

We have shown that $Pr(\underline{x}/\underline{a}_\alpha) \subseteq X_a(\underline{x})/\underline{a}_\alpha \cup Y_a(\underline{x})$, completing the proof that $Pr(\underline{x}/\underline{a}_\alpha) = X_a(\underline{x})/\underline{a}_\alpha \cup Y_a(\underline{x})$. We next show that the union is disjoint.

If $\underline{u} \in X_a(\underline{x})/\underline{a}_\alpha \cap Y_a(\underline{x})$, then $\underline{a}_\alpha.\underline{u} \in X_a(\underline{x})$ and $\underline{u} \in Y_a(\underline{x})$. It follows that $\underline{u}, \underline{a}_\alpha \leq \underline{x}$ and, as $\underline{u} \in Y_a(\underline{x})$, $\underline{a}_\alpha \not\leq \underline{u}$, and so by 10.2.1, $\underline{a}_\alpha$ *ind* $\underline{u}$. But then,

$\underline{u}.\underline{a}_\alpha = \underline{a}_\alpha.\underline{u} = \underline{a}_\alpha.\underline{v}.\underline{b}_\alpha$, and so $a = b$, as $\underline{a}_\alpha.\underline{u} \in Pr(A_\alpha{}^*)$. And now we have a contradiction, for from $\underline{a}_\alpha$ *ind* $\underline{u}$ and $\underline{u} = \underline{v}.\underline{b}_\alpha$, we deduce that $\underline{a}_\alpha$ *ind* $\underline{b}_\alpha$. Thus, $X_a(\underline{x})/\underline{a}_\alpha \cap Y_a(\underline{x}) = \varnothing$.

For (3), we note that by (10.3), if $\underline{u}_1, \underline{u}_2 \in X_a(\underline{x})$, then $\underline{u}_1 = \underline{u}_2 \Leftrightarrow \underline{u}_1/\underline{a}_\alpha = \underline{u}_2/\underline{a}_\alpha$ so the mapping $\underline{u} \mapsto \underline{u}/\underline{a}_\alpha$ establishes a bijection between $X_a(\underline{x})$ and $X_a(\underline{x})/\underline{a}_\alpha$, so $|X_a(\underline{x})| = |X_a(\underline{x})/\underline{a}_\alpha|$. Using (1) and (2):

$$|Pr(\underline{x}/\underline{a}_\alpha)| = |X_a(\underline{x})/\underline{a}_\alpha| + |Y_a(\underline{x})| = |X_a(\underline{x})| + |Y_a(\underline{x})| = |Pr(\underline{x})| - 1$$

By induction, $|Pr(\underline{x}/\underline{a}_\alpha)| = |\underline{x}/\underline{a}_\alpha|$. Hence, by 9.2.11 (2) and (3)

$$|Pr(\underline{x})| = |Pr(\underline{x}/\underline{a}_\alpha)| + 1 = |\underline{x}/\underline{a}_\alpha| + |\underline{a}_\alpha| = |\underline{x}|$$

■

12.2.7. LEMMA. If L is left-closed in $A_\alpha{}^*$, then $B_{L,A}$ is λ-reduced.

PROOF. Suppose $\underline{x}_1, \underline{x}_2 \in L$ such that $Pr(\underline{x}_1) \equiv Pr(\underline{x}_2)$ via a bijection ϕ. We shall prove that $\underline{x}_1 = \underline{x}_2$, which then gives $Pr(\underline{x}_1) = Pr(\underline{x}_2)$. The claim is obvious if $\underline{x}_1 = \underline{\Omega}_\alpha$, so suppose $\underline{x}_1 \neq \underline{\Omega}_\alpha$ and let $a \in A$ such that $\underline{a}_\alpha \leq \underline{x}_1$. The idea is to show that $Pr(\underline{x}_1/\underline{a}_\alpha) \equiv Pr(\underline{x}_2/\underline{a}_\alpha)$ and appeal to induction on $|\underline{x}_1|$.

It is clear that $\underline{a}_\alpha \in Pr(\underline{x}_1)$ and $\lambda_L(\underline{a}_\alpha) = a$. Let $\underline{u} = \phi(\underline{a}_\alpha)$, then $a = \lambda_L(\underline{a}_\alpha) = \lambda_L(\varphi(\underline{a}_\alpha)) = \lambda_L(\underline{u})$, which means that $\underline{u} = \underline{v}.\underline{a}_\alpha$, some $\underline{v} \in A_\alpha{}^*$. If $\underline{v} \neq \underline{\Omega}_\alpha$, then there exists $b \in A$ such that $\underline{b}_\alpha \leq \underline{v}$. Thus, $\underline{b}_\alpha \in Pr(\underline{x}_2)$ and $\underline{b}_\alpha < \underline{u}$. But this entails that $\phi^{-1}(\underline{b}_\alpha) < \phi^{-1}(\underline{u}) = \underline{a}_\alpha$, which is impossible. Thus $\underline{u} = \underline{a}_\alpha$.

It follows that $\underline{a}_\alpha \leq \underline{x}_2$ and that $\phi(\underline{a}_\alpha) = \underline{a}_\alpha$.

Next, if $\underline{u} \in Pr(\underline{x}_1)$, then $\underline{a}_\alpha \leq \underline{u} \Leftrightarrow \phi(\underline{a}_\alpha) \leq \phi(\underline{u}) \Leftrightarrow \underline{a}_\alpha \leq \phi(\underline{u})$. Thus, in the terminology of 12.2.6,

$$\phi(X_a(\underline{x}_1)) \subseteq X_a(\underline{x}_2) \wedge \phi(Y_a(\underline{x}_1)) \subseteq Y_a(\underline{x}_2) \tag{12.2}$$

But, ϕ is a bijection, and thus

$$Pr(\underline{x}_2) = \phi(Pr(\underline{x}_1)) = \phi(X_a(\underline{x}_1) \cup Y_a(\underline{x}_1)) \subseteq X_a(\underline{x}_2) \cup Y_a(\underline{x}_2) = Pr(\underline{x}_2).$$

using (12.2). From this, and the fact that all unions are disjoint, it easily follows that $\phi(X_a(\underline{x}_1)) = X_a(\underline{x}_2) \wedge \phi(Y_a(\underline{x}_1)) = Y_a(\underline{x}_2)$.

Now $X_a(\underline{x}_1)/\underline{a}_\alpha \cap Y_a(\underline{x}_1) = \varnothing = X_a(\underline{x}_2)/\underline{a}_\alpha \cap Y_a(\underline{x}_2)$, by 12.2.6, so there is a map $\psi: X_a(\underline{x}_1)/\underline{a}_\alpha \cap Y_a(\underline{x}_1) \to X_a(\underline{x}_2)/\underline{a}_\alpha$, given by

$$\psi(\underline{u}) = \begin{cases} \phi(\underline{u}) \text{ if } \underline{u} \in Y_a(\underline{x}_1) \\ \phi(\underline{a}_\alpha . \underline{u})/\underline{a}_\alpha \text{ if } \underline{u} \in X_a(\underline{x}_1/\underline{a}_\alpha) \end{cases}$$

which is bijective, by (10.3). And if $\underline{u} \in Pr(\underline{x}_1)$, then $\lambda_L(\underline{u}) = \lambda_L(\underline{u}/\underline{a}_\alpha)$ and so $\lambda_L(\underline{u}) = \lambda_L(\psi(\underline{u}))$.

To complete the proof that $Pr(\underline{x}_1/\underline{a}_\alpha) \equiv Pr(\underline{x}_2/\underline{a}_\alpha)$, we show that ψ preserves $\leq$ in both directions. There are four cases to consider.

Case 1: $\underline{u}, \underline{u}' \in X_a(\underline{x}_1)/\underline{a}_\alpha$, then

$$\underline{u} \leq \underline{u}' \Leftrightarrow \underline{a}_\alpha . \underline{u} \leq \underline{a}_\alpha . \underline{u}' \Leftrightarrow \phi(\underline{a}_\alpha . \underline{u}) \leq \phi(\underline{a}_\alpha . \underline{u}') \Leftrightarrow \phi(\underline{a}_\alpha . \underline{u})/\underline{a}_\alpha \leq \phi(\underline{a}_\alpha . \underline{u}')/\underline{a}_\alpha$$

and so $\underline{u} \leq \underline{u}' \Leftrightarrow \psi(\underline{u}) \leq \psi(\underline{u}')$.

Case 2: $\underline{u}, \underline{u}' \in Y(\underline{x}_1)$, then $\underline{u} \leq \underline{u}' \Leftrightarrow \phi(\underline{u}) \leq \phi(\underline{u}') \Leftrightarrow \psi(\underline{u}) \leq \psi(\underline{u}')$.

Case 3: $\underline{u} \in X_a(\underline{x}_1)/\underline{a}_\alpha \wedge \underline{u}' \in Y_a(\underline{x}_1)$. If $\underline{u} \leq \underline{u}'$, then $\underline{a}_\alpha, \underline{u}' \leq \underline{x}_1$ with $\underline{a}_\alpha \not\leq \underline{u}'$ and so by 10.2.1, $\underline{a}_\alpha$ *ind* $\underline{u}'$ and hence $\underline{u}$ *ind* $\underline{a}_\alpha$ (since $\underline{u} \leq \underline{u}'$). But we have already seen, in the proof of 12.2.6 that $\underline{a}_\alpha \; ind \; \underline{u} \Rightarrow \underline{a}_\alpha . \underline{u} \notin Pr(A_\alpha{}^*)$ which contradicts $\underline{u} \in X_a(\underline{x}_1)/\underline{a}_\alpha$. Thus this case cannot arise.

Case 4: $\underline{u}' \in X_a(\underline{x}_1)/\underline{a}_\alpha \wedge \underline{u} \in Y_a(\underline{x}_1)$. If $\underline{u} \leq \underline{u}'$, then $\underline{a}_\alpha . \underline{u}', \underline{u} \leq \underline{x}_1$ with $\underline{a}_\alpha \not\leq \underline{u}$. So $\underline{u}$ *ind* $\underline{a}_\alpha$, by 10.2.1, and by 10.2.2, $\underline{u} \leq \underline{u}' \Leftrightarrow \underline{u} \leq \underline{a}_\alpha . \underline{u}'$. Also, $\phi(\underline{a}_\alpha . \underline{u}'), \phi(\underline{u}) \leq \underline{x}_2$ and $\underline{a}_\alpha \not\leq \phi(\underline{u})$, so $\phi(\underline{u}) \leq \varphi(\underline{a}_\alpha . \underline{u}') \Leftrightarrow \phi(\underline{u}) \leq \varphi(\underline{a}_\alpha . \underline{u}')/\underline{a}_\alpha$. Thus

$$\underline{u} \leq \underline{u}' \Leftrightarrow \underline{u} \leq \underline{a}_\alpha . \underline{u}' \Leftrightarrow \phi(\underline{u}) \leq \phi(\underline{a}_\alpha . \underline{u}') \Leftrightarrow \phi(\underline{u}) \leq \phi(\underline{a}_\alpha . \underline{u}')/\underline{a}_\alpha \Leftrightarrow \psi(\underline{u}) \leq \psi(\underline{u}').$$

Thus, $Pr(\underline{x}_1/\underline{a}_\alpha) \equiv Pr(\underline{x}_2/\underline{a}_\alpha)$. By induction, $\underline{x}_1/\underline{a}_\alpha = \underline{x}_2/\underline{a}_\alpha$ and hence $\underline{x}_1 = \underline{x}_2$.
■

We shall show that the properties of $B_{L,A}$ that we identified in lemmas 12.2.2, 12.2.4 and 12.2.7 are characteristic of such behavioural presentations. Let us give a name to the class in question.

12.2.8. DEFINITION. Let B be a behavioural presentation and ι an independence relation on E, then B will be said to be ι-*linguistic* $\Leftrightarrow$

1• B is discrete and asynchronous;
2• $\forall o_1, o_2 \in O: o_1 \; co \; o_2 \Rightarrow \lambda(o_1) \; \iota \; \lambda(o_2)$;
3• $\forall o_1, o_2 \in O: o_1 \; unsep \; o_2 \Rightarrow (\lambda(o_1) \; \iota \; \lambda(o_2) \Rightarrow o_1 \; co \; o_2)$;
4• B is λ-reduced.

We thus have the following characterisation of behavioural presentations which derive from left-closed trace languages.

12.2.9. THEOREM. Suppose L is left-closed in A_α^*, then $B_{L,A}$ is an ι_α-linguistic behavioural presentation.

■

To complete the chapter, we show that all left-closed trace or vector languages are images of ι-linguistic behavioural presentations.

12.2.10. DEFINITION. If B is a discrete, asynchronous behavioural presentation, then for each $p \in P$ define $\mathfrak{I}(p)$ to be the set of all total orders $D = (p, \leq_D)$ such that $\rightarrow \subseteq \leq_D$.

Define $\rho_B(p)$ to be the set of all strings $\lambda(o_1)\cdots\lambda(o_n) \in E^*$ such that $p = \{o_1, \cdots, o_n\}$ and $o_1 <_D \cdots <_D o_n$ for some $D \in \mathfrak{I}(p)$. $\rho_B(p)$ is called the set of all *sequential interleavings* of p.

12.2.11 THEOREM. If B is an ι-linguistic behavioural presentation, then for each $p \in P$, $\rho_B(p) \in E_\iota^*$

PROOF. Let $p \in P$ and $x \in \rho_B(p)$, then there exists $(p, \leq_x) \in \mathfrak{I}(p)$ such that $p = \{o_1, \cdots, o_n\}$, with $o_1 <_x \cdots <_x o_n$ and $x = \lambda(o_1)\cdots\lambda(o_n)$. If $y \in E^*$ such that $y \equiv_\iota^{(1)} x$, then for some $1 \leq i < n$, $y = \lambda(o_1')\cdots\lambda(o_n')$ and $\lambda(o_i')\ \iota\ \lambda(o_{i+1}')$, where

$$o_k' = \begin{cases} o_{i+1}' \text{ if } k = i \\ o_i \text{ if } k = i+1 \\ o_k \text{ otherwise} \end{cases}$$

Since $o_i, o_{i+1} \in p$, we cannot have, $\neg o_i \# o_{i+1}$ and since $o_i <_x o_{i+1}$, $o_{i+1} \not< o_i$, so $o_{i+1} < o \Rightarrow o \not< o_i$. And we cannot have $o_i < o < o_{i+1}$, since this would give the contradiction $o_i <_x o <_x o_{i+1}$. Hence, $o_i < o \Rightarrow o \not< o_{i+1}$. Thus, o_i *unsep* o_{i+1}. But $\lambda(o_{i+1}) = \lambda(o_i')\ \iota\ \lambda(o_{i+1}') = \lambda(o_i)$ so o_i *co* o_{i+1}. Now, if the total order $\leq_y$ is given by $o_1' <_y \cdots <_y o_n'$, then $\leq_y = (\leq_x - \{(o_i, o_{i+1})\}) \cup \{(o_{i+1}, o_i)\}$. Since o_i *co* o_{i+1}, it follows that $(o_i, o_{i+1}), (o_{i+1}, o_i) \notin \rightarrow$ and hence $\rightarrow \subseteq \leq_y$. Thus $(p, \leq_y) \in \mathfrak{I}(p)$ and so $y \in \rho_B(p)$. Since $\equiv_\iota$ is the reflexive, transitive closure of, $\equiv_\iota^{(1)}$, we may conclude that

$$\forall p \in P\ \forall x \in \rho_B(p): x_\iota \subseteq \rho_B(p) \tag{12.3}$$

In view of (12.3), in order to complete the proof, it will suffice to show that

$$\forall p \in P\ \forall x, y \in \rho_B(p): x \equiv_\iota y \tag{12.4}$$

So, suppose that $x, y \in \rho_B(p)$, so that there exists $(p, \leq_x), (p, \leq_y) \in \Im(p)$ such that

$$p = \{o_1, \cdots, o_n\}, \text{ with } o_1 <_x \cdots <_x o_n \text{ and } x = \lambda(o_1) \cdots \lambda(o_n);$$
$$p = \{o'_1, \cdots, o'_n\}, \text{ with } o_1 <_x \cdots <_x o_n \text{ and } y = \lambda(o'_1) \cdots \lambda(o'_n).$$

First, we show that

$$o_1 \cdots o_n \equiv_{co} o'_1 \cdots o'_n \tag{12.5}$$

(recall that co is an independence relation on O^*). We argue by induction on $n - |u \wedge v|$, where $u = o_1 \cdots o_n$, $v = o'_1 \cdots o'_n$ and $|u \wedge v|$ denotes the longest common prefix of u and v. If $n - |u \wedge v| = 0$ then $u = v$, and so certainly $u \equiv_{co} v$.

Suppose that $n - |u \wedge v| > 0$. We shall show that there exists $(p, \leq_w) \in \Im(p)$ such that $|u \wedge v| < |w \wedge v|$ and $u \equiv_{co} w$, where $w = \tilde{o}_1 \cdots \tilde{o}_n$, $p = \{\tilde{o}_1, \cdots, \tilde{o}_n\}$, with $\tilde{o}_1 <_w \cdots <_w \tilde{o}_n$. By induction, $w \equiv_{co} v$ and hence $u \equiv_{co} v$, by transitivity of $\equiv_{co}$.

Let $o_1 \cdots o_k = o'_1 \cdots o'_k = u \wedge v$, so $o'_{k+1} \neq o_i$ for $1 \leq i \leq k+1$. Thus $o_l = o'_{k+1}$ for some $k+1 < l \leq n$. Suppose that $k+1 < j \leq n$, then $o_j <_x o_l$, so $o_l \not\rightarrow o_j$, by 12.2.10, and $\neg o_l \# o_j$ since $o_l, o_j \in p$ and so either $o_j \rightarrow o_l$, in which case $o_j < o_l$ or $o_j \not\rightarrow o_l$, in which case $o_j \; co \; o_l$, by 2.3.6. But, if $o_j < o_l$, then $o_j <_y o_l = o'_{k+1}$ and so $o_j \in \{o'_1, \cdots, o'_k\} = \{o_1, \cdots, o_k\}$, a contradiction. Thus $o_j \; co \; o_l$. It follows that if we define $\leq_w$ by $\tilde{o}_1 \leq_w \cdots \leq_w \tilde{o}_n$, where

$$\tilde{o}_i = \begin{cases} o_l \text{ if } i = k+1 \\ o_{k-1} \text{ if } k+1 < i < l \\ o_i \text{ otherwise} \end{cases}$$

then $(p, \leq_w) \in \Im(p)$ and $|u \wedge v| < |w \wedge v|$ where $w = \tilde{o}_1 \cdots \tilde{o}_n$. Since we may obtain w from u by permuting adjacent elements satisfying $o_j \; co \; o_l$, we have $u \equiv_{co} w$. This concludes the proof of (12.5).

Finally, by 12.2.8. (2),

$$o_1 \cdots o_n \equiv^{(1)}_{co} o'_1 \cdots o'_n \Rightarrow \lambda(o_1) \cdots \lambda(o_n) \equiv^{(1)}_{\iota} \lambda(o'_1) \cdots \lambda(o'_n)$$

and since $\equiv_\iota$ is the reflexive, transitive closure of, $\equiv_\iota^{(1)}$ and $\equiv_{co}$ is the reflexive, transitive closure of, $\equiv_{co}^{(1)}$ we have, using (12.5)

$$x = \lambda(o_1) \cdots \lambda(o_n) \equiv_\iota \lambda(o'_1) \cdots \lambda(o'_n) = y$$

completing the proof.

■

12.2.12 THEOREM. If B is an ι-linguistic behavioural presentation, then $\rho_B(P) \subseteq E_\iota^*$ is left-closed and $\rho_B : P \to \rho_B(P)$ is a poset isomorphism.

PROOF. $\rho_B(P) \subseteq E_\iota^*$ by 12.2.11. Considered as a map $\rho_B : P \to \rho_B(P)$, ρ_B is surjective. To show that ρ_B is a poset isomorphism it suffices to show that

$$\forall p_1, p_2 \in P : p_1 \subseteq p_2 \Leftrightarrow \rho_B(p_1) \leq_\iota \rho_B(p_2) \tag{12.7}$$

for (12.7) also entails that ρ_B is injective.

First, suppose that $p_1 \subseteq p_2$ and let $(p_1, \leq_1) \in \Im(p_1)$, where $p_1 = \{o_1, \cdots, o_m\}$ and $o_1 <_1 \cdots <_1 o_m$. Let $p_2 - p_1 = \{o_{m+1}, \cdots, o_n\}$, where $o_i < o_j \Rightarrow i < j$ when $i, j \in \{m+1, \cdots, n\}$. Define $<_2$ by $o_i <_2 o_j \Leftrightarrow i < j$. $<_2$ is a total order on p_2. Suppose that $i, j \in \{1, \cdots, n\}$ and $o_i < o_j$. If $j \leq m$, then $o_j \in p_1$ and so $o_i \in p_1$ and so $i < j$. If $j > m$, then either $i > m$, in which case $i < j$ by hypothesis, or $i \leq m$ in which case $i < j$, as $j > m$. Thus, $(p_2, \leq_2) \in \Im(p_2)$ and so

$$\rho_B(p_1) = (\lambda(o_1) \cdots \lambda(o_m))_\iota \leq_\iota (\lambda(o_1) \cdots \lambda(o_n))_\iota = \rho_B(p_2).$$

Next, let $\rho_B(p_1) \leq_\iota \rho_B(p_2)$, then there exists $(p_1, \leq_1) \in \Im(p_1)$ and $(p_2, \leq_2) \in \Im(p_2)$ such that $p_1 = \{o_1, \cdots, o_m\}$, $o_1 <_1 \cdots <_1 o_m$, $p_2 = \{o'_1, \cdots, o'_n\}$, $o'_1 <_2 \cdots <_2 o'_n$ and $\lambda(o_1) \cdots \lambda(o_m) \leq \lambda(o'_1) \cdots \lambda(o'_n)$. We prove by contradiction that $p_1 \subseteq p_2$

If $p_1 \not\subseteq p_2$, then there exists $k < m$, such that $o_i = o'_i$ when $i \leq k$, but $o_{k+1} \neq o'_{k+1}$. Let $\tilde{p}_1 = \{o_1, \cdots, o_k, o_{k+1}\}$ and $\tilde{p}_2 = \{o_1, \cdots, o_k, o'_{k+1}\}$. We shall show that $\tilde{p}_1, \tilde{p}_2 \in P$ with $\tilde{p}_1 \equiv \tilde{p}_2$. By the λ-reduced property, we may conclude that $\tilde{p}_1 = \tilde{p}_2$ and hence that $o_{k+1} = o'_{k+1}$, the desired contradiction.

Certainly, $\tilde{p}_1 \subseteq p_1$. Suppose that $o_i \in \tilde{p}_1$, $o' \in p_1$ such that $o' \to o_i$, then $o' \leq_1 o_i$ and so $o' \in \tilde{p}_1$. We have shown that $\tilde{p}_1 \lhd p_1$. Similarly, $\tilde{p}_2 \lhd p_2$. By 3.3.14 and 12.2.8 (1), $\tilde{p}_1, \tilde{p}_2 \in P$

Next, define $\phi : \tilde{p}_1 \to \tilde{p}_2$ by $\phi(o_i) = o_i$ if $i \leq k$ and $\phi(o_{k+1}) = o'_{k+1}$. ϕ is clearly bijective, so in order to prove that $\phi : \tilde{p}_1 \equiv \tilde{p}_2$, it suffices to show:

$$\forall j \leq k : o_j < o_{k+1} \Leftrightarrow o_j < o'_{k+1} \tag{12.8}$$

Suppose $o_j < o_{k+1}$, then $o'_{k+1} \not\leq_2 o_j$ and so, since $o_j, o_{k+1} \in p_1$ and B is asynchronous, either $o_j < o'_{k+1}$ or $o_j \; co \; o'_{k+1}$. We assume the latter and derive a contradiction.

If $o_j \; co \; o'_{k+1}$, then by 12.2.8 (2), $\lambda(o_j) \iota \lambda(o'_{k+1})$. But $\lambda(o_{k+1}) = \lambda(o'_{k+1})$, and so $\lambda(o_j) \iota \lambda(o_{k+1})$ and since $o_j < o_{k+1}$, we do not have $o_j \; unsep \; o_{k+1}$, by 12.2.8 (3). Let $j \le l \le k+1$ be maximal with respect to $o_j < o_l < o_{k+1}$; by construction $o_l \; unsep \; o_{k+1}$. We cannot have $o_l < o'_{k+1}$, since this would mean $o_j < o_l < o'_{k+1}$, contradicting $o_j \; co \; o'_{k+1}$. Since $o_l <_2 o'_{k+1}$, we cannot have $o'_{k+1} < o_l$ either, and so since $o_j, o'_{k+1} \in p_2$, $o_l \; co \; o'_{k+1}$, by 2.3.6. By 12.2.8 (2), $\lambda(o_l) \iota \lambda(o'_{k+1})$. But $\lambda(o_{k+1}) = \lambda(o'_{k+1})$, and so $\lambda(o_l) \iota \lambda(o_{k+1})$. Since $o_l \; unsep \; o_{k+1}$, $o_l \; co \; o_{k+1}$, the required contradiction. This completes the proof that $\phi: \tilde{p}_1 \equiv \tilde{p}_2$.

Finally, we prove that $\rho_B(P)$ is left-closed. Let $x_\iota = \rho_B(p)$, $p \in P$, and $y \in E^*$, with $y \le x$. By 12.2.10, $x = \lambda(o_1) \cdots \lambda(o_n)$, where $p = \{o_1, \cdots, o_n\}$ and $o_1 <_x \cdots <_x o_n$, some $(p, \le_x) \in \Im(p)$. If $|y| = m$, then $y = \lambda(o_1) \cdots \lambda(o_m)$ so that if we can show that $p' = \{o_1, \cdots, o_m\} \lhd p$ and that $(p', \le_y) \in \Im(p)$, where $o_1 <_y \cdots <_y o_m$, then $p' \in P$, by 3.3.14 and 12.2.8 (1) and $\rho_B(p') = y_\iota$. We will then have established that

$$y_\iota \le \rho_B(p) \Rightarrow y_\iota \in \rho_B(P).$$

which concludes the proof. Now, certainly $p' \subseteq p$ and if $o_i \in p$ and $o_j \in p'$ such that $o_i \to o_j$, then $o_i \le_x o_j$ and so $i \le j$, and hence $o_i \in p'$. We have established that $p' \lhd p$.

Next suppose that $o_i, o_j \in p'$ with $o_i \to o_j$, then $o_i, o_j \in p$ with $o_i \to o_j$ so $o_i \le_x o_j$ and $i \le j$ and we have shown that $(p', \le_y) \in \Im(p)$.

■

Since $L = \rho_\iota(P)$ is a left-closed trace language, we may construct a behavioural presentation $B_{L,E}$. It should not come as a great surprise that there is some relationship between B and $B_{L,E}$. The relationship is one of isomorphism.

12.2.13. DEFINITION. Let B_1 and B_2 be behavioural presentations, then a *label isomorphism*, from B_1 to B_2 is a triple $f = (f_O, f_P, f_E)$ such that $f_O: O_1 \to O_2$, $f_P: P_1 \to P_2$ and $f_E: E_1 \to E_2$ are bijections and

1• $\forall o \in O_1: f_E(\lambda_1(o)) = \lambda_2(f_O(o))$;

2• $\forall p \in P_1: f_P(p) = \{f_O(o) | o \in p\}$.

We write $f: B_1 \equiv_E B_2$ if f is a label isomorphism from B_1 to B_2 and we write $B_1 \equiv_E B_2$ and say that B_1 and B_2 are *label isomorphic* if there exists a label isomorphism from B_1 to B_2.

An *isomorphism* from B_1 to B_2 is a pair $f=(f_O,f_P)$ such that $f_O:O_1 \to O_2$ and $f_P:P_1 \to P_2$ are bijections such that $f=(f_O,f_P,Id_E)$ is a label isomorphism, where Id_E is the identity function on E_1.
We write $f:B_1 \equiv B_2$ if f is an isomorphism from B_1 to B_2 and we write $B_1 \equiv B_2$ and say that B_1 and B_2 are *isomorphic* if there exists an isomorphism from B_1 to B_2. Note that if B_1 and B_2 are isomorphic, then $E_1 = E_2$.

Two behavioural presentations are label isomorphic if they are identical except for the names of occurrences and their events.

Two behavioural presentations are isomorphic if they are identical except for the names of occurrences.

If $f_P(p_1)=p_2$, then p_1 and p_2 describe exactly the same orderings of event occurrences, as the next result shows.

12.2.14. LEMMA. If B_1 and B_2 are behavioural presentations and f is a label isomorphism or an isomorphism, from B_1 to B_2, then

1• $\forall o,\tilde{o} \in O_1: o \to_1 \tilde{o} \Leftrightarrow f_O(o) \to_2 f_O(\tilde{o})$;
2• $\forall o,\tilde{o} \in O_1: o \#_1 \tilde{o} \Leftrightarrow f_O(o) \#_2 f_O(\tilde{o})$.

PROOF. Arguing directly:

$$\begin{aligned} o \to_1 \tilde{o} &\Leftrightarrow (\forall p_1 \in P_1: o \in p_1 \Rightarrow \tilde{o} \in p_1), \text{ by 2.3.1 (2)} \\ &\Leftrightarrow (\forall p_1 \in P_1: f_O(o) \in f_P(p_1) \Rightarrow f_O(\tilde{o}) \in f_P(p_1)), \text{ by 12.2.13} \\ &\Leftrightarrow (\forall p_2 \in P_2: f_O(o) \in p_2 \Rightarrow f_O(\tilde{o}) \in p_2), \text{ as } f_P \text{ is surjective} \\ &\Leftrightarrow f_O(o) \to_2 f_O(\tilde{o}), \text{ by 2.3.1 (2)} \end{aligned}$$

$$\begin{aligned} o \#_1 \tilde{o} &\Leftrightarrow (\forall p_1 \in P_1: \tilde{o} \in p_1 \Rightarrow o \notin p_1), \text{ by 2.3.1 (1)} \\ &\Leftrightarrow (\forall p_1 \in P_1: f_O(\tilde{o}) \in f_P(p_1) \Rightarrow f_O(o) \notin f_P(p_1)), \text{ by 12.2.13} \\ &\Leftrightarrow (\forall p_2 \in P_2: f_O(\tilde{o}) \in p_2 \Rightarrow f_O(o) \notin p_2), \text{ as } f_P \text{ is surjective} \\ &\Leftrightarrow f_O(o) \#_2 f_O(\tilde{o}), \text{ by 2.3.1 (1)} \end{aligned}$$

■

If B_1 and B_2 are asynchronous and left-closed, then f_O is actually determined by f_P as the next result shows.

12.2.15. PROPOSITION. If B_1 and B_2 are asynchronous and left-closed and π is a poset isomorphism from $(P_1,\subseteq)$ to $(P_2,\subseteq)$, then there exists a bijection $\omega: O_1 \to O_2$, such that

$$\forall p \in P_1: \pi(p) = \{\omega(o) \mid o \in p\}.$$

PROOF. Since π is a poset isomorphism it must map $Pr(B_1)$ bijectively onto $Pr(B_2)$. By 4.3.2, if $o_1 \in O_1$, then there exists unique $o_2 \in O_2$ such that $\pi(\downarrow_1 o_1) = \downarrow_2 o_2$. Define $\omega(o_1) = o_2$. And now, for all $p \in P_1$:

$$\begin{aligned} o_1 \in p &\Leftrightarrow \downarrow_1 o_1 \subseteq p, \text{ by 3.3.10 (5)} \\ &\Leftrightarrow \pi(\downarrow_1 o_1) \subseteq \pi(p), \text{ by hypothesis} \\ &\Leftrightarrow \downarrow_2 \omega(o_1) \subseteq \pi(p), \text{ by definition of } \omega \\ &\Leftrightarrow \omega(o_1) \in \pi(p), \text{ by 3.3.9 and 2.3.2 (3)} \end{aligned}$$

■

12.2.16. PROPOSITION. If B is an ι-linguistic behavioural presentation and $L = \rho_\iota(P)$, then $B \equiv B_{L,E}$.

PROOF. Define $\pi: P \to P_L$ by $\pi(p) = Pr(\rho_B(p))$, then π surjective, by 12.1.1, while

$$\begin{aligned} p_1 \subseteq p_2 &\Leftrightarrow \rho_B(p_1) \leq_\iota \rho_B(p_2) \text{ by 12.2.12} \\ &\Leftrightarrow Pr(\rho_B(p_1)) \subseteq Pr(\rho_B(p_2)), \text{ by 10.4.5} \end{aligned}$$

so π is a poset isomorphism By 12.2.15, there is a bijection $\omega: O \to O_L$ satisfying $\pi(p) = \{\omega(o) \mid o \in p\}$, and we recall that ω is defined in such a way that $\downarrow_L \omega(o) = \pi(\downarrow o)$. Since $\lambda(o) = \lambda_L(\rho_\iota(\downarrow o))$, $\lambda(o) = \lambda_L(\omega(o))$. Also the set of events of $B_{L,E}$ is the same as that of B, namely E. Thus $B \equiv B_{L,E}$.

■

There is a similar relationship between left-closed trace languages and the behavioural presentations they determine.

12.2.17. PROPOSITION. Suppose L is left-closed in $A_\iota{}^*$, then

$$\forall X \in L: \rho_B(Pr(X)) = X.$$

where $B = B_{L,A}$. Consequently, $L = \rho_B(P_L)$.

PROOF. Suppose that $a^1 \cdots a^n \in X$. Let $X_1 = a^1{}_\iota$ and for $i > 1$ let $X_i = X_{i-1}.a^i{}_\iota$. Let $U_1 = X_1$ and for each $i > 1$ define $U_i = pr(U_{i-1}, a_i)$. Since $|Pr(X)| = n$, by (3) of 12.2.6, and the U_i are clearly distinct, it follows that $Pr(X) = \{U_1, \cdots, U_n\}$.

We show that the ordering $\leq_X$ given by $U_1 <_X \cdots <_X U_n$ satisfies $\to \subseteq \leq_X$.

Suppose that $i < j \leq n$ and let $U_i = V_i.a^i{}_\iota$ and $U_j = V_j.a^j{}_\iota$ with $a^i, a^j \in A$. By 10.3.3, $X_i = V_i.W_i.a^i{}_\iota$ where $a^i{}_\iota \; ind \; W_i$, and $X_j = V_j.W_j.a^j{}_\iota$, where $a^j{}_\iota \; ind \; W_j$. If $V_j < U_i$, then $V_j < V_i$, by 12.2.1. Let $W = V_i / V_j$. By construction,

$X_j = V_j.a^j{}_\iota.W.W_i.a^i{}_\iota$, so $W_j = W.W_i.a^i{}_\iota$. But this means that $a^j{}_\iota$ *ind* $W.W_i.a^i{}_\iota$ and from that we may deduce that $a^j \neq a^i$ and that $U_i = V_j.a^j{}_\iota.W.a^i{}_\iota = V_j.W.a^i{}_\iota.a^j{}_\iota$, contradicting the primality of U_i.

Thus, if $i<j$, then $U_i < U_j$ or $U_i \; co_L \; U_j$. It follows that if $U_i < U_j$, then $j \not\leq i$, for $i=j \Rightarrow U_i = U_j$ and $j<i \Rightarrow U_j < U_i \vee U_j \; co_L \; U_i$.

Thus, $(Pr(X), \leq_x) \in \Im(Pr(X), \rightarrow)$ so

$$\rho_B(Pr(X)) = (\lambda(U_1)\cdots\lambda(U_n))_\iota = (a_1 \cdots a_n)_\iota = X.$$

■

We have finally accumulated all the material we need for the main result of this section, which sums up the relationships that we have ascertained between trace/vector languages and behavioural presentations. First, let us say that B is *linguistic* $\Leftrightarrow$ B is ι-linguistic for some independence relation ι. A linguistic behavioural presentation could be ι-linguistic for several ι. However, we note that $\rho_B(P)$ does not, in fact, depend on the ι in question, but only on $\rightarrow$ and λ.

12.2.18. THEOREM. If L is left-closed in $A_\iota{}^*$, then $B_{L,A}$ is linguistic. Conversely, if B is ι-linguistic, then there exists L left-closed in $E_\iota{}^*$ such that $B \equiv B_{L,E}$. Furthermore, if B_1 and B_2 are linguistic behavioural presentations, then

$$B_1 \equiv B_2 \Leftrightarrow E_1 = E_2 \wedge \rho_{B_1}(P_1) = \rho_{B_2}(P_2)$$

PROOF. If L is left-closed in $A_\iota{}^*$, then $B_{L,A}$ is linguistic, by 12.2.9. If B is ι-linguistic, then $L = \rho_B(P)$ is left-closed in $E_\iota{}^*$, by 12.2.12, and $B \equiv B_{L,A}$ by 12.2.16.

Suppose finally, that B_1 and B_2 are linguistic behavioural presentations. If $E_1 = E_2$ and $\rho_{B_1}(P_1) = \rho_{B_2}(P_2)$ then $B_1 \equiv B_{\rho_{B_1}(P_1),E_1} = B_{\rho_{B_2}(P_2),E_2} \equiv B_2$.

Conversely, if $f: B_1 \equiv B_2$, then $E_1 = E_2$ and for any $p \in P_1$, if $(p, \leq_1) \in \Im(p)$, with $p = \{o_1, \cdots, o_n\}$ and $o_1 <_1 \cdots <_1 o_n$, then by 12.2.14, $(f_P(p), \leq_2) \in \Im(f_P(p))$, where $f_O(o_1) <_2 \cdots <_2 f_O(o_n)$ and hence

$$\rho_B(p) = (\lambda_1(o_1)\cdots\lambda_1(o_k))_\iota = (\lambda_2(f_O(o_1))\cdots\lambda_2(f_O(o_k)))_\iota = \rho_B(f_P(p))$$

and so $\rho_{B_1}(P_1) = \rho_{B_2}(P_2)$.

■

12.3. Strong Forward Stability Revisited.

In chapters 7 and 11, we defined a notion of strong forward stability as a property of behavioural presentations and of left-closed vector languages. In this section, we show that the two concepts are equivalent. We also look at the trace language versions of two related properties.

12.3.1. DEFINITION. Suppose that L is left-closed in A_ι^*, then

1• L is *determinate* $\Leftrightarrow \forall u_\iota, v_\iota \in Pr(L)\, \exists x_\iota \in L: u_\iota, v_\iota \leq x_\iota$;

2• L is *forward stable* $\Leftrightarrow$
$\forall x \in A^*\, \forall a,b \in A: x_\iota.a_\iota, x_\iota.b_\iota \in L \wedge a\,\iota\,b \Rightarrow x_\iota.a_\iota.b_\iota \in L$;

3• L is *strongly forward stable* $\Leftrightarrow$
$\forall x \in A^*\, \forall a,b \in A: x_\iota.a_\iota, x_\iota.b_\iota \in L \wedge a \neq b \Rightarrow a\,\iota\,b \wedge x_\iota.a_\iota.b_\iota \in L$.

We now show that a left-closed trace language L satisfies one of the above properties if and only if $B_{L,A}$ satisfies the corresponding property of behavioural presentations. First, we need a result which relates the derivation structure of $B_{L,A}$, with the structure of L

12.3.2. LEMMA. If L is left-closed in A_ι^* and $u_\iota \in O_L$, with $\lambda_L(u_\iota) = a$, and $x_\iota, y_\iota \in L$ then

$$Pr(x_\iota) \succ_L^{u_\iota} Pr(y_\iota) \Leftrightarrow y_\iota = x_\iota.a_\iota \wedge u_\iota = pr(x_\iota, a).$$

PROOF. Since $Pr(x_\iota) \succ_L^{u_\iota} Pr(y_\iota)$, it follows that $Pr(x_\iota) \subseteq Pr(y_\iota)$, so that $x_\iota = \sqcup Pr(x_\iota) \leq \sqcup Pr(y_\iota) = y_\iota$, by 10.4.5.

If $x_\iota = a^1{}_\iota \cdots a^m{}_\iota$ and $y_\iota = a^1{}_\iota \cdots a^n{}_\iota$, define $u^i{}_\iota = a^1{}_\iota \cdots a^{i-1}{}_\iota$, $i = 1, \cdots, n$. By 10.3.4, $Pr(x_\iota) = \{pr(u^i{}_\iota, a^i) \mid i = 1, \cdots, m\}$ and $Pr(y_\iota) = \{pr(u^i{}_\iota, a^i) \mid i = 1, \cdots, n\}$. But, $Pr(x_\iota) \succ_L^{u_\iota} Pr(y_\iota)$ and so $Pr(y_\iota) - Pr(x_\iota) = \{u_\iota\}$, from which we deduce that $u_\iota = pr(u^n{}_\iota, a^n)$ and so $n = m+1$, $a = \lambda_L(pr(u^n{}_\iota, a^n)) = a^n$, whence $y_\iota = x_\iota.a_\iota$, and $u_\iota = pr(u^n{}_\iota, a^n{}_\iota) = pr(x_\iota, a)$.

Conversely, if $y_\iota = x_\iota.a_\iota$ and $u_\iota = pr(u^n{}_\iota, a^n{}_\iota) = pr(x_\iota, a)$, then $Pr(x_\iota) \subseteq Pr(y_\iota)$ and by 10.3.4, $Pr(y_\iota) - Pr(x_\iota) = \{u_\iota\}$, so as $B_{L,A}$ is asynchronous, by 12.2.2, $Pr(x_\iota) \succ_L^{u_\iota} Pr(y_\iota)$, by 3.3.1.

∎

12.3.3. PROPOSITION. If L is left-closed in A_ι^*, then L is determinate $\Leftrightarrow B_{L,A}$ is determinate.

PROOF. Suppose that L is determinate and let $u_\iota, v_\iota \in O_L$, then $u_\iota, v_\iota \in Pr(L)$ and so by 12.3.1 (1), there exists $x_\iota \in L$ such that $u_\iota, v_\iota \leq x_\iota$. Thus,

$u_\iota, v_\iota \in Pr(x_\iota) \in P_L$ and so $\neg u_\iota \#_L v_\iota$, by 2.3.1 (1). Since this holds for all $u_\iota, v_\iota \in O_L$, $B_{L,A}$ is determinate, by 2.3.10. Conversely, suppose that $B_{L,A}$ is determinate and let $u_\iota, v_\iota \in Pr(L)$, then $u_\iota, v_\iota \in O_L$, and so $\neg u_\iota \#_L v_\iota$, which means that $u_\iota, v_\iota \in Pr(x_\iota)$, some $Pr(x_\iota) \in P_L$, that is $u_\iota, v_\iota \leq x_\iota$, some $x_\iota \in L$, and L is determinate.

■

12.3.4. PROPOSITION. If L is left-closed in A_ι^*, then L is forward stable $\Leftrightarrow$ $B_{L,A}$ is forward stable.

PROOF. Suppose that L is forward stable and let $x_\iota, y_\iota, z_\iota \in L$ and $u_\iota, v_\iota \in O_L$ such that $u_\iota \; co \; v_\iota$, $Pr(x_\iota) \succ_L^{u_\iota} Pr(y_\iota)$ and $Pr(x_\iota) \succ_L^{v_\iota} Pr(z_\iota)$. Then by 12.3.2, $y_\iota = x_\iota.a_\iota$, $z_\iota = x_\iota.b_\iota$, $u_\iota = pr(x_\iota, a)$ and $v_\iota = pr(x_\iota, b)$ where $a = \lambda_L(u_\iota)$ and $b = \lambda_L(v_\iota)$. Since $B_{L,A}$ is ι-linguistic, $a \, \iota \, b$. Thus, $x_\iota.a_\iota, x_\iota.b_\iota \in L$ with $a \, \iota \, b$ and so $x_\iota.a_\iota.b_\iota \in L$, by forward stability. Since $a \, \iota \, b$, $a_\iota \; ind_\iota b_\iota$, by 9.2.9. Therefore $x_\iota.a_\iota.b_\iota = x_\iota.a_\iota \sqcup x_\iota.b_\iota$, by 10.2.5 (3). So $y_\iota \sqcup z_\iota \in L$ and accordingly

$$Pr(y_\iota) \cup Pr(z_\iota) = Pr(y_\iota \sqcup z_\iota) \in P_L$$

establishing that $B_{L,A}$ is forward stable.

Conversely, suppose that $B_{L,A}$ is forward stable and suppose that $x_\iota.a_\iota, x_\iota.b_\iota \in L$ with $a \, \iota \, b$. By 12.3.2, $Pr(x_\iota) \succ_L^{u_\iota} Pr(y_\iota)$ and $Pr(x_\iota) \succ_L^{v_\iota} Pr(z_\iota)$, where $y_\iota = x_\iota.a_\iota$, $z_\iota = x_\iota.b_\iota$, $u_\iota = pr(x_\iota, a)$ and $v_\iota = pr(y_\iota, a)$. Suppose that $w_\iota \in Pr(\dot{L})$ such that $w_\iota < u_\iota$, then $w_\iota \in Pr(y_\iota) - \{w_\iota\} = Pr(x_\iota)$. Therefore, $v_\iota \nless w_\iota$, since otherwise, $v_\iota \in Pr(x_\iota)$, contradicting $Pr(x_\iota) \succ_L^{v_\iota} Pr(z_\iota)$. We have shown that $u_\iota \; unsep \; v_\iota$. Since $a \, \iota \, b$, $u_\iota \; co \; v_\iota$. By forward stability of $B_{L,A}$, $Pr(x_\iota.a_\iota) \cup Pr(x_\iota.b_\iota) \in P_L$. Hence,

$$x_\iota.a_\iota \sqcup x_\iota.b_\iota = \sqcup(Pr(x_\iota.a_\iota) \cup Pr(x_\iota.b_\iota)) \in L \cdot$$

But again we may use 10.2.5 to show $x_\iota.a_\iota.b_\iota = x_\iota.a_\iota \sqcup x_\iota.b_\iota$.

■

12.3.5. PROPOSITION. If L is left-closed in A_ι^*, then L is strongly forward stable $\Leftrightarrow$ $B_{L,A}$ is strongly forward stable.

PROOF. Suppose that L is strongly forward stable and let $x_\iota, y_\iota, z_\iota \in L$ and $u_\iota, v_\iota \in O_L$ such that $Pr(x_\iota) \succ_L^{u_\iota} Pr(y_\iota)$ and $Pr(x_\iota) \succ_L^{v_\iota} Pr(z_\iota)$. By 12.3.2, $y_\iota = x_\iota.a_\iota$, $z_\iota = x_\iota.b_\iota$, $u_\iota = pr(x_\iota, a)$ and $v_\iota = pr(y_\iota, a)$ where $a = \lambda_L(u_\iota)$ and $b = \lambda_L(v_\iota)$. Thus, $x_\iota.a_\iota, x_\iota.b_\iota \in L$. If $a = b$, then $y_\iota = z_\iota$ and so

$$Pr(y_\iota) \sqcup Pr(z_\iota) = Pr(y_\iota) \in P_L.$$

Otherwise $a \neq b$ and so $a \iota b$ and $x_\iota.a_\iota.b_\iota \in L$, by strong forward stability. Hence, a_ι *ind* b_ι. Thus $x_\iota.a_\iota.b_\iota = x_\iota.a_\iota \sqcup x_\iota.b_\iota$, by 10.2.5. So $y_\iota \sqcup z_\iota \in L$ and:

$$Pr(y_\iota) \cup Pr(z_\iota) = Pr(y_\iota \sqcup z_\iota) \in P_L$$

establishing that $B_{L,A}$ is strongly forward stable.

Conversely, suppose that $B_{L,A}$ is strongly forward stable and let $x_\iota.a_\iota, x_\iota.b_\iota \in L$ with $a \neq b$, then $Pr(x_\iota) \succ_L^{u_\iota} Pr(y_\iota)$ and $Pr(x_\iota) \succ_L^{v_\iota} Pr(z_\iota)$, where $y_\iota = x_\iota.a_\iota$, $z_\iota = x_\iota.b_\iota$, $u_\iota = pr(x_\iota, a)$ and $v_\iota = pr(y_\iota, a)$, by 12.3.2. Thus, $Pr(x_\iota.a_\iota) \cup Pr(x_\iota.b_\iota) \in P_L$, by strong forward stability of $B_{L,A}$, Therefore, for some $y_\iota \in L$, $Pr(x_\iota.a_\iota) \cup Pr(x_\iota.b_\iota) = Pr(y_\iota)$ and now if $w_\iota \in L$, then

$$\begin{aligned} y_\iota \leq w_\iota &\Leftrightarrow Pr(y_\iota) \subseteq Pr(w_\iota), \text{ by 10.3.1} \\ &\Leftrightarrow Pr(x_\iota.a_\iota) \cup Pr(x_\iota.b_\iota) \subseteq Pr(w_\iota) \\ &\Leftrightarrow Pr(x_\iota.a_\iota), Pr(x_\iota.b_\iota) \subseteq Pr(w_\iota) \\ &\Leftrightarrow \sqcup Pr(x_\iota.a_\iota), \sqcup Pr(x_\iota.b_\iota) \leq \sqcup Pr(w_\iota) \\ &\Leftrightarrow x_\iota.a_\iota, x_\iota.b_\iota \leq w_\iota, \text{ by 10.4.5} \end{aligned}$$

Thus, $y_\iota = x_\iota.a_\iota \sqcup x_\iota.b_\iota$, and as $a \neq b$, by 10.2.8, a_ι *ind* b_ι and $x_\iota.a_\iota.b_\iota = y_\iota \in L$. ■

12.3.6. COROLLARY. Suppose that L is left-closed in A_ι^*, then L is strongly forward stable $\Leftrightarrow$ L is forward stable and determinate.

PROOF. L is strongly forward stable $\Leftrightarrow$ $B_{L,A}$ is strongly forward stable, by 12.3.5 $\Leftrightarrow$ $B_{L,A}$ is forward stable and determinate, by 7.4.7 $\Leftrightarrow$ L is forward stable and determinate, by 12.3.3 and 12.3.4

■

Notes.

The results in section 12.2 of this chapter were obtained by the author for a related model involving sets of labelled partially ordered sets [38, 42].

Chapter 13

Unambiguous Transition Systems

> We represent the time-sequence by a line progressing to infinity, in which the manifold constitutes a series of one dimension only; and we reason from the properties of this line to all the properties of time, with this one exception, that while the parts of the line are simultaneous the parts of time are always successive.
>
> IMMANUEL KANT
> Critique of Pure Reason

13.1. Introduction.

Apart from chapter 8, we have concentrated on objects such as behavioural presentations, event structures and vector and trace languages, whose purpose is to model the set of all possible behaviours of some system. In chapter 8, we gave our first illustration of the use of such objects in providing a descriptive formalism, computation graphs, with a behavioural semantics. This consisted of a mapping, pictured in figure 13.1. which associated each computation graphs G with a behavioural presentation B_G.

In general, a semantics based on behavioural presentations would consist of such a mapping, with the objects constituting the descriptive notation occupying the left-hand oval.

Figure 13.1.

As we have explained, a central feature of the approach taken in this book is the use of automata, called hybrid transition systems, to mediate in the definition of semantic mappings, as pictured in figure 13.2, so that each object *Ob* in the descriptive formalism is associated with an initialised hybrid transition system, $H_{Ob}{}^{*}$.

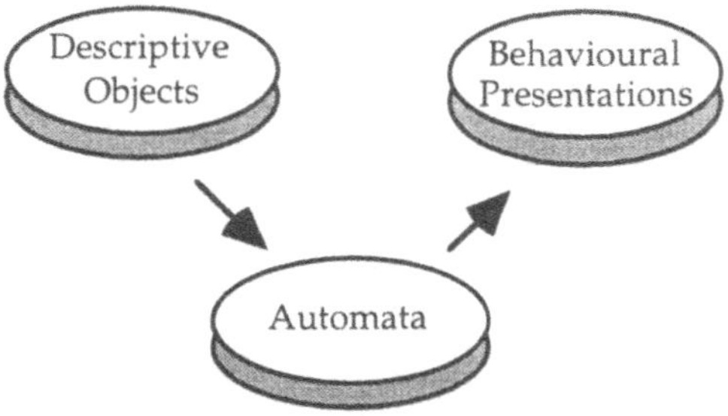

Figure 13.2.

As we shall show, each initialised hybrid transition system in turn determines a discrete behavioural presentation. In fact, every discrete behavioural presentation is determined, up to isomorphism, by an initialised hybrid transition system.

We shall examine a number of the descriptive notations in the course of this book and show how they may be interpreted in terms of hybrid transition systems. In this way, we illustrate our claim that Parallelism Theory is capable of being unified at the semantic level.[1]

We begin with a special case; discrete, sequential systems. We do this for several reasons.

1• It presents the arguments in the setting of sequential systems. State machines, formal languages and the connections between them are familiar.
2• It establishes a place for the very important idea of a *transition system* within the general theory.
3• It brings into focus the problem of *ambiguity*[2] in a relatively simple context. Roughly speaking in an ambiguous transition system, a given *action sequence*[3] from a given initial state may actually correspond to more than one distinct *derivation sequence* [4] Hence, the same action sequence may describe two distinct behaviours; in the general case, action sequences does not provide an adequate behavioural semantics for transition systems in general.

Perhaps the most important reason is the following;

---

[1]Readers acquainted with theories of parallelism such as Petri Nets or CCS will see immediately that a given marked net or system of CCS equations determines a transition system. However, we wish to model concurrency and/or simultaneity explicitly.
[2]Otherwise, 'non-determinism'.
[3]See 13.2.2.
[4]See 13.2.2.

4• Many semantic models are equivalent to transition systems. The relationship between the corresponding semantics to the non-interleaving semantics presented in this book becomes very simple; just throw away information, namely that of the concurrency structure.

13.2. Transition Systems.

A transition system is essentially a partial (possibly infinite-state) automaton. Formally, we have the following definition.

13.2.1. DEFINITION. A *transition system* is a triple $T=(Q,A,\rightarrow)$ where

1• Q is a set of (global) *states;*
2• A is a set of *actions;*
3• $\rightarrow\subseteq Q\times A\times Q$ is the *transition relation.*

An *initialised transition system* is a pair (T,q), where $T=(Q,A,\rightarrow)$ is a transition system and $q\in Q$.

We write $q_1 \rightarrow^a q_2$ to indicate that $(q_1,a,q_2)\in\rightarrow$ and $q_1 \rightarrow^a$ to indicate that $q_1 \rightarrow^a q_2$ some q_2

As in the case of behavioural presentations, we make a convention that the elements in the triple corresponding to a subscripted or otherwise embellished transition system inherit the subscript or embellishment. Thus, for example, $T_B=(Q_B,A_B,\rightarrow_B)$ and $T^\infty=(Q^\infty,A^\infty,\rightarrow^\infty)$.

Transition systems are most conveniently pictured as directed graphs whose nodes represent, and are labelled by, the names of states and where there is a directed arc labelled a from a node labelled q_1 to a node labelled q_2 if and only if $q_1 \rightarrow^a q_2$.

Figure 13.3 is a picture of a transition system describing the behaviour of a vending machine designed by RIP-U-OFF plc.

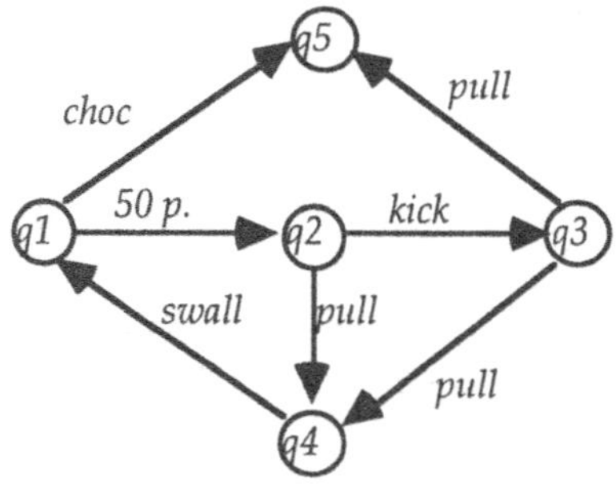

Figure 13.3.

The event names are abbreviations as follows

- *50p.*: the customer puts a 50 pence coin in the slot:
- *pull:* the customer pulls the drawer:
- *swal:* the machine makes a derisive noise as it swallows the customer's money;
- *kick*: the customer gives the machine a good kick;
- *choc*: the customer takes a bar of chocolate from the drawer.

A transition system describes, in a very low-level way, the actions that some system may perform and how the performance of these actions transforms the system's (global) internal state.

Formally, we have the notion of a derivation sequence. A derivation sequence is simply a string of actions and states derived by tracing a contiguous path of directed arcs through the graph from that initial state. We also have a notion of a set of action sequences of the transition system from some initial state. An action sequence is a sequence of actions obtained from a derivation sequence by throwing away the state names.

13.2.2. DEFINITION. Let T be a transition system. A *derivation sequences* of T is either $q \in Q$ or a sequence $x = q_0 a_1 q_1 \cdots a_n q_n$, where $q_i \in Q$ and $a_i \in A$, each i, such that for each $i > 0$, $q_{i-1} \rightarrow^{a_i} q_i$. Denote the set of derivation sequences of T by $Der(T)$.

If $x = q \in Q$, then define $head(x) = end(x) = q$ and $act(x) = \Omega$. Otherwise, define $head(x) = q_0$, $end(x) = q_n$ and $act(x) = a_1 \cdots a_n$

If $q \in Q$, then the set of derivation sequences of T from q, denoted $Der(T,q)$, is defined to be $\{x \in Der(T) \mid head(x) = q\}$

If $q \in Q$, then the set of finite *action sequences* of T from q, denoted by $L(T,q)$, is defined to be $\{act(x) \mid x \in Der(T,q)\}$. The set of all infinite action sequences of T from q, denoted by $L^{\omega}(T,q)$, is defined to be $\{x \in A^{\omega} \mid \forall u \in A^*: u \leq x \Rightarrow u \in L(T,q)\}$. The set of all action sequences of T from q, denoted by $L^{\infty}(T,q)$, is defined to be $L(T,q) \cup L^{\omega}(T,q)$

We also extend the transition relation to action sequences, by defining

$$q_1 \rightarrow^{u} q_2 \Leftrightarrow \exists x \in Der(T,q_1): act(x) = u \wedge end(x) = q_2$$

Note that if $q \in Q$, then $q \rightarrow^{\Omega} q$ and if $a \in A$, $x \in A^*$, and $q_1, q_2 \in Q$, then

$$q_1 \rightarrow^{x.a} q_2 \Leftrightarrow \exists \tilde{q} \in Q: q_1 \rightarrow^{x} \tilde{q} \wedge \tilde{q} \rightarrow^{a} q_2.$$

Thus, in the example system of figure 13.1, we have a derivation sequence

$$x_1 = q1.50p.q2.pull.q4.swal.q1.50p.q2.pull.q4.swal.q1$$

describing the acquisition by RIP-U-OFF plc. of £1 clear profit and also a derivation sequence

$$q1.50p.q2.kick.q3.pull.q4.swal.q1.50p.q2.kick.q3.pull.q5.choc.q1$$

which shows that persistence pays off in the end, at the price of an increase in urban violence. Both the above sequences belong to $Der(T,q1)$, and

$$act(x_1) = 50p.pull.swal.50p.pull.swal \text{ and } q1 \rightarrow^{50p.pull.swal.50p.pull.swal} q1$$

The following construction will enable us to dispense with unnecessary structure.

13.2.3. DEFINITION. If (T,q) is a initialised transition system, then we define the set of states reachable from q by

$$R(T,q) = \{q' \in Q \mid \exists x \in A^*: q \rightarrow^x q'\}$$

and the reduction T of with respect to q by

$$r_q(T) = (Q', A, \rightarrow \cap (Q' \times A \times Q'))$$

where $Q' = R(T,q)$.

Immediately, we have:

$$Der(r_q(T),q) = Der(T,q) \tag{13.1}$$
$$L(r_q(T),q) = L(T,q) \tag{13.2}$$
$$L^\omega(r_q(T),q) = L^\omega(T,q) \tag{13.3}$$
$$L^\infty(r_q(T),q) = L^\infty(T,q) \tag{13.4}$$

Consider the action sequence $50p.kick.pull \in L(T,q1)$ for the transition system of figure 13.1. This action sequence corresponds both to the derivation sequence $q1.50p.q2.kick.q3.pull.q4$ and $q1.50p.q2.kick.q3.pull.q5$.

We can see from this that the action sequence alone does not tell us all we need to know about the state of the system; for example, whether or not I will get a chocolate. For this reason, derivation sequences are to be preferred to action sequences in the case of *ambiguous* transition systems such as that in the example.

This being the case, it may seem strange that we shall be working mainly with action sequences. The reason is that it is easier to generalise action sequences (to Mazurkiewicz traces) in the non-sequential case than it is to generalise derivation sequences.

Any behavioural presentation determines a transition system - the states of the system are the elements of P and transitions are steps, as given in definition 3.3.1. Formally, we have the following construction.

13.2.4. DEFINITION. If B is an asynchronous behavioural presentation, then define $T_B = (P, E, \rightarrow_B)$, where if $e \in E$, and $p_1, p_2 \in P$, then

$$p_1 \rightarrow_B^e p_2 \Leftrightarrow \exists o \in O : p_1 \succ^o p_2 \wedge \lambda(o) = e$$

Of course, T_B is a transition system.

Figure 13.4. pictures a behavioural presentation B. T_B is represented on its right.

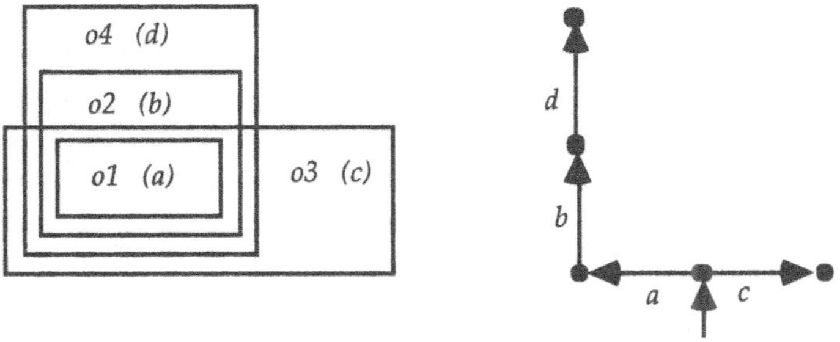

Figure 13.4.

A behavioural presentation has its own transition structure, and so may be considered a transition system in disguise, so to speak. One might well ask why, in that case, one should bother with transition systems at all. The answer is that a transition system is generally a more convenient way of representing the (interleaved) behaviour of a system than a behavioural presentation. For example a transition system may be finite while a corresponding behavioural presentation may not.

As we have observed, for a discrete, sequential behavioural presentation B, T_B has its own behavioural semantics; from a given initial state, p, we may construct the language of action sequences $L(T_B, p)$. B itself has an obvious 'initial state', namely $\emptyset$, and B describes the behaviours possible to some system from that initial state. What, then, is the relationship between B and $L(T_B, \emptyset)$? In one direction, it is relatively straightforward. As we shall show in 13.2.7, the function $\tilde{\rho}_B$ of 13.2.6 below maps the set of points of a behavioural presentation onto $L(T_B, \emptyset)$.

First, as we shall be dealing almost exclusively with discrete, sequential behavioural presentations in this and the following chapter, we establish some of their basic properties in the following result.

13.2.5. LEMMA. If B is a discrete, sequential behavioural presentation and $p \in P$ then:

1• p is totally ordered by $\rightarrow$;
2• If $p = \{o_1, \cdots, o_n\}$ with $o_1 < \cdots < o_n$, then

$$p' \lhd p \Leftrightarrow \exists i : p' = \{o_1, \cdots, o_i\}.$$

PROOF. (1) Let $o, o' \in p$. We cannot have $o \# o'$, by 2.3.1 (1). Also, since B is sequential, we cannot have $o \; co \; o'$, by 2.3.9 (1). Hence, by 2.3.6, either $o \rightarrow o'$ or $o' \rightarrow o$.
(2) Let $p = \{o_1, \cdots, o_n\}$ with $o_1 < \cdots < o_n$ and $p' \lhd p$. Let $i = max(\{j \mid o_j \in p'\})$. then by definition, if $j > i$ then $o_j \notin p'$ and if $j \leq i$, then $o_i \in p'$, $o_j \in p$, $o_j \rightarrow o_i$ and $p' \lhd p$, so $o_j \in p'$, by 3.3.7 (2). Thus, $o_j \in p' \Leftrightarrow j \leq i$, that is $p' = \{o_1, \cdots, o_i\}$.
Conversely, suppose that $p' = \{o_1, \cdots, o_i\}$, then $p' \subseteq p$, giving 3.3.7 (1). If $o_i \in p'$, $o_j \in p$, $o_j \rightarrow o_i$, then since $o_1 < \cdots < o_n$, $j \leq i$, that is, $o_j \in p'$, and we have established 3.3.7 (2). Thus, $p' \lhd p$

■

13.2.6 DEFINITION. Let B be a discrete, sequential behavioural presentation and let $p \in P$. If $p = \varnothing$ then $\tilde{\rho}_B(p) = \Omega$. Otherwise, by 13.2.5 (1), for some $n \in \omega$, $p = \{o_1, \cdots, o_n\}$ with $o_1 < \cdots < o_n$ and we define $\tilde{\rho}_B(p) = \lambda(o_1) \cdots \lambda(o_n)$.

Note that if B is a discrete, sequential behavioural presentation and $p \in P$ with $p = \{o_1, \cdots, o_n\}$ and $o_1 < \cdots < o_n$, then by 12.2.10, $\Im(p) = \{(p, \leq)\}$ and so $\rho_B(p) = \{\tilde{\rho}_B(p)\}$.

For the behavioural presentation of figure 13.4:

$$\tilde{\rho}_B(\varnothing) = \Omega$$
$$\tilde{\rho}_B(\{o_1\}) = a$$
$$\tilde{\rho}_B(\{o_1, o_2\}) = ab$$
$$\tilde{\rho}_B(\{o_1, o_3\}) = ac$$
$$\tilde{\rho}_B(\{o_1, o_2, o_4\}) = abd$$

while $L(T_B, \varnothing) = \{\Omega, a, ab, ac, abd\}$. Thus, in this example, $L(T_B, \varnothing) = \tilde{\rho}_B(P)$. This is always the case.

13.2.7. PROPOSITION. If B is a discrete, sequential behavioural presentation then

$$\forall p \in P : \varnothing \rightarrow_B^x p \Leftrightarrow x = \tilde{\rho}_B(p).$$

and hence $L(T_B, \varnothing) = \rho_B(P)$.

PROOF. First, suppose that $x \in L(T_B, \varnothing)$ so that $\varnothing \rightarrow_B^x p$, some $p \in P$. By 13.2.4, $\varnothing \succ^{o_1} p_1 \succ^{o_2} \cdots \succ^{o_n} p_n = p$ with $x = \lambda(o_1) \cdots \lambda(o_n)$. By 3.3.1, we must have $p_i = \{o_1, \cdots, o_i\}$, each i and in particular, $p = \{o_1, \cdots, o_n\}$. Now, for each i, j with $1 \le i < j \le n$, $o_i \in p_i$ but $o_j \notin p_i$, so $o_j \nrightarrow o_i$ and hence, $o_i < o_j$, by 13.2.5 (1). Thus, $o_1 < \cdots < o_n$ and so $\rho_B(p) = \lambda(o_1) \cdots \lambda(o_n) = x$.

This shows that if $p \in P$, then $\varnothing \rightarrow_B^x p \Rightarrow x = \tilde{\rho}_B(p)$, Let us show that the opposite implication holds.

Suppose $p = \{o_1, \cdots, o_n\}$ with $o_1 < \cdots < o_n$. Now, $p_i = \{o_1, \cdots, o_i\}$ is left-closed in p for each i, by 13.2.5 (2), and so $p_i \in P$ for each i, since B is discrete and hence left-closed, by 3.3.14. Thus, $\varnothing \succ^{o_1} p_1 \succ^{o_2} \cdots \succ^{o_n} p_n = p$ in B, so $\varnothing \rightarrow_B^{\lambda(o_1)} p_1 \rightarrow_B^{\lambda(o_2)} \cdots \rightarrow_B^{\lambda(o_n)} p_n = p$ in T_B, that is $\varnothing \rightarrow_B^x p$. This shows that if $p \in P$, then $x = \tilde{\rho}_B(p) \Rightarrow \varnothing \rightarrow_B^x p$ giving the reverse implication.

■

We shall find it useful to cite some additional properties of $\tilde{\rho}_B$.

13.2.8. LEMMA. If B is a discrete, sequential behavioural presentation and $p_1, p_2 \in P$, then

1• $p_1 \subseteq p_2 \Rightarrow \tilde{\rho}_B(p_1) \le \tilde{\rho}_B(p_2)$;
2• If in addition $\tilde{\rho}_B$ is injective, then $\tilde{\rho}_B(p_1) \le \tilde{\rho}_B(p_2) \Rightarrow p_1 \subseteq p_2$.

PROOF. (1) By 13.2.5 (1), $p_2 = \{o_1, \cdots, o_n\}$ with $o_1 < \cdots < o_n$, some $n \in \omega$ If $p_1 \subseteq p_2$, then $p_1 = \downarrow p_1 \lhd \downarrow p_2 = p_2$, by 3.3.10 (2) and (4), so $p_1 = \{o_1, \cdots, o_m\}$, some $m \le n$, by 13.2.5 (2). Thus,

$$\tilde{\rho}_B(p_1) = \lambda(o_1) \cdots \lambda(o_m) \le \lambda(o_1) \cdots \lambda(o_n) = \tilde{\rho}_B(p_2).$$

(2) Suppose that $\tilde{\rho}_B$ is injective and let $\tilde{\rho}_B(p_1) \le \tilde{\rho}_B(p_2)$ and suppose that $|p_1| = m$ and $p_2 = \{o_1, \cdots, o_n\}$. Let $\tilde{p}_1 = \{o_1, \cdots, o_m\}$, then $\tilde{p}_1 \subseteq p_2$, $\tilde{\rho}_B(\tilde{p}_1) \le \tilde{\rho}_B(p_2)$ and $\tilde{\rho}_B(p_1) \le \tilde{\rho}_B(p_2)$ and $|\tilde{\rho}_B(\tilde{p}_1)| = |\tilde{\rho}_B(p_1)|$. Hence $\tilde{\rho}_B(p_1) = \tilde{\rho}_B(\tilde{p}_1)$ and thus $p_1 = \tilde{p}_1$, since $\tilde{\rho}_B$ is injective. But we know that $\tilde{p}_1 \subseteq p_2$. Thus $p_1 \subseteq p_2$.

■

This establishes a relationship between B and $L(T_B, \varnothing)$, but what kind of relationship is it? We may obtain the latter from the former, but not the former from the latter, as the following example shows. In it, we exhibit two behavioural presentations B_1 and B_2 which are distinct, but which satisfy $\tilde{\rho}_B(P_1) = \tilde{\rho}_B(P_2)$.

13.2.9. EXAMPLE. Define behavioural presentations B_1 and B_2 as follows.

$O_1 = \{o_1, o_2, \cdots\}$
$P_1 = \{p_1, p_2, \cdots\}$ where for each $n \in \omega$, $p_n = \{o_1, \cdots, o_{n-1}, o_n\}$
$E_1 = \{e_1, e_2, \cdots\}$
for each $n \in \omega$, for each $\lambda_1(o_n) = e_n$

$O_2 = O_1 \cup \tilde{O}_1$, where $\tilde{O}_1 = \{\tilde{o}_1, o_2, \cdots\}$ and $O_1 \cap \tilde{O}_1 = \varnothing$
$P_2 = P_1 \cup \{\tilde{p}_1, \tilde{p}_2, \cdots\}$ where for each $n \in \omega$, $\tilde{p}_n = \{o_1, \cdots, o_{n-1}, \tilde{o}_n\}$
$E_2 = E_1$
for each $n \in \omega$, $\lambda_2(o_n) = \lambda_2(\tilde{o}_n) = e_n$

Their transition systems are partially pictured in figure 13.5.

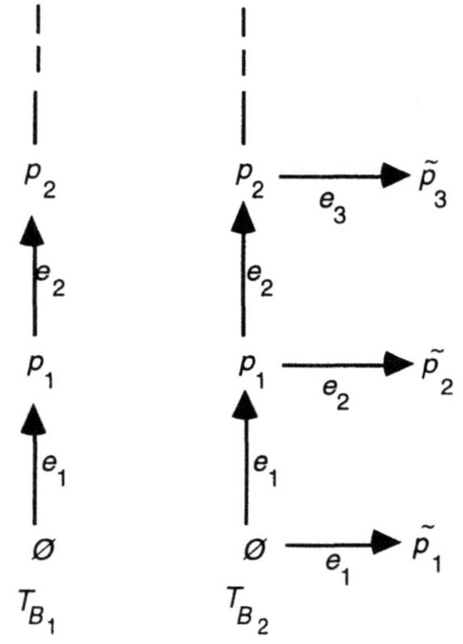

Figure 13.5

Now $\tilde{\rho}_{B_1}(P_1) = \tilde{\rho}_{B_2}(P_2) = \{e_1 \cdots e_n \mid n \in \omega\}$, but B_1 and B_2 are not in any sense equivalent. For every $o \in O_1$, there exists $o' \in O_1$ such that $o <_1 o'$ but if $\tilde{o} \in O_2 - O_1$, there is no $o' \in O_2$ such that $\tilde{o} <_2 o'$. B_1 describes a non-terminating system, while B_2 describes a system that may terminate in an infinite number of ways.

It appears that sequential behavioural presentations may not be distinguished simply by the languages of action sequences associated with them. The problem is that in the case of B_2, the mapping ρ_{B_2} is not injective.

This is because B_2 has a property akin to ambiguity in phrase structure grammars or non-determinism in automata theory.

This marks a limitation of formal languages as representatives of sequential behaviour. Essentially the same limitation was described in the interleaving semantics for the COSY language [33] presented in [48]. Milner, in chapter 1 of [11] makes much the same point - almost with the same example as that in [48], indeed, but he goes on to develop a solution to the problem which underlies much of the technical development of his approach. This involves the construction of equivalences based on the state/transition structure of systems, rather than merely the string languages which may be derived from such structure. We shall consider one of these relations, namely strong equivalence, in section 13.4.

Our approach is to use behavioural presentations as semantics for transition systems.

Initially, we shall consider the class of behavioural presentations for which such problems do not arise. This is precisely the class of behavioural presentations/systems for which a representation of behaviour by a string language is adequate. In Milner's taxonomy, these are the *strongly determinate* systems. We shall describe them as unambiguous.

13.3. Properties of Unambiguous Transition Systems.

13.3.1. DEFINITION. A transition system T is *unambiguous* $\Leftrightarrow$

$$\forall q, q_1, q_2 \in Q \, \forall a \in A: (q \to^a q_1 \wedge q \to^a q_2 \Rightarrow q_1 = q_2).$$

If B is an asynchronous behavioural presentation, then B is unambiguous $\Leftrightarrow$

$$\forall p, p_1, p_2 \in P \, \forall o_1, o_2 \in O: p \succ^{o_1} p_1 \wedge p \succ^{o_2} p_2 \wedge \lambda(o_1) = \lambda(o_2) \Rightarrow p_1 = p_2$$

Equivalently, B is unambiguous $\Leftrightarrow$

$$\forall p, p_1, p_2 \in P \, \forall o_1, o_2 \in O: p \succ^{o_1} p_1 \wedge p \succ^{o_2} p_2 \wedge \lambda(o_1) = \lambda(o_2) \Rightarrow o_1 = o_2$$

We note some immediate consequences of the definition.

13.3.2. LEMMA. Suppose that B is a discrete, sequential behavioural presentation and T is an unambiguous transition system, then

1• B is unambiguous $\Leftrightarrow$ T_B is unambiguous.

2• If $q \in Q$, then

$\forall q, q_1, q_2 \in Q \forall x \in A^*: (q \rightarrow^x q_1 \wedge q \rightarrow^x q_2 \Rightarrow q_1 = q_2)$.

PROOF. (1) By 13.2.4 and 13.3.1.
(2) By 13.3.1 and induction on $|x|$.

■

13.3.3. PROPOSITION. Suppose that B is an unambiguous, discrete, sequential behavioural presentation, then the map $\tilde{\rho}_B$ is a poset isomorphism from $(P, \subseteq)$ to $(\tilde{\rho}_B(P), \leq)$.

PROOF. In view of 13.2.8 and the fact that $\tilde{\rho}_B$ is surjective by definition, it suffices to show that $\tilde{\rho}_B$ is injective. Suppose that $p_1, p_2 \in P$ such that $\tilde{\rho}_B(p_1) = \tilde{\rho}_B(p_2) = x$, say. Then $\varnothing \rightarrow_B{}^x p_1$ and $\varnothing \rightarrow_B{}^x p_2$, by 13.2.7, and so $p_1 = p_2$, by 13.3.2.

■

The following result allows us to apply material from chapter 12 to unambiguous, sequential behavioural presentations.

13.3.4. PROPOSITION. Suppose that B is a behavioural presentation, then B is discrete, sequential and unambiguous $\Leftrightarrow$ B is $\varnothing$-linguistic.

PROOF. Of course, $\varnothing$ is an independence relation.

Suppose first that B is discrete, sequential and unambiguous, then certainly B is discrete and asynchronous, by hypothesis and by 2.3.9, giving 12.2.8 (1), while 12.2.8 (2) and (3) hold trivially. Finally, suppose that $p_1, p_2 \in P$ such that $p_1 \equiv p_2$ via a bijection ϕ. By 13.2.5 (2), $p_1 = \{o_1, \cdots, o_k\}$ and $p_2 = \{o_1', \cdots, o_k'\}$ such that $o_1 < \cdots < o_k$, $o_1' < \cdots < o_k'$ and $\phi(o_i) = o_i'$, each i, so that $\lambda(o_i) = \lambda(o_i')$ each i, by 12.2.13 (1). If $p_1 \neq p_2$, then there exists $1 \leq m \leq k$ such that $o_m \neq o_m'$ and $o_i = o_i'$, each $i < m$. Let $p = \{o_i \mid i < m\}$, $p_1' = \{o_i \mid i \leq m\}$ and $p_2' = \{o_i' \mid i \leq m\}$. By 13.2.5 (2), $p \lhd p_1' \lhd p_1$ and $p \lhd p_2' \lhd p_2$. Since B is discrete, $p, p_1', p_2' \in P$, by 3.3.14. But $p \succ^{o_m} p_1'$ and $p \succ^{o_m'} p_2'$, by 3.3.1, with $o_m \neq o_m'$ and $\lambda(o_m) = \lambda(o_m')$, contradicting unambiguity. Hence, $p_1 = p_2$ and B is λ-reduced.

Conversely, suppose that B is Ø-linguistic, then B is discrete by 12.2.8 (1) and sequential, by 12.2.8 (2). Suppose that $p, p_1, p_2 \in P$ and $o_1, o_2 \in O$ with $p \succ^{o_1} p_1$, $p \succ^{o_2} p_2$ and $\lambda(o_1) = \lambda(o_2)$. Since B is sequential and $\lambda(o_1) = \lambda(o_2)$, the function $\phi: p_1 \rightarrow p_2$ given by

$$\phi(o) = \begin{cases} o_2 \text{ if } o = o_1 \\ o \text{ otherwise} \end{cases}$$

is an isomorphism, and so $p_1 = p_2$, since B is λ-reduced. Therefore $\{o_1\} = p_1 - p = p_2 - p = \{o_2\}$, showing that B is unambiguous.

■

With 13.3.4, the constructions and results of chapter 12 become available; in particular, the construction of an Ø-linguistic behavioural presentation from a left-closed trace language with independence relation Ø. But a trace language with independence relation Ø is just a string language in disguise. So, we are considering languages L which are left-closed in the usual sense that

$$\forall x \in L : y \leq x \Rightarrow y \in L \tag{13.5}$$

Observe that if (T,q) is an initialised transition system, then $L(T,q)$ is left-closed, since if $q \to^{x} \tilde{q}$ and $x = y.u$, then there must exist $q' \in Q$ such that $q \to^{y} q' \to^{u} \tilde{q}$, by 13.2.2.

13.3.5. DEFINITION. If $L \subseteq E^*$, define $B_{L,E} = (O_L, P_L, E, \lambda_L)$ by

1• $O_L = L - \{\Omega\}$;
2• $P_L = \{p_x \mid x \in L\}$, where $p_x = \{u \in O_L \mid u \leq x\}$;
3• If $u \in E^*$ and $e \in E$ such that $u.e \in L$, then $\lambda_L(u.e) = e$.

The reader might care to refer back to 2.2.1, which deals with the possible behaviours of a tossed coin, to see that the behavioural presentation given in that example is $B_{E^*,E}$, where $E = \{H,T\}$.

Let us first observe that as Ø is an independence relation, we may form a trace monoid, $E_\varnothing{}^* = \{\{x\} \mid x \in E^*\}$. It is trivial that the map $x \mapsto \{x\}$ is both a monoid and poset isomorphism from $f_O(u) = \{u\}$ to $E_\varnothing{}^*$

13.3.6. THEOREM. If L is left-closed in E^*, then $B_{L,E}$ is a discrete, unambiguous, sequential behavioural presentation. Conversely, if B is a discrete, unambiguous, sequential behavioural presentation, then there exists L left-closed in E^*, by 3.3.1, such that $B \equiv B_{L,E}$. Furthermore, if B_1 and B_2 are linguistic behavioural presentations, then

$$B_1 \equiv B_2 \Leftrightarrow E_1 = E_2 \wedge \tilde{\rho}_{B_1}(P_1) = \tilde{\rho}_{B_2}(P_2)$$

PROOF. Suppose first that L is left-closed in E^*. Define $L_\varnothing = \{\{x\} \mid x \in L\}$, then $L_\varnothing$ is left-closed in $E_\varnothing{}^*$ and so $B_{L_\varnothing,E}$ is a Ø-linguistic behavioural presentation, by 12.2.18, and so $B_{L_\varnothing,E}$. is a discrete, unambiguous, sequential behavioural presentation, by 13.3.4. But the functions given by $f_O(u) = \{u\}$

and $f_P(p) = \{\{u\} \mid u \in p\}$ clearly define an isomorphism from $B_{L,E}$ to $B_{L_\varnothing,E}$ and so $B_{L,E}$ is a discrete, unambiguous, sequential behavioural presentation.

Conversely, suppose that B is a discrete, unambiguous, sequential behavioural presentation, then $B_{L_\varnothing,E}$ is a Ø-linguistic, by 13.3.4, and so by 12.2.18, there exists $\tilde{L}$ left-closed in $E_\varnothing{}^*$ such that $B \equiv B_{\tilde{L},E}$. But clearly, $\tilde{L} = L_\varnothing$, where $L = \{x \mid \{x\} \in \tilde{L}\}$ and so $B \equiv B_{L_\varnothing,E} \equiv B_{L,E}$.

Finally, by 12.2.18

$$B_1 \equiv B_2 \Leftrightarrow E_1 = E_2 \wedge \rho_{B_1}(P_1) = \rho_{B_2}(P_2) \Leftrightarrow E_1 = E_2 \wedge \tilde{\rho}_{B_1}(P_1) = \tilde{\rho}_{B_2}(P_2)$$

■

13.3.7. PROPOSITION. If L is left-closed in E^* and $B = B_{L,E}$, then $\tilde{\rho}_B(P) = L$.

PROOF. Let $L_\varnothing = \{\{x\} \mid x \in L\}$, then $L_\varnothing$ is left-closed in $E_\varnothing{}^*$ and so if $B' = B_{L_\varnothing,E}$, then $\rho_{B'}(P') = L_\varnothing$, by 12.2.17, and hence $\tilde{\rho}_{B'}(P') = L$. But, as we saw in the proof of 13.3.6, $B \equiv B'$, and so by 13.3.6, $\tilde{\rho}_B(P) = \tilde{\rho}_{B'}(P') = L$.

■

13.4. Bisimulation.

Let us pause to take our bearings. If T is a transition system and $q \in Q$, then T determines a left-closed language $L(T,q)$, which in turn determines a sequential, discrete, unambiguous behavioural presentation, $B_{L(T,q),A}$. which may be characterised up to isomorphism as the unique sequential, discrete, unambiguous behavioural presentation B such that $\rho_B(P) = L(T,q)$.

Conversely, if B is a sequential, discrete, unambiguous behavioural presentation, then we can construct an unambiguous transition system T_B and we know that $\rho_B(P) = L(T_B,\varnothing)$.

Thus, up to isomorphism, every unambiguous transition system together with an initial state determines a unique sequential, discrete, unambiguous behavioural presentation and every sequential, discrete, unambiguous behavioural presentation is determined by a unambiguous transition system together with an initial state.

There remains the question: what is the relationship between transition systems plus initial states that 'accept' the same behavioural presentations. It is immediate that they have the same languages of execution sequences. We may give an alternative characterisation which relates the transition systems structurally and gives us a fine excuse to introduce the notion of a bisimulation into our discussion

13.4.1. DEFINITION. Suppose (T_1,q_1) and (T_2,q_2) are initialised transition systems. A *strong bisimulation* from (T_1,q_1) to (T_2,q_2) is a relation $R \subseteq Q_1 \times Q_2$ such that $q_1 \, R \, q_2$ and that whenever $q_1' \, R \, q_2'$, $e \in E_1 \cup E_2$, $\tilde{q}_1 \in Q_1$ and $\tilde{q}_2 \in Q_2$, then

1• $q_1' \to_1^e \tilde{q}_1 \Rightarrow \exists \hat{q}_2 \in Q_2 : q_2' \to_2^e \hat{q}_2 \wedge \tilde{q}_1 \, R \, \hat{q}_2$;
2• $q_2' \to_2^e \tilde{q}_2 \Rightarrow \exists \hat{q}_1 \in Q_1 : q_1' \to_1^e \hat{q}_1 \wedge \hat{q}_1 \, R \, \tilde{q}_2$.

We shall say that (T_1,q_1) and (T_2,q_2) are *strongly equivalent* and write $(T_1,q_1) \sim (T_2,q_2) \Leftrightarrow A_1 = A_2$ and there exists a strong bisimulation R from (T_1,q_1) to (T_2,q_2).

Intuitively, if $(T_1,q_1) \sim (T_2,q_2)$ then whatever T_1 can do from its initial state, T_2 may also do (and vice versa), that is if $q_1' \to_1^e \tilde{q}_1$ then $q_2' \to_2^e \tilde{q}_2$ some $\tilde{q}_2$. Furthermore, $\tilde{q}_2$ may be chosen so that $(T_1,\tilde{q}_1) \sim (T_2,\tilde{q}_2)$, that is, they may continue to simulate each other after the first step.

$\sim$ is reflexive, symmetric and transitive. To see this, note that (a) the identity on Q is a bisimulation establishing $(T,q) \sim (T,q)$, (b) If R is a strong bisimulation from (T_1,q_1) to (T_2,q_2) then

$$R^{-1} = \{(\tilde{q}_2,\tilde{q}_1) \mid \tilde{q}_1 \, R \, \tilde{q}_2\}$$

is a strong bisimulation from (T_2,q_2) to (T_1,q_1) and (c) if R_1 is a strong bisimulation from (T_1,q_1) to (T_2,q_2) and R_2 is a strong bisimulation from (T_2,q_2) to (T_3,q_3) then

$$R_1 \circ R_2 = \{(\tilde{q}_1,\tilde{q}_3) \in Q_1 \times Q_3 \mid \exists \tilde{q}_2 \in Q_2 : \tilde{q}_1 \, R_1 \, \tilde{q}_2 \wedge \tilde{q}_2 \, R_2 \, \tilde{q}_3\}$$

is a bisimulation from (T_1,q_1) to (T_3,q_3).

13.4.2. LEMMA. If (T_1,q_1) and (T_2,q_2) are initialised transition systems, then

1• $(T_1,q_1) \sim (T_2,q_2) \Rightarrow L(T_1,q_1) = L(T_2,q_2)$;
2• $L(T_1,q_1) = L(T_2,q_2) \Rightarrow (T_1,q_1) \sim (T_2,q_2)$ if T_1 and T_2 are unambiguous.

PROOF. By induction on the length of x, using 13.4.1 (1), if R is a strong bisimulation from (T_1,q_1) to (T_2,q_2), then

$$\forall x \in L(T_1,q_1) : (q_1 \to_1^x \tilde{q}_1 \Rightarrow \exists \tilde{q}_2 \in Q_2 : q_2 \to_2^x \tilde{q}_2 \wedge \tilde{q}_1 \, R \, \tilde{q}_2)$$

which entails that $L(T_1,q_1) \subseteq L(T_2,q_2)$. Since $(T_2,q_2) \sim (T_1,q_1)$, by symmetry of $\sim$, an analogous argument gives the converse inclusion.

Suppose that T_1 and T_2 are unambiguous and that $L(T_1,q_1)=L(T_2,q_2)$. Define a relation $R \subseteq Q_1 \times Q_2$, by

$$\tilde{q}_1 R \tilde{q}_2 \Leftrightarrow \exists x \in L(T_1,q_1): q_1 \to_1^x \tilde{q}_1 \wedge q_2 \to_2^x \tilde{q}_2.$$

As $q_1 \to_1^\Omega q_1$ and $q_2 \to_2^\Omega q_2$, $q_1 R q_2$.

Next, let $\tilde{q}_1 R \tilde{q}_2$ and suppose that $\hat{q}_1 \in Q_1$ and $e \in E_1$ such that $\tilde{q}_1 \to_1^e \hat{q}_1$. By definition of R, there exists $x \in L(T_1,q_1)$ such that $q_1 \to_1^x \tilde{q}_1$ and $q_2 \to_2^x \tilde{q}_2$. As $\tilde{q}_1 \to_1^e \hat{q}_1$, $q_1 \to_1^{x.e} \hat{q}_1$, so $x.e \in L(T_1,q_1)$. Since $L(T_1,q_1)=L(T_2,q_2)$, $x.e \in L(T_2,q_2)$ and so for some $\bar{q}_2, \hat{q}_2 \in Q_2$ $q_2 \to_2^x \bar{q}_2 \to_2^e \hat{q}_2$. But T_2 is unambiguous, and therefore, by 13.3.2 (2), $\bar{q}_2 = \tilde{q}_2$. Thus, $\tilde{q}_2 \to_2^e \hat{q}_2$. Since $q_1 \to_1^{x.e} \hat{q}_1$ and $q_2 \to_2^{x.e} \hat{q}_2$, $\hat{q}_1 R \hat{q}_2$. This shows that 13.4.1 (1) holds. 13.4.1 (2) follows by symmetry.

■

We may sum up our findings about unambiguous sequential systems as follows. Let **USB** denote the class whose elements are the $\equiv$-equivalence classes of unambiguous, sequential, discrete behavioural presentations. Write $[\![B]\!]_\equiv$ for the $\equiv$-class of B.

Let **UTS** denote the class whose elements are the $\sim$-equivalence classes of unambiguous, initialised transition systems. Write $[\![T,q]\!]_\sim$ for the $\sim$class of (T,q).

13.4.3. THEOREM. There is a bijective mapping

$$Beh_u : \mathbf{UTS} \to \mathbf{UBP}$$

such that

$$Beh_u([\![T,q]\!]_\sim) = [\![B]\!]_\equiv \Leftrightarrow \tilde{\rho}_B(P) = L(T,q).$$

Furthermore, $Beh_u([\![T,q]\!]_\sim) = [\![B]\!]_\equiv \Leftrightarrow (T_B,\emptyset) \in [\![T,q]\!]_\sim \Leftrightarrow B \equiv B_{L(T,q),A}$.

PROOF. First, we show that Beh_u is well-defined.

Suppose that (T_1,q_1) and (T_2,q_2) are unambiguous, initialised transition systems, then $L(T_1,q_1)$ and $L(T_2,q_2)$ are both left-closed, so by 13.3.6 and 13.3.7 there exists unambiguous, sequential, discrete behavioural presentations B_1 and B_2 such that $\tilde{\rho}_{B_1}(P_1)=L(T_1,q_1)$ and $\tilde{\rho}_{B_2}(P_2)=L(T_2,q_2)$. If $[\![T_1,q_1]\!]_\sim = [\![T_2,q_2]\!]_\sim$, then $(T_1,q_1) \sim (T_2,q_2)$ and so $L(T_1,q_1)=L(T_2,q_2)$ and $A_1 = A_2$, by 13.4.2 (1) and 13.4.1. But then, $\tilde{\rho}_{B_1}(P_1)=\tilde{\rho}_{B_2}(P_2)$ and $A_1 = A_2$ and so $B_1 \equiv B_2$, by 13.3.6.

Beh_U is surjective. For if B is an unambiguous, sequential, discrete behavioural presentation, then $\tilde{\rho}_B(P) = L(T_B, \varnothing)$, by 13.2.7, and hence $Beh_U([\![T_B, \varnothing]\!]_\sim) = [\![B]\!]_\equiv$.

Beh_U is injective. For if $Beh_U([\![T_1, q_1]\!]_\sim) = [\![B]\!]_\equiv = Beh_U([\![T_2, q_2]\!]_\sim)$ then $L(T_1, q_1) = \rho_B(P) = L(T_2, q_2)$ and $A_1 = A_2$ and hence $(T_1, q_1) \sim (T_2, q_2)$, by 13.4.2 (2).

Finally,

$$\begin{aligned}(T_B, \varnothing) \in [\![T, q]\!] &\Leftrightarrow (T_B, \varnothing) \sim (T, q)\\ &\Leftrightarrow L(T_B, \varnothing) = L(T, q), \text{ by 13.4.2}\\ &\Leftrightarrow \tilde{\rho}_B(P) = L(T, q), \text{ by 13.2.7.}\end{aligned}$$

The third characterisation is equivalent to the first, since if $B' = B_{L(T,q),A}$

$$\begin{aligned}\tilde{\rho}_B(P) = L(T, q) &\Leftrightarrow \tilde{\rho}_B(P) = \rho_{B'}(P')\text{ , by 13.3.7}\\ &\Leftrightarrow B \equiv B', \text{ by 13.3.6.}\end{aligned}$$

■

13.5. Computation Graphs Revisited.

We may use transition systems to define the node-set sequence semantics of the computation graphs we examined in chapter 8. In the following definition $[D \to \omega]$ is the set of all functions from D to the finite ordinals. $r_q(T)$ for an initialised transition system (T, q) is defined in 13.2.3.

13.5.1. DEFINITION. If G is a computation graph, then we define

$$T_G = r_A([D \to \omega], \Sigma_G, \to_G)$$

where $\underline{q}_1 \to_G^X \underline{q}_2 \Leftrightarrow$

1• $\forall d \in \bullet X : T(d) \leq \underline{q}_1(d);$

2• $\forall d \in D : \underline{q}_2(d) = \underline{q}_1(d) + U(X, d) - W(X, d)$

and

$$U(X, d) = \begin{cases} U(d) \text{ if } d \in X\bullet \\ 0 \text{ otherwise}\end{cases}$$

$$W(X, d) = \begin{cases} W(d) \text{ if } d \in \bullet X \\ 0 \text{ otherwise}\end{cases}$$

It is immediate that T_G is an unambiguous transition system.

Note that it may be constructed iteratively, beginning with state A, as we illustrate in figure 13.7. with the computation graph of figure 13.6. In the figure, the expression next to each arc is $d{:}(A(d), U(d), W(d), T(d))$.

In the figure, the vector (a_1, a_2, a_3, a_4) represents the state $\underline{q}$ with $\underline{q}(d) = a_d$, each d.

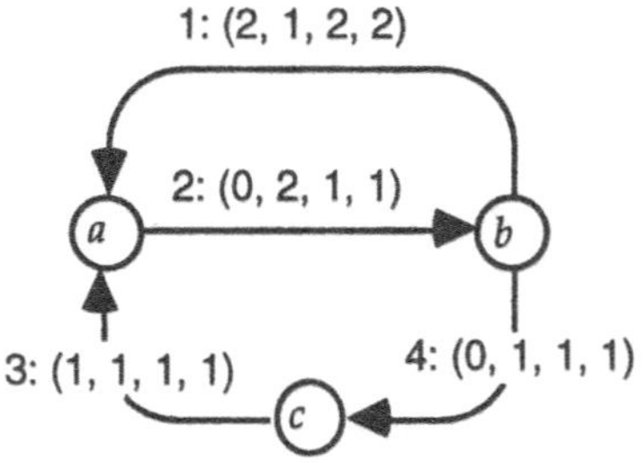

Figure 13.6.

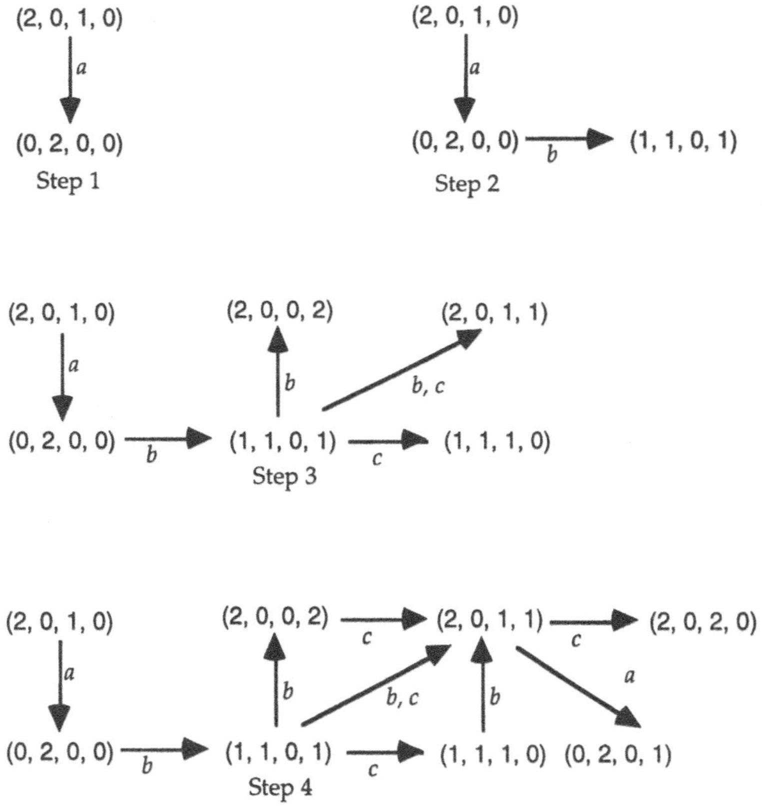

Figure 13.7.

We now relate the indigenous semantics of computation graphs with that given by action sequences of (T_G, A). We start with a lemma.

13.5.2. LEMMA. If G is a computation graph, then

$$\forall \mathbf{S} \in \Sigma_G^*: A \rightarrow_G^{\mathbf{S}} \underline{q} \Rightarrow (\forall d \in D: \underline{q}(d) = \chi(\mathbf{S}, d, |\mathbf{S}|)),$$

PROOF. We argue by induction on $n = |\mathbf{S}|$. If $n = 0$, then $\mathbf{S} = \Omega$ and $\underline{q} = A$. By 8.2.2, $\chi(\Omega, d, 0) = A(d)$. Otherwise, $\mathbf{S} = \mathbf{S}'.X$, some $\mathbf{S}' \in \Sigma_G^*$ and $X \in \wp(V)$. Thus, $A \rightarrow_G^{\mathbf{S}'} \underline{q}' \rightarrow_G^{X} \underline{q}$ and by induction, $\underline{q}'(d) = \chi(\mathbf{S}', d, |\mathbf{S}'|)$, all $d \in D$. And now, by 8.2.2

$$\begin{aligned}\chi(\mathbf{S}, d, |\mathbf{S}|) &= \chi(\mathbf{S}', d, |\mathbf{S}'|) + U(X, d) - W(X, d)\\ &= \underline{q}'(d) + U(X, d) - W(X, d) = \underline{q}(d).\end{aligned}$$

■

13.5.3. PROPOSITION. If G is a computation graph, then

$$Px(G) = L^{\infty}(T_G, A)$$

PROOF. We prove by induction on $n = |\mathbf{S}|$ that $\mathbf{S} \in Px(G) \cap \Sigma_G^* \Leftrightarrow A \rightarrow_G^{\mathbf{S}}$, that is $Px(G) \cap \Sigma_G^* = L(T_G, A)$, from which the claim of the proposition follows.

The case $n = 0$ is clear. Otherwise, $\mathbf{S} = \mathbf{S}'.X$, some $\mathbf{S}' \in \Sigma_G^*$ and $X \in \wp(V)$. Suppose that $\mathbf{S} \in Px(G) \cap \Sigma_G^*$, then $\mathbf{S}' \in Px(G) \cap \Sigma_G^*$ and so by induction, $A \rightarrow_G^{\mathbf{S}'} \underline{q}'$, say. By 13.5.2, $\underline{q}'(d) = \chi(\mathbf{S}', d, |\mathbf{S}'|)$, all $d \in D$, and so by 8.2.3, $T(d) \leq \underline{q}'(d)$, all $d \in \bullet X$. Hence, $\underline{q}' \rightarrow_G^{X}$ and so $A \rightarrow_G^{\mathbf{S}'.X}$. Conversely, if $A \rightarrow_G^{\mathbf{S}'.X}$, then $A \rightarrow_G^{\mathbf{S}'} \underline{q}'$, say, and $\underline{q}' \rightarrow_G^{X}$. By induction, $\mathbf{S}' \in Px(G) \cap \Sigma_G^*$ and by 13.5.2, $T(d) \leq \underline{q}'(d)$, all $d \in \bullet X$, so by 8.2.3, $\mathbf{S} \in Px(G) \cap \Sigma_G^*$.

■

We conclude this section by showing how transition systems may be used to decide whether or not the queues in a computation graph may accumulate an unbounded number of items.

In step 4 of figure 13.7 we have constructed a state $(2,0,1,1)$, in which every queue contains a number of items greater than or equal to the number on the queue initially. Furthermore, the number of items on queue 4 has actually increased.

This means that the sequence *abbc* may be repeated an arbitrary number of times, and after each repetition, queue 4 will contain one more item than the previous one. Thus, the queue may accumulate an arbitrary number of items and the transition system is, in consequence, infinite. This illustrates a decision procedure for boundedness:

13.5.4. THEOREM ([9]). T_G is infinite if and only if there exists $\underline{q}_1, \underline{q}_2 \in Q_G$ and $x \in \wp(V)^*$ such that

1• $\underline{q}_1 \to_G^x \underline{q}_2$;
2• $\forall d \in D: \underline{q}_1(d) \leq \underline{q}_2(d)$;
3• $\exists d \in D: \underline{q}_1(d) < \underline{q}_2(d)$.

PROOF. (Sketch) If the conditions hold, then we have derivations $\underline{q}_1 \to_G^x \underline{q}_2 \to_G^x \underline{q}_3 \cdots \to_G^x \underline{q}_n$ with $\underline{q}_{n+1}(d) = \underline{q}_n(d) + (\underline{q}_2(d) - \underline{q}_1(d))$, each $d \in D$. If d satisfies (3), then $\underline{q}_{n+1}(d) > \underline{q}_n(d)$, so the $\underline{q}_n$ are distinct, and Q_G is infinite. Conversely, if Q_G is infinite then by Konig's Lemma, there exists $\underline{q}_1 \to_G^{x_1} \underline{q}_2 \cdots \to_G^{x_n} \underline{q}_{n+1} \cdots$ such that the $\underline{q}_n$ are distinct. By a property of sequences of finite vectors of natural numbers, there exists $i < j$ such that $\underline{q}_i(d) \leq \underline{q}_j(d)$, each $d \in D$ and $\underline{q}_i(d) < \underline{q}_j(d)$ some $d \in D$.

■

This result may be used to determine boundedness as, in the construction illustrated in figure 13.7, after a finite number of steps either no new states will be introduced, in which case the system is bounded, or vectors $\underline{q}_1$ and $\underline{q}_2$ will be constructed satisfying (1), (2) and (3) of 13.5.4, in which case the system is unbounded.

Notes.

Transition systems were proposed by Keller in [21] as a universal semantic model for parallelism. While we hesitate to accept this claim, it is certainly true that transition systems provide a general setting for describing methods for examining those properties of systems which are preserved under interleaving of non-sequential behaviour. For example, [21] uses them to present a general notion of invariance, while Kwong uses transition systems in [49] to discuss general notions of property-preserving transformations of systems and in [50] to investigate the phenomenon of livelock. Sifakis uses transition systems in [51] to describe strategies for classifying and analysing termination properties of systems.

Where transition systems fail is in handling phenomena in which there is an uncomfortable propinquity between concurrency and non-determinism -

phenomena such as '*confusion*' in General Net Theory. Since transition systems present an interleaving representation of the behaviour of non-sequential system, they are incapable of making a distinction between the two and are therefore inadequate for the treatment of such properties.

Confusion is a phenomenon whereby a conflict between two occurrences may be created or destroyed by the occurrence of a third. One possible consequence of the presence of confusion is shown in the following. Consider the unambiguous transition system T pictured in figure 13.8.

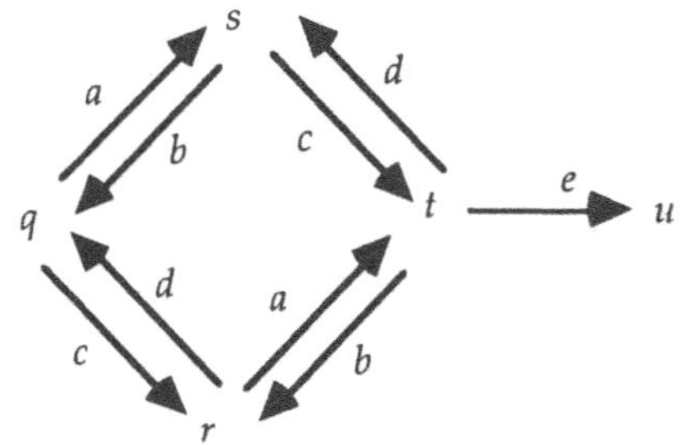

Figure 13.8.

The infinite sequences $(abcd)^\omega$ and $(acbd)^\omega$ are both action sequences of (T,q). In the former, the state t is never entered; in the former it is entered an infinite number of times. Thus, in the former, the transition e is never possible, while in the latter it is possible infinitely often, but never taken. In such circumstances, one might be inclined to say that the former sequence is *fair* to e and that the latter is not.

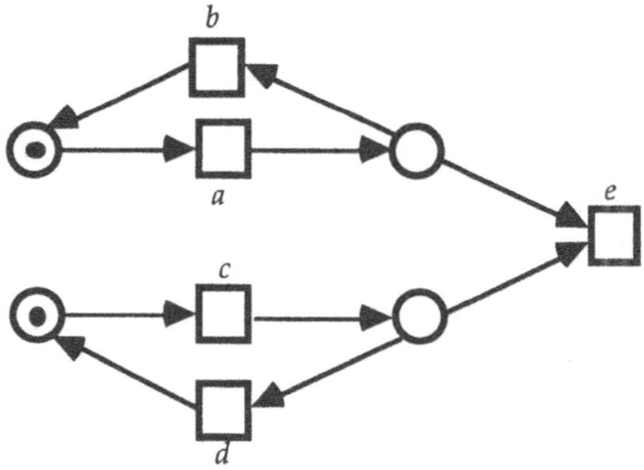

Figure 13.9.

However, this transition system arises from the Petri Net of figure 13.9[5] and in this net, the two sequences actually represent the same behaviour.

[5]See chapter 16.

The apparently meaningful definition now turns out to be ambiguous. The culprit is the subnet of figure 13.10, which exhibits *asymmetric confusion*. In a behaviour in which only a and c it does not make sense to talk of a conflict (between b and e) being resolved or not, because a and c are concurrent.

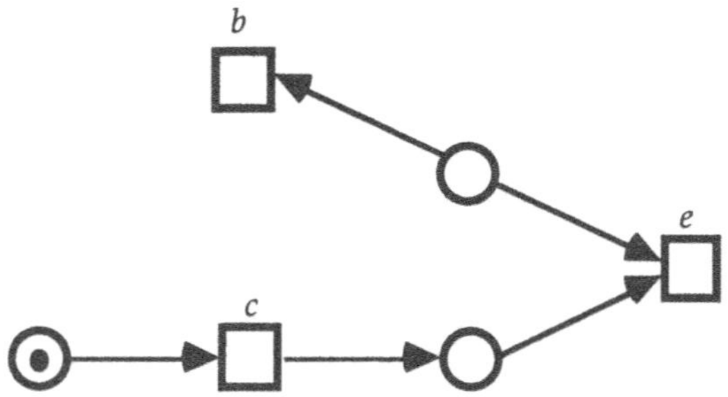

Figure 13.10.

For a more detailed discussion of this and other issues, see [52].

Transition systems are perfectly adequate for the discussion of systems properties which depend only on the reachability of states, such as absence or presence of deadlock.

The notion of strong equivalence was introduced by Milner in [11]. It is one of a number of equivalence relations that he constructed to describe behavioural equivalences between systems described in his CCS formalism. Such systems could be ambiguous (in our sense) and therefore equality of languages of execution sequences is too coarse a relation to serve as a means of expressing equivalence. Bisimulations were introduced by Park [37] in order to reformulate the definition of strong equivalence.

The notion of strong equivalence presented in 13.4.1 differs from that of [3] in that we make an explicit additional requirement that the sets of events are the same. This is to ensure that the same is true for the corresponding behavioural presentations, which is one of the requirements for them to be isomorphic. However, this requirement is automatically satisfied in the case of CCS, as we shall see in chapter 27.

Chapter 14
General Transition Systems

'Ambiguity' itself can mean an indecision as to what you mean, an intention to mean several things, a probability that one or other or both of two things have been meant, and the fact that a statement has several meanings.

WILLIAM EMPSON
Seven Types of Ambiguity

14.1. Introduction.

Let us now turn to the general case. We have seen, in 13.2.4, that sequential behavioural presentations determine transition systems. However, to obtain an isomorphic copy of the behavioural presentation back from the transition system via its language of execution sequences is not possible in general, as 13.2.9 illustrates.

The problem is that there may be an action sequence which corresponds to more than one derivation. In 13.2.9, for instance, the action sequence e_1 in T_{B_2} arises in two ways, either through the execution of the occurrence o_1, corresponding to the transition $\varnothing \rightarrow_{B_2}^{e_1} p_1$, or through the execution of the occurrence $\tilde{o}_1$, corresponding to the transition $\varnothing \rightarrow_{B_2}^{e_1} \tilde{p}_1$.

This ambiguity is removed, however, if we include information about states. We may then distinguish between 'the e_1 that lands us in state p_1' and 'the e_1 that lands us in state $\tilde{p}_1$'. By attaching the name of the state to the name of the event, we remove the ambiguity.

This will allow us to exploit the material of chapter 13. We shall also see that the notion of strong congruence plays a more important rôle in the presence of ambiguity. In the unambiguous case, it is equivalent to equality of languages of action sequences, as we saw in 13.4.2. In the general case, it is a way of distinguishing between transition systems whose behavioural presentations are structurally distinct but whose corresponding languages are equal.

14.2. Behaviour of Transition Systems.

14.2.1. DEFINITION. Let T be a transition system. Define T_∇ to be $(Q, A_\nabla, \rightarrow_\nabla)$, where

1• $A_\nabla = \{(a,q) \in A \times Q \mid \exists q' \in Q: q' \rightarrow^a q\}$;

2• $q_1 \rightarrow_\nabla^{(a,\tilde{q})} q_2 \Leftrightarrow \tilde{q} = q_2 \wedge q_1 \rightarrow^a q_2$.

It is immediate that T_∇ is an unambiguous transition system and

$$(a_1,q_1)\cdots(a_n,q_n) \in L(T_\nabla,q) \Leftrightarrow q \rightarrow^{a_1} q_1 \rightarrow^{a_2} \cdots \rightarrow^{a_n} q_n$$

Thus, $L(T_\nabla,q)$ contains the additional information about the route through the transition system taken by an execution sequence.

By 13.2.2, $(a_1,q_1)\cdots(a_n,q_n) \in L(T_\nabla,q)$ if and only if $qa_1q_1\cdots a_nq_n \in Der(T,q)$.

As T_∇ is unambiguous, we may apply the constructions of chapter 13, obtaining a behavioural presentation from $L(T_\nabla,q)$ in the usual way.

14.2.2. DEFINITION. If T is a transition system and $q \in Q$, then we define $B_{(T,q)} = (O_{(T,q)}, P_{(T,q)}, A, \lambda_{(T,q)})$, where

1• $O_{(T,q)} = L(T_\nabla,q) - \{\Omega\}$;

2• $P_{(T,q)} = \{p_x \mid x \in L(T_\nabla,q)\}$, where $p_x = \{u \in O_{(T,q)} \mid u \leq x\}$, each $x \in L(T_\nabla,q)$;

3• If $(a,\tilde{q}) \in A_\nabla$ and $u.(a,\tilde{q}) \in O_{(T,q)}$, then $\lambda_{(T,q)}(u.(a,\tilde{q})) = a$.

The argument of 13.3 goes through with hardly any alteration and we leave the reader to satisfy himself of herself of the validity of the following.

14.2.3. LEMMA. T is a transition system and $q \in Q$, then $B_{(T,q)}$ is a discrete, sequential behavioural presentation.

■

If T is unambiguous, then each element of $L(T,q)$ determines a unique element of $L(T_\nabla,q)$, that is to say, there is a bijection $f: L(T_\nabla,q) \rightarrow L(T,q)$ such that $f((a_1,q_1)\cdots(a_n,q_n)) = a_1\cdots a_n$. Indeed, by 13.2.2, $a_1\cdots a_n \in L(T,q)$ if and only if there is a derivation sequence $qa_1q_1\cdots a_nq_n$ in T if and only if $(a_1,q_1)\cdots(a_n,q_n) \in L(T_\nabla,q)$, so f is surjective. Injectivity follows from the following property.

$$\forall x,y \in L(T_\nabla,q): x \leq y \Leftrightarrow f(x) \leq f(y) \tag{14.1}$$

for then

$$f(x)=f(y)\Rightarrow f(x)\leq f(y)\wedge f(y)\leq f(x)\Rightarrow x\leq y\wedge y\leq x\Rightarrow x=y.$$

The $\Rightarrow$ implication of (14.1) is trivial. For the $\Leftarrow$ implication, suppose that $x,y\in L(T_{\nabla},q)$ such that $f(x)\leq f(y)$, then $x=(a_1,q_1)\cdots(a_m,q_m)$ and $y=(a_1,q_1')\cdots(a_n,q_n')$ with $m\leq n$. An easy induction on m shows that $q_i'=q_i$, all $i\leq m$.

We use f to construct mappings from the various sets associated with $B_{(T,q)}$ and $B_{L(T,q),A}$ as follows,

- $g_O:O_{(T,q)}\rightarrow O_{L(T,q),A}$, defined by $g_O(u)=f(u)$;
- $g_P:P_{(T,q)}\rightarrow P_{L(T,q),A}$, defined by $g_O(p_x)=p_{f(x)}$.

As $f:L(T_{\nabla},q)\rightarrow L(T,q)$ is a poset isomorphism, g_P is surjective. Also, using (14.1), we have

$$p_x\subseteq p_y\Leftrightarrow x\leq y\Leftrightarrow f(x)\leq f(y)\Leftrightarrow p_{f(x)}\subseteq p_{f(y)}\Leftrightarrow g_P(p_x)\subseteq g_P(p_y)$$

so g_P is a poset isomorphism. Also, if $u=(a_1,q_1)\cdots(a_m,q_m)\in O_{(T,q)}$, then it is immediate that $g_P(\downarrow_{(T,q)}u)=\downarrow_{L(T,q)}f(u)$ and $\lambda_{(T,q)}(u)=\lambda_{L(T,q)}(f(u))$. Applying 12.2.15, we obtain:

14.2.4. LEMMA. If T is an unambiguous transition system and $q\in Q$ then $g:B_{(T,q)}\equiv B_{L(T,q),A}$, where $g=(g_O,g_P)$.

■

Thus, our new construction coincides - up to isomorphism - with our construction in the unambiguous case, as it certainly should.

To conclude, we should establish a generalisation of 13.4.3. Let us begin by formally defining what it means for a transition system to have a behavioural presentation as its behavioural semantics.

14.2.5 DEFINITION. Let (T,q) be an initialised transition system and let B be a discrete, sequential behavioural presentation, then (T,q) will be said to *accept* $B\Leftrightarrow B\equiv B_{(T,q)}$.

We should expect that any sequential behavioural presentation should be accepted by at least one pair (T,q), namely the pair $(T_B,\emptyset)$ constructed from B. Let us establish that desired isomorphism exists.

Let $p\in P$. If $p\neq\emptyset$, then as B is discrete and asynchronous, by 3.2.5 and 3.3.3, we may find $o_1,\cdots,o_n\in O$ and $p_1,\cdots,p_n\in P$ such that

$$\varnothing \succ^{o_1} p_1 \succ^{o_2} \cdots \succ^{o_n} p_n = p$$

and hence, by 13.2.4 and 14.2.2,

$$x(p) = (\lambda(o_1), p_1) \cdots (\lambda(o_n), p_n) \in L(T_\nabla, \varnothing)$$

Define $f_P(p) = p_{x(p)}$ and $f_P(\varnothing) = \varnothing$. If $o \in O$, define $f_O(o) = x(\downarrow o)$. Let us show that $f: B \equiv \tilde{B}$, where $f = (f_O, f_P)$ and $\tilde{B} = B_{(T,\varnothing)}$. First, we observe that from the definition of $x(p)$, $x(p_1) = x(p_2) \Rightarrow p_1 = p_2$, so that by 14.2.2 (2), $p_{x(p_1)} = p_{x(p_2)} \Rightarrow p_1 = p_2$ and f_P is injective. We have seen that every element of $\tilde{P}$ is of the form $p_{x(p)}$, some $p \in P$ and hence f_P is surjective.

Let $o_1, o_2 \in O$, then using the facts that $x(p_1) = x(p_2) \Rightarrow p_1 = p_2$ and that B is asynchronous:

$$f_O(o_1) = f_O(o_2) \Rightarrow x(\downarrow o_1) = x(\downarrow o_2) \Rightarrow \downarrow o_1 = \downarrow o_2 \Rightarrow o_1 = o_2$$

and so f_O is injective. If $\tilde{o} \in \tilde{O}$, then $\tilde{o} = (\lambda(o_1), p_1) \cdots (\lambda(o_n), p_n) \in L(T_\nabla, \varnothing)$, by 14.4.2 (1). Since p_n is totally ordered, $p_n = \downarrow o_n$ and so $\tilde{o} = f_O(o_n)$, and f_O is surjective.

If $p \in P$, then using 13.2.5

$$o \in p \Leftrightarrow \downarrow o \lhd p \Leftrightarrow x(\downarrow o) \leq x(p) \Leftrightarrow f_O(o) \in f_P(p).$$

Finally, if $o \in O$, then $f_O(o)$ is of the form $(\lambda(o_1), p_1) \cdots (\lambda(o_n), p_n)$ with $o_n = o$, and so by 14.4.2 (3), $\tilde{\lambda}(f_O(o)) = \lambda(o)$.

We have proved the following:

14.2.6. PROPOSITION. Let B be a discrete, sequential behavioural presentation, then $(T_B, \varnothing)$ accepts B.

■

Since (T,q) accepts $B_{(T,q)}$ by definition, it follows that, up to isomorphism, every initialised transition system determines a unique discrete, sequential behavioural presentation and every discrete, sequential behavioural presentation is determined by a initialised transition system.

To establish the generalisation of 13.4.3, we need to know the circumstances under which two transition systems with initial state accept the same behavioural presentations.

Alas, it turns out that strong equivalence is not strong enough to distinguish between non-isomorphic behavioural presentations in some cases.

We may illustrate this with a simple example. Let T_1 and T_2 be the transition systems pictured in figure 14.1. It is easy to check that the relation $R=\{(q_1,q_2),(q',q'),(q',q'')\}$ is a strong bisimulation from (T_1,q_1) to (T_2,q_2).

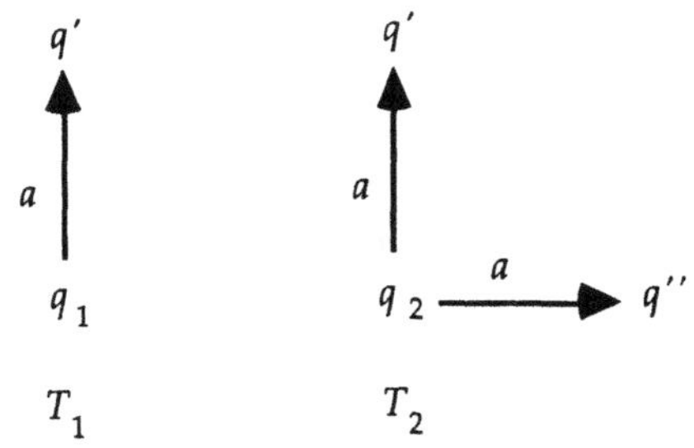

Figure 14.1.

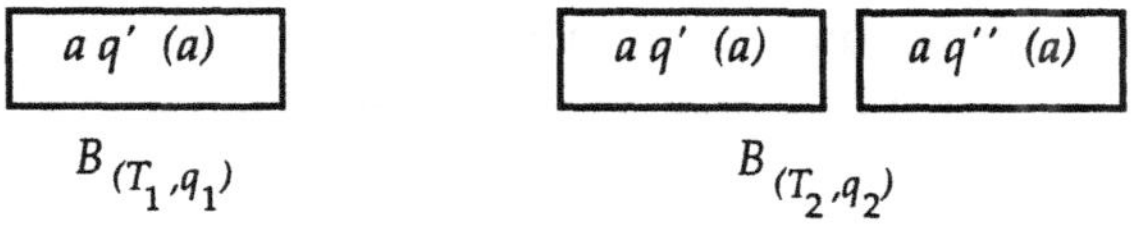

Figure 14.2

However, the behavioural presentations, $B_{(T_1,q_1)}$ and $B_{(T_2,q_2)}$, pictured in figure 14.2, are clearly not isomorphic to each other.

At this point, it seems that there is a choice of positions.

- Since the definition of isomorphism is clearly the 'natural' one, the notion of bisimulation is too weak to capture structural identity of system behaviour and we therefore need an alternative and stronger version.
- The notion of isomorphism may be 'natural' from a purely formal point of view, but the example above shows that it is too strong. Intuitively, T_1 and T_2 describe the same system, which executes a and then stops.

Rather than trying to arbitrate between these two positions we will therefore develop the theory relating to both of them. The reader may then use whichever of them they prefer.

14.3. Strong Equivalence.

We begin by formulating the definition of strong bisimulation, presented in 13.3.4.1, in terms of behavioural presentations.

14.3.1. DEFINITION. Suppose B_1 and B_2 are behavioural presentations. A *strong bisimulation* from B_1 to B_2 is a relation $R \subseteq P_1 \times P_2$ such that $\varnothing\, R\, \varnothing$ and that whenever $p_1\, R\, p_2$, then

$$p_1 \succ_1^{o_1} \tilde{p}_1 \Rightarrow \exists o_2 \in O_2\, \exists \tilde{p}_2 \in P_2 : p_2 \succ_2^{o_2} \tilde{p}_2 \wedge \lambda_1(o_1) = \lambda_2(o_2) \wedge \tilde{p}_1\, R\, \tilde{p}_2;$$
$$p_2 \succ_2^{o_2} \tilde{p}_2 \Rightarrow \exists o_1 \in O_1\, \exists \tilde{p}_1 \in P_1 : p_1 \succ_1^{o_1} \tilde{p}_1 \wedge \lambda_1(o_1) = \lambda_2(o_2) \wedge \tilde{p}_1\, R\, \tilde{p}_2.$$

We shall say that B_1 and B_2 are *strongly equivalent* and write $B_1 \sim B_2 \Leftrightarrow$ there exists a strong bisimulation from B_1 to B_2 and $E_1 = E_2$.

Clearly, $\sim$ is reflexive, symmetric and transitive. As expected:

14.3.2. LEMMA. If B_1 and B_2 are discrete sequential behavioural presentations then $(T_{B_1}, \varnothing) \sim (T_{B_2}, \varnothing) \Leftrightarrow B_1 \sim B_2$.

PROOF. A simple induction argument shows that R is a strong bisimulation from $(T_{B_1}, \varnothing)$ to $(T_{B_2}, \varnothing) \Leftrightarrow R$ is a strong bisimulation from B_1 to B_2.

■

To sum up. Let $\mathbf{SB}_{\sim}$ denote the class whose elements are the $\sim$-equivalence classes of sequential, discrete behavioural presentations. Write $[\![B]\!]_{\sim}$ for the $\sim$-class of B. Let $\mathbf{TS}_{\sim}$ denote the class whose elements are the $\sim$-equivalence classes of transition systems with initial states. Write $[\![T, q]\!]_{\sim}$ for the $\sim$-class of (T, q).

14.3.3. THEOREM. There is a bijective mapping

$$Beh_T : \mathbf{TS}_{\sim} \to \mathbf{SB}_{\sim}$$

given by $Beh_T([\![T, q]\!]_{\sim}) = [\![B_{(T,q)}]\!]_{\sim}$.

PROOF. Beh_T is well defined and injective, by 14.3.2. It is surjective, by 14.2.6.

■

14.4. Strict Equivalence.

Now, we move on to a relation between transition systems which corresponds to that of isomorphism between behavioural presentations.

First, for $q \in Q$, define

$$|q\rangle = \{(a,q') \in A \times Q \mid q \to^a q'\}.$$

Thus, $|q\rangle \subseteq A_\nabla$, each $q \in Q$. We will use the usual subscript convention. Thus, if T_1 and T_2 are transition systems and $q \in Q_1 \cap Q_2$ then

$$|q\rangle_i = \{(a,q') \in A_i \times Q_i \mid q \to_i^a q'\}.$$

14.4.1. DEFINITION. Let (T_1, q_1) and (T_2, q_2) be initialised transition systems. A *strict equivalence* from (T_1, q_1) to (T_2, q_2) is a relation $R \subseteq Q_1 \times Q_2$ such that

1• $q_1 \, R \, q_2$;
2• If $\tilde{q}_1 \, R \, \tilde{q}_2$, then there is a bijection $f_{\tilde{q}_1} : |\tilde{q}_1\rangle \to |\tilde{q}_2\rangle$ such that if $(a, \hat{q}_1) \in |\tilde{q}_1\rangle$, then $f_{\tilde{q}_1}(a, \hat{q}_1) = (b, \hat{q}_2) \Rightarrow a = b \wedge \hat{q}_1 \, R \, \hat{q}_2$.

We will say that (T_1, q_1) and (T_2, q_2) are *strictly equivalent* $\Leftrightarrow$ $A_1 = A_2$ and there is a strict equivalence from (T_1, q_1) to (T_2, q_2). Write $(T_1, q_1) \equiv (T_2, q_2)$. It is elementary that $\equiv$ is an equivalence relation on the class of all initialised transition systems.

14.4.2. LEMMA. If (T_1, q_1) and (T_2, q_2) are initialised transition systems, then

1• $(T_1, q_1) \equiv (T_2, q_2) \Rightarrow (T_1, q_1) \sim (T_2, q_2)$;
2• $(T_1, q_1) \sim (T_2, q_2) \Rightarrow (T_1, q_1) \equiv (T_2, q_2)$ if T_1 and T_2 are unambiguous.

PROOF. Comparing 13.4.1 and 14.4.1 shows that if R is a strict equivalence from (T_1, q_1) to (T_2, q_2), then R is a strong bisimulation from (T_1, q_1) to (T_2, q_2).

Suppose that T_1 and T_2 are unambiguous and R is a strong bisimulation from (T_1, q_1) to (T_2, q_2). We show that R is a strict equivalence from (T_1, q_1) to (T_2, q_2).

Indeed, if $\tilde{q}_1 \, R \, \tilde{q}_2$, and $(a, \hat{q}_1) \in |\tilde{q}_1\rangle_1$, then $\tilde{q}_1 \to_1^a \hat{q}_1$, so by 13.4.1, there exists $\hat{q}_2 \in Q_2$ such that $\tilde{q}_2 \to_2^a \hat{q}_2$ and $\hat{q}_1 \, R \, \hat{q}_2$. As T_2 is unambiguous, $\hat{q}_2 \in Q_2$ is unique, by 13.3.1 and so we may define $f_{\tilde{q}_1}(a, \hat{q}_1) = (a, \hat{q}_2)$. As T_1 is unambiguous, $f_{\tilde{q}_1}$ is injective. If $(a, \hat{q}_2) \in |\tilde{q}_2\rangle_2$, then $\tilde{q}_2 \to_2^a \hat{q}_2$, so by 13.4.1, there exists $\hat{q}_1 \in Q_1$ such that $\tilde{q}_1 \to_1^a \hat{q}_1$ and $\hat{q}_1 \, R \, \hat{q}_2$, so $f_{\tilde{q}_1}(a, \hat{q}_1) = (a, \hat{q}_2)$ and $f_{\tilde{q}_1}$ is surjective.

■

Suppose $f{:}B_1 \equiv B_2$, then we may define a strict equivalence R from $(T_{B_1},\varnothing)$ to $(T_{B_2},\varnothing)$ by defining $p\,R\,p' \Leftrightarrow f_P(p)=p'$ and if $(e,p')\in|p\rangle_{B_1}$, then $f_p(e,p')=(e,f_P(p'))$. We note that f_p has the required property since

$$\begin{aligned} p \rightarrow_{B_1}^{e} p' &\Leftrightarrow \exists o \in O_1 : p \succ^{o} p' \wedge \lambda_1(o)=e \\ &\Leftrightarrow \exists o \in O_1 : f_P(p) \rightarrow_{B_2}^{\lambda_2(f_O(o))} f_P(p') \wedge \lambda_2(f_O(o))=e \\ &\Leftrightarrow f_P(p) \rightarrow_{B_2}^{e} f_P(p') \end{aligned}$$

Thus,

$$B_1 \equiv B_2 \Rightarrow (T_{B_1},\varnothing) \equiv (T_{B_2},\varnothing) \tag{14.2}$$

Next, suppose that $(T_1,\varnothing)\equiv(T_2,\varnothing)$ with strict equivalence R. First note that this allows us to define a function $\beta{:}L(T_{1\nabla},\varnothing)\rightarrow(T_{2\nabla},\varnothing)$ by

$$\beta((e_1,q_1)\cdots(e_n,q_n)) = f_q(e_1.q_1).f_{q_1}(e_2,q_2)\cdots f_{q_{n-1}}(e_n,q_n) \tag{14.3}$$

β is a bijection (since all the f_{q_i} are) and we may therefore define bijections $f_P{:}P_{(T_1,q_1)} \rightarrow P_{(T_2,q_2)}$ and $f_O{:}O_{(T_1,q_1)} \rightarrow O_{(T_2,q_2)}$ by $f_P(p_x)=p_{\beta(x)}$ and $f_O(u)=\beta(u)$. It is clear from equation (14.3) that $\lambda_{(T_1,q_1)}(u)=\lambda_{(T_2,q_2)}(f_O(u))$, giving requirement (1) of 12.2.13. It is also clear that

$$u \in p_x \Leftrightarrow u \le x \Leftrightarrow \beta(u) \le \beta(x) \Leftrightarrow f_O(u) \in f_P(p_x)$$

giving requirement (2) of 12.2.13. Thus

$$(T_1,q_1)\equiv(T_2,q_2) \Rightarrow B_{(T_1,q_1)} \equiv B_{(T_2,q_2)} \tag{14.4}$$

In particular, if $(T_1,\varnothing)\equiv(T_2,\varnothing)$, then $B_{(T_{B_1},\varnothing)} \equiv B_{(T_{B_2},\varnothing)}$ But $B_1 \equiv B_{(T_{B_1},\varnothing)}$ and $B_{(T_{B_2},\varnothing)} \equiv B_2$, by 14.2.5, and so we have

$$(T_{B_1},\varnothing)\equiv(T_{B_2},\varnothing) \Rightarrow B_1 \equiv B_2 \tag{14.5}$$

To sum up:

14.4.3. LEMMA. If B_1 and B_2 are discrete sequential behavioural presentations then $(T_{B_1},\varnothing)\equiv(T_{B_2},\varnothing) \Leftrightarrow B_1 \equiv B_2$.

PROOF. By (14.2) and (14.5).

■

Next, suppose that $B_{(T_1,q_1)} \equiv B_{(T_2,q_2)}$. By 14.2.2, $B_{(T_i,q_i)} \equiv B_{L(T_{i\nabla},q_i)}$, each i. Since each $T_{i\nabla}$ is unambiguous, $\rho(P_{L(T_{1\nabla},q_1)}) = \rho(P_{L(T_{2\nabla},q_2)})$, by 13.3.6 and so by 13.3.7, $L(T_{1\nabla},q_1) = L(T_{2\nabla},q_2)$ and so by 13.4.2 (2), $(T_{1\nabla},q_1) \sim (T_{2\nabla},q_2)$, and so by 14.4.2 (2), $(T_{1\nabla},q_1) \equiv (T_{2\nabla},q_2)$. Let R be a strict equivalence from $(T_{1\nabla},q_1)$ to $(T_{2\nabla},q_2)$; we show that it is also a strict equivalence from (T_1,q_1) to (T_2,q_2).

Certainly, $q_1 R q_2$. Next, suppose that $\tilde{q}_1 R \tilde{q}_2$, then by 14.2.1 (2), $(a,\hat{q}_1) \in [\tilde{q}_1\rangle_1 \Leftrightarrow ((a,\hat{q}_1),\hat{q}_1) \in [\tilde{q}_1\rangle_{1\nabla}$, and therefore $f^{\nabla}_{\tilde{q}_1} : |\tilde{q}_1\rangle_{1\nabla} \to |\tilde{q}_2\rangle_{2\nabla}$ is a bijection satisfying 14.4.1 (2), and $f_{\tilde{q}_1} : |\tilde{q}_1\rangle_1 \to |\tilde{q}_2\rangle_2$, defined by

$$f_{\tilde{q}_1}(a,\hat{q}_1) = (b,\hat{q}_2) \Leftrightarrow f^{\nabla}_{\tilde{q}_1}((a,\hat{q}_1),\hat{q}_1) = ((b,\hat{q}_2),\hat{q}_2)$$

is also a bijection satisfying 14.4.1 (2), and we have proved that

$$B_{(T_1,q_1)} \equiv B_{(T_2,q_2)} \Rightarrow (T_1,q_1) \equiv (T_2,q_2) \tag{14.6}$$

To sum up:

14.4.4. LEMMA. If (T_1,q_1) and (T_2,q_2) are initialised transition systems, then

$$(T_1,q_1) \equiv (T_2,q_2) \Leftrightarrow B_{(T_1,q_1)} \equiv B_{(T_2,q_2)}.$$

PROOF. By (14.4) and (14.6).

■

Finally, we have a result analogous to 14.3.3 for the case of strict equivalence. Let **SB** denote the class whose elements are all the $\equiv$-equivalence classes of sequential, discrete behavioural presentations. Write $[\![B]\!]$ for the $\equiv$-class of the sequential, discrete behavioural presentation B. Let **TS** denote the class whose elements are all the $\equiv$-equivalence classes of transition systems with initial states. Write $[\![T,q]\!]$ for the $\equiv$-class of the transition system with initial state (T,q).

14.4.5. THEOREM. There is a bijective mapping

$$Beh_T : \mathbf{TS} \to \mathbf{SB}$$

such that $B \in Beh_T([\![T,q]\!]) \Leftrightarrow (T,q)$ accepts B.

PROOF. Define Beh_T by $Beh_T([\![T,q]\!]) = [\![B_{(T,q)}]\!]$. By 14.2.5:

$$B \in Beh_T([\![T,q]\!]) \Leftrightarrow B \equiv B_{(T,q)} \Leftrightarrow (T,q) \text{ accepts } B$$

Since $(T_B, \varnothing)$ accepts B, by 14,2,6, Beh_T, is surjective.
Finally, if (T_1, q_1) and (T_2, q_2) are initialised transition systems, then

$$\begin{aligned} Beh_T([\![T_1, q_1]\!]) = Beh_T([\![T_2, q_2]\!]) &\Rightarrow [\![B_{(T_1,q_1)}]\!] = [\![B_{(T_2,q_2)}]\!] \\ &\Rightarrow B_{(T_1,q_1)} \equiv B_{(T_2,q_2)} \\ &\Rightarrow (T_1, q_1) \equiv (T_2, q_2), \text{ by 14.4.4} \\ &\Rightarrow [\![T_1, q_1]\!] = [\![T_2, q_2]\!] \end{aligned}$$

so Beh_T, is injective.

■

Chapter 15
Asynchronous Transition Systems

Oh Diamond! Diamond! thou little knowest the mischief done.
ISAAC NEWTON
Allegedly, after his dog had overturned a candle which set fire to some of Newton's papers.

15.1. Introduction.

We now seek to extend the work of the preceding two chapters to a larger class of behavioural presentations. We begin by considering those which are discrete and asynchronous.

Where do we start? It seems a good idea to see how sequential behavioural presentations relate to the automata that accept them. The relevant construction is given in 13.2.4, which indeed applies to arbitrary discrete, asynchronous behavioural presentations.

Unfortunately, it is clear that this construction is not adequate to capture all the structure of an asynchronous behavioural presentation as it contains no information relating to concurrency. On the other hand, we have already seen, in 7.5.1, how concurrency interacts with the derivation structure of asynchronous behavioural presentations in what we called the 'lozenge property' or 'diamond property'.

However, the lozenge property refers not to the transitions $p_1 \rightarrow_B^e p_2$ of 13.2.4, but steps $p_1 \succ^o p_2$ of 3.3.1. We shall therefore begin by constructing our automata with a transition structure based directly on the derivation structure of the behavioural presentation.

This gives us an automaton. The next question is: how do such automata relate to behavioural presentations? The answer is simple: each such initialised automaton together determines a left-closed trace language. And we have seen in chapter 12 that every left-closed trace language determines a linguistic behavioural presentation.

15.2. The Automata.

With 7.5.1 in mind, we make the following definition:

15.2.1. DEFINITION. An ι-*asynchronous transition system* is quadruple $C=(Q,A,\to,\iota)$ where

1• $T(C)=(Q,A,\to)$ is an unambiguous transition system;
2• ι is an independence relation;
3• If $q_1,q_2,q_3 \in Q$ and $a,b \in A$ such that $q_1 \to^a q_2 \to^b q_3$ and $a\,\iota\,b$, then there exists $\tilde{q}_2 \in Q$ such that $q_1 \to^b \tilde{q}_2 \to^a q_3$.

As we have said, condition (3) is will be referred to as the '*lozenge rule*' or the *diamond rule,* because it entails the existence of diamond shapes in the transition system, as we saw in the example of figure 1.3.

As automata, ι-asynchronous transition systems determine languages, in this case, trace languages.

15.2.2. DEFINITION. Let (C,q) be an initialised ι-asynchronous transition system. Define $L(C,q)=\{x_\iota \mid x \in L(T(C),q)\}$.

We state some elementary properties of $L(C,q)$

15.2.3. LEMMA. If (C,q) is an initialised ι-asynchronous transition system, then

1• $u \in L(T(C),q) \wedge x \equiv_\iota^{(1)} u \Rightarrow x \in L(T(C),q)$;
2• $q \to^x q_1 \wedge q \to^y q_2 \wedge x \equiv_\iota y \Rightarrow q_1 = q_2$.

PROOF. (1) In view of 9.2.1, it suffices to show that if $w \in L(T(C),q)$ and $w \equiv_\iota^{(1)} x$, then $x \in L(T(C),q)$. Suppose that $v_1,v_2 \in A^*$ and $a,b \in A$ such that $w=v_1.a.b.v_2$, $x=v_1.b.a.v_2$ with $a\,\iota\,b$. By 15.2.2, there exists $q_1,q_2,\tilde{q},q' \in Q$ such that $q \to^{v_1} q_1 \to^a q_2 \to^b \tilde{q} \to^{v_2} q'$. By the lozenge rule, there exists $q_3 \in Q$ such that $q \to^{v_1} q_1 \to^b q_3 \to^a \tilde{q} \to^{v_2} q'$. Therefore, $x=v_1.b.a.v_2 \in L(T(C),q)$.
(2) In view of 9.2.1, it suffices to show that if $q \to^{u.a.b.v} q_1$ and $q \to^{u.b.a.v} q_2$ with $a,b \in A$ then $q_1=q_2$. So suppose that $q \to^u q' \to^a q_1' \to^b q_1'' \to^v q_1$ and $q \to^u q' \to^b q_2' \to^a q_2'' \to^v q_2$. By the lozenge rule, there exists $\tilde{q} \in Q$ such that $q \to^u q' \to^b \tilde{q} \to^a q_1''$. By unambiguity, $q_1''=q_2''$ and now $q_1''=q_2'' \to^v q_2$ and $q_1'' \to^v q_1$. By unambiguity, $q_1=q_2$.

■

15.2.3 (2) allows us to define unambiguously

$$q \rightarrow^{x_\iota} q' \Leftrightarrow q \rightarrow_x q' \tag{15.1}$$

15.2.4. PROPOSITION. If (C,q) is an initialised ι-asynchronous transition system, then, then $L(C,q)$ is left-closed.

PROOF. Let $x \in L(T(C),q)$ and $y_\iota \in A_\iota{}^*$ such that $y_\iota \leq_\iota x_\iota$ By 9.2.2 (2), there exists $z \in A^*$ such that $y.z \equiv_\iota x$. By 15.2.3, $y.z \in L(T(C),q)$, so there exists $\tilde{q},q' \in Q$ such that $q \rightarrow^y \tilde{q} \rightarrow^z q'$ and in particular, $q \rightarrow^y \tilde{q}$ and so $y \in L(T(C),q)$ and so $y_\iota \in L(C,q)$.

■

15.2.5. COROLLARY. If (C,q) is an initialised ι-asynchronous transition system, then $L(C,q)$ is prime algebraic and consistently complete.

PROOF. By 10.4.5 and 15.2.4.

■

We know from section 12.1, that we may construct a behavioural presentation from $L(C,q)$.

15.2.6. DEFINITION. If (C,q) is an initialised ι-asynchronous transition system, then we define $B_{(C,q)} = B_{L(C,q),A}$.

Note that by 12.2.9, $B_{(C,q)}$ is ι-linguistic.

15.2.7. DEFINITION. If (C,q) is an initialised ι-asynchronous transition system, then we shall say that (C,q) *accepts* $B \Leftrightarrow B \equiv B_{(C,q)}$.

We know that every initialised ι-asynchronous transition system accepts a ι-linguistic behavioural presentation. Is every ι-linguistic behavioural presentation accepted by some initialised ι-asynchronous transition system? We might seek an answer to such a question by examining ways of constructing automata from behavioural presentations as we did in chapter 13.

15.2.8. DEFINITION. If B is a discrete, asynchronous behavioural presentation, then we define $\overline{C_B} = (P,O,\succ,co)$.

If B is discrete and asynchronous, then the transition system $(P,O,\succ)$ satisfies the lozenge rule, by 7.5.1. We also know that co is an independence relation. Finally, if $p \succ^{o} p_1$ and $p \succ^{o} p_2$, then $p_1 = p \cup \{o\} = p_2$, so the transition system is also ambiguous. Hence:-

15.2.9. PROPOSITION. If B is a discrete, asynchronous behavioural presentation, then $\overline{C_B}$ is a *co*-asynchronous transition system.

■

Note that in going from B to $\overline{C_B}$, we lose information about labels, which effectively prevents us from reconstructing B up to isomorphism from $\overline{C_B}$. In order to be able to do that, we must add extra structure to $\overline{C_B}$, giving what we shall call a hybrid transition system.[1] However, there is another construction, which works for ι-linguistic behavioural presentations, and as these are the type of behavioural presentation determined by asynchronous transition systems, this construction is not without relevance, as we shall see.

15.2.10. DEFINITION. Suppose B is a ι-linguistic behavioural presentation Define $C_B = (L, E, \rightarrow, \iota)$, where $L = \rho_B(P)$ and $\rightarrow \subseteq L \times E \times L$ is defined by

$$(x_\iota, e, y_\iota) \in \rightarrow \Leftrightarrow y_\iota = x_\iota . e_\iota .$$

As usual, write $x_\iota \rightarrow^{e} y_\iota$ for $(x_\iota, e, y_\iota) \in \rightarrow$.

15.2.11. PROPOSITION. If B is a ι-linguistic behavioural presentation, then C_B is an ι-asynchronous transition system and (C_B, Ω_ι) accepts B.

PROOF. $T(C_B)$ is unambiguous, since by definition of $\rightarrow$

$$x_\iota \rightarrow^{e} y_\iota \wedge x_\iota \rightarrow^{e} y'_\iota \Rightarrow y_\iota = x_\iota . e_\iota = y'_\iota .$$

Next, if, $x_\iota \rightarrow^{e} y_\iota \rightarrow^{e'} y'_\iota$ with $e \,\iota\, e'$, then $x.e.e' \equiv_\iota^{(1)} x.e'.e$, and so $x_\iota . e_\iota . e'_\iota = x_\iota . e'_\iota . e_\iota$ But by definition of $\rightarrow$, $y'_\iota = y_\iota . e_\iota = x_\iota . e_\iota . e'_\iota$ and therefore $x_\iota . e'_\iota \in \rho_B(P)$, since $\rho_B(P)$ is left-closed, by, 2.2.12 and again by definition of $\rightarrow$, $x_\iota \rightarrow^{e'} \tilde{y}_\iota \rightarrow^{e} y'_\iota$, where $\tilde{y}_\iota = x_\iota . e'_\iota$.

A simple induction argument shows that if $z \in E^*$ and $y_\iota \in L(C_B, \Omega_\iota)$ then $\Omega_\iota \rightarrow_B^{z_\iota} y_\iota \Leftrightarrow z_\iota = y_\iota$. Hence, $L(C_B, \Omega_\iota) = \rho_B(P)$. But $B \equiv B_{\rho_B(P),E}$, by 12.2.16. Hence, $B \equiv B_{\rho_B(P),E} = B_{L(C_B,\Omega_\iota),E} = B_{(C_B,\Omega_\iota)}$ and so (C_B, Ω_ι) accepts B.

■

---

[1]See chapter 22.

15.3. Equivalences.

We now know that every ι-linguistic behavioural presentation is accepted by some ι-asynchronous transition system with initial state and that, conversely, every ι-asynchronous transition system with initial state accepts some ι-linguistic behavioural presentation. We are on the heels of another 'semantics theorem', analogous to 13.4.3, 14.3.3 and 14.4.5. First, we need to define an appropriate equivalence between initialised ι-asynchronous transition systems.

Let (C_1,q_1) and (C_2,q_2) be initialised asynchronous transition systems. These give us two unambiguous transition systems with initial states, namely $(T(C_1),q_1)$ and $(T(C_2),q_2)$. Now, in chapters 13 and 14 we defined two relations on transition systems, namely strong equivalence (13.4.1) and strict equivalence (14.4.1), but we also noted, in 14.4.2, that these two relations coincided for unambiguous transition systems such as $(T(C_1),q_1)$ and $(T(C_2),q_2)$. By 14.4.2 (2) and 13.4.2 (2), if $(T(C_1),q_1)\not\equiv(T(C_2),q_2)$, then $L(T(C_1),q_1)\neq L(T(C_2),q_2)$, which means that $L(C_1,q_1)\neq L(C_2,q_2)$, since by 15.2.3 (1), for any asynchronous transition system with initial state, (C,q),

$$L(T(C),q)=\bigcup_{x_\iota\in L(C,q)} x_\iota \tag{15.2}$$

Thus, if $(T(C_1),q_1)\not\equiv(T(C_2),q_2)$, then (C_1,q_1) and (C_2,q_2) cannot accept the same behavioural presentations. We therefore shall require in the definition of our equivalence that $L(T(C_1),q_1)=L(T(C_2),q_2)$.

Of course, in order that $L(C_1,q_1)=L(C_2,q_2)$ it is not sufficient that $L(T(C_1),q_1)=L(T(C_2),q_2)$, since there is also the concurrency structure to be taken into account. However, when comparing the concurrency structures of (C_1,q_1) and (C_2,q_2), we are only interested in those pairs of events which can be reached from the initial states. So, if (C,q) is an initialised asynchronous transition system, then we define $\iota_{(C,q)}$ to be the smallest independence relation such that for each $x\in L(T(C),q)$, $x_\iota=x_{\iota_{(C,q)}}$. Formally:

$$a\,\iota_{(C,q)}\,b\Leftrightarrow a\,\iota\,b\wedge(\exists x\in A^*:x.a.b\in L(T(C),q)) \tag{15.3}$$

We are only interested in concurrency in so far as it relates to pairs of transitions which are reachable and which may go concurrently. This leads us to make the following definition.

15.3.1. DEFINITION. Let (C_1,q_1) and (C_2,q_2) be initialised asynchronous transition systems. We say that (C_1,q_1) and (C_2,q_2) are *strictly equivalent* and write $(C_1,q_1)\equiv(C_2,q_2)\Leftrightarrow$

1• $L(T(C_1),q_1)=L(T(C_2),q_2)$;
2• $\iota_{(C_1,q_1)}=\iota_{(C_2,q_2)}$;
3• $A_1=A_2$.

It is clear that $\equiv$ is an equivalence relation on the class of all initialised asynchronous transition systems

The next result shows that our equivalence has the right consequences.

15.3.2. PROPOSITION. If (C_1,q_1) and (C_2,q_2) be initialised asynchronous transition systems, then the following are equivalent.

1• $(C_1,q_1)\equiv(C_2,q_2)$;
2• $A_1=A_2 \wedge L(C_1,q_1)=L(C_2,q_2)$;
3• $B_{(C_1,q_1)}=B_{(C_2,q_2)}$.

PROOF. (1) $\Rightarrow$ (2) By 15.3.1 (2) $L(T(C_1),q_1)=L(T(C_2),q_2)$. If $x,y\in L(T(C_1),q_1)$, then $x\equiv_{\iota_1} y \Leftrightarrow x\equiv_{\iota_{(C_1,q_1)}} y \Leftrightarrow x\equiv_{\iota_{(C_2,q_2)}} y \Leftrightarrow x\equiv_{\iota_2} y$. Thus if $x\in L(T(C_1),q_1)$, then $x_{\iota_1}=x_{\iota_2}$. But now, using 15.2.3:

$$x_{\iota_1}\in L(C_1,q_1)\Leftrightarrow x\in L(T(C_1),q_1)\Leftrightarrow x\in L(T(C_2),q_2)\Leftrightarrow x_{\iota_1}\in L(C_2,q_2).$$

(2) $\Leftrightarrow$ (3) By 15.2.6.
(2) $\Rightarrow$ (1) Let $L(C_1,q_1)=L(C_2,q_2)$ and $A_1=A_2$. 15.3.1 (1) holds by (15.2). If $a\,\iota_{(C_1,q_1)}\,b$, then there exists $x\in A_1^*$ such that $x.a.b\in L(T(C_1),q_1)$ and $a\,\iota_1\,b$, Therefore $(x.b.a)_{\iota_1}=(x.a.b)_{\iota_1}\in L(C_1,q_1)$ and so $(x.b.a)_{\iota_2}=(x.a.b)_{\iota_2}\in L(C_2,q_2)$. But now $x.a.b\in L(T(C_2),q_2)$ and by 9.3.2 and 9.3.3, $(a.b)_{\iota_2}=(b.a)_{\iota_2}$. Hence $a\,\iota_2\,b$ and thus $a\,\iota_{(C_2,q_2)}\,b$. We have shown that $\iota_{(C_1,q_1)}\subseteq\iota_{(C_2,q_2)}$. By symmetry, $\iota_{(C_2,q_2)}\subseteq\iota_{(C_1,q_1)}$, giving 15.3.1 (2). Since $A_1=A_2$ by hypothesis, 15.3.(3) holds.

■

We may now state the asynchronous analogue of 14.4.5. Let $\mathbf{LBP}_\iota$ denote the class whose elements are all the $\equiv$-equivalence classes of ι-linguistic, discrete behavioural presentations. Write $[\![B]\!]$ for the $\equiv$-class of B. Let $\mathbf{ATS}_\iota$ denote the class whose elements are all the $\equiv$-equivalence classes of ι-asynchronous transition systems with initial states. Write $[\![C,q]\!]$ for the $\equiv$-class of (C,q).

15.3.3. THEOREM. There is a bijective mapping

$$Beh_A:\mathbf{ATS}^*_\iota\rightarrow\mathbf{LBP}_\iota$$

such that $Beh_A([\![C,q]\!]) = [\![B]\!] \Leftrightarrow (C,q)$ accepts B.

PROOF. Beh_A is well defined since if $(C_1,q_1) \equiv (C_2,q_2)$ and (C_1,q_1) accepts B_1 and (C_2,q_2) accepts B_2, then by 15.2.7 and 15.3.2, $B_1 \equiv B_{(C_1,q_1)} \equiv B_{(C_2,q_2)} \equiv B_2$.

Beh_A is injective since if $Beh_A([\![C_1,q_1]\!]) = Beh_A([\![C_2,q_2]\!])$, then there exists B_1 and B_2 such that (C_1,q_1) accepts B_1 and (C_2,q_2) accepts B_2, with $B_1 \equiv B_2$. By 15.2.7, $B_{(C_1,q_1)} \equiv B_1 \equiv B_2 \equiv B_{(C_2,q_2)}$ and so by 15.3.2, $(C_1,q_1) \equiv (C_2,q_2)$. Hence,

$$Beh_A([\![C_1,q_1]\!]) = Beh_A([\![C_2,q_2]\!]) \Rightarrow [\![C_1,q_1]\!] = [\![C_2,q_2]\!].$$

Beh_A is surjective since $Beh_A([\![C_B,\Omega_\iota]\!]) = [\![B]\!]$, by 15.2.11.

∎

Notes.

The diamond rule is related to concurrency, the assertion of the existence of which forms part of a destructive critique of Newton's synchronous view of space-time. Hence, the quotation at the head of this chapter.

Asynchronous transition systems are due to the author [39, 42] and were subsequently reinvented in [43].

Subsequently, other automata-based models of concurrency have been proposed, for example, by Winskel in [55]. A discussion of the relationships between these appears in [56].

Chapter 16

Condition/Event Systems: Process and Occurrence Net Semantics

> The aim of this paper...is to exhibit one finitary combinatorial model of event topology which is in close correspondence with the models of modern physics, is capable of describing total information flow, and has proved superior to some conventional models both in construction and in analysis of systems of complex organisation.
>
> C. A. PETRI [8]

16.1. Introduction.

The theory of Petri Nets was one of the first general theories of concurrent systems. Nets were introduced by C.A. Petri in his doctoral thesis in 1962 [7]. Since then, Net Theory has been widely studied and is arguably one of the most developed of all the theories of parallelism.

Net Theory, at least as practised by Petri and his associates, is built around a non-interleaving view of parallelism. Concurrency is a key idea.

That there is such an emphasis on the importance of a non-interleaving view of parallelism may come as a surprise to readers who are familiar mainly with texts such as [57] or [58] in which nets are provided with an interleaving semantics in terms of so-called *firing sequences*. The authors of these books are largely concerned with the application of Net Theory to the analysis of systems - so-called *token mathematics* - to discover, for example, whether or not the system deadlocks or requires the generation or consumption of an unbounded number of resources. As long as the property in question is definable in terms of reachable global states - deadlock is an example of such a property - then there is no harm in adopting an interleaving approach. Matters become more complicated if the property in question is sensitive to possible confusion[1] or involves infinite behaviour.

---

[1]'See notes to chapter 13.

In this chapter, we describe net-based semantics for nets at the conceptual level of atomic conditions and events, first in terms of sets of nets representing executions of the target net from some initial (global) state and next in terms of a single occurrence nets (in the sense of [20]) representing all possible evolutions of the system.

In chapter 17, we offer a semantics based on asynchronous transition systems, which allows us to associate a given condition/event system with a linguistic behavioural presentation. We then show that both net semantics may be derived from the behavioural presentation semantics. This indicates that our behavioural presentation semantics is consistent with both the 'official' semantics and that using occurrence nets, a requirement set out in the preface.

16.2. Net Theory, C/E Systems and Process Semantics.

Let us begin by saying what a net is.

16.2.1. DEFINITION. A *net* N is a triple $(S,T;F)$ such that

1• S is a set of S-*elements* - also known as *places;*
2• T is a set of T-*elements* - also known as *transitions;*
3• $F \subseteq (S \times T) \cup (T \times S)$ is the *flow relation*, satisfying

a• $S \cap T = \varnothing$ and $S \cup T \neq \varnothing$;
b• $dom(F) \cup ran(F) = S \cup T$[2]

There is a well-known graphical representation of nets in which S-elements are drawn as circles, T-elements are drawn as squares and a directed arc is drawn from the representation of an element $x \in S \cup T$ to that of $y \in S \cup T$ if $x\ F\ y$.

Figure 16.1 illustrates a net N with

$S = \{s_1, s_2, s_3, s_4, s_5\}$
$T = \{t_1, t_2, t_3, t_4\}$

$s_1\ F\ t_1$	$t_1\ F\ s_2$	$s_5\ F\ t_1$	$s_2\ F\ t_2$	$t_2\ F\ s_1$
$s_3\ F\ t_3$	$t_3\ F\ s_4$	$t_3\ F\ s_5$	$s_4\ F\ t_4$	$t_4\ F\ s_3$

[2]If $R \subseteq A \times B$, then the *domain* of R is the set $dom(R) = \{a \in A \mid \exists b \in B: a\ R\ b\}$ and the *range* of R is the set $ran(R) = \{b \in B \mid \exists a \in A: a\ R\ b\}$.

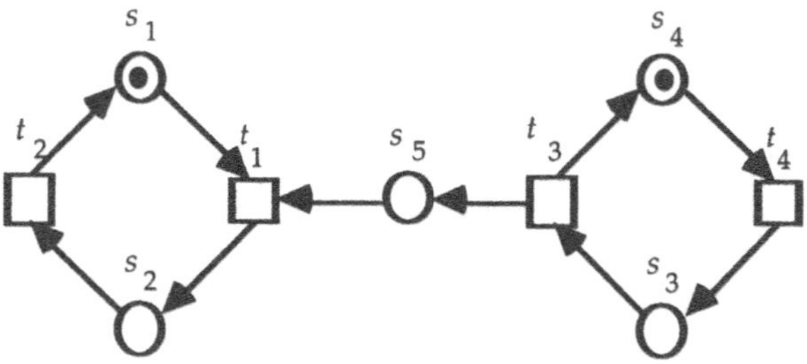

Figure 16.1.

We explain the significance of the dots shortly.
We shall find it convenient to define, for $x \in S \cup T$

$$\bullet x = \{y \in S \cup T \mid y\ F\ x\}$$
$$x \bullet = \{y \in S \cup T \mid x\ F\ y\}$$

If there are several nets in question, then we shall identify them in phrases such as '$\bullet\, x$ in N_1'.

T-elements represent the actions of a system, things that it does. S-elements are components of distributed states; that is, the state of a system described by a net is an aggregate of the current states of its component S-elements. Graphically, these are exhibited by the presence of tokens of some form in the S-elements, as in the case of s_1 and s_4 of figure 16.1. The dynamics of a net is defined in terms of changes to the local states of the S-elements consequent on the activity - or *firing*[3] - of the T-elements. The fact that these changes are local establishes an independence between certain events and consequent concurrency between their firings.

Any system may be described in a number of levels of detail. A processor executing a non-interactive program, absorbs the input data, churns away at it and then - hopefully - produces a result. A description on this low level of detail (high level of abstraction) involves no more than two or three actions and (by implication) no more than two or three states. But these actions are complex, structured from simpler ones; input and output of individual data items, executions of simple program statements.

Likewise, the states of the system represent the complex of local states in the memory of the processor. One could describe the execution of the program at a more detailed level; listing the sequence of executions of simple program statements and the intermediate assignments of values to all its variables. One could go further to the level of compiled op codes or microcode, to the operation of the processor in terms of logic gates or transistors.

---

[3]The term is suggestive of the firing of neurones in the brain.

In all cases, however, the behaviour of the system may be regarded as an example of the dynamics of a representation involving state and action elements, linked by pre-condition and post-condition relations which determine how items of 'data' flow through the system as actions occur.

In a like manner, there are several classes of net model occupying different levels of abstraction, but ultimately related in the sense that each instance of each model may be given an interpretation in terms of a basic net where S-elements and T-elements represent, respectively, atomic proposition and atomic events. On this basic level, where the net models are called *condition/event systems*, each atomic action requires, in order to fire, that atomic propositions belonging to a specific set (the preconditions of the event) hold simultaneously; on firing, these all cease to hold and propositions from another specific set (the postconditions of the event) begin to hold. The flow relation shows which conditions are related to which events. Single tokens residing in S-elements indicate instantaneous holdings of conditions.

On the next level up, that of *place/transition nets*, S-elements represent resources (the term is used loosely and generally) and the number of tokens on the S-element indicates how many instances of that resource are currently available. T-elements represent actions and the effect of an action is to consume a certain number of resources of various kinds and to produce certain number of various kinds of resource in return. Again, the flow relation, augmented by numerical weights indicate how many of each resource a given action consumes or produces. In chapter 23, we examine a model of parallel computation based on nets on this level.

At a yet higher level, that of *predicate/transition nets* [59, 60], S-elements are associated with predicates and tokens have structure or 'colours' [61] so that the presence of several tokens in a place indicates the existence in the system of that number of individuals currently satisfying that predicate. The firing of a T-element depends of the ability of tokens in its input places to satisfy a predicate associated with that T-element, using a binding mechanism which determines ways in which individuals may be consumed and produced by that T-element.

In moving from condition/event systems to predicate/transition nets, there is a progressive increase in abstraction: the same system may be described using smaller and smaller nets. However, it is a requirement of General Net Theory that any net model should possess an interpretation in terms of the basic condition/event model. In that way, net models acquire a semantics either directly, in the case of condition/event systems, or indirectly by their having a formal interpretation in terms of condition/event systems.[4]

---

[4]We emphasise that there is no 'canonical' interpretation. The principle is that any system described by a higher level net must have an equivalent description on the condition/event level and that the two representation be formally related.

In this chapter, we look at the basic interpretation. First, we need:

16.2.2. DEFINITION. Let $N=(S,T;F)$ be a net and suppose that $c \subseteq S$, then we say that $t \in T$ has *concession* at c or is *enabled* at c and write $c[t\rangle \Leftrightarrow \bullet t \subseteq c \wedge t\bullet \cap c = \varnothing$.

Regarding S-elements as atomic propositions, then $\bullet t$ represents the set of preconditions of t and $t\bullet$ represents the set of postconditions of t. If c is a set of propositions which are true at some stage in the evolution of the system represented by N, then $\bullet t \subseteq c$ says that the preconditions of t hold at that stage. Briefly, all the preconditions of t hold in the case c. Given that the effect of an atomic event t is to cause the preconditions $\bullet t$ to cease to hold and the postconditions $t\bullet$ to begin to hold, then in order that t may occur in case c, we must not only have $\bullet t \subseteq c$ but also $t\bullet \cap c = \varnothing$, since if $p \in t\bullet \cap c$, then p cannot become true as a result of an occurrence of t- it is true already.

16.2.3. DEFINITION. ([23]). A *condition/event system* (C/E system) is a quadruple $\Sigma = (B,E;F,C)$, where

- $N_\Sigma = (B,E;F)$ is a *simple net*, that is a net for which

$$\forall x,y \in B \cup E{:}\bullet x = \bullet y \wedge x\bullet = y\bullet \Rightarrow x = y$$

The elements of B are called *conditions*. The elements of E are called *events*.

1• $C \subseteq \wp(B)$ is a non-empty set of *cases*;

2• $\forall e \in E\, \exists c \in C{:}c[e\rangle$.

3• C is an equivalence class of the full reachability relation R, where

(a) $R = (R_1 \cup R_1^{-1})^*$;[5]

(b) $\forall X,Y \in \wp(B){:}X\,R_1\,Y \Leftrightarrow (\exists G \subseteq E{:}G \neq \varnothing \wedge X[G\rangle Y)$

(c) $\forall X,Y \in \wp(B){:}X\,[G\rangle\,Y \Leftrightarrow (Y - X = \bigcup_{e\in G} e\bullet) \wedge (X - Y = \bigcup_{e\in G} \bullet e)$
$\wedge(\forall e_1,e_2 \in G{:}e_1 \neq e_2 \Rightarrow (\bullet e_1 \cup e_1\bullet) \cap (\bullet e_2 \cup e_2\bullet) = \varnothing)$.

Note that if $X[G\rangle Y$ and $e \in G$, then by (3c), $\bullet e \subseteq X - Y \subseteq X$ and $e\bullet \cap X \subseteq (X-Y) \cap X = \varnothing$, so $X[e\rangle$.

---

[5]If $R \subseteq X \times X$ then $x\,R^{-1}\,y \Leftrightarrow y\,R\,x$ and $x\,R^*\,y \Leftrightarrow x = y$ or there exists $x_1,\dots,x_n$ such that $x = x_1 R \cdots R x_n = y$.

Suppose that $e\bullet \cap \bullet e \neq \varnothing$, then if $c \in C$ with $\bullet e \subseteq c$ then $\bullet e \cap e\bullet \subseteq c \cap e\bullet$ and so $e\bullet \cap c \neq \varnothing$, giving $\neg c\, [e>$. But this contradicts 16.2.3 (2). Hence for every $e \in E$, $e\bullet \cap \bullet e = \varnothing$. We shall use this property of C/E systems repeatedly in what follows.

Since C is an equivalence class, it is determined by any of its elements. Thus, a C/E system may be represented diagrammatically by a picture of its underlying net together with some indication of the elements of one of its cases. The standard way of doing the latter is to place a single token or marker into each condition of the chosen case as in figure 16.1.

We now turn to the dynamics of C/E systems. Any piece of behaviour of a C/E system will consist of a partial ordering of condition holdings and transition firings. The partial order determines a flow relation on the holdings and firings. A piece of behaviour therefore determines a net whose T-elements represent firings of T-elements of the original net and whose S-elements represent holding of the S-elements of the original net. This behaviour net is labelled to show which firings are of which T-elements and which holdings are of which S-elements. The result is a *process* of the net. We need some definitions.

16.2.4. DEFINITION. $N=(S,T;F)$ is a *Marked-graph Net* (MG Net) $\Leftrightarrow$ $\forall s \in S: |s\bullet| \leq 1 \wedge |\bullet s| \leq 1$.

MG nets contain no *conflicts*, conditions with more than one output event or more than one input event. Figure 16.1 represents an MG net.

16.2.5. DEFINITION. $N=(S,T;F)$ is a *acyclic* $(S \cup T, F^*)$ is a partially ordered set

Figure 16.2, later, pictures an acyclic net; essentially, an acyclic net is one containing no cycles. The net is also an MG net, and hence a process net, from the following.

16.2.6. DEFINITION. $N=(S,T;F)$ is a *Process Net* $\Leftrightarrow$ N is an acyclic MG Net and $|T| < \infty$.

If N is an acyclic MG net, then define

$min(N) = \{x \in S \cup T \mid \bullet x = \varnothing\}$ and
$max(N) = \{x \in S \cup T \mid x\bullet = \varnothing\}$.

16.2.7. DEFINITION. If $N=(S,T;F)$ is a net and $c \subseteq S$, then a *process* of N from c is a pair $(\hat{N},\lambda)$, where $\hat{N}=(\hat{S},\hat{T};\hat{F})$ is a process net and $\lambda: \hat{S} \cup \hat{T} \to S \cup T$ satisfies

1• $\lambda(\hat{S}) \subseteq S$ and $\lambda(\hat{T}) \subseteq T$;

2• $\forall t \in T: \lambda(\bullet t) = \bullet\lambda(t) \wedge \lambda(t\bullet) = \lambda(t)\bullet$

3• $\forall s_1, s_2 \in \hat{S}: \lambda(s_1) = \lambda(s_2) \Rightarrow s_1\ F^*\ s_2 \vee s_2\ F^*\ s_1$;

4• $\forall x \in min(N): (x \in T \Rightarrow \bullet\lambda(x) = \varnothing) \wedge (x \in S \Rightarrow \lambda(x) \in c)$.

Intuitively, $(\hat{N}, \lambda)$ describes a non-sequential execution of N starting from case $c \subseteq S$. Each $\hat{s} \in \hat{S}$ represents a single holding of condition $\lambda(\hat{s})$ and each $\hat{t} \in \hat{T}$ represents a single occurrence of the event $\lambda(\hat{t})$. If $\hat{s}\ \hat{F}\ \hat{t}$, then condition $\lambda(\hat{s})$ ceases to hold on the occurrence of the event $\lambda(\hat{t})$. If $\hat{t}\ \hat{F}\ \hat{s}$, then condition $\lambda(\hat{s})$ begins to hold on the firing of $\lambda(\hat{t})$.

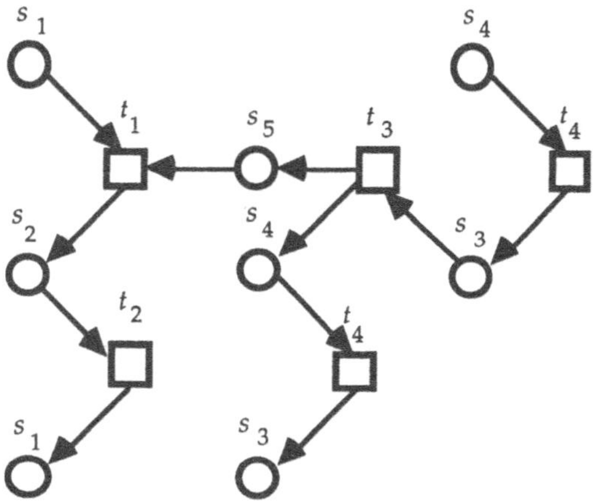

Figure 16.2

Figure 16.2 pictures a process of the net of figure 16.1 from the initial case $\{s_1, s_4\}$. Each S-element and T-element is labelled by its image under λ.

Of course, there may be several processes describing the same execution - take a process and consistently rename conditions and events in it. However, all such processes will be reciprocally isomorphic.

16.2.8. DEFINITION. Let $N_i = (S_i, T_i; F_i)$, $i = 1, 2$, be nets. An *isomorphism* from N_1 to N_2 is a bijective function $\phi: S_1 \cup T_1 \rightarrow S_2 \cup T_2$ such that

1• $\phi(S_1) = S_2$ and $\phi(T_1) = T_2$;

2• $\forall x, y \in S_1 \cup T_1: x\ F_1\ y \Leftrightarrow \phi(x)\ F_1\ \phi(y)$.

Write $\phi: N_1 \equiv N_2$ to indicate that ϕ is an isomorphism from N_1 to N_2. Write $N_1 \equiv N_2$ to indicate that $\phi: N_1 \equiv N_2$, for some ϕ.

Thus, if we draw two pictures of isomorphic labelled nets and erase the names of places and transitions, leaving just the labels, then the resulting pictures will be (structurally) identical.

16.2.9. DEFINITION. Let $P_i = (N_i, \lambda_i)$, $i = 1,2$, be processes of the net $N = (S,T;F)$ from $c \subseteq S$. An isomorphism from P_1 to P_2 is an isomorphism $\phi: N_1 \equiv N_2$ such that

$$\forall x \in S_1 \cup T_1 : \lambda_1(x) = \lambda_2(\phi(x)).$$

Given a C/E system, $\Sigma = (S,T;F,C)$, the net-based behavioural semantics of Σ may be described to be the map that sends each $c \in C$ to the class $Proc(N_\Sigma, c)/\equiv$, where $Proc(N_\Sigma, c)$ denotes the class of all processes of N_Σ from c and $Proc(N_\Sigma, c)/\equiv$ denotes the class of all isomorphism classes of $Proc(N_\Sigma, c)$.

In chapter 17 we shall show how we may construct representatives of each isomorphism class from a time point in a behavioural presentation determined by an asynchronous transition system constructed from Σ.

16.3. Occurrence Net Semantics.

A set of processes may be used to give a complete description of the behaviours of a C/E system. An alternative representation is proposed in [20], from which we cite the following

> 'It has been argued that the process level semantics of a transition net is the class of causal nets[6] it unfolds into, where all the choices associated with such an unfolding are made by the environment. However, from a computational point of view, we would prefer to deal with conflicts in one semantic object.'

We describes the objects in question below in 16.3.1. First, some notation. Suppose $t_1, t_2 \in T$. Define

$$t_1 \#_N^{(1)} t_2 \Leftrightarrow t_1 \neq t_2 \wedge {}^\bullet t_1 \cap {}^\bullet t_2 \neq \emptyset$$
$$t_1 \#_N t_2 \Leftrightarrow \exists t_1', t_2' \in T : t_1' F^* t_1 \wedge t_2' F^* t_2 \wedge t_1' \#_N^{(1)} t_2'.$$

16.3.1. DEFINITION. A net $N = (S,T;F)$ is an *occurrence net* $\Leftrightarrow$

[6] That is, processes.

1• F^* is a partial order;

2• $\forall s \in S: |\bullet s| \leq 1$;

3• $\#_N$ is irreflexive.

Occurrence nets may be thought of as asynchronous generalisations of tree-like objects used in branching-time models of systems or of the synchronisation trees described by Milner in [11]. Note that we do not require that $|s\bullet| \leq 1$ for $s \in S$, that is, forward branching is possible.

In order to define occurrence nets of C/E systems, we need to define an order relation on.

16.3.2. DEFINITION. If N_1 and N_2 are nets, then N_1 is a *prefix* of N_2 $\Leftrightarrow$

1• $S_1 \subseteq S_2$, $T_1 \subseteq T_2$, and $F_1 = F_2 \cap ((S_1 \times T_1) \cup (T_1 \times S_1))$;

2• $\forall x \in S_1 \cup T_1\ \forall y \in S_2 \cup T_2 : y\ F_2\ x \Rightarrow y \in S_1 \cup T_1$

Write $N_1 \leq N_2$ to indicate that N_1 is a prefix of N_2.

Suppose $\tilde{N}$ is an occurrence net and let $s_1, s_2 \in \tilde{S}$. Define $s_1 \#_{\tilde{N}} s_2$ $\Leftrightarrow$ whenever $\overline{N} \leq \tilde{N}$ and $\overline{N}$ is a process net, then $\{s_1, s_2\} \not\subseteq \overline{S}$.

16.3.3. DEFINITION. Suppose that $\Sigma = (S, T; F, C)$ is a C/E system and $c \in C$, then an occurrence net of Σ from c is a pair $(\tilde{N}, \tilde{\lambda})$ such that

1• $\tilde{N}$ is an occurrence net;

2• $\tilde{\lambda} : \tilde{S} \cup \tilde{T} \to S_\Sigma \cup T_\Sigma$;

3• $\tilde{\lambda}(\tilde{S}) \subseteq S_\Sigma$ and $\tilde{\lambda}(\tilde{T}) \subseteq T_\Sigma$;

4• $\forall t \in \tilde{T} : \tilde{\lambda}(t\bullet) = \tilde{\lambda}(t)\bullet \wedge \bullet\tilde{\lambda}(t) = \bullet\tilde{\lambda}(t)$;

5• $\forall s_1, s_2 \in \tilde{S} : \tilde{\lambda}(s_1) = \tilde{\lambda}(s_2) \Rightarrow s_1\ \tilde{F}^*\ s_2 \vee s_2\ \tilde{F}^*\ s_1 \vee s_1 \#_{\tilde{N}} s_2$;

6• $\forall x \in min(\tilde{N}) : (x \in \tilde{T} \Rightarrow \bullet x = \emptyset) \wedge (x \in \tilde{S} \Rightarrow \bullet\tilde{\lambda}(x) \in c)$.

One of the stated aims of [20] was to reconcile Petri's theory of concurrency with the domain theory associated with Scott's theory of computation. The link between the two is the concept of event structure that we met in chapter 4.

If $N = (S, T; F)$ is an occurrence net. Define $Es(N) = (T, F^* \cap (T \times T), \#_N)$. It is trivial that

16.3.4. PROPOSITION. If N is an occurrence net then $Es(N)$ is an event structure.

■

Finally, we show how processes may be extracted from occurrence nets.

16.3.5. PROPOSITION. If $(\tilde{N},\tilde{\lambda})$ an occurrence net of Σ from c and $\overline{N} \leq \tilde{N}$ satisfies

1• $|\overline{T}| < \infty$;

2• If $\forall t \in \overline{T}\ \forall s \in \tilde{S}: t\ \tilde{F}\ s \Rightarrow t\ \overline{F}\ s$;

3• $\forall t_1, t_2 \in \overline{T}: \neg t_1 \#_{\tilde{N}}\ t_2$.

then $(\overline{N},\overline{\lambda})$ is a process of Σ from c, where $\forall x \in \overline{S} \cup \overline{T}: \overline{\lambda}(x) = \tilde{\lambda}(x)$.

PROOF. If $s \in \overline{S}$, then $t\ \overline{F}\ s \Rightarrow t\ \tilde{F}\ s$, by 16.3.2 (1), and so $|\bullet s| \leq 1$ in $\overline{N}$ as $|\bullet s| \leq 1$ in $\tilde{N}$, by 16.3.1 (2). If $s\ \overline{F}\ t_1$ and $s\ \overline{F}\ t_2$ with $t_1 \neq t_2$ then $s\ \tilde{F}\ t_1$ and $s\ \tilde{F}\ t_2$ with $t_1 \neq t_2$, by 16.3.2 (2) and so $t_1 \#_{\tilde{N}}\ t_2$, contradicting (3). Hence $|s\bullet| \leq 1$ in $\overline{N}$ and we have shown that is an MG net.

Since $\tilde{F}^*$ is a partial order, by 16.3.1 (1), and $\overline{F}^* \subseteq \tilde{F}^*$, $\overline{F}^*$ is a partial order and so $\overline{N}$ is acyclic, by 16.2.5. Since $|\overline{T}| < \infty$, by (1), $\overline{N}$ is a process net, by 16.2.6.

16.2.7 (1) follows from 16.3.3 (2) and 16.3.2 (1).

Let $t \in \overline{T}$, then $\bullet t$ in $\overline{N}$ equals $\bullet t$ in $\tilde{N}$, by 16.3.2 (2) and $t\bullet$ in $\overline{N}$ equals $t\bullet$ in $\tilde{N}$, by (2), so 16.2.7 (2) follows from 16.3.3 (4).

If $s_1, s_2 \in \overline{S}$, then $s_1 \#_{\tilde{N}}\ s_2$ cannot hold, so 16.2.7 (3) follows from 16.3.3 (5).

Finally, suppose that $x \in min(\overline{N})$, then $x \in min(\tilde{N})$. If $x \in \overline{T}$ then $x \in \tilde{T}$ and so $\bullet x = \varnothing$ in $\tilde{N}$, by 16.3.3 (6), and so $\bullet x = \varnothing$ in $\overline{N}$, by 16.3.2. If $x \in \overline{S}$ then $x \in \tilde{S}$ and so $\tilde{\lambda}(x) \in c$, by 16.3.3 (6), and so $\overline{\lambda}(x) \in c$, by definition of $\overline{\lambda}$ ■

Notes.

There is an extensive literature on Net Theory. For aspects of the semantics of C/E systems, there are comprehensive accounts in [18, 23, 60]. Our treatment of processes is largely taken from [60] with one slight modification.

Chapter 17

Condition/Event Systems: Asynchronous Transition System Semantics

17.1. Introduction.

In this chapter, we apply our theory of behavioural presentations and asynchronous transition systems to the C/E systems that we considered in the previous chapter. For a given C/E system $\Sigma=(S,T;F,C)$, we construct an asynchronous transition system C_Σ with states C. For any case $c \in C$, we may therefore construct a left-closed trace language $L(C_\Sigma,c)$, using 15.2.2 and from this, a linguistic behavioural presentation $B_{(\Sigma,c)} = B_{(C_\Sigma,c)}$, by 15.2.6. Therefore, we have a semantic function which maps Σ to $B_{(\Sigma,c)}$.

This is all very well, but C/E systems have a semantics already, if not two. What, we may ask is the relationship between this behavioural presentation semantics and that involving process nets and occurrence nets that we have already met.

Roughly speaking, the answer is as follows. Each process of Σ from c determines a labelled partial order of its T-elements; it transpires that this partial order is label-isomorphic to some $p \in P_{(\Sigma,c)}$, considering the latter as partially ordered by $\rightarrow_{(\Sigma,c)}$ and labelled by $\lambda_{(\Sigma,c)}$.

Conversely, given $p \in P_{(\Sigma,c)}$, we may regard its elements as occurrences of T-elements of Σ and we may then use the structure of N_Σ to augment p with S-elements giving rise to a process $Proc_\Sigma(p)$ of Σ from c. We shall show that every process of Σ from c is isomorphic to exactly one of the $Proc_\Sigma(p)$. Thus, our construction allows us to construct a complete set of representatives of the isomorphism classes of processes of Σ from c. Not only that, but there is an additional advantage that it preserves ordering; there is an ordering $\leq$ on processes such that if $p_1, p_2 \in P_{(\Sigma,c)}$, then $p_1 \subseteq p_2 \Leftrightarrow Proc_\Sigma(p_1) \leq Proc_\Sigma(p_2)$.

Furthermore, we may form the union of the $Proc_\Sigma(p)$ and obtain an occurrence net of Σ from c. Thus, among other things, our construction serves several purposes:

•1 It reconciles the various semantics by showing that each may be recovered from any other;
•2 It allows us to construct a canonical process/occurrence nets semantics.

17.2. Asynchronous Transition System Semantics.

We begin by constructing a mapping from C/E systems to asynchronous transition systems. For the transition rule, we use that given in 16.2.3, defined there for sets of transitions. If $\Sigma=(S,T;F,C)$ is a C/E system, $c_1,c_2 \subseteq S$ and $t \in T$, then we define:

$$c_1\,[t\rangle\, c_2 \Leftrightarrow c_1\,[t\rangle \wedge c_2 = (c_1 - {\bullet}t) \cup t{\bullet} \tag{17.1}$$

where $c_1\,[t\rangle$ is given in 16.2.2. This is sometimes known as the *'safe firing rule'*. Pictorially, in passing from case c_1 to case c_2 one removes a token from each precondition of t and places a token on each postcondition of t.

17.2.1. DEFINITION. Let $\Sigma=(S,T;F,C)$ be a C/E system. Define $C_\Sigma=(C,T,\rightarrow,\iota)$, where

1• If $c_1,c_2 \in C$ and $t \in T$, then $c_1 \rightarrow^t c_2 \Leftrightarrow c_1\,[t\rangle\, c_2$;
2• If $t_1,t_2 \in T$, then $t_1\,\iota\,t_2 \Leftrightarrow ({\bullet}t_1 \cup t_1{\bullet}) \cap ({\bullet}t_2 \cup t_2{\bullet}) = \varnothing$.

17.2.2. PROPOSITION. If $\Sigma=(S,T;F,C)$ is a C/E system, then C_Σ is an asynchronous transition system.

PROOF. $c \rightarrow^t c_1 \wedge c \rightarrow^t c_2 \Rightarrow c\,[t\rangle\, c_1 \wedge c[t\rangle\, c_2 \Rightarrow c_1 = (c_1 - {\bullet}t) \cup t{\bullet} = c_2$, so $T(C_\Sigma)$ is unambiguous.

It is trivial that ι is an independence relation.

It remains to establish that the 'lozenge rule' holds. We must show that if $c_1,c_2,c_3 \in C$ and $t_1,t_2 \in T$, then

$$c_1\,[t_1\rangle\, c_2 \wedge c_2\,[t_2\rangle\, c_3 \wedge t_1\,\iota\,t_2 \Rightarrow (\exists \tilde{c}_2 \in C{:}\, c_1\,[t_2\rangle\, \tilde{c}_2 \wedge \tilde{c}_2\,[t_1\rangle\, c_3.$$

Let $s \in {\bullet}t_2$, then as $c_2\,[t_2\rangle$, by (17.1), $s \in c_2 = (c_1 - {\bullet}t_1) \cup t_1{\bullet}$ by 16.2.2. As $t_1\,\iota\,t_2$, $s \notin t_1{\bullet}$, by 17.2.1 (2), so $s \in ((c_1 - {\bullet}t_1) \cup t_1{\bullet}) - t_1{\bullet} \subseteq c_1$. We have proved that ${\bullet}t_2 \subseteq c_1$.

Next suppose that $s \in c_1 \cap t_2\bullet$, then $s \notin \bullet t_1$, by 17.2.1 (2), as $t_1 \iota t_2$. Hence, $s \in c_1 - \bullet t_1 \subseteq (c_1 - \bullet t_1) \cup t_1\bullet = c_2$. Since $s \in t_2\bullet$, it follows that $s \in c_2 \cap t_2\bullet$. But, $c_2 \cap t_2\bullet = \varnothing$, by 16.2.2, because $c_2\, [t_2\rangle$. From this contradiction, we deduce that $c_1 \cap t_2\bullet = \varnothing$. And, as $\bullet t_2 \subseteq c_1$, it follows that $c_1\, [t_2\rangle$.

Thus, if we define $\tilde{c}_2 = (c_1 - \bullet t_2) \cup t_2\bullet$, then $c_1\, [t_2\rangle\, \tilde{c}_2$, by (17.1). Note that $\tilde{c}_2 \in C$, by 16.2.3.

Suppose that $s \in \bullet t_1$, then $s \in c_1$, since $c_1\, [t_1\rangle$. Also $s \notin \bullet t_2$, by 17.2.1 (2), because $t_1 \iota t_2$. Hence $s \in c_1 - \bullet t_2 \subseteq \tilde{c}_2$. and we have shown that $\bullet t_1 \subseteq \tilde{c}_2$.

Next, suppose that $s \in \tilde{c}_2 \cap t_1\bullet$. Since $t_1 \iota t_2$and $s \in t_1\bullet$, $s \notin t_2\bullet$ and so $s \in \tilde{c}_2 - t_2\bullet = ((c_1 - \bullet t_2) \cup t_2\bullet) - t_2\bullet \subseteq c_1$. Since $s \in t_1\bullet$, it follows that $s \in c_1 \cap t_1\bullet$. But $c_1 \cap t_1\bullet = \varnothing$ by 16.2.2, because $c_1\, [t_1\rangle$. From this contradiction, we deduce that $c_2 \cap t_1\bullet = \varnothing$. And, as $\bullet t_1 \subseteq \tilde{c}_2$, it follows that $\tilde{c}_2\, [t_1\rangle$.

Finally, note that since $(\bullet t_1 \cup t_1\bullet) \cap (\bullet t_2 \cup t_2\bullet) = \varnothing$,

$$\begin{aligned}(\tilde{c}_2 - \bullet t_1) \cup t_1\bullet &= (((c_1 - \bullet t_2) \cup t_2\bullet) - \bullet t_1) \cup t_1\bullet \\ &= (((c_1 - \bullet t_1) \cup t_1\bullet) - \bullet t_2) \cup t_2 = (c_2 - \bullet t_2) \cup t_2\bullet = c_3\end{aligned}$$

and so $\tilde{c}_2\, [t_1 > c_3$.

■

C_Σ has another property, corresponding to forward stability in behavioural presentations and left-closed trace languages.

17.2.3. DEFINITION. An asynchronous transition system C is forward stable $\Leftrightarrow$ for all $q, q_1, q_2 \in Q$ and $a, b \in A$

$$q \to^a q_1 \wedge q \to^b q_2 \wedge a \iota b \Rightarrow \exists q' \in Q : q_2 \to^a q' \wedge q_1 \to^b q'.$$

It is immediate from 12.3.1 (2) that C is forward stable if and only if $L(C,q)$ is forward stable for every $q \in Q$.

17.2.4. PROPOSITION. Let $\Sigma = (S,T;F,C)$ be a C/E system, then C_Σ is forward stable.

PROOF. Suppose that $c, c_1, c_2 \in C$ and $t_1, t_2 \in T$ such that $c\, [t_1\rangle\, c_1$, $c\, [t_2\rangle\, c_2$ and $t_1 \iota t_2$, then $\bullet t_1, \bullet t_2 \subseteq c$, by (17.1) and 16.2.2, and $\bullet t_1 \cap \bullet t_2 = \varnothing$, by 17.2.1 (2). Thus, $\bullet t_2 \subseteq (c - \bullet t_1) \cup t_1\bullet = c_1$. Also, $c \cap t_1\bullet = c \cap t_2\bullet = \varnothing$, by (17.1) and 16.2.2, and $t_1\bullet \cap t_2\bullet = \varnothing$, by 17.2.1 (2), and so $t_2\bullet \cap c_1 = t_2\bullet \cap ((c - \bullet t_1) \cup t_1\bullet) = \varnothing$. Since $\bullet\, t_2 \subseteq c_1$, it follows that $c_1\, [t_2\rangle$. By symmetry, $c_2\, [t_1\rangle$.

Finally, since $(\bullet t_1 \cup t_1\bullet) \cap (\bullet t_2 \cup t_2\bullet) = \varnothing$,

$$\begin{aligned}(c_1 - \bullet t_2) \cup t_2\bullet &= (((c - \bullet t_1) \cup t_1\bullet) - \bullet t_2) \cup t_2\bullet \\ &= (((c - \bullet t_2) \cup t_2\bullet) - \bullet t_1) \cup t_1\bullet = (c_2 - \bullet t_1) \cup t_1\bullet\end{aligned}$$

and so if we define $\tilde{c} = (c_1 - \bullet t_2) \cup t_2\bullet$, then $c_1\, [t_2\rangle\, \tilde{c}$ and $c_2\, [t_1\rangle\, \tilde{c}$.

■

The construction of 17.2.1 is illustrated in figure 17.1, which shows the asynchronous transition system determined by the C/E system determined by the marked net of figure 16.1. In the figure the node marked $i_1 \cdots i_k$ represents the case $\{s_{i_1}, \cdots, s_{i_k}\}$. Essentially, the transition system consists of four oriented rectangles connected at their edges.

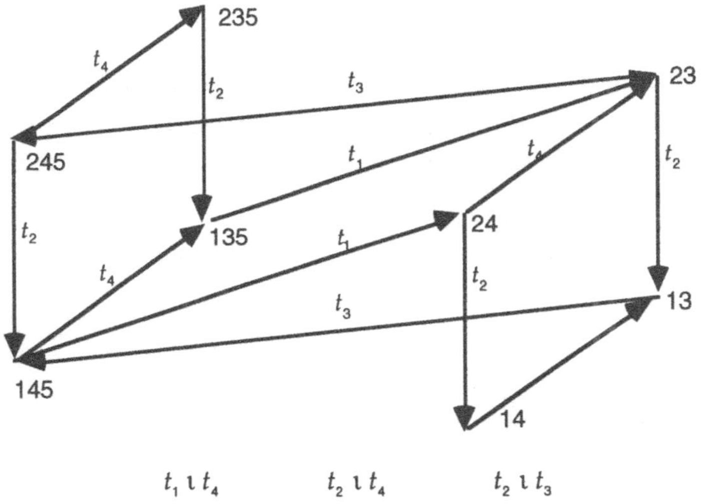

Figure 17.1.

17.3. Comparison of the Semantics: Preliminaries.

A C/E system Σ determines a forward stable asynchronous transition system $C_\Sigma = (C, T, \rightarrow, \iota)$ and hence, for any $c \in C$, an ι-linguistic behavioural presentation, which we shall denote by $B_{(\Sigma,c)}$. $B_{(\Sigma,c)}$ represents a class of partial orders of transition firings. $Proc(N_\Sigma, c)$, on the other hand, is a class of partial orders of transition firings and condition holdings. From $Proc(N_\Sigma, c)$, we may obtain a set of partial orders of transition firings very easily - just throw away the conditions.

Suppose $\tilde{P} = ((\tilde{S}, \tilde{T}; \tilde{F}), \tilde{\lambda}) \in Proc(N_\Sigma, c)$, then $(\tilde{S} \cup \tilde{T}, \tilde{F}^*)$ is a partially ordered set, by 16.2.6, and so if we define

$$po_T(\tilde{P}) = (\tilde{T}, F^* \cap (\tilde{T} \times \tilde{T}), \tilde{\lambda} \mid_{\tilde{T}})$$

then $po_T(\tilde{P})$ is a labelled poset. Define

$$Proc_T(\Sigma, c) = \{po_T(\tilde{P}) \mid \tilde{P} \in Proc(N_\Sigma, c)\}.$$

Now, each $p \in P_{(\Sigma,c)}$ may also be regarded as a labelled poset[1] since it inherits a partial order from $\rightarrow_{(\Sigma,c)}$ and a labelling function from $\lambda_{(\Sigma,c)}$. Thus, we may compare the 'official' semantics of C/E systems with our asynchronous transition system/linguistic behavioural presentation semantics by comparing two classes of partial orders labelled by T, namely $Proc_T(\Sigma, c)$ and $P_{(\Sigma,c)}$.

For, example if $c = \{s_1, s_4\}$ in the examples of figures 16.1 and 17.1, then $L(C_\Sigma, c)$ contains the trace $(t_4 t_3 t_4 t_1 t_2)_\iota$, which determines the labelled poset of figure 17.2. This is clearly isomorphic to $po_T(\tilde{P})$, where $\tilde{P} \in Proc(N_\Sigma, c)$ is that member of the isomorphism class of processes pictured in figure 16.2.

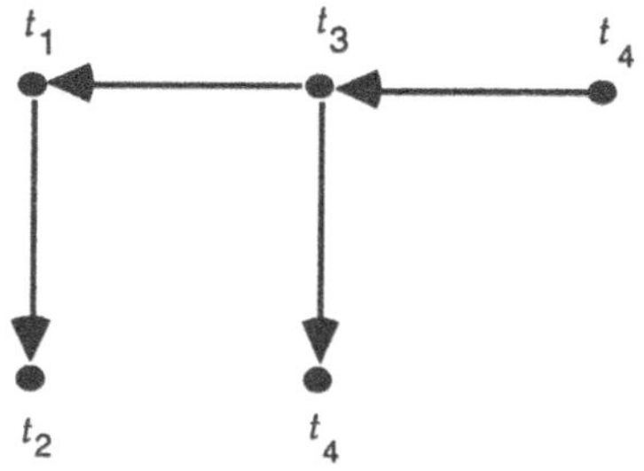

Figure 17.2.

We shall establish such a connection for a particular class of C/E systems, those which are contact-free. $\Sigma = (S, T; F, C)$ is *contact-free* $\Leftrightarrow$

$$\forall c \in C \; \forall t \in T: {}^\bullet t \subseteq c \Rightarrow t^\bullet \cap c = \varnothing$$

It follows that for a contact free system,

$$\forall c \in C \; \forall t \in T: {}^\bullet t \subseteq c \Leftrightarrow c[t\rangle$$

The reason why we restrict ourselves to contact-free C/E systems is that C/E systems without this property fail to have a reasonable process semantics. Let us consider a simple example, which is pictured in figure 17.3.

[1]In the sequel we shall identify p with this labelled poset.

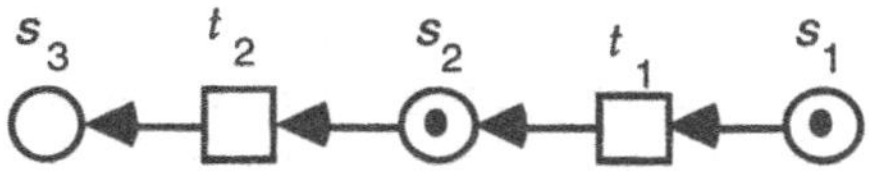

Figure 17.3.

Note that $t_1\, F^*\, t_2$ but that t_2 happens first. The reason is that there is a 'contact situation', that is $\bullet t_1 \subseteq \{s_1, s_2\}$ and $t_1\bullet \cap \{s_1, s_2\} = \varnothing$, and hence t_1 cannot fire until the firing of t_2 removes the token from s_2. Because the dynamic structure of this system is somewhat at odds with its static structure, there is no process of this C/E system corresponding to the behaviour $t_2 t_1$. There is no place which is an input place of t_1 and an output place of t_2 which would establish the correct ordering.

This is a little disappointing, (but only for Net Theorists). We experience no such difficulty in handling C/E systems with contact.[2]

However, if $\Sigma = (S, T; F, C)$ is a contact-free C/E system, $c \in C$, then we may show the following:

F1: Every $p \in P_{(\Sigma,c)}$, with $p \neq \varnothing$, considered as a labelled poset, equals $po_T(\tilde{P})$, for some $\tilde{P} \in Proc(N_\Sigma, c)$. In other words, every $p \in P_{(\Sigma,c)}$ describes the partial ordering of transition firings of some process of N_Σ from c.

F2: For every $P_1 \in Proc(N_\Sigma, c)$, there exists $P_2 \in Proc(N_\Sigma, c)$ such that $P_1 \equiv P_2$ and $T_2 \in P_{(\Sigma,c)}$. In other words, the partial order of transition firings of each process of N_Σ from c is described, up to isomorphism, by some $p \in P_{(\Sigma,c)}$

F1 and **F2** describe the basics of the relationships between the two semantics. Essentially, each may be recovered from the other. Note, however, that in going from $B_{(\Sigma,c)}$ to $Proc(N_\Sigma, c)$ we need to refer to N_Σ. Diagrammatically:

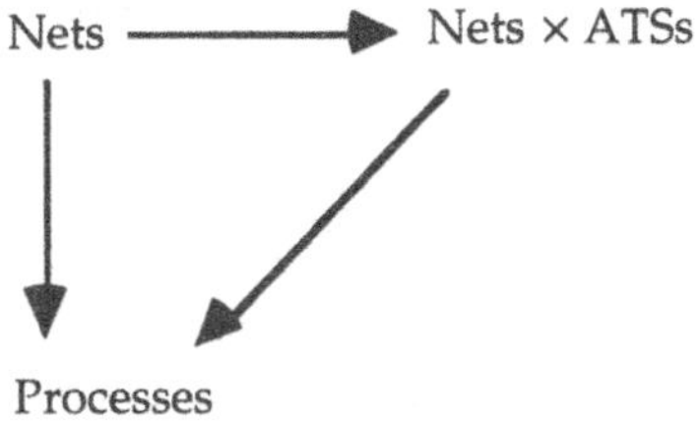

[2]It should be pointed out that the technique of place-complementation, which we discuss in chapter 21, always gives rise to a contact free C/E system having exactly the same behaviour with respect to transition firings. This is how process semantics are defined for arbitrary C/E systems in [23].

We shall look at **F1** first, but before that, we adopt the following notation, which will avoid an uncomfortable plethora of subscripts. As long as the C/E system Σ is clear from context, we shall write $\underline{u}$ for $u_{\iota_{N_\Sigma}}$, $u \in T_\Sigma{}^*$.

17.4. From Behavioural Presentations to Processes ...

We begin with a construction that takes a non-empty point in $P_{(\Sigma,c)}$ and turns it into a process of Σ from a case c the element of the point become T-elements and by inserting S-elements having one of the following forms. An S-element of the form (Ω, s) represents the initial holding of a condition s which is a precondition for the occurrence of some element of the point. An S-element of the form $(\underline{u}, s)$ represents the initial holding of a condition s as a consequence of the occurrence $\underline{u}$.

17.4.1. DEFINITION. Let $\Sigma = (S,T;F,C)$ be a C/E system and $c \in C$. If $p \in P_{(\Sigma,c)}$, with $p \neq \varnothing$, then let $Proc_\Sigma(p) = ((\tilde{S},\tilde{T};\tilde{F}),\tilde{\lambda})$, where:

1• $\tilde{T} = p;$

2• $\tilde{S}_\Omega = \{(\Omega, s) \in \{\Omega\} \times c \mid \exists \underline{u} \in p : s \in \bullet\lambda_{(\Sigma,c)}(\underline{u})\};$

3• $\tilde{S}_+ = \{(\underline{u}, s) \in p \times S \mid s \in \lambda_{(\Sigma,c)}(\underline{u})\bullet\};$

4• $\tilde{S} = \tilde{S}_\Omega \cup \tilde{S}_+;$

5• $\underline{u}\, F\, (\underline{u}', s) \Leftrightarrow \underline{u} = \underline{u}';$

6• $(\underline{u}, s)\, \tilde{F}\, \underline{u}' \Leftrightarrow \underline{u} < \underline{u}' \wedge s \in \bullet\lambda_{(\Sigma,c)}(\underline{u}')$
$\wedge (\forall \underline{u}'' \in p : \underline{u} < \underline{u}'' < \underline{u}' \Rightarrow s \notin \bullet\lambda_{(\Sigma,c)}(\underline{u}''));$

7• $\tilde{\lambda}(\underline{u}) = \lambda_{(\Sigma,c)}(\underline{u})$ and $\tilde{\lambda}(\underline{u}, s) = s$.

This construction is illustrated in figure 17.4, using the point $\underline{t}_4\underline{t}_3\underline{t}_4\underline{t}_1\underline{t}_2$ pictured in figure 17.2. Note that the resulting net is isomorphic to that of figure 17.1.

The same construction applied to the totally ordered set $\{\underline{t}_2, \underline{t}_2\underline{t}_1\}$ describing the behaviour of the net of figure 17.3 gives the net of figure 17.5. The reader may be able to see why this net does not describe the behaviour properly.

We shall show if $\Sigma = (S,T;F,C)$ is contact-free, then $Proc_\Sigma(p)$ is indeed a process of N_Σ from $c \in C$, providing $p \neq \varnothing$.

17.4.2. LEMMA. Suppose that $\Sigma = (S,T;F,C)$ is a C/E system and $c \in C$, then if $p \in P_{(\Sigma,c)}$ with $p \neq \varnothing$, and $Proc_\Sigma(p) = (\tilde{N},\tilde{\lambda})$, then $\tilde{N}$ is a net.

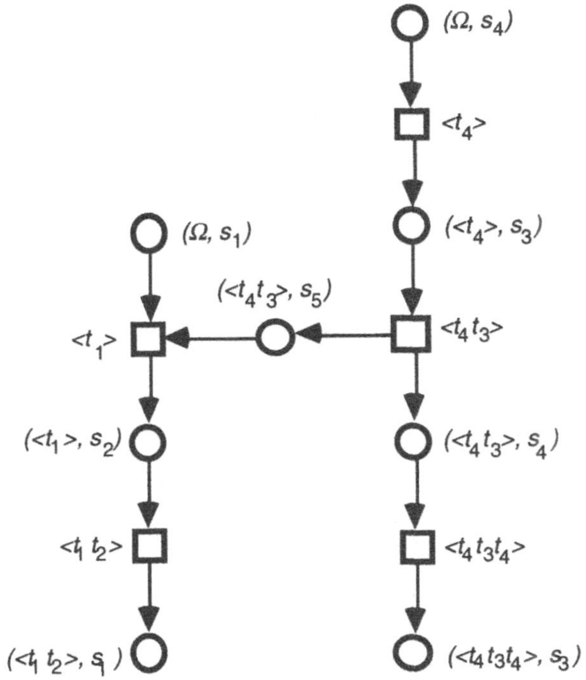

Figure 17.4.

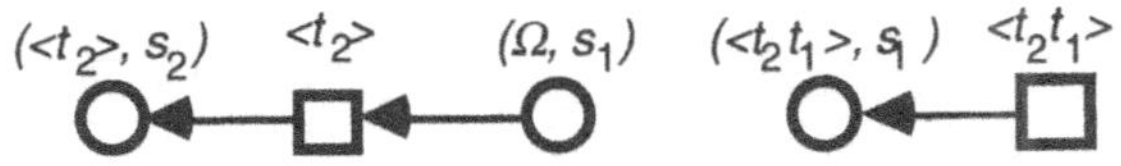

Figure 17.5.

PROOF. We check the conditions of 16.2.1. As the elements of $\tilde{S}$ are ordered pairs and the elements of $\tilde{T}$ are sets of sequences, $\tilde{S} \cap \tilde{T} = \varnothing$ and as $\tilde{T} = p \neq \varnothing$, $\tilde{S} \cup \tilde{T} \neq \varnothing$, so that (1) holds. We check (2). If $(\Omega, s) \in \tilde{S}_\Omega$ then by 17.4.1 (2), $s \in c \cap \bullet\lambda_{(\Sigma,c)}(\underline{u})$, some $\underline{u} \in p$. If we select $\underline{u}' \in p$ to be minimal with respect to this property, then $(\Omega, s)\, \tilde{F}\, \underline{u}'$, by 17.4.1 (6). Thus $\tilde{S}_\Omega \subseteq dom(\tilde{F})$. If $(\underline{u}, s) \in \tilde{S}_+$, then $\underline{u}\, \tilde{F}\, (\underline{u}, s)$, by 17.4.1 (5). Thus, $\tilde{S}_+ \subseteq ran(\tilde{F})$ and therefore $\tilde{S} \subseteq dom(\tilde{F}) \cup ran(\tilde{F})$.

If $\underline{u} \in \tilde{T}$ and $\lambda_{(\Sigma,c)}(\underline{u})\bullet \neq \varnothing$, then $\underline{u}\, \tilde{F}\, (\underline{u}, s)$, for $s \in \lambda_{(\Sigma,c)}(\underline{u})\bullet$, by 17.4.1 (5), and so $\underline{u} \in dom(\tilde{F})$. If $\underline{u} \in \tilde{T}$ and $\lambda_{(\Sigma,c)}(\underline{u})\bullet = \varnothing$, then by 16.2.1 (2), we must have $\bullet\lambda_{(\Sigma,c)}(\underline{u}) \neq \varnothing$. Suppose $s \in \bullet\lambda_{(\Sigma,c)}(\underline{u})$. If there exists $\underline{u}' \in p$ such that $\underline{u}' < \underline{u}$ and $(\underline{u}', s) \in \tilde{S}$, then for a maximal such $\underline{u}'$, $(\underline{u}', s)\, \tilde{F}\, \underline{u}$, by 17.4.1 (6) and hence $\underline{u} \in ran(\tilde{F})$. If no such $\underline{u'}$ exists, then for all $\underline{u'} \in p$ such that $\underline{u'} < \underline{u}$, $(\underline{u'}, s) \notin \tilde{S}$, and so $s \notin \lambda_{(\Sigma,c)}(\underline{u'})\bullet$, by 17.4.1 (3). Thus, if $c \rightarrow^{\underline{u'}} c_1$ and $\underline{u'} < \underline{u}$, then $s \in c_1 \Rightarrow s \in c$. So, if $\underline{u} = \underline{u'}.\underline{t}$, then $c \rightarrow^{\underline{u'}} c_1 \rightarrow^{t} \tilde{c}$. Since $s \in \bullet\lambda_{(\Sigma,c)}(\underline{u})$, it follows that $s \in c_1$ and hence $s \in c$. Thus $(\Omega, s) \in \tilde{S}_\Omega$, by 17.4.1 (2), and it is clear that $(\Omega, s)\, \tilde{F}\, \underline{u}$. Therefore, in this case also, $\underline{u} \in ran(\tilde{F})$.

We have shown that $\tilde{S} \cup \tilde{T} \subseteq dom(\tilde{F}) \cup ran(\tilde{F})$. Since the opposite inclusion always holds, the proof is complete.

■

17.4.3. LEMMA. Let $\Sigma = (S,T;F,C)$ be a contact-free C/E system and $c \in C$, and let $p \in P_{(\Sigma,c)}$ such that $p \neq \varnothing$. Let $Proc_\Sigma(p) = (\tilde{N}, \tilde{\lambda})$, and let $t_1, t_2 \in T$ and $u_1, u_2 \in T^*$ such that $\underline{w}_1 = \underline{u}_1.\underline{t}_1, \underline{w}_2 = \underline{w}_1.\underline{u}_2.\underline{t}_2 \in p$, then

$$\underline{t}_1\ ind_\iota\ \underline{u}_2 \Rightarrow \exists \tilde{s} \in \tilde{S}: \underline{w}_1\ \tilde{F}\ \tilde{s}\ \tilde{F}\ \underline{w}_2.$$[3]

PROOF. Let $\underline{w}_1, \underline{w}_2 \in p$ and $\underline{t}_1\ ind_\iota\ \underline{u}_2$, then, $\neg t_1\ \iota\ t_2$, for otherwise $\underline{t}_1 \neq \underline{t}_2$ and $\underline{u}_1.\underline{t}_1.\underline{u}_2.\underline{t}_2 = \underline{u}_1.\underline{u}_2.\underline{t}_2.\underline{t}_1$, contradicting the primality of $\underline{w}_2$. Thus, by 17.2.1 (2):

$$(\bullet t_1 \cup t_1\bullet) \cap (\bullet t_2 \cup t_2\bullet) \neq \varnothing \tag{17.2}$$

As $\underline{u}_1.\underline{u}_2.\underline{t}_1.\underline{t}_2 = \underline{w}_2 \in L(C_\Sigma, c)$, $c \rightarrow_\Sigma^{u_1.u_2} c_1 \rightarrow_\Sigma^{t_1} c_2 \rightarrow_\Sigma^{t_2} \tilde{c}$, some $c_1, c_2, \tilde{c} \in C$. By definition of $\rightarrow_\Sigma$, $t_1\bullet \subseteq c_2$ and $t_2\bullet \cap c_2 = \varnothing$, so $t_1\bullet \cap t_2\bullet = \varnothing$. Likewise, $\bullet\, t_2 \subseteq c_2$ and $\bullet\, t_1 \cap c_2 = \varnothing$, so $\bullet\, t_1 \cap \bullet t_2 = \varnothing$. Thus, from (17.2),

$$(\bullet t_1 \cap t_2\bullet) \cup (\bullet t_2 \cap t_1\bullet) \neq \varnothing \tag{17.3}$$

We prove by contradiction that $t_1\bullet \cap \bullet\, t_2 \neq \varnothing$.

Suppose that $t_1\bullet \cap \bullet\, t_2 = \varnothing$, then by (17.3), $\bullet t_1 \cap t_2\bullet \neq \varnothing$. Since $\bullet\, t_1 \subseteq c_1$, it follows that $t_2\bullet \cap c_1 \neq \varnothing$, and since $\bullet t_2 \subseteq c_2$ and $t_1\bullet \cap \bullet\, t_2 = \varnothing$, $\bullet t_2 \subseteq c_2 - t_1\bullet \subseteq (c_2 - t_1\bullet) \cup \bullet\, t_1 = c_1$. But now $\bullet t_2 \subseteq c_1$ and $t_2\bullet \cap c_1 \neq \varnothing$ and this contradicts the assumption that Σ is contact free. Thus $t_1\bullet \cap \bullet\, t_2 \neq \varnothing$, as required.

Let $s \in t_1\bullet \cap \bullet\, t_2$. We shall show that $\underline{w}_1\ \tilde{F}\ \tilde{s}\ \tilde{F}\ \underline{w}_2$, where $\tilde{s} = (\underline{w}_1, s)$.

Certainly $\underline{w}_1\ \tilde{F}\ \tilde{s}$, by 17.4.1 (5).

Finally, by choice of s, $s \in \bullet t_2 = \bullet\, \tilde{\lambda}(\underline{w}_2)$. Suppose that $\underline{w} \in p$ such that $\underline{w}_1 < \underline{w} < \underline{w}_2$, so that $\underline{w} = \underline{w}_1.\underline{u}.\underline{t}$, some $\underline{u}.\underline{t} \leq \underline{u}_2$. If $s \in \bullet\tilde{\lambda}(\underline{w})$, then

---

[3]Note that this is not the case for the example of figures 17.3 and 17.5 and so this is where we would expect to evoke the hypothesis of contact-freeness.

$\bullet\tilde{\lambda}(\underline{w}) \cap \bullet\tilde{\lambda}(\underline{w}_1) \neq \varnothing$, that is $\bullet t \cap \bullet t_1 \neq \varnothing$, so $\neg t \,\iota\, t_1$, from which it follows that $\neg \underline{t}_1 \, ind_\iota \, \underline{u}.\underline{t}$, so $\neg \underline{t}_1 \, ind_\iota \, \underline{u}_2$. But $\underline{t}_1 \, ind_\iota \, \underline{u}_2$ by hypothesis. From this contradiction we deduce that if $\underline{w} \in p$ and $\underline{w}_1 < \underline{w} < \underline{w}_2$ then $s \notin \bullet\tilde{\lambda}(\underline{w})$ and accordingly, $\tilde{s} \, \tilde{F} \, \underline{w}_2$ as required, by 17.4.1 (6).

■

17.4.4. PROPOSITION. Let $\Sigma = (S, T; F, C)$ be a contact-free C/E system, $c \in C$, and let $p \in P_{(\Sigma,c)}$ such that $p \neq \varnothing$, then $Proc_\Sigma(p)$ is a process of N_Σ from c

PROOF. Let $(\tilde{N}, \tilde{\lambda}) = Proc_\Sigma(p)$. First, we show that $\tilde{N}$ is an MG net. To begin with, we shall find it useful to observe that:

$$\forall \underline{u}_1, \underline{u}_2 \in p : \neg\tilde{\lambda}(\underline{u}_1) \,\iota_{(\Sigma,c)}\, \tilde{\lambda}(\underline{u}_2) \Rightarrow \underline{u}_1 \leq \underline{u}_2 \vee \underline{u}_2 \leq \underline{u}_1 \tag{17.4}$$

For $B_{(\Sigma,c)}$ is $\iota_{(\Sigma,c)}$-linguistic, by 12.2.9, so if $\neg\tilde{\lambda}(\underline{u}_1) \,\iota_{(\Sigma,c)}\, \tilde{\lambda}(\underline{u}_2)$ then $\neg \underline{u}_1 \, co_{(\Sigma,c)} \, \underline{u}_2$, by 12.2.8 (2). Since $\underline{u}_1, \underline{u}_2 \in p$, we cannot have $\underline{u}_1 \# \underline{u}_2$ and since $B_{(\Sigma,c)}$ is asynchronous, we cannot have $\underline{u}_1 \approx \underline{u}_2$ unless $\underline{u}_1 = \underline{u}_2$. Hence, by 2.3.8 and (12.1), either $\underline{u}_1 \leq \underline{u}_2$ or $\underline{u}_2 \leq \underline{u}_1$, proving (17.4).

Now we show that $|(\underline{u}, s)\bullet| \leq 1$. Let $(\underline{u}, s) \, \tilde{F} \, \underline{u}_1$ and $(\underline{u}, s) \, \tilde{F} \, \underline{u}_2$. By 17.4.1, $\underline{u} \leq \underline{u}_1$ and $\underline{u} \leq \underline{u}_2$ with $s \in \bullet\tilde{\lambda}(\underline{u}_1) \cap \bullet\tilde{\lambda}(\underline{u}_2)$. Since $\bullet\tilde{\lambda}(\underline{u}_1) \cap \bullet\tilde{\lambda}(\underline{u}_2) \neq \varnothing$, it follows, by 17.2.1 (2), that $\neg\tilde{\lambda}(\underline{u}_1) \,\iota\, \tilde{\lambda}(\underline{u}_2)$, that is $\neg\tilde{\lambda}(\underline{u}_1) \,\iota_{(\Sigma,c)}\, \tilde{\lambda}(\underline{u}_2)$, and so $\underline{u}_1 \leq \underline{u}_2$ or $\underline{u}_2 \leq \underline{u}_1$, by (17.4). If $\underline{u}_1 < \underline{u}_2$, then $\underline{u} < \underline{u}_1 < \underline{u}_2$ and $(\underline{u}, s) \, \tilde{F} \, \underline{u}_2$. By 17.4.1 (6), $s \notin \bullet\tilde{\lambda}(\underline{u}_1)$, a contradiction. Thus $\underline{u}_1 \not< \underline{u}_2$. Likewise $\underline{u}_2 \not< \underline{u}_1$. Hence, $\underline{u}_1 = \underline{u}_2$.

Since $\underline{u}_1 \, \tilde{F} \, (\underline{u}, s)$ and $\underline{u}_2 \, \tilde{F} \, (\underline{u}, s)$ entail $\underline{u}_1 = \underline{u} = \underline{u}_2$, by 17.4.1 (5), it follows that also $|\bullet(\underline{u}, s)| \leq 1$. Hence, $\tilde{N}$ is an MG net.

Next, we show that $\tilde{N}$ is acyclic. In fact, we shall first show that

$$\forall \underline{u}_1, \underline{u}_2 \in p : \underline{u}_1 \, \tilde{F}^* \, \underline{u}_2 \Leftrightarrow \underline{u}_1 \leq \underline{u}_2 \tag{17.5}$$

That $\tilde{N}$ is acyclic follows from (17.5). For if $x \, \tilde{F}^* \, y$ and $y \, \tilde{F}^* \, x$, with $x \neq y$, then we cannot have $x \, \tilde{F} \, y$ and $y \, \tilde{F} \, x$, since this would entail $\bullet x \cap x\bullet \neq \varnothing$ and $\bullet y \cap y\bullet \neq \varnothing$, with either $x \in T$ or $y \in T$, which we know to be impossible in a C/E system. It follows that there are two distinct transitions $\underline{u}_1$ and $\underline{u}_2$ in the 'cycle' $x \, \tilde{F} \, x_1 \, \tilde{F} \cdots x_{m-1} \, \tilde{F} \, x_m = y \, \tilde{F} \, y_1 \, \tilde{F} \cdots y_{n-1} \, \tilde{F} \, y_n = x$, so $\underline{u}_1 \, \tilde{F}^* \, \underline{u}_2$ and $\underline{u}_2 \, \tilde{F}^* \, \underline{u}_1$. But, by (17.5), this entails that $\underline{u}_1 \leq \underline{u}_2$ and $\underline{u}_2 \leq \underline{u}_1$, giving $\underline{u}_1 = \underline{u}_2$, a contradiction. Therefore, $x \tilde{F}^* y \wedge y \tilde{F}^* x \Rightarrow x = y$ and $\tilde{N}$ is acyclic.

We prove (17.5). One direction of the implication is easy. Indeed, if $(\underline{u},s) \tilde{F} \underline{u}'$, then $\underline{u} < \underline{u}'$, by 17.4.1 (6), and if $\underline{u} \tilde{F} (\underline{u}',s)$, then $\underline{u} = \underline{u}'$ by 17.4.1 (5), and so by transitivity of F^*,

$$\forall \underline{u}_1, \underline{u}_2 \in p : \underline{u}_1 \; F^* \; \underline{u}_2 \Rightarrow \underline{u}_1 \leq \underline{u}_2$$

Suppose $\underline{u}_1, \underline{u}_2 \in p$ such that $\underline{u}_1 < \underline{u}_2$, then $\underline{u}_1 = \underline{u.t}$ and $\underline{u}_2 = \underline{u.t.u'.t'}$, some $u, u' \in T^*$ and $t, t' \in T$. If $\underline{u'} \, ind_\iota \, \underline{t}$, then $\underline{u.t} \tilde{F}^* \underline{u.t.u'.t'}$ by 17.4.3. Otherwise, we may find $u'' \in T^*$ and $t'' \in T$ such that $\underline{u''} \, ind_\iota \, \underline{t}$, $\underline{u''.t''} < \underline{u'.t'}$ and $\underline{u.t.u''.t''} \in p$. By 17.4.3, $\underline{u.t} \tilde{F}^* \underline{u.t.u''.t''}$ so by induction on $|\underline{u}_2 / \underline{u}_1|$, $\underline{u.t.u''.t''} \tilde{F}^* \underline{u.t.u'.t'}$ and so $\underline{u.t} \tilde{F}^* \underline{u.t.u'.t'}$ by transitivity of F^*. This completes the proof of (17.5) and hence of the acyclicity of $\tilde{N}$.

$\tilde{T}$ is finite, since $\tilde{T} = Pr(\underline{x})$, for some $x \in T^*$, and by 12.2.6 (3), $|Pr(\underline{x})| = |\underline{x}| < \infty$.

To conclude, we show that $\tilde{N}$ is a process of N_Σ from c.

Checking the conditions of 16.2.7, we note that $\tilde{\lambda}(\tilde{S}) \subseteq S_\Sigma$ and that $\tilde{\lambda}(\tilde{T}) \subseteq T_\Sigma$, giving 16.2.7 (1).

If $\underline{u} \in p$, then $\tilde{\lambda}(\underline{u}\bullet) = \tilde{\lambda}(\{(\underline{u},s) \mid s \in \tilde{\lambda}(\underline{u})\bullet\}) = \tilde{\lambda}(\underline{u})\bullet$, giving 16.2.7 (2),

If $s \in \bullet\tilde{\lambda}(\underline{u})$, then as we have seen in the proof of 17.4.2, $(\underline{v},s) \in \bullet\underline{u}$ for some $\underline{v} < \underline{u}$. So $s = \tilde{\lambda}(\underline{v},s) \in \tilde{\lambda}(\bullet\underline{u})$. Conversely, if $s \in \tilde{\lambda}(\bullet\underline{u})$, then $(\underline{v},s) \, F \, \underline{u}$, for some $\underline{v} < \underline{u}$ such that $s \in \bullet\tilde{\lambda}(\underline{u})$. Thus, $\bullet\tilde{\lambda}(\underline{u}) = \tilde{\lambda}(\bullet\underline{u})$.

For 16.2.7 (3), suppose $s_1, s_2 \in \tilde{S}$ such that $\tilde{\lambda}(s_1) = \tilde{\lambda}(s_2)$. We thus have $\underline{u}_1, \underline{u}_2 \in p$ such that $s_1 = (\underline{u}_1, s)$ and $s_2 = (\underline{u}_2, s)$, where $s = \tilde{\lambda}(s_1)$. Since $s \in \tilde{\lambda}(\underline{u}_1)\bullet \cap \tilde{\lambda}(\underline{u}_2)\bullet$, by 17.4.1 (5), it follows that $\neg\tilde{\lambda}(\underline{u}_1) \, \iota \, \tilde{\lambda}(\underline{u}_2)$, by 17.2.1 (2). By (17.4), either $\underline{u}_1 \leq \underline{u}_2$ or $\underline{u}_2 \leq \underline{u}_1$. Suppose without loss of generality that $\underline{u}_1 \leq \underline{u}_2$. If $\underline{u}_1 = \underline{u}_2$, then $s_1 = s_2$ so, trivially, $s_1 \tilde{F}^* s_2$. Otherwise $\underline{u}_1 < \underline{u}_2$, so $\underline{u}_1 = \underline{u.t}$ and $\underline{u}_2 = \underline{u.t.u'.t'}$, some $u, u' \in T^*$ and $t, t' \in T$, where $t = \tilde{\lambda}(\underline{u}_1)$ and $t' = \tilde{\lambda}(\underline{u}_2)$. Now, $\underline{u.t.u'.t'} \in L(C_\Sigma, c)$ and so $c \rightarrow_\Sigma^{\underline{u}} c_1 \rightarrow_\Sigma^{\underline{t}} c_2 \rightarrow_\Sigma^{\underline{u'}} c_3 \rightarrow_\Sigma^{\underline{t'}} c'$, some $c_1, c_2, c_3, c' \in C$. Since $s \in \tilde{\lambda}(\underline{u}_1)\bullet \cap \tilde{\lambda}(\underline{u}_2)\bullet = t\bullet \cap t'\bullet$, it follows that $s \in c_2 \cap c'$ and $s \notin c_3$. Therefore, there exists $\underline{u''} \in T_\iota^*$ and $t'' \in T$ such that $\underline{u''.t''} \leq \underline{u'}$ and $s \in \bullet t''$. Let $\underline{w} \in p$ be minimal with respect to the property that $\underline{u}_1 < \underline{w} < \underline{u}_2$ and $s \in \bullet\tilde{\lambda}(\underline{w})$, then by 17.4.1 (6), $(\underline{u}_1, s) \tilde{F} \underline{w}$. Since $\underline{w} < \underline{u}_2$,

$\underline{w}\ \tilde{F}^* \underline{u}_2$, by (17.3). By 17.4.1 (5), $\underline{u}_2\ \tilde{F}\,(\underline{u}_2, s)$. Thus, by transitivity of F^*, $s_1 = (\underline{u}_1, s)\ \hat{\tilde{F}}^*(\underline{u}_2, s) = s_2$.

Finally, if $\underline{u} \in p$ such that $\underline{u} \in min(\tilde{N})$, then $\bullet\tilde{\lambda}(\underline{u}) = \tilde{\lambda}(\bullet\underline{u}) = \varnothing$ and hence $\bullet\,\underline{u} = \varnothing$. On the other hand, if $s \in \tilde{S}$ such that if $s \in min(\tilde{N})$, then $s \in \tilde{S}_\Omega$ and so $\tilde{\lambda}(s) \in c$. This establishes 16.2.7 (4).

■

Thus, $Proc_\Sigma$ is a mapping from $P_{(\Sigma,c)} - \{\varnothing\}$ to $Proc(N_\Sigma, c)$ satisfying $po_T(Proc_\Sigma(p)) = p$. This mapping preserves other structure as well, because we can impose a partial order on $Proc(N_\Sigma, c)$, and $Proc_\Sigma$ turns out to be a poset morphism. The ordering on processes is based on the relation on nets defined in 16.3.2. We begin by establishing a property of this relation.

17.4.5. LEMMA. $\le$ is a partial order on the class of all nets.

PROOF. Reflexivity is obvious. If $N_1 \le N_2$ and $N_2 \le N_1$, then $S_1 \subseteq S_2 \subseteq S_1$ and so $S_1 = S_2$. Likewise $T_1 = T_2$. Therefore, $F_1 = F_2 \cap ((S_2 \times T_2) \cup (T_2 \times S_2)) = F_2$ and so $\le$ is antisymmetric. Finally, if $N_1 \le N_2$ and $N_2 \le N_3$, then $S_1 \subseteq S_2 \subseteq S_3$ and so $S_1 \subseteq S_3$. Likewise $T_1 \subseteq T_3$. Hence:

$$\begin{aligned} &F_3 \cap ((S_1 \times T_1) \cup (T_1 \times S_1)) \\ &\quad = (F_3 \cap ((S_2 \times T_2) \cup (T_2 \times S_2))) \cap ((S_1 \times T_1) \cup (T_1 \times S_1)) \\ &\quad = F_2 \cap ((S_1 \times T_1) \cup (T_1 \times S_1)) = F_1 \end{aligned}$$

And if $x \in S_1 \cup T_1$ and $y \in S_3 \cup T_3$ such that $y\ F_3\ x$, then as $x \in S_2 \cup T_2$ and $N_2 \le N_3$, we have $y \in S_2 \cup T_2$, Thus, $(y, x) \in F_3 \cap ((S_2 \times T_2) \cup (T_2 \times S_2)) = F_2$. Hence, $x \in S_1 \cup T_1$ and $y \in S_2 \cup T_2$ such that $y\ F_2\ x$. As $N_1 \le N_2$, $y \in S_1 \cup T_1$ and we have, completing the proof that $\le$ is transitive.

■

17.4.6. DEFINITION. Suppose $P_1, P_2 \in Proc(N, c)$, where $P_1 = (N_1, \lambda_1)$ and $P_2 = (N_2, \lambda_2)$, then P_1 is a *prefix* of P_2 $\Leftrightarrow$

1• $N_1 \le N_2$;

2• $\forall x \in S_1 \cup T_1 : \lambda_2(x) = \lambda_1(x)$.

We write $P_1 \le P_2$ to indicate that P_1 is a prefix of P_2.

Intuitively, if P_1 and P_2 are processes of (N, c), then $P_1 \le P_2$ means that P_1 represents an initial part of the behaviour described by P_2. It is clear, from 17.4.5, that $\leq$ is a partial order on $Proc(N,c)$. It is also clear that if $P_1 \leq P_2$ and $\phi_2 : P_2 \equiv P_2'$, with $P_2' \in Proc(N,c)$ then ϕ_2 cuts down to an isomorphism $\phi_1 : P_1 \equiv P_1'$ with $P_1' \in Proc(N,c)$ and $P_1' \leq P_2'$. It is also easy enough to prove that if $P_1 \leq P_2$ and $\phi_1 : P_1 \equiv P_1'$, then there exists $P_2' \in Proc(N,c)$ such that $P_1' \leq P_2'$ and such that ϕ_1 extends to an isomorphism $\phi_2 : P_2 \equiv P_2'$ - one adjoins to P_1' a copy of that part of P_2 lying outside of P_1 and defines ϕ_2 to be the identity on that part.

It follows from this that if we define a relation $\leq$ on $Proc(N,c)/\equiv$ by

$$\mathbf{P}_1 \leq \mathbf{P}_2 \Leftrightarrow \exists P_1 \in \mathbf{P}_1\ \exists P_2 \in \mathbf{P}_2 : P_1 \leq P_2$$

then $\leq$ is a partial order on $Proc(N,c)/\equiv$

Indeed if $\mathbf{P}_1 \leq \mathbf{P}_2$ and $\mathbf{P}_2 \leq \mathbf{P}_1$ then there exists $\tilde{P}_1, \overline{P}_1 \in \mathbf{P}_1$, and $\tilde{P}_2, \overline{P}_2 \in \mathbf{P}_2$ such that $\tilde{P}_1 \leq \tilde{P}_2$ and $\overline{P}_2 \leq \overline{P}_1$. As $\tilde{P}_2 \equiv \overline{P}_2$, we can find $P_1' \equiv \overline{P}_1$ such that $\tilde{P}_2 \leq P_1'$. Since $\tilde{P}_1 \leq P_1'$ and $P_1' \equiv \tilde{P}_1$ and P_1', and $\tilde{P}_1$ are finite, we conclude that $P_1' = \tilde{P}_1$. This means that $\tilde{P}_1 \leq \tilde{P}_2 \leq P_1' = \tilde{P}_1$. $\tilde{P}_1 = \tilde{P}_2$, and accordingly $\mathbf{P}_1 = \mathbf{P}_2$, now follows, giving antisymmetry.

We shall now prove **F1.** Actually, we shall show more than this; we may compare order structure as well.

17.4.7. PROPOSITION. If $\Sigma = (S,T;F,C)$ is a contact-free C/E system and $c \in C$, then there exists an injective mapping $Proc_\Sigma : P_{(\Sigma,c)} \to Proc(N_\Sigma, c)$ such that

1• $\forall p \in P_{(\Sigma,c)} - \{\varnothing\} : po_T(Proc_\Sigma(p)) = p;$

2• $\forall p_1, p_2 \in P_{(\Sigma,c)} : p_1 \subseteq p_2 \Leftrightarrow Proc(p_1) \leq Proc(p_2).$

PROOF. By 17.4.4, $(\tilde{N}, \tilde{\lambda}) = Proc_\Sigma(p)$ is a process of (N,c). By construction, $\tilde{T} = p$. Also, if $\underline{u}_1, \underline{u}_2 \in p$ then $\underline{u}_1\ \tilde{F}^*\ \underline{u}_2 \Leftrightarrow \underline{u}_1 \leq \underline{u}_2$ by (17.5) $\Leftrightarrow \underline{u}_1 \to \underline{u}_2$ in $B_{(\Sigma,c)}$, by (12.1). Hence $po_T(Proc_\Sigma(p)) = p$.

Suppose that $(\tilde{N}_1, \tilde{\lambda}_1) = Proc_\Sigma(p_1) \leq Proc_\Sigma(p_2) = (\tilde{N}_2, \tilde{\lambda}_2)$, then by 17.4.1 (1), $p_1 = \tilde{T}_1 \subseteq \tilde{T}_2 = p_2$. Let $p_1 \subseteq p_2$. It is clear from the construction of the $Proc_\Sigma(p_i)$ that 16.3.2 (1) and 17.4.6 (2) hold.

Let us show that 16.3.2 (2) holds. Suppose $x \in \tilde{S}_1 \cup \tilde{T}_1$, $y \in \tilde{S}_2 \cup \tilde{T}_2$ such that $y\ \tilde{F}_2\ x$. We must prove that $y \in \tilde{S}_1 \cup \tilde{T}_1$.

First, let $y \in \tilde{T}_2$, then $y = \underline{u} \in p_2$ and $x = (\underline{u}, s)$, some $s \in \tilde{\lambda}_2(\underline{u})^\bullet = \tilde{\lambda}_1(\underline{u})^\bullet$. Since $x \in \tilde{S}_1$, it follows that $\underline{u} \in p_1$.

Next suppose that $y \in \tilde{S}_2$, then $y = (\underline{u}, s)$, where $\underline{u} \in p_2$, and $x = \underline{u}'$, which is minimal with respect to $\underline{u} < \underline{u}'$ and $s \in \bullet\tilde{\lambda}_2(\underline{u}')$. As $\underline{u}' \in p_1$, it follows that $\underline{u} \in p_1$. This is because $\underline{u} \to \underline{u}'$, by (12.1) and $\underline{u}' \in p_1$. Hence, $y = (\underline{u}, s) \in p_1 \times S$. Since, in addition, $s \in \bullet\tilde{\lambda}_2(\underline{u}') = \bullet\tilde{\lambda}_1(\underline{u}')$, it follows that $y = (\underline{u}, s) \in \tilde{S}_1$.

■

The reader may, perhaps, be thumbing on ahead for the usual theorem that says that $Proc(N,c)/\equiv$ is prime algebraic and consistently complete. Alas, such a reader will look in vain. Since empty nets are not allowed, by 16.2.1 (1), $Proc(N,c)/\equiv$ will not generally possess a bottom element. However, we may easily get round this. Define

$$Proc_{\perp}(N,c) = Proc(N,c) \cup \{\perp\},$$

where $\perp$ is not any element of $Proc(N,c)$, and extend $\leq$ to $Proc_{\perp}(N,c)$ by defining $\perp \leq \tilde{P}$ for all $\tilde{P} \in Proc_{\perp}(N,c)$. We also extend $Proc_{\Sigma}$ to the whole of $P_{(\Sigma,c)}$ by defining $Proc_{\Sigma}(\varnothing) = \perp$. The anticipatory reader is now advised not to thumb past 17.5.5.

17.5. ...And Back Again.

Now, we look at the converse, **F2**. Suppose that $\tilde{P} \in Proc_{\perp}(N,c)$, then we must show that there exists $p \in P_{(\Sigma,c)}$ such that $\tilde{P} \in Proc_{\Sigma}(p)/\equiv$. p will be of the form $Pr(\underline{x})$, some $\underline{x} \in L(C_{\Sigma},c)$, so we ask ourselves whether there are any $\underline{x} \in L(C_{\Sigma},c)$ specifically related to $\tilde{P}$.

Consider the process $\tilde{P}$ of the C/E system Σ of figure 16.1 which is pictured in figure 16.2. We know that $\tilde{P}$ is isomorphic to $Proc_{\Sigma}(Pr(\underline{t}_4\underline{t}_3\underline{t}_4\underline{t}_1\underline{t}_2))$, pictured in figure 17.4. What is the relationship between $\underline{t}_4\underline{t}_3\underline{t}_4\underline{t}_1\underline{t}_2$ and $\tilde{P}$? The answer is that $\underline{t}_4\underline{t}_3\underline{t}_4\underline{t}_1\underline{t}_2$ is the set of all sequential interleavings of the partial order $po_T(\tilde{P})$. Looking back to 12.2.17, we see that traces may arise from all the sequential interleavings of a labelled partial order. We may apply the same trick to $po_T(\tilde{P})$, as follows

Let $\tau(\tilde{P})$ denote the set of all sequences $t_1 \cdots t_n$ where $\tilde{T} = \{t_1, \cdots, t_n\}$ and the total order $t_1 < \cdots < t_n$ satisfies $\tilde{F}^* \cap (\tilde{T} \times \tilde{T}) \subseteq \leq$. Define

$$\rho(\tilde{P}) = \{\tilde{\lambda}(t_1) \cdots \tilde{\lambda}(t_n) \mid t_1 \cdots t_n \in \tau(\tilde{P})\}.$$

This, as it will transpire, is the $\underline{x}$ we are looking for. First, recall that if $(X_1,\leq_1,\lambda_1)$ and $(X_2,\leq_2,\lambda_2)$ are labelled posets, then a label isomorphism is a poset isomorphism $\phi: X_1 \to X_2$ such that for every $x \in X_1$, $\lambda_2(\phi(x)) = \lambda_1(x)$. Write $(X_1,\leq_1,\lambda_1) \equiv (X_2,\leq_2,\lambda_2)$ to indicate that $(X_1,\leq_1,\lambda_1)$ and $(X_2,\leq_2,\lambda_2)$ are label isomorphic.

17.5.1. PROPOSITION. Suppose that $\Sigma = (S,T;F,C)$ is a contact-free C/E system and $c \in C$, then if $\tilde{P} = (\tilde{N},\tilde{\lambda}) \in Proc(N_\Sigma,c)$, then $\rho(\tilde{P}) \in L(C_\Sigma,c)$.

PROOF. Suppose $x = t_1 \cdots t_n \in \tau(\tilde{P})$, with $a_i = \tilde{\lambda}(t_i)$, each i.

Let us begin by showing that $a_1 \cdots a_n \in L(T(C_\Sigma),c)$, that is, that there exists $c_1 \in C$ such that $c \to^{a_1} c_1 \cdots c_n \to^{a_n} c_{n+1}$.

A prefix $t_1 \cdots t_i \leq x$ represents a preliminary part of the behaviour $\tilde{P}$. Its maximal S-elements represent the last conditions in N_Σ to become true as a consequence of the occurrences of the t_j. These, together with those elements of c which have not been falsified by the occurrence of any t_j, are the conditions which hold after $t_1 \cdots t_i$. This suggests that we should be able to construct c_i from $\tilde{P}$ as follows. If $T_i = \{t_1,\cdots,t_i\}$, then

$$c_i = \tilde{\lambda}(T_i \bullet - \bullet T_i) \cup (c - \tilde{\lambda}(\bullet T_i)) \tag{17.6}$$

We argue by induction on n that if $a_i = \tilde{\lambda}(t_i)$, each i then

$$x = t_1 \cdots t_n \in \tau(\tilde{P}) \Rightarrow c \to^{a_1} c_1 \cdots c_{n-1} \to^{a_n} c_n \tag{17.7}$$

For $n = 1$, we observe that $t_1 \in min(po_T(\tilde{P}))$ and hence $\bullet t_1 \subseteq min(\tilde{N}) \cap \tilde{S}$. Thus, $\bullet a_1 = \tilde{\lambda}(\bullet t_1) \subseteq \tilde{\lambda}(min(\tilde{N}) \cap \tilde{S}) \subseteq c$, by 16.2.7 (4). Since Σ is contact-free, $a_1 \bullet \cap c = \varnothing$. Thus $c\,[a_1\rangle$.

By (17.6), $c_1 = \tilde{\lambda}(t_1 \bullet - \bullet t_1) \cup (c - \tilde{\lambda}(\bullet t_1)) = (a_1 \bullet - \bullet a_1) \cup (c - \bullet a_1)$, and since $a_1 \bullet \cap \bullet a_1 = \varnothing$, $a_1 \bullet - \bullet a_1 = a_1 \bullet$. Thus, $c_1 = (c - \bullet a_1) \cup a_1 \bullet$ so $c\,[a_1\rangle\,c_1$, by (17.1), that is, $c \to^{a_1} c_1$.

Now for the induction step. Suppose that $s \in \bullet a_{i+1}$, then $s = \tilde{\lambda}(s')$, some $s' \in \bullet t_{i+1}$, by 16.2.7 (2). We observe first of all that $s' \notin \bullet T_i$, for otherwise, $s' \in \bullet t_j$ some $j \leq i$, so $s' \in \bullet t_j \cap \bullet t_{i+1}$, contradicting the fact that $\tilde{N}$ is an MG net. We have two cases.

If $s' \in T_i \bullet$, then $s' \in T_i \bullet - \bullet T_i$, because $s' \notin \bullet T_i$. Therefore, by (17.6), $s = \tilde{\lambda}(s') \in \tilde{\lambda}(T_i \bullet - \bullet T_i) \subseteq c_i$.

On the other hand, if $s' \notin T_i\bullet$ then $s' \in min(\tilde{N}) \cap \tilde{S}$. By 16.2.7 (4), $s = \tilde{\lambda}(s') \in c$. We show that $\tilde{\lambda}(s') \notin \tilde{\lambda}(\bullet T_i)$. If $s'' \in \bullet T_i$ with $\tilde{\lambda}(s'') = \tilde{\lambda}(s')$, then $s' \neq s''$, because $s' \notin \bullet T_i$ and $s'' \in \bullet T_i$, and by 16.2.7 (3), either $s' \tilde{F}^* s''$ or $s'' \tilde{F}^* s'$. If $s' \tilde{F}^* s''$, then $s' \tilde{F}\, t\, \tilde{F}^* s'' \tilde{F}\, t'$ for some $t, t' \in \tilde{T}$, and by the MG property, $t = t_{i+1}$ and $t' = t_j$, some $j \leq i$. But then, $t_{i+1} \tilde{F}^* t_j$, contradicting the hypothesis that $t_1 \cdots t_n \in \tau(\tilde{P})$. Thus $s'' \tilde{F}^* s'$. But, $s' \in min(\tilde{N}) \cap \tilde{S}$, so $s'' \tilde{F}^* s'$ is only possible if $s' = s''$, which we know to be false. Thus, $\tilde{\lambda}(s') \notin \tilde{\lambda}(\bullet T_i)$ and so in this case, $s = \tilde{\lambda}(s') \in (c - \tilde{\lambda}(\bullet T_i)) \subseteq c_i$.

We have shown that $\bullet a_{i+1} \subseteq c_i$. Since Σ is contact-free, $a_{i+1}\bullet \cap c_i = \varnothing$ and hence $c_i\, [a_{i+1}\rangle$. We shall now prove that $c_i\, [a_{i+1}\rangle\, c_{i+1}$.

Suppose that $s \in a_{i+1}\bullet$, then $s' = \tilde{\lambda}(s)$, for some $s' \in t_{i+1}\bullet$, by 16.2.7 (2). If $s' \in \bullet T_{i+1}$, then either $s' \in \bullet t_{i+1}$ - giving the contradiction $\bullet t_{i+1} \cap t_{i+1}\bullet \neq \varnothing$ - or $s' \in \bullet t_j$, some $j < i$ - giving $t_{i+1} \tilde{F}\, s' \tilde{F}\, t_j$, contradicting the hypothesis that $t_1 \cdots t_n \in \tau(\tilde{P})$. Thus, $s' \in T_{i+1}\bullet - \bullet T_{i+1}$ and so $s = \tilde{\lambda}(s') \in \tilde{\lambda}(T_{i+1}\bullet - \bullet T_{i+1}) \subseteq c_{i+1}$. We have proved that $a_{i+1}\bullet \subseteq c_{i+1}$. Since Σ is contact-free, $a_{i+1}\bullet \cap c_i = \varnothing$ and hence $a_{i+1}\bullet \subseteq c_{i+1} - c_i$.

Next, suppose $s \in c_{i+1} - c_i$. We show that $s \in a_{i+1}\bullet$.

Define $X = \{s' \in \bullet T_{i+1} \cup T_{i+1}\bullet \mid \tilde{\lambda}(s') = s\}$. We know that $X \neq \varnothing$. If $X \subseteq \bullet T_{i+1}$, then $X \cap (T_{i+1}\bullet - \bullet T_{i+1}) = \varnothing$ so $\tilde{\lambda}(s') \notin \tilde{\lambda}(T_{i+1}\bullet - \bullet T_{i+1})$ and $\tilde{\lambda}(s') \notin c - \tilde{\lambda}(\bullet T_{i+1})$, for each $s' \in X$. But, this means that $s \notin c_{i+1}$, a contradiction.

Thus, the set $Y = \{s' \in X \mid s' \notin \bullet T_{i+1}\}$ is non-empty. If $Y \subseteq T_i\bullet$, then $Y \subseteq T_i\bullet - \bullet T_{i+1} \subseteq T_i\bullet - \bullet T_i$ so $\{s\} = \tilde{\lambda}(Y) \subseteq c_i$, contradicting $s \in c_{i+1} - c_i$. Hence, there exists $s' \in Y$ with $s' \notin T_i\bullet$. Since $s' \in Y$, $s' \notin \bullet T_{i+1}$ and so, as $s' \in X$, $s' \in (\bullet T_{i+1} \cup T_{i+1}\bullet) - \bullet T_{i+1} \subseteq T_{i+1}\bullet$. Therefore, $s' \in T_{i+1}\bullet - T_i\bullet \subseteq t_{i+1}\bullet$ and so, $s = \tilde{\lambda}(s') \in \tilde{\lambda}(t_{i+1}\bullet) = a_{i+1}\bullet$. We have proved that $c_{i+1} - c_i \subseteq a_{i+1}\bullet$ and so $a_{i+1}\bullet = c_{i+1} - c_i$. As we know that $c_i\, [a_{i+1}\rangle$, it follows that $c_i\, [a_{i+1}\rangle\, c_{i+1}$, completing the proof of the induction step. We have established (17.7).

Next, we shall show that

$$y \in \rho(\tilde{P}) \Rightarrow y_\iota \subseteq \rho(\tilde{P}) \tag{17.8}$$

By 9.2.1, it suffices to show that if $y \in \rho(\tilde{P})$ and $y' \in \tilde{T}^*$ with $y \equiv_\iota^{(1)} y'$, then $y' \in \rho(\tilde{P})$. Now, $y = a_1 \cdots a_n = \tilde{\lambda}(x)$, where $x = t_1 \cdots t_n \in \tau(\tilde{P})$. As $y \equiv_\iota^{(1)} y'$, for some k, $y' = a_1 \cdots a_k a_{k+2} a_{k+1} a_{k+3} \cdots a_n$ with $a_{k+1}\, \iota\, a_{k+2}$. Let $x' = t_1 \cdots t_k t_{k+2} t_{k+1} t_{k+3} \cdots t_n$ so that $y' = \tilde{\lambda}(x')$.

We show by contradiction that $\neg t_{k+1}\, \tilde{F}^*\, t_{k+2}$.

Suppose that $t_{k+1}\, \tilde{F}^*\, t_{k+2}$, then either $t_{k+1}{}^\bullet \cap {}^\bullet t_{k+2} \neq \varnothing$ or there exists $t \neq t_{k+1}, t_{k+2}$ such that $t_{k+1}\, \tilde{F}^*\, t\, \tilde{F}^*\, t_{k+2}$ In the latter case, $t_{k+1} < t < t_{k+2}$ for every total order extending $\tilde{F}^*$, which would entail $x \notin \tau(\tilde{P})$. In the former, $a_{k+1}{}^\bullet \cap {}^\bullet a_{k+2} \neq \varnothing$, by 16.2.7 (2), and hence $\neg a_{k+1}\, \iota\, a_{k+2}$, by 17.2.1 (2), a contradiction. Thus $\neg t_{k+1}\, \tilde{F}^*\, t_{k+2}$. From this, it follows that if $t_1 \cdots t_n \in \tau(\tilde{P})$, then $t_1 \cdots t_k t_{k+2} t_{k+1} t_{k+3} \cdots t_n \in \tau(\tilde{P})$. Thus $y' = \tilde{\lambda}(x') \in \rho(\tilde{P})$, proving (17.8).

Next, suppose that $y, y' \in \rho(\tilde{P})$. We shall show that $y \equiv_\iota y'$. We have $x = t_1 \cdots t_n \in \tau(\tilde{P})$ and $x' = t'_1 \cdots t'_n \in \tau(\tilde{P})$ such that $y = \tilde{\lambda}(x)$ and $y' = \tilde{\lambda}(x')$. Also, suppose $\tilde{\lambda}(t_k) = a_k$ and $\tilde{\lambda}(t'_k) = a'_k$ for each k.

We argue by induction on $n - |x \wedge x'|$.[4] If $n - |x \wedge x'| = 0$ then $x = x'$ and hence $y = y'$ so that $y \equiv_\iota y'$. Next, let $n - |x \wedge x'| > 0$ and let $u = t_1 \cdots t_k = t'_1 \cdots t'_k = x \wedge x'$. $\{t_1, \cdots, t_n\} = \{t'_1, \cdots, t'_n\} = \tilde{T}$ and so $t_{k+1} = t'_l$, some $l > k+1$. As in an earlier part of this argument, we may show that if $l > i > k+1$, then $\neg t_i\, \tilde{F}^*\, t_l$, since otherwise, we would have $t_{k+1} < t_i < t'_l$ for every total order extending $\tilde{F}^*$, which would entail $x' \notin \tau(\tilde{P})$. Likewise, for $l > i > k+1$, $\neg t_l\, \tilde{F}^*\, t_i$, otherwise $x \notin \tau(\tilde{P})$. Thus if $x'' = t_1 \cdots t_k t_l t_{k+1} \cdots t_{l-1} t_{l+1} \cdots t_n$, then $x'' \in \tau(\tilde{P})$. Since $\neg t_i\, \tilde{F}^*\, t_l$ and $\neg t_l\, \tilde{F}^*\, t_i$, for $l > i > k+1$, $t_i\, \iota\, t_l$ for $l > i > k+1$, so $a_i\, \iota\, a_l$ for $l > i > k+1$. Thus $\tilde{\lambda}(x'') \equiv_\iota y$. But $|x'' \wedge x'| > |x \wedge x'|$, so by induction, $\tilde{\lambda}(x'') \equiv_\iota y'$. Hence $y \equiv_\iota y'$. Together with (17.8), this gives

$$y \in \rho(\tilde{P}) \Rightarrow y_\iota = \rho(\tilde{P}).$$

Now, we know from (17.7), that if $y \in \rho(\tilde{P})$ then $y \in L(T(C_\Sigma), c)$, and therefore $\rho(\tilde{P}) = y_\iota \in L(C_\Sigma, c)$.

For the remainder of the proof, suppose $t \in \tilde{T}$ and let $\downarrow t = \{t' \in \tilde{T} \mid t'\, \tilde{F}^*\, t\}$. Define $\rho(\downarrow t)$ in the usual way.

The reader may check that $Pr(\rho(\tilde{P})) = \{\downarrow t \mid t \in \tilde{T}\}$ and that the mapping $\phi : Pr(\rho(\tilde{P})) \to po_T(\tilde{P})$, given by $\phi(Pr(\rho(\downarrow t))) = t$ is a bijection satisfying $t_1\, \tilde{F}^*\, t_2 \Leftrightarrow \downarrow t_1 \subseteq \downarrow t_2 \Leftrightarrow \rho(\downarrow t_1) \leq \rho(\downarrow t_2)$.

■

[4]Recall that $x \wedge x'$ denotes the longest common prefix of x and x'.

In order to relate $\tilde{P}$ to $Proc_\Sigma(Pr(\rho(\tilde{P})))$, we need the following lemma, which shows that a label isomorphism from $po_T(\tilde{P}_1)$ to $po_T(\tilde{P}_2)$ extends to a process isomorphism from $\tilde{P}_1$ to $\tilde{P}_2$.

17.5.2. LEMMA. Suppose that $\Sigma=(S,T;F,C)$ is a contact-free C/E system, $c\in C$ and $\tilde{P}_1,\tilde{P}_2\in Proc(N_\Sigma,c)$, then $po_T(\tilde{P}_1)\equiv po_T(\tilde{P}_2)\Rightarrow \tilde{P}_1\equiv\tilde{P}_2$.

PROOF. Suppose $\varphi{:}\,po_T(\tilde{P}_1)\equiv po_T(\tilde{P}_2)$ We must extend φ to a suitable mapping $\phi{:}\,\tilde{S}_1\cup\tilde{T}_1\to\tilde{S}_2\cup\tilde{T}_2$. We argue by induction on $|\tilde{T}_1|=|\tilde{T}_2|$. Recall that this number will be finite, by 16.2.6.

If $|\tilde{T}_1|=1$, then there exists t_1 and t_2 such that $\tilde{T}_1=\{t_1\}$ and $\tilde{T}_2=\{t_2\}$ whence $\varphi(t_1)=t_2$ so $\tilde{\lambda}_1(t_1)=\tilde{\lambda}_2(t_2)$. Thus, $\tilde{\lambda}_1(\bullet t_1)=\bullet\tilde{\lambda}_1(t_1)=\bullet\tilde{\lambda}_2(t_2)=\tilde{\lambda}_2(\bullet t_2)$. Likewise, $\tilde{\lambda}_1(t_1\bullet)=\tilde{\lambda}_2(t_2\bullet)$.

By 16.2.7 (3), and the fact that in a C/E system, $\bullet t_1\cap t_1\bullet=\varnothing$, $\tilde{\lambda}_1$ determines bijections from $t_1\bullet$ to $\tilde{\lambda}_1(t_1\bullet)$ and from $\bullet t_1$ to $\tilde{\lambda}_1(\bullet t_1)$ whose union is a bijection from $\tilde{S}_1=t_1\bullet\cup\bullet t_1$. (16.2.1 (b)) to $\tilde{\lambda}_1(t_1\bullet)\cup\tilde{\lambda}_1(\bullet t_1)$. Likewise, for t_2. So if we define $\hat{\phi}{:}\,\tilde{S}_1\to\tilde{S}_2$ by $\hat{\phi}(s_1)=s_2\Leftrightarrow\tilde{\lambda}_1(s_1)=\tilde{\lambda}_2(s_2)$, then $\hat{\phi}$ is a composition of bijections and is therefore a bijection. It is easy to see that

$$s_1\,\tilde{F}_1\,t_1\Leftrightarrow\tilde{\lambda}_1(s_1)\,F_\Sigma\,\tilde{\lambda}_1(t_1)\Leftrightarrow\tilde{\lambda}_2(\hat{\phi}(s_1))\,F_\Sigma\,\tilde{\lambda}_2(t_2)\Leftrightarrow\hat{\phi}(s_1)\,\tilde{F}_2\,t_2.$$

Likewise, $t_1\,\tilde{F}_1\,s_1\Leftrightarrow t_2\,\tilde{F}_2\,\hat{\phi}(s_1)$. Therefore, if we define $\phi{:}\,\tilde{S}_1\cup\tilde{T}_1\to\tilde{S}_2\cup\tilde{T}_2$ by $\phi=\varphi\cup\hat{\phi}$, it is clear that $\phi{:}\,\tilde{P}_1\equiv\tilde{P}_2$.

Next, suppose that $|\tilde{T}_1|>1$. Let t_1 be maximal in $po_T(\tilde{P}_1)$. Note that $\phi(t_1)=t_2$ is maximal in $po_T(\tilde{P}_2)$.

For $i=1,2$, we define $\tilde{P}_i\setminus t_i=(\tilde{N}'_i,\tilde{\lambda}'_i)$, where $\tilde{S}'_1=\tilde{S}_1-t_1\bullet$, $\tilde{T}'_1=\tilde{T}_1-\{t_1\}$, $\tilde{F}'_i=\tilde{F}_i\cap((\tilde{S}'_i\times\tilde{T}'_i)\cup(\tilde{T}'_i\times\tilde{S}'_i))$ and $\tilde{\lambda}'_i=\tilde{\lambda}_i|\tilde{S}'_i\cup\tilde{T}'_i$. Since t_1 is maximal in $po_T(\tilde{P}_1)$, then $\tilde{P}_1\setminus t_1$ may easily be shown to be a process of N_Σ from c.

Clearly, φ cuts down to a label isomorphism $\varphi'{:}\,po_T(\tilde{P}_1\setminus t_1)\equiv po_T(\tilde{P}_2\setminus t_2)$. $|\tilde{T}'_1|<|\tilde{T}_1|$, so by induction, φ' extends to a process isomorphism $\phi'{:}\,\tilde{P}_1\setminus t_1\equiv\tilde{P}_2\setminus t_2$. We now extend ϕ' to $\tilde{P}_1\equiv\tilde{P}_2$ by defining $\phi(t_1)=t_2$ and for $s\in t_1\bullet$, defining $\phi(s)=s'\Leftrightarrow s'\in t_2\bullet\wedge\tilde{\lambda}_1(s)=\tilde{\lambda}_2(s')$. The rest of the proof is routine.

■

17.5.3 PROPOSITION. Suppose that $\Sigma = (S,T;F,C)$ is a contact-free C/E system, $c \in C$ and $\tilde{P} \in Proc(N_\Sigma, c)$ then $\tilde{P} \equiv Proc_\Sigma(Pr(\rho(\tilde{P})))$.

PROOF. Since $\underline{\Omega} \neq \rho(\tilde{P}) \in L(C_\Sigma, c)$, by 17.5.1, $\varnothing \neq Pr(\rho(\tilde{P})) \in B_{(C_\Sigma,c)}$, by 12.1.1 and so $Proc_\Sigma(Pr(\rho(\tilde{P}))) \in Proc(N_\Sigma, c)$, by 17.4.4. By 17.5.2, it suffices to show that $po_T(\tilde{P}) \equiv po_T(Proc_\Sigma(Pr(\rho(\tilde{P}))))$.

Now, $po_T(Proc_\Sigma(Pr(\rho(\tilde{P})))) = (Pr(\rho(\tilde{P})), \leq, \lambda)$, where $\leq$ is the usual ordering on traces, by (17.5) and if $u_\iota.t_\iota \in Pr(\rho(\tilde{P}))$, $t \in \tilde{T}$, then $\lambda(u_\iota.t_\iota) = t$, by 17.4.1 (7).

We shall prove that if $t_1, \cdots, t_n, t_1', \cdots, t_n' \in \tilde{T}$ such that $t_1 \cdots t_n, t_1' \cdots t_n' \in \tau(\tilde{P})$ and $\tilde{\lambda}(t_1) \cdots \tilde{\lambda}(t_n) = \tilde{\lambda}(t_1') \cdots \tilde{\lambda}(t_n') \in \rho(\tilde{P})$, then $t_i' = t_i'$, each i. Suppose not, then there exists j such that $t_j \neq t_j'$ and that $t_i = t_i'$ for $i < j$, then $t_j = t_k'$, some $k > j$, and so by definition of $\tau(\tilde{P})$, $\neg t_k' \tilde{F}^* t_j'$ and so $\neg t_j \tilde{F}^* t_j'$. By symmetry, $\neg t_j' \tilde{F}^* t_j$.

By hypothesis $\tilde{\lambda}(t_j) = \tilde{\lambda}(t_j') = t$, say. Now, by 16.2.1 (b), either $\bullet t \neq \varnothing$ or $t\bullet \neq \varnothing$.

Suppose first that $\bullet t \neq \varnothing$ and let $s \in \bullet t$, then by 16.2.7 (2) there exists $s_j \in \bullet t_j$ and $s_j' \in \bullet t_j'$ such that $\tilde{\lambda}(s_j) = \tilde{\lambda}(s_j') = s$. By 16.2.7 (3), either $s_j \tilde{F}^* s_j'$ or $s_j' \tilde{F}^* s_j$. Without loss of generality, assume the former. $s_j' \neq s_j$, since otherwise we would have $s_j \in \bullet t_j \cap \bullet t_j'$ with $t_j \neq t_j'$ contradicting the MG property. Thus, there exists $\tilde{t} \in \tilde{T}$ such that $s_j \tilde{F} \tilde{t} \tilde{F}^* s_j' F t_j'$. And now $s_j \in \bullet \tilde{t} \cap \bullet t_j$, so that $\tilde{t} = t_j$, by the MG property, and we have a contradiction to $\neg t_j \tilde{F}^* t_j'$. The case $t\bullet \neq \varnothing$ is similar. This completes the proof of our claim.

Now suppose that $u_\iota \in Pr(\rho(\tilde{P}))$, then there exists $v, w \in T^*$ and $t \in T$ such that $u_\iota = v_\iota.t_\iota$ and $v.t.w \in \rho(\tilde{P})$. By the foregoing, there exists unique $\tilde{v}, \tilde{w} \in \tilde{T}^*$ and $\tilde{t} \in \tilde{T}$ such that $\tilde{v}.\tilde{t}.\tilde{w} \in \tau(\tilde{P})$, $\tilde{\lambda}(\tilde{v}) = v$, $\tilde{\lambda}(\tilde{w}) = w$ and $\tilde{\lambda}(\tilde{t}) = t$. Define $\phi(u_\iota.t_\iota) = t$, so that $\phi: Pr(\rho(\tilde{P})) \to \tilde{T}$. Immediately, $\tilde{\lambda}(\phi(u_\iota)) = \tilde{\lambda}(\tilde{t}) = t = \lambda(u_\iota)$ it only remains to show that ϕ is a poset isomorphism. It will actually suffice to show that if $u_\iota, u_\iota' \in Pr(\rho(\tilde{P}))$, then

$$u_\iota \leq u_\iota' \Leftrightarrow \phi(u_\iota) \tilde{F}^* \phi(u_\iota) \qquad (17.9)$$

For then, ϕ is injective and since $|Pr(\rho(\tilde{P}))| = |\rho(\tilde{P})| = |\tilde{P}|$, by 12.2.6 (3), ϕ is surjective and hence bijective and so, by (17.9), a poset isomorphism.

Suppose that $u_1, u_1' \in Pr(\rho(\tilde{P}))$, then $u_1 = v_1.t_1$ and $u_1' = v_1'.t_1'$, with $t, t' \in T$. If $\neg\phi(u_1)\,\tilde{F}^*\,\phi(u_1')$, then there exists $\tilde{w}_1, \tilde{w}_2, \tilde{w}_3 \in \tilde{T}^*$ such that $\tilde{w}_1.\phi(u_1').\tilde{w}_2.\phi(u_1).\tilde{w}_3 \in \tau(\tilde{P})$, so that $\tilde{\lambda}(\tilde{w}_1).t'.\tilde{\lambda}(\tilde{w}_2).t.\tilde{\lambda}(\tilde{w}_3) \in \rho(\tilde{P})$ and so it cannot be the case that $w_1.t_1' \leq \rho(\tilde{P}) \Rightarrow \#_t\, w > 0$. Hence, $u_1 \not\leq u_1'$, using 12.2.1. We have shown that $u_1 \leq u_1' \Rightarrow \phi(u_1)\,\tilde{F}^*\,\phi(u_1')$. Conversely, suppose that $\phi(u_1)\,\tilde{F}^*\,\phi(u_1')$, then if $u_1' = v_1'.t_1'$, then there exists $\tilde{w} \in \tilde{T}^*$ such that $u' = \tilde{\lambda}(\tilde{w})$ and since $\phi(u_1)\,\tilde{F}^*\,\phi(u_1')$, $\#_{\phi(u_1)}\,\tilde{w} > 0$. Hence there exists $\tilde{v}, \tilde{v}' \in \tilde{T}^*$ such that $\tilde{w} = \tilde{v}.\phi(u_1).\tilde{v}'$, and now $u_1 = pr(\tilde{\lambda}(v)_1, t) \leq u_1'$.

■

We are now in a position to prove a converse to 17.4.7.

17.5.4. PROPOSITION. Suppose that $\Sigma = (S, T; F, C)$ is a contact-free C/E system and $c \in C$, then the function $f_\Sigma : P_{(\Sigma,c)} \rightarrow Proc_\perp(N_\Sigma, c)/\!\equiv$ defined by

$$f_\Sigma(p) = [Proc_\Sigma(p)]_\equiv$$

is a poset isomorphism, where $[Proc_\Sigma(p)]_\equiv$ denotes the $\equiv$-class of $Proc_\Sigma(p)$.

PROOF. We know from 17.4.7 that $f_\Sigma(P_{(\Sigma,c)}) \subseteq Proc_\perp(N_\Sigma, c)/\!\equiv$.

If $f_\Sigma(p_1) = f_\Sigma(p_2)$, then $Proc_\Sigma(p_1) \equiv Proc_\Sigma(p_2)$ and hence, by 17.4.7 (1) and 17.5.1, $p_1 = po_T(Proc_\Sigma(p_1)) \equiv po_T(Proc_\Sigma(p_2)) = p_2$. Since $B_{(\Sigma,c)}$ is linguistic, it is $\lambda_{(\Sigma,c)}$-reduced[5] and so as $p_1 \equiv p_2$, $p_1 = p_2$. Thus f_Σ is injective.

Next, suppose $\tilde{P} \in Proc_\perp(N_\Sigma, c)$. If $\tilde{P} = \perp$, then $f_\Sigma(\emptyset) = [\tilde{P}]_\equiv$. Otherwise, $\tilde{P} \equiv Proc_\Sigma(Pr(\rho(\tilde{P})))$ by 17.5.3. Thus, $f_\Sigma(p) = [\tilde{P}]_\equiv$, where $p = Pr(\rho(\tilde{P})) \in P_{(\Sigma,c)}$, and we may conclude that f_Σ is surjective.

That f_Σ is a poset isomorphism now follows easily from 17.4.7 (2).

■

17.5.5. COROLLARY. Suppose that $\Sigma = (S, T; F, C)$ is a contact-free C/E system and $c \in C$, then $(Proc_\perp(N_\Sigma, c)/\!\equiv, \leq)$ is prime algebraic and consistently complete.

■

Thus, the set $\{Proc_\Sigma(p) \mid p \in P_{(\Sigma,c)}\}$ completely represents all the processes of N_Σ from c. Moreover, it completely reflects the order structure of

---

[5]See 12.2.8.

$Proc(N_\Sigma,c)$, so that processes may be related directly, rather by way of equivalent representations, since

$$[Proc_\Sigma(p_1)]_\equiv \leq [Proc_\Sigma(p_2)]_\equiv \Leftrightarrow Proc_\Sigma(p_1) \leq Proc_\Sigma(p_2) \Leftrightarrow p_1 \subseteq p_2.$$

The first equivalence derives from the fact that (a) if $(\tilde{N},\tilde{\lambda}) \in Proc_\Sigma(p)$, then $\tilde{T} = p$, by 17.4.1 (1), which gives the $\Rightarrow$ implication, and (b) the definition of $\leq$ on $\equiv$-classes, which gives the $\Leftarrow$ implication. The other equivalence is given by 17.4.7. (2)

We may in fact combine the elements $Proc_\Sigma(P_{(\Sigma,c)})$ into a single net to give an occurrence net of Σ from c, a case in Σ. Define $Occ(\Sigma,c) = (\tilde{N},\tilde{\lambda})$, where

$$\tilde{S} = \bigcup_{P' \in Proc_\Sigma(P_{(\Sigma,c)})} S' \qquad \tilde{T} = \bigcup_{P' \in Proc_\Sigma(P_{(\Sigma,c)})} T' \qquad \tilde{F} = \bigcup_{P' \in Proc_\Sigma(P_{(\Sigma,c)})} F' \qquad \tilde{\lambda} = \bigcup_{P' \in Proc_\Sigma(P_{(\Sigma,c)})} \lambda'$$

From the definitions involved in the $Proc_\Sigma$ construction, we may describe the components of $\tilde{N}$ in more detail. First, note that $\tilde{T} = O_{(\Sigma,c)}$ and that $\tilde{S} = \tilde{S}_\Omega \cup \tilde{S}_+$, where

$$\tilde{S}_\Omega = \{(\Omega,s) \in \{\Omega\} \times c \mid \exists \underline{u} \in O_{(\Sigma,c)} : s \in \bullet\lambda_{(\Sigma,c)}(\underline{u})\}$$
$$\tilde{S}_+ = \{(\underline{u},s) \in O_{(\Sigma,c)} \times S \mid s \in \lambda_{(\Sigma,c)}(\underline{u})\bullet\}$$

For the flow relation, we have,

$$\underline{u}\,\tilde{F}\,(\underline{u}',s) \Leftrightarrow \underline{u} = \underline{u}'$$
$$(\underline{u},s)\,\tilde{F}\,\underline{u}' \Leftrightarrow \underline{u} < \underline{u}' \wedge s \in \bullet\lambda_{(\Sigma,c)}(\underline{u}') \wedge (\forall \underline{u}'' \in \tilde{T} : \underline{u} < \underline{u}'' < \underline{u}' \Rightarrow s \notin \bullet\lambda_{(\Sigma,c)}(\underline{u}''))$$

and for the labelling function,

$$\tilde{\lambda}(\underline{u}) = \lambda_{(\Sigma,c)}(\underline{u}) \text{ and } \tilde{\lambda}(\underline{u},s) = s$$

With these observations, we may now prove:

17.5.6. PROPOSITION. If $\Sigma = (S,T;F,C)$ is a C/E system, $c \in C$ and $(\tilde{N},\tilde{\lambda}) = Occ(\Sigma,c)$ then $\tilde{N}$ is an occurrence net.

PROOF. The proof that $\tilde{N}$ is a net is exactly the same as that of 17.4.2. For acyclicity of $\tilde{N}$, note first of all that, using (17.5), if

$$\underline{u}_1 \tilde{F}^* \underline{u}_2 \Leftrightarrow (\exists \hat{P} \in Proc_\Sigma(P_{(\Sigma,c)}): \underline{u}_1 \hat{F}^* \underline{u}_2) \Leftrightarrow \underline{u}_1 \le \underline{u}_2$$

Thus $\tilde{F}^*$ is a partial order on $\tilde{T}$. The argument that it is consequently a partial order on $\tilde{S} \cup \tilde{T}$ is the same as in the proof of 17.4.4. Thus 16.3.1 (1) holds for $\tilde{N}$.

For 16.3.1 (2), $\underline{u}_1 \tilde{F}(\underline{u},s) \wedge \underline{u}_2 \tilde{F}(\underline{u},s) \Rightarrow \underline{u}_1 = \underline{u} = \underline{u}_2$, and so $|{}^\bullet(\underline{u},s)| \le 1$.

Finally, we prove that $\tilde{N}$ satisfies 16.3.1 (3). Note that the statement that $\#_{\tilde{N}}$ is irreflexive is equivalent to the statement

$$\forall \underline{u}, \underline{u}_1, \underline{u}_2 \in \tilde{T}: \underline{u}_1 \tilde{F}^* \underline{u} \wedge \underline{u}_2 \tilde{F}^* \underline{u} \Rightarrow \neg \underline{u}_1 \#_{\tilde{N}}^{(1)} \underline{u}_2.$$

By (17.5), this is equivalent to

$$\forall \underline{u}, \underline{u}_1, \underline{u}_2 \in \tilde{T}: \underline{u}_1 \le \underline{u} \wedge \underline{u}_2 \le \underline{u} \Rightarrow \neg \underline{u}_1 \#_{\tilde{N}}^{(1)} \underline{u}_2.$$

Suppose $p = Pr(\underline{u})$ and let $(\overline{N}, \overline{\lambda}) = Proc_\Sigma(p)$. If $\underline{u}_1, \underline{u}_2 \in \tilde{T}$ with $\underline{u}_1, \underline{u}_2 \le \underline{u}$, then $\underline{u}_1, \underline{u}_2 \in \overline{T}$ and it is clear that for each i, $s \overline{F} \underline{u}_i \Leftrightarrow s \tilde{F} \underline{u}_i$ Thus, if $\underline{u}_1 \#_{\tilde{N}}^{(1)} \underline{u}_2$ then $\underline{u}_1 \#_{\overline{N}}^{(1)} \underline{u}_2$. But this is impossible, since $\overline{N}$ is an MG net.

■

17.5.7. PROPOSITION. Suppose that $\Sigma = (S,T;F,C)$ is a C/E system and $c \in C$ then $Occ(\Sigma, c)$ is an occurrence net of N_Σ from c.

PROOF. 16.3.2 (1) is proved in 17.5.6. (2) and (3) are clear. For 16.3.2 (4), we note that if $P_1, P_2 \in Proc_\Sigma(P_{(\Sigma,c)})$, then

$$\forall x, y \in \tilde{S} \cup \tilde{T}: x \tilde{F}_1 y \Leftrightarrow x \tilde{F}_2 y.$$

For example, $\underline{u} \tilde{F}_1 (\underline{u}', s) \Leftrightarrow \underline{u} = \underline{u}' \Leftrightarrow \underline{u} \tilde{F}_2 (\underline{u}', s)$. 16.3.2 (4) now follows from 16.2.7 (2) and 17.4.4.

Now suppose $\tilde{\lambda}(s_1) = \tilde{\lambda}(s_2)$. If $\neg s_1 \#_{\tilde{N}} s_2$, then $s_1, s_2 \in S'$, some $(N', \lambda') \le (\tilde{N}, \tilde{\lambda})$ such that $(N', \lambda') \in Proc(N_\Sigma, c)$. It now follows that $s_1 \tilde{F}^* s_2$ or $s_2 \tilde{F}^* s_1$, by 16.2.7 (3). This gives 16.3.2 (5). The proof that $(\tilde{N}, \tilde{\lambda})$ satisfies 16.3.2 (6) is entirely analogous to that of the corresponding part of 17.4.4.

■

If $\Sigma = (S,T;F,C)$ is a contact free C/E system and $c \in C$, then it is associated with two semantic objects, namely the labelled event structure

$(Es(\tilde{N}),\tilde{\lambda})$, where $(\tilde{N},\tilde{\lambda}) = Occ(\Sigma,c)$, and the behavioural presentation, $B_{(\Sigma,c)}$. We know from chapter 4 that there are relationships between (unlabelled) event structures and unlabelled behavioural presentations. These relationships extend quite easily to labelled objects

Recall from chapter 4 that every event structure W determines a (closed) unlabelled behavioural presentation, $\Lambda(W)$. Now, $Es(Occ(\Sigma,c))$ actually comes with a labelling function, $\tilde{\lambda} \mid \tilde{T}$. Thus, it determines a labelled behavioural presentation. If $\Lambda(Es(Occ(\Sigma,c))) = (O,P)$, then $\tilde{\lambda} \mid \tilde{T}$ gives us a labelled behavioural presentation $(O,P,\tilde{T},\tilde{\lambda} \mid \tilde{T})$. In general, if $W = (O,\leq,\#)$ is an event structure and $\lambda : O \rightarrow E$, then we may define $\Lambda(W,\lambda)$ to be the behavioural presentation (O,P,E,λ), where $(O,P) = \Lambda(W)$.

Define $Les(\Sigma,c) = (Es(\tilde{N}),\tilde{\lambda})$, where $(\tilde{N},\tilde{\lambda}) = Occ(\Sigma,c)$. It is clear that $\Lambda(Les(\Sigma,c))$ is a closed behavioural presentation. In fact we may show that it is the closure of the behavioural presentation $B_{(\Sigma,c)}$.

17.5.8. PROPOSITION. If $\Sigma = (S,T;F,C)$ is a C/E system and $c \in C$, then

$$\Lambda(Les(\Sigma,c)) = \overline{B_{(\Sigma,c)}}.$$

PROOF. Let $\overline{B_{(\Sigma,c)}} = (\overline{O},\overline{P},\overline{E},\overline{\lambda})$. First, we note that if $(\tilde{N},\tilde{\lambda}) = Occ(\Sigma,c)$, then $Les(\Sigma,c) = (Es(\tilde{N}),\tilde{\lambda}) = ((\tilde{T},\leq,\tilde{\#}),\tilde{\lambda})$, where $\tilde{T} = O_{(\Sigma,c)}$, $\leq$ is the usual ordering on traces, $\tilde{\lambda} = \lambda_{(\Sigma,c)}$ and

$$\underline{u}_1 \,\tilde{\#}\, \underline{u}_2 \Leftrightarrow \exists \underline{u}'_1,\underline{u}'_2 \in O_{(\Sigma,c)} : \underline{u}'_1 \neq \underline{u}'_2 \wedge \underline{u}'_1 \leq \underline{u}_1 \wedge \underline{u}'_2 \leq \underline{u}_2 \wedge {}^\bullet\underline{u}'_1 \cap {}^\bullet\underline{u}'_2 \neq \varnothing.$$

Thus, if $\Lambda(Les(\Sigma,c)) = B'$, then

- $O' = \tilde{T} = O_{(\Sigma,c)} = \overline{O}$
- $E' = T_\Sigma = E_{(\Sigma,c)} = \overline{E}$ and
- $\lambda' = \lambda_{(\Sigma,c)} = \overline{\lambda}$.

Thus, to complete the proof, we need to show that $P' = \overline{P}$.

If $p \in \overline{P}$, then certainly $p \leq O'$. Suppose $\underline{u}_1,\underline{u}_2 \in p$. We shall show that $\neg \underline{u}_1 \,\tilde{\#}\, \underline{u}_2$ and that hence $p \in P'$. Now since $p \in \overline{P}$, then p is $\#_{(\Sigma,c)}$-conflict-free in $O_{(\Sigma,c)}$ and so $\underline{u}'_1,\underline{u}'_2 \in p'$, some $p' \in P_{(\Sigma,c)}$. If $\underline{u}_1 \,\tilde{\#}\, \underline{u}_2$, then ${}^\bullet\underline{u}'_1 \cap {}^\bullet\underline{u}'_2 \neq \varnothing$, for some $\underline{u}'_1 \leq \underline{u}_1$, $\underline{u}'_2 \leq \underline{u}_2$ such that $\underline{u}'_1 \neq \underline{u}'_2$. But this means that ${}^\bullet\underline{u}'_1 \cap {}^\bullet\underline{u}'_2 \neq \varnothing$, for $\underline{u}'_1,\underline{u}'_2 \in p'$, contradicting the fact that $Proc_\Sigma(p')$ is an MG net. Hence, $\neg \underline{u}_1 \,\tilde{\#}\, \underline{u}_2$ so $p \in P'$. Thus, $\overline{P} \subseteq P'$.

Next, let $p \in P'$. If $p_1, p_2 \in P'$ such that $p_1, p_2 \subseteq p$, then $p_1 \sqcup p_2 \in P'$ and $p_1 \sqcup p_2 \subseteq p$, since P' is consistently complete. Thus $Pr(p)$ is pairwise consistent in $P_{(\Sigma,c)}$ and hence $\bigsqcup Pr(p)$ exists and belongs to $\overline{P}$. But, by prime algebraicity, $\bigsqcup Pr(p) = p$. Thus, $p \in \overline{P}$ and so $P' \subseteq \overline{P}$.

■

To recap, we have a number of behavioural semantics for a contact-free C/E system, using: classes of process nets, occurrence nets, behavioural presentations and event structures. All of these are equivalent, in the sense that any may be constructed from the others. All of them, in fact, may be obtained using the asynchronous transition system C_Σ.

Chapter 18
Loosely Coupled Systems

> We are here concerned with a new formalism of restrictions on processes in distributed systems parts, called loosely coupled systems. This is a model which syntactically is based only on mutual exclusion of states.
>
> A. MAGGIOLO-SCHETTINI ET AL. [63]

18.1. Introduction.

The next model we examine is pleasantly simple and quite closely related to Nets on the condition/event level.

In fact, we shall show in chapter 21 that every loosely coupled system with initial case may be associated with a net with initial case which map to the same initialised asynchronous transition system. In other words, the loosely coupled system model may be embedded into the net model.

In its simplest form, a loosely coupled system [63, 64] consists of a set of conditions (called phases in [63]) which is partitioned into blocks (called parts in [63]). Conceptually, each part corresponds to a sequential subsystem; at any time precisely one of its conditions holds. In the simplest model and with no further constraint, the parts may change from state to state arbitrarily. The loosely coupled system becomes a system by virtue of a mechanism by which the activity of the parts is constrained.

What is unusual and interesting about this model is that co-ordination is not enforced by synchronisation on events, as is usually the case, but through a relation of mutual exclusion on conditions. Thus, in the simplest form of the model, loosely coupled systems model systems by abstracting away from everything except synchronisation required to enforce mutual exclusion on critical sections.

The extension to the model described in [63] allows one to model systems of finite state machines co-ordinated by the same synchrononisation mechanism. We provide both models with a semantics based on asynchronous transition systems.

18.2. Pure Loosely Coupled Systems.

The following definition is a modified version of that in [63].

18.2.1. DEFINITION. A *loosely coupled system* is a triple $L = (B, P, K)$ where

- B is a finite set of *conditions;*
- P is a partition of B into *parts;*
- K is a symmetric, irreflexive relation (of *mutual exclusion*) on B satisfying

$$\forall p \in P\ \forall b_1, b_2 \in p : b_1 \neq b_2 \Rightarrow b_1\ K\ b_2 \qquad (18.1)$$

We adopt the usual convention about components of loosely coupled systems inheriting subscripts or whatever. For example, L^2 is the triple (B^2, P^2, K^2).

Loosely coupled systems may be represented graphically. Conditions are represented by nodes, parts distinguished by surrounding sets of nodes with closed curves and the mutual exclusion relation is indicated by drawing lines between nodes of different parts which are in that relation. Figure 18.1 shows a system with

$$B = \{1, 2, 3, 4, 5\}$$
$$P = \{\{1, 2, 3\}, \{4, 5\}\}$$
$$K = \{(2,5), (5,2), (4,5), (5,4)\} \cup \{(a,b) \in \{1,2,3\} \times \{1,2,3\} \mid a \neq b\}$$

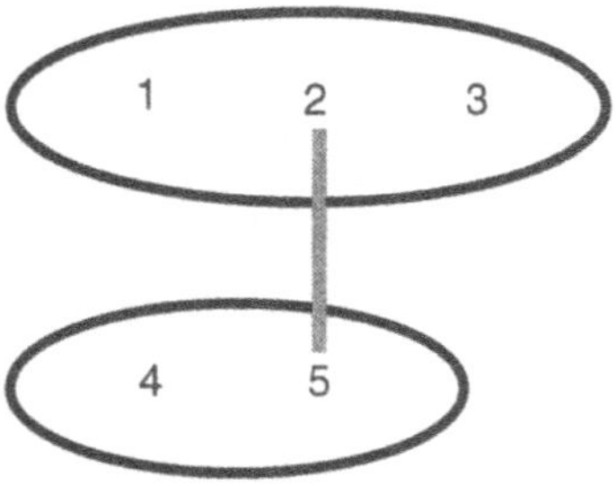

Figure 18.1

Global states of L are collections of conditions, one from each part.

18.2.2. DEFINITION. Let L be a loosely coupled system, then $c \subseteq B$ is a *maximal configuration* of $L \Leftrightarrow$

$$\forall p \in P : |c \cap p| = 1.$$

Thus a maximal configuration is a global state such that each of the sequential components (the parts of the system) is in a specific state. Of course, not all maximal configurations are allowed. If $b_1 \, K \, b_2$ then b_1 and b_2 are not allowed to hold concurrently. A permissible maximal configuration is called a case, as for C/E systems.

c is a *case* of $L \Leftrightarrow c$ is a maximal configuration and $\forall b_1, b_2 \in c: \neg b_1 \, K \, b_2$.

We shall write Q_L to denote the set of all cases of L.

For example, for figure 18.1, we have

$$Q_L = \{\{1,4\},\{1,5\},\{2,4\},\{3,4\},\{3,5\}\}.$$

Elementary events of the system are single state changes. Each event a is therefore uniquely determined by a pair of conditions which we denote $\bullet a$ and $a\bullet$, where $\bullet a$ and $a\bullet$ are distinct and belong to the same part. The interpretation is that the event a has precisely the effect of changing the local state of its associated part from $\bullet a$ to $a\bullet$.

18.2.3. DEFINITION. Let L be a loosely coupled system, then the set of elementary events of L, denoted A_L is defined

$$A_L = \{(b_1,b_2) \in B \times B \mid b_1 \neq b_2 \wedge \exists p \in P: b_1, b_2 \in p\} \tag{18.2}$$

If $a = (b_1,b_2) \in A_L$, then we define $\bullet a = b_1$ and $a\bullet = b_2$. If $b \in B$, then we define

$$\bullet b = \{a \in A_L \mid b = a\bullet\} \text{ and } b\bullet = \{a \in A_L \mid b = \bullet a\}.$$

We may now define a transition in L. It is simply a change of state of one of its parts.

18.2.4. DEFINITION. The relation $\rightarrow_L \subseteq Q_L \times A_L \times Q_L$ is defined as follows. (As usual, we write $c_1 \rightarrow_L^a c_2$ for $(c_1,a,c_2) \in \rightarrow_L$).

$$c_1 \rightarrow_L^a c_2 \Leftrightarrow c_1 - c_2 = \{\bullet a\} \wedge c_2 - c_1 = \{a\bullet\} \tag{18.3}$$

$T_L = (Q_L, A_L, \rightarrow_L)$ is clearly a transition system. That of figure 18.1 is pictured in figure 18.2. Here, in order to simplify the diagram, we are using the fact, which follows immediately from 18.2.3 and 18.2.4, that

$$c_1 \rightarrow_L^{(b_1,b_2)} c_2 \Leftrightarrow c_1 - c_2 = \{b_1\} \wedge c_2 - c_1 = \{b_2\} \Leftrightarrow c_2 \rightarrow_L^{(b_2,b_1)} c_1$$

Each arc is a double arrow representing the state change in both directions. The arcs are not labelled; the elementary events related to each may be deduced from the diagram. For example, the arc connecting $\{1,4\}$ to $\{2,4\}$ represents the fact that $\{1,4\} \rightarrow_L^{(1,2)} \{2,4\}$ and $\{2,4\} \rightarrow_L^{(2,1)} \{1,4\}$.

It remains to construct an independence relation on A_L. In [63] the authors define a_1 to be concurrent with a_2 if and only if $\neg a_1\bullet\ K\ a_2\bullet$. This is reasonable enough. If a_1 and a_2 had concurrent occurrences, then there would be a 'case' of L, corresponding to a point in the potential behaviour of the system, containing $a_1\bullet$ and $a_2\bullet$, which would not be possible if $a_1\bullet\ K\ a_2\bullet$.

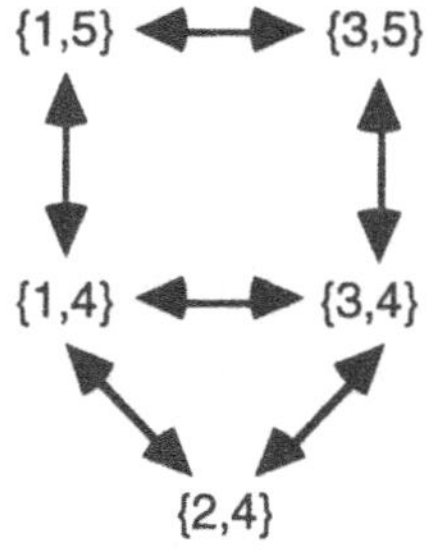

Figure 18.2

However, this definition has its problems. If K relates either of $\bullet a_1$ and $a_1\bullet$ to either of $\bullet a_2$ and $a_2\bullet$, then by a similar argument, if they were concurrent, then there would be a 'case' which would violate the mutual exclusion constraint. Technically, with the relation of [63], $(Q_L, A_L, \rightarrow_L)$ does not become an asynchronous transition system. Consider the example of figure 18.1. Let $a_1 = (2,1)$ and let $a_2 = (4,5)$. Since $(1,5) \notin K$, a_1 is concurrent with a_2 according to [63]. However, we may check from figure 18.2 that

$$\{2,4\} \rightarrow_L^{a_1} \{1,4\} \rightarrow_L^{a_2} \{1,5\}$$

but that for no $c \in Q_L$ do we have

$$\{2,4\} \rightarrow_L^{a_2} c$$

since c would have to be $\{2,5\}$ and $\{2,5\} \in K$. Thus, with this definition of concurrency, the transition system would not be asynchronous. Accordingly, we reject the definition of concurrency given in [63] as failing to satisfy the properties required of an independence relation and offer the following alternative, (which works).

18.2.5. DEFINITION. Suppose $a_1, a_2 \in A_L$. Define

$$a_1\ \iota_L\ a_2 \Leftrightarrow \forall b_1 \in \{\bullet a_1, a_1\bullet\}\ \forall b_2 \in \{\bullet a_2, a_2\bullet\}: \neg b_1\ K\ b_2 \qquad (18.4)$$

With this definition, we have our non-interleaving semantics.

18.2.6. PROPOSITION. Let L be a loosely coupled system and let $C_L = (Q_L, A_L, \rightarrow_L, \iota_L)$, then C_L is an asynchronous transition system. Furthermore, C_L is forward stable.

PROOF. By (18.3),

$$c \rightarrow_L^a c_1 \wedge c \rightarrow_L^a c_2 \Rightarrow c_1 = (c - \bullet a) \cup \{a \bullet\} = c_2$$

and so $T(C_L)$ is unambiguous.

Let $c_1 \rightarrow_L^{a_1} c_2 \rightarrow_L^{a_2} c_3$ with $a_1 \; \iota_L \; a_2$. and let $p_1, p_2 \in P$ such that $\bullet a_i \in p_i$, each i then $p_1 \neq p_2$, by (18.1) and (18.4). Define $\hat{c}_2 = (c_1 - \{\bullet a_2\}) \cup \{a_2 \bullet\}$. We show that $\hat{c}_2 \in Q_L$. First, $\hat{c}_2$ is a maximal configuration, since for each $p \in P$,

$$\hat{c}_2 \cap p = \begin{cases} c_1 \cap p \text{ if } p \neq p_2 \\ a_2 \bullet \text{ otherwise} \end{cases}$$

that is, $|\hat{c}_2 \cap p| = 1$, each $p \in P$. Now suppose that there exists $b_1, b_2 \in \hat{c}_2$ such that $b_1 \, K \, b_2$. We shall derive a contradiction, thereby showing that $\hat{c}_2$ is a case.

If $b_1 \, K \, b_2$, then we cannot have $\{b_1, b_2\} \subseteq c_1$, as $c_1 \in Q_L$, so one of the b_i must belong to $\hat{c}_2 - c_1 = \{a_2 \bullet\}$. Suppose, without loss of generality that $b_2 = a_2 \bullet$. We thus have $b_1 \, K \, a_2 \bullet$. To obtain the contradiction, we show that $b_1 \in c_3$. Since $c_2 \rightarrow_L^{a_2} c_3$, $a_2 \bullet \in c_3$, and so we would have $b_1, a_2 \bullet \in c_3$, contradicting $b_1 \, K \, a_2 \bullet$.

We first note $b_1 \neq b_2$, since K is irreflexive, and so $b_1 \in \hat{c}_2 - \{a_2 \bullet\} \subseteq c_1$, and so if $b_1 \notin c_3$, then $b_1 \in c_1 - c_3 = \{\bullet a_1, \bullet a_2\}$. But if $b_1 = \bullet a_1$, then $\bullet a_1 \, K \, a_2 \bullet$ contradicting $a_1 \; \iota_L \; a_2$, while if $b_1 = \bullet a_2$, then $\bullet a_2, a_2 \bullet \in \hat{c}_2$ which means that $|\hat{c}_2 \cap p_2| > 1$, contradicting $\hat{c}_2$ a maximal configuration. This establishes that $b_1 \in c_3$ which concludes the proof that $\hat{c}_2 \in Q_L$.

The reader may easily check that $c_1 - \hat{c}_2 = \{\bullet a_2\}$, $\hat{c}_2 - c_1 = \{a_2 \bullet\}$, $\hat{c}_2 - c_3 = \{\bullet a_1\}$ and $c_3 - \hat{c}_2 = \{a_1 \bullet\}$, so that $c_1 \rightarrow_L^{a_2} \hat{c}_2 \rightarrow_L^{a_1} c_3$, as required.

Finally, we show that C_L is forward stable.

Let $c \rightarrow_L^{a_1} c_1$ and $c \rightarrow_L^{a_2} c_2$ with $a_1 \; \iota_L \; a_2$. and let $p_1, p_2 \in P$ such that $\bullet a_i \in p_i$, each i then $p_1 \neq p_2$, by (18.1) and (18.4). Define $c' = (c_1 - \{\bullet a_2\}) \cup \{a_2 \bullet\}$. We show that $c' \in Q_L$. First, c' is a maximal configuration, since for each $p \in P$,

$$c' \cap p = \begin{cases} c_1 \cap p \text{ if } p \neq p_2 \\ a_2 \bullet \text{ otherwise} \end{cases}$$

that is, $|c' \cap p| = 1$, each $p \in P$. Now suppose that there exists $b_1, b_2 \in c'$ such that $b_1 \, K \, b_2$. We shall derive a contradiction, thereby showing that c' is a case.

If $b_1 \, K \, b_2$, then we cannot have $\{b_1, b_2\} \subseteq c_1$, as $c_1 \in Q_L$, so one of the b_i must belong to $c' - c_1 = \{a_2\bullet\}$. Suppose, without loss of generality that $b_2 = a_2\bullet$. We thus have $b_1 \, K \, a_2\bullet$. To obtain the contradiction, we show that $b_1 \in c_2$. Since $c \rightarrow_L^{a_2} c_2$, $a_2\bullet \in c_2$, and so we would have $b_1, a_2\bullet \in c_2$, contradicting $b_1 \, K \, a_2\bullet$.

We first note $b_1 \neq a_2\bullet$, since K is irreflexive, and so $b_1 \in c' - \{a_2\bullet\} \subseteq c_1$, and so if $b_1 \notin c_2$, then $b_1 \in c_1 - c_2 = \{a_1\bullet\}$. But this means that $a_1\bullet \, K \, a_2\bullet$ contradicting $a_1 \, \iota_L \, a_2$. This establishes that $b_1 \in c_2$ which concludes the proof that $c' \in Q_L$.

By definition of c', $c_1 \rightarrow_L^{a_2} c'$, and since $\bullet a_1$, $a_1\bullet$, $\bullet a_2$ and $a_2\bullet$ are distinct, by 18.2.3 and 18.2.5, $c' = (c - \{\bullet a_1, \bullet a_2\}) \cup \{a_1\bullet, a_2\bullet\} = (c_2 - \{\bullet a_1\}) \cup \{a_1\bullet\}$, and so $c_2 \rightarrow_L^{a_1} c'$.

■

18.2.6 allows us to construct, for any loosely coupled system L and initial state $c \in Q_L$, an asynchronous behavioural presentation, $B_{(C_L, c)}$. This gives the model a non-interleaving semantics.

18.3. Loosely Coupled Systems with Transitions.

The model described in 18.2.1 abstracts away from everything except the idea of independent sequential processes synchronised by mutual exclusion. [63] deals with an extension of the model in which the parts of a loosely coupled system are equipped with a transition structure.

18.3.1. DEFINITION. A *loosely coupled system with a transition relation* (LCST) is a pair $S = (L, R)$ where L is a loosely coupled system and $R \subseteq A_L$.

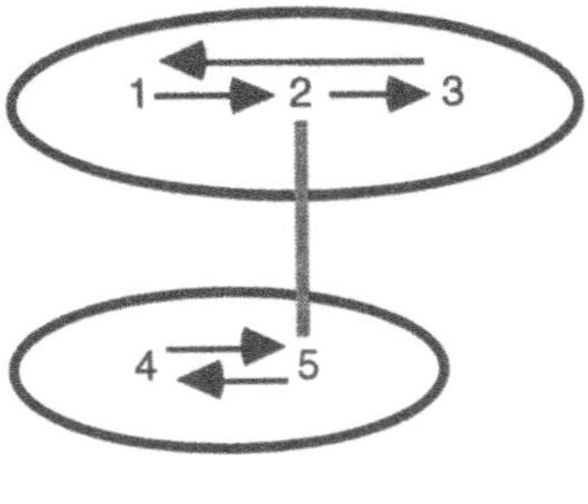

Figure 18.3

R says which local state changes are permitted.

Figure 18.3 shows a picture of a LCST derived from the example of figure 18.1. Here $R=\{a,b,c,d,e\}$, where $a=(1,2)$, $b=(2,3)$, $c=(3,1)$, $d=(4,5)$ and $e=(5,4)$.

The non-interleaving semantics of S may readily be obtained from that of L. We simply remove some of the transitions!

18.3.2. DEFINITION. If $S=(L,R)$ is a LCST, then $C_S=(Q_S,A_S,\rightarrow_S,\iota_S)$ where

1• $Q_S=Q_L$;

2• $A_S=R$;

3• $\rightarrow_S=\rightarrow_L\cap(Q_L\times R\times Q_S)$;

4• $\iota_S=\iota_L\cap(R\times R)$.

It is clear that C_S is unambiguous (because C_L is). Also note that conditions 18.3.2 (3) and (4) guarantee that if $c_1\rightarrow_S^{a_1}c_2\rightarrow_S^{a_2}c_3$ with $a_1\,\iota_S\,a_2$ then $c_1\rightarrow_L^{a_1}c_2\rightarrow_L^{a_2}c_3$ with $a_1\,\iota_L\,a_2$ so by 18.2.6, there exists $\hat{c}_2\in Q_L$ such that $c_1\rightarrow_L^{a_2}\hat{c}_2\rightarrow_L^{a_1}c_3$ and as $a_1,a_2\in R$, this entails that there exists $\hat{c}_2\in Q_L$ such that $c_1\rightarrow_S^{a_2}\hat{c}_2\rightarrow_S^{a_1}c_3$.

Thus, C_S is an ι_S asynchronous transition system. A similar argument proves forward stability. We have:

18.3.3. PROPOSITION. If S is a LCST, then C_S is an asynchronous transition system. Furthermore, C_S is forward stable.

■

Figure 18.4 shows the asynchronous transition system derived from the LCST of figure 18.3. In the figure, both $c\,\iota_S\,d$ and $c\,\iota_S\,e$.

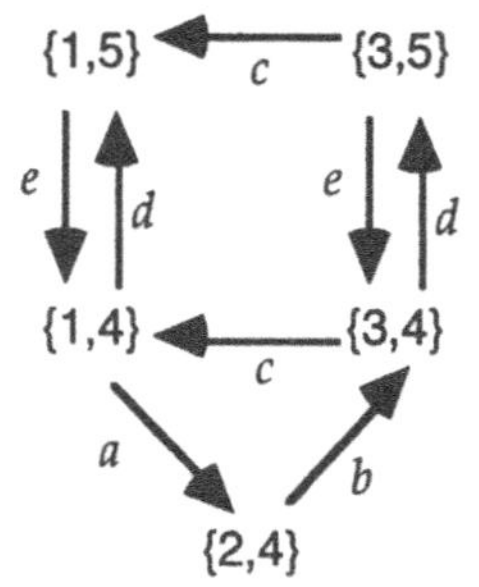

Figure 18.4.

Chapter 19
Categorical Connections

> Category theory asks of every type of Mathematical object: "What are the morphisms?"
>
> S. MAC LANE [67]

19.1. Introduction.

In various parts of this book, we have established relationships between behavioural presentations, trace and vector languages and asynchronous transition systems. We have also discovered relations between objects of the latter kinds corresponding to isomorphism of behavioural presentations.

However, there are useful structural relationships between objects other than isomorphism. For example, in chapter 21, we present conditions under which a given initialised asynchronous transition system may be generated (up to isomorphism) by a Petri Net. It turns out that the condition is that the initialised asynchronous transition system in question may be embedded in a 'concurrent product' of transition systems of a particular kind.

The embedding is an example of a morphism. A class of objects together with their morphisms form a category.

For the sake of readers who have successfully avoided Category Theory, a category is an abstraction from the idea of the class of objects of a mathematical theory (for example, groups, vector spaces, topological spaces) together with their structure-preserving maps (for example, group homomorphisms, linear transformations, continuous functions).

In this chapter, we construct categories for asynchronous transition systems with and without initial states, left-closed trace languages and discrete behavioural presentations and establish the formal relationships between these categories.

A *category* thus consists of a class $\underline{\underline{O}}$ of *objects* and a class of $\underline{\underline{A}}$ of *arrows* or *morphisms*. Each morphism f has a *domain*, $dom(f) \in \underline{\underline{O}}$ and a *codomain*, $cod(f) \in \underline{\underline{O}}$. We write $f{:}d \rightarrow c$ to denote that f is a morphism with domain

d and codomain c. There is a partial composition of morphisms. If $f, g \in \underline{\underline{A}}$ and $cod(f) = dom(g)$, then the composition of f and g is defined and denoted by $g \circ f$. It satisfies $dom(g \circ f) = dom(f)$ and $cod(g \circ f) = cod(g)$. Each object **o** is associated with a unique *identity morphism*, denoted by Id_o, which is defined to have $dom(Id_o) = cod(Id_o) = \mathbf{o}$. We also require that if $f, g, h \in \underline{\underline{A}}$, then

$$\forall f \in \underline{\underline{A}}: dom(f) = \mathbf{o} \Rightarrow f \circ Id_o = f \wedge cod(f) = \mathbf{o} \Rightarrow Id_o \circ f = f;$$
$$\forall f, g, h \in \underline{\underline{A}}: cod(f) = dom(g) \wedge cod(g) = dom(h) \Rightarrow (h \circ g) \circ f = h \circ (g \circ f)$$

We shall follow [67] in using the same symbol for the class of objects and of morphisms. If $\underline{\underline{C}}$ denotes a category, then the assertion

$$\mathbf{o} \in \underline{\underline{C}}$$

indicates that **o** is an object in $\underline{\underline{C}}$, whereas the assertion

$$f \ in \ \underline{\underline{C}}$$

indicates that f is a morphism in $\underline{\underline{C}}$.

$f: d \to c$ is an *isomorphism* in a category $\underline{\underline{C}} \Leftrightarrow$

$$\exists g: c \to d: g \circ f = Id_d \wedge f \circ g = Id_c.$$

The morphism g may easily be shown to be unique. It is called the *inverse* of f. If there exists an isomorphism $f: d \to c$, then we shall say that d is *isomorphic* to c, and write $d \equiv c$. It may easily be shown that identity morphisms, inverses of isomorphisms and composites of isomorphisms are also isomorphisms and that accordingly, $\equiv$ is an equivalence relation.

19.2. Asynchronous Transition Systems.

A category of asynchronous transition systems has been defined and studied in M. A. Bednarcyk's doctoral thesis [43].

19.2.1. DEFINITION. Suppose C_1 and C_2 are asynchronous transition systems, then a *morphism* from C_1 to C_2 is a pair $f = (f_Q, f_A)$ where $f_Q: Q_1 \to Q_2$, $f_A: A_1 \to A_2 \cup \{\Omega\}$, and if $a, b \in A_1$ and $q, q' \in Q_1$, then

1• $f_A(a) \neq \Omega \neq f_A(b) \Rightarrow (a \, \iota_1 \, b \Rightarrow f_A(a) \, \iota_2 \, f_A(b));$

2• $f_A(a) \neq \Omega \Rightarrow (q \to_1^a q' \Rightarrow f_Q(q) \to_2^{f_A(a)} f_Q(q'))$;

3• $f_A(a) = \Omega \Rightarrow (q \to_1^a q' \Rightarrow f_Q(q) = f_Q(q'))$.

In such a case, we define $dom(f) = C_1$ and $codom(f) = C_2$.

Write $f: C_1 \to C_2$ to denote that f is a morphism from C_1 to C_2.

In the sequel we use the subscript convention of 2.1.2, so that, for example, if f_2 is a morphism, then f_2 is a pair $((f_2)_Q, (f_2)_A)$.

19.2.1 is illustrated in figure 19.1.

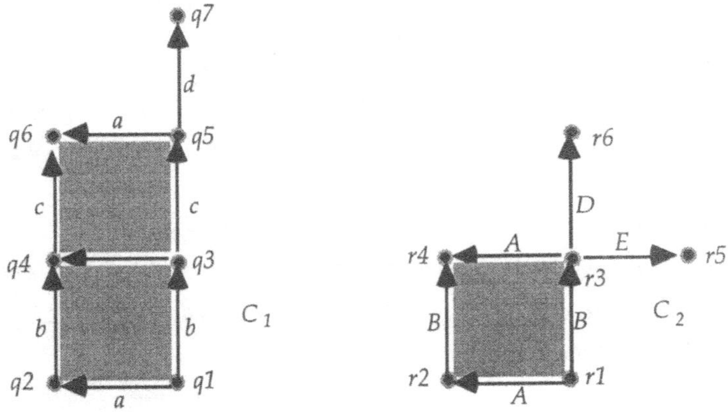

$f_Q(q1) = r1$ $f_Q(q2) = r2$ $f_Q(q3) = r3$ $f_Q(q4) = r4$

$f_Q(q5) = r3$ $f_Q(q6) = r4$ $f_Q(q7) = r6$

$f_A(a) = A$ $f_A(b) = B$ $f_A(c) = \Omega$ $f_A(d) = D$

Figure 19.1.

Thus, a morphism relates the graphical structures of C_1 and C_2 by either

1• sending arcs to arcs in a way that respects incidence of arcs to nodes and the labelling functions or

2• collapsing an arc and its incident nodes to a single node when the label of the arc maps to Ω.

We also need a definition of composition of morphisms and of identity morphisms.

19.2.2. DEFINITION. If $f: C_1 \to C_2$ and $g: C_2 \to C_3$, then we define $g \circ f = (g_Q \circ f_Q, g_A * f_A)$ where $g_Q \circ f_Q$ is composition of functions and

$$(g_A * f_A)(a) = \begin{cases} g_A(f_A(a)) \text{ if } f_A(a) \neq \Omega \\ \Omega \text{ otherwise} \end{cases} \tag{19.1}$$

If C is an asynchronous transition system, then we define $Id_C = (Id_Q, Inc_A)$, where Id_Q is the identity function of Q and $Inc_A : A \to A \cup \{\Omega\}$ is the inclusion mapping.

19.2.3. PROPOSITION. If **ATS** denotes the class of all asynchronous transition systems then **ATS** is a category with morphisms as in 19.2.1 and composition and identities as in 19.2.2.

PROOF. We show that the composition of morphisms is a morphisms. The rest of the proof merely involves checking that the definition are satisfied.

Suppose that $f: C_1 \to C_2$ and $g: C_2 \to C_3$, then $g \circ f$ is defined.

Let $a, b \in A_1$ and suppose that $(g \circ f)_A(a) \neq \Omega \neq (g \circ f)_A(b)$ and $a \,\iota_1\, b$, then $f_A(a) \neq \Omega \neq f_A(b)$, by (19.1), so $f_A(a) \,\iota_2\, f_A(b)$, by 19.2.1. (1), and since $f_A(a) \neq \Omega \neq f_A(b)$ and $f_A(a) \,\iota_2\, f_A(b)$, $g_A(f_A(a)) \,\iota_3\, g_A(f_A(a))$, by 19.2.1 (1), that is, $(g \circ f)_A(a) \,\iota_3\, (g \circ f)_A(b)$.

Next, suppose that $a \in A_1$ and $q_1, q_2 \in Q_1$ such that $q_1 \to_1^a q_2$.

If $(g \circ f)_A(a) \neq \Omega$, then $f_A(a) \neq \Omega$ so that $f_Q(q_1) \to_2^{f_A(a)} f_Q(q_2)$, by 19.2.1 (2) and since $(g \circ f)_A(a) \neq \Omega$, $g_Q(f_Q(q_1)) \to_3^{g_A(f_A(a))} g_Q(f_Q(q_2))$, by 19.2.1 (2), that is, $(g \circ f)_Q(q_1) \to_3^{(g \circ f)_A(a)} (g \circ f)(q_2)$.

Finally, suppose that $(g \circ f)_A(a) = \Omega$. If $f_A(a) = \Omega$, then $f_Q(q_1) = f_Q(q_2)$, by 19.2.1 (3) so $(g \circ f)_Q(q_1) = g_Q(f_Q(q_1)) = g_Q(f_Q(q_2)) = (g \circ f)_Q(q_2)$. If $f_A(a) \neq \Omega$, then $f_Q(q_1) \to_2^{f_A(a)} f_Q(q_2)$, by 19.2.1 (2), and $g_A(f_A(a)) = (g \circ f)_A(a) = \Omega$. Thus, $(g \circ f)_Q(q_1) = g_Q(f_Q(q_1)) = g_Q(f_Q(q_2)) = (g \circ f)_Q(q_2)$, by 19.2.1 (3).

■

19.2.4. PROPOSITION. Suppose $f: C_1 \to C_2$, then f is an isomorphism $\Leftrightarrow$

1• f_Q is bijective;
2• f_A is injective and $f_A(A_1) = A_2$;
3• $\forall a, b \in A_1 : a \,\iota_1\, b \Leftrightarrow f_A(a) \,\iota_2\, f_A(b)$;
4• $\forall a \in A_1 \forall q, q' \in Q_1 : q \to_1^a q' \Leftrightarrow f_Q(q) \to_1^{f_A(a)} f_Q(q')$".

PROOF. Let $g = (f_Q^{-1}, g_A)$, where g_A is given by $g_A(a) = a' \Leftrightarrow f_A(a') = a$. It is straightforward to show that g is a morphism and the inverse of f and that conversely if f has an inverse then it must satisfy (1) to (4).

■

Next, we turn to initialised asynchronous transition systems. Note that we only consider cases in which every part of the asynchronous transition system is 'used'. We do not wish to have to consider irrelevant structure.

19.2.5. DEFINITION. We denote by **ATS*** the class of all initialised asynchronous transition systems. (C,q) such that

1• $Q = R(T(C),q)$;

2• $\forall a \in A\ \exists q' \in R(T(C),q): q' \rightarrow^{a}$;

3• $\forall a,b \in A: a\,\iota\,b \Rightarrow \exists q' \in R(T(C),q): q' \rightarrow^{ab}$.

A *morphism* from (C_1,q_1) to (C_2,q_2) is a morphism $f:C_1 \rightarrow C_2$ such that $f_Q(q_1) = q_2$. Write $f:(C_1,q_1) \rightarrow (C_2,q_2)$ to indicate that f is a morphism from (C_1,q_1) to (C_2,q_2)

Identity morphisms and composition of morphisms is defined as in **ATS**.

As an almost immediate consequence of 19.2.3, we have:

19.2.6. PROPOSITION. **ATS*** is a category. f is an isomorphism in **ATS*** $\Leftrightarrow$ f is an isomorphism in **ATS**.

■

19.3. Trace Languages.

We have seen that the class of all initialised transition systems may be made into a category. Let us next see what we can do with trace languages.

By 15.2.2, each $(C,q) \in$ **ATS*** determines a left-closed trace language $L(C,q)$. Let us see how morphisms carry over to left-closed trace languages.

19.3.1. LEMMA. If $f:(C,q) \rightarrow (C',q')$, then f_A determines a monoid morphism $f_A{}^*: A_\iota{}^* \rightarrow A'_{\iota'}{}^*$ satisfying $f_A{}^*(x_\iota) = f_A(x)_{\iota'}$.

Furthermore if $q_1 \rightarrow^{x_\iota} q_2$, then $f_Q(q_1) \rightarrow^{f_A{}^*(x_\iota)} f_Q(q_2)$.

In particular, $f_A{}^*(L(C,q)) = L(C',q')$.

PROOF. By 19.2.1 (1), if $a,b \in A$ and $u,v \in A^*$, then

$$u.a.b.v \equiv_\iota^{(1)} u.b.a.v \Rightarrow f_A(u).f_A(a).f_A(b).f_A(v) \equiv_{\iota'}^{(1)} f_A(u).f_A(b).f_A(a).f_A(v)$$

Note that this is true if either or both of $f_A(a) = \Omega$ or $f_A(b) = \Omega$ apply.

Since $\equiv_\iota$ is the transitive closure of $\equiv_\iota^{(1)}$ and $\equiv_{\iota'}$ is the transitive closure of $\equiv_{\iota'}^{(1)}$, then we may deduce that

$$\forall x,y \in A^*: x \equiv_\iota y \Rightarrow f_A(x) \equiv_{\iota'} f_A(y).$$

Thus, $f_A{}^*$ is well defined and is clearly a monoid morphism. The claim that if $q_1 \to^{x_\iota} q_2$, then $f_Q(q_1) \to^{f_A{}^*(x_\iota)} f_Q(q_2)$ follows by induction on $|x|$ using 19.2.1 (2) and (3).

■

Now that we are beginning to compare trace languages, we need to be a little more detailed about their *sorts*, that is to say, the alphabets A and independence relations ι of the trace monoid $A_\iota{}^*$ of which the languages are considered to lie. We would like morphisms of the trace languages to be the type of function described in 19.3.1. But the descriptions of such functions make explicit reference to domains, ranges and independence relations. Following [43], we put left-closed trace languages into a category in the following form.

19.3.2. DEFINITION. Let **LCTL** denote the class of all triples (A, ι, L), where A is a set, ι is an independence relation on A and $L \subseteq A_\iota{}^*$ is left-closed.

We shall use the usual subscript convention. Thus, $D_\$ = (A_\$, \iota_\$, L_\$)$

If $D_1, D_2 \in$ **LCTL**, then a morphism from D_1 to D_2 is a function $f: A_1 \to A_2 \cup \{\Omega\}$ such that

1• $\forall a, b \in A_1 : f(a) \neq \Omega \neq f(b) \Rightarrow (a \, \iota_1 \, b \Rightarrow f(a) \, \iota_2 \, f(b));$

2• $\forall \underline{x} \in L_1 \; \forall y \in \underline{x} \; \exists \underline{z} \in L_2 : f(y) \in \underline{z}.$

Write $f: D_1 \to D_2$ to denote that f is a morphism from D_1 to D_2

Composition of morphisms is defined as in (19.1). If $D \in$ **LCTL**, then the identity morphism is defined by $Id_D = inc_A$.

19.3.3. PROPOSITION. **LCTL** is a category.

PROOF. We show that the composition of morphisms is a morphism. The remaining properties of a category may be easily verified. Let $f: D_1 \to D_2$ and $g: D_2 \to D_3$. The proof that $g \circ f$ satisfies (1) of 19.3.2 is exactly the same as (1) of 19.2.3. Suppose that $\underline{x} \in L_1$ and that $x' \in \underline{x}$, then by (2) of 19.3.2, there exists $\underline{y} \in L_2$ such that $f(x') \in \underline{y}$ and hence by 19.3.2 again, there exists $\underline{z} \in L_3$ such that $g(f(x')) \in \underline{z}$.

■

The following fact will come in useful in the sequel. As in the proof of 19.3.1, if $f: D_1 \to D_2$, then f determines a function $f^*: (A_1)_{\iota_1}{}^* \to (A_2)_{\iota_2}{}^*$, which satisfies $f^*(x_{\iota_1}) = f(x)_{\iota_2}$. f^* is a monoid morphism and also a poset morphism, This allows us to reformulate 19.3.2 (2) as:

2'• $f^*(L_1) \subseteq L_2$.

19.3.4. LEMMA. If $f: D \to \hat{D}$, then f is an isomorphism $\Leftrightarrow$

1• f is injective and $f(A) = \hat{A}$;

2• $f^*(L) = \hat{L}$;

3• $\forall a, b \in A: a \iota b \Leftrightarrow f(a) \hat{\iota} f(b)$.

PROOF. The conditions ensure that we may define a function $g: \hat{A} \to A \cup \{\Omega\}$ satisfying $g(\hat{a}) = a \Leftrightarrow f(a) = \hat{a}$. It is easy to show that g is a morphism with $g*f = inc_A$ and $f*g = inc_{\hat{A}}$ as functions, that is $g \circ f = Id_D$ and $f \circ g = Id_{\hat{D}}$ as morphisms. Hence, f is an isomorphism.

Conversely, suppose f is an isomorphism with inverse g. If $\hat{a} \in \hat{A}$, then $f(g(\hat{a})) = \hat{a}$ and so $\hat{A} \subseteq f(A)$. If $a \in A$, then as $g(f(a)) = a$, we cannot have $f(a) = \Omega$. Hence $\hat{A} = f(A)$. f is injective by simple set theory.

As $f^*(L) \subseteq \hat{L}$ and $g^*(\hat{L}) \subseteq L$, it follows that $\hat{L} = f^*(g^*(\hat{L})) \subseteq f^*(L) \subseteq \hat{L}$ and hence $f^*(L) = \hat{L}$. Finally, if $a, b \in A$, then, by 19.3.2 (1) $a \iota b \Rightarrow f(a) \hat{\iota} f(b)$ and $f(a) \hat{\iota} f(b) \Rightarrow g(f(a)) \iota g(f(b)) \Rightarrow a \iota b$.

■

We now have two categories **ATS*** and **LCTL** which we know to be related through the mapping that sends (C, q) to $(A, \iota, L(C, q))$. We also know, following 19.3.1, that morphisms in **ATS*** are related to morphisms in **LCTL**. The relationship is functorial.

A *functor* is a kind of morphism of categories. More formally, if **C** and **D** are categories, then a functor $\mathbf{F}: \mathbf{C} \to \mathbf{D}$ consists of a pair of maps (both called **F**) mapping the objects of **C** to the objects of **D** and the morphisms of **C** to the morphisms of **D**, in such a way that if $f: \mathbf{c} \to \mathbf{c}'$ in **C**, then $\mathbf{F}(f): \mathbf{F}(\mathbf{c}) \to \mathbf{F}(\mathbf{c}')$ in **D** and

1• $\mathbf{F}(Id_{\mathbf{c}}) = Id_{\mathbf{F}(\mathbf{c})}$, all $\mathbf{c} \in \mathbf{C}$;

2• If $f \circ g$ is defined, then so is $\mathbf{F}(f) \circ \mathbf{F}(g)$ and $\mathbf{F}(f \circ g) = \mathbf{F}(f) \circ \mathbf{F}(g)$.

19.3.5. PROPOSITION. The pair of mappings that sends $(C, q) \in$ **ATS*** to $(A, \iota, L(C, q)) \in$ **LCTL** and the morphism f *in* **ATS*** to f_A *in* **LCTL** is a functor $\mathbf{L}$: **ATS*** $\to$ **LCTL**.

PROOF. If $(C,q),(C',q') \in \underline{\underline{\mathbf{ATS}^*}}$ and $f:(C,q)\to(C',q')$, then f_A is a morphism, since 19.3.2 (1) holds because of 19.2.1 (1) and 19.3.2 (2) holds because of 19.3.1.

If $(C,q) \in \underline{\underline{\mathbf{ATS}^*}}$, then $\mathbf{L}(Id_{(C,q)}) = inc_A = Id_{(A,\iota,\mathbf{L}(C,q))} = Id_{\mathbf{L}(C,q)}$.

Finally, if f,g *in* $\underline{\underline{\mathbf{ATS}^*}}$ and $dom(g)=cod(f)$, then

$$dom(\mathbf{L}(g)) = dom(g_A) = cod(f_A) = cod(\mathbf{L}(f))$$

and

$$\mathbf{L}(g\circ f) = (g\circ f)_A = g_A * f_A = \mathbf{L}(g)\circ\mathbf{L}(f).$$

■

Now, not only may we transform objects and morphisms from $\underline{\underline{\mathbf{ATS}^*}}$ into objects and morphisms from $\underline{\underline{\mathbf{LCTL}}}$, we may also go in the opposite direction.

19.3.6. DEFINITION. If $D \in \underline{\underline{\mathbf{LCTL}}}$, then define

$$\mathbf{A}(D) = ((L, A, \to_L, \iota), \Omega_\iota)$$

where

$$x_\iota \to^a y_\iota \Leftrightarrow y_\iota = x_\iota . a_\iota$$

and if $f:D\to\hat{D}$, then $\mathbf{A}(f)=(f_Q,f)$, where $f_Q:L\to\hat{L}$ is given by $f_Q(x_\iota)=f^*(x_\iota)$. f_Q is well defined because of 19.3.2 (2').

19.3.7. PROPOSITION. $\mathbf{A}$ is a functor $\mathbf{A}:\underline{\underline{\mathbf{LCTL}}}\to\underline{\underline{\mathbf{ATS}^*}}$.

PROOF. We first show that $(L,A,\to_L,\iota)$ is an asynchronous transition system. It is unambiguous because

$$x_\iota \to_L^a y_\iota \wedge x_\iota \to_L^a y'_\iota \Rightarrow y_\iota = x_\iota . a_\iota = y'_\iota$$

and it satisfies the lozenge rule because if $x_\iota \to_L^a y_\iota \to_L^b z_\iota$ and $a\,\iota\,b$, then $z_\iota = y_\iota . b_\iota = x_\iota . a_\iota . b_\iota = x_\iota . b_\iota . a_\iota$ and so $x_\iota \to_L^b x_\iota . b_\iota \to_L^a z_\iota$.

If $f:D\to\hat{D}$, then $\mathbf{A}(f):\mathbf{A}(D)\to\mathbf{A}(\hat{D})$ is a morphism because

$$x_\iota \to_L^a y_\iota \Rightarrow y_\iota = (x.a)_\iota \Rightarrow f^*(y_\iota) = f^*(x_\iota).f^*(a_\iota)$$

$$\Rightarrow \begin{cases} f^*(x_\iota) \rightarrow_L^{f(a)} f^*(y_\iota) \text{ if } f(a) \neq \Omega \\ f^*(x_\iota) = f^*(y_\iota) \text{ otherwise} \end{cases}$$

which shows that 19.2.1 (2) and (3) hold. 19.2.1 (1) holds because of 19.3.2 (1).

If $D \in \underline{\underline{\mathrm{LCTL}}}$, then $\mathbf{A}(Id_D) = \mathbf{A}(inc_A) = (Id_L, inc_A) = Id_{\mathbf{A}(D)}$

Finally, it takes a simple computation to show that if $f \circ g$ is defined then so is $\mathbf{A}(f) \circ \mathbf{A}(g)$ and $\mathbf{A}(f \circ g) = \mathbf{A}(f) \circ \mathbf{A}(g)$.

■

Given the close relationship between asynchronous transition systems and left-closed trace languages, it is reasonable to ask what relationship there is between **A** and **L** The answer is that they are mutually *adjoint*.

Generally, if $\mathbf{F}:\underline{\underline{\mathrm{C}}} \rightarrow \underline{\underline{\mathrm{D}}}$ and $\mathbf{G}:\underline{\underline{\mathrm{D}}} \rightarrow \underline{\underline{\mathrm{C}}}$ are functors, then **F** is a left adjoint to **G** and **G** is a right adjoint to **F** if and only if for each $\mathbf{c} \in \underline{\underline{\mathrm{C}}}$, there exists a morphism $h_c:\mathbf{c} \rightarrow \mathbf{GF}(\mathbf{c})$ such that for any $\mathbf{d} \in \underline{\underline{\mathrm{D}}}$ and morphism $f:\mathbf{c} \rightarrow \mathbf{G}(\mathbf{d})$, there exists a unique morphism $g:\mathbf{F}(\mathbf{c}) \rightarrow \mathbf{d}$ such that the diagram of figure 19.2 commutes, that is, $f = \mathbf{G}(g) \circ h_c$.

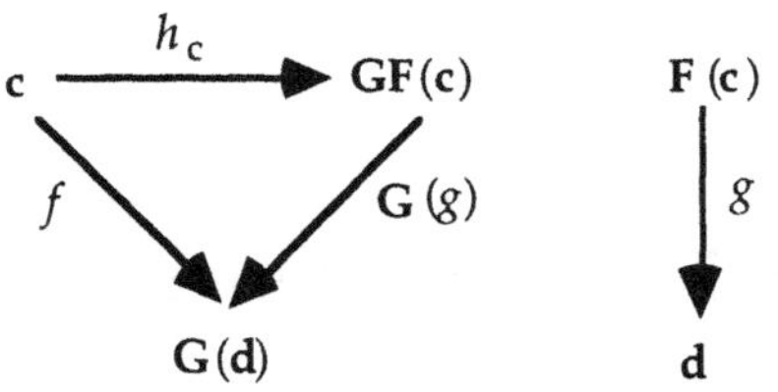

Figure 19.2.

19.3.8. PROPOSITION. **A** is left adjoint to **L**.

PROOF. If $D \in \underline{\underline{\mathrm{LCTL}}}$, then by induction on $|x|$, $\Omega_\iota \rightarrow_L^x y_\iota \Leftrightarrow x_\iota = y_\iota$, so $L(\mathbf{A}(D)) = L$ and hence $\mathbf{LA}(D)) = D$. Define $h_D = Id_D$.

We shall show that if $f:D \rightarrow \mathbf{L}(\hat{C},\hat{q})$ then there exists unique $g:\mathbf{A}(D) \rightarrow (\hat{C},\hat{q})$ such that $\mathbf{L}(g) = f$ and consequently $\mathbf{L}(g) \circ h_D = f$.

Define $g_A = f$. We know that $f:A \rightarrow \hat{A} \cup \{\Omega\}$ determines a mapping $f^*:L \rightarrow L(\hat{C},\hat{q})$. Suppose that $x_\iota \in L$, then by unambiguity, there exists a unique $\tilde{q} \in \hat{Q}$ such that $\hat{q} \,\hat{\rightarrow}^{f^*(x_\iota)}\, \tilde{q}$. Define $g_Q(x_\iota) = \tilde{q}$.

g_A satisfies 19.2.1 (1) since f does, by 19.3.2 (1). Suppose that $x_\iota \rightarrow_L^a y_\iota$, then $y_\iota = x_\iota.a_\iota$, $\hat{q} \,\hat{\rightarrow}^{f^*(x_\iota)}\, g_Q(x_\iota)$ and $\hat{q} \,\hat{\rightarrow}^{f^*(x_\iota.a_\iota)}\, g_Q(y_\iota)$. If $f(a) = \Omega$, then $f^*(x_\iota) = f^*(y_\iota)$ and so $g_Q(x_\iota) = g_Q(y_\iota)$, by unambiguity. On the other hand, if $f(a) \neq \Omega$, then $\hat{q} \hat{\rightarrow}^{f^*(x_{1,1})} g_Q(x_1) \hat{\rightarrow}^{f^*(a_1)} g_Q(y_1)$, so $g_Q(x_1) \hat{\rightarrow}^{f^*(a_1)} g_Q(y_1)$. Thus g_Q satisfies 19.2.1 (2) and (3).

Thus g is a morphism. g also satisfies $\mathbf{L}(g) = g_A = f$. It remains to be shown that g is unique.

Suppose g' also satisfies the conditions of the proposition. Let $x_1 \in L$. Since g' is a morphism, $\Omega_1 \rightarrow_L^{x_1} x_1 \Rightarrow \hat{q} \hat{\rightarrow}^{f^*(x_1)} g'_Q(x_1)$. But $\hat{q} \hat{\rightarrow}^{f^*(x_1)} g_Q(x_1)$. Thus $g'_Q(x_1) = g_Q(x_1)$, by unambiguity. We have shown that $g'_Q = g_Q$. As we have observed, $\mathbf{L}(g') = g'_A = f$. Thus $g'_A = g_A$ and g is indeed unique.

■

Effectively, 19.3.8 says that the structure of $\underline{\underline{\mathbf{LCTL}}}$ is determined by that of $\underline{\underline{\mathbf{ATS^*}}}$. Indeed if $D_1, D_2 \in \underline{\underline{\mathbf{ATS^*}}}$ and $f: D_1 \rightarrow D_2$, then $f: D_1 \rightarrow \mathbf{L}(C_2^*)$, where $C_2^* = \mathbf{A}(D_2)$, as $D_2 = \mathbf{LA}(D_2)$. Thus, there exists a morphism $g: \mathbf{A}(D_1) \rightarrow C_2^*$ such that $f = \mathbf{L}(g)$. In other words, there exists a morphism $f: D_1 \rightarrow D_2$ if and only if there exists a morphism $g: \mathbf{A}(D_1) \rightarrow \mathbf{A}(D_2)$ such that $f = \mathbf{L}(g)$. This extends the view that all left-closed trace languages come from initialised asynchronous transition systems to the wider view that all left-closed trace language morphisms come from initialised asynchronous transition system morphisms.

19.4. Discrete Behavioural Presentations.

Now let us turn to behavioural presentations. Given a $D \in \underline{\underline{\mathbf{ATS^*}}}$, we can construct a linguistic behavioural presentation $\mathbf{B}(D) = B_{L,A}$, using 12.1.1. This gives us the object part of a functor $\mathbf{B}$ from $\underline{\underline{\mathbf{ATS^*}}}$ into the class of linguistic behavioural presentations. If $D_1, D_2 \in \underline{\underline{\mathbf{ATS^*}}}$ and $f: D_1 \rightarrow D_2$, what can we extract which gives something that looks like a morphism $\mathbf{B}(f)$ from B_{L_1,A_1} to B_{L_2,A_2}? Bear in mind that we already have a notion of isomorphism[1], so that $\mathbf{B}(f)$ ought to be some form of generalisation of this concept. In particular, it will be a triple of functions between points, occurrences and events.

The latter is straightforward. Define $\mathbf{B}(f)_E: A_1 \rightarrow A_2 \cup \{\Omega\}$ by $\mathbf{B}(f)_E = f$.

It is also not hard to find a mapping from P_{L_1} to P_{L_2}. Recall that a typical element of P_{L_i} is a set of primes $Pr(x_{1_i})$, $x_{1_i} \in L_i$. We may define $\mathbf{B}(f)_P: P_{L_1} \rightarrow P_{L_2}$, by $\mathbf{B}(f)_P(Pr(x_{1_1})) = Pr(f^*(x_{1_1}))$. As f is a poset morphism, $\mathbf{B}(f)_P$ is also a poset morphism from P_{L_1} to P_{L_2}.

---

[1] See 12.2.13.

Turning to O_{L_1} and O_{L_2}, we do have a mapping defined on O_{L_1}, since $O_{L_1} \subseteq Pr(L_1) \subseteq L_1$. The mapping is defined by $f^*(x_{\iota_1}) = f(x)_{\iota_2}$. However, f^* does not necessarily map a prime to a prime. For example, if $a, b \in A$ such that $a \iota b$, then the function $Id_A{}^*: A^* \to A_\iota{}^*$ sends the prime ab to the non-prime $(ab)_\iota = (ba)_\iota$.

If $u_{\iota_1}.a_{\iota_1}$ is prime, with $a \in A_1$, then $f^*(u_{\iota_1}.a_{\iota_1}) = f^*(u_{\iota_1}).f^*(a_{\iota_1})$. If $f(a) \neq \Omega$, then we may construct a prime in L_2, namely $pr(f^*(u_{\iota_1}), f(a))$.[2] Furthermore, if $u_{\iota_1}.a_{\iota_1} \in Pr(x_{\iota_1})$, then $f^*(u_{\iota_1}.a_{\iota_1}) \leq f^*(x_{\iota_1})$ and so

$$pr(f^*(u_{\iota_1}), f(a)) \in Pr(f^*(x_{\iota_1})) = \mathbf{B}(f)_P)(Pr(x_{\iota_1}))$$

and so it seems promising to map $u_{\iota_1}.a_{\iota_1}$ to $pr(f^*(u_{\iota_1}), f(a))$. If $f(a) = \Omega$, then there isn't anything that cries out to be the image of $u_{\iota_1}.a_{\iota_1}$. Accordingly, we don't map $u_{\iota_1}.a_{\iota_1}$ to anything in that case. This leads us to define for primes $u_{\iota_1}.a_{\iota_1}$, $a \in A_1$:

$$\mathbf{B}(f)_O(u_{\iota_1}.a_{\iota_1}) = \begin{cases} pr(f^*(u_{\iota_1}), f(a)) \text{ if } f(a) \neq \Omega \\ \text{undefined, otherwise} \end{cases}$$

Let us examine this mapping a little more closely. We already know from the argument above that

$$\mathbf{B}(f)_O(Pr(x_{\iota_1})) \subseteq \mathbf{B}(f)_P(Pr(x_{\iota_1})) \tag{19.3}$$

What about the converse inclusion? Suppose $v_{\iota_2}.b_{\iota_2} \in Pr(y_{\iota_2})$, where $y = f(x)$. Let $X = \{a \in A_1 \mid f(a) = b\}$. We observe that $\#_b y > 0$ and $y = f(x)$ and so $X \neq \emptyset$ and $\#_b y = \sum_{a \in X} \#_a x$. Let $n = \#_b v.b$. Let $u.a \leq x$ such that $\sum_{a' \in X} \#_{a'} u.a = n$ and $a \in X$. Let $w_{\iota_1}.a_{\iota_1} = pr(u_{\iota_1}, a)$. We shall show that $\mathbf{B}(f)_O(w_{\iota_1}.a_{\iota_1}) = v_{\iota_2}.b_{\iota_2}$.

By (10.10), $u_{\iota_1} = w_{\iota_1}.u'_{\iota_1}$, where u'_{ι_1} ind_{ι_1} a_{ι_1}. Since $\neg a' \iota_1 a''$, for every $a', a'' \in X$ - otherwise, by irreflexivity of ι_2, 19.3.2 (1) is violated - it follows that $\sum_{a' \in X} \#_{a'} u' = 0$ and that therefore $\sum_{a' \in X} \#_{a'} w.a = \sum_{a' \in X} \#_{a'} v.a = n$.

Now $f(a) \neq \Omega$ because $f(a) = b$, and so $\mathbf{B}(f)_O(w_{\iota_1}.a_{\iota_1})$ is defined and equal to $pr(f^*(w_{\iota_1}), f(a)) = pr(f^*(w_{\iota_1}), b)$ Let $v'_{\iota_2}.b_{\iota_2} = pr(f^*(w_{\iota_1}), b)$ As above, we argue that $\#_b v'_{\iota_2}.b_{\iota_2} = \#_b f^*(w_{\iota_1}).b_{\iota_2} = \sum_{a' \in X} w_{\iota_1}.a_{\iota_1} = n$.

---

[2] See (10.10). Note hat in this argument we are referring to traces while using concepts and results developed for vectors. This is justified by 9.3.3, which allows us to define for traces $\underline{u}$ and $\underline{v}$, $\underline{u}$ ind_ι $\underline{v} \Leftrightarrow \phi_\alpha{}^*(\underline{u})$ ind_α $\phi_\alpha{}^*(\underline{v})$, where $\iota = \iota_\alpha$

Thus, $v_{\iota_2}.b_{\iota_2}, v'_{\iota_2}.b_{\iota_2} \in Pr(y_{\iota_2})$ with $\#_b v_{\iota_2}.b_{\iota_2} = \#_b v'_{\iota_2}.b_{\iota_2}$. We shall show that this entails $v_{\iota_2}.b_{\iota_2} = v'_{\iota_2}.b_{\iota_2}$. From this it will follow that $\mathbf{B}(f)_O(w_{\iota_1}.a_{\iota_1}) = v_{\iota_1}.b_{\iota_2}$. This, together with (19.3) gives

$$\mathbf{B}(f)_O(Pr(x_{\iota_1})) = \mathbf{B}(f)_P(Pr(x_{\iota_1})) \tag{19.4}$$

Let $v_{\iota_2}.b_{\iota_2}, v'_{\iota_2}.b_{\iota_2} \in Pr(y_{\iota_2})$ with $\#_b v_{\iota_2}.b_{\iota_2} = \#_b v'_{\iota_2}.b_{\iota_2}$. Since $v_{\iota_2}.b_{\iota_2}, v'_{\iota_2}.b_{\iota_2} \leq y_{\iota_2}$, $v_{\iota_2}.b_{\iota_2} \sqcup v'_{\iota_2}.b_{\iota_2}$ exists, by 10.2,4. By 10.2.8, $v_{\iota_2}.b_{\iota_2} \sqcup v'_{\iota_2}.b_{\iota_2} = \hat{v}_{\iota_2}.z_{\iota_2}.z'_{\iota_2}$, where $\hat{v}_{\iota_2} = v_{\iota_2}.b_{\iota_2} \sqcap v'_{\iota_2}.b_{\iota_2}$. $z_{\iota_2} = (v_{\iota_2}.b_{\iota_2})/\hat{v}_{\iota_2}$, $z'_{\iota_2} = (v'_{\iota_2}.b_{\iota_2})/\hat{v}_{\iota_2}$ and $z_{\iota_2}\ ind_{\iota_2}\ z'_{\iota_2}$.

If $z_{\iota_2} \neq \Omega_{\iota_2}$, then $z_{\iota_2} = z''_{\iota_2}.c_{\iota_2}$, some $c \in A_1$ so $v_{\iota_2}.b_{\iota_2} = \hat{v}_{\iota_2}.z''_{\iota_2}.c_{\iota_2}$ and hence $b = c$, by primality. Similarly, if $z'_{\iota_2} \neq \Omega_{\iota_2}$, then z'_{ι_2} is of the form $z''_{\iota_2}.b_{\iota_2}$. Since $z_{\iota_2}\ ind_{\iota_2}\ z'_{\iota_2}$, we thus cannot have both $z_{\iota_2} \neq \Omega_{\iota_2}$ and $z'_{\iota_2} \neq \Omega_{\iota_2}$.

Suppose without loss of generality that $z'_{\iota_2} = \Omega_{\iota_2}$. If $z_{\iota_2} \neq \Omega_{\iota_2}$, then $v'_{\iota_2}.b_{\iota_2} = \hat{v}_{\iota_2} \leq v_{\iota_2}.b_{\iota_2}$ and so by 12.2.1, $v'_{\iota_2}.b_{\iota_2} \leq v_{\iota_2}$. But then $\#_b v_{\iota_2}.b_{\iota_2} < \#_b v'_{\iota_2}.b_{\iota_2}$, a contradiction. Thus, $z_{\iota_2} = \Omega_{\iota_2}$ and therefore $v_{\iota_2}.b_{\iota_2} = \hat{v}_{\iota_2} = v'_{\iota_2}.b_{\iota_2}$.

The above argument suggests the following definition, which also entails that the triple $\mathbf{B}(f) = (\mathbf{B}(f)_O, \mathbf{B}(f)_P, \mathbf{B}(f)_E)$ is a morphism of behavioural presentations and hence the morphism part of a functor from the category **ATS*** to a category of behavioural presentations.

19.4.1. DEFINITION. Let **BP** denote the class of all behavioural presentations. If $B_1, B_2 \in \mathbf{BP}$, then a morphism from B_1 to B_2 is a triple $f = (f_O, f_P, f_E)$, where

1• $f_O : O_1 \to O_2$ is a partial function;
2• $f_P : P_1 \to P_2$ is a function satisfying

$$\forall p \in P_1 : f_P(p) = \{f_O(o) \mid o \in p\};$$

3• $f_E : E_1 \to E_2 \cup \{\Omega\}$ is a function satisfying

$$\forall o \in dom(f_O) : \lambda_2(f_O(o)) = f_E(\lambda_1(o)).$$

Write $f : B_1 \to B_2$ to indicate that f is a morphism from B_1 to B_2.

In (2), the set on the right hand side of the equation is the set of all $f_O(o)$ such that $o \in p$ and $f_O(o)$ is defined.

Composition of morphisms is componentwise composition of partial functions, that is $f \circ g = (f_O \circ g_O, f_P \circ g_P, f_E * g_E)$, where $f_P \circ g_P$ is normal function composition $f_E * g_E$ is given in (19.1) and $(f_P \circ g_P)(o) = f_P(g_P(o))$ if it is defined. If $B \in \mathbf{BP}$, then we define $Id_B = (Id_O, Id_P, inc_E)$.

The proof of the following is routine.

19.4.2. PROPOSITION. **BP** is a category. ■

$f_P(\varnothing)=\varnothing \quad f_P(\{a\})=\{A\} \quad f_P(\{b\})=\{B\} \quad f_P(\{a,b\})=\{A,B\}$

$f_P(\{a,b,c\})=\{A,B\} \quad f_P(\{b,c,d\})=\{B,D\}$

$f_O(a)=A \quad f_O(b)=B \quad f_O(d)=D \quad f_O(c)$ undefinied

$f_E(a)=A \quad f_E(b)=B \quad f_E(d)=D \quad f_E(c)=\Omega$

Figure 19.3.

19.4.1 is illustrated in figure 19.3. In each case, it is assumed that $E=O$ and $\lambda = Id_O$.

Within the context of asynchronous systems, we are particularly interested in linguistic behavioural presentations. The main difference between this and the general case is that linguistic behavioural presentations respect independence relations. This must be reflected in the way that morphisms handle independence relations, prompting (1) of the following definition.

19.4.3. DEFINITION. Let **LBP** denote the class of all pairs (B,ι), where B is an ι-linguistic behavioural presentation. If $(B_1,\iota_1),(B_2,\iota_2)\in$ **LBP**, then a *morphism* from (B_1,ι_1) to (B_2,ι_2) in **LBP** is a morphism $f:B_1 \to B_2$ in **BP** satisfying the additional properties that

1• $\forall e,e' \in E_1 : e\,\iota_1\,e' \wedge f_E(e) \neq \Omega \neq f_E(e') \Rightarrow f_E(e)\,\iota_2\,f_E(e')$;

2• $\forall o,o' \in dom(f_O) : o \neq o' \wedge f_O(o) = f_O(o') \Rightarrow o \,\#_1\, o'$.

Write $f:(B_1,\iota_1)\to(B_2,\iota_2)$ to indicate that f is a morphism from (B_1,ι_1) to (B_2,ι_2).

Composition of morphisms and identity morphisms are as in **BP**.

19.4.4. PROPOSITION. **LBP** is a category.

PROOF. We show that condition (2) is preserved under composition; the rest is straightforward verification.

Suppose that $f:(B_1,\iota_1)\to(B_2,\iota_2)$ and $g:(B_2,\iota_2)\to(B_3,\iota_3)$ and suppose that $o,o'\in dom((g\circ f)_O)$ such that $o\neq o'$ and $(g\circ f)_O(o)=(g\circ f)_O(o')$. We must show that $o\,\#_1\,o'$.

First, let $\neg o\,\#_1\,o'$, then $o,o'\in p$, some $p\in P_1$, so $f_O(o),f_O(o')\in f_P(p)\in P_2$, by 19.4.1 (2), that is $\neg f_O(o)\,\#_2\,f_O(o')$. We have proved:

$$f_O(o)\,\#_2\,f_O(o')\Rightarrow o\,\#_1\,o' \tag{19.6}$$

As $o,o'\in dom((g\circ f)_O)$, $o,o'\in dom(f_O)$ so if $f_O(o)=f_O(o')$, then $o\,\#_1\,o'$. On the other hand, $f_O(o),f_O(o')\in dom(g_O)$ and $g_O(f_O(o))=g_O(f_O(o'))$ and hence if $f_O(o)\neq f_O(o')$, then $f_O(o)\,\#_2\,f_O(o')$, and consequently $o\,\#_1\,o'$, by (19.6).

■

Turning to isomorphisms, the reader may recall that we have used this term quite a lot already. Fortunately, our previous use is consistent with the categorical notion, as the reader may care to verify.

19.4.5. LEMMA. Let $B_1,B_2\in\underline{\underline{\mathbf{BP}}}$ and $f:B_1\to B_2$, then f is an isomorphism $\Leftrightarrow$ f is a label isomorphism in the sense of 12.2.13.

■

Let us now return to the discussion preceding definition 19.4.1. This was an attempt to argue our way to a definition of morphism in $\underline{\underline{\mathbf{BP}}}$, given that we wished morphisms in $\underline{\underline{\mathbf{BP}}}$ to reflect those in $\underline{\underline{\mathbf{LCTL}}}$. As a consequence, we have a functor between the two categories.

19.4.6. DEFINITION. Suppose $D,D_1,D_2\in\underline{\underline{\mathbf{LCTL}}}$. Define $\mathbf{B}(D)=(B_{L,A},\iota)$. If $f:D_1\to D_2$, then define $\mathbf{B}(f)=(\mathbf{B}(f)_O,\mathbf{B}(f)_P,\mathbf{B}(f)_E)$ by

$$\mathbf{B}(f)_O(u_{\iota_1}.a_{\iota_1})=\begin{cases}pr(f^*(u_{\iota_1}),f(a)) \text{ if } f(a)\neq\Omega\\ \text{undefined, otherwise}\end{cases}$$
$$\mathbf{B}(f)_P(Pr(x_{\iota_1}))=Pr(f^*(x_{\iota_1}))$$
$$\mathbf{B}(f)_E=f$$

19.4.7 PROPOSITION. $\mathbf{B}$ is a functor, $\mathbf{B}:\underline{\underline{\mathbf{LCTL}}}\to\underline{\underline{\mathbf{LBP}}}$.

PROOF. We know that if $D\in\underline{\underline{\mathbf{LCTL}}}$, then $\mathbf{B}(D)\in\underline{\underline{\mathbf{LBP}}}$. Suppose that $f:D_1\to D_2$. We must show that $\mathbf{B}(f):\mathbf{B}(D_1)\to\mathbf{B}(D_2)$.

19.4.1 (1) and (2) have already been established in the argument preceding that definition. For 19.4.1 (3), if $u_{\iota_1}.a_{\iota_1}\in Pr(L_1)$ and $f(a)\neq\Omega$ then $\mathbf{B}(f)_O(u_{\iota_1}.a_{\iota_1})=pr(f^*(u_{\iota_1}),f(a))$.

But $\lambda_1(u_{\iota_1}.a_{\iota_1}) = a$ and $\lambda_2(pr(f^*(u_{\iota_1}), f(a)) = f(a)$ and so

$$\lambda_2(\mathbf{B}_O(f)(u_{\iota_1}.a_{\iota_1})) = f(a)) = \mathbf{B}_E(f)(a) = \mathbf{B}_E(f)(\lambda_1(u_{\iota_1}.a_{\iota_1}))$$

Thus, $\mathbf{B}(f)$ *in* $\underline{\underline{\mathbf{BP}}}$.

$\mathbf{B}(f)$ satisfies 19.4.3 (1) by 19.3.2 (1).

For (2), let $u_{\iota_1}.a_{\iota_1}, v_{\iota_1}.b_{\iota_1} \in dom(\mathbf{B}(f)_O)$ such that $\neg u_{\iota_1}.a_{\iota_1} \#_{(L_1,A_1)} v_{\iota_1}.b_{\iota_1}$. We show by contradiction that $u_{\iota_1}.a_{\iota_1} \neq v_{\iota_1}.b_{\iota_1} \Rightarrow \mathbf{B}(f)_O(u_{\iota_1}.a_{\iota_1}) \neq \mathbf{B}(f)_O(v_{\iota_1}.b_{\iota_1})$. Since $\neg u_{\iota_1}.a_{\iota_1} \#_{(L_1,A_1)} v_{\iota_1}.b_{\iota_1}$, $u_{\iota_1}.a_{\iota_1}, v_{\iota_1}.b_{\iota_1} \in Pr(x_{\iota_1})$, some $x_{\iota_1} \in L_1$ and so $u_{\iota_1}.a_{\iota_1}, v_{\iota_1}.b_{\iota_1} \in Pr(x_{\iota_1}) \cap dom(\mathbf{B}(f)_O)$. Now, by (19.4) and 19.4.6

$$Pr(f^*(x_{\iota_1})) = \{\mathbf{B}(f)_O(u_{\iota_1}.a_{\iota_1}) \mid u_{\iota_1}.a_{\iota_1} \in Pr(x_{\iota_1}) \wedge f(a) \neq \Omega\},$$

Therefore, if $u_{\iota_1}.a_{\iota_1} \neq v_{\iota_1}.b_{\iota_1}$ and $\mathbf{B}(f)_O(u_{\iota_1}.a_{\iota_1}) = \mathbf{B}(f)_O(v_{\iota_1}.b_{\iota_1})$ then $|Pr(f^*(x_{\iota_1}))| < \sum_{f(a)\neq\Omega} \#_a x$. But, $\sum_{f(a)\neq\Omega} \#_a x = |f(x)|$ and $|Pr(f^*(x_{\iota_1}))| = |f(x)|$, by 12.2.6 (3), the required contradiction.

It is a simple matter to check that $\mathbf{B}(Id_D) = Id_{\mathbf{B}(D)}$.

Finally, we show that composition of morphisms is preserved. Let $f: D_1 \to D_2$ and $g: D_2 \to D_3$. For each i, let $(B_i, \iota_i) = \mathbf{B}(D_i)$.

Suppose that $p \in P_1$, then $p = Pr(x_{\iota_1})$, for some $x_{\iota_1} \in L_1$. Calculating,

$$\begin{aligned}\mathbf{B}(g)_P \circ \mathbf{B}(f)_P(p) &= \mathbf{B}(g)_P(Pr(f^*(x_{\iota_1})) = Pr(g^*(f^*(x_{\iota_1})) \\ &= Pr((g \circ f)^*(x_{\iota_1})) = \mathbf{B}(g \circ f)_P(p)\end{aligned}$$

Thus, $\mathbf{B}(g)_P \circ \mathbf{B}(f)_P = \mathbf{B}(g \circ f)_P$.

We also have that for all $e \in E_1$, $(\mathbf{B}(g)_E * \mathbf{B}(f)_E)(e) = (g * f)(e) = \mathbf{B}(g \circ f)_E(e)$ and therefore $\mathbf{B}(g)_E \circ \mathbf{B}(f)_E = \mathbf{B}(g \circ f)_E$.

Finally, we show $\mathbf{B}(g)_O \circ \mathbf{B}(f)_O = \mathbf{B}(g \circ f)_O$. The proof hinges on the following fact. If $u_{\iota_1}.a_{\iota_1} \in Pr(x_{\iota_1})$ and $f(a) = b \neq \Omega$ and $n = \sum_{f(a')=b} \#_{a'} u_{\iota_1}.a_{\iota_1}$, then $\mathbf{B}(f)_O(u_{\iota_1}.a_{\iota_1})$ is the unique $v_{\iota_2}.b_{\iota_2} \in Pr(f^*(x_{\iota_1}))$ such that $\#_b v_{\iota_2}.b_{\iota_2} = n$.

Let us prove this. First, recall from the discussion preceding 19.4.1 that

$$v_{\iota_2}.b_{\iota_2}, v'_{\iota_2}.b_{\iota_2} \in Pr(y_{\iota_2}) \Rightarrow (\#_b v_{\iota_2}.b_{\iota_2} = \#_b v'_{\iota_2}.b_{\iota_2} \Rightarrow v_{\iota_2}.b_{\iota_2} = v'_{\iota_2}.b_{\iota_2})$$

so that $v_{\iota_2}.b_{\iota_2}$ will be unique if it exists. Let $w_{\iota_2} = f^*(u_{\iota_1})$, then $\mathbf{B}(f)_O(u_{\iota_1}.a_{\iota_1}) = v''_{\iota_2}.b_{\iota_2}$ where $v''_{\iota_2} \leq w_{\iota_2}$ and w_{ι_2}/v_{ι_2} ind_{ι_2} b_{ι_2}. From the latter, it is clear that $\#_b w_{\iota_2}/v_{\iota_2} = 0$ so $\#_b w_{\iota_2} = \#_b v_{\iota_2}$. But, $\#_b w_{\iota_2} = \sum_{f(a')=b} \#_{a'} u_{\iota_1} = n-1$. Thus $\#_b v''_{\iota_2}.b_{\iota_2} = n$. Uniqueness entails that any $v''_{\iota_2}.b_{\iota_2}$ with this property equals $\mathbf{B}(f)_O(u_{\iota_1}.a_{\iota_1})$.

A simple calculation now shows that if $u_{\iota_1}.a_{\iota_1} \in Pr(x_{\iota_1})$ then $(\mathbf{B}(g)_O \circ \mathbf{B}(f)_O)(u_{\iota_1}.a_{\iota_1})$ and $\mathbf{B}(g \circ f)_O(u_{\iota_1}.a_{\iota_1})$ are both equal to the unique prime $w_{\iota_3}.c_{\iota_3} = g^*(f^*(x_{\iota_1}))$ such that $\#_c w_{\iota_3}.c_{\iota_3} = \sum_{g(f(a'))=c} \#_{a'} u_{\iota_1}.a_{\iota_1}$.

■

We also have a map going in the other direction. Suppose $(B,\iota) \in \underline{\underline{\mathbf{LBP}}}$, then we define $\rho(B,\iota) = (E,\iota,\rho_B(P))$. By 12.2.12, $\rho(B,\iota) \in \underline{\underline{\mathbf{LCTL}}}$. If f *in* $\underline{\underline{\mathbf{LBP}}}$, then we define $\rho(f) = f_E$. We shall show that $\rho(f)$ is a morphism in $\underline{\underline{\mathbf{LCTL}}}$. We note first of all that condition 19.3.2 (1) holds because of 19.4.3 (1).

Suppose that $f:(B_1,\iota_1) \to (B_2,\iota_2)$ and let $p \in P_1$ and suppose that $\rho_{B_1}(p) = (e_1 \cdots e_n)_{\iota_1}$, so that we have $p = \{o_1 \cdots o_n\}$ and strict total ordering $\tilde{<}_1$ on p, which extends $\to_1$, satisfying $o_1 \tilde{<}_1 \cdots \tilde{<}_1 o_n$ and $e_i = \lambda_1(o_i)$, each i. As $\tilde{<}_1$ extends $\to_1$, it follows that $\{o_1 \cdots o_i\} \lhd p$. and hence that $\{o_1,\cdots,o_i\} \in P_1$ for each i since B_1 is discrete. Therefore, $f_P(\{o_1,\cdots,o_i\}) \in P_2$, each i. Now, suppose that $o_{j_1},\cdots,o_{j_m}$ are those elements of p such that $o_{j_1} \tilde{<}_1 \cdots \tilde{<}_1 o_{j_m}$ and that $f_O(o_{j_i})$ is defined, each i. By 19.4.1 (2) and (3), $f_P(p) = f_P(\{o_{j_1},\cdots,o_{j_m}\})$ and $f_E(e_1 \cdots e_n) = \lambda_2(f_O(o_{j_1}) \cdots f_O(o_{j_m}))$.

If $f_O(o_{j_k}) \to_2 f_O(o_{j_l})$, then $f_O(o_{j_l}) \in f_O(\{o_1,\cdots,o_{j_l}\})$, and therefore $f_O(o_{j_k}) \in f_O(\{o_1,\cdots,o_{j_l}\})$ and hence $j_k \le j_l$. By 19.4.3 (2), there is a strict total ordering $f_O(o_{j_1}) \tilde{<}_2 \cdots \tilde{<}_2 f_O(o_{j_m})$ of $f_P(p)$ and we have seen that it extends $\to_2$. Thus

$$\rho_{B_2}(f_P(p)) = (\lambda_2(f_O(o_{j_1})) \cdots \lambda_2(f_O(o_{j_m})))_{\iota_2} = f(e_1 \cdots e_n)_{\iota_2}.$$

We have shown that

$$\forall p \in P_1 : \rho(f)^*(\rho_{B_1}(p)) = \rho_{B_2}(f_P(p)) \tag{19.8}$$

(19.8) shows that $\rho(f)^*(\rho_{B_1}(P_1)) \subseteq \rho_{B_2}(P_2)$, so $\rho(f)$ satisfies 19.3.2 (2). By 19.4.3 (1), $\rho(f)$ also satisfies 19.3.2 (1) Thus $\rho(f):\rho(B_1,\iota_1) \to \rho(B_2,\iota_2)$.

It is easy to check that $\rho(Id_{(B,\iota)}) = Id_{\rho(B,\iota)}$ and that $\rho(f \circ g) = \rho(f) \circ \rho(g)$, so

19.4.8. PROPOSITION. ρ is a functor, $\rho : \underline{\underline{\mathbf{LBP}}} \to \underline{\underline{\mathbf{LCTL}}}$.

■

Again, we know that the objects in $\underline{\underline{\mathbf{LBP}}}$ are intimately related to those in $\underline{\underline{\mathbf{LCTL}}}$. Indeed, it should not come as too much of a shock to the reader to learn that ρ is left adjoint to B and B is left adjoint to ρ.

Actually, more than this is true. The categories $\underline{\underline{\mathbf{LBP}}}$ and $\underline{\underline{\mathbf{LCTL}}}$ are *equivalent* in the sense in which we now define.

If $\mathbf{F},\mathbf{G}:\underline{\underline{\mathbf{C}}}\to\underline{\underline{\mathbf{D}}}$, are functors then a *natural transformation*, $\tau:\mathbf{F}\to\mathbf{G}$ is a function which assigns each $c\in\underline{\underline{\mathbf{C}}}$ with a morphism $\tau_c:\mathbf{F}(c)\to\mathbf{G}(c)$ in such a way that for each morphism $f:c\to d$, in $\underline{\underline{\mathbf{C}}}$ the diagram of figure 19.4 commutes.

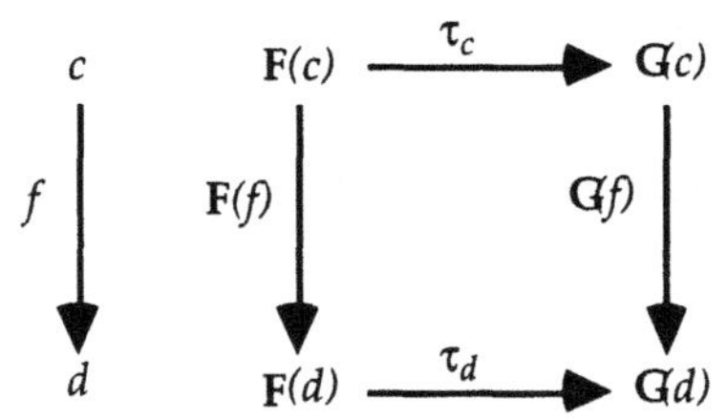

Figure 19.4.

If each τ_c is an isomorphism, then τ is a *natural equivalence*. We shall write $\mathbf{F}\equiv\mathbf{G}$ to indicate that there is a natural equivalence $\tau:\mathbf{F}\to\mathbf{G}$.

$\mathbf{F}:\underline{\underline{\mathbf{C}}}\to\underline{\underline{\mathbf{D}}}$ is an *equivalence of categories* $\Leftrightarrow$ there exists a functor $\mathbf{G}:\underline{\underline{\mathbf{D}}}\to\underline{\underline{\mathbf{C}}}$ such that $\mathbf{FG}\equiv\mathbf{1}_{\underline{\mathbf{D}}}$ and $\mathbf{GF}\equiv\mathbf{1}_{\underline{\mathbf{C}}}$. Here $\mathbf{1}_{\underline{\mathbf{C}}}$ is the *identity functor* $\mathbf{1}_{\underline{\mathbf{C}}}:\underline{\underline{\mathbf{C}}}\to\underline{\underline{\mathbf{C}}}$ which satisfies $\mathbf{1}_{\underline{\mathbf{C}}}(c)=c$ and $\mathbf{1}_{\underline{\mathbf{C}}}(f)=f$, for every $c\in\underline{\underline{\mathbf{C}}}$ and f *in* $\underline{\underline{\mathbf{C}}}$. It is clear that equivalence of categories is an equivalence relation. We shall refer to $\mathbf{G}$ as the *inverse* of $\mathbf{F}$.

We observe that if we have an equivalence of categories $\mathbf{F}:\underline{\underline{\mathbf{C}}}\to\underline{\underline{\mathbf{D}}}$ with inverse $\mathbf{G}$, then $\mathbf{F}$ is left adjoint to $\mathbf{G}$ and vice versa. Indeed, if $\tau:\mathbf{1}_{\underline{\mathbf{D}}}\to\mathbf{GF}$ is a natural equivalence and if $c\in\underline{\underline{\mathbf{C}}}$, $d\in\underline{\underline{\mathbf{D}}}$ and $f:c\to\mathbf{G}(d)$, then we may define $h_c=\tau_c$ and $g:\mathbf{F}(c)\to d$ by $g=\tau_c^{-1}\circ\mathbf{F}(f)$.

19.4.9. THEOREM. ρ is an equivalence of categories with inverse $\mathbf{B}$. In particular, ρ is left adjoint to $\mathbf{B}$ and vice versa.

PROOF. If $D\in\underline{\underline{\mathbf{LCTL}}}$, then $\rho\mathbf{B}(D)=\rho(B_{L,A})=(A,\iota,\rho_{B_{L,A}}(P_{L,A}))=(A,\iota,L)=D$, by 12.2.17. Likewise, if f *in* $\underline{\underline{\mathbf{LCTL}}}$, then $\rho\mathbf{B}(f)(e)=\mathbf{B}(f)_E(e)=e$. Thus, $\rho\mathbf{B}=1_{\underline{\mathbf{LCTL}}}$, and therefore $\rho\mathbf{B}\equiv 1_{\underline{\mathbf{LCTL}}}$.

Now let us look at $\mathbf{B}\rho$. Let $(B,\iota)\in\underline{\underline{\mathbf{LBP}}}$, then $\mathbf{B}\rho(B,\iota)=(B',\iota')$, where, $O'=Pr(\rho_B(P))$, $P'=\{Pr(\underline{x})\mid\underline{x}\in\rho_B(P)\}$, $E'=E$ and $\lambda'(u_\iota.e_\iota)=e$. Define

$$(\tau_{(B,\iota)})_O(o)=\rho_B(\downarrow o)$$
$$(\tau_{(B,\iota)})_P(p)=Pr(\rho_B(p))$$

$(\tau_{(B,\iota)})_E(e) = e$

Let us first prove that $\tau_{(B,\iota)}$ is a morphism from (B,ι) to (B',ι').

First, we note that $\rho_B(\downarrow o) \in \rho_B(Pr(P)) = Pr(\rho_B(P)) = O'$, since ρ_B is an isomorphism from P to $\rho_B(P)$ by 12.2.12. Thus 19.4.1 (1) holds. For 19.4.1 (2), we note that since ρ_B is an isomorphism, $\rho_B(\downarrow o) \leq \rho_B(p) \Leftrightarrow o \in p$ and that since $\downarrow o$ is prime, by 4.3.2, $\underline{u} \in Pr(\rho_B(p)) \Leftrightarrow \underline{u} = \rho_B(\downarrow o)$, some $o \in p$ and so

$$\begin{aligned} \underline{u} \in (\tau_{(B,\iota)})_O(p) &\Leftrightarrow \underline{u} = \rho_B(\downarrow o), \text{ some } o \in p \\ &\Leftrightarrow \underline{u} \in Pr(\rho_B(p)) \\ &\Leftrightarrow \underline{u} \in (\tau_{(B,\iota)})_P(p). \end{aligned}$$

If $\lambda(o) = e$ then $\rho_B(\downarrow o)$ must be of the form $u_\iota.e_\iota$ and so $\lambda'(\rho_B(\downarrow o)) = \lambda(o)$ and we have

$$\lambda'((\tau_{(B,\iota)})_O(o)) = \lambda'(\rho_B(\downarrow o)) = \lambda(o) = (\tau_{(B,\iota)})_E(\lambda(o))$$

Thus, $\tau_{(B,\iota)}$ also satisfies 19.4.1 (3) and is a morphism. It is an isomorphism because ρ_B is a poset isomorphism.

Now suppose $f:(B_1,\iota_1) \to (B_2,\iota_2)$. In order to show that τ is a natural transformation, we must show that $\mathbf{B}\rho(f) \circ \tau_{(B_1,\iota_1)} = \tau_{(B_2,\iota_2)} \circ f$. This reduces to proving several identities.

Evaluating both sides of the equation gives

$$(\tau_{(B_2,\iota_2)} \circ f)_O(o) = (\tau_{(B_2,\iota_2)})_O(f_O(o)) = \rho_{B_2}(\downarrow f(o))$$

and

$$(\mathbf{B}\rho(f) \circ \tau_{(B_1,\iota_1)})_O(o) = \mathbf{B}\rho(f)_O((\tau_{(B_1,\iota_1)})_O(o)) = pr(f_E{}^*(u_{\iota_1}), f_E(e))$$

where $\rho_{B_1}(\downarrow o) = u_{\iota_1}.e_{\iota_1}$. This means that we must establish that

$$\rho_{B_1}(\downarrow o) = u_{\iota_1}.e_{\iota_1} \Leftrightarrow \rho_{B_2}(\downarrow f(o)) = pr(f_E{}^*(u_{\iota_1}), f_E(e)) \tag{19.8}$$

where it is to be understood that if either side of the equation is defined then so is the other. This is indeed the case, since, by 19.4.1 (3), $\lambda_2(f_O(o)) = f_E(\lambda_1(o)) = f_E(e)$ and therefore $f_O(o)$ is defined $\Leftrightarrow$ $f_E(e)$ is defined $\Leftrightarrow pr(f_E{}^*(u_{\iota_1}), f_E(e))$ is defined.

We also have

$$(\tau_{(B_2,\iota_2)} \circ f)_P(p) = (\tau_{(B_2,\iota_2)})_P(f_P(p)) = Pr(\rho_{B_2}(f_P(p)))$$

and

$$(\mathbf{B}\rho(f) \circ \tau_{(B_1,\iota_1)})_P(p) = \mathbf{B}\rho(f)_P(Pr(\rho_{B_1}(p)) = Pr(f_E^*(\rho_{B_1}(p)))$$

But, by (19.7), $\rho_{B_2}(f_P(p)) = \rho(f)(\rho_{B_1}(p)) = f_E^*(\rho_{B_1}(p))$. Thus,

$$(\mathbf{B}\rho(f) \circ \tau_{(B_1,\iota_1)})_P = (\tau_{(B_2,\iota_2)} \circ f)_P$$

Finally,

$$(\tau_{(B_2,\iota_2)} \circ f)_E(e) = (\tau_{(B_2,\iota_2)})_E(f_E(e)) = f_E(e)$$

and

$$(\mathbf{B}\rho(f) \circ \tau_{(B_1,\iota_1)})_E(e) = \mathbf{B}\rho(f)_E((\tau_{(B_1,\iota_1)})_E(e) = \mathbf{B}\rho(f)_E(e) = f_E(e)$$

Thus,

$$(\mathbf{B}\rho(f) \circ \tau_{(B_1,\iota_1)})_E = (\tau_{(B_2,\iota_2)} \circ f)_E$$

It remains to prove $(\mathbf{B}\rho(f) \circ \tau_{(B_1,\iota_1)})_O = (\tau_{(B_2,\iota_2)} \circ f)_O$ which, as we have seen, follows from (19.8). Suppose $\rho_{B_1}(\downarrow o) = u_{\iota_1}.e_{\iota_1}$.

First, we observe that if $o_1 \in O_1$ then $\downarrow f_O(o_1) \subseteq f_P(\downarrow o_1)$. Indeed, if $o_2 \in \downarrow f_O(o)$, then as $f_O(o_1) \in f_P(\downarrow o_1)$, by 19.4.1 (2), and $o_2 \to_2 f_O(o)$, it follows that $o_2 \in f_P(\downarrow o_1)$. Since ρ_{B_2} is a poset morphism,

$$\rho_{B_2}(\downarrow f_O(o_1)) \leq \rho_{B_2}(f_P(\downarrow o_1))$$

By (19.7), $\rho_{B_2}(f_P(\downarrow o_1)) = \rho(f)(f_P(\downarrow o_1)) = f_E^*(\rho_{B_1}(\downarrow o_1)) = f_E^*(u_{\iota_1}).e'_{\iota_2}$, where $e' = f_E(e)$. Clearly, $\rho_{B_2}(f_P(\downarrow o_1))$ is of the form $v_{\iota_2}.e''_{\iota_2}$, where $e'' = \lambda_2(f_O(o_1))$. But $\lambda_2(f_O(o_1)) = f_E(\lambda_1(o_1)) = f_E(e) = e'$ and therefore $\rho_{B_2}(f_P(\downarrow o_1)) = v_{\iota_2}.e'_{\iota_2}$. Furthermore $v_{\iota_2}.e'_{\iota_2} \in Pr(\rho_{B_2}(P_2))$, since ρ_{B_2} maps primes to primes.

As $\downarrow f_O(o_1) \subseteq f_P(\downarrow o_1)$, it follows that $\downarrow f_O(o_1) = \{o' \in f_P(\downarrow o_1) \mid o' \to_2 f_O(o_1)\}$, so that if $o' \in f_P(\downarrow o_1) - \downarrow f_O(o_1)$, then $o' \, co_2 \, f_O(o_1)$. By the ι_2-linguistic property, $\lambda_2(o') \, \iota_2 \, \lambda_2(f_O(o)) = e'$. Thus $\lambda_2(o') \neq e'$ and so

$$\#_{e'} \rho_{B_2}(\downarrow f_O(o)) = \#_{e'} \rho_{B_2}(f_P(\downarrow o))$$

Therefore, the proof will be complete if we can show the following

$$\left.\begin{array}{l} v_{\iota_2}.e'_{\iota_2} \le x_{\iota_2}.e'_{\iota_2} \\ \#_{e'} v_{\iota_2}.e'_{\iota_2} = \#_{e'} x_{\iota_2}.e'_{\iota_2} \\ v_{\iota_2}.e'_{\iota_2} \in Pr(\rho_{B_2}(P_2)) \end{array}\right\} \Rightarrow v_{\iota_2}.e'_{\iota_2} = pr(x_{\iota_2}, e')$$

Let $w_{\iota_2} = x_{\iota_2}.e'_{\iota_2}/v_{\iota_2}.e'_{\iota_2}$. Since $\#_{e'} v_{\iota_2}.e'_{\iota_2} = \#_{e'} x_{\iota_2}.e'_{\iota_2}$, we must have $\#_{e'} w_{\iota_2} = 0$. We may argue that w_{ι_2} ind_{ι_2} e_{ι_2}. It follows that $x_{\iota_2}.e'_{\iota_2} = v_{\iota_2}.e'_{\iota_2}.w_{\iota_2} = v_{\iota_2}.w_{\iota_2}.e'_{\iota_2}$ and that hence $x_{\iota_2} = v_{\iota_2}.w_{\iota_2}$. Therefore $v_{\iota_2} \le x_{\iota_2}$ and $x_{\iota_2}/v_{\iota_2} = w_{\iota_2}$ ind_{ι_2} e_{ι_2} By 10.3.3, $v_{\iota_2}.e'_{\iota_2} = pr(x_{\iota_2}, e')$.

■

Notes.

While we have taken [43] as a starting point, our treatment differes from Bednarcztk's in an important respect. He relates trace languages not to linguistic behavioural presentations but to event structures whose primes are finite sets. Such event structures are closures of discrete, unlabelled behavioural presentations, which, equipped with an identity function on occurrences, constitute a subclass of the class of linguistic behavioural presentations. Our results are therefore stronger.

[43] goes into much more detail generally about categorical aspects of asynchrony theory, however.

Chapter 20
Categorical Applications

> A landlady may be reduced to her lowest terms by a sequence of propositions.
>
> STEPHEN LEACOCK
> Boarding House Geometry.

20.1. Introduction.

In this chapter we look at a number of constructions involving initialised asynchronous transition systems. Some are instances of familiar constructions in algebra and automata theory, such as forming quotients and reduction. The other is associated with a constructor which crops up in a number of specification notations.

We approach these constructions from a categorical point of view; the category in question being a subcategory of **ATS** which we call **SATS**. The objects of **SATS** are those of **ATS** and if $f:C_1 \to C_2$ then f *in* **SATS** if and only if $A_2 \subseteq A_1$ and

$$f_A(a) = \begin{cases} a \text{ if } a \in A_2 \\ \Omega \text{ otherwise} \end{cases} \tag{20.1}$$

20.2. Reduction Morphisms.

One useful kind of morphism is the reduction morphism.

20.2.1. DEFINITION. $f:C_1 \to C_2$ is a *reduction morphism* $\Leftrightarrow$ f *in* **SATS** and

1• f_Q is surjective;

2• $(A_1, \iota_1) = (A_2, \iota_2)$;

3• $\forall a \in A_1 \forall q_1 \in Q_1 \forall q_2 \in Q_2 : f_Q(q_1) \to_2^a q_2 \Leftrightarrow \exists q_2' \in f_Q^{-1}(q_2) : q_1 \to_1^a q_2'$.

From this definition and induction on $|x|$, we may deduce that if $x \in A_1^*$, $q_1 \in Q_1$ and $q_2 \in Q_2$ then

$$f_Q(q_1) \to_2^x q_2 \Leftrightarrow \exists q_2' \in f_Q^{-1}(q_2) : q_1 \to_1^x q_2' \tag{20.2}$$

and in particular, if $q_1 \in Q_1$, then $L(T(C_1), q_1) = L(T(C_2), f_Q(q_1))$. Since $\iota_1 = \iota_2$, it follows, by 15.2.2, that $L(C_1, q_1) = L(C_2, f_Q(q_1))$. C_2 is thus a 'no less efficient' generator of its trace languages than C_1 since by 20.2.1 (1) it has at most the same number of states as C_1. Reduction morphisms are closely related to what we shall call *Q-congruences* and *kernels*.

20.2.2. DEFINITION. If $f: C_1 \to C_2$, then we define the *kernel* of f to be the relation $Ker(f) \subseteq Q_1 \times Q_1$ given by

$$q_1 \; Ker(f) \; q_2 \Leftrightarrow f_Q(q_1) = f_Q(q_2).$$

20.2.3. DEFINITION. If C is an asynchronous transition system, then a Q-congruence is an equivalence relation $\equiv \subseteq Q \times Q$ satisfying:

1• $\forall q, q' \in Q : q \equiv q' \Leftrightarrow L(T(C), q) = L(T(C), q')$;
2• $\forall q_1, q_2, \hat{q}_1, \hat{q}_2 \in Q : q_1 \equiv q_2 \wedge q_1 \to^a \hat{q}_1 \wedge q_2 \to^a \hat{q}_2 \Rightarrow \hat{q}_1 \equiv \hat{q}_2$.

Let $q/\equiv$ denote the $\equiv$-class of $q \in Q$ and let $Q_\equiv = \{q/\equiv \mid q \in Q\}$. We may impose a transition structure on $Q_\equiv$ as follows

$$q_1/\equiv \; \to_\equiv^a q_2/\equiv \Leftrightarrow \exists q_1' \in q_1/\equiv \exists q_2' \in q_2/\equiv : q_1' \to^a q_2' \tag{20.3}$$

and define $C_\equiv = (Q_\equiv, A, \to_\equiv, \iota)$.

Note that the identity relation on Q, Id_Q, is a Q-congruence and $C_{Id_Q} \equiv C$.

20.2.4. LEMMA. If $f: C_1 \to C_2$ is a reduction, then $Ker(f)$ is a Q-congruence.

PROOF. $Ker(f)$ is clearly an equivalence relation. For (1), we note that from (20.2), if $q \; Ker(f) \; q'$ then $f_Q(q) = f_Q(q')$ and so

$$L(T(C_1), q) = L(T(C_2), f_Q(q)) = L(T(C_2), f_Q(q')) = L(T(C_1), q').$$

Finally, if $q_1 \; Ker(f) \; q_2$, $q_1 \to_1^a \hat{q}_1$ and $q_2 \to_1^a \hat{q}_2$, then by (20.1), $f_A(a) = a$, and so by 19.2.1 (2), $f_Q(q_1) \to_2^a f_Q(\hat{q}_1)$ and $f_Q(q_2) \to_2^a f_Q(\hat{q}_2)$. But $q_1 \; Ker(f) \; q_2$, that is $f_Q(q_1) = f_Q(q_2)$, so, as $T(C_2)$ is unambiguous, by 15.2.1 (1), $f_Q(\hat{q}_1) = f_Q(\hat{q}_2)$, that is $\hat{q}_1 \; Ker(f) \; \hat{q}_2$, and we have proved (2).

■

20.2.5. PROPOSITION. If C is an asynchronous transition system and $\equiv$ is a Q-congruence on C, then $C_\equiv$ is an asynchronous transition system and $f_\equiv = (f/\equiv, inc_A)$ is a reduction, $f_\equiv : C \to C/\equiv$, where $f/\equiv$ is given by $(f/\equiv)(q) = q/\equiv$. Furthermore, $Ker(f_\equiv) = \equiv$

PROOF. We check the conditions of 15.2.1.

If $q/\equiv \to_\equiv^a q_1/\equiv$ and $q/\equiv \to_\equiv^a q_2/\equiv$, then by (20.3), there exists $q', q'' \in q/\equiv$, $q_1' \in q_1/\equiv$ and $q_2' \in q_2/\equiv$ such that $q' \to^a q_1'$ and $q'' \to^a q_2'$. But $q' \equiv q''$, so by 20.2.3 (2), $q_1' \equiv q_2'$, that is $q_1'/\equiv = q_2'/\equiv$, so $T(C_\equiv)$ is unambiguous.

Next, suppose that $q/\equiv \to_\equiv^a q_1/\equiv \to_\equiv^b q_2/\equiv$ and $a \iota b$, then by (20.3), there exists $q' \in q/\equiv$, $q_1', q_1'' \in q_1/\equiv$ and $q_2'' \in q_2/\equiv$ such that $q' \to^a q_1'$ and $q_1'' \to^b q_2''$. But $q_1' \equiv q_1''$ and so by 20.2.3 (1), $L(C, q_1') = L(C, q_1'')$, and since $b_\iota \in L(C, q_1'')$, there must exist $q_2' \in Q$ such that $q_1' \to^b q_2'$. By the lozenge rule, there exists $\tilde{q}_1 \in Q$ such that $q' \to^b \tilde{q}_1 \to^a q_2'$. By (20.3), $q/\equiv = q'/\equiv \to_\equiv^a \tilde{q}_1/\equiv \to_\equiv^b q_2'/\equiv$. But $q_1' \equiv q_1''$, $q_1' \to^b q_2'$ and $q_1'' \to^b q_2''$, and so $q_2' \equiv q_2''$, by 20.2.3 (2). Hence, $q/\equiv \to_\equiv^a \tilde{q}_1/\equiv \to_\equiv^b q_2''/\equiv = q_2/\equiv$ and the lozenge rule holds for $C_\equiv$.

The conditions of 19.2.1 and 20.2.1 (1) and (2) follow immediately from the definition of $C_\equiv$ and $f_\equiv$.

Next, suppose that $q \in Q$ and $\hat{q}/\equiv \in Q_\equiv$. If $f_\equiv(q) \to_\equiv^a \hat{q}/\equiv$, then $q/\equiv \to_\equiv^a \hat{q}/\equiv$ and so by (20.3) there exists $q' \in q/\equiv$ such that $q' \to^a \hat{q}'$. Again, using 20.2.3, there exists $\hat{q}'' \in Q$ such that $q \to^a \hat{q}''$ and $\hat{q}' \equiv \hat{q}''$, that is $\hat{q}'' \in f_\equiv^{-1}(\hat{q})$. On the other hand, if $\hat{q}'' \in f_\equiv^{-1}(\hat{q})$ such that $q \to^a \hat{q}''$, then by (20.4), $q/\equiv \to_\equiv^a \hat{q}''/\equiv$, that is $f_\equiv(q) \to_\equiv^a \hat{q}/\equiv$. Hence, 20.2.1 (3) holds. Finally

$$q_1 \; Ker(f_\equiv) \; q_2 \Leftrightarrow (f/\equiv)(q_1) = (f/\equiv)(q_2) \Leftrightarrow q_1/\equiv = q_2/\equiv \Leftrightarrow q_1 \equiv q_2$$

and so $Ker(f_\equiv) = \equiv$.

■

We have the usual interaction between kernels, congruences and reduction morphisms.

20.2.6. PROPOSITION. Suppose that $f : C \to \hat{C}$ is a reduction and that $\equiv$ is a Q-congruence on C such that $Ker(f) \subseteq \equiv$, then there exists a reduction $g : \hat{C} \to C_\equiv$ such that $g \circ f = f_\equiv$.

PROOF. Suppose $\hat{q} \in \hat{Q}$, then as f_Q is surjective, by 20.2.1 (1), there exists $q \in Q$ such that $f_Q(q) = \hat{q}$. We wish to define $g_Q(\hat{q}) = q/\equiv$. To show that g_Q is well-defined, suppose that it is also the case that $f_Q(q') = \hat{q}$, then $f_Q(q') = f_Q(q)$ that is $q \; Ker \; q'$ so $q \equiv q'$, as $Ker(f) \subseteq \equiv$. Let $g = (g_Q, inc_A)$. We check that $g: \hat{C} \to C_{\equiv}$.

$(A, \iota) = (\hat{A}, \hat{\iota})$, by 20.2.1 (2), as f is a reduction, and $(A, \iota) = (A_{\equiv}, \iota_{\equiv})$, by 20.2.3, so $(\hat{A}, \hat{\iota}) = (A_{\equiv}, \iota_{\equiv})$. Thus, 20.2.1 (2) holds for g. 19.2.1 (1) follows.

We prove that 19.2.1 (2) holds. Suppose that $\hat{q}_1, \hat{q}_2 \in \hat{Q}$ and $a \in A$ such that $\hat{q}_1 \hat{\to}^a \hat{q}_2$. Since f_Q is surjective, by 20.2.1 (1), there exists $q_1 \in Q$ such that $f_Q(q_1) = \hat{q}_1$. We thus have $f_Q(q_1) \hat{\to}^a \hat{q}_2$, and so by 20.2.1 (3), there exists $q_2 \in f_Q^{-1}(\hat{q}_2)$ such that $q_1 \to^a q_2$. By 20.2.3, $q_1/\equiv \to_{\equiv}^a q_2/\equiv$. But for each i, $f_Q(q_i) = \hat{q}_i$ and so $g_Q(\hat{q}_i) = q_i/\equiv$, that is $g_Q(\hat{q}_1) \to_{\equiv}^a g_Q(\hat{q}_2)$, as required.

19.2.1 (3) holds trivially, and we have established that g is a morphism.

Since, $A = A_{\equiv}$, by 20.2.3, and $g_A = inc_A$, g *in* **SATS**, by (20.1).

If $q/\equiv \in Q_{\equiv}$, then $q/\equiv = g_Q(f_Q(q))$. This establishes the surjectivity of g_Q, giving 20.2.1 (1). It also shows that $g_Q \circ f_Q = f/\equiv$ and so, since $inc_A * inc_A = inc_A$, we have

$$g \circ f = (g_Q \circ f_Q, inc_A * inc_A) = (f/\equiv, inc_A) = f_{\equiv}$$

We have already seen that 20.2.1 (2) holds. To complete the proof, we show that 20.2.1 (3) holds. Suppose therefore that $a \in A$, $\hat{q}_1 \in \hat{Q}$ and $q_2 \in Q$ such that $g_Q(\hat{q}_1) \to_{\equiv}^a q_2/\equiv$. By definition of g_Q, there exists $q_1 \in Q$ such that $f_Q(q_1) = \hat{q}_1$ and $g_Q(\hat{q}_1) = q_1/\equiv$. By (20.3), there exists $q_1' \in q_1/\equiv$ and $q_2' \in q_2/\equiv$ such that $q_1' \to^a q_2'$. Hence, $f_Q(q_1') \hat{\to}^a f_Q(q_2')$, by 19.2.1 (2). But $f_Q(q_1) = \hat{q}_1$ and $g_Q(f_Q(q_2)) = q_2''/\equiv = q_2/\equiv$, and hence there exists $\hat{q}_2 = f_Q(q_2) \in g_Q^{-1}(q_2/\equiv)$ such that $\hat{q}_1 \hat{\to}^a \hat{q}_2$.

■

We shortly prove an example of one of the classical isomorphism theorems. First, we prove a lemma establishing a relationship between reduction morphisms and isomorphisms in **SATS**.

20.2.7. LEMMA. If $f: C \to \hat{C}$ is a reduction, then f is an isomorphism in **SATS** $\Leftrightarrow$ f_Q is injective.

PROOF. $\Rightarrow$ follows from 19.2.4 (1)

$\Leftarrow$: We check the conditions of 19.2.4. Conditions (2) and (3) follow by virtue of 20.2.1 (2), since f is a reduction, as does (1), given that f_Q is surjective. As f_Q is injective by hypothesis f_Q is bijective, giving (1).

Finally, let $a \in A$ and $q_1, q_2 \in Q$. As f is a morphism and $f_A(a) = a \neq \Omega$, $q_1 \to^a q_2 \Rightarrow f_Q(q_1) \hat{\to}^{f_A(a)} f_Q(q_2)$, by 19.2.1 (2).

Conversely, suppose that $f_Q(q_1) \hat{\to}^{f_A(a)} f_Q(q_2)$, that is $f_Q(q_1) \hat{\to}^a f_Q(q_2)$. By 20.2.1 (3), there exists $q_2' \in f_Q^{-1}(f_Q(q_2))$ such that $q_1 \to^a q_2'$. But f_Q is injective, so $q_2' = q_2$. We have shown that $f_Q(q_1) \hat{\to}^{f_A(a)} f_Q(q_2) \Rightarrow q_1 \to^a q_2$, thereby completing the proof that 19.2.4 (4) holds.

■

20.2.8. THEOREM. If $f: C \to \hat{C}$ is a reduction morphism, then $\hat{C}$ is isomorphic to $C_{Ker(f)}$ in **SATS***.

PROOF. By 20.2.4 and 20.2.6, there exists a reduction morphism $g: \hat{C} \to C_{Ker(f)}$ such that $g \circ f = f_{Ker(f)}$. In view of 20.2.7 it suffices to show that g_Q is injective.

If $\hat{q}_1, \hat{q}_2 \in \hat{Q}$ such that $g_Q(\hat{q}_1) = g_Q(\hat{q}_2)$, then there exists $q_1, q_2 \in Q$ such that for each i, $f_Q(q_i) = \hat{q}_i$ and $g_Q(\hat{q}_i) = q_i/\equiv$. Therefore, $q_1/Ker(f) = q_2/Ker(f)$ and so $\hat{q}_1 = f_Q(q_1) = f_Q(q_2) = \hat{q}_2$.

■

Each asynchronous transition system has a 'largest' Q-congruence; the idea is familiar from automata theory. Suppose that C is an asynchronous transition system and define a relation $\equiv_{red}$ on Q by

$$q \equiv_{red} q' \Leftrightarrow L(T(C), q) = L(T(C), q') \tag{20.4}$$

$\equiv_{red}$ is clearly an equivalence relation which satisfies 20.2.3 (1) by (20.4). Let $Q_{red} = \{q/\equiv_{red} \mid q \in Q\}$.

We show that $\equiv_{red}$ is a Q-congruence. Let $q_1, q_2 \in Q$ such that $q_1 \equiv_{red} q_2$. If $q_1 \to^a \hat{q}_1$, then by definition of $\equiv_{red}$, $q_2 \to^a \hat{q}_2$, some ,$\hat{q}_2 \in q_2/\equiv_{red}$, and so by (20.4)

$$x \in L(T(C), \hat{q}_1) \Rightarrow a.x \in L(T(C), q_1) \Rightarrow a.x \in L(T(C), q_2)$$

and hence there exists $q, \bar{q}_2 \in Q$ such that $q_2 \to^a \bar{q}_2 \to^x q$. But $T(C)$ is unambiguous, by 15.2.1 (1), and so $\bar{q}_2 = \hat{q}_2$. Thus, $x \in L(T(C), \hat{q}_2)$, and we have shown that $L(T(C), \hat{q}_1) \subseteq L(T(C), \hat{q}_2)$. The reverse inclusion follows by symmetry. We have shown that $\hat{q}_1 \equiv_{red} \hat{q}_2$. This completes the proof of the first part of the following:

20.2.9. LEMMA. If C is an asynchronous transition system, then $\equiv_{red}$ is a Q-congruence. Furthermore, if $\equiv$ is a Q-congruence, then $\equiv \subseteq \equiv_{red}$.

PROOF. If $\equiv$ is a Q-congruence and $q_1 \equiv q_2$, then $L(T(C),q_1) = L(T(C),q_2)$, by 20.2.3 (1), that is $q_1 \equiv_{red} q_2$.

■

20.2.10. PROPOSITION. If C is an asynchronous transition system and $\equiv = \equiv_{red}$, then $f_{\equiv}:C \to C_{\equiv}$ is a reduction.

Furthermore, if $f:C \to \hat{C}$ is a reduction morphism, then there exists a unique reduction morphism $g:\hat{C} \to C_{\equiv}$ such that $g \circ f = f_{\equiv}$.

PROOF. That $f_{\equiv}:C \to C_{\equiv}$ is a reduction follows from 20.2.5 and 20.2.9.

Suppose that $f:C \to \hat{C}$ is a reduction morphism, then as $Ker(f) \subseteq \equiv_{red}$, by 20.2.9, there exists a reduction morphism $g:\hat{C} \to C_{\equiv}$ such that $g \circ f = f_{\equiv}$, by 20.2.6. We prove uniqueness. Suppose that $g':\hat{C} \to C_{\equiv}$ is a reduction morphism such that $g' \circ f = f_{\equiv}$ and let $\hat{q} \in \hat{Q}$, then since f_Q is surjective, there exists $q \in Q$, such that $f_Q(q) = \hat{q}$. But now,

$$g_Q(\hat{q}) = g_Q(f_Q(q)) = f_{\equiv}(q) = g'_Q(f_Q(q)) = g'_Q(\hat{q})$$

■

We define f *in* **ATS*** to be a reduction morphism if it is a reduction morphism in **ATS**. The following result indicates one of a number of connections between reduction morphisms and strict equivalence.

20.2.11. LEMMA. Suppose that $(C_1,q_1),(C_2,q_2) \in$ **ATS***, then if there exists a reduction morphism $f:(C_1,q_1) \to (C_2,q_2)$, then $(C_1,q_1) \equiv (C_2,q_2)$.

PROOF. We have noted in the discussion following (20.2) that if $f:C_1 \to C_2$ is a reduction morphism, then $L(C_1,q_1) = L(C_2,f_Q(q_1))$. Since $f_Q(q_1) = q_2$, it follows that $L(C_1,q_1) = L(C_2,q_2)$ and since $A_1 = A_2$, by 20.2.1 (2), $(C_1,q_1) \equiv (C_2,q_2)$, by 15.3.2.

■

If $(C,q) \in$ **ATS***, we define $(C,q)_{red} = (C_{\equiv_{red}}, q/\equiv_{red})$. As, $f_{\equiv_{red}}:C \to C_{\equiv_{red}}$ is a reduction morphism such that $(f_{\equiv_{red}})_Q(q) = q/\equiv_{red}$, if follows that $f_{\equiv_{red}}$ is also a reduction $f_{\equiv_{red}}:(C,q) \to (C,q)_{red}$, in **ATS***.

20.2.12. COROLLARY. If $(C,q) \in$ **ATS***, then $(C,q) \equiv (C,q)_{red}$.

PROOF. By 20.2.11.

■

20.2.13. PROPOSITION. If f *in* **SATS** is an isomorphism, then f is a reduction morphism.

PROOF. 20.2.1 (1) follows from 19.2.4 (1). 20.2.1 (2) follows from 19.2.4(2) and (3) and (20.1). 20.2.1 (3) follows from 19.2.4 (4).

■

The final result of this section relates strict equivalence to reductions and isomorphisms. We write $(C_1,q_1)\approx(C_2,q_2)$ to indicate that there is an isomorphism f *in* **SATS** from C_1 to C_2 such that $f_Q(q_1)=q_2$.

20.2.14. PROPOSITION. If $(C_1,q_1),(C_2,q_2)\in$ **ATS***, then $(C_1,q_1)\equiv(C_2,q_2)\Leftrightarrow(C_1,q_1)_{red}\approx(C_2,q_2)_{red}$.

PROOF. If $(C_1,q_1)_{red}\approx(C_2,q_2)_{red}$, then $(C_1,q_1)_{red}\equiv(C_2,q_2)_{red}$ by 20.2.11 and 20.2.13. As $(C_i,q_i)\equiv(C_i,q_i)_{red}$, each i, by 20.2.12, $(C_1,q_1)\equiv(C_2,q_2)$, since $\equiv$ is transitive.

Conversely, suppose that $(C_1,q_1)\equiv(C_2,q_2)$ and define $R\subseteq Q_1\times Q_2$ by

$$q_1'\,R\,q_2'\Leftrightarrow\exists x\in A_1^*:q_1\rightarrow_1^x q_1'\wedge q_2\rightarrow_2^x q_2'$$

Let $q_1'\in Q_1$, then by 19.2.5, there exists $x\in A_1^*$ such that $q_1\rightarrow_1^x q_1'$ and so, by 15.3.1 (1), there exists $q_2'\in Q_2$ such that $q_2\rightarrow_2^x q_2'$. Thus,

$$\forall q_1'\in Q_1\,\exists q_2'\in Q_2:q_1'\,R\,q_2' \tag{20.5}$$

Let $q_1'\in Q_1$ and $q_2',q_2''\in Q_2$ with $q_1'\,R\,q_2'$ and $q_1'\,R\,q_2''$. We show that $q_2'\equiv_{red}q_2''$. Let $q_2'\rightarrow_2^y$ and let $x',x''\in A_1^*$ such that $q_1\rightarrow_1^{x'}q_1'$, $q_2\rightarrow_2^{x'}q_2'$, $q_1\rightarrow_1^{x''}q_1'$ and $q_2\rightarrow_2^{x''}q_2''$. Then $q_2\rightarrow_2^{x'.y}$ and so by 15.3.1 (1), $q_1\rightarrow_1^{x'.y}$ and by unambiguity of $T(C_1)$, $q_1\rightarrow_1^{x'}q_1'\rightarrow_1^y$. Thus $q_1\rightarrow_1^{x''}q_1'\rightarrow_1^y$ and so by 15.3.1 (1), $q_2\rightarrow_2^{x''.y}$, and by unambiguity of $T(C_2)$, $q_2\rightarrow_2^{x''}q_2''\rightarrow_2^y$. We have shown that $L(T(C_2),q_2')\subseteq L(T(C_2),q_2'')$. The converse inclusion follows by symmetry. We may therefore define a function $h_Q:Q_1\rightarrow Q_2/\equiv_{red}$ by

$$h(q_1')=q_2'/\equiv_{red}\Leftrightarrow q_1'\,R\,q_2' \tag{20.6}$$

We claim that the pair $h=(h_Q,inc_{A_1})$ is a reduction morphism $h:(C_1,q_1)\rightarrow(C_1,q_1)_{red}$ such that $Ker(h)=\equiv_{red}$. $(C_1,q_1)_{red}\approx(C_2,q_2)_{red}$ follows by 20.2.8.

First, we note that by 19.2.5 and 15.3.1, $(A_1,\iota_1)=(A_2,\iota_2)$. Thus if $a,b\in A_1$, with $a\,\iota_1\,b$ then $h_A(a),h_A(b)>\Omega$ and $a\,\iota_2\,b$, that is $h_A(a)\,\iota_2\,h_A(b)$, since $h_A=inc_{A_1}$. Thus, h satisfies 19.2.1 (1) as well as 20.2.1 (2).

Next, suppose that $a\in A_1$ and $q_1',q_1''\in Q_1$ such that $q_1'\rightarrow_1^a q_1''$, then $h_A(a)\neq\Omega$ and h satisfies 19.2.1 (3) vacuously. Let $q_2',q_2''\in Q_2$ such that $q_1'\,R\,q_2'$ and $q_1''R\,q_2''$ and let $x',x''\in A_1{}^*$ such that $q_1\rightarrow_1^{x'} q_1'$, $q_2\rightarrow_2^{x'} q_2'$, then $q_1\rightarrow_1^{x'.a} q_1''$ and so $q_2\rightarrow_2^{x'.a}$, and by unambiguity of $T(C_2)$, $q_2\rightarrow_2^{x'} q_2'\rightarrow_2^{a} q_2''$. Hence, $q_2'/{\equiv_{red}}\rightarrow^a q_2''/{\equiv_{red}}$, by (20.3), that is $h_Q(q_2')\rightarrow^a h_Q(q_2'')$ and h satisfies 19.2.1 (3). We have proved that h *in* **ATS**.

Since $q_1\,R\,q_2$, $h_Q(q_1)=q_2/{\equiv_{red}}$, and h *in* **ATS***.

We have observed that if $q_1'\in Q_1$, then there exists $q_2'\in Q_2$ such that $q_1'\,R\,q_2'$. By symmetry, if $q_2'\in Q_2$, then there exists $q_1'\in Q_1$ such that $q_1'\,R\,q_2'$ and h_Q is surjective, giving 20.2.1 (1).

Next, suppose that $h_Q(q_1')\rightarrow^a q_2''/{\equiv_{red}}$, then there exists $q_2'\in Q_2$ such that $q_1'\,R\,q_2'$ and $q_2'\rightarrow_2^a q_2''$. As in previous arguments, there exists $q_1''\in Q_1$ such that $q'\rightarrow_1^a q''$ and $q_1''R\,q_2''$. But $q_1''\in h_Q^{-1}(q_2''/{\equiv_{red}})$, and so 20.2.1 (3) holds, and we have shown that h is a reduction.

Finally, we show that $Ker(h)={\equiv_{red}}$, Since $Ker(h)$ is a Q-congruence, by 20.2.4, $Ker(h)\subseteq{\equiv_{red}}$, by 20.2.9. On the other hand, let $q_1',q_1''\in Q_1$ with $q_1'\equiv_{red} q_1''$. By (20.6), we must show that $h_Q(q_1')=h_Q(q_1'')$, that is, if $q_1'\,R\,q_2'$ and $q_1''R\,q_2''$, then $q_2'\equiv_{red} q_2''$. Suppose, then that $x\in L(T(C_2),q_2')$. By (20.5), there exists $y\in A_1{}^*$ such that $q_1\rightarrow_1^y q_1'$ and $q_2\rightarrow_2^y q_2'$. Hence, $y.x\in L(T(C_2),q_2')$ and so $y.x\in L(T(C_1),q_1')$, by 15.3.1 (1). Therefore, $q_1\rightarrow_1^y \hat{q}_1\rightarrow_1^x$, some $\hat{q}_1\in Q_1$. By unambiguity of $T(C_1)$, $\hat{q}_1=q_1'$. Thus, $x\in L(T(C_1),q_1')$, and since $q_1'\equiv_{red} q_1''$, $x\in L(T(C_1),q_1'')$, by (20.4). By (20.5), there exists $y\in A_1{}^*$ such that $q_1\rightarrow_1^y q_1''$ and $q_2\rightarrow_2^y q_2''$. Hence, $y.x\in L(T(C_1),q_1'')$ and so $y.x\in L(T(C_2),q_2'')$, by 15.3.1 (1). Hence, $q_2\rightarrow_2^y \hat{q}_2\rightarrow_2^x$, some $\hat{q}_2\in Q_2$. By unambiguity of $T(C_2)$, $\hat{q}_2=q_2''$. Thus, $x\in L(T(C_2),q_2'')$. We have shown that $L(T(C_2),q_2')\subseteq L(T(C_2),q_2'')$. The reverse inclusion follows by symmetry.

■

20.3. Products.

If $(c_i)_{i\in I}$ is an indexed family of objects of a category $\underline{\mathbf{C}}$, then a *product* of the family is an object c together with a family of morphisms $\pi_i:c\rightarrow c_i$, called the *projections* of the product, with the property that given any family of morphisms $f_i:d\rightarrow c_i$ there exists a unique morphism $h:d\rightarrow c$ such that for each i, the following diagram commutes.

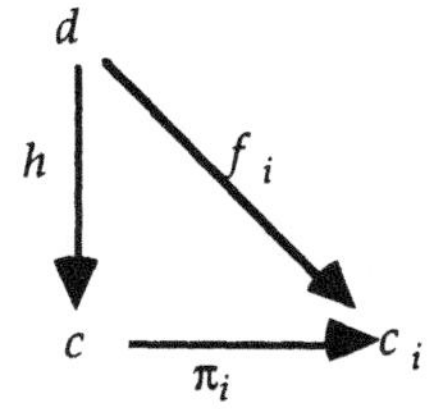

The definition ensures that the objects of any two products of the same family are isomorphic.

We show that products in **SATS** are related to an operator on string languages which has cropped up in various contexts. For example, in the interleaving semantics for unambiguous path expressions[1] described in [48], the string language associated with a set of path expressions is obtained, by the use of this operator, from those corresponding to the individual paths. The same operator is used in the trace semantics to TCSP ([2]).[2]

We define versions of this operator for the categories **LCTL** and **SATS** and show that it gives rise to products in the latter category.

20.3.1. DEFINITION. Let $(A_i)_{i\in I}$ be a family of sets and let $(\iota_i)_{i\in I}$ be a family of independence relations, where $\iota_i \subseteq A_i \times A_i$, each i Define, $\underset{i\in I}{||}(A_i,\iota_i) = (A,\iota)$ where $A = \bigcup_{i\in I} A_i$ and for all $a,b \in A$:

$$a\,\iota\,b \Leftrightarrow (\forall i \in I{:}a,b \in A_i \Rightarrow a\,\iota_i\,b).$$

Define the *projection functions* $\hat{\pi}_i{:}A^* \to A_i{}^*$ as follows.

$$\hat{\pi}_i(\Omega) = \Omega \tag{20.7}$$

and if $a \in A$ and $u,v \in A^*$ then

$$\hat{\pi}_i(a) = \begin{cases} a \text{ if } a \in A_i \\ \Omega \text{ otherwise} \end{cases} \tag{20.8}$$

$$\hat{\pi}_i(u.v) = \hat{\pi}_i(u).\hat{\pi}_i(v) \tag{20.9}$$

Next, suppose that $x,y \in A^*$ such that $x \equiv_\iota^{(1)} y$, then there exists $u,v \in A^*$ and $a,b \in A$ such that $x = u.a.b.v$, $y = u.b.a.v$ and $a\,\iota\,b$. Let $i \in I$. If $\{a,b\} \not\subseteq A_i$, then by (20.8) and (20.9), $\hat{\pi}_i(a).\hat{\pi}_i(b) = \hat{\pi}_i(b).\hat{\pi}_i(a)$ and so

[1]See chapter 25.
[2]See chapter 28.

$\hat{\pi}_i(u.a.b.v) = \hat{\pi}_i(u.b.a.v)$. If $\{a,b\} \subseteq A_i$, then $a\ \iota_i\ b$, by 20.3.1 (2), and by (20.8) and (20.9),

$$\hat{\pi}_i(u.a.b.v) = \hat{\pi}_i(u).a.b.\hat{\pi}_i(v) \equiv_{\iota_i}^{(1)} \hat{\pi}_i(u).b.a.\hat{\pi}_i(v) = \hat{\pi}_i(u.b.a.v).$$

Since $\equiv_\iota$ is the reflexive, transitive closure of $\equiv_\iota^{(1)}$ and since the $\equiv_{\iota_i}$ are the reflexive, transitive closures of the $\equiv_{\iota_i}^{(1)}$, we deduce that for all $i \in I$:

$$\forall x,y \in A^*: x \equiv_\iota y \Rightarrow \hat{\pi}_i(x) \equiv_{\iota_i} \hat{\pi}_i(y) \tag{20.10}$$

which allows us to define trace projection functions $\pi_i : A_\iota^* \to (A_i)_{\iota_i}^*$ by

$$\pi_i(x_\iota) = \hat{\pi}_i(x)_{\iota_i}$$

each $x \in A^*$. We observe that these functions satisfy the trace analogues of (20.7), (20.8) and (20.9).

20.3.2. DEFINITION. If $(D_i)_{i\in I}$ is a family of elements of **LCTL**, we define $\|_{i\in I} D_i = (A, \iota, L)$, where

1• $(A,\iota) = \|_{i\in I}(A_i, \iota_i)$;
2• $L = \{x_\iota \in A_\iota^* \mid \forall i \in I: \pi_i(x_\iota) \in L_i\}$.

20.3.3. PROPOSITION. If $(D_i)_{i\in I}$ is a family of elements of **LCTL**, then $\|_{i\in I} D_i \in$ **LCTL** and for each $i \in I$ the function $\hat{\pi}_i : A \to A_i \cup \{\Omega\}$ is a morphism $\|_{i\in I} D_j \to D_i$.

PROOF. Let $D = \|_{i\in I} D_i$. Suppose $x_\iota \in L$ and $y_\iota \in A_\iota^*$ such that $y_\iota \leq x_\iota$, then by 9.2.2 there exists $z \in A^*$ such that $y.z \equiv_\iota x$, that is to say, $y_\iota.z_\iota = x_\iota$, so $\pi_i(y_\iota).\pi_i(z_\iota) = \pi_i(x_\iota)$, that is $\pi_i(y_\iota) \leq \pi_i(x_\iota)$. As $\pi_i(x_\iota) \in L_i$ and $D_i \in$ **LCTL**, each i, $\pi_i(y_\iota) \in L_i$. By 20.3.2 (2), $y_\iota \in L$. We have shown that $D \subseteq A_\iota^*$ is left-closed.

Checking that each $\hat{\pi}_i$ is a morphism from D to D_i, we see that 19.3.2 (1) follows from 20.3.1 and that 19.3.2 (2') follows from 20.3.2 (2).

■

We pause to look at the special case of parallel composition in which the independence relations of the component languages are all trivial. In such a case, these languages are structurally identical to ordinary string languages and we have a corresponding construction.

20.3.4. DEFINITION. Let $\mathcal{A} = (A_i)_{i \in I}$ be a family of sets, then we define $\coprod_{i \in I} A_i = (A, \iota)$, where

1• $A = \bigcup_{i \in I} A_i$;

2• $\forall a, b \in A: a \,\iota\, b \Leftrightarrow \forall i \in I: \{a, b\} \not\subseteq A_i$.

We observe that $\iota = \iota_{\mathcal{A}}$,[3] whence ι is an independence relation.[4]

Next, suppose that $x, y \in A^*$ such that $x \equiv_\iota^{(1)} y$, then there exists $u, v \in A^*$ and $a, b \in A$ such that $x = u.a.b.v$, $y = u.b.a.v$ and $a \,\iota\, b$. Thus, for each $i \in I$, $\{a, b\} \not\subseteq A_i$, and so using (20.9) we see that $\hat{\pi}_i(a).\hat{\pi}_i(b) = \hat{\pi}_i(b).\hat{\pi}_i(a)$ and so $\hat{\pi}_i(u.a.b.v) = \hat{\pi}_i(u.b.a.v)$. Since $\equiv_\iota$ is the reflexive, transitive closure of $\equiv_\iota^{(1)}$, we deduce that for all $i \in I$:

$$\forall x, y \in A^*: x \equiv_\iota y \Rightarrow \hat{\pi}_i(x) = \hat{\pi}_i(y) \tag{20.11}$$

which allows us to define projection functions $\pi_i: A_\iota^* \to A_i^*$ by

$$\pi_i(x_\iota) = \hat{\pi}_i(x)$$

each $x \in A^*$.

20.3.5. DEFINITION. Let $(A_i, L_i)_{i \in I}$ be a family of sorted string languages. Define, $(A, \iota, L) = \coprod_{i \in I}^{\tau} (A_i, L_i)$ where

1• $(A, \iota) = \coprod_{i \in I} A_i$;

2• $\forall x \in A^*: x_\iota \in L \Leftrightarrow (\forall i \in I: \pi_i(x) \in L_i)$.

There is a direct relationship between this construction and the parallel composition operator of section 10.5.

20.3.6. PROPOSITION. Let $\mathcal{L} = (A_i, L_i)_{i \in I}$ be a family of sorted string languages, then if $(A, \iota, L) = \coprod_{i \in I}^{\tau} (A_i, L_i)$, then $\| \mathcal{L} = \phi_\alpha^*(L)$, where $\phi_\alpha^*: A_\iota^* \to A_\alpha^*$ is the function whose existence is asserted by 9.3.3.

PROOF. Referring to the proof of 9.3.3, we see that for all $x \in A^*$, $\phi_\alpha^*(x_\iota) = \pi_\alpha(x)$, where $(\pi_\alpha(x))(i) = \pi_i(x)$. Hence, if $\underline{x} \in \| \mathcal{L}$, then there exists $x_\iota \in L$ such that for all $i \in I$, $\underline{x}(i) = \pi_i(x_\iota)$, and $\underline{x}(i) \in L_i$, by 20.3.5.

---

[3]See 10.5.1

[4]See 9.2.4

Conversely, suppose that $\underline{x}(i) \in L_i$, all $i \in I$, and let $x_\iota = \phi_\alpha^{*-1}(\underline{x})$, then for each $i \in I$, $\pi_i(x_\iota) = \underline{x}(i) \in L_i$, and so $\underline{x} \in \| \mathcal{L}$.

■

20.3.7. COROLLARY. Let $\mathcal{L} = (A_i, L_i)_{i\in I}$ be a family of sorted string languages then

1• If each L_i is left-closed in A_i^* then $\|^{\tau}_{i\in I}(A_i, L_i)$ is left-closed in $(\|_{i\in I} A_i)^*$;

2• If $X \subseteq \|^{\tau}_{i\in I}(A_i, L_i)$ and $\bigsqcup X$ exists, then $\bigsqcup X \in \|^{\tau}_{i\in I}(A_i, L_i)$.

PROOF. By 10.5.3 and 20.3.6.

■

We now consider an equivalent construction for asynchronous transition systems.

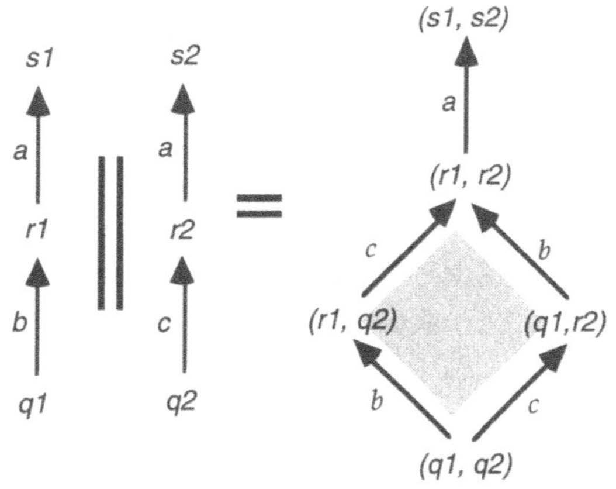

Figure 20.1.

20.3.8. DEFINITION Suppose $(C_i)_{i\in I}$ is an indexed family of asynchronous transition systems. We define their parallel composite, denoted by $\|_{i\in I} C_i$ to be the tuple $(Q, A, \rightarrow, \iota)$ where

1• Q is the set of functions $\underline{q}: I \rightarrow \bigcup_{i\in I} Q_i$ such that $\underline{q}(i) \in Q_i$, each $i \in I$

2• $(A, \iota) = \|_{i\in I} A_i$;

3• $\underline{q}_1 \rightarrow^a \underline{q}_2 \Leftrightarrow \forall i \in I: (\underline{q}_1(i) \rightarrow^a \underline{q}_2(i)) \vee (a \notin A_i \wedge \underline{q}_1(i) = \underline{q}_2(i))$.

If $I = \{1, \cdots, n\}$, then we shall write $C_1 \| \cdots \| C_n$ for $\|_{i\in I} C_i$

If $(C_i, q_i)_{i\in I}$ is a family of initialised asynchronous transition systems, we define

$$\underset{i\in I}{||}(C_i, q_i) = r_{\underline{q}}(\underset{i\in I}{||} C_i, \underline{q})$$

where $\underline{q}(i) = q_i$, each $i \in I$.

20.3.8 is illustrated in figure 20.1, where it is assumed that $A_1 = \{a,b\}$, $A_2 = \{a,c\}$ and ι_1 and ι_2 are both trivial.

20.3.9. PROPOSITION. If $(C_i)_{i\in I}$ is an indexed family of asynchronous transition systems, then $\underset{i\in I}{||} C_i$ is an asynchronous transition system. If each of the C_i are forward stable, then so is $\underset{i\in I}{||} C_i$.

PROOF. Let $C = \underset{i\in I}{||} C_i$. $T(C)$ is obviously a transition system.

Suppose $\underline{q} \to^a \underline{q}_1$ and $\underline{q} \to^a \underline{q}_2$. If $a \in A_i$ then $\underline{q}(i) \to_i^a \underline{q}_1(i)$ and $\underline{q}(i) \to_i^a \underline{q}_2(i)$, and so $\underline{q}_1(i) = \underline{q}_2(i)$, as each $T(C_i)$ is unambiguous. Otherwise, $\underline{q}_1(i) = \underline{q}(i) = \underline{q}_2(i)$. Thus $\underline{q}_1 = \underline{q}_2$ and so $T(C)$ is unambiguous.

Next, suppose $\underline{q} \to^a \underline{q}_1$ and $\underline{q}_1 \to^b \underline{q}_2$, where $a \,\iota\, b$. If $a,b \in A_i$, then $\underline{q}(i) \to_i^a \underline{q}_1(i)$ and $\underline{q}_1(i) \to_i^b \underline{q}_2(i)$, by 20.3.8 (3), and $a \,\iota_i\, b$, by 20.3.1. Hence $\underline{q}(i) \to_i^b q_i$ and $q_i \to_i^a \underline{q}_2(i)$, some $q_i \in Q_i$. Define

$$\underline{q}'_1(i) = \begin{cases} q_i \text{ if } a,b \in A_i \\ \underline{q}(i) \text{ if } b \notin A_i \\ \underline{q}_2(i) \text{ if } a \notin A_i \end{cases}$$

Note that if $a \notin A_i$ and $b \notin A_i$, then $\underline{q}(i) = \underline{q}_1(i) = \underline{q}_2(i)$, so $\underline{q}'_1$ is well defined.

If $a \in A_i$, then either $b \in A_i$, in which case $\underline{q}'_1(i) = q_i \to_i^a \underline{q}_2(i)$, or $b \notin A_i$, in which case $\underline{q}'_1(i) = \underline{q}(i) \to_i^a \underline{q}_1(i) = \underline{q}_2(i)$. If $a \notin A_i$, then $\underline{q}'_1(i) = \underline{q}_2(i)$. We have shown that $\underline{q}'_1 \to^a \underline{q}_2$.

If $b \in A_i$, then either $a \in A_i$, in which case $\underline{q}(i) \to_i^b q_i = \underline{q}'_1(i)$, or $a \notin A_i$, in which case $\underline{q}(i) = \underline{q}_1(i) \to_i^b \underline{q}_2(i) = \underline{q}'_1(i)$. If $b \notin A_i$, then $\underline{q}(i) = \underline{q}'_1(i)$. We have shown that $\underline{q} \to^b \underline{q}'_1$.

Thus, the lozenge rule holds and so $\underset{i\in I}{||} C_i$ is an asynchronous transition system.

The proof that $\underset{i\in I}{\|} C_i$ is forward stable if the C_i are forward stable is similar and will be omitted.

■

Finally, we return to the categorical side of things.

20.3.10. PROPOSITION. Suppose that $(C_i)_{i\in I}$ is a family of asynchronous transition systems, and $C = \underset{i\in I}{\|} C_i$, then $(C, \underline{q})$ is the object of a product with projections $\pi^i : C \to C_i$ given by

$$\pi^i{}_Q(\underline{q}) = \underline{q}(i) \tag{20.12}$$

$$\pi^i{}_A(\underline{a}) = \begin{cases} a \text{ if } a \in A_i \\ \Omega \text{ otherwise} \end{cases} \tag{20.13}$$

PROOF. First, we show that the π^i are morphisms in **SATS**. Suppose that $a, b \in A$ such that $\pi^i{}_A(a), \pi^i{}_A(b) > \Omega$ and $a \iota b$, then by (20.1), $a, b \in A_i$ and so by 20.3.4 and (20.13) $\pi^i{}_A(a) \iota_i \pi^i{}_A(b)$, giving 19.2.1 (1). Next, suppose that $a \in A$ and $\underline{q}_1, \underline{q}_2 \in Q$ such that $\underline{q}_1 \to^a \underline{q}_2$. If $\pi^i{}_A(a) = \Omega$, then $a \notin A_i$ and so by 20.9.3 (3) and (20.12) $\pi^i{}_Q(\underline{q}_1) = \pi^i{}_Q(\underline{q}_2)$, giving 19.2.1 (3). If $\pi^i{}_A(a) \neq \Omega$, then $a \in A_i$ and so by 20.9.3 (3), (20.12) and (20.13) $\pi^i{}_Q(\underline{q}_1) \to_i^{\pi^i{}_A(a)} \pi^i{}_Q(\underline{q}_2)$, giving 19.2.1 (2). Thus, the π^i are morphisms in **ATS**. By (20.1) and (20.13), the π^i are morphisms in **SATS**.

Next, suppose that $(f^i)_{i\in I}$ is a family of morphisms $f^i : \hat{C} \to C_i, q_i$. We define functions $h_Q : \hat{Q} \to Q$ and $h_A : \hat{A} \to A$ by

$$(h_Q(\hat{q}'))(i) = f^i(\hat{q}') \tag{20.14}$$

$$h_A(a) = \begin{cases} a \text{ if } f^i{}_A(a) = a, \text{ some } i \in I \\ \Omega \text{ otherwise} \end{cases} \tag{20.15}$$

Suppose that $a, b \in \hat{A}$ with $h_A(a), h_A(b) > \Omega$ and $a \,\hat{\iota}\, b$. If $a, b \in A_i$, then by 20.3.1, $f^i{}_A(a) = a$ and $f^i{}_A(b) = b$, and so by 19.2.1 (1), $a \iota_i b$. Hence, by 20.3.8 $a \iota b$, that is $h_A(a) \iota h_A(b)$, (20.15). Thus, h satisfies 19.2.1 (1). Next, let $a \in \hat{A}$ and $\hat{q}_1, \hat{q}_2 \in \hat{Q}$ such that $\hat{q}_1 \hat{\to}^a \hat{q}_2$. If $h_A(a) = \Omega$, then $a \in A_i$, all i, and consequently $f^i{}_A(a) = \Omega$ by (20.1). By 19.2.1 (2) $f^i{}_Q(\hat{q}_1) = f^i{}_Q(\hat{q}_2)$, each i, that is $h_Q(\hat{q}_1) = h_Q(\hat{q}_2)$, by (20.14). Thus, h satisfies 19.2.1 (2). If $h_A(a) \neq \Omega$, then for each i such that $a \in A_i$, $f^i{}_Q(\hat{q}_1) \to^{f^i{}_A(a)} f^i{}_Q(\hat{q}_2)$, by 19.2.1 (3), that is

$(h_Q(\hat{q}_1))(i) \to^a (h_Q(\hat{q}_2))(i)$, by (20.14) and (20.15). Otherwise, we have $h_Q(\hat{q}_1) = h_Q(\hat{q}_2)$, as we have seen. Thus, by 20.3.8 (3), $h_Q(\hat{q}_1) \to^{h_A(a)} h_Q(\hat{q}_2)$ and h satisfies 19.2.1 (3). We have proved that h *in* **ATS**. By construction, h *in* **SATS**.

If $\hat{q}' \in \hat{Q}$, then for each i, $f^i{}_Q(\hat{q}') = (h_Q(\hat{q}'))(i) = \pi^i(h_Q(\hat{q}'))$. Let $a \in \hat{A}$ and $i \in I$. If $a \in A_i$, then $h_A(a) = f^i{}_A(a) = \pi^i{}_A(a) = a$ so $f^i{}_A(a) = (\pi^i{}_A * h_A)(a)$, while if $a \in A_i$, then $h_A(a) = \pi^i{}_A(a) = \Omega$ and so $f^i{}_A(a) = (\pi^i{}_A * h_A)(a)$ in this case also. We have shown that $f^i = \pi^i \circ h$, all $i \in I$.

Finally suppose that $h' : \hat{C} \to C$ such that $f^i = \pi^i \circ h'$, then for each $i \in I$

$$(h'_Q(\hat{q}'))(i) = \pi^i{}_Q(h'_Q(\hat{q}')) = f^i{}_Q(\hat{q}') = \pi^i{}_Q(h_Q(\hat{q}')) = (h_Q(\hat{q}'))(i)$$

so $h'_Q = h_Q$. Finally, $h'_A = h_A$ by (20.1).

■

The relationship between parallel composition of languages and of asynchronous transition systems is what one might expect.

20.3.11. PROPOSITION. Suppose that $(C_i, q_i)_{i \in I}$ is a family of initialised asynchronous transition systems, and $(C, \underline{q}) = \|_{i \in I}(C_i, q_i)$, then

$$\mathbf{L}(C, \underline{q}) = \|_{i \in I} \mathbf{L}(C_i, q_i).$$[5]

PROOF. $\mathbf{L}(C, \underline{q}) = (A, \iota, L(C, \underline{q}))$ and $\mathbf{L}(C_i, q_i) = (A_i, \iota_i, L(C_i, q_i))$, each i so if $(A', \iota', L') = \|_{i \in I} \mathbf{L}(C_i, q_i)$, then $(A', \iota') = \|_{i \in I}(A_i, \iota_i) = (A, \iota)$. By 20.3.8 (3) and induction:

$$\forall x \in A^* : (\underline{q} \to^x \underline{\hat{q}} \Leftrightarrow \forall i \in I : q_i \to_i^{\hat{\pi}_i(x)} \underline{\hat{q}}(i))$$

That is

$$\forall x \in A^* : (x \in L(T(C), \underline{q}) \Leftrightarrow \forall i \in I : \hat{\pi}_i(x) \in L(T(C_i), q_i)).$$

Hence

$$\forall x \in A^* : (x_\iota \in L(C, \underline{q}) \Leftrightarrow \forall i \in I : \pi_i(x_\iota) \in L(C_i, q_i))$$

which, in view of 20.3.2 (2) gives $L(C, \underline{q}) = L'$

■

---

[5]See 19.3.5.

Chapter 21
An Application: Realisation by Nets

The reason why so few marriages are happy, is, because young ladies spend their time in making nets, not in making cages.
JONATHAN SWIFT
Thoughts on Various Subjects

21.1. Introduction.

In chapter 17, we treated the semantics of C/E systems within our general framework, putting us in a position to ask: how expressive is the C/E system model? What kind of systems may be modelled - or as we shall say realised - by C/E systems? We shall make things easier for ourselves, for the moment, by asking a more general question: what systems may be realised by a net using the dynamics of a C/E system?[1]

Clearly, we must restrict ourselves to systems whose behaviour may be described by asynchronous transition systems; that is the kind of behaviour nets have at this level. We may rephrase the question in the form: which asynchronous transition systems may be implemented on a net? Given a net, N, we define

$$C_N = (\wp(S), T, \rightarrow_N, \iota_N), \text{ where } c \rightarrow_N{}^t\, c' \Leftrightarrow c[t\rangle c' \tag{21.1}$$

The proofs of propositions 17.2.2 and 17.2.3 go through without any alteration. C_N is a forward stable asynchronous transition system. Now we shall explain what we mean by a system **S** being capable of being realised by a net; the initialised, asynchronous transition system associated with **S** may be embedded into the initialised, asynchronous transition system of some net. As a consequence, the two automata determine identical trace languages and hence identical behavioural presentations.

---

[1]That is, we allow nets to be non-simple but we use (1.17) as the 'firing rule'.

In section 21.2 we establish necessary and sufficient conditions for realisation by a net, which use the parallel composition operator described in chapter 20.

In section 21.3, we show that every loosely coupled system with transitions can be realised by a net. The theory also shows us how a given loosely coupled system with transitions can be transformed into a net which realises it.

21.2. Conditions for Realisation.

We begin by defining the notion of an embedding.

21.2.1. DEFINITION. Let C_1 and C_2 be asynchronous transition systems. An *embedding* of C_1 into C_2 is a morphism $f:C_1 \to C_2$ such that f_Q is injective and if $q_1 \in Q_1$, $q_2 \in Q_2$ and $a,b \in A_1$, then

1• $f_A(a) = a$;

2• $a \iota_2 b \Rightarrow a \iota_1 b$;

3• $f_Q(q_1) \to_2^a q_2 \Rightarrow \exists q_1' \in f_Q^{-1}(q_2): q_1 \to_1^a q_1'$.

An easy induction argument allows us to extend (3) to the following:

4• $\forall x \in A_1^*: (f_Q(q_1) \to_2^x q_2 \Rightarrow \exists q_1' \in f_Q^{-1}(q_2): q_1 \to_1^x q_1')$

It is easy enough to show that the composition of embeddings is an embedding. We also note that an isomorphism f is an embedding $\Leftrightarrow$ f_A is the inclusion mapping. Isomorphisms which are also embeddings, we shall call *embedding isomorphisms.*

The idea is that if $f:C_1 \to C_2$ is an embedding then every behaviour possible to C_1 is also possible to C_2. Indeed, since f is a morphism, we have $L(C_1,q_1) = f_A^*(L(C_1,q_1)) \subseteq L(C_2,f_Q(q_1))$, by 19.3.1.

Conversely, suppose $x \in L(T(C_2),f_Q(q_1))$. Let $y \in L(T(C_1),q_1)$ be such that $y \le x$ and that if $y' \in L(T(C_1),q_1)$ and $y' \le x$ then $y' \le y$. If $y < x$, then $f_Q(q_1) \to_2^y q_2 \to_2^a q_2'$, some $q_2, q_2' \in Q_2$, where $a \in A_1$ and $y.a \le x$. But now by 21.2.1 (3) and (4), there exists $q_2' \in f_Q^{-1}(q_2)$ and $q_1'' \in f_Q^{-1}(q_2')$ such that $q_1 \to_1^y q_1' \to_1^a q_1''$. So $y.a \in L(C_1,q_1)$ and $y.a \le x$, contradicting maximality of y. Hence $y = x$.

We have shown that $L(T(C_2),f_Q(q_1)) \subseteq L(T(C_1),q_1)$. As $a \iota_2 b \Leftrightarrow a \iota_1 b$, all $a,b \in A_1$, by 20.2.1 (1), and 21.2.1 (3), it follows that $L(C_2,f_Q(q_1)) \subseteq L(C_1,q_1)$. To sum up:

21.2.2. LEMMA. If $f: C_1 \to C_2$ is an embedding, then

$$\forall q_1 \in Q_1 : L(C_1, q_1) = L(C_2, f_Q(q_1)).$$

■

Thus, if there is an embedding of C_1 into C_2, then every pattern of event occurrences possible in C_1 from some initial state is also possible in C_2 from some initial state. Anything C_1 can do, C_2 can do also.

21.2.3. DEFINITION. We shall say that an asynchronous transition system C may be *realised by a net* $\Leftrightarrow$ N there exists an embedding $f: C \to C_N$.

So, which asynchronous transition systems may be realised by nets? The main points of the argument are as follows:

1• Every net has a kind of 'decomposition' into so-called 'state-machine' nets, which are equivalent to transition systems.
2• The parallel composite of these transition systems is the asynchronous transition system of the net with which we started.
3• Thus, if an asynchronous transition system may be embedded into an asynchronous transition system which is the composite of transition systems, then it may be realised by a net.[2]
4• Conversely, any family of transition systems may be combined to form a net whose asynchronous transition system is their parallel composite.
5• Thus, an asynchronous transition system may be realised by a net if and only if it may be embedded into an asynchronous transition system which is the composite of transition systems.

We begin by applying the construction of 20.3.8 to unambiguous transition systems; an unambiguous transition system $T = (Q, A, \to)$ may be regarded as an asynchronous transition system in which $\iota = \varnothing$.[3] We shall make this identification in the sequel, especially in 21.2.4 which follows. Note that in this case, the independence relation of the composite equals ι, where $\|_{i \in I} A_i = (A, \iota)$.

21.2.4. DEFINITION. A *parallel decomposition* of an asynchronous transition system C is a pair $(\mathbf{T}, f)$, where $\mathbf{T}$ is a family of non-empty, unambiguous transition systems and $f: C \to \|_{i \in I} T_i$ is an embedding.

---

[2]For example, we prove below that any asynchronous transition system that provides a non-interleaving semantics for a loosely coupled system has an embedding into an asynchronous transition system which may be thus decomposed. Hence any loosely coupled system may be realised by a net.
[3] See 22.3.4.

A transition system T will be said to be *singly labelled* $\Leftrightarrow$

$$q_1 \to^{a_1} \hat{q}_1 \wedge q_2 \to^{a_2} \hat{q}_2 \wedge a_1 = a_2 \Rightarrow (q_1, \hat{q}_1) = (q_2, \hat{q}_2).$$

A parallel decomposition $(\mathbf{T}, f)$ will be said to be *singular* $\Leftrightarrow$ each T_i is singly labelled.

The notion 'singly labelled' needs a few words of explanation. The idea is that a singly labelled transition system may be regarded as a net in which each T-element t satisfies $|\bullet t| \leq 1$ and $|t \bullet| \leq 1$. Such nets are known as *state machine nets*. Any state machine net may also be regarded as a singly labelled transition system. Now, if C has a singular parallel decomposition, then we may break C down into state machine nets. These may then be reconstituted into a net according to the composition of 21.2.5 below. 21.2.6 will then allow us to conclude that C may be embedded into the asynchronous transition system of this composite net. This will enable us to begin a structural characterisation of asynchronous transition systems which may be realised by nets. Let us begin with composition of nets.

21.2.5. DEFINITION. Let $(N_i)_{i \in I}$ be a family of nets. We define its *parallel composite*, written $\underset{i \in I}{||} N_i$, to be the net $N = (S, T; F)$ where

1• $S = \bigcup_{i \in I} (S_i \times \{i\})$;

2• $T = \bigcup_{i \in I} T_i$;

3• $\forall (s,i) \in S \; \forall t \in T: ((s,i) \, F \, t \Leftrightarrow s \, F_i \, t) \wedge (t \, F \, (s,i) \Leftrightarrow t \, F_i \, s)$.

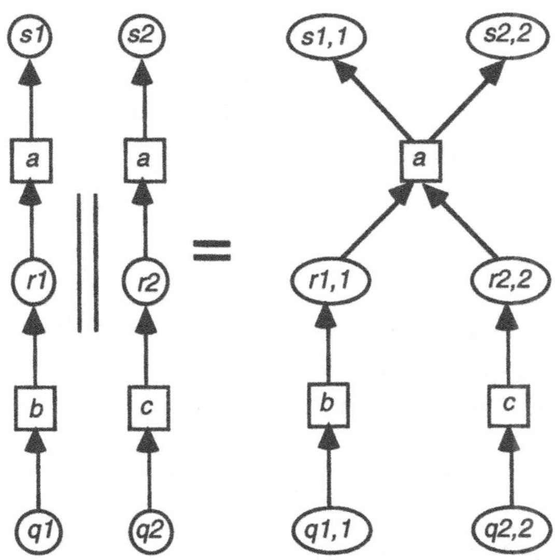

Figure 21.1

If $I=\{1,\cdots,n\}$ then we shall write $N_1 \| \cdots \| N_n$ for $\|_{i\in I} N_i$.

Figure 21.2 illustrates the construction. Pictorially, we make N from N_1 and N_2 by forming the disjoint union of their sets of S-elements, the union of the set of T-elements, and we connect the copy of an element of S_1 (respectively S_2) to an element of T if and only if the two elements are so connected in N_1 (respectively N_2).

Note that the nets N_1 and N_2 are state machine nets and correspond to the two transition systems T_1 and T_2 of figure 20.1 in the sense that each T_i is isomorphic to C_{N_i}. It is also the case that $C_{N_1 \| N_2}$ is isomorphic to $T_1 \| T_2 \equiv C_{N_1} \| C_{N_2}$ of figure 20.1. This is typical.

21.2.6. LEMMA. If $(N_i)_{i\in I}$ is a family of nets and $N = \|_{i\in I} N_i$, then there is an embedding isomorphism

$$f: \|_{i\in I} C_{N_i} \to C_N.$$

PROOF. Let $C_i = C_{N_i}$, each $i \in I$ and let $\hat{C} = \|_{i\in I} C_i$.

By 20.3.9, $\hat{C}$ is the set of all function $\underline{q}: I \to \bigcup_{i\in I} Q_i$ such that $\underline{q}(i) \in Q_i$, each $i \in I$. By (21.1), $Q_i = \wp(S_i)$ and by (21.1) and 21.2.5 (1), $Q_N = \wp(\bigcup_{i\in I}(S_i \times \{i\}))$. Therefore, we may define a function $f_Q: \hat{Q} \to Q_N$ by:

$$f_Q(\underline{q}) = \bigcup_{i\in I}(\underline{q}(i) \times \{i\})$$

The function $h_Q: Q_N \to \hat{Q}$ given by $(h_Q(X))(i) = \{s \in S_i \mid (s,i) \in X\}$ may be seen to be inverse to f_Q, which is consequently a bijection.

By 20.3.8 (2), 21.2.5 (2) and (21.1)

$$\hat{A} = \bigcup_{i\in I} T_i = A_N.$$

We may therefore define $f_A: \hat{A} \to A_N$ by $f_A(a) = a$. Define $f = (f_Q, f_A)$. We prove that f is an embedding isomorphism.

First, note that if $t_1, t_2 \in T$ and $(s,i) \in S$, then by 21.2.5 (3):

$$(s,i) \in (\bullet t_1 \cup t_1\bullet) \cap (\bullet t_2 \cup t_2\bullet) \text{ in } N \Leftrightarrow s \in (\bullet t_1 \cup t_1\bullet) \cap (\bullet t_2 \cup t_2\bullet) \text{ in } N_i$$

and hence by 17.2.1 (2) and 20.3.8 (2)

$$t_1 \iota_N t_2 \Leftrightarrow (\forall i \in I: t_1, t_2 \in T_i \Rightarrow t_1 \iota_i t_2) \Leftrightarrow t_1 \hat{\iota}\, t_2$$

Next, suppose that $\underline{q} \hat{\rightarrow}^a \underline{q}'$. If $(s,i) \in \bullet a$ in $\hat{N}$, then $s \in \bullet a$ in N_i by 21.2.5 (3). By 20.3.8 (3), $s \in \underline{q}(i)$, so that $(s,i) \in \underline{q}(i) \times \{i\} \subseteq f_Q(\underline{q})$. This shows that $\bullet a \subseteq f_Q(\underline{q})$ in N. If $(s,i) \in a\bullet \cap f_Q(\underline{q}')$ in N then $s \in a\bullet \cap \underline{q}'(i)$ in N_i a contradiction to $\underline{q}(i) \rightarrow_i^a \underline{q}'(i)$, in view of 20.3.8 (3) and (21.1). We have shown that

$$\underline{q} \hat{\rightarrow}^a \underline{q}' \Rightarrow f_Q(\underline{q})\,[a\rangle\, f_Q(\underline{q}') \Rightarrow f_Q(\underline{q}) \rightarrow_N^a f_Q(\underline{q}')$$

For the converse implication, we note that

$$\begin{aligned} &\bigcup_{i \in I} (\underline{q}(i) \times \{i\})\,[a\rangle \bigcup_{i \in I} (\underline{q}'(i) \times \{i\}) \\ &\quad \Rightarrow ((a \in A_i \Rightarrow \underline{q}(i) \rightarrow_i^a \underline{q}'(i)) \wedge (a \notin A_i \Rightarrow \underline{q}(i) = \underline{q}'(i))) \\ &\quad \Rightarrow \underline{q} \hat{\rightarrow}^a \underline{q}' \end{aligned}$$

which completes the proof.

■

21.2.7. COROLLARY. If C has a singular parallel decomposition, then C may be realised by a net.

PROOF. Suppose $f: C \rightarrow \|_{i \in I} T_i$, where each T_i is a singly labelled transition system. By suitably renaming we may ensure that $Q_i \cap A_i = \varnothing$, each $i \in I$ so that if we define $N_i = (Q_i, A_i; F_i)$, where $F_i \subseteq (A_i \times Q_i) \cup (Q_i \times A_i)$ satisfies:

$$(x,y) \in F \Leftrightarrow \exists q \in Q: x \rightarrow^y q \vee q \rightarrow^x y,$$

then N_i is a net.

Since T_i is singly labelled, we have an embedding, $h_i: T_i \rightarrow C_{N_i}$, regarding T_i as an asynchronous transition system with $\iota_i = \varnothing$, given by $(h_i)_Q(q) = \{q\}$ and $(h_i)_A(a) = a$.

It is now not hard to show that the pair $h = (h_Q, h_A)$ given by $(h_Q(\underline{q}))(i) = \{\underline{q}(i)\}$ and $h_A(a) = a$ is an embedding $h: \|_{i \in I} T_i \rightarrow \|_{i \in I} C_i$.

Let $N = \|_{i \in I} N_i$. By 21.2.6, there is an embedding isomorphism $g: \|_{i \in I} C_i \rightarrow C_N$. Hence, we have a morphism $g \circ h \circ f: C \rightarrow C_N$ and since g, h and f are embeddings, then so is $g \circ h \circ f$.

■

Now, we shall establish the converse. This depends, as we have said, on the fact that nets may be decomposed into nets describing sequential behaviour. The decomposition will be familiar to net theorists as so-called *place complementation*, and is at the heart of the following construction.

21.2.8. DEFINITION. Let N be a net. Define the *parallel decomposition* of N to be the family of transition systems $decomp(N) = (T_s)_{s\in S}$, where for all $s \in S$:

1• $Q_s = \{(s,0),(s,1)\}$;

2• $A_s = \bullet s \cup s\bullet$;

3• $(s,0) \rightarrow_s^t (s,1) \Leftrightarrow s\ F\ t \wedge \neg t\ F\ s$

$(s,1) \rightarrow_s^t (s,0) \Leftrightarrow \neg s\ F\ t \wedge t\ F\ s,$

$(s,0) \rightarrow_s^t (s,0) \Leftrightarrow s\ F\ t \wedge t\ F\ s$

$(s,1) \rightarrow_s^t (s,1)$ for no $t \in T$

The construction is illustrated in figure 21.2.

We remark that the construction involves duplicating each place s to give places $(s,0)$ and $(s,1)$ - hence the name 'place complementation'. $(s,0)$ has the same input and output transitions as s, as does $(s,1)$ but with the arrows reversed. If we think of T_s as a net (which we can, because it is clearly singly labelled), then it arises by adding a place complementary to s with the same incidence but with arrows reversed and keeping only those transitions incident to s.

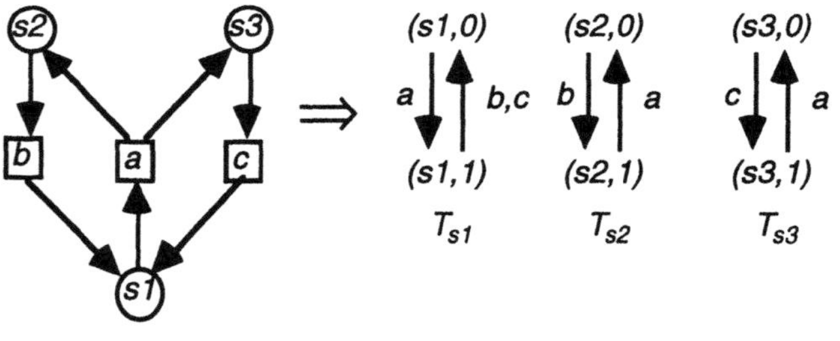

Figure 21.2

We use the parallel decomposition of a net to establish that its corresponding asynchronous transition system has a singular parallel decomposition, as we show in the next lemma.

21.2.10. LEMMA Let N be a net. If $decomp(N) = (T_s)_{s\in S}$, then there is an embedding $f: C_N \rightarrow \|_{s\in S} T_s$. Furthermore, each T_s is singly labelled. That is, C_N has a singular parallel decomposition.

PROOF. Define $f = (f_Q, f_A)$, where for all $c \in \wp(S)$

$$f_Q(c) = \begin{cases} (s,0) \text{ if } s \in c \\ (s,1) \text{ if } s \notin c \end{cases}$$

and for all $t \in T$, $f_A(t) = t$. The rest of the proof is a straightforward verification.

■

We may now give a characterisation of systems which may be realised by a net.

21.2.11. THEOREM. Let C be an asynchronous transition system, then C may be realised by a net $\Leftrightarrow$ C has a singular parallel decomposition.

PROOF. We have shown sufficiency in corollary 21.2.7.

Suppose C may be realised by a net N, so that we have an embedding $f: C \to C_N$. By 21.2.10, there is an embedding $g: C_N \to \parallel_{s \in S} T_s$. Thus, the composite is an embedding $g \circ f: C \to \parallel_{s \in S} T_s$, and this is clearly a parallel decomposition. It is singular by 21.2.10.

■

21.3. Representation of Loosely Coupled Systems.

At the beginning of chapter 18, we remarked that the loosely coupled systems model was closely related to nets. Indeed, we have a similar set-up; global states are sets of local states and elementary events consist of local changes to states which are determined by the events themselves. In fact, if S is a LCST[4] and Σ is a C/E system, then the underlying transition systems of both C_S and C_Σ are examples of case-transition systems satisfying the *extensionality principle*. (See, for example, [23])

It is interesting to compare the two models more closely. In view of the previous section, we shall find it most useful to approach the comparison by looking at the decomposition of C_S into sequential components.

If we look back to the parallel decomposition of nets, we see that as far as the dynamics of the system is concerned, each transition system in the decomposition constrains some set of events of the system to occur only in certain sequences and that the behaviour of the system is precisely that allowed by a conjunction of these constraints. In a LCST, there are two kinds of sequential constraint:

1• That enforced by the mutual exclusion relation, K;

---

[4]That is, a loosely coupled system with transitions.

2• That determined by the transition structure of each individual part[5].

To each of these constraints, we may associate a transition system, as follows. Let $S=(L,R)$ be a LCST where $L=(B,P,K)$ is a loosely coupled system.

Parts. If $p \in P$, then define $T_p=(p,A_p,\rightarrow_p)$, where

$$A_p=\{a \in A_L \mid a\bullet \in p\} \tag{21.2}$$
$$b_1 \rightarrow_p b_2 \Leftrightarrow a=(b_1,b_2) \wedge a \in R \tag{21.3}$$

Note that if $a \in A_p$, then $\bullet a \in p$, by (21.2) and so $a\bullet \in p$, by (21.3), 18.3.1 and 18.2.3. Thus

$$\forall p \in P: a \in A_p \Rightarrow \bullet a, a\bullet \in p \tag{21.4}$$

Exclusion. If $k=(b_1,b_2) \in K$, then for any case c, we have exactly three possibilities.

(K1) $\{b_1,b_2\} \cap c=\varnothing$: here, k is 'invoked' upon the occurrence of some $a \in \bullet b_1 \cap \bullet b_2$. If $a \in \bullet b_1$ occurs, then we reach a new case c', giving a second possibility
(K2) $b_1 \in c'$: here, no event of $\bullet b_2$ may occur until some event of $b_1\bullet$ has occurred - otherwise, we would reach a 'case' containing both b_1 and b_2, violating the mutual exclusion constraint. After some event of $b_1\bullet$ has occurred, we are back in case (K1)
(K3) The third possibility, $b_2 \in c$, is analogous to (K2).

Let $Q_k=\{q_k^0,q_k^1,q_k^2\}$. The states of Q_k correspond to the three cases (K1), (K2) and (K3) above. We also have actions

$$A_k=\bullet b_1 \cup \bullet b_2 \cup b_1\bullet \cup b_2\bullet \tag{21.5}$$

and we define

$$q_k^i \rightarrow_k^a q_k^j \Leftrightarrow (i=0 \wedge j \neq 0 \wedge a \in \bullet b_j) \vee (i \neq 0 \wedge j=0 \wedge a \in b_i\bullet) \tag{21.6}$$

Finally, define $T_k=(Q_k,A_k,\rightarrow_k)$.

For example, the transition system corresponding to the pair $(2,5)$ of figure 18.3 is pictured in figure 21.3.

---

[5]See 18.2.1.

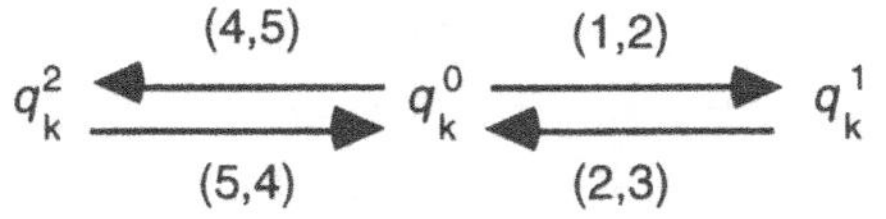

Figure 21.3

If $(b_1,b_2) \in K$ and $b_1,b_2 \in p$, some $p \in P$, then the fact that T_p ensures that these two states be mutually exclusive means that we need not add T_k to our set of constraints. Define

$$K^- = \{(b_1,b_2) \in K \mid \forall p \in P: \{b_1,b_2\} \not\subseteq p\} \tag{21.7}$$

$(T_i)_{i \in P \cup K^-}$ is an indexed family of transition systems and hence determines an asynchronous transition system.

$$C^S = \underset{i \in P \cup K^-}{\|} T_i \tag{21.8}$$

C^S expresses the conjunction of the constraints associated with each T_i.

In chapter 18, we also associated a LCST S with an asynchronous transition system C_S.[6] We shall show that there is an embedding f from C_S to C^S. Let us begin by defining f_Q.

Suppose $c \in Q_S$, then we may define a state $f_Q(c) \in Q^S$ as follows. Recall that the states of Q^S are vectors with a co-ordinate for each part $p \in P$ and for each pair $(b_1,b_2) \in K^-$.

If $p \in P$, then $|c \cap p| = 1$, by 18.2.2 and we define

$$(f_Q(c))(p) = b \Leftrightarrow c \cap p = \{b\} \tag{21.9}$$

If $k = (b_1,b_2) \in K^-$, then $|c \cap \{b_1,b_2\}| \leq 1$, by 18.2.2, and we define

$$(f_Q(c))(k) = \begin{cases} q_k^0 \text{ if } c \cap \{b_1,b_2\} = \emptyset \\ q_k^i \text{ if } c \cap \{b_1,b_2\} = \{b_i\} \end{cases} \tag{21.10}$$

For the rest of the definition of f we observe that $A^S = R = A_S$, so we may define $f_A : A_S \to A^S \cup \{\Omega\}$, by $f_A(a) = a$.

C^S for the LCST of figure 18.3 is pictured in figure 21.4. The states of the transition system are represented as tuples in the following way: if

[6]See 18.3.2.

$\underline{q}(\{1,2,3\}) = x_1$, $\underline{q}(\{4,5\}) = x_2$, $\underline{q}(2,5) = q^{x_3}_{(2,5)}$ and $\underline{q}(5,2) = q^{x_4}_{(5,2)}$ then we represent $\underline{q}$ by the tuple (x_1, x_2, x_3, x_4). We also have $c\,\iota^S\,d$ and $c\,\iota^S\,e$

Note the connected component contained in the shaded rectangle. This is isomorphic to the asynchronous transition system pictured in figure 18.4. We note that in this isomorphism:

$f_Q(\{1,5\}) = (1,5,2,1)$ $\quad f_Q(\{3,5\} = (3,5,1,2)$ $\quad f_Q(\{1,4\}) = (1,4,0,0)$

$f_Q(\{3,4\} = (3,4,1,2)$ $\quad f_Q(\{2,4\} = (2,4,1,2)$

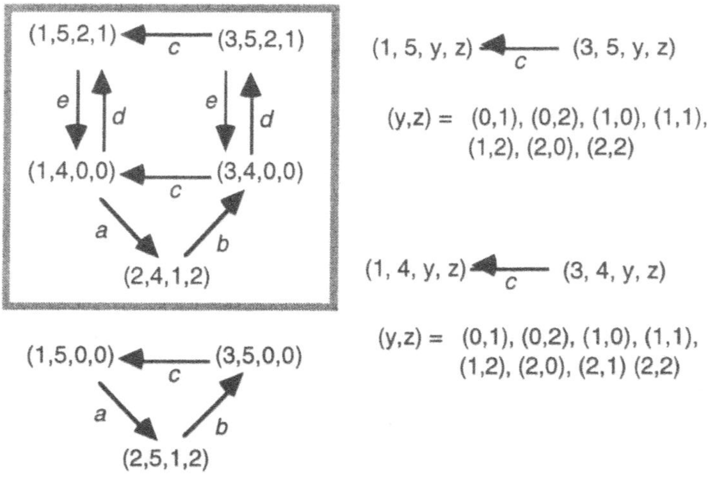

The remaining 31 nodes are incident to no arcs.

Figure 21.4

21.3.1. PROPOSITION. The pair $f = (f_Q, f_A)$ is an embedding from C_S to C^S.

PROOF. First, note that $A^S = A_S$ and that, since $f_A(a) = a$, by definition, f_A satisfies 21.2.1 (1). By 18.2.5 and 20.3.8,

$$\neg a\,\iota_S\,b \Leftrightarrow (\exists k \in K^- : k \in \{\bullet a, a\bullet\} \times \{\bullet b, b\bullet\}) \vee (\exists p \in P : a, b \in A_p)$$
$$\Leftrightarrow \exists i \in P \cup K^- : \{a,b\} \subseteq A_i \Leftrightarrow \neg a\,\iota^S\,b$$

so that $\iota^S = \iota_S$. In particular, f_A satisfies 19.2.1 (1) and 21.2.1 (2).

In view of 19.2.1 (2) and (3) and the fact that $f_A(a) = a$, all $a \in A_S$, in order to complete the proof that f is a morphism, it suffices to show that if $c_1, c_2 \in Q_S$ and $a \in A_S$ such that $c_1 \rightarrow_S{}^a c_2$, then $f_Q(c_1) \rightarrow^{S\,a} f_Q(c_2)$.

By (18.2), 18.2.1, 18.3.1 and 18.3.2, there is a unique $p \in P$ such that $a \in A_p$. By 18.2.4, $c_2 = (c_1 - \{\bullet a\}) \cup \{a\bullet\}$ and by (21.3), $\bullet a \rightarrow_p{}^a a\bullet$. Since $\{\bullet a\} = c_1 \cap p$

and $\{a\bullet\} = c_2 \cap p$, $(f_Q(c_1))(p) = \bullet a \rightarrow_p^a a\bullet = (f_Q(c_2))(p)$. If $a \notin A_p$, then $c_1 \cap p = c_2 \cap p$ and hence, $(f_Q(c_1))(p) = (f_Q(c_2))(p)$.

Next, suppose $k = (b_1, b_2) \in K^-$ such that $a \in A_k$, then by (21.5) either $a \in b_i\bullet$ or $a \in \bullet b_i$, some $i \in \{1,2\}$.

If $a \in b_i\bullet$, then $b_i \in c_1$ and $b_i \notin c_2$, by 18.2.4. Hence, $(f_Q(c_1))(k) = q_k^i$, by (21.10) and $b_{3-i} \notin c_1$, by 18.2.2. If $b_{3-i} \in c_2$, then as $c_2 = (c_1 - \{\bullet a\}) \cup \{a\bullet\}$, we would have to have $b_{3-i} = a\bullet$ so that $\{b_1, b_2\} \subseteq p$, where $a \in A_p$, contradicting the definition of K^-. Thus, $\{b_1, b_2\} \cap c_2 = \varnothing$ and hence $(f_Q(c_2))(k) = q_k^0$. Since $a \in b_i\bullet$, $q_k^i \rightarrow_k^a q_k^0$, by (21.6), that is to say $(f_Q(c_1))(k) \rightarrow_k^a (f_Q(c_2))(k)$.

If $a \in \bullet b_i$, then $b_i \notin c_1$ and $b_i \in c_2$ by 18.2.4. Hence, $(f_Q(c_2))(k) = q_k^i$, by (21.10) and $b_{3-i} \notin c_2$, by 18.2.2. If $b_{3-i} \in c_1$. then as $c_2 = (c_1 - \{\bullet a\}) \cup \{a\bullet\}$, we would have to have $b_{3-i} = \bullet a$, so that $\{b_1, b_2\} \subseteq p$, where $a \in A_p$, contradicting the definition of K^-. Thus, $\{b_1, b_2\} \cap c_1 = \varnothing$ and so $(f_Q(c_1)) = q_k^0$. Since $a \in \bullet b_i$, we have $q_k^0 \rightarrow_k^a q_k^i$, by (21.6) that is $(f_Q(c_1))(k) \rightarrow_k^a (f_Q(c_2))(k)$.

If $k \in K^-$ such that $a \notin A_k$, then clearly $(f_Q(c_1))(k) = (f_Q(c_2))(k)$.

By 20.3.9, $f_Q(c_1) \rightarrow^{S\,a} f_Q(c_2)$, completing the proof that f is a morphism.

Now we show that f is an embedding morphism. Suppose that $c_1, c_2 \in Q_S$ such that $f_Q(c_1) = f_Q(c_2)$, then by (21.9)

$$c_1 = \bigcup_{p \in P}(c_1 \cap p) = \bigcup_{p \in P}\{(f_Q(c_1))(p)\} = \bigcup_{p \in P}\{(f_Q(c_2))(p)\} = \bigcup_{p \in P}(c_2 \cap p) = c_2$$

and so f_Q is injective. We have already seen that 21.2.1 (1) and (2) hold. We prove 21.2.1 (3).

Let $c \in Q_S$ with $f_Q(c) \rightarrow^{S\,a} \underline{q}$. If $a \in A_p$, then $(f_Q(c))(p) \rightarrow_p^a \underline{q}(p)$, by (21.8) and 20.3.9 (3), and so $(f_Q(c))(p) = \bullet a$ and $\underline{q}(p) = a\bullet$, by (21.2). Define

$$c' = (c - \{\bullet a\}) \cup \{a\bullet\} \tag{21.11}$$

We shall show that $c' \in Q_S$, $c \rightarrow_S^a c'$ and $f_Q(c') = \underline{q}$.

First, we show that c' is a case. If $p \in P$ then

$$p \cap c' = \begin{cases} a\bullet \text{ if } a \in A_p \\ p \cap c \text{ otherwise} \end{cases}$$

so that $|p \cap c'| = 1$ and therefore c' is a maximal configuration.

Let $b_1, b_2 \in c'$ such that $b_1 \, K \, b_2$; we prove a contradiction. Since $c \in Q_S$, we cannot have $b_1, b_2 \in c$. We may therefore assume without loss of generality that $b_1 \in c$ and $b_2 \notin c$. It follows that $b_2 \in c' - c$, that is $b_2 = a\bullet$.

Hence, $a \in A_{(b_1,b_2)}$, by (21.5). As $b_1 \in c$ and $b_2 \notin c$, $(f_Q(c))(b_1,b_2) = q^1_{(b_1,b_2)}$, by (21.10). As $f_Q(c) \to^{Sa} \underline{q}$ and $a \in A_{(b_1,b_2)}$, $q^1_{(b_1,b_2)} = (f_Q(c))(b_1,b_2) \to_{(b_1,b_2)}{}^a \underline{q}(b_1,b_2)$, by 20.3.9 (3). Thus, $\underline{q}(b_1,b_2) = q^0_{(b_1,b_2)}$, by (21.6). But $q^1_{(b_1,b_2)} \to_{(b_1,b_2)}{}^a q^0_{(b_1,b_2)}$ implies $a \in b_1\bullet$, by (21.6) and so $b_1 = \bullet a$. But, $\bullet a \notin c'$, by (21.11). This contradiction to $b_1, b_2 \in c'$ shows that for no $b_1, b_2 \in c'$ do we have $b_1 \, K \, b_2$ and accordingly, since c' is a maximal configuration, $c' \in Q_S$.

By (21.11), $c - c' = \{\bullet a\}$ and $c' - c = \{a\bullet\}$ and hence $c \to_S{}^a c'$.

Finally, we show that $f_Q(c') = \underline{q}$.

Let $p \in P$. If $a \notin A_p$, then $\underline{q}(p) = (f_Q(c))(p)$, by 20.3.9 (3), and $p \cap c = p \cap c'$. by (21.11) so $(f_Q(c))(p) = (f_Q(c'))(p)$. Thus, $(f_Q(c'))(p) = \underline{q}(p)$. If $a \in A_p$, then $(f_Q(c))(p) \to_p{}^a \underline{q}(p)$, so $\underline{q}(p) = a\bullet$. But $p \cap c' = \{a\bullet\}$, so $(f_Q(c'))(p) = \underline{q}(p)$ in this case also, by (21.9). Therefore

$$\forall p \in P: (f_Q(c'))(p) = \underline{q}(p) \tag{21.12}$$

Finally, suppose $(b_1,b_2) = k \in K^-$. If $a \notin A_k$, then $\underline{q}(k) = (f_Q(c))(k)$, by 20.3.9 (3). Also, $\{b_1,b_2\} \cap c = \varnothing \Leftrightarrow \{b_1,b_2\} \cap c' = \varnothing$ and $b_i \in c \Leftrightarrow b_i \in c'$ and so, by (21.10), $(f_Q(c'))(k) = (f_Q(c))(k) = \underline{q}(k)$. On the other hand, if $a \in A_k$, then $(f_Q(c))(k) \to_k{}^a \underline{q}(k)$, by 20.3.9 (3) and so if $(f_Q(c))(k) = q^0_k$, then $\underline{q}(k) = q^i_k$, where $a \in \bullet b_i$. But $c \to_S{}^a c'$, and so $b_i \in c'$ and hence $(f_Q(c'))(k) = q^i_k = \underline{q}(k)$. The argument in the case $(f_Q(c))(k) = q^i_k$, for $i \neq 0$ is entirely analogous. Thus

$$\forall k \in P: (f_Q(c'))(k) = \underline{q}(k)$$

which together with (21.12) gives $f_Q(c') = \underline{q}$.

■

Thus, C_S has a parallel decomposition. With 21.2.11 in mind, it seems reasonable to ask whether this decomposition is singular. Happily, it is.

21.3.2. LEMMA. Let $\mathbf{T} = (T_i)_{i \in P \cup K^-}$, then $(\mathbf{T}, f)$ is singular, where f is the embedding of 21.3.1.

PROOF. If $p \in P$, then

$$b_1 \to_p{}^a b_2 \wedge b_1' \to_p{}^a b_2' \Rightarrow b_1 = \bullet a = b_1' \wedge b_2 = a\bullet = b_2'$$

so T_p is singly labelled.

Next, suppose $k=(b_1,b_2)\in K^-$ and suppose that $q_k^i \to_k^a q_k^l$ and $q_k^j \to_k^a q_k^m$. If $i\neq j$, then suppose without loss of generality that $i=0$ and $j\neq 0$. By (21.6), $l\neq 0$, $a\in \bullet b_l$, $m=0$ and $a\in b_i\bullet$. Hence, $b_l, b_j\in T_p$, where $a\in A_p$, by (21.4) and hence $b_l=b_j$. For otherwise, we would have $(b_1,b_2)\in K$ and $b_1,b_2\in T_p$, contradicting $(b_1,b_2)\in K^-$. Thus $q_k^0 \to_k^a q_k^j$ and $q_k^j \to_k^a q_k^0$. But now $\bullet a = b_j = a\bullet$, by (21.6), contradicting 18.2.3 and 18.3.1.

Hence, $i=j$. If $i=0$, then by (21.6), $l,m\neq 0$, $a\in \bullet b_l$ and $a\in \bullet b_m$.. If $l\neq m$ then $a\in \bullet b_1, \bullet b_2$ giving $b_1,b_2\in T_p$, where $a\in A_p$, by (21.4), contradicting (21.7). If $i\neq 0$, then $l=0=m$, by (21.6). In each case $q_k^i=q_k^j$ and $q_k^l=q_k^m$, so T_k is also singly labelled and we are done.

■

As a corollary, we therefore have the following.

21.3.3. THEOREM. Every LCST (and in particular, every loosely coupled system) may be realised by a net.

PROOF. From 21.2.11, 21.3.1 and 21.3.2.

■

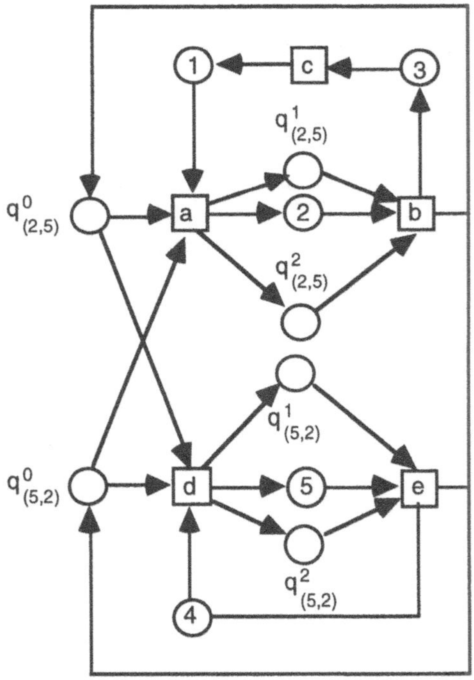

Figure 21.5

As an illustration of 21.3.3, the net corresponding to the example of figure 18.3 is shown in figure 21.5.

Notes.

It has long been known that Nets have 'State machine decompositions'; this has given rise to characterisations of the trace languages arising from nets. The material on realisation of nets by asynchronous transition systems is thus an old idea in new clothes.

If we simplify the net of figure 21.5, by identifying all pairs of places having the same sets of input transitions and of output transitions, then we obtain the net of figure 21.6, below. Readers familiar with Net Theory will recognise this net as a solution to a mutual exclusion problem, in which conditions 2 and 5 are to be prevented from holding concurrently. q_0 acts as a binary semaphore with a and d as 'P' operations and b and e as 'V' operations. This reflects the fact that the two states are required by the K relation to be mutually exclusive.

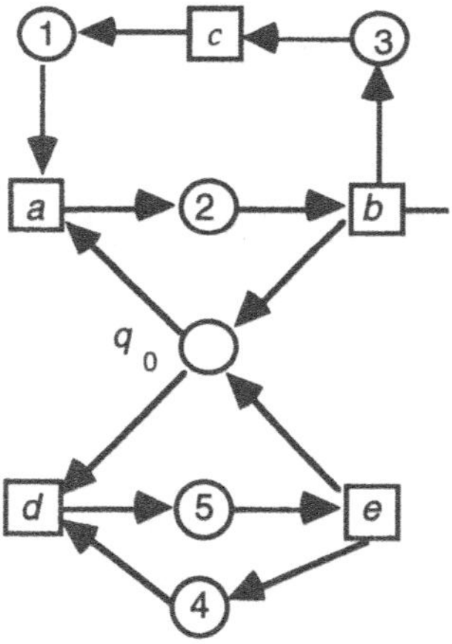

Figure 21.6

Chapter 22
Hybrid Transition Systems

> Hybrids from two species which are very difficult to cross and which rarely produce any offspring, are generally very sterile; but the parallelism between the difficulty of making the first cross, and the sterility of the hybrids thus produced - two classes of facts which are generally confounded together - is by no means strict.
>
> CHARLES DARWIN
> The Origin of Species

22.1. Introduction.

So far we have largely concerned ourselves with systems whose behaviours could be represented by trace and vector languages or equivalently by linguistic behavioural presentations. Such systems may be modelled at the system level by initialised asynchronous transition systems.

However, it is clear that not all systems may be modelled in this way. Indeed, we may immediately establish shortcomings of such models:

1• They cannot cope with non-trivial simultaneity.

2• They cannot cope with systems which are unambiguous.

3• They cannot deal with systems in which two specific actions may sometimes be capable of concurrent activation and sometimes not, owing, say, to possible absence or presence of sufficient shared resources. For example, whether or not my wife and I can concurrently quaff our bedtime cocoa depends on whether or not both mugs have been washed up.

The first point need no further amplification. The second is of particular significance because ambiguous systems arise naturally from process algebras. We shall come back to this point later.

As a further illustration of the third point, we note that in the behavioural presentation corresponding to the Karp and Miller Computation graph of figure 22.1, which is pictured in figure 22.2, of the two execution of b, the first is concurrent with a, while the second is not, even though they are not separated. Hence, the corresponding behavioural presentation is not linguistic and can therefore not be accepted by an initialised asynchronous transition system.

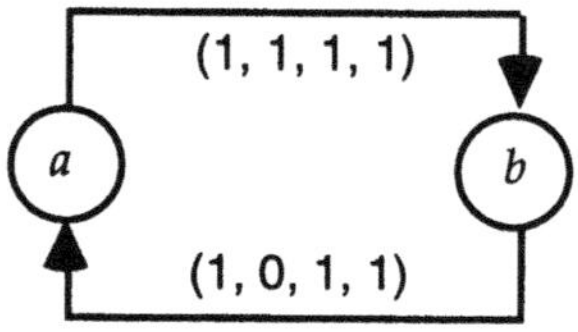

Figure 22.1

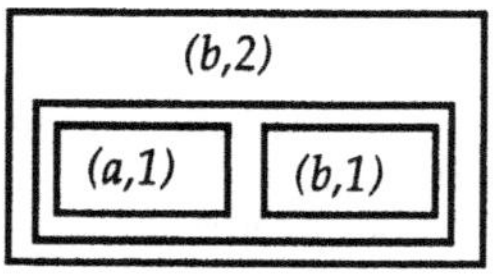

Figure 22.2.

The automata that we describe in this chapter, the hybrid transition systems, cope with all three of these inadequacies. Simply speaking, such an automaton consists of an asynchronous transition system with an extra layer of labelling.

In section 22.2, we formally define the model and show how a hybrid transition system with initial state determines a discrete behavioural presentation, and vice versa, and we establish semantics theorems corresponding to 14.3.3 and 14.4.5 for transition systems and 15.3.3 for asynchronous transition systems.

In section 22.3, we examine the relationships between the classes of automata that we have considered in this book.

22.2. The Model.

22.2.1. DEFINITION. A *hybrid transition system* is a 6-tuple $H=(Q,A,\rightarrow,\iota,E,\mu)$, where

1• $C(H)=(Q,A,\rightarrow,\iota)$ is an ι-asynchronous transition system;
2• $\mu : A \rightarrow \mathcal{B}(E)$.

where $\mathcal{B}(E)$ denotes the set of *bags* over E. A bag over E is a function

$\mathscr{b}: E \to \omega$.

The idea is that $\mathscr{b}$ is a collection having $\mathscr{b}(e)$ instances of the item $e \in E$. This is how we deal with simultaneity. $\mu(a)$ represents a class of simultaneous occurrences in which $(\mu(a))(e)$ instances of the event $e \in E$ occur.

In the course of constructing a behavioural presentation from a hybrid transition systems with initial state, we shall find ourselves with an ordered pair (B,μ), where $B=(O,P,A,\lambda)$ is a (linguistic) behavioural presentation and $\mu: A \to \mathcal{B}(E)$. We then apply the following construction, which replaces a single 'occurrence' $a \in A$ by a simultaneity class $\varepsilon_\mu(a)$.

22.2.2. DEFINITION. A *bagged behavioural presentation* is a pair (B,μ), where

1• $B=(O,P,A,\lambda)$ is a behavioural presentation;
2• $\mu: A \to \mathcal{B}(E)$.

If (B,μ) is a bagged behavioural presentation, then for each $a \in A$,define:

$$\varepsilon_\mu(o) = \{(o,e,i) \mid 1 \le i \le (\mu(\lambda(o)))(e)\}$$

and if $X \subseteq O$ and $\mathcal{X} \subseteq \wp(O)$ then define

$$\varepsilon_\mu(X) = \bigcup_{o \in X} \varepsilon_\mu(o)$$
$$\varepsilon_\mu(\mathcal{X}) = \{\varepsilon_\mu(X) \mid X \in \mathcal{X}\}$$

Finally, let $\varepsilon(B,\mu) = (\varepsilon_\mu(O), \varepsilon_\mu(P), E, \lambda_\mu)$, where if $(o,e,i) \in \varepsilon_\mu(O)$, then

$$\lambda_\mu(o,e,i) = e.$$

22.2.3. DEFINITION. If (H,q) is an initialised hybrid transition system, then define

$$B_{(H,q)} = \varepsilon(B_{(C(H),q)}, \mu).$$

Of course, $(B_{(C(H),q)}, \mu)$ is a bagged behavioural presentation, and so $\varepsilon(B_{(C(H),q)}, \mu)$ is well defined.

22.2.4. LEMMA. If (H,q) is an initialised hybrid transition system, then $B_{(H,q)}$ is a discrete behavioural presentation.

PROOF. If (H,q) is an initialised hybrid transition system, then $\tilde{B} = B_{(H,q)}$ is of the form $\varepsilon(B,\mu)$, where B is discrete. By 22.2.2

$$\bigcup_{\tilde{p}\in\tilde{P}} \tilde{p} = \bigcup_{p\in P} \varepsilon(p) = \varepsilon(\bigcup_{p\in P} p) = \varepsilon(O) = \tilde{O}$$

and so $\tilde{B}$ is a behavioural presentation.

Since ε_μ is clearly a poset isomorphism from P to $\tilde{P}$ and B satisfies the ACC and DCC, by 3.2.5, so does $\tilde{B}$, and since $\varnothing \in P$, by 3.2.5, $\varnothing = \varepsilon_\mu(\varnothing) \in \tilde{P}$. Finally, suppose that $\tilde{p}_1, \tilde{p}_2 \in \tilde{P}$ with $\tilde{p}_1 \subseteq \tilde{p}_2$ and $\tilde{o}_1, \tilde{o}_2 \in \tilde{p}_2 - \tilde{p}_1$ such that $\tilde{o}_2 \not\twoheadrightarrow \tilde{o}_1$. By 22.2.3, there exists $p_1, p_2 \in P$, $o_1, o_2 \in O$, $e_1, e_2 \in E$ and $j_1, j_2 \in \omega$ such that for each $\varepsilon_\mu(p_i) = \tilde{p}_i$ and $\tilde{o}_i = (o_i, e_i, j_i)$. By 22.2.3, if $o \in O$ and $p \in P$, then $o \in p \Leftrightarrow \varepsilon_\mu(o) \subseteq \varepsilon_\mu(p)$, from which it follows that $p_1 \subseteq p_2$, $o_1, o_2 \in p_2 - p_1$ and $o_2 \not\twoheadrightarrow o_1$. By 3.2.5 and 3.2.3, there exists $p \in P$ such that $p_1 \subseteq p \subseteq p_2$, $o_1 \in p$ and $o_2 \notin p$. Hence, there exists $\tilde{p} \in \tilde{P}$, namely $\varepsilon_\mu(p)$, such that $\tilde{p}_1 \subseteq \tilde{p} \subseteq \tilde{p}_2$, $\tilde{o}_1 \in \tilde{p}$ and $\tilde{o}_2 \notin \tilde{p}$. Hence, $\tilde{B}$ satisfies the ACC and DCC, has a bottom element and is replete, that is $\tilde{B}$ is discrete.

■

Figure 22.3 shows a hybrid transition system, H. In this example, the only mutually independent transitions are a and b. and the labelling function is given by $\mu(a) = e_1.e_2$, $\mu(b) = e_2.e_3$, $\mu(c) = e_2$ and $\mu(d) = e_1.e_3$. Here an expression $a_1^{r_1} \cdots a_n^{r_n}$ represents a bag $\mathscr{b}$ in which

$$\mathscr{b}(a) = \begin{cases} r_i \text{ if } a = a_i \\ 0 \text{ otherwise} \end{cases}$$

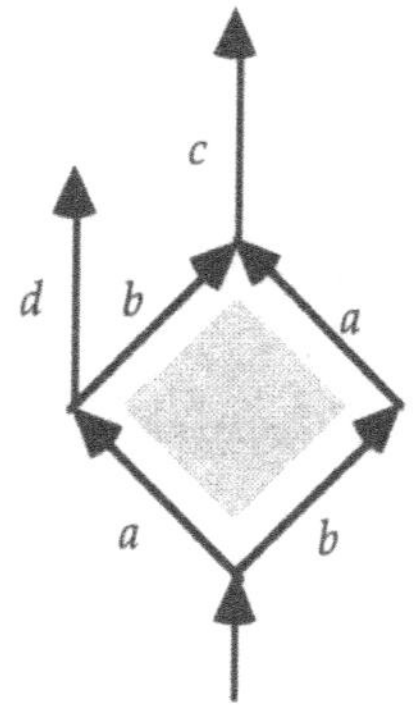

Figure 22.3

The trace language $L(C(H),q)$ is as follows

$$\{\{\Omega\},\{a\},\{b\},\{ab,ba\},\{abc,bac\},\{ad\}\}$$

The complete primes of this language are the following.

$$o_1=\{a\} \qquad o_2=\{b\} \qquad o_3=\{abc,bac\} \qquad o_4=\{ad\}$$

and so we get the behavioural presentation $B_{(C(H),q)}$ pictured in figure 22.4.

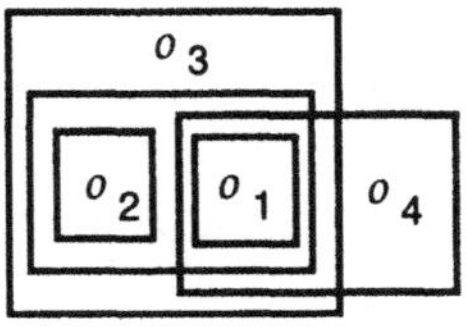

$\lambda(o_1)=a$ $\quad$ $\lambda(o_2)=b$ $\quad$ $\lambda(o_3)=c$ $\quad$ $\lambda(o_4)=d$

$\mu(a)=e_1.e_2$ $\quad$ $\mu(b)=e_2.e_3$ $\quad$ $\mu(c)=e_2$ $\quad$ $\mu(d)=e_1.e_3$

Figure 22.4.

Finally we use the ε expansion technique to obtain the discrete behavioural presentation of figure 22.5.

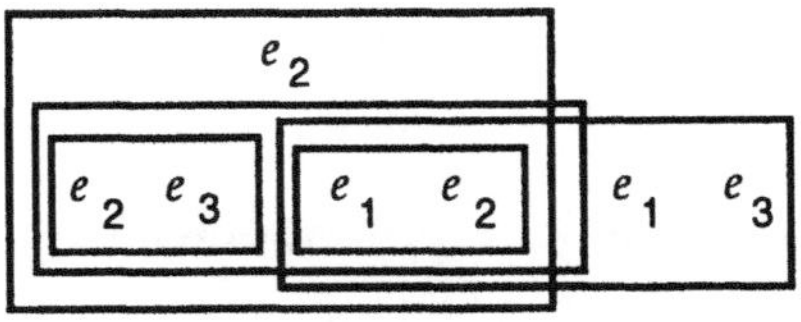

Figure 22.5

Actually, figure 22.5 doesn't completely describe $B_{(H,q)}$. We have omitted the names of the occurrences. As these will turn out to include such entities as $(\{b\},2,e_2)$ we may possibly be excused from cluttering up the diagram. However, figure 22.5 does represent something. Any behavioural presentation isomorphic to $B_{(H,q)}$ would have exactly the same picture associated with it, since it would only differ from $B_{(H,q)}$ in the names of occurrences. The figure represents the members of the *isomorphism class* of behavioural presentations which are accepted by (H,q).

22.2.5. DEFINITION. We shall say that *(H,q) accepts B* $\Leftrightarrow B\equiv B_{(H,q)}$.

Our construction ensures that every initialised hybrid transition system accepts a discrete behavioural presentation. Is every discrete behavioural presentation accepted by some initialised transition system?

We have seen in chapter 12 that this is the case when the behavioural presentation in question is ι-linguistic. If B is a ι-linguistic behavioural presentation, then there is an ι-asynchronous transition system C_B such that (C_B, Ω_ι) accepts B.[1] The following construction allows us to exploit this result for hybrid transition systems subject to the following finiteness condition.

We shall define a behavioural presentation B to be *bounded* $\Leftrightarrow$

$$\forall o \in O\ \forall e \in E: (\mu_B(o/\approx))(e) < \infty \tag{22.1}$$

22.2.6. DEFINITION. If B is a discrete behavioural presentation, then define:

1• $P/\approx = \{p/\approx \mid p \in P\}$;

2• $B/\approx = (O/\approx, P/\approx, O/\approx, Id_{O/\approx})$,

and if B is bounded, then define

3• $\mu_B : O/\approx \to \mathcal{B}(E)$ by

$$(\mu_B(o/\approx))(e) = \left|\{o \in o/\approx \mid \lambda(o) = e\}\right|.$$

We shall find it useful to collect together some properties of this construction.

22.2.7. LEMMA. Suppose that B is a discrete behavioural presentation and that $o, o_1, o_2 \in O$, $p \in P$ and $X, Y \subseteq O$, then

1• $o \in p \Leftrightarrow o/\approx \in p/\approx$

2• $o_1 \to o_2 \Leftrightarrow o_1/\approx \to_{\approx} o_2/\approx$;

3• $o_1 \# o_2 \Leftrightarrow o_1/\approx \#_{\approx} o_2/\approx$;

4• $o_1\ co\ o_2 \Leftrightarrow o_1/\approx co_{\approx}\ o_2/\approx$;

5• $X \lhd Y \Rightarrow X/\approx \lhd Y/\approx$;

6• $X/\approx \lhd Y/\approx \Rightarrow \downarrow X \lhd \downarrow Y$.

PROOF. (1) $o \in p \Rightarrow o/\approx \in p/\approx$, by 3.3.4. Suppose $o/\approx \in p/\approx$, then $o \approx o'$, some $o' \in p$, and so $o \to o'$, by 2.3.4, and hence $o \in p$, by 2.3.1 (2).

---

[1] See 15.2.10 and 15.2.11.

(2) Suppose that $o_1 \to o_2$. If $p_{\approx} \in P_{\approx}$, such that $o_2/{\approx} \in p_{\approx}$, then $p_{\approx} = p/{\approx}$, some $p \in P$. Thus, $o_2 \in p$, by (1), so $o_1 \to o$, since $o_1 \to o_2$, so $o_1 \in p$, by 2.3.1 (2), so $o_1/{\approx} \in p/{\approx} = p_{\approx}$. We have shown that $o_2/{\approx} \in p_{\approx} \Rightarrow o_1/{\approx} \in p_{\approx}$, that is $o_1/{\approx} \to_{\approx} o_2/{\approx}$.

Conversely, suppose that $o_1/{\approx} \to_{\approx} o_2/{\approx}$. If $o_2 \in p$, then $o_2/{\approx} \in p/{\approx} \in P_{\approx}$ and hence $o_1/{\approx} \in p/{\approx}$, by 2.3.1 (2), so $o_1 \in p$, by (1). We have shown that $o_2 \in p_{\approx} \Rightarrow o_1 \in p_{\approx}$, that is $o_1 \to o_2$.

(3) By 2.3.1 (1), $\neg o_1 \# o_2$ if and only if there exists $p \in P$ such that $o_1, o_2 \in p$ if and only if there exists $p_{\approx} \in P_{\approx}$, namely $p/{\approx}$, such that $o_1/{\approx}, o_2/{\approx} \in p/{\approx}$, by (1), if and only $\neg o_1/{\approx} \#_{\approx} o_2/{\approx}$.

(4) By (2), (3) and 2.3.6.

(5) Suppose $X \lhd Y$, then $X \subseteq Y$, by 3.3.7, and so $X/{\approx} \subseteq Y/{\approx}$, by 3.3.4. If $o_2/{\approx} \in X/{\approx}$ and $o_1/{\approx} \in Y/{\approx}$ such that $o_1/{\approx} \to_{\approx} o_2/{\approx}$ then $o_1 \to o_2$, by (1), and there exists $o_2' \in Y$ such that $o_2 \approx o_2'$. Hence, $o_1 \to o_2'$, by 2.3.4, and so $o_1 \in X$, by 3.3.7, and so $o_1/{\approx} \in X/{\approx}$. By 3.3.7, $X/{\approx} \lhd Y/{\approx}$.

(6) Suppose $X/{\approx} \lhd Y/{\approx}$ and let $o \in {\downarrow}X$, then by 3.3.4, there exists $o' \in X$ such that $o \to o'$. Now, $o'/{\approx} \in X/{\approx}$ and so $o'/{\approx} \in Y/{\approx}$, by 3.3.7, and so there exists $o'' \in Y$ such that $o' \approx o''$ and so $o' \to o''$, by 2.3.4. Thus, $o \to o''$ and consequently, $o \in {\downarrow}Y$, by 3.3.7. We have shown that ${\downarrow}X \subseteq {\downarrow}Y$.

Next, suppose that $o_2 \in {\downarrow}X$ and $o_1 \in {\downarrow}Y$ such that $o_1 \to o_2$, then $o_1/{\approx} \to_{\approx} o_2/{\approx}$ by (1). Since $o_2 \in {\downarrow}X$ there exists $o_2' \in X$ such that $o_1 \to o_2'$, by 3.3.7 and 2.3.4, so $o_2/{\approx} \to_{\approx} o_2'/{\approx}$. Thus, $o_2'/{\approx} \in X/{\approx}$ and $o_1/{\approx} \to_{\approx} o_2'/{\approx}$, from which we deduce $o_1/{\approx} \in X/{\approx}$, by 2.3.1 (2). Thus, there exists $o_1' \in X$ such that $o_1 \to o_1'$, that is $o_1 \in {\downarrow}X$.

■

The link between arbitrary discrete behavioural presentations and linguistic behavioural presentations is established in the following result.

22.2.8. PROPOSITION. If B is a discrete behavioural presentation, then

1• $B/{\approx}$ is a $co/{\approx}$-linguistic behavioural presentation;
2• If B is bounded then $(B/{\approx}, \mu_B)$ is a bagged behavioural presentation.

PROOF. We note that $co/{\approx}$ is an independence relation.

It is trivial, following 3.3.4 that B is $\approx$-finitary if and only if $B/{\approx}$ is $\approx$-finitary. Let us show that $B/{\approx}$ is left-closed.

Let $p_{\approx} \in P_{\approx}$ and $X \lhd p_{\approx}$, then $p_{\approx} = p/{\approx}$ and $X = Y/{\approx}$, some $p \in P$ and $Y \subseteq O$. By 22.2.7 (6) and 3.3.10 (4), ${\downarrow}Y \lhd p$. Hence, ${\downarrow}Y \in P$, by 3.3.14, and so $({\downarrow}Y)/{\approx} \in P_{\approx}$. We show that $({\downarrow}Y)/{\approx} = X$.

As $Y \subseteq \downarrow Y$, $X = Y/\approx \subseteq (\downarrow Y)/\approx$. Conversely, if $o/\approx \in (\downarrow Y)/\approx$, then there exists $o' \in O$ and $o'' \in Y$ such that $o \approx o'$ and $o' \to o''$. Hence, by (2), $o/\approx = o'/\approx \to_{\approx} o''/\approx$. But $o''/\approx \in Y/\approx = X$ and as $\downarrow Y \lhd p$, $(\downarrow Y)/\approx \subseteq p/\approx = p_{\approx}$, so $o/\approx \in p_{\approx}$ Hence, $o/\approx \in X$, since $X \lhd p_{\approx}$. Thus $(\downarrow Y)/\approx \subseteq X$.

As $B_{\approx}$ is $\approx$-finitary and left-closed, it is discrete, by 3.3.14. By 22.2.7 (2) and 2.3.4, $o_1/\approx \approx_{\approx} o_2/\approx \Leftrightarrow o_1 \approx o_2$, and so $B/\approx$ is asynchronous. This gives 12.2.8 (1). (2) and (3) hold trivially since the labelling function is the identity, by 22.2.6 (3).

Finally, suppose that $p_1, p_2 \in P_{\approx}$ such that $p_1 \equiv p_2$, then there exists a bijection $\phi: p_1 \to p_2$ such that if $o \in p_1$, then $o = \lambda_{\approx}(o) = \lambda(\phi(o)) = \phi(o)$. So $p_1 = p_2$ and $B/\approx$ is λ-reduced.

(2) Immediate.

■

22.2.9. DEFINITION. If B is a bounded, discrete behavioural presentation, then we define $H_B = (Q, A, \to, \iota, E, \mu)$, where

1• $(Q, A, \to, \iota) = C_{B/\approx}$;

2• $\mu = \mu_B$.

Before we state our next result, we observe that if $(C, q) = (C_{B/\approx}, \Omega_{co/\approx})$ then by 15.2.10, the states of C are traces, elements of the left-closed trace language $\rho_{B/\approx}(P/\approx) \subseteq (O/\approx)_{co/\approx}{}^*$. Thus, $\Omega_{co/\approx}{}^*$ is a state of C. By 22.2.9 (1), $(H, q) = (H_B, \Omega_{co/\approx})$ has the same set of states as C. So $\Omega_{co/\approx}{}^*$ is also a state of H.

22.2.10. COROLLARY. Suppose that B is a bounded, discrete behavioural presentation and suppose that $(C, q) = (C_{B/\approx}, \Omega_{co/\approx})$ and $(H, q) = (H_B, \Omega_{co/\approx})$, then

1• H is a hybrid transition system;

2• $(B_{(C,q')}, \mu_B)$ is a bounded, bagged behavioural presentation;

3• $B_{(H,q)} = \varepsilon(B_{(C,q)}, \mu_B)$.

PROOF. (1) $B/\approx$ is a linguistic behavioural presentation, by 22.2.8, and so $C(H) = C$ is an asynchronous transition system, by 15.2.11. Hence, H is a hybrid transition system, by 22.2.1.

(2) If $B = (O, P, E, \lambda)$, then by 15.2.10 and 22.2.6, C is of the form $(Q, O/\approx, \to, \iota)$ and so $B_{(C,q)}$ is of the form $(O', P', O/\approx, \lambda')$, by 12.1.1. But the domain of μ_B is $O/\approx$, by 22.2.6. Thus $(B_{(C,q')}, \mu_B)$ is a bagged behavioural presentation. It is bounded because B is bounded.

(3) By 22.2.9 (1), $C(H) = C_{B/\approx}$ and so $(C(H),q) = (C_{B/\approx},q)$. (3) now follows from (2) and 22.2.3.

■

We shall show that if B is a bounded, discrete behavioural presentation and $(H,q) = (H_B,\Omega_{co/\approx})$, then (H,q) accepts B. By 22.2.5, we must show that $B \equiv B_{(H,q)}$, which is the same thing as $B \equiv \varepsilon(B_{(C,q)},\mu_B)$, by 22.2.10 (2), where $(C,q) = (C_{B/\approx},\Omega_{co/\approx})$.

We know that $B/\approx$ is a $co_{\approx}$-linguistic behavioural presentation, by 22.2.8 (1), and so by 15.2.11, C is an asynchronous transition system such that (C,q) accepts $B/\approx$, that is $B/\approx \equiv B_{(C,q)}$, by 15.2.7. Our approach will be to 'lift' the isomorphism $B/\approx \equiv B_{(C,q)}$ to an isomorphism $B = \varepsilon(B_{(C,q)},\mu_B)$. We begin by noting that $(B/\approx,\mu_B)$ is itself a bagged behavioural presentation.

22.2.11. LEMMA. Suppose that B is a bounded, discrete behavioural presentation and let $(C,q) = (C_{B/\approx},\Omega_{co/\approx})$, then

$$\varepsilon(B/\approx,\mu_B) \equiv \varepsilon(B_{(C,q)},\mu_B).$$

PROOF. Let $B' = B_{(C,q)}$, $B'' = B/\approx$, $\tilde{B}'' = \varepsilon(B'',\mu_B)$ and $\tilde{B}' = \varepsilon(B',\mu_B)$. By 15.2.7 and 15.2.11, there exists an isomorphism $f\colon B'' \equiv B'$. By 12.2.13 (1) and 22.2.6 (2), if $o \in O''$, then $o = \lambda''(o) = \lambda'(f_O(o))$. Since f_O is bijective, we may define a bijection $\tilde{f}_O\colon \varepsilon_{\mu_B}(O'') \to \varepsilon_{\mu_B}(O')$ by $\tilde{f}_O(o,e,i) = (f_O(o),e,i)$. By 22.2.2 and the definition of $\tilde{f}_O$

$$\tilde{\lambda}''(o,e,i) = e = \lambda'(f_O(o),e,i) = \tilde{\lambda}'(\tilde{f}_O(o,e,i))$$

so that 12.2.13 (1) holds.

We also have a bijection $f_P\colon P'' \to P'$ and so we may define a bijection $\tilde{f}_P\colon \varepsilon_{\mu_B}(P'') \to \varepsilon_{\mu_B}(P')$, by $\tilde{f}_P(\varepsilon_{\mu_B}(p)) \to \varepsilon_{\mu_B}(f_P(p))$. $\tilde{f}_P$ is well defined as it is an immediate consequence of 22.2.2 that if $\varepsilon_{\mu_B}(p_1) = \varepsilon_{\mu_B}(p_2)$, then $p_1 = p_2$. Finally,

$$\begin{aligned}
\tilde{f}_P(\varepsilon_{\mu_B}(p)) &= \varepsilon_{\mu_B}(f_P(p)), \text{by definition of } \tilde{f}_P \\
&= \varepsilon_{\mu_B}(\{f_O(o) \mid o \in p\}), \text{by 12.2.13 (2)} \\
&= \bigcup_{o \in p} \varepsilon_{\mu_B}(f_O(o)), \text{by 22.2.2} \\
&= \bigcup_{o \in p} \tilde{f}_O(\varepsilon_{\mu_B}(o)), \text{by definition of } \tilde{f}_O.
\end{aligned}$$

$$= \{\tilde{f}_O(o') \mid o' \in \varepsilon_{\mu_B}(p)\}, \text{ by 22.2.2.}$$

So 12.2.13 (2) holds.

■

Now for the last link in the chain

22.2.12. LEMMA. Suppose that B is a bounded, discrete behavioural presentation, then

$$B \equiv \varepsilon(B/\approx, \mu_B).$$

PROOF. Let $B' = B/\approx$ and $B'' = \varepsilon(B', \mu_B)$.

If $X \in O'$, then both X and $\varepsilon_{\mu_B}(X)$ contain the same number of elements labelled by a given event e, namely $(\mu_B(X))(e)$, and so for each such set X we may define a bijection $h_X : X \to \varepsilon_{\mu_B}(X)$ with the property that if $o \in X$, then $\lambda(o) = \lambda''(h_X(o))$. By 22.2.2, the sets $\varepsilon_{\mu_B}(X)$ partition $\varepsilon_{\mu_B}(O)$ and so we may define a bijection $f_O : O \to O''$ so that if $o \in X$, then $f_Q(o) = h_X(o)$. This mapping consequently satisfies condition 12.2.13 (1).

If $p \in P$, then we define $f_P(p) = \varepsilon_{\mu_B}(p/\approx)$. It is clear that $f_P : P \to P''$ and is surjective. If $p_1, p_2 \in P$, then

$$\varepsilon_{\mu_B}(p_1/\approx) = \varepsilon_{\mu_B}(p_2/\approx) \Rightarrow p_1/\approx = p_2/\approx \Rightarrow p_1 = \bigcup_{X \in p_1/\approx} X = \bigcup_{X \in p_2/\approx} X = p_2$$

so f_P is injective and hence bijective.

Finally:

$$f_P(p) = \varepsilon_{\mu_B}(p/\approx) = \bigcup_{X \in p/\approx} \varepsilon_{\mu_B}(X) = \bigcup_{X \in p/\approx} f_O(X) = \{f_O(o) \mid o \in p\}.$$

so 12.2.13 (2), also holds.

■

We are now assured that every bounded, discrete behavioural presentation is accepted by at least one initialised hybrid transition system.

22.2.13. PROPOSITION. If B is a bounded, discrete behavioural presentation, then H_B is a hybrid transition system and $(H_B, \Omega_{co/\approx})$ accepts B.

PROOF. By 22.2.10, 22.2.11 and 22.2.12.

■

We are again on the trail of a semantics theorem. This means that we need to define some kind of equivalence. Things are a little more complex than in the asynchronous case however because in order that (H_1, q_1) and (H_2, q_2) accept the same $\equiv$ class of behavioural presentations, it is not necessary that $(C(H_1), q_1) \equiv (C(H_2), q_2)$. In particular, we need not have $\iota_{(C(H_1), q_1)} = \iota_{(C(H_2), q_2)}$, since the transitions involved are 'hidden' when the ε operator is applied.

Basically, we need a relation on states, $R \subseteq Q_1 \times Q_2$ such that whenever $q_1' \; R \; q_2'$, then apart from the names of transitions, H_1 and H_2 look the same at these two states. For example, if there are three transitions at q_1' labelled $a.b^3.d$, then there should be three transitions at q_2' labelled $a.b^3.d$. In other words, there should be a bijection between transitions from q_1' to transitions from q_2' which preserve labelling. We also need to respect the concurrency structure.

22.2.14. DEFINITION. Let (H_1, q_1) and (H_2, q_2) be initialised hybrid transition system. A *strict equivalence* from (H_1, q_1) to (H_2, q_2) is a pair (R, F), where $R \subseteq Q_1 \times Q_2$ and F is a family of bijections indexed by R, such that if $(q, q') \in R$, then $f_{(q,q')} : |q\rangle_1 \to |q'\rangle_2$ [2] and

1• $q_1 \; R \; q_2$;

2• $f_{(q,q')}(a, \tilde{q}) = (b, \tilde{q}') \Rightarrow \mu_1(a) = \mu_2(b) \wedge \tilde{q} \; R \; \tilde{q}'$;

3• $(f_{(q,q')}(a, \tilde{q}) = (a', \tilde{q}') \wedge (f_{(\tilde{q},\tilde{q}')}(b, \bar{q}) = (b', \bar{q}')) \Rightarrow (a \; \iota_1 \; b \Leftrightarrow a' \; \iota_2 \; b')$

4• $(f_{(q,q')}(a, \tilde{q}) = (a', \tilde{q}') \wedge f_{(\tilde{q},\tilde{q}')}(b, \bar{q}) = (b', \bar{q}') \wedge f_{(q,q')}(b, \hat{q}) = (b'', \hat{q}')$
$\wedge f_{(\hat{q},\hat{q}')}(a, \breve{q}) = (a'', \breve{q}') \wedge a \; \iota_1 \; b) \Rightarrow a' = a'' \wedge b' = b''$.

We will say that (H_1, q_1) is *strictly equivalent* to (H_2, q_2) $\Leftrightarrow$ there is a strict equivalence from (H_1, q_1) to (H_2, q_2). Write $(H_1, q_1) \equiv (H_2, q_2)$ to indicate that (H_1, q_1) is strictly equivalent to (H_2, q_2).

To avoid unreadable subscripts, we shall write $(q, a, \tilde{q}) \; F \; (q', a', \tilde{q}')$ to indicate that $q \; R \; q'$ and $f_{(q,q')}(a, \tilde{q}) = (a', \tilde{q}')$.

$(q, a, \tilde{q}) \; F \; (q', a', \tilde{q}')$ means that under a local bijection, the transition $q \to_1^a \tilde{q}$ in H_1 corresponds to the transition $q' \to_2^{a'} \tilde{q}'$ in H_2. (2) says that the two transitions must have the same label. (3) says that if the transitions $q \to_1^a \tilde{q} \to_1^b \bar{q}$ in H_1 corresponds to the transition $q' \to_2^{a'} \tilde{q}' \to_2^{b'} \bar{q}'$ in H_2, then $a \; \iota_1 \; b \Leftrightarrow a' \; \iota_2 \; b'$, so that locally, concurrency is preserved. Finally, if $q \to_1^a \tilde{q} \to_1^b \bar{q}$ in H_1 corresponds to the transition $q' \to_2^{a'} \tilde{q}' \to_2^{b'} \bar{q}'$ in H_2 and $q \to_1^b \hat{q} \to_1^a \bar{q}$ in H_1 corresponds to the transition $q' \to_2^{b''} \hat{q}' \to_2^{a''} \breve{q}'$ in H_2, with $a \; \iota_1 \; b$, then $a' = a''$ and $b' = b''$; lozrenges are not 'split open'.

---

[2]See the introduction to section 14.4.

It is not hard to show that $\equiv$ is an equivalence relation on the class of all initialised hybrid transition systems. Reflexivity is obvious. If (R,F) is a strict equivalence from (H_1,q_1) to (H_2,q_2), then (R^{-1},F^{-1}) is a strict equivalence from (H_2,q_2) to (H_1,q_1) where $F^{-1}=\{f_X^{-1} \mid X\in R\}$. Finally, if (R_1,F_1) is a strict equivalence from (H_1,q_1) to (H_2,q_2) and (R_2,F_2) is a strict equivalence from (H_2,q_2) to (H_3,q_3), then $(R_1\circ R_2,F_1\circ F_2)$ is a strict equivalence from (H_1,q_1) to (H_3,q_3), where

$$F_1\circ F_2=\{f_2\circ f_1 \mid f_1\in F_1\wedge f_2\in F_2\wedge dom(f_2)=range(f_1)\}.$$

We shall prove that $(H_1,q_1)\equiv(H_2,q_2)\Leftrightarrow B_{(H_1,q_1)}\equiv B_{(H_2,q_2)}$.

First, we tackle the forward implication. The idea is to use the strict equivalence from (H_1,q_1) to (H_2,q_2) to set up a bijection between $L(C(H_1),q_1)$ and $L(C(H_2),q_2)$ which induces an isomorphism between $B_{(C(H_1),q_1)}$ and $B_{(C(H_2),q_2)}$, and consequently between $B_{(H_1,q_1)}$ and $B_{(H_2,q_2)}$.

22.2.15. PROPOSITION. If (H_1,q_1) and (H_2,q_2) are initialised hybrid transition systems then $(H_1,q_1)\equiv(H_2,q_2)\Rightarrow B_{(H_1,q_1)}\equiv B_{(H_2,q_2)}$.

PROOF. Suppose that $(H_1,q_1)\equiv(H_2,q_2)$ with strict equivalence (R,F). For $i=1,2$, define $T_i=T(C(H_i))$ and $L_i=L(C(H_i),q_i)$. We shall show that there is a bijection $\phi:L(T_1,q_1)\to L(T_2,q_2)$ such that

$$\forall x,y\in L(T_1,q_1): x\equiv_{\iota_1} y\Leftrightarrow \phi(x)\equiv_{\iota_2}\phi(y) \tag{22.2}$$

Thus, there exists a bijection $\Phi:L_1\to L_2$, given by $\Phi(x_{\iota_1})=\phi(x)_{\iota_2}$. From Φ we shall construct an isomorphism from $B_{(H_1,q_1)}$ to $B_{(H_2,q_2)}$.

Let $x=a_1\cdots a_n\in L(T_1,q_1)$. As T_1 is unambiguous, there exists unique $q_1^i\in Q_1$, with $q_1=q_1^0\xrightarrow{a_1}_1 q_1^1\xrightarrow{a_2}_1\cdots\xrightarrow{a_n}_1 q_1^n$. If $q_2^0=q_2$, then $q_1^0\,R\,q_2^0$, by 22.2.14 (1). By induction on n, for each i, $1\le i\le n$, there exists $(q_2^{i-1},b_i,q_2^i)\in\to_2$ such that $(q_1^{i-1},a_i,q_1^i)\,F\,(q_2^{i-1},b_i,q_2^i)$. Thus $q_2=q_2^0\xrightarrow{b_1}_2 q_2^1\xrightarrow{b_2}_2\cdots\xrightarrow{b_n}_2 q_2^n$.

Let $\phi(x)=b_1\cdots b_n$. It is easy to show, by induction on n and using the fact that the elements of F are bijections, that ϕ determines a bijection $L(T_1,q_1)\cap A_1^n\to L(T_2,q_2)\cap A_2^n$ and hence that $\phi:L(T_1,q_1)\to L(T_2,q_2)$ is bijective. It is also clear from the definition of ϕ that it is a poset isomorphism. We establish that

$$\forall x,y\in L(T_1,q_1): x\equiv_{\iota_1}^{(1)} y\Leftrightarrow \phi(x)\equiv_{\iota_2}^{(1)}\phi(y)$$

from which (22.2) follows.

If $x \equiv_{\iota_1}^{(1)} y$, then we may write $x = u.a.b.v$ and $y = u.b.a.v$, where $a, b \in A_1$, $u, v \in A_1^*$ and $a\,\iota_1\,b$, and so we have $q_1 \to_1^u q_1' \to_1^a q_1'' \to_1^b q_1''' \to_1^v$ and $q_1 \to_1^u q_1' \to_1^b \tilde{q}_1'' \to_1^a q_1''' \to_1^v$. As above, $q_2 \to_2^{u'} q_2' \to_2^{a'} q_2'' \to_2^{b'} q_2''' \to_2^{v'}$ and $q_2 \to_2^{u'} q_2' \to_2^{b''} \tilde{q}_2'' \to_2^{a''} \tilde{q}_2''' \to_2^{v''}$, where $\phi(x) = u'.a'.b'.v'$, $\phi(y) = u'.b''.a''.v''$, $(q_1', a, q_1'')\,F\,(q_2', a', q_2'')$, $(q_1'', b, q_1''')\,F\,(q_2'', b', q_2''')$, $(q_1', b, \tilde{q}_1'')\,F\,(q_2', b'', \tilde{q}_2'')$ and $(\tilde{q}_1'', a, \tilde{q}_1''')\,F\,(\tilde{q}_2'', a'', \tilde{q}_2''')$. Thus, $a'\,\iota_2\,b'$, by 22.2.14 (3) and by 22.2.14 (4), $a' = a''$ and $b' = b''$. Since $a'\,\iota_2\,b'$ and $T(C(H_2))$ is unambiguous, $q_2''' = \tilde{q}_2'''$ and consequently, $v' = v''$. We have shown that $a'\,\iota_2\,b'$, $\phi(x) = u'.a'.b'.v'$, $\phi(y) = u'.b'.a'.v'$, that is $\phi(x) \equiv_{\iota_2}^{(1)} \phi(y)$.

Thus, we may define $\Phi: L_1 \to L_2$ as above and note that Φ is a poset isomorphism. It follows that the restriction of Φ to $O_{L_1} = Pr(L_1)$, which we shall call g_O, is a poset isomorphism $g_O: O_{L_1} \to O_{L_2}$. Φ also determines a poset isomorphism $g_P: P_{L_1} \to P_{L_2}$, defined by $f_P(Pr(x_{\iota_1})) = Pr(\Phi(x_{\iota_1}))$. (Recall that by prime algebraicity, the map that takes x_{ι_1} to $Pr(x_{\iota_1})$ is a bijection from L_1 to P_{L_1}).

We show that g_O and g_P determine an isomorphism $f: B_{(H_1,q_1)} \equiv B_{(H_2,q_2)}$.

First, if $\underline{u} \in Pr(L_1)$ and $\underline{x} \in L_1$

$$g_O(\underline{u}) \in g_P(Pr(\underline{x})) \Leftrightarrow \Phi(\underline{u}) \le \Phi(\underline{x}) \Leftrightarrow \underline{u} \le \underline{x} \Leftrightarrow \underline{u} \in Pr(\underline{x}).$$

Thus

$$\forall p \in P_{L_1}: g_P(p) = \{g_O(o) \mid o \in p\} \tag{22.3}$$

Next, note that if $\phi(x.a) = y.b$, with $a \in A_1$ and $b \in A_2$, then by 22.2.14 (2), $\mu_1(a) = \mu_2(b)$.

Thus, if $x_{\iota_1}.a_{\iota_1} \in Pr(L_1)$ and $y_{\iota_2}.b_{\iota_2} = g_O(x_{\iota_1}.a_{\iota_1})$, with $a \in A_1$ and $b \in A_2$, then $\mu_2(b) = \mu_2(\lambda_{L_2}(y_{\iota_2}.b_{\iota_2})) = \mu_2(\lambda_{L_2}(g_O(x_{\iota_1}.a_{\iota_1})))$ and so

$$\mu_1(\lambda_{L_1}(x_{\iota_1}.a_{\iota_1})) = \mu_1(a) = \mu_2(b) = \mu_2(\lambda_{L_2}(g_O(x_{\iota_1}.a_{\iota_2}))).$$

Thus

$$\forall o \in O_{L_1}: \mu_1(\lambda_{L_1}(o)) = \mu_2(\lambda_{L_2}(g_O(o))) \tag{22.4}$$

By (22.4), for each $\underline{u} \in O_{L_1}$ there exists a bijection $h_{\underline{u}}: \varepsilon_{\mu_1}(\underline{u}) \to \varepsilon_{\mu_2}(g_O(\underline{u}))$ satisfying $\lambda_{\mu_2}(h_o(o')) = \lambda_{\mu_1}(o')$, all $o' \in \varepsilon_{\mu_1}(\underline{u})$ and, as in the proof of 22.2.12, we may define a bijection $f_O: \varepsilon_{\mu_1}(O_{L_1}) \to \varepsilon_{\mu_2}(O_{L_2})$ by $f_O(o') = h_{\underline{u}}(o')$ if $o' \in \varepsilon_{\mu_1}(\underline{u})$. Note that $\lambda_{\mu_2}(f_O(o')) = \lambda_{\mu_1}(o')$, all $o' \in \varepsilon_{\mu_1}(O_{L_1})$.

We may also define a bijection $f_P:\varepsilon_{\mu_1}(P_{L_1}) \rightarrow \varepsilon_{\mu_2}(P_{L_2})$ by $f_P(\varepsilon_{\mu_1}(p)) = \varepsilon_{\mu_2}(g_O(p))$. And now:

$$\begin{aligned} f_P(\varepsilon_{\mu_1}(p)) &= \varepsilon_{\mu_2}(g_P(p)), \text{by definition of } f_P \\ &= \varepsilon_{\mu_2}(\{g_O(o) \mid o \in p\}), \text{by (22.3)} \\ &= \bigcup_{o\in p} \varepsilon_{\mu_1}(g_O(o)) \\ &= \bigcup_{o\in p} f_O(\varepsilon_{\mu_1}(o)), \text{by definition of } f_O. \\ &= \{f_O(o') \mid o' \in \varepsilon_{\mu_1}(p)\}. \end{aligned}$$

It is now clear that $f:\varepsilon(B_{(C_1,q_1)},\mu_1) \equiv \varepsilon(B_{(C_2,q_2)},\mu_2)$, that is $f:B_{(H_1,q_1)} \equiv B_{(H_2,q_2)}$

■

Now let us show that the converse holds. We begin by defining a relation between states of initialised hybrid transition systems whose corresponding behavioural presentations are isomorphic.

22.2.16. DEFINITION. If $f:B_{(H_1,q_1)} \equiv B_{(H_2,q_2)}$ then for all $q_1' \in Q_1$ and $q_2' \in Q_2$, we define $q_1' \, R_f \, q_2' \Leftrightarrow$ there exists $x_{\iota_1} \in L(C(H_1),q_1)$ and $y_{\iota_2} \in L(C(H_2),q_2)$ such that

1• $q_1 \rightarrow_1^{x_{\iota_1}} q_1'$ and $q_2 \rightarrow_2^{y_{\iota_2}} q_2'$;
2• $f_P(\varepsilon_{\mu_1}(Pr(x_{\iota_1}))) = \varepsilon_{\mu_2}(Pr(y_{\iota_2}))$.

We note that $\Omega_{\iota_2} \in L(C(H_1),q_1)$ and $\Omega_{\iota_1} \in L(C(H_2),q_2)$ and that $q_1 \rightarrow_1^{\Omega_{\iota_1}} q_1$ and $q_2 \rightarrow_2^{\Omega_{\iota_2}} q_2$. Finally,

$$f_P(\varepsilon_{\mu_1}(Pr(\Omega_{\iota_1}))) = f_P(\varnothing) = \varnothing = \varepsilon_{\mu_2}(Pr(\Omega_{\iota_2}))$$

and we have demonstrated:

$$f:B_{(H_1,q_1)} \equiv B_{(H_2,q_2)} \Rightarrow q_1 \, R_f \, q_2 \tag{22.5}$$

Thus, 22.2.14 (1) holds. We now address ourselves to the construction of the functions F_f.

Suppose that $q_1' \, R_f \, q_2'$ and $(a,q_1'') \in |q_1'\rangle_1$. We have $(x_1)_{\iota_1} \in L(C(H_1),q_1)$ such that $q_1 \rightarrow_1^{(x_1)_{\iota_1}} q_1'$ and so $q_1 \rightarrow_1^{(x_1.a)_{\iota_1}} q_1''$, giving $(x_1.a)_{\iota_1} \in L(C(H_1),q_1)$. By 12.3.2, $Pr((x_1)_{\iota_1}) \succ^{u_{\iota_1}} Pr((x_1.a)_{\iota_1})$, where $u_{\iota_1} = pr((x_1)_{\iota_1},a)$. We shall show that

$$\varepsilon_{\mu_1}(Pr((x_1)_{\iota_1})) \succ^{\varepsilon_{\mu_1}(u_{\iota_1})} \varepsilon_{\mu_1}(Pr((x_1.a)_{\iota_1})) \tag{22.6}$$

from which we argue that

$$f_P(\varepsilon_{\mu_1}(Pr((x_1)_{\iota_1}))) \succ^{f_O(\varepsilon_{\mu_1}(u_{\iota_1}))} f_P(\varepsilon_{\mu_1}(Pr((x_1.a)_{\iota_1}))) \tag{22.7}$$

Since $q_1' \, R_f \, q_2'$, there exists $(x_2)_{\iota_2} \in L(C(H_2), q_2)$ such that $q_2 \rightarrow_2^{(x_2)_{\iota_2}} q_2'$ and $f_P(\varepsilon_{\mu_1}(Pr((x_1)_{\iota_1}))) = \varepsilon_{\mu_2}(Pr((x_2)_{\iota_2}))$. From (22.7), we deduce that there exists $y_{\iota_2} \in L(C(H_2), q_2)$ such that

$$\varepsilon_{\mu_2}(Pr((x_2)_{\iota_2})) \succ^{f_O(\varepsilon_{\mu_1}(u_{\iota_1}))} \varepsilon_{\mu_2}(Pr(y_{\iota_2})) \tag{22.8}$$

$$f_P(\varepsilon_{\mu_1}(Pr((x_1.a)_{\iota_1}))) = \varepsilon_{\mu_2}(Pr(y_{\iota_2})) \tag{22.9}$$

We shall then argue that there exists unique $(u_2)_{\iota_2} \in Pr(L(C(H_2), q_2)$ such that

$$Pr((x_2)_{\iota_2}) \succ^{(u_2)_{\iota_2}} Pr(y_{\iota_2}) \tag{22.10}$$

By 12.3.2, there exists unique, $b \in A_2$ such that $y_{\iota_2} = (x_2.b)_{\iota_2}$ and $(u_2)_{\iota_2} = pr((x_2)_{\iota_2}, b)$, so there exists $\hat{q}_2', q_2'' \in Q_2$ such that $q_2 \rightarrow_2^{x_2} \hat{q}_2' \rightarrow_2^{b} q''$. Since $T(C(H_2))$ is unambiguous, $\hat{q}_2' = q_2'$ and so $q_2' \rightarrow_2^{b} q_2''$, that is $(b, q_2'') \in |q_2'\rangle_2$. As $T(C(H_2))$ is unambiguous, q_2'' is unique. We may therefore define $f_{(q_1', q_2')}(a, q_1'') = (b, q_2'')$ and note that $f_{(q_1', q_2')} : |q_1'\rangle_1 \rightarrow |q_2'\rangle_2$.

Let us pause to establish the above claims. To deduce (22.6), we need the following:

22.2.17. LEMMA. If (B, μ) is a bagged behavioural presentation such that B is asynchronous, then

1• If $o \in O$, then $\varepsilon_\mu(o)$ is a $\approx$-class in $\varepsilon(B, \mu)$;

2• If $p_1, p_2 \in P$ and $o \in O$, then $p_1 \succ^o p_2 \Rightarrow \varepsilon_\mu(p_1) \succ^{\varepsilon_\mu(o)} \varepsilon_\mu(p_2)$.

PROOF. (1) If $o_1, o_2 \in \varepsilon_\mu(o)$, then $o_1 \in \varepsilon_\mu(p) \Leftrightarrow o \in p \Leftrightarrow o_2 \in \varepsilon_\mu(p)$, for all $p \in P$, so $o_1 \approx o_2$ in $\varepsilon(B, \mu)$. If $o_1 \in \varepsilon_\mu(o)$ and $o_2 \in \varepsilon_\mu(o')$ such that $o \neq o'$, then $o \not\approx o'$ as B is asynchronous, and so there exists $p \in P$ such that $\{o, o'\} \not\subseteq p$. It easily follows that $\{o_1, o_2\} \not\subseteq \varepsilon_\mu(p)$ and hence $o_1 \not\approx o_2$.

(2) If $p_1 \succ^o p_2$, then $p_1 \subseteq p_2$ and $p_2 - p_1 = \{o\}$ and therefore $\varepsilon_\mu(p_1) \subseteq \varepsilon_\mu(p_2)$ and $\varepsilon_\mu(p_2) - \varepsilon_\mu(p_1) = \varepsilon_\mu(o)$. Since $\varepsilon_\mu(o)$ is a $\approx$-class in $\varepsilon(B, \mu)$, $\varepsilon_\mu(p_1) \succ^{\varepsilon_\mu(o)} \varepsilon_\mu(p_2)$.

■

To deduce (22.7) from (22.6), we need the following:

22.2.18. LEMMA. If B_1 and B_2 are behavioural presentation and $f:B_1 \equiv B_2$, then if $o_1, o_2 \in O_1$, $p_1, p_2 \in P$ and $X \in O/\approx$:

1• $o_1 \approx_1 o_2 \Leftrightarrow f_O(o_1) \approx_2 f_O(o_2)$;
2• If $p_1 \succ_1^X p_2$, then $f_P(p_1) \succ_2^{f_O(X)} f_P(p_2)$.

PROOF (1) Suppose that $o_1, o_2 \in O_1$, then

$$\begin{aligned} o_1 \approx_1 o_2 &\Leftrightarrow (\forall p \in P_1 : o_1 \in p \Leftrightarrow o_2 \in p) \\ &\Leftrightarrow (\forall p \in P_1 : f_O(o_1) \in f_P(p) \Leftrightarrow f_O(o_2) \in f_P(p)) \\ &\Leftrightarrow (\forall p \in P_2 : f_O(o_1) \in p \Leftrightarrow f_O(o_2) \in p) \Leftrightarrow f_O(o_1) \approx_2 f_O(o_2). \end{aligned}$$

(2) If $p_1 \succ^X p_2$, then $p_1 \subseteq p_2$ and $p_2 - p_1 = X$ and so $f_P(p_1) \subseteq f_P(p_2)$ and $f_P(p_2) - f_P(p_1) = f_O(X)$. Since $f_O(X)$ is a $\approx$-class in B_2, $f_P(p_1) \succ_2^{f_O(X)} f_P(p_2)$.
■

(22.8) and (22.9) follow directly from (22.7) and the definition of $B_{(H_2,q_2)}$. To deduce (22.10) from (22.8), we need the following lemma.

22.2.19. LEMMA. Suppose that (B_1, μ_1) and (B_2, μ_2) are bagged behavioural presentation, such that B_1 and B_2 are asynchronous, and let $f: \varepsilon(B_1, \mu_1) \equiv \varepsilon(B_2, \mu_2)$, then for each $o_1 \in O_1$, there exists unique $o_2 \in O_2$ such that $f_O(\varepsilon_{\mu_1}(o_1)) = \varepsilon_{\mu_2}(o_2)$.

PROOF. Since $\varepsilon_{\mu_1}(o_1)$ is a $\approx$-class of (B_1, μ_1), by 22.2.17 (1), $f_O(\varepsilon_{\mu_1}(o_1))$ is a $\approx$-class of (B_2, μ_2), by 22.2.18 (1). It suffices to show that every $\approx$-class of (B_2, μ_2) is of the form $\varepsilon_{\mu_2}(o_2)$. But every such set is a $\approx$-class of (B_2, μ_2), by 22.2.17, (1) and these sets partition $\varepsilon_{\mu_2}(O_2)$, so they must be the complete set of $\approx$-classes of (B_2, μ_2).
■

To summarise our findings so far, we have:

22.2.20. COROLLARY. Suppose that (H_1, q_1) and (H_2, q_2) are initialised hybrid transition systems such that $f: B_{(H_1,q_1)} \equiv B_{(H_2,q_2)}$. If $x_1 \in A_1^*$ and $a \in A_1$ such that $q_1 \rightarrow_1^{x_1} q_1' \rightarrow_1^{a} q_1''$ and $x_2 \in A_2^*$ such that $q_2 \rightarrow_2^{x_2} q_2'$ and $f_P(\varepsilon_{\mu_1}(Pr((x_1)_{\iota_1}))) = \varepsilon_{\mu_2}(Pr((x_2)_{\iota_2}))$, then there exists unique $b \in A_2$ and $q_2'' \in Q_2$ such that

1• $q_2' \to_1^b q_2''$;
2• $\mu_1(a) = \mu_2(b)$;
3• $f_P(\varepsilon_{\mu_1}(Pr((x_1.a)_{\iota_1}))) = \varepsilon_{\mu_2}(Pr((x_2.b)_{\iota_2}))$;
4• $q_1'' R_f q_2''$.

■

22.2.21. PROPOSITION. If (H_1, q_1) and (H_2, q_2) are initialised hybrid transition systems then

$$B_{(H_1,q_1)} \equiv B_{(H_2,q_2)} \Rightarrow (H_1, q_1) \equiv (H_2, q_2).$$

PROOF. Suppose $f: B_{(H_1,q_1)} \equiv B_{(H_2,q_2)}$. We prove that (R_f, F_f) is a strict equivalence. We know from (22.5) that (R_f, F_f) satisfies 22.2.14 (1).

If $q_1' R q_2'$, then by 12.3.2, f_P determines a bijection between the two sets

$$\{\varepsilon_{\mu_1}(Pr((x_1.a_1)_{\iota_1})) \mid (x_1.a_1)_{\iota_1} \in L(C(H_1), q_1)\} \text{ and}$$
$$\{\varepsilon_{\mu_2}(Pr((x_2.a_2)_{\iota_2})) \mid (x_2.a_2)_{\iota_2} \in L(C(H_2), q_2)\}$$

and so $f_{(q_1',q_2')}$ is a bijection.

Suppose next, that $(a, q_1'') \in |q'\rangle_1$ and $f_{(q_1',q_2')}(a, q_1'') = (b, q_2'')$, then by (22.9), $f_P(\varepsilon_{\mu_1}(Pr((x_1.a)_{\iota_1}))) = \varepsilon_{\mu_2}(Pr((x_2.b)_{\iota_2}))$ and since $q_1 \to_1^{x_1.a} q_1''$ and $q_2 \to_2^{x_2.b} q_2''$, it follows that $q_1'' R_f q_2''$.

By comparing (22.8) and (22.9), we may deduce

$$f_O(\varepsilon_{\mu_1}(u_{\iota_1})) = \varepsilon_{\mu_2}((u_2)_{\iota_2}) \tag{22.11}$$

from which it follows that

$$\mu_1(a) = \mu_1(\lambda_1(u_{\iota_1})) = \mu_2(\lambda_2(u_{\iota_2})) = \mu_2(b) \tag{22.12}$$

Thus, (R_f, F_f) satisfies 22.2.14 (2). We establish 22.2.14 (3).

Suppose $(a, q_1'') \in |q'\rangle_1$ and $(a', q_1''') \in |q_1''\rangle_1$, such that $a \,\iota_1\, a'$ then

$$q_1 \to_1^x q_1' \to_1^a q_1'' \to_1^{a'} q_1'''.$$

Since $C(H_1)$ is ι_1-asynchronous, there exists $\tilde{q}_1 \in Q_1$ such that

$$q_1 \to_1^x q_1' \to_1^{a'} \tilde{q}_1 \to_1^a q_1'''.$$

Let $(q_1', a, q_1'') F_f (q_2', b, q_2'')$, $(q_1'', a', q_1''') F_f (q_2'', b', q_2''')$, $(q_1', a', \tilde{q}_1) F_f (q_2', \tilde{b}, \tilde{q}_2)$ and $(\tilde{q}_1, a, q_1''') F_f (\tilde{q}_2, \bar{b}, \bar{q}_2)$, then

$$q_2 \to_2^{y} q_2' \to_2^{b} q_2'' \to_2^{b'} \breve{q}_2$$

and

$$q_2 \to_2^{y} q_2' \to_2^{\tilde{b}} \tilde{q}_2 \to_2^{\bar{b}} \bar{q}_2$$

satisfying

$$f_P(\varepsilon_{\mu_1}(Pr((x.a.a')_{\iota_1}))) = \varepsilon_{\mu_2}(Pr((y.b.b')_{\iota_2})) \tag{22.13}$$
$$f_P(\varepsilon_{\mu_1}(Pr((x.a'.a)_{\iota_1}))) = \varepsilon_{\mu_2}(Pr((y.\tilde{b}.\bar{b})_{\iota_2})) \tag{22.14}$$

Now, $a\,\iota_1\,a'$, and so $(x.a.a')_{\iota_1} = (x.a'.a)_{\iota_1}$. Therefore by (22.13) and (22.14), $\varepsilon_{\mu_2}(Pr((y.b.b')_{\iota_2})) = \varepsilon_{\mu_2}(Pr((y.\tilde{b}.\bar{b})_{\iota_2}))$ and hence $Pr((y.b.b')_{\iota_2}) = Pr((y.\tilde{b}.\bar{b})_{\iota_2})$. Therefore,

$$(y.b.b')_{\iota_2} = \bigsqcup Pr((y.b.b')_{\iota_2}) = \bigsqcup Pr((y.\tilde{b}.\tilde{b})_{\iota_2}) = (y.\tilde{b}.\tilde{b})_{\iota_2}.$$

and so by 12.3.2 and 12.3.3,

$$(b.b')_{\iota_2} = (y.b.b')_{\iota_2}/y_{\iota_2} = (y.\tilde{b}.\bar{b})_{\iota_2}/y_{\iota_2} = (\tilde{b}.\bar{b})_{\iota_2}.$$

Thus, $b.b' \equiv_{\iota_2} \tilde{b}.\bar{b}$ and therefore either (i) $b = \tilde{b}$ and $b' = \bar{b}$ or (ii) $b = \bar{b}$, $b' = \tilde{b}$ and $b\,\iota_2\,b'$. Now, $a \neq a'$, since ι_1 is irreflexive. Since $f_{(\tilde{q}_1, \tilde{q}_2)}$ is bijective and $(q_1', a, q_1'')\,F_f\,(q_2', b, q_2'')$ and $(q_1', a', \tilde{q}_1)\,F_f\,(q_2', \tilde{b}, \tilde{q}_2)$, it follows that $(b, q_2'') \neq (\tilde{b}, \tilde{q}_2)$ Since $T(C(H_2))$ is unambiguous, $b = \tilde{b} \Rightarrow (b, q_2'') = (\tilde{b}, \tilde{q}_2)$. Thus $b \neq \tilde{b}$ and so case (ii) holds. In particular, $b\,\iota_2\,b'$ as required. We have proved that $a\,\iota_1\,a' \Rightarrow b\,\iota_2\,b'$. The converse implication follows by symmetry.

In the course of the proof of 22.2.14 (3), we have also demonstrated 22.2.14 (4).

■

We now present the last of our 'semantics theorems'. Let **DBP** denote the class whose elements are all the $\equiv$-equivalence classes of bounded, discrete behavioural presentations. Write $[\![B]\!]$ for the $\equiv$-class of B. Let **HTS*** denote the class whose elements are all the $\equiv$-equivalence classes of initialised hybrid transition systems. Write $[\![H,q]\!]$ for the $\equiv$-class of (H,q).

22.2.22. THEOREM. There is a bijective mapping

$$Beh_H : \mathbf{HTS^*} \to \mathbf{DBP}$$

such that $Beh_H([\![H,q]\!]) = [\![B]\!] \Leftrightarrow (H,q)$ accepts B.

PROOF. Suppose that (H,q) is an initialised hybrid transition system, then $B_{(H,q)}$ is a discrete behavioural presentation by 22.2.4.

Suppose that $(H_1,q_1) \equiv (H_2,q_2)$ and that (H_1,q_1) accepts B_1 and (H_2,q_2) accepts B_2. By 22.2.5, $B_{(H_1,q_1)} \equiv B_1$ and $B_{(H_2,q_2)} \equiv B_2$. By 22.2.15 and 22.2.21, $(H_1,q_1) \equiv (H_2,q_2) \Leftrightarrow B_{(H_1,q_1)} \equiv B_{(H_2,q_2)}$. Hence, $B_1 \equiv B_{(H_1,q_1)} \equiv B_{(H_2,q_2)} \equiv B_2$. It follows that Beh_H is well defined.

Beh_H is injective since if $Beh_H([\![H_1,q_1]\!]) = Beh_H([\![H_2,q_2]\!])$, then there exists B_1 and B_2 such that (H_1,q_1) accepts B_1 and (H_2,q_2) accepts B_2 with $B_1 \equiv B_2$. $B_{(H_1,q_1)} \equiv B_1 \equiv B_2 \equiv B_{(H_2,q_2)}$, by 22.24, and $B_{(H_1,q_1)} \equiv B_{(H_2,q_2)} \Rightarrow (H_1,q_1) \equiv (H_2,q_2)$, by 22.2.21. Beh_H is surjective since, by 22.2.13, if B is a bounded, discrete behavioural presentation, then $Beh_H([\![H_B,\emptyset]\!] = [\![B]\!]$.

∎

22.3. Classes of Hybrid Transition Systems.

There is one further task we need to perform in order to complete our construction of automata for parallelism. We have built up semantic machinery for three kinds of transition systems: sequential, asynchronous and hybrid. However, the corresponding classes of behavioural presentations are not disjoint. Indeed sequential behavioural presentations are special cases of (in certain cases) asynchronous behavioural presentations, while discrete, asynchronous behavioural presentations are special cases of discrete behavioural presentations. We shall show that these inclusions may be formalised as embeddings of the corresponding ≡-classes which commute with the semantic functions Beh.

First, we observe that if $f: B_1 \equiv B_2$, then B_1 is determinate (respectively, unambiguous, sequential, synchronous, ι-linguistic, discrete) $\Leftrightarrow$ B_2 is determinate (respectively, unambiguous, sequential, synchronous, ι-linguistic, discrete). This is because all five of these properties are defined in terms of $\rightarrow$ and # and labelling, all of which are preserved under isomorphism. For example, if $o_1, o_2 \in O_1$, then

$$o_1 \#_1 o_2 \Leftrightarrow f_O(o_1) \#_2 f_O(o_2)$$

so that B_1 is determinate $\Leftrightarrow$ B_2 is determinate.

Thus, if any member of $[\![B]\!]$ has one or more of the above properties, then they all have. With this in mind, we name the classes of interest.

22.3.1. DEFINITION. 1• **UBP** denotes the class whose elements are all the ≡-equivalence classes of unambiguous, sequential, discrete behavioural presentations.

2• **SB** denotes the class whose elements are all the ≡-equivalence classes of sequential, discrete behavioural presentations.

3• **LBP** denotes the class whose elements are all the ≡-equivalence classes of linguistic, discrete behavioural presentations.

4• **DBP** denotes the class whose elements are all the ≡-equivalence classes of discrete behavioural presentations.

Likewise, we name the ≡-classes of initialised transition systems.

22.3.2. DEFINITION. 1• **UTS*** denotes the class whose elements are all the ≡-equivalence classes of initialised, unambiguous transition systems. Write $[\![T,q]\!]_{UTS}$ for the ≡-equivalence class of $(T,q) \in \mathbf{UTS}^*$.

2• **TS*** denotes the class whose elements are all the ≡-equivalence classes of initialised transition systems. Write $[\![T,q]\!]_{TS}$ for the ≡-equivalence class of $(T,q) \in \mathbf{TS}^*$.

3• **ATS*** denotes the class whose elements are all the ≡-equivalence classes of initialised, asynchronous transition systems Write $[\![C,q]\!]_{ATS}$ for the ≡-equivalence class of $(C,q) \in \mathbf{ATS}^*$.

4• **HTS*** denotes the class whose elements are all the ≡-equivalence classes of initialised, hybrid transition systems. Write $[\![H,q]\!]_{HTS}$ for the ≡-equivalence class of $(H,q) \in \mathbf{HTS}^*$.

$$\begin{array}{ccc} \mathbf{UTS}^* & \xrightarrow{Beh_U} & \mathbf{UBP} \\ {\scriptstyle inc_{SH}} \downarrow & & {\scriptstyle inc_{BUT}} \downarrow \\ \mathbf{TS}^* & \xrightarrow{Beh_T} & \mathbf{SB} \end{array}$$

Figure 22.6.

22.3.3. THEOREM. There are injective functions $inc_{UT}:\mathbf{UTS}^* \to \mathbf{TS}^*$ and $inc_{BUT}:\mathbf{UBP} \to \mathbf{SB}$ such that

1• If $(T,q) \in \mathbf{UTS}^*$, then $inc_{UT}([\![T,q]\!]_{UTS}) = [\![T,q]\!]_{TS}$;

2• If $B \in \mathbf{UBP}$, then $inc_{BUT}([\![B]\!]) = [\![B]\!]$;

3• The diagram of figure 22.6 commutes:

PROOF. First, observe that $inc_{UT}:\mathbf{UTS}^* \to \mathbf{TS}^*$ is well defined and injective. To show that the diagram commutes, we note that if $(T,q) \in \mathbf{UTS}^*$ then

$$Beh_T(inc_{UT}([\![T,q]\!]_{UTS})) = Beh_T([\![T,q]\!]_{TS}) = [\![B_{(T,q)}]\!] \tag{22.15}$$

and by 13.3.12 and the definition of inc_{BUT}, if $(T,q) \in \mathbf{UTS}^*$, then

$$inc_{BUT}(Beh_U([\![T,q]\!]_{UTS})) = Beh_U([\![T,q]\!]_{UTS}) = [\![B_{L(T,q)}]\!] \tag{22.16}$$

But $B_{(T,q)} \equiv B_{L(T,q)}$, by 14.2.4, that is $[\![B_{(T,q)}]\!] = [\![B_{L(T,q)}]\!]$, and so, using (22.15) and (22.16), if $(T,q) \in \mathbf{UTS}^*$, then

$$Beh_T(inc_{UT}([\![T,q]\!]_{UTS})) = inc_{BUT}(Beh_U([\![T,q]\!]_{UTS})).$$

■

Our next result embeds the class of unambiguous transition systems into the class of asynchronous transition systems. The embedding is effected using the following function. If $U = (Q, A, \rightarrow) \in \mathbf{UTS}$, define

$$\varpi(U) = (Q, A, \rightarrow, \emptyset).$$

22.3.4. THEOREM. There are injections $inc_{UA}: \mathbf{UTS}^* \rightarrow \mathbf{ATS}^*$ and $inc_{BUA}: \mathbf{UBP} \rightarrow \mathbf{LBP}$ such that

1• If $(U,q) \in \mathbf{UTS}^*$, then $inc_{SH}([\![U,q]\!]_{UTS}) = [\![\varpi(U),q]\!]_{ATS}$;
2• If $B \in \mathbf{UBP}$, the $inc_{BUA}([\![B]\!]) = [\![B]\!]$;
3• The diagram of figure 22.7 commutes.

$$\begin{array}{ccc} \mathbf{UTS}^* & \xrightarrow{Beh_U} & \mathbf{USBP} \\ {\scriptstyle inc_{SH}} \downarrow & & {\scriptstyle inc_{BSH}} \downarrow \\ \mathbf{ATS}^* & \xrightarrow{Beh_A} & \mathbf{LBP} \end{array}$$

Figure 22.7

PROOF. We first observe that $L(\varpi(U),q) = \{\{x\} \mid x \in L(T(\varpi(U)),q)\}$, for all $(U,q) \in \mathbf{UTS}^*$. But $T(\varpi(U)) = U$ and so $L(\varpi(U),q) = \{\{x\} \mid x \in L(U,q)\}$. Therefore, $L(U_1,q_1) = L(U_2,q_2) \Leftrightarrow L(\varpi(U_1),q_1) = L(\varpi(U_2),q_2)$ for all $(U_1,q_1),(U_2,q_2) \in \mathbf{UTS}^*$.

Suppose that $(U_1,q_1),(U_2,q_2) \in \mathbf{UTS}^*$ and $(U_1,q_1) \equiv (U_2,q_2)$, then $(U_1,q_1) \sim (U_2,q_2)$, by 14.4.2 (2) and so $L(U_1,q_1) = L(U_2,q_2)$, by 13.4.2 (1) and so $L(\varpi(U_1),q_1) = L(\varpi(U_2),q_2)$ and so $(\varpi(U_1),q_1) \equiv (\varpi(U_2),q_2)$ by 15.3.2. This entails that the mapping $inc_{UA}: \mathbf{UTS}^* \rightarrow \mathbf{ATS}^*$ is well-defined.

Next, let $(U_1,q_1),(U_2,q_2) \in \mathbf{UTS}^*$ with $inc_{UA}([\![U_1,q_1]\!]) = inc_{UA}([\![U_2,q_2]\!])$, so that $(\varpi(U)_1,q_1) \equiv (\varpi(U_2,q_2)$, then $L(\varpi(U_1),q_1) = L(\varpi(U_2),q_2)$, by 15.3.2, and so $L(U_1,q_1) = L(U_2,q_2)$ and so $(U_1,q_1) \sim (U_2,q_2)$, by 13.4.2, and so $(U_1,q_1) \equiv (U_2,q_2)$, by 14.4.2 (2). Hence, inc_{UA} is injective.

Let $(U,q) \in \mathbf{UTS^*}$. By 13.4.3, $Beh_U[\![U,q]\!] = [\![B]\!] \Leftrightarrow B \equiv B_{L(U,q)}$ and by 15.3.3, $Beh_A[\![\varpi(U),q]\!] = [\![B]\!] \Leftrightarrow B \equiv B_{L(\varpi(U),q)}$.

If $\{x\} \in L(\varpi(U),q)$, then $\{x\}$ is prime $\Leftrightarrow$ x is prime $\Leftrightarrow$ $x > \Omega$. Thus, there is a bijection $f_O : O_{L(U,q)} \to O_{L(\varpi(U),q)}$ given by $f_O(x) = \{x\}$, and a bijection $f_P : P_{L(U,q)} \to P_{L(\varpi(U),q)}$ given by $f_P(Pr(x)) = Pr(\{x\})$. It is trivial to check that f_O and f_P determine an isomorphism, $B_{L(U,q)} \equiv B_{L(\varpi(U),q),A}$. Hence,

$$\begin{aligned} Beh_A(inc_{UA}([\![U,q]\!]_{UTS})) = Beh_A([\![\varpi(U),q]\!]) &= [\![B_{(\varpi(U),q)}]\!] \\ &= [\![B_{L(U,q)}]\!] = Beh_U([\![U,q]\!]_{UTS}) \\ &= inc_{BUA}(Beh_T([\![U,q]\!]_{UTS})) \end{aligned}$$

■

Finally, we consider embedding the class of asynchronous transition systems into the class of hybrid transition systems. If $C = (Q,A,\to,\iota)$, then let $\gamma(C) = (Q,A,\to,\iota,A,\mu)$, where μ is the canonical embedding of A into $\mathcal{B}(A)$.

22.3.5. THEOREM. There are injections $inc_{AH} : \mathbf{ATS^*} \to \mathbf{HTS^*}$ and $inc_{BAH} : \mathbf{LBP} \to \mathbf{DBP}$ such that

1• If $(C,q) \in \mathbf{ATS^*}$, then $inc_{AH}([\![C,q]\!]_{ATS}) = [\![\gamma(C),q]\!]_{HTS}$;
2• If $B \in \mathbf{LBP}$, the $inc_{BAH}([\![B]\!]) = [\![B]\!]$;
3• The diagram of figure 22.8 commutes:

$$\begin{array}{ccc} \mathbf{ATS^*} & \xrightarrow{Beh_A} & \mathbf{LBP} \\ {\scriptstyle inc_{AH}} \downarrow & & {\scriptstyle inc_{BAH}} \downarrow \\ \mathbf{HTS^*} & \xrightarrow{Beh_H} & \mathbf{DBP} \end{array}$$

Figure 22.8

PROOF. Suppose that $(C_1,q_1),(C_2,q_2) \in \mathbf{ATS^*}$ such that $(C_1,q_1) \equiv (C_2,q_2)$, then $L(C_1,q_1) = L(C_2,q_2)$, by 15.3.2, and so we may define a relation $R \subseteq Q_1 \times Q_2$ by $q_1' \, R \, q_2' \Leftrightarrow \exists \underline{x} \in L(C_1,q_1) : q_1 \to_1^{\underline{x}} q_1' \wedge q_2 \to_2^{\underline{x}} q_2'$ and if $q_1' \, R \, q_2'$, then we define $f_{(q_1',q_2')}(a,q_1'') = (a,q_2'') \Leftrightarrow q_2' \to_2^{a} q_2''$, for all $(a,q_1'') \in |q_1'\rangle_1$. it is not hard to check that (R,F) is a strict equivalence from $(\gamma(C_1),q_1)$ to $(\gamma(C_2),q_2)$, so that $(\gamma(C_1),q_1) \equiv (\gamma(C_2),q_2)$ and so, $inc_{AH} : \mathbf{ATS^*} \to \mathbf{HTS^*}$ is well defined.

Let $(C_1,q_1),(C_2,q_2) \in \mathbf{ATS^*}$ such that $inc_{AH}([\![C_1,q_1]\!]_{ATS}) = inc_{AH}([\![C_2,q_2]\!]_{ATS})$, then $(\gamma(C_1),q_1) \equiv (\gamma(C_2),q_2)$, so there exists a strict equivalence (R,F) from $(\gamma(C_1),q_1)$ to $(\gamma(C_2),q_2)$. As in the proof of 22.2.15, there exists a poset isomorphism $\Phi: L(C(\gamma(C_1)), q_1) \rightarrow L(C(\gamma(C_2)), q_2)$ such that $\mu_1(\underline{x}) = \mu_2(\Phi(\underline{x}))$ for all $\underline{x} \in L(C(\gamma(C_1)), q_1)$. But $C(\gamma(C_i)) = C_i$, and μ_i is the identity, each i. Hence, $L(C_1, q_1) = L(C_2, q_2)$, and so $(C_1, q_1) \equiv (C_2, q_2)$, by 15.3.2.

Finally, using (22.17),

$$\begin{aligned} Beh_H(inc_{AH}([\![C,q]\!]_{ATS})) &= Beh_H([\![\gamma(C),q]\!]_{HTS}) = [\![B_{(\gamma(C),q)}]\!] \\ &= [\![B_{(C,q)}]\!] = Beh_A([\![C,q]\!]_{ATS}) = inc_{BAH}(Beh_A([\![C,q]\!]_{ATS})). \end{aligned}$$

■

At last we have a class of automata capable of modelling any system whose behaviour may be described by a bounded discrete behavioural presentation. In fact, boundedness is a restriction that may be lifted providing we are prepared to use transfinite bags. This extension is discussed in the notes at the end of this chapter.

In the chapters that follow, we shall use hybrid transition systems to provide non-interleaving semantics for a number of notations.

Notes.

Hybrid transition systems and their semantics was first presented in [68].

The additional structure which goes to make an asynchronous transition system into a hybrid transition system permits us to handle ambiguity. It is possible to handle ambiguity more directly. In [52], M. Z. Kwiatowska relaxes the ambiguity constraint and provides a semantics for the extended class of asynchronous transition systems. The construction is rather similar to that of chapter 14 for general transition systems.

The following construction extends the hybrid transition system model to deal with general discrete behavioural presentations. Let $\aleph$ be a cardinal number, then $\mathcal{B}_\aleph(E)$ denotes the set of all functions from E to the set of cardinal numbers less than $\aleph$. We then define a hybrid transition system to be a 6-tuple $H = (Q, A, \rightarrow, \iota, E, \mu)$, where

1• $C(H) = (Q, A, \rightarrow, \iota)$ is an ι-asynchronous transition system;

2• $\mu: A \rightarrow \mathcal{B}_\aleph(E)$ for some cardinal number $\aleph$.

Chapter 23
The Keller Parallel Computation Model

> We will deal with two models. The 'conceptual model'[1] is highly abstract. In fact, it is abstract enough to include almost every existing parallel program model as a special case
>
> R. M. KELLER [21]

23.1. Introduction.

In chapters 16 and 17, we looked at nets on the condition/event level, but we observed that there is a hierarchy of net models, in which condition/event nets are at the lowest level. We also observed that nets belonging to higher levels of the hierarchy are required to have an interpretation on the condition/event level.

This does not mean that to each net in higher levels, with appropriate initial state, there exists a unique (up to isomorphism) C/E system and unique 'net morphism' providing the interpretation. Rather, depending on the system which is being modelled, there should exist such a C/E system and interpreting morphism, reflecting the assumption that all discrete systems may be modelled by C/E systems.

If this is the case, then it is unclear how, in general, we can give a unique non-interleaving semantics to higher-level nets directly. However, we can do something of the sort for particular net-based models in which the type of system being modelled is of a specific kind.

We have chosen to examine a model of parallel computation described by Keller in [21], which he uses to illustrate his transition-system-based approach to the formal verification of parallel programs.

This model has a number of aspects, which we shall first describe informally:

---

[1]This 'conceptual model' is that of the transition system discussed in chapters 13 and 14.

1• Events are associated with guarded assignments to global variables. That is, each occurrence is an execution of a statement of the form

when $Pr(\underline{x})$ **do** $\underline{x} \leftarrow F(\underline{x})$

where $\underline{x}$ represents a vector of n variables $x_1, \cdots, x_n$ - which for the sake of simplicity we shall assume to be of type real - Pr is a predicate $Pr: \mathbb{R}^n \rightarrow \mathbb{B}$[2] and F is a function $F: \mathbb{R}^n \rightarrow \mathbb{R}^n$. If Pr evaluates to true on the current values v_i of the variables x_i, then the assignment $\underline{x} \leftarrow F(\underline{x})$ resets x_i to $F(\underline{v})_i$.

2• Control flow through the system is described using a Petri Net with multiple arcs.[3] Each T-element is labelled by a guarded assignment.

3• At any particular stage in the execution of a parallel program, each S-element will contain a number of tokens. These may be thought of as local program counters. Each global variable will also have some value, which we may represent by the appropriate co-ordinate of an n-tuple $\underline{v}$.

Let us explain the dynamics of a parallel program. A T-element t may fire, that is to say execute, providing that t has concession under the current marking - that is, each of its input S-elements S contains sufficient tokens - and if its associated predicate Pr_t evaluates to true on $\underline{v}$. If this is the case, then the marking changes according to a particular firing rule and the values of the variables are modified to $\underline{v}' = F_t(\underline{v})$.

Let us now formalise all this. The presentation we give is different from but equivalent to that of [21] - the main difference is that we make explicit the underlying Net structure and allow nets to have multiple arcs.

23.2. The Model and its Transition System Semantics.

23.2.1. DEFINITION. A *net with multiple arcs* is a triple $N = (S, T, \Phi)$, where;

1• S is a non-empty set of S-elements, known as *places*;
2• T is a non-empty set of T-elements known as *transitions*;
3• $\Phi: (S \times T) \cup (T \times S) \rightarrow \omega$ is the *incidence function*,

satisfying $S \cap T = \emptyset$.

---

[2] $\mathbb{B}$ denotes the set of Boolean values.
[3] See definition 23.2.1.

Pictorially, places are represented by circles and transitions by squares and there are $\Phi(x,y)$ arcs drawn from the node x to the node y.

We will say that a net $N=(S,T,\Phi)$ is finite if and only if S and T are finite.

For the rest of this chapter, by 'net' we mean 'net with multiple arcs'.

23.2.2. DEFINITION. A *parallel program* is a triple $Prog=(N,n,\lambda)$, where

1• $N(Prog)=N$ is a finite net;
2• $n\in\omega$ is the number of variables in the system;
3• $\lambda: T\rightarrow[\mathbb{R}^n\rightarrow\mathbb{B}]\times[\mathbb{R}^n\rightarrow\mathbb{R}^n]$.

that is to say, $\lambda(t)=(Pr_t,F_t)$, where Pr_t is a predicate defined on the set of all possible values of the n variables and F_t is a function $F_t:\mathbb{R}^n\rightarrow\mathbb{R}^n$.

Figure 23.1 is a pictorial representation of a computation graph. We adopt the usual conventions for representing nets. The guarded assignments associated with the transitions are as follows:

$t1$: **when true do** $x_1:=x_1\times x_2$
$t2$: **when true do** $x_3:=x_3+1$
$t3$: **when** $x_1<0.001\times x_3$ **do** $x_4:=x_4+x_1/x_3$

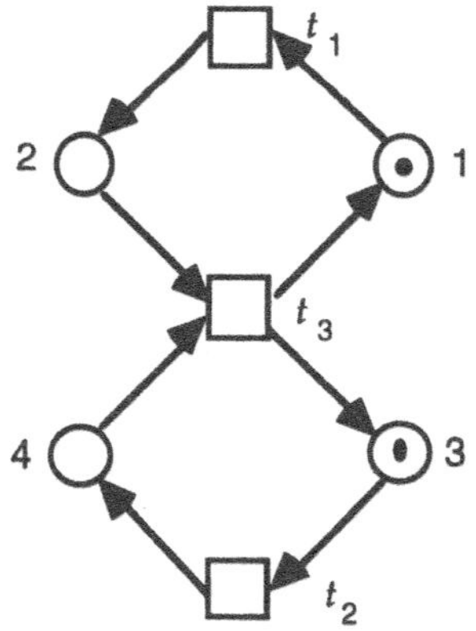

Figure 23.1.

We are using the familiar assignment statement notation to describe the functions associated with the transitions. This is to save us the trouble of writing, for example

$$F_{t_1}(x)_i=\begin{cases}x_1\times x_2 \text{ if } i=1\\ x_i \text{ otherwise}\end{cases}$$

In terms of notation to be introduced later, $dom(F_{t_1}) = \{x_1, x_2\}$ and $range(F_{t_1}) = \{x_1\}$.

23.2.3. DEFINITION. If $N = (S, T, \Phi)$ is a net, then a *marking* of N is a function $M: S \to \omega$ $M(s)$ is the number of tokens on the place $\mathbf{s}$. $Mark(N)$ denote the set of all markings of N. A *configuration* of a parallel program $Prog = (N, n, \lambda)$ is a pair $(M, \underline{v})$, where $M \in Mark(N)$ and $\underline{v} \in \mathbb{R}^n$. Let $Conf_{Prog}$ denote the set of all configurations of $Prog$.

Next we consider how transitions may become active in certain configurations.

23.2.4. DEFINITION. If $N = (S, T, \Phi)$ is a net, $M_1, M_2 \in Mark(N)$ and $t \in T$, then $M_1\ [t\rangle_N\ M_2 \Leftrightarrow$

1• $\forall s \in S: M_1(s) \geq \Phi(s, t)$ - we shall abbreviate this condition to $M_1\ [t\rangle_N$;
2• $\forall s \in S: M_2(s) = M_1(s) + \Phi(t, s) - \Phi(s, t)$.

Note that (1) entails that if $M_1\ [t\rangle_N$, then there exists $M_2 \in Mark(N)$ such that $M_1\ [t\rangle_N\ M_2$.

If $Prog = (N, n, \lambda)$ is a parallel program, $\underline{v}_1, \underline{v}_2 \in \mathbb{R}^n$ and $t \in T$, then $\underline{v}_1 \to^t{}_{Prog}\ \underline{v}_2 \Leftrightarrow$

1• $Pr(\underline{v}_1) = \textbf{true}$;
2• $\underline{v}_2 = F_t(\underline{v}_1)$.

Now that we have described the control flow imposed by the underlying net and the effect on the state space of variable values by the firing of transitions, we may put the two together to give a transition system interpretation.

23.2.5. DEFINITION. If $Prog = (N, n, \lambda)$ is a parallel program, then we define $T_{Prog} = (Conf_{Prog}, T, \to_{Prog})$ where

$$(M_1, \underline{v}_1) \to_{Prog}^t (M_2, \underline{v}_2) \Leftrightarrow M_1\ [t\rangle_N\ M_2 \wedge \underline{v}_1 \to^t{}_{Prog}\ \underline{v}_2.$$

Of course, T_{Prog} is a transition system. It is easy to see that it is unambiguous since, by definition 23.2.4, M_2 is uniquely determined by M_1 and t, and $\underline{v}_2$ is uniquely determined by $\underline{v}_1$ and F_t. T_{Prog} determines an interleaving representation of the behaviour of $Prog$ as a set of sequences of executions of the guarded commands associated with transition firings.

Figure 23.2 shows a fragment of the transition system associated with the example of figure 23.1 in which initially $(x_1, x_2, x_3, x_4) = (1, 0.5, 0, 1)$.

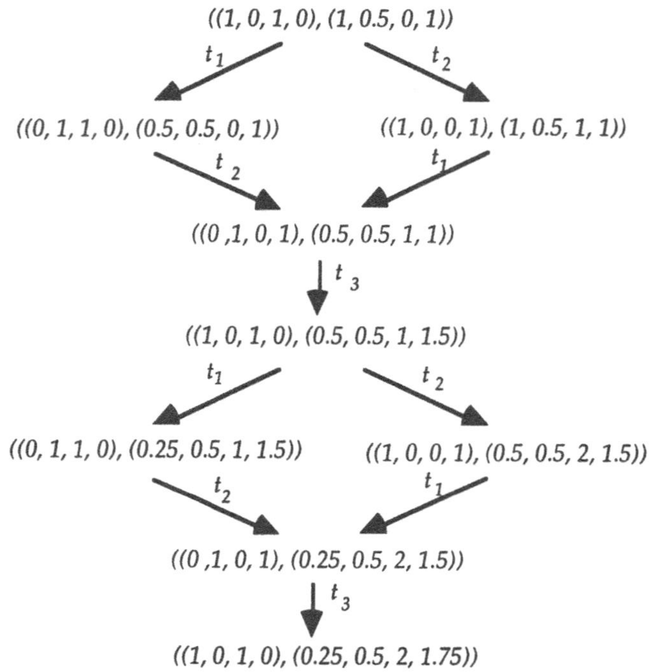

Figure 23.2

23.3. A Hybrid Transition System Semantics for General Nets.

We now embark upon the provision of a non-interleaving semantics for parallel programs. We do this in two stages:

1• We construct a hybrid transition semantics for marked nets;
2• We combine this with an asynchronous transition system semantics for the language of guarded commands.

Consider the net of figure 23.3. Its associated transition system is given in figure 23.4.

The question is, how do we associate concurrency, if at all, with the behaviour of this net? There are certainly a few 'lozenges'. For example, there is a lozenge involving transitions a and c in the top left hand corner, and involving transitions b and c in the top right hand corner. Inspection of the net shows this to have a plausible interpretation: transitions a and c

certainly have enough tokens in the initial marking for them to fire concurrently and transitions b and c have enough tokens between them to fire concurrently from the marking $(0,2,1)$.

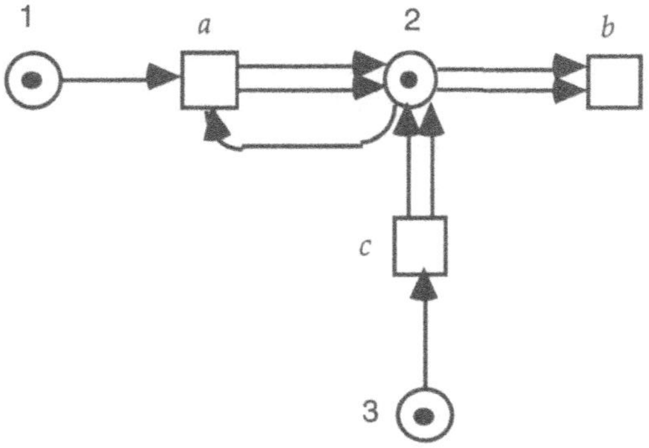

Figure 23.3

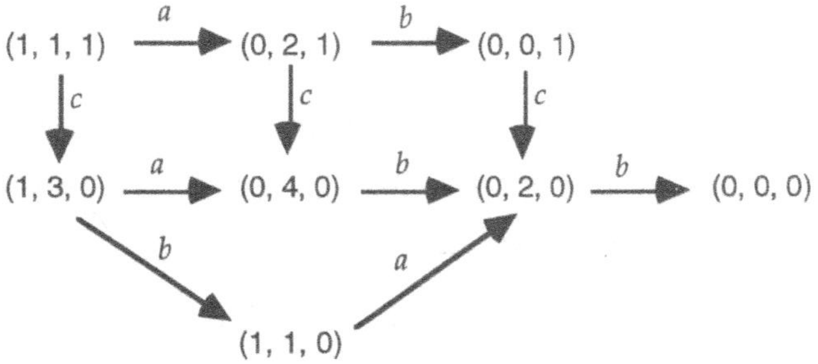

Figure 23.4.

The problem arises from the fact that at the bottom of the transition system, we have a lozenge involving a and b, and there are enough tokens in the marking $(1,3,0)$ for them to fire concurrently. But if we try to construct a hybrid transition system to reflect this we end up with an underlying 'asynchronous transition system' which fails in fact to be so, for the following reason:

1• In order that we have concurrency in the top left hand corner, the transition from $(1,1,1)$ to $(0,2,1)$ must be the same as that from $(1,3,0)$ to $(0,4,0)$. Call it t_1;

2• In order that we have concurrency in the top right hand corner, the transition from $(0,2,1)$ to $(0,0,1)$ must be the same as that from $(0,4,0)$ to $(0,2,0)$. Call it t_2;

3• In order that we have concurrency at the bottom of the diagram, we must have $t_1 \iota t_2$;

4• But then $(1,1,1) \rightarrow^{t_1} (0,2,1,) \rightarrow^{t_2} (0,0,1)$ and $t_1 \iota t_2$ but for no marking M do we have $(1,1,1) \rightarrow^{t_2} M \rightarrow^{t_1} (0,0,1)$, in violation of the definition of an asynchronous transition system.

We may resolve this problem if we think of tokens as *abstract resources* that are being consumed and produced by the firing of transitions. After the firing of c, place 2 contains three tokens, one that was there initially and another two produced by c. In the subsequent firing of a, one of two things can happen: the token from place 2 consumed by this firing is either the one that was there in the initial marking or is one of the two produced by c. In the first case, the firings of a and c may take place concurrently; in the latter case, the firing of a must wait for the firing of c.

Let us see how we can exploit this observation. Recall that $\mathcal{B}(A)$ denotes the set of bags over A.[4] The states of our projected hybrid transition system will be those functions $\underline{q}: S \rightarrow \mathcal{B}(T \cup \{\bot\})$, where $\bot$ is some element presumed not to belong to T, satisfying:

$$\forall s \in S\ \forall t \in T: (\underline{q}(s))(t) > 0 \Rightarrow \Phi(t,s) > 0 \tag{23.1}$$

Write Q_N for this set. The state $\underline{q}$ represents a marking in which place s contains $(\underline{q}(s))(t)$ tokens deposited there by t - we shall refer to these as tokens of type t - and $(\underline{q}(s))(\bot)$ tokens remaining from the initial marking - we shall refer to these as tokens of type $\bot$.

The firing of a transition will consume tokens of various kinds from each place. Let us define A_N to be the set of all pairs $(t,\underline{m}) \in T \times Q_N$ such that

$$\forall s \in S: \textstyle\sum_{x \in T \cup \{\bot\}} (\underline{m}(s))(x) = \Phi(s,t) \tag{23.2}$$

The idea is that $(t,\underline{m})$ is the instance of t which removes $(\underline{m}(s))(x)$ tokens of type x from place s. (23.2) guarantees that $(t,\underline{m})$ will consume $\Phi(s,t)$ tokens overall. In order for $(t,\underline{m})$ to fire, there must be enough tokens of each type. Suppose $\underline{q} \in Q_N$ and $(t,\underline{m}) \in A_N$. We define

$$\underline{q}\,[(t,\underline{m})\rangle \Leftrightarrow \underline{m} \le \underline{q} \tag{23.3}$$

where for $\underline{m}, \underline{n} \in Q_N$,

$$\underline{m} \le \underline{n} \Leftrightarrow \forall s \in S\ \forall x \in T \cup \{\bot\}: (\underline{m}(s))(x) \le (\underline{n}(s))(x).$$

---

[4]See section 22.2.

(23.3) defines *concession*. It is possible to fire $(t,\underline{m})$ at state $\underline{q}$ providing each place s contains enough tokens of the appropriate kind. The firing rule may now be stated. If $\underline{q}_1,\underline{q}_2 \in Q_N$ and $(t,\underline{m}) \in A_N$, then

$$\underline{q}_1 \rightarrow_N^{(t,\underline{m})} \underline{q}_2 \Leftrightarrow \underline{q}_1 [(t,\underline{m})\rangle \wedge \underline{q}_2 = \mathcal{T}(\underline{q}_1,t,\underline{m})$$

where if $\underline{q} \in Q_N$, $s \in S$ and $x \in T \cup \{\bot\}$ then

$$((\mathcal{T}(\underline{q},t,\underline{m}))(s))(x) = \begin{cases} (\underline{q}(s))(x) - (\underline{m}(s))(x) \text{ if } t \neq x \\ (\underline{q}(s))(t) + \Phi(t,s) - (\underline{m}(s))(t) \text{ if } t = x \end{cases} \tag{23.4}$$

so that if $t \neq x$, then no tokens of type t are created but $(\underline{m}(s))(x)$ tokens of type x are removed from each place s, whereas if $t = x$, then in addition, $\Phi(t,s)$ tokens of type t are placed on s.

It is easy to check that if $\underline{q} \in Q_N$, $s \in S$ and $x \in T \cup \{\bot\}$ then $\mathcal{T}(\underline{q},t,\underline{m})$ satisfies (23.1) and therefore belongs to Q_N. We therefore have a transition system $(Q_N, A_N, \rightarrow_N)$.

Two transitions $(t_1,\underline{m}_1),(t_2,\underline{m}_2) \in A_N$ are certainly capable of occurring concurrently when neither depends on the firing of the other in order to obtain concession. For an independence relation, we define:

$$(t_1,\underline{m}_1)\ \iota_N\ (t_2,\underline{m}_2) \Leftrightarrow (t_1,\underline{m}_1) \neq (t_2,\underline{m}_2) \wedge (t_1,\underline{m}_1)\ i_N\ (t_2,\underline{m}_2) \tag{23.5}$$

where

$$(t_1,\underline{m}_1)\ i_N\ (t_2,\underline{m}_2) \Leftrightarrow \forall s \in S : (\underline{m}_1(s))(t_2) = (\underline{m}_2(s))(t_1) = 0 \tag{23.6}$$

In words, $(t_1,\underline{m}_1)\ i_N\ (t_2,\underline{m}_2)$ says that $(t_1,\underline{m}_1)$ consumes no tokens on s put there by the firing of $(t_2,\underline{m}_2)$ and vice versa, that is, these two transitions do not attempt to concurrently change the number of tokens of any type given in any given place.

Finally, the labelling function, $\mu_N : A_N \rightarrow \mathcal{B}(T)$, is given by

$$(\mu_N(t,\underline{m}))(t') = \begin{cases} 1 \text{ if } t = t' \\ 0 \text{ otherwise} \end{cases} \tag{23.7}$$

23.3.1. DEFINITION. If N is a net, then define $H_N = (Q_N, A_N, \rightarrow_N, \iota_N, T, \mu_N)$.

It is clear that $T(C(H_N))$ is unambiguous, for

$$\underline{q} \rightarrow_N^{(t,\underline{m})} \underline{q}_1 \wedge \underline{q} \rightarrow_N^{(t,\underline{m})} \underline{q}_2 \Rightarrow \underline{q}_1 = \mathcal{T}(\underline{q},t,\underline{m}) = \underline{q}_2.$$

Suppose that $\underline{q}_1 \rightarrow_N^{(t_1,\underline{m}_1)} \underline{q}_2 \rightarrow_N^{(t_2,\underline{m}_2)} \underline{q}_3$ with $(t_1,\underline{m}_1)\ \iota_N\ (t_2,\underline{m}_2)$. Let $s \in S$ and $x \in T \cup \{\bot\}$. By (23.3), $(\underline{q}_2(s))(x) \geq (\underline{m}_2(s))(x)$. If $x \neq t_1$, then by (23.4), $(\underline{q}_2(s))(x) = (\underline{q}_1(s))(x) - (\underline{m}_1(s))(x)$ and so

$$(\underline{q}_1(s))(x) \geq (\underline{m}_1(s))(x) + (\underline{m}_2(s))(x) \geq (\underline{m}_2(s))(x),$$

whereas if $x = t_1$, then $(\underline{q}_1(s))(x) \geq 0 = (\underline{m}_2(s))(x)$, by (23.5) and (23.6). Thus by (23.3), $\underline{q}_1\ [(t_2,\underline{m}_2)\rangle_N$, and so $\underline{q}_1 \rightarrow_N^{(t_2,\underline{m}_2)} \underline{q}'_2$, where $\underline{q}'_2 = \mathcal{T}(\underline{q}_1, t_2, \underline{m}_2)$.

We know that $(\underline{q}_1(s))(t) \geq (\underline{m}_1(s))(t) + (\underline{m}_2(s))(t)$ when $x \neq t_1$, so when $x \neq t_1, t_2$, $(\underline{q}'_2(s))(x) \geq (\underline{m}_1(s))(x)$, by (23.4). Also, $(\underline{q}'_2(s))(t_2) \geq 0 = (\underline{m}_1(s))(t_2)$, because $(t_1,\underline{m}_1)\ \iota_N\ (t_2,\underline{m}_2)$. The remaining case is $x = t_1 \neq t_2$. In this case,

$$\begin{aligned}(\underline{q}'_2(s))(t_1) &= (\underline{q}_1(s))(t_1) + \Phi(t_1,s) - (\underline{m}_2(s))(t_1)\\ &= (\underline{q}_2(s))(t_1) + (\underline{m}_1(s))(t_1) + \Phi(t_1,s) - (\underline{m}_2(s))(t_1) \geq (\underline{m}_1(s))(t_1)\end{aligned}$$

since $(\underline{q}_2(s))(t_1) \geq (\underline{m}_2(s))(t_1)$, by (23.3). We have shown that $\underline{q}'_2\ [(t_1,\underline{m}_1)\rangle_N$, from which it follows that $\underline{q}'_2 \rightarrow_N^{(t_1,\underline{m}_1)} \underline{q}'_3$, where $\underline{q}'_3 = \mathcal{T}(\underline{q}'_2, t_1, \underline{m}_1)$.

Now, using (23.4), a calculation shows that

$$\underline{q}'_3 = \mathcal{T}(\mathcal{T}(\underline{q}_1, t_2, \underline{m}_2), t_1, \underline{m}_1) = \mathcal{T}(\mathcal{T}(\underline{q}_1, t_1, \underline{m}_1), t_2, \underline{m}_2) = \underline{q}_3.$$

Thus the 'lozenge rule' holds in $C(H_N)$, and we have:

23.3.2. PROPOSITION. If N is a net then H_N is a hybrid transition system. ■

We now extend the construction to marked nets.

23.3.3. DEFINITION. Suppose that (N,M) is a marked net. Define $H_{(N,M)}{}^* = (H_N, \underline{q}_M)$ where

$$(\underline{q}_M(s))(x) = \begin{cases} M(s) \text{ if } x = \bot \\ 0 \text{ otherwise} \end{cases}.$$

The part of the hybrid transition system corresponding to figure 23.3 reachable from its initial marking is given in figure 23.5.

The diagram is to be read as follows. A state $\underline{q}$ is represented by a quadruple $(u_1, u_2, u_3, u_\bot)$, where u_x represents the bag $\underline{q}(x)$ as follows. If

$u_x = f_a f_b f_c f_\perp$, then $(\underline{q}(x))(y) = f_y$ if f_y is a number and $(\underline{q}(x))(y) = 0$ otherwise.

A label xi represents the transition $(x, \underline{m}_i)$ where:

$\underline{m}_1 = (///1, ///1, ////)$ $\quad$ $\underline{m}_2 = (////, 2///, ////)$ $\quad$ $\underline{m}_3 = (////, ////, ///1)$

$\underline{m}_4 = ////, //2/, ////)$ $\quad$ $\underline{m}_5 = (////, //11/, ////)$ $\quad$ $\underline{m}_6 = (////, 1/1/, ////)$

$\underline{m}_7 = (///1, //1/, ////)$

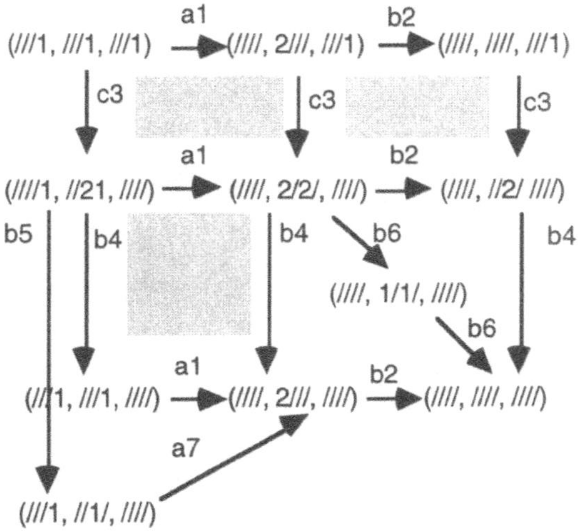

Figure 23.5.

Before we finish the construction of hybrid transition system semantics for parallel programs, we make the observation that the hybrid transition systems arising from general nets are *asynchronous*, in the sense of the following definition.

23.3.4. DEFINITION. Let H be a hybrid transition system, then H is *asynchronous* providing

$$\sum_{e \in E} (\mu(a))(e) = 1, \text{ for each } a \in A.$$

For asynchronous hybrid transition systems, we shall abuse notation and write $\mu(a) = e$ when $(\mu(a))(e) = 1$. It is clear that such hybrid transition systems give rise to behavioural presentations which are themselves asynchronous. It should be pointed out that almost all the theories that we shall be examining give rise to asynchronous hybrid transition systems.

23.4. Hybrid Transition System Semantics for Parallel Programs.

Now that we have a hybrid transition semantics for marked nets, we may move on to the second part of the programme and give a semantics for parallel programs.

Let $Prog$ be a parallel program. We wish to construct a hybrid transition system corresponding to $Prog$. First, states correspond to states of $H_{N(Prog)}$, reflecting the control flow structure of $Prog$, and allocations of values to the variables of $Prog$. So

$$Q_{Prog} = Q_{N(Prog)} \times \mathbb{R}^n \tag{23.8}$$

Transitions are transitions of $H_{N(Prog)}$

$$A_{Prog} = A_{N(Prog)} \tag{23.9}$$

The transition structure reflects that of both $H_{N(Prog)}$ and the guarded commands.

$$(\underline{q}_1, \underline{v}_1) \rightarrow_{Prog}{}^{a} (\underline{q}_2, \underline{v}_2) \Leftrightarrow \underline{q}_1 \rightarrow_{N(Prog)}{}^{a} \underline{q}_2 \Leftrightarrow \underline{v}_1 \rightarrow_{Prog}{}^{t} \underline{v}_2 \tag{23.10}$$

where $\mu_{N(Prog)}(a) = t$.

The events of the transition system are the events of $H_{N(Prog)}$ and the labelling function is also that of $H_{N(Prog)}$.

$$E_{Prog} = T \tag{23.11}$$
$$\mu_{Prog} = \mu_{N(Prog)} \tag{23.12}$$

Finally, we come to the independence relation. We shall assume that transitions are atomic in the sense that during the execution of the command **when** $Pr(\underline{x})$ **do** $\underline{x} \leftarrow F(\underline{x})$ there are no concurrent modifications of the pertinent variables. This entails a constraint on parallel execution; two transitions should not be allowed to fire concurrently if either changes the values of significant input variables of the other or both try to rewrite the same output variable.

Let us formalise this. First, if $\underline{v} \in \mathbb{R}^n$, $1 \le i \le n$ and $u \in \mathbb{R}$, let $\underline{v}[i \backslash u]$ denote the vector obtained from $\underline{v}$ by replacing its ith co-ordinate by u. Let Pr be a predicate. If $Pr = \textbf{true}$, then we define $dom(Pr) = \varnothing$. Otherwise, we define

$$dom(Pr) = \{x_i \mid \exists \underline{v} \in \mathbb{R}^n\ \exists v, v' \in \mathbb{R} : Pr(\underline{v}[i \setminus v]) \neq Pr(\underline{v}[i \setminus v'])\}.$$

then $dom(Pr)$ is the set of variables upon whose values $Pr(\underline{v})$ depends. In our example, $dom(Pr_{t_1}) = \emptyset$ and $dom(Pr_{t_3}) = \{x_1, x_3\}$. It is clear that if $\underline{v}_1, \underline{v}_2 \in \mathbb{R}^n$, then

$$(\forall i \in dom(Pr) : (\underline{v}_1)_i = (\underline{v}_2)_i) \Rightarrow Pr(\underline{v}_1) = Pr(\underline{v}_2) \tag{23.13}$$

Similarly, let F be a function. If F is the identity, then we define $dom(F) = \emptyset$. Otherwise, we define

$$dom(F) = \{x_i \mid \exists \underline{v} \in \mathbb{R}^n\ \exists v, v' \in \mathbb{R} : F(\underline{v}[i \setminus v]) \neq F(\underline{v}[i \setminus v'])\}.$$

then $dom(F)$ is the set of variables upon whose values $F(\underline{v})$ depends. In the case of our example, $dom(F_{t_3}) = \{x_1, x_3, x_4\}$. Again, we have

$$(\forall i \in dom(F) : (\underline{v}_1)_i = (\underline{v}_2)_i) \Rightarrow F(\underline{v}_1) = F(\underline{v}_2) \tag{23.14}$$

Finally, if F is the identity, then we define $range(F) = \emptyset$. Otherwise, we define

$$range(F) = \{x_i \mid \exists \underline{v} \in \mathbb{R}^n : F(\underline{v})_i \neq \underline{v}_i\} \tag{23.15}$$

then $range(F)$ is the set of variables whose values may be altered by execution of the assignment $\underline{x} \leftarrow F(\underline{x})$. In our example, $range(F_{t_3}) = \{x_4\}$.

Intuitively, two transitions t and t' may execute concurrently if

1• The underlying marked net permits it;
2• Execution of t does not alter the values of variables in the domains of $Pr_{t'}$ or $F_{t'}$ - likewise with their roles reversed;
3• F_t and $F_{t'}$ do not alter the values of the same variable - in particular, they cannot be equal unless they are both the identity function on $\mathbb{R}^n$.

23.4.1. DEFINITION. Suppose $t, t' \in T$. Define $t \leftrightarrow_{Prog} t' \Leftrightarrow t \neq t'$ and

1• $(dom(F_t) \cup dom(Pr_t)) \cap range(F_{t'}) = \emptyset$;
2• $(dom(F_{t'}) \cup dom(Pr_{t'})) \cap range(F_t) = \emptyset$;
3• $range(F_t) \cap range(F_{t'}) = \emptyset$.

For instance, in the case of the example of figure 23.1, we have $t_1 \leftrightarrow_{Prog} t_2$ but $t_1 \not\leftrightarrow_{Prog} t_3$ and $t_2 \not\leftrightarrow_{Prog} t_3$. Let us now define the independence relation.

$$a\, \iota_{Prog}\, b \Leftrightarrow a\, \iota_{N(Prog)}\, b \wedge \mu_{Prog}(a) \leftrightarrow_{Prog} \mu_{Prog}(b) \tag{23.16}$$

We would like to show that the structure

$$H_{Prog} = (Q_{Prog}, A_{Prog}, \rightarrow_{Prog}, \iota_{Prog}, T, \mu_{Prog})$$

is a hybrid transition system. The main problem is, as always, the 'lozenge rule'. This is our next goal. We begin with an observation.

$t_1 \leftrightarrow_{Prog} t_2$ is to be interpreted to mean that all else being equal, the assignments associated with the transitions t_1 and t_2 may take place concurrently. If our construction reflects our intuition - that the two assignments do not 'interfere' with each other - then it should not matter in which order the corresponding functions F_{t_1} and F_{t_2} are applied to any vector, if they were to be applied serially. That this is the case is shown by the next result.

23.4.2. LEMMA. $t_1 \leftrightarrow_{Prog} t_2 \Rightarrow \forall \underline{v} \in \mathbb{R}^n : F_{t_1}(F_{t_2}(\underline{v})) = F_{t_2}(F_{t_1}(\underline{v}))$.

PROOF. We shall show that if $t_1 \leftrightarrow_{Prog} t_2$ then

$$F_{t_1}(F_{t_2}(\underline{v}))_i = \begin{cases} F_{t_2}(\underline{v})_i \text{ if } i \in range(F_{t_2}) \\ F_{t_1}(\underline{v})_i \text{ if } i \in range(F_{t_1}) \\ \underline{v}_i \text{ otherewise} \end{cases} \tag{23.17}$$

Since $\leftrightarrow_{Prog}$ is clearly symmetric, we have $t_2 \leftrightarrow_{Prog} t_1$, and so by (23.17),

$$F_{t_2}(F_{t_1}(\underline{v}))_i = \begin{cases} F_{t_1}(\underline{v})_i \text{ if } i \in range(F_{t_1}) \\ F_{t_2}(\underline{v})_i \text{ if } i \in range(F_{t_2}) \\ \underline{v}_i \text{ otherewise} \end{cases}$$

so $F_{t_1}(F_{t_2}(\underline{v}))_i = F_{t_2}(F_{t_1}(\underline{v}))_i$, each i, that is, $F_{t_1}(F_{t_2}(\underline{v})) = F_{t_2}(F_{t_1}(\underline{v}))$.

It remains to prove (23.17). Let $\underline{v}_1 = F_{t_2}(\underline{v})$ and $\underline{v}_2 = F_{t_1}(\underline{v}_1)$. If $i \in range(F_{t_2})$, then $i \notin range(F_{t_1})$ by 23.4.1 (3), so $(\underline{v}_2)_i = (\underline{v}_1)_i$ by (23.15), that is $F_{t_1}(F_{t_2}(\underline{v}))_i = F_{t_2}(\underline{v})_i$. If $i \notin range(F_{t_1}) \cup range(F_{t_2})$ then $(\underline{v}_2)_i = (\underline{v}_1)_i = (\underline{v})_i$, by (23.15), that is $F_{t_1}(F_{t_2}(\underline{v}))_i = (\underline{v})_i$. Finally, suppose that $i \in range(F_{t_1})$. If $j \in dom(F_{t_1})$, then $j \notin dom(F_{t_2})$ by 23.4.1 (1) and so $(\underline{v}_1)_j = (\underline{v})_j$. By (23.14), $F_{t_1}(\underline{v}_1) = F_{t_1}(\underline{v})$ and hence $F_{t_1}(F_{t_2}(\underline{v}))_i = F_{t_1}(\underline{v})_i$.

■

We now show that guarded commands obey the lozenge rule.

23.4.3. LEMMA. Let $\underline{v},\underline{v}_1,\underline{v}_2 \in \mathbb{R}^n$ and let $t_1,t_2 \in T$ such that $t_1 \leftrightarrow_{Prog} t_2$, then:

$$\underline{v} \rightarrow_{Prog}^{t_1} \underline{v}_1 \rightarrow_{Prog}^{t_2} \underline{v}_2 \Rightarrow \exists \underline{v}'_1 \in \mathbb{R}^n : \underline{v} \rightarrow_{Prog}^{t_2} \underline{v}'_1 \rightarrow_{Prog}^{t_1} \underline{v}_2 .$$

PROOF. Consider $Pr_{t_2}(\underline{v})$. As $dom(Pr_{t_2}) \cap range(F_{t_1}) = \emptyset$, by 23.4.1 (2), $\underline{v}_i = (\underline{v}_1)_i$, all $i \in dom(Pr_{t_2})$, by (23.15). Therefore, $Pr_{t_2}(\underline{v}) = Pr_{t_2}(\underline{v}_1) = \textbf{true}$, so $\underline{v} \rightarrow_{Prog}^{t_2} \underline{v}'_1$, where $\underline{v}'_1 = F_{t_2}(\underline{v})$, by 23.2.5. Now, as $dom(Pr_{t_1}) \cap range(F_{t_2}) = \emptyset$, by 23.4.1 (1), for all $i \in dom(Pr_{t_1})$, $\underline{v}_i = (\underline{v}'_1)_i$. Hence, $Pr_{t_1}(\underline{v}'_1) = Pr_{t_1}(\underline{v}) = \textbf{true}$, and thus $\underline{v}'_1 \rightarrow^{t_1}{}_{Prog} \underline{v}'_2$, where $\underline{v}'_2 = F_{t_1}(\underline{v}'_1)$.

It remains to show that $\underline{v}'_2 = \underline{v}_2$, that is that $F_{t_1}(F_{t_2}(\underline{v})) = F_{t_2}(F_{t_1}(\underline{v}))$. But this has been established in 23.4.2.

■

Let us collect our results together.

23.4.4. PROPOSITION. H_{Prog} is a hybrid transition system.

PROOF. First, if $(\underline{q},\underline{v}) \rightarrow_{Prog}^{a} (\underline{q}_1,\underline{v}_1)$ and $(\underline{q},\underline{v}) \rightarrow_{Prog}^{a} (\underline{q}_2,\underline{v}_2)$, then by (23.10), $\underline{q} \rightarrow_{N(Prog)}^{a} \underline{q}_1$, $\underline{q} \rightarrow_{N(Prog)}^{a} \underline{q}_2$, $\underline{v} \rightarrow_{Prog}^{a} \underline{v}_1$ and $\underline{v} \rightarrow_{Prog}^{a} \underline{v}_2$. From the first two, we obtain $\underline{q}_1 = \underline{q}_2$, since $T(C(H_{N(Prog)}))$ is unambiguous, by 23.3.2, and from the latter two we obtain $\underline{v}_1 = F_t(\underline{v}) = \underline{v}_2$, where $\mu_{Prog}(a) = t$. Hence $(\underline{q}_1,\underline{v}_1) = (\underline{q}_2,\underline{v}_2)$, showing that $T(C(H_{Prog}))$ is unambiguous.

Next let $(\underline{q},\underline{v}) \rightarrow_{Prog}^{a} (\underline{q}_1,\underline{v}_1) \rightarrow_{Prog}^{b} (\underline{q}_2,\underline{v}_2)$, with $a\,\iota_{Prog}\,b$, then $\underline{q} \rightarrow_{N(Prog)}^{a} \underline{q}_1 \rightarrow_{N(Prog)}^{b} \underline{q}_2$ and $\underline{v} \rightarrow_{N(Prog)}^{a} \underline{v}_1 \rightarrow_{N(Prog)}^{b} \underline{v}_2$, with $a\,\iota_{N(Prog)}\,b$ and $a \leftrightarrow_{Prog} b$, by (23.10) and (23.16). As $C(H_{N(Prog)})$ is an asynchronous transition system, by 23.3.2, there exists $\underline{q}'_1 \in Q_{N(Prog)}$ such that $\underline{q} \rightarrow_{N(Prog)}^{b} \underline{q}'_1 \rightarrow_{N(Prog)}^{a} \underline{q}_2$ and by 23.4.3, there exists $\underline{v}'_1 \in \mathbb{R}^n$ such that $\underline{v} \rightarrow_{Prog}^{b} \underline{v}'_1 \rightarrow_{Prog}^{a} \underline{v}_2$. By (23.10), $(\underline{q},\underline{v}) \rightarrow_{Prog}^{b} (\underline{q}'_1,\underline{v}'_1) \rightarrow_{Prog}^{a} (\underline{q}_2,\underline{v}_2)$. Hence the 'lozenge rule' holds in $C(H_{Prog})$ and we are done.

■

23.4.5. DEFINITION. Let $Prog$ be a parallel program with initial configuration $(M_0,\underline{v}_0)$, then we define

$$\rho[\![Prog,(M_0,\underline{v}_0)]\!] = (H_{Prog},(M_0,\underline{v}_0)).$$

23.5. Comparison of the Semantics.

It is reasonable to ask, what is the connection between the non-interleaving semantics we have so laboriously constructed, and that of Keller?

Let us look at the question more generally. Suppose we have a descriptive formalism giving rise to a class of objects $\mathbf{O}$, which has a transition system semantics in the form of a map $\sigma : \mathbf{O} \to \mathbf{TS}^*$. Suppose also that we provide it with a semantics in terms of initialised hybrid transition systems via a map $\rho : \mathbf{O} \to \mathbf{HTS}^*$. We would like to be able to justify our semantics somehow. We shall assume that we are talking only about asynchronous systems - as these are the only kind of systems that we consider in what immediately follows. The manner of our justification will be as follows:

1• $\sigma[\![\mathbf{o}]\!]$ is an interleaved version of $\rho[\![\mathbf{o}]\!]$, as we shall explain;

2• The additional concurrency structure possessed by $\rho[\![\mathbf{o}]\!]$ is consistent with that suggested informally in the descriptive formalism.

(2) must involve the waving of hands and allusions to persuasive syntax. For the former, we introduce a definition.

23.5.1. DEFINITION. Let H be an asynchronous hybrid transition system. The *interleaving* of H is the transition system $T_H = (Q, E, \to_H)$, where

$$q_1 \to_H^e q_2 \Leftrightarrow \exists a \in A : q_1 \to^a q_2 \wedge \mu(a) = e.$$

We may now formalise (1) above as a requirement that if $\rho[\![\mathbf{o}]\!] = (H, q)$, then $\sigma[\![\mathbf{o}]\!] \sim (T_H, q)$. We now show that this holds in the case of Keller parallel programs.

23.5.2. PROPOSITION. Let $Prog$ be a parallel program with initial configuration $(M_0, \underline{v}_0)$ and $(H, q) = \rho[\![(Prog, (M_0, \underline{v}_0))]\!]$, then

$$(T_{Prog}, (M_0, \underline{v}_0)) \sim (T_H, q).$$

PROOF. It is simple to check that the relation $R \subseteq Conf_{Prog} \times Q$ given by

$$(M, \underline{v})\, R\, (\underline{q}, \underline{v}') \Leftrightarrow \underline{v} = \underline{v}' \wedge \forall s \in S : \sum_{t \in T \cup \{\bot\}} (q(s))(\underline{t}) = M(s)$$

is a strong bisimulation.

■

Notes.

In [69], Goltz and Reisig provide a process-based net semantics for Petri Nets in which it is possible for several instances of the same transition to fire concurrently. In this respect, they differ from our approach, in which we obviously don't want two concurrent evaluations of the same function. They also differ in an important respect. Their semantics describes the behaviour of nets in terms of individual processes, rather than of a single object which contains them all, like a behavioural presentation or an occurrence net in the sense of [20]. This means that their approach would be unsuitable in situations where one is interested in relationships between distinct behaviours. For example, we can define a system to be deadlock-free if for each $p \in P$, there exists $p' \in P$ such that $p \subset p'$, and we need some kind of notion of strict prefix relation between behaviours. However, their paper describes some interesting connections between safeness (no place may acquire more than some fixed number of tokens) and the Net Theoretic property of K-density. (See, for example [29]).

The constructions of section 23.3 effectively turn the underlying net of a parallel program into a coloured Petri Net [61], in which tokens have different 'colours' and transitions require a certain number of tokens of each 'colour' in order to fire.

Chapter 24
From Path Expressions to TCSP

> The COSY notation ... is based on the recognition that human knowledge is expressed in language, and that knowledge about dynamic systems can be effectively expressed in terms of the behaviours of automata interacting with each other and their environment which model the relevant aspects of the systems.
>
> R. JANICKI AND P. E. LAUER [66]

24.1. Introduction.

Path expressions were originally introduced by R. H. Campbell and A. N. Habermann [67] as a synchronisation mechanism for programs consisting of parallel processes synchronising on shared data objects. The idea is that, to avoid problems arising from attempts to concurrently access shared data objects, such objects only allow access to their data via certain externally available procedures or operations. Paths determine restrictions on the order in which such operations may be applied. The concept is evidently an extension of C. A. R. Hoare's monitors [68]. and thereby is associated with one of the two major paradigms of parallel computation, that whereby process synchronisation is achieved via properly protected shared variables. This is in contrast with a second paradigm; synchronisation via communication, as exemplified by CCS and CSP. Ironically, we shall see later that path/process programs may be syntactically transformed into TCSP programs.

A path expression specifies a further constraint on the order in which these operations may be applied. For example, a buffer frame with the two operations *write* (write into the frame) and *read* (read from the frame) may be augmented by the path expression

path *write; read* **end**.

This has the effect of forcing writes and reads to take place in alternation, beginning with a write - thereby ensuring that the contents of the frame are never overwritten before they have been read. A data object might contain several path expressions; their constraints are to apply together.

The presence of several paths in the same object entails the possibility of deadlock. For example, an object augmented by the paths

path *a; b* **end** **path** *b; a* **end**

would never allow operations *a* and *b* to take place at all; the first occurrence of either would be allowed by one path but disallowed by the other. Deadlocks may arise in all sorts of ways - the above example is the simplest possible - and hence there is a need to determine at some stage whether a given program has such properties.

In order to make possible the formal analysis of systems of processes synchronising on shared data objects augmented by path expressions, P. E. Lauer and R. H. Campbell devised an abstract notation [15] in which data objects are represented solely by collections of paths and processes are represented by regular expressions, defining the order in which the processes would make use of the operations provided by the data objects. This permits a description of the system purely from the point of view of its synchronisation structure. They then equipped this notation with a formal semantics in terms of marked, labelled Petri Nets; at the time, Net Theory had the richest theory of deadlock in concurrent systems and Lauer and Campbell wished to exploit this fact.

In the following years, the path-process notation gradually emerged as a specification notation in its own right. With the additional carapace of a macro notation for the definition of repetitive constructs, the path/process language became absorbed into the COSY (COncurrent SYstems) notation [33].

We shall concentrate on the path/process notation without macros.[1] First, we give the syntax of the notation and its semantics in terms of marked, labelled nets. We observe that such nets are associated with certain hybrid transition systems having equivalent behaviour. We then note that the hybrid transition systems may be constructed directly from transition systems associated with each path and process, using two combinators. This allows us to reformulate the syntax of the notation using these combinators to express the way in which the component paths/processes (renamed 'flow expressions') are to be combined - in the standard notation the way in which sequential subsystems are combined is indicated by their titles; paths are combined together differently from processes. The notation is again reformulated so that the flow expressions are defined by systems of

[1]COSY programs written in the macro notation translate to path/process programs without macros, so we need not consider macros as far as the semantics of the notation is concerned.

equations. We now have something which looks very like TCSP [2] and we suggest that, firstly, the interleaving semantics of this sublanguage coincides with that of TCSP and, secondly, that the additional structure in our semantics to express concurrency is consistent with the amount of concurrency suggested by the syntax of TCSP.

Again, we have used our general semantic model to compare two formalisms. One conclusion we may draw from this exercise is that the path/process notation occupies a very interesting place among the family of concurrency theories, in that it may be seen to form a bridge between Net Theory and Process Algebra.

24.2. Syntax and Marked, Labelled Net Semantics.

We begin by giving the syntax of the path/process notation.

24.2.1. DEFINITION. A *path/process program* is a terminal string generated from the non-terminal < *program* > via the following rewrite rules.

< *program* >::= **begin** [< *process* > | < *path* >]$^+$ **end**
< *process* >::= **process** < *sequence* > **end**
< *path* >::= **path** < *sequence* > **end**
< *sequence* >::= [< *orelement* >;] * | < *orelement* >
< *orelement* >::= [< *element* >,] * | < *element* >
< *element* >::= < *event* > | (< *sequence* >)

Terms between angled brackets represent non-terminal symbols. | indicates syntactic alternatives. $[x]^+$ indicates one or more repetitions of x and $[x]^*$ indicates zero or more repetitions of x. We assume a set of event names E and rules which allows one to rewrite < *event* > by elements of E.

As is usual in such cases, the plethora of parentheses is a notational embarrassment and so we establish a binding precedence on the combinators as follows: ',' binds more strongly than ';'.

The intuition behind the notation is as follows. Each process describes a sequential agent which is capable of performing the actions named in it in the order suggested by the regular expression it contains. For example

process $a; b, c$ **end**

must begin by performing a, after which - ';' is a sequencing operator - it has the choice - ',' is a mutual exclusion operator - of performing b or of performing c. After this, it starts from the beginning again with a - the process loops forever. On the other hand

process $(a;b),c$ **end**

represents a process which either performs a followed by b or performs c, and may do either repeatedly.

Several processes together operate independently of each other. Thus, two processes may perform operations with the same names at the same time. Synchronisation is provided by the path expressions. The path expression

path $a;b$ **end**

says that the action a and the action b must always occur in that order iteratively and that there can only be one execution of either at a time. For example, the path program

begin
  process $P;$ *crit sec* 1; V **end**
  process $P;$ *crit sec* 2; V **end** (24.1)
  path $P;V$ **end**
end

enforces mutual exclusion between the actions *crit sec* 1 and *crit sec* 2 via a binary semaphore. Only one process may initially perform the P action and the other process cannot progress until the path allows another P action to take place - and a V action must proceed it, which may only be performed by the process which has just performed the P action.

The semantics of the path/process notation is given in terms of marked, labelled nets.

24.2.2. DEFINITION. A *labelled net* is a tuple $LN=(N,E,\lambda)$, where $N=(S,T;F)$ is a net, in the sense of 16.2.1, E is a set of events and $\lambda:T\rightarrow E$ is a labelling function.

A *marked, labelled net* is a tuple $MN=(N,E,\lambda,c)$, where $LN(MN)=(N,E,\lambda)$ is a labelled net and $c\subseteq S$.

We shall use LN (respectively, MN) for labelled nets (respectively marked, labelled nets) with the usual subscript conventions. Thus, MN_i is a marked labelled net $((S_i,T_i;F_i),E_i,\lambda_i,c_i)$.

If the net part of MN is contact-free, then it may be given a process semantics as in chapter 16. A *process* of MN from c is a pair $(N',\lambda\circ\lambda')$, where (N',λ') is a process of N from c. That is, a process of MN from c displays the order in which transitions representing instances of elements of E have occurred. It is possible to show that the marked, labelled net arising from a path/process program is contact-free provising each path and process in the program contains at least one semicolon, and hence the sublanguage of such path/process programs may be given a process-net-based or occurrence-net-based semantics.

The marked, labelled net determined by a path/process program is constructed using structural recursion on the syntax of programs. We shall present it pictorially, after the manner of [46], which involves a slight modification to the semantics given in [15].

We begin by associating each path or process with a net. With **path** *s* **end** or **process** *s* **end**, we associate a net as pictured in figure 24.1.

Figure 24.1.

Nets produced during the construction will be labelled by terms made up of events and the terminal symbols. Each term has a *type*, in the sense that it parses as deriving from one of the non-terminals <*sequence*>, <*orelement*>, <*element*> and <*event*>.

A term of type <*sequence*>, will either be of the form $o_1;\cdots;o_n$, where each o_i is of type <*orelement*>, or will be of the form o, where o is of type <*orelement*>. In the former case, figure 24.2 depicts the rewrite rule for a T element labelled by a term of type <*sequence*>.[2]

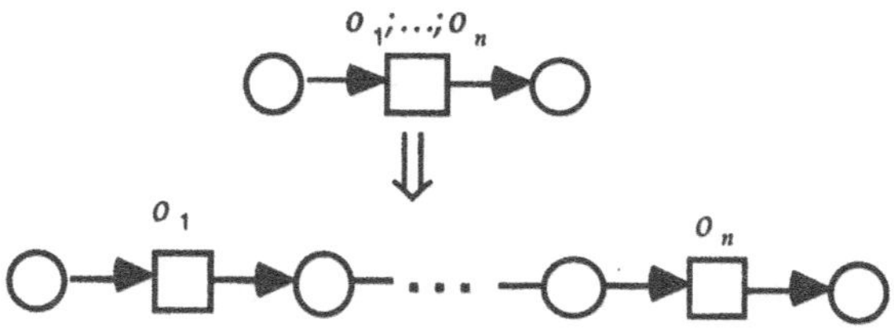

Figure 24.2.

A term of type <*orelement*>, will either be of the form $e_1,\cdots,e_n$, where each e_i is of type <*element*>, or will be of the form e, where $e \in E$. In the former case, figure 24.3, depicts the rewrite rule for a T element labelled by a term of type <*orelement*>.

A term of type <*element*>, will either be of the form (<*sequence*>), or will be of the form e, where $e \in E$. In the former case, figure 24.4 depicts the appropriate rewrite rule.

[2]In this and the following rules, all tokens are preserved; that is, if there is a token in the input place to the T element on the left hand side of the derivation arrow, then there will be an input place to the T element labelled o_1.

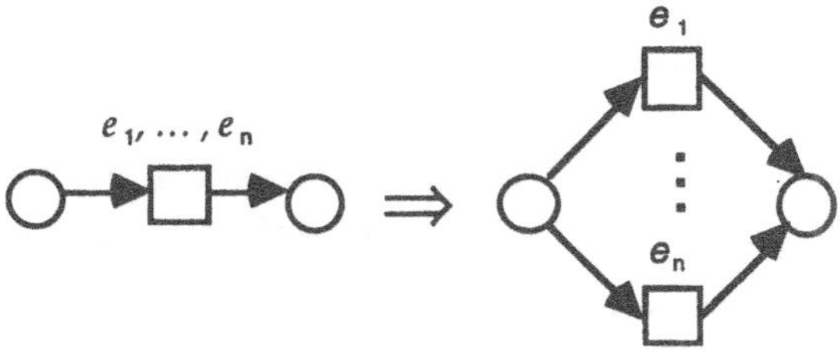

Figure 24.3.

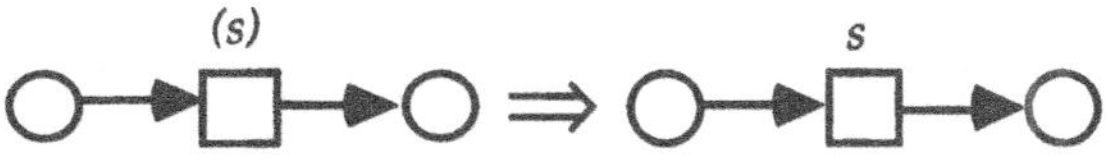

Figure 24.4.

These rules are applied until it is no longer possible to do so. This can only be the case if each T element is labelled by e, some $e \in E$. The resulting net will have exactly one place marked and each transition will be labelled by some element of E. It is thus a marked labelled net. Denote by $\sigma[\![P]\!]$, the marked labelled net associated with the path/process P.[3]

Now we compose the paths and processes together. This involves two combinators, one for combining nets deriving from processes, which we denote by '|||' and one for composing the combined process nets and the path nets together, which is a form of the '||' operator of chapter 21 and will be given the same symbol.

24.2.3. DEFINITION. Let $(LN_i)_{i \in I}$ be an indexed family of labelled nets, where $LN_i = (N_i, E_i, \lambda_i)$, each $i \in I$. Define $\underset{i \in I}{||} LN_i$ to be the tuple $LN = (N, E, \lambda)$, where $N = (S, T; F)$ and

1• $S = \bigcup_{i \in I} (S_i \times \{i\})$;

2• $E = \bigcup_{i \in I} E_i$;

3• T is the set of all functions $\underline{t}: I \to \bigcup_{i \in I} T_i \cup \{\Omega\}$ such that

(a) $\forall i \in I: \underline{t}(i) \in T_i \cup \{\Omega\}$;

(b) $\exists i \in I: \underline{t}(i) \neq \Omega$;

(c) $\exists e \in E\, \forall \in I: (e \in E_i \Rightarrow \underline{t}(i) \neq \Omega \wedge \lambda_i(\underline{t}(i)) = e)$;

[3]Strictly speaking, $\sigma[\![P]\!]$ represents a class of somorphic marked, labelled nets, as we have not explicitly defined sets of S and T elements.

4• $\underline{t}\, F\,(s,i) \Leftrightarrow \underline{t}(i)\, F_i\, s$ and $(s,i)\, F\, \underline{t} \Leftrightarrow s\, F_i\, \underline{t}(i)$;

5• $\lambda(\underline{t}) = e \Leftrightarrow \exists i \in I : \underline{t}(i) \neq \Omega \wedge \lambda_i(\underline{t}(i)) = e$.

If $(MN_i)_{i \in I}$ is an indexed family of marked, labelled nets, where $(MN_i) = (N_i, E_i, \lambda_i, c_i)$, each $i \in I$, then we define $\underset{i \in I}{\|} MN_i$ to be the tuple $MN = (N, E, \lambda, c)$, where $(N, E, \lambda) = \underset{i \in I}{\|} LM(MN_i)$ and

6• $c = \bigcup_{i \in I} c_i \times \{i\}$.

If $I = \{1, \cdots, n\}$, then we shall use an infix notation

$$XN_1 \| \cdots \| XN_n = \underset{i \in I}{\|} XN_i$$

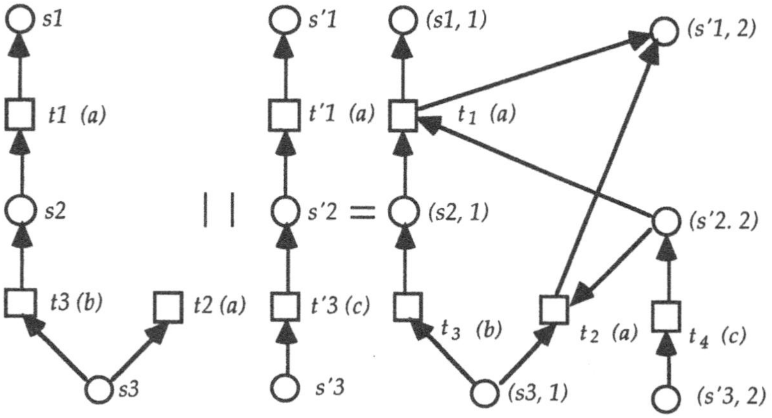

Figure 24.5.

Pictorially, N is constructed as follows. First, make disjoint copies of the sets S_i. Next, for each $e \in E$, let $J_e \in \{i \in I \mid e \in \lambda_i(T_i)\}$. For each selection of elements t_j, one from each T_j such that $j \in J_e$ and $\lambda_j(t_j) = e$, we construct a transition labelled by e whose inputs (respectively outputs) are those S elements (s, j), $j \in J_e$, such that $s\, F_j\, t_j$ (respectively $t_j\, F_j\, s$). The transition is technically a vector whose j component is t_j if $j \in J_e$ and Ω otherwise. The construction is illustrated in figure 24.5.

In the figure, $t_1 = (t1, t'1)$, $t_2 = (t2, t'1)$, $t_3 = (t3, \Omega)$ and $.t_4 = (\Omega, t'3)$

The effect is to force the component marked nets to synchronise T elements from the component nets with the same label. The process combinator, on the other hand, produces a system in which each process object runs independently of its fellows.

24.2.4. DEFINITION. Let $(LN_i)_{i\in I}$ be an indexed family of labelled nets, where $LN_i = (N_i, E_i, \lambda_i)$, each $i \in I$. Define $\mathop{|||}_{i\in I} LN_i$ to be the tuple $LN = (N, E, \lambda)$, where $N = (S, T; F)$ and

1• $S = \bigcup_{i\in I}(S_i \times \{i\});$

2• $E = \bigcup_{i\in I} E_i;$

3• $T = \bigcup_{i\in I}(T_i \times \{i\});$

4• $(x,i)\, F\, (y,j) \Leftrightarrow i = j \wedge x\, F_i\, y$

5• $\lambda(t,i) = \lambda_i(t);$

If $(MN_i)_{i\in I}$ is an indexed family of marked, labelled nets, where $(MN_i) = (N_i, E_i, \lambda_i, c_i)$, each $i \in I$, then we define $\mathop{|||}_{i\in I} MN_i$ to be the tuple $MN = (N, E, \lambda, c)$, where $(N, E, \lambda) = \mathop{|||}_{i\in I} LM(MN_i)$ and

6• $c = \bigcup_{i\in I} c_i \times \{i\}.$

If $I = \{1, \cdots, n\}$, then we shall use an infix notation

$$XN_1 \mathbin{|||} \cdots\cdots \mathbin{|||} XN_n = \mathop{|||}_{i\in I} XN_i$$

The ||| operator simply constructs a net consisting of disjoint copies of the MN_i.

We may now give the net semantics for the path/process notation.

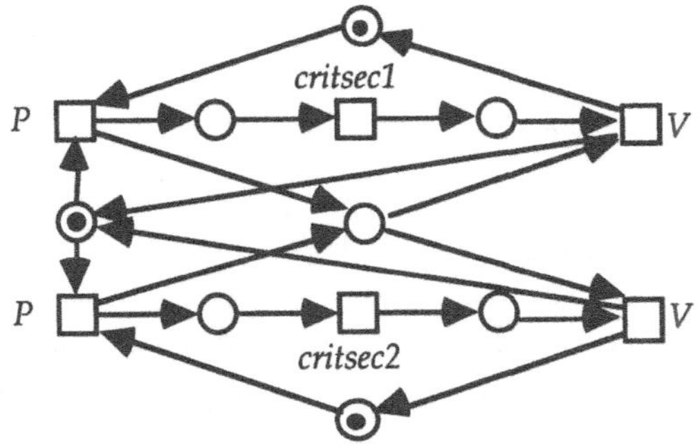

Figure 24.6.

24.2.5. DEFINITION. Let $P =$ **begin** $P_1 \cdots P_n$ **end**. Define

$$I_{path}(P) = \{i \mid P_i \text{ is a path}\}$$

$I_{proc}(P) = \{i \mid P_i \text{ is a process}\}$

The marked, labelled net of P, denoted by $\sigma[\![P]\!]$ is given by

$$\sigma[\![P]\!] = (\mathop{||}_{i \in I_{path}(P)} \sigma[\![P_i]\!]) \,||\, (\mathop{|||}_{i \in I_{proc}(P)} \sigma[\![P_i]\!])$$

Figure 24.6 shows $\sigma[\![P]\!]$ for the program (24.1).

24.3. Hybrid Transition System Semantics.

We shall now equip the path/process notation with a semantics in terms of initialised hybrid transition systems. This comes in three stages:

(H1) Construction of an initialised hybrid transition system for an individual path or process;
(H2) Combining hybrid transition systems together with a version of the $||$ operator - this deals with hybrid transition systems arising from paths and their combination with that arising from processes in (H3);
(H3) Combining hybrid transition systems together with a version of the $|||$ operator - this deals with hybrid transition systems arising from processes.

For (H1), we take the soft option. If P is a path or process then recall that $\sigma[\![P]\!]$ denotes marked labelled net determined from P by the net semantics of section 24.1

24.3.1. DEFINITION. Let P be a path or process and suppose that $\sigma[\![P]\!] = ((S,T;F),E,\lambda,c)$. Define $\rho[\![P]\!] = (S,T,\rightarrow,\varnothing,E,\mu),c)$ where

1• $s \rightarrow^t s' \Leftrightarrow s\,F\,t \wedge t\,F\,s'$;

2• $(\mu(t))(e) = \begin{cases} 1 \text{ if } \lambda(t) = e \\ 0 \text{ otherwise} \end{cases}$

$\rho[\![P]\!]$ is an asynchronous, hybrid transition system with initial state. INdeed, the lozenge rule holds trivially, while unambiguity follows from the fact (an invariant under the construction of section 24.2), that $\rho[\![P]\!]$ is a state machine net, that is, each transition has at most one input and at most one output place. Hence, if $s \rightarrow^t s_1$ and $s \rightarrow^t s_2$, then $t\,F\,s_1$ and $t\,F\,s_2$ and therefore $s_1 = s_2$.

Next, we define the effect of the $||$ operator as promised in (H2).

24.3.2. DEFINITION. Let $(H_i)_{i\in I}$ be an indexed family of hybrid transition systems. We define $\|_{i\in I} H_i$ to be a tuple $(Q, A, \rightarrow, \iota, E, \mu)$, where

1• Q is the set of all functions $\underline{q}: I \rightarrow \bigcup_{i\in I} Q_i$ such that $\underline{q}(i) \in Q_i$, each i

2• $E = \bigcup_{i\in I} E_i$;

3• A is the set of all functions $\underline{a}: I \rightarrow \bigcup_{i\in I} A_i \cup \{\Omega\}$ such that

(a) $\forall i \in I: \underline{a}(i) \in A_i \cup \{\Omega\}$;

(b) $\exists i \in I: \underline{a}(i) \neq \Omega$;

(c) $\exists \mathit{b} \in \mathcal{B}(E)\ \forall i \in I: \mathit{b} \in \mu_i(A_i) \Rightarrow \mu_i(\underline{a}(i)) = \mathit{b}$;

4• $$\underline{q}_1 \rightarrow^{\underline{a}} \underline{q}_2 \Leftrightarrow \forall i \in I: (\underline{a}(i) \neq \Omega \Rightarrow \underline{q}_1(i) \rightarrow_i^{\underline{a}(i)} \underline{q}_2(i)) \wedge (\underline{a}(i) = \Omega \Rightarrow \underline{q}_1(i) = \underline{q}_2(i));$$

5• $\underline{a}\ \iota\ \underline{b} \Leftrightarrow \forall i \in I: \underline{a}(i), \underline{b}(i) \in A_i \Rightarrow \underline{a}(i)\ \iota_i\ \underline{b}(i)$;

6• $\mu(\underline{a}) = \mathit{b} \Leftrightarrow \exists i \in I: \underline{a}(i) \neq \Omega \wedge \mu_i(\underline{a}(i)) = \mathit{b}$;

If $(H_i, q_i)_{i\in I}$ is an indexed family of hybrid transition systems with initial state, then we define $\|_{i\in I} (H_i, q_i) = (\|_{i\in I} H_i, \underline{q})$ where

7• $\forall i \in I: \underline{q}(i) = q_i$.

24.3.2 is rather long, so a few words about it are in order. It is perhaps worth comparing it with the definition of parallel composition of asynchronous transition systems given in 20.3.9. The states are the same in each case; a state of $\|_{i\in I} H_i$ is a vector showing the current 'local states' of the constituent transition systems. Transitions are also vectors of transitions, showing which transitions of the constituent transition systems are combined. The difference is that for asynchronous transition systems, transitions are merged when they have the same name, whereas in the hybrid case transitions are merged if they have the same label under μ.

We also note that the construction of 24.3.2 generalises that of 20.3.9 in the following sense.

Recall that we may regard asynchronous transition systems as special cases of hybrid transition systems.[4] If $C = (Q, A, \rightarrow, \iota)$, is an asynchronous transition system then $\gamma(C) = (Q, A, \rightarrow, \iota, A, \mu)$ is a hybrid transition system, where μ is the canonical embedding of A into $\mathcal{B}(A)$. It is straightforward to prove that if $(C_i, q_i)_{i\in I}$ is an indexed family of asynchronous transition systems with initial state, then

---

[4] See the discussion preceding 22.3.5.

$$\underset{i\in I}{||}(\gamma(C_i),q_i)=\gamma(\underset{i\in I}{||}(C_i,q_i))$$

Of course, before making these remarks we should ascertain that || does what it's supposed to do. Here is the result.

24.3.3. LEMMA. If $(H_i,q_i)_{i\in I}$ is a family of initialised hybrid transition systems then $\underset{i\in I}{||}(H_i,q_i)$ is an initialised hybrid transition system.

PROOF. Let $(H,q)=\underset{i\in I}{||}(H_i,q_i)$.

First, suppose that $\underline{q}\rightarrow^{\underline{a}}\underline{q}_1$ and $\underline{q}\rightarrow^{\underline{a}}\underline{q}_2$. If $\underline{a}(i)=\Omega$, then by 24.3.2 (4), $\underline{q}_1(i)=\underline{q}(i)=\underline{q}_2(i)$. On the other hand, if $\underline{a}(i)\neq\Omega$, then $\underline{q}(i)\rightarrow_i^{\underline{a}(i)}\underline{q}_1(i)$ and $\underline{q}(i)\rightarrow_i^{\underline{a}(i)}\underline{q}_2(i)$, by 24.3.2 (4). As each $T(C(H_i))$ is unambiguous, $\underline{q}_1(i)=\underline{q}_2(i)$. Thus $\underline{q}_1=\underline{q}_2$ and hence $T(C(H))$ is unambiguous.

Next, suppose that $\underline{q}_1\rightarrow^{\underline{a}}\underline{q}_2$ and $\underline{q}_2\rightarrow^{\underline{b}}\underline{q}_3$, with $\underline{a}\,\iota\,\underline{b}$. If $\underline{a}(i),\underline{b}(i)\neq\Omega$, then $\underline{q}_1(i)\rightarrow_i^{\underline{a}(i)}\underline{q}_2(i)\rightarrow_i^{\underline{b}(i)}\underline{q}_3(i)$ and $\underline{a}(i)\,\iota_i\,\underline{b}(i)$, by 24.3.2 (4) and (5), and so there exists $q_i'\in Q_i$ such that $\underline{q}_1(i)\rightarrow_i^{\underline{b}(i)}q_2'\rightarrow_i^{\underline{a}(i)}\underline{q}_3(i)$. Define

$$\underline{q}'_2(i)=\begin{cases}q_2' \text{ if } \underline{a}(i),\underline{b}(i)\neq\Omega\\ \underline{q}_1(i) \text{ if } \underline{b}(i)=\Omega\\ \underline{q}_3(i) \text{ if } \underline{a}(i)=\Omega\end{cases}$$

Note that if $\underline{a}(i)=\Omega$ and $\underline{b}(i)=\Omega$, then $\underline{q}_1(i)=\underline{q}_2(i)=\underline{q}_3(i)$, by 24.3.2 (4), and so $\underline{q}'_2$ is well defined.

If $\underline{a}(i)\neq\Omega$, then either $\underline{b}(i)\neq\Omega$, so that $\underline{q}'_2(i)=q_2'\rightarrow_i^{\underline{a}(i)}\underline{q}_3(i)$ or $\underline{b}(i)=\Omega$, so that $\underline{q}'_2(i)=\underline{q}_1(i)\rightarrow_i^{\underline{a}(i)}\underline{q}_2(i)=\underline{q}_3(i)$. If $\underline{a}(i)=\Omega$, then $\underline{q}'_2(i)=\underline{q}_3(i)$. We have shown that $\underline{q}'_2\rightarrow^{\underline{a}}\underline{q}_3$

If $\underline{b}(i)\neq\Omega$, then either $\underline{a}(i)\neq\Omega$, so that $\underline{q}_1(i)\rightarrow_i^{\underline{b}(i)}q_2'=\underline{q}'_2(i)$ or $\underline{a}(i)=\Omega$, so that $\underline{q}_1(i)=\underline{q}_2(i)\rightarrow_i^{\underline{b}(i)}\underline{q}_3(i)=\underline{q}'_2(i)$. If $\underline{b}(i)=\Omega$, then $\underline{q}'_2(i)=\underline{q}_1(i)$. We have shown that $\underline{q}_1\rightarrow^{\underline{b}}\underline{q}'_2$.

Thus, $C(H)$ is an ι-asynchronous transition system, and we are done. ■

24.3.4. DEFINITION. Let $(H_i)_{i\in I}$ be an indexed family of hybrid transition systems. We define $\underset{i\in I}{|||}H_i$ to be a tuple $(Q,A,\rightarrow,\iota,E,\mu)$, where

1• Q is the set of all functions $\underline{q}: I \to \bigcup_{i \in I} Q_i$ such that $\underline{q}(i) \in Q_i$, each i

2• $E = \bigcup_{i \in I} E_i$;

3• $A = \bigcup_{i \in I} (A_i \times \{i\})$;

4• $\underline{q}_1 \to^{(a,i)} \underline{q}_2 \Leftrightarrow (\underline{q}_1(i) \to_i^a \underline{q}_2(i)) \wedge (j \neq i \Rightarrow \underline{q}_1(i) = \underline{q}_2(i))$;

5• $(a,i) \iota (b,j) \Leftrightarrow (i \neq j) \vee (i = j \wedge a \iota_i b)$;

6• $\mu(a,i) = \mu_i(a)$;

If $(H_i, q_i)_{i \in I}$ is a family of initialised hybrid transition systems, then we define $|||_{i \in I} (H_i, q_i) = (|||_{i \in I} H_i, \underline{q})$, where

7• $\forall i \in I: \underline{q}(i) = q_i$.

It is useful to compare 24.3.4 with 24.3.2. Again, states are vectors of 'local states', but this time, transitions are local to constituent transition systems. The transition (a,i) represents the ith transition system performing an a transition, independently of anything that the other transition systems are doing. Thus, in (4), only the ith co-ordinate of the state vector changes as a consequence of the execution of i. Transitions arising from distinct constituents are always reciprocally independent.

The proof of the following lemma is very similar to that of 24.3.3 and is omitted.

24.3.5. LEMMA. If $(H_i, q_i)_{i \in I}$ is a family of hybrid transition systems with initial state, then $|||_{i \in I} (H_i, q_i)$ is a hybrid transition system with initial state. ■

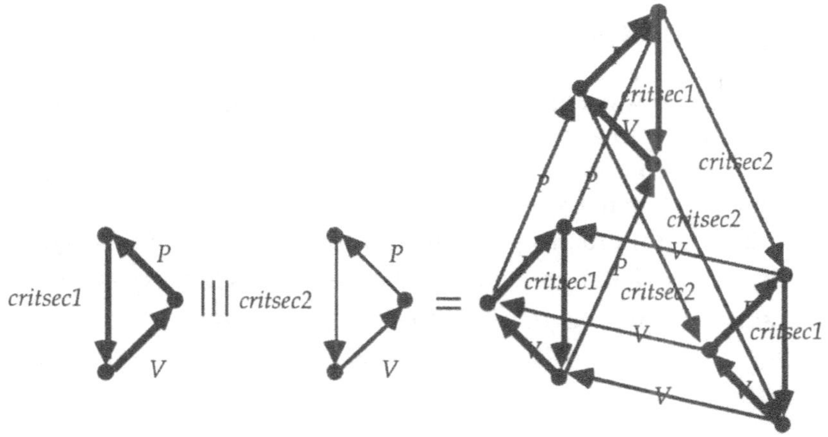

Figure 24.7

Combining the semantics for individual paths with the two combinators, we arrive at a hybrid transition system semantics.

24.3.6. DEFINITION. If $P = \textbf{begin}\ P_1 \cdots P_n\ \textbf{end}$ then define

$$\rho[\![P]\!] = (\mathop{||}_{i \in I_{path}(P)} \rho[\![P_i]\!]) \,||\, (\mathop{|||}_{i \in I_{process}(P)} \rho[\![P_i]\!])$$

where $I_{path}(P)$ and $I_{process}(P)$ are defined as in 24.2.5.

We illustrate the construction in figure 24.7, which shows the combination of the transition systems associated with the two processes of example (24.1), and figure 24.8, which shows the result of adjoining the path of that program.

The independence relation satisfies $(x,i)\,\iota\,(y,j) \Leftrightarrow i \neq j$

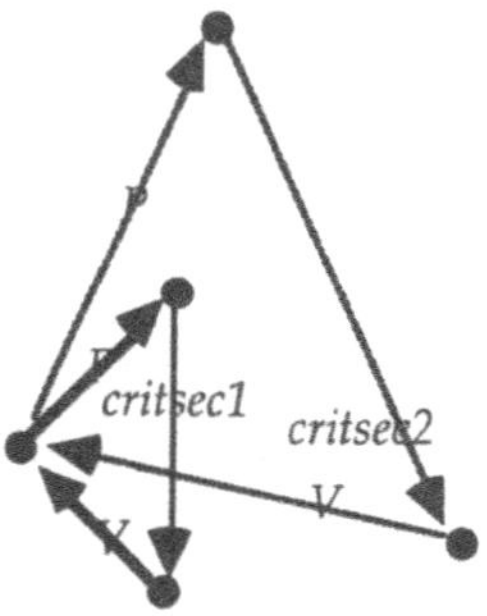

Figure 24.8.

24.4. Comparison of the Semantics.

We now have two ways of assigning a meaning to a path/process program P. The function σ maps P to a marked labelled net and the function ρ maps P to a hybrid transition system with initial state. We now consider the relationship between these two semantics.

Let P be a path/process program and let $(N,E,\lambda,c) = MN = \sigma[\![P]\!]$. By (21.1), the net N determines an asynchronous transition system C_N with initial state c, where $C_N = (\wp(S),T,\rightarrow,\iota_N)$[5] and $c_1 \rightarrow' c_2 \Leftrightarrow c_1\,[t\rangle\,c_2$. Regarding λ as a map, $\lambda{:}T \rightarrow \mathcal{B}(E)$,we obtain a hybrid transition system, $H_{MN} = (\wp(S),T,\rightarrow,\iota_N,E,\lambda)$. Let $H_{MN}{}^* = (H_{MN},c)$.

---

[5]See 17.2.1.

We shall examine the relationship between $H_{MN}{}^*$ and $\rho[\![P]\!]$.

Recall from chapter 21 the notion of an embedding morphism between asynchronous transition systems (21.2.1) and that if $f{:}C_1 \to C_2$ is an embedding morphism then for every $q \in Q_1$, $L(C_1,q) = L(C_2,f_Q(q))$ (21.2.2). We shall extend this notion to hybrid transition systems.

24.4.1. DEFINITION. Suppose that H_1 and H_2 are hybrid transition systems. An embedding $f{:}C(H_1) \to C(H_2)$ is an embedding from H_1 to H_2 $\Leftrightarrow$ $\mu_1(a) = \mu_2(a)$, all $a \in A_1$.

If $H_1{}^* = (H_1,q_1)$ and $H_2{}^* = (H_2,q_2)$ then f is an embedding from H_1^* to $H_2{}^*$ $\Leftrightarrow f$ is an embedding from H_1 to H_2 such that $f_Q(q_1) = q_2$.

Write $f{:}H_1^* \Rightarrow H_2^*$ to indicate that f is an embedding from H_1^* to H_2^* .

If $f{:}H_1^* \Rightarrow H_2^*$, then by 21.2.2, $L(C(H_1),q_1) = L(C(H_2),q_2)$ and since $\mu_1(a) = \mu_2(a)$, all $a \in A_1$, it follows that $B_{H_1^*} = B_{H_2^*}$. and so we have (1) of the following result. (2) follows from (1) and 20.2.19.

24.4.2. PROPOSITION. Suppose that $f{:}H_1^* \Rightarrow H_2^*$, then

1• $B_{H_1^*} = B_{H_2^*}$;

2• $H_1^* \equiv H_2^*$.

■

Let us first consider the case of an individual path or process, P. By construction, $\rho[\![P]\!] = (S,T,\to,\varnothing,E,\lambda),c)$ where $\sigma[\![P]\!] = ((S,T;F),E,\lambda,c)$ and $s_1 \to' s_2 \Leftrightarrow s_1\ F\ t \wedge t\ F\ s_2$. By definition $H_{\sigma[\![P]\!]}{}^* = ((\wp(S),T,\to,E,\lambda),c)$ in such a way that $s_1 \to' s_2$ in $\sigma[\![P]\!] \Leftrightarrow \{s_1\} \to' \{s_2\}$ in $H_{\sigma[\![P]\!]}{}^*$. It is now clear that the pair $h_P = ((h_P)_Q,(h_P)_A)$ defined by

$$(h_P)_Q(s) = \{s\} \text{ and } (h_P)_A(t) = t$$

is an embedding $h_P{:}\rho[\![P]\!] \Rightarrow H_{\sigma[\![P]\!]}{}^*$. We next show that embeddings are preserved by composition.

24.4.3. LEMMA. Suppose that $(H_i^*)_{i\in I}$ and $(\tilde{H}_i^*)_{i\in I}$ are families of initialised hybrid transition systems and suppose that for each $i \in I$,

i• There is an embedding $f_i{:}H_i^* \Rightarrow \tilde{H}_i^*$;

ii• $E_i = \tilde{E}_i$,

then there exists morphisms:

1• $f: \coprod_{i \in I} H_i^* \Rightarrow \coprod_{i \in I} \tilde{H}_i^*$;

2• $f: \Vert\!| _{i \in I} H_i^* \Rightarrow \Vert\!|_{i \in I} \tilde{H}_i^*$.

PROOF. (1) Let $H^* = \coprod_{i \in I} H_i^*$ and $\tilde{H}^* = \coprod_{i \in I} \tilde{H}_i^*$. By (i) we may define a function $f_Q : Q \to \tilde{Q}$ by

$$(f_Q(\underline{q}))(i) = (f_i)_Q(\underline{q}(i)) \tag{24.2}$$

If $\underline{a} \in A$, then by 24.3.2 (2), $\underline{a}: I \to \bigcup_{i \in I} A_i \cup \{\Omega\}$ such that (a) $\underline{a}(i) \in A_i \cup \{\Omega\}$, all $i \in I$ and hence $\underline{a}(i) = (f_i)_A(\underline{a}(i)) \in \tilde{A}_i \cup \{\Omega\}$, since the f_i are embeddings, and by (b) there exists $e \in E$ such that $e \in E_i \Leftrightarrow \underline{a}(i) \neq \Omega \wedge \mu_i(\underline{a}(i)) = e$ and hence, using (ii), there exists $e \in E$ such that $e \in \tilde{E}_i \Leftrightarrow \underline{a}(i) \neq \Omega \wedge \tilde{\mu}_i(\underline{a}(i))$. Hence, $A \subseteq \tilde{A}$ and so we can define $f_A : A \to \tilde{A}$ by $f_A(\underline{a}) = \underline{a}$.

It easily follows from 24.3.2 (6) and 24.4.1 that $\mu(\underline{a}) = \tilde{\mu}(\underline{a})$. Hence, it only remains to check the conditions of 21.2.1. To check that f is an embedding morphism, it suffices to show that f_Q is injective, that 21.2.1 (1) and (3) hold and that

(a) $\underline{q}_1 \to^{\underline{a}} \underline{q}_2$ in $H \Rightarrow f_Q(\underline{q}_1) \to^{\underline{a}} f_Q(\underline{q}_2)$ in $\tilde{H}$;
(b) if $\underline{a}, \underline{b} \in A$, then $\underline{a}\, \iota\, \underline{b} \Leftrightarrow \underline{a}\, \tilde{\iota}\, \underline{b}$.

(a) follows from the fact that if $\underline{a}(i) \neq \Omega$ and $\underline{q}_1(i) \to_i^{\underline{a}(i)} \underline{q}_2(i)$ in H_i then $(f_Q(\underline{q}_1))(i) = (f_i)_Q(\underline{q}_1(i)) \to_i^{\underline{a}(i)} (f_i)_Q(\underline{q}_2(i)) = (f_Q(\underline{q}_2))(i)$ in $\tilde{H}_i$.

(b) is clear from the fact that $\underline{a}(i)\, \iota_i\, \underline{b}(i) \Leftrightarrow \underline{a}(i)\, \tilde{\iota}_i\, \underline{b}(i)$, where both are defined, and 24.3.2 (5).

Since each $(f_i)_Q$ is injective, the injectivity of f_Q easily follows. $f_A(\underline{a}) = \underline{a}$ holds by definition, giving 21.2.1 (3). To complete the proof, suppose that $f_Q(\underline{q}_1) \to^{\underline{a}} \tilde{\underline{q}}_2$ in $\tilde{H}$. By 24.3.2 (4), we have two cases:

(a) $\underline{a}(i) \neq \Omega$ and $(f_Q(\underline{q}_1))(i) \to_i^{\underline{a}(i)} \tilde{\underline{q}}_2(i)$ in $\tilde{H}_i$. Then by definition of f_Q, $(f_i)_Q(\underline{q}_1(i)) \to_i^{\underline{a}(i)} \tilde{\underline{q}}_2(i)$, and since f_i is an embedding, there exists $q_i \in Q_i$ such that $\underline{q}_1(i) \to_i^{\underline{a}(i)} q_i$ in H_i..
(b) $\underline{a}(i) = \Omega$ and $(f_Q(\underline{q}_1))(i) = \tilde{\underline{q}}_2(i)$.

Thus, if we define

$$\underline{q}_2(i) = \begin{cases} q_i \text{ if } \underline{a}(i) \neq \Omega \\ \underline{q}_1(i) \text{ otherwise} \end{cases}$$

the $f_Q(\underline{q}_2) = \tilde{q}_2$, by definition of f_Q, and $\underline{q}_1 \rightarrow^{\underline{a}} \underline{q}_2$ in H, by 24.3.2 (4).

(2) We define f_Q as in (24.2) and $f_A(a,i) = (a,i)$.

The proof that f is an embedding is analogous to (1).

■

24.4.4. PROPOSITION. Suppose that $(MN_i)_{i \in I}$ is an indexed family of marked, labelled nets, and let $MN = \|_{i \in I} MN_i$ then there exists an embedding isomorphism $f: H_{MN}{}^* \Rightarrow \|_{i \in I} H_{MN_i}{}^*$.

PROOF. Let $(MN_i) = (N_i, E_i, \lambda_i, c_i)$, where $N_i = (S_i, T_i; F_i)$, each $i \in I$ and let $MN = (N, E, \lambda, c)$, where $N = (S, T; F)$.

Let $(H, \underline{q}_0) = H_{MN}{}^*$, let $(H_i, c_i) = H_{MN_i}{}^*$ each $i \in I$ and let $(\tilde{H}, \tilde{q}_0) = \|_{i \in I} (H_i, c_i)$. We are required to construct an embedding isomorphism $f: (H, \underline{q}_0) \Rightarrow (\tilde{H}, \tilde{q}_0)$.

We note first of all that $A = T$ and that $A_i = T_i$, each $i \in I$. A comparison of 24.2.3 (3) and 24.3.2 (3) shows that $A = \tilde{A}$. Thus, we may take f_A to be the inclusion mapping.

Next, let $\underline{a}, \underline{b} \in A$. We show that $\underline{a}\, \iota\, \underline{b} \Leftrightarrow \underline{a}\, \tilde{\iota}\, \underline{b}$.

If $\underline{a}\, \iota\, \underline{b}$, then $\underline{a}\; co_N\; \underline{b}$ and so $(\bullet\underline{a} \cup \underline{a}\bullet) \cap (\bullet\underline{b} \cup \underline{b}\bullet) = \varnothing$, by 17.2.1 (2). Therefore, if $\underline{a}(i) \neq \Omega$ and $\underline{b}(i) \neq \Omega$, then by 24.2.3 (4), we have $(\bullet\underline{a}(i) \cup \underline{a}(i)\bullet) \cap (\bullet\underline{b}(i) \cup \underline{b}(i)\bullet) = \varnothing$ in N_i and hence $\underline{a}(i)\; co_{N_i}\; \underline{b}(i)$ and hence $\underline{a}(i)\; \iota_i\; \underline{b}(i)$, by 17.2.1 (2). By 24.3.2 (5), $\underline{a}\, \tilde{\iota}\, \underline{b}$.

If $\neg \underline{a}\, \iota\, \underline{b}$, then $(\bullet\underline{a} \cup \underline{a}\bullet) \cap (\bullet\underline{b} \cup \underline{b}\bullet) \neq \varnothing$, by 17.2.1 (2). Suppose, without loss of generality, that $(s,i) \in \underline{a}\bullet \cap \bullet\underline{b}$, then by 24.2.3 (4), $\underline{a}(i) \neq \Omega$, $\underline{b}(i) \neq \Omega$ and $s \in \underline{a}(i)\bullet \cap \bullet\, \underline{b}(i)$. Thus $\neg \underline{a}(i)\; \iota_i\; \underline{b}(i)$ by 17.2.1, and so by 24.3.2 (5), $\neg \underline{a}\, \tilde{\iota}\, \underline{b}$. We have shown that $\underline{a}\, \iota\, \underline{b} \Leftrightarrow \underline{a}\, \tilde{\iota}\, \underline{b}$.

By 24.2.3 (3) and (5), the definition of $H_{MN}{}^*$ given at the beginning of this section and 24.3.2 (3) and (6), $\mu = \tilde{\mu}$.

Next, let $\underline{q} \in Q$, so that $\underline{q} \subseteq S$. Recall that $S = \bigcup_{i \in I} S_i \times \{i\}$ so that if we define

$$(f_Q(\underline{q}))(i) = \{s \in S_i \mid (s,i) \in \underline{q}\} \tag{24.3}$$

then as $(f_Q(\underline{q}))(i) \subseteq S_i$ and $Q_i = \wp(S_i)$, each $i \in I$, $f_Q: Q \rightarrow \tilde{Q}$. It is clearly injective. It is also surjective, for if $\tilde{q} \in \tilde{Q}$, then $f_Q(\underline{q}) = \tilde{q}$, where

$\underline{q} = \{(s,i) \in S \mid s \in \underline{q}(i)\}$

Now $\underline{q}_0 = \bigcup_{i \in I} c_i \times \{i\}$, by 24.2.3 (6), and so $(f_Q(\underline{q}_0))(i) = c_i$, each $i \in I$. By 24.3.2 (7), $f_Q(\underline{q}_0) = \underline{q}'_0$.

Finally, we need to show that $\underline{q}_1 \rightarrow^{\underline{a}} \underline{q}_2 \Leftrightarrow f_Q(\underline{q}_1) \rightarrow^{\underline{a}} f_Q(\underline{q}_2)$.

Now $\underline{q}_1 \rightarrow^{\underline{a}} \underline{q}_2 \Leftrightarrow \underline{q}_1 [\underline{a}\rangle_N \underline{q}_2 \Leftrightarrow \bullet\underline{a} \subseteq \underline{q}_1 \wedge \underline{q}_2 = (\underline{q}_1 - \bullet\underline{a}) \cup \underline{a}\bullet$. By 24.2.3 (4), $\bullet\underline{a} \subseteq \underline{q}_1 \Leftrightarrow \forall i \in I : (\underline{a}(i) \neq \Omega \Rightarrow \bullet\underline{a}(i) \subseteq f_Q(\underline{q}_1))$. Also, if $\underline{a}(i) = \Omega$, then again by 24.2.3, (4) $\underline{q}_1 \cap (S_i \times \{i\}) = \underline{q}_2 \cap (S_i \times \{i\})$ so $(f_Q(\underline{q}_1))(i) = (f_Q(\underline{q}_2))(i)$. If $\underline{a}(i) \neq \Omega$, then $(s,i) \in (\underline{q}_1 - \bullet\underline{a}) \cup \underline{a}\bullet \Leftrightarrow s \in ((f_Q(\underline{q}_1))(i) - \bullet\underline{a}(i)) \cup \underline{a}(i)\bullet$. Thus

$$\begin{aligned} \underline{q}_1 \rightarrow^{\underline{a}} \underline{q}_2 \Leftrightarrow (\underline{a}(i) = \Omega \Rightarrow (f_Q(\underline{q}_1))(i) = (f_Q(\underline{q}_2))(i)) \\ \wedge (\underline{a}(i) \neq \Omega \Rightarrow (f_Q(\underline{q}_1))(i) \rightarrow^{\underline{a}(i)} (f_Q(\underline{q}_2))(i)). \end{aligned}$$

and we are done.

■

24.4.5. PROPOSITION. Suppose that $(MN_i)_{i \in I}$ is an indexed family of marked, labelled nets, and let $MN = \Vert\!|_{i \in I} MN_i$ then there exists an embedding isomorphism $f : H_{MN}{}^* \Rightarrow \Vert\!|_{i \in I} H_{MN_i}{}^*$.

PROOF. Suppose $\underline{q}$ is a state of $H_{MN}{}^*$, then by the definition at the beginning of section 24.4, we have $\underline{q} \subseteq S$, where S is the set of states of $\Vert\!|_{i \in I} H_{MN_i}{}^*$. Define $\pi_i(\underline{q}) = \{s \in S_i \mid (s,i) \in \underline{q}\}$. $\pi_i(\underline{q}) \subseteq S_i$, the set of places of MN_i. By 24.3.4 (1), we may define $(f_Q(\underline{q}))(i) = \pi_i(\underline{q})$.

By definition, the transitions of $H_{MN}{}^*$ are the T elements of $\Vert\!|_{i \in I} MN_i{}^*$. By 24.2.4 (3) and 24.3.4 (3), this set is equal to the set of transitions of $\Vert\!|_{i \in I} H_{MN_i}{}^*$. Hence, we may legitimately define f_A to be the inclusion map. The proof that $f = (f_Q, f_A)$ is an embedding isomorphism is straightforward.

■

We may now show that the direct and the marked-net-based semantics are equivalent.

24.4.6. PROPOSITION. If P is a path/process program, then $\rho[\![P]\!] \equiv H_{\sigma[\![P]\!]}{}^*$.

PROOF. We write $H \Rightarrow \tilde{H}$ to indicate that there is an embedding from H to $\tilde{H}$. Let $P_{path} = \Vert_{i \in I_{path}(P)} \sigma[\![P_i]\!]$ and $P_{proc} = \Vert\!|_{i \in I_{proc}(P)} \sigma[\![P_i]\!]$.

We know that $\rho[\![P_i]\!] \equiv H_{\sigma[\![P_i]\!]}{}^*$, each $i \in I$ and so by 24.4.3 (1) and 24.4.4

$$\mathop{||}_{i \in I_{path}(P)} \rho[\![P_i]\!] \Rightarrow \mathop{||}_{i \in I_{path}(P)} H_{\sigma[\![P_i]\!]}{}^* \Rightarrow H_{\sigma[\![P_{path}]\!]}$$

and by 24.4.3 (2) and 24.4.5

$$* \mathop{|||}_{i \in I_{proc}(P)} \rho[\![P_i]\!] \Rightarrow \mathop{|||}_{i \in I_{proc}(P)} H_{\sigma[\![P_i]\!]}{}^* \Rightarrow H_{\sigma[\![P_{proc}]\!]}{}^*$$

and so, by 24.3.6, 24.4.3 (1) and 24.4.4

$$\rho[\![P]\!] = (\mathop{|||}_{i \in I_{proc}(P)} \rho[\![P_i]\!]) || (\mathop{||}_{i \in I_{path}(P)} \rho[\![P_i]\!]) \Rightarrow H_{\sigma[\![P_{proc}]\!]}{}^* || H_{\sigma[\![P_{path}]\!]}{}^* \Rightarrow H_{\sigma[\![P_{proc} \,||\, P_{path}]\!]} = H_{\sigma[\![P]\!]}{}^*$$

Since the composition of embeddings is clearly an embedding, it follows that $\rho[\![P]\!] \Rightarrow H_{\sigma[\![P]\!])}{}^*$ and so $\rho[\![P]\!] \equiv H_{\sigma[\![P]\!])}{}^*$, by 24.4.2 (2).

■

24.5. Embedding COSY Into TCSP.

Let us return to definition 24.3.6. The expression on the right hand side is the composite of a family of hybrid transition systems and such compositions may be applied to arbitrary families of hybrid transition systems, not merely those arising from path/process programs. Nor are we obliged to combine the transition systems in the form given by 24.3.6. There is nothing to stop us nesting applications of the combinators.

This suggests an extension of the path/process notation. The only thing that distinguishes paths from processes is the way in which they are combined, so we need not make any distinction between them if we can express this distinction explicitly using two combinators corresponding to || and |||. Let us provisionally call a flow expression, an expression of the form

flow s **end**

where s may be derived from the non-terminal, $<sequence>$ using the production rules of definition 24.2.1. We may achieve a specification language at least as powerful as the path/process notation in the following.

24.5.1. DEFINITION. A program in the *flow notation* is a terminal string generated from the non-terminal $<system>$ via the following rewrite rules.

$<system>::=<flow>|([<system>||]^*)|([<system>|||]^*)$
$<flow>::=$ **flow** $<sequence>$ **end**

$< sequence >::= [< orelement >;]^* \mid < orelement >$
$< orelement >::= [< element >,]^* \mid < element >$
$< element >::=< event > \mid (< sequence >)$

The semantics of a flow program may be given recursively. First, if $P = \textbf{flow}\ s\ \textbf{end}$, then we define $\rho[\![P]\!] = \rho[\![\tilde{P}]\!]$ where $\tilde{P} = \textbf{path}\ s\ \textbf{end}$ and

$\rho[\![P_1 \parallel \cdots \parallel P_m]\!] = \rho[\![P_1]\!] \parallel \cdots \parallel \rho[\![P_m]\!]$
$\rho[\![P_1 \mid\mid\mid \cdots \mid\mid\mid P_n]\!] = \rho[\![P_1]\!] \mid\mid\mid \cdots \mid\mid\mid \rho[\![P_n]\!]$

Suppose we have a path/process program P given by

begin
  process s_1 **end** $\cdots$ **process** s_n **end**
  path s_{n+1} **end** $\cdots$ **path** s_{n+m} **end**
end

then we may construct a flow program $\phi(P)$

$((\textbf{flow}\ s_1\ \textbf{end} \mid\mid\mid \cdots \mid\mid\mid \textbf{flow}\ s_n\ \textbf{end})$
$\quad \parallel (\textbf{flow}\ s_{n+1}\ \textbf{end} \parallel \cdots \parallel \textbf{flow}\ s_{n+m}\ \textbf{end})$

and by definition 24.3.6, $\rho[\![P]\!] = \rho[\![\phi(P)]\!]$.

For example, the path/process program of (24.1) may be translated into the flow program having exactly the same meaning.[6]

$$((\textbf{flow}\ P; crit\ sec\ 1; V\ \textbf{end} \mid\mid\mid \textbf{flow}\ P; crit\ sec\ 1; V\ \textbf{end}) \parallel \textbf{flow}\ P; V\ \textbf{end}) \tag{24.4}$$

Thus, the flow notation can do everything the path/process notation of definition 24.2.1 can do but has the possible advantage of making the form of the combination of sequential subsystems explicit rather than by using the path/process dichotomy.

We may massage the notation still further. One objection to the path/process notation is that paths and processes are always cyclic. This means that there are certain sequential systems which cannot be described by a single path or process, namely those whose behaviour is not described by a regular expression of the form $(s)^*$. Things are not quite as bad as this, since we can construct a path/process program having such behaviour as

[6]That is, if F denotes the flow program of (24.4) and P denotes the path/process program of (24.1) then $\rho[\![F]\!] = \rho[\![P]\!]$.

follows. If a and b are two events not occurring in s, then the following path program does the trick.

$$\textbf{begin path}\ (s); a; b\ \textbf{end path}\ b; a\ \textbf{end end} \tag{24.5}$$

Informally, once one has reached the end of s, one must stop, since the second of the two paths prevents a from occurring next.

It is a matter of taste, perhaps, but (24.5) seems to me a rather roundabout way of saying something simple. Of course, there is nothing to stop us from avoiding this problem in the flow expression language by recasting the semantics - although in order to retain 'parity' with the path/process notation we would have to introduce the Kleene star into the flow notation.[7]

We shall take a different approach, which allows an interesting comparison with process algebra. Consider a single path or process, P. This gives rise to a hybrid transition system H with initial state q_0, say. Each state q of H may be completely described by an equation of the form:

$$P_q = (e_1 \rightarrow P_{q_1} \sqcap \cdots \sqcap e_n \rightarrow P_{q_n})$$

where

$$\{(e_i, q_i) \mid i = 1, \cdots, n\} = \{(e, q') \mid \exists a \in A : q \rightarrow q' \wedge \mu(a) = e\}$$

and the transition structure of H is completely described, up to isomorphism, by a set of such equations, one for each state.

First of all, it may be seen that the notation is beginning to look less like the path process notation and more like a sublanguage of 'Theoretical CSP' [2]. In fact, our choice of symbolism was not made in ignorance of the fact that $\parallel$ and $\parallel\!\parallel$ are symbols for parallel combinators in TCSP and we may argue that they mean exactly the same in both notations, once one has equipped TCSP with some formal notion of concurrency.

Nevertheless, the new notation is capable of expressing any path/process program. Let us take example (24.1).

$$\begin{aligned}
&PROC1 = (P \rightarrow (crit\,sec1 \rightarrow (V \rightarrow PROC1)))\\
&PROC2 = (P \rightarrow (crit\,sec2 \rightarrow (V \rightarrow PROC2)))\\
&SEM = (P \rightarrow (V \rightarrow SEM))\\
&SYS = (PROC1 \parallel\!\parallel PROC2) \parallel SEM
\end{aligned} \tag{24.6}$$

In the 'guarded command' notation, it rewrites to (24.6) or using the recursion construct of [2], to (24.7).

[7] There is a Kleene star is in the full path/process notation; we have not used it for the sake of simplicity.

$$SYS = [(\mu X.(P \rightarrow (crit\,sec\,1 \rightarrow (V \rightarrow X))))$$
$$\qquad ||| ((\mu X.(P \rightarrow (crit\,sec\,2 \rightarrow (V \rightarrow X)))))] \qquad (24.7)$$
$$|| (\mu X.(P \rightarrow (V \rightarrow X)))$$

Of course, all we have done is to effect a syntactic transformation of (24.1), via (24.4). The fact that the result looks like a TCSP program does not imply that it has the same meaning in TCSP as it does as a flow program. What we would like to do is to show that if F is a flow program and $\chi(F)$ is the derived TCSP program, then $\rho[\![F]\!] = \sigma[\![\chi(F)]\!]$, where σ is the semantics function for TCSP.

Now we are faced by two problems. The first is that the semantics for path programs presented in this chapter is not fully formal; we have presented it using pictures, as in Lauer and Campbell's original paper [15]. A proper proof that $\rho[\![F]\!] = \sigma[\![\chi(F)]\!]$ would involve an induction over the structure of F and for that we would need to give a formal, recursive definition of ρ.

The second problem concerns the semantics of TCSP. The theories we have looked at so far admit a formal notion of concurrency, even though, in the case of loosely coupled systems, we have ventured to disagree with it. With TCSP, we have none of that; Hoare explicitly refrains from modelling contemporaneity of occurrences.

We face this problem also with CCS, and our approach there is to provide the language with a non-interleaving semantics and show that if $\rho[\![Exp]\!]$ is the hybrid transition systems associated with the expression Exp and $\sigma[\![Exp]\!]$ is the transition system associated with Exp by Milner's semantics. then $\rho[\![Exp]\!] \sim T_{\sigma[\![Exp]\!]}$, where $T_{\sigma[\![Exp]\!]}$ is defined in 24.5.1. In other words, Milner's semantics may be recovered from ours, up to ~, if we throw away the concurrency.

We shall do the same for TCSP, or at least for a sublanguage of it, in chapter 28. As for the relationship between TCSP and the path/process notation, we merely remark that if γ is the mapping that sends hybrid transition systems into the semantic domain of TCSP, then it suffices to prove $\rho[\![F]\!] = \gamma(\sigma[\![\chi(F)]\!])$ for individual flow expressions. The general case, then follows from the fact, which we *shall* prove, that γ preserves both parallel combination operators.

Notes.

Our concerns in this chapter have been with semantics. It should therefore be mentioned that there exists a substantial body of results which assist in the analysis of COSY programs. This material is largely collected together in [70].

As an historical note, the vector representation of behaviours described in chapter 9 was originally devised by the author to give a semantics to path programs ([25]). In fact, the path process notation was the author's first taste of concurrency theory.

Ironically, the author's first contribution to the study of path/process programs was to propose a direct *interleaving* semantics for the notation.

Chapter 25
The Nivat/Arnold Process Model

> Clearly, the specification of a system of concurrent processes can be divided into two parts; (i) specification of all possible behaviour of each individual process; (ii) specification of the interference between the behaviours of these individual processes
>
> A. ARNOLD [26]

25.1. Introduction.

The process model proposed by M. Nivat in [37] somewhat resembles the path notation in the sense that systems are conceived of consisting of a number of finite-state sequential processes which synchronise on actions. It differs from the path/process notation in a number of ways. For example, the synchronisation mechanism adopted by Nivat is more general than that of the path/process notation in that, instead of only synchronising on actions with the same name, actions with different names may synchronise. On the other hand, processes are represented by the sets of sequences of actions they may execute, so Nivat's formalism cannot describe non-deterministic (ambiguous) processes.

Nivat's semantics bears a strong resemblance to the vector representation of traces we met in chapter 9 and this has led A. Arnold to claim that the latter is a special case of the former. In a purely formal sense (that is, neglecting the intended meaning of the formalisms) this is no doubt true, but there are differences. In the Nivat model, a behaviour is represented by a vector of sequences whose i th co-ordinate describes the sequence of actions performed by the i th process of the system in the course of that behaviour.

However, from that vector one cannot, in general, deduce the order in which actions from different processes have occurred, unlike the vectors of chapter 9.

We shall see how we may equip the model with a semantics in terms of hybrid transition systems. We shall do so with the intention of being able to recover the Nivat/Arnold semantics from out own, in line with our general position that we are obliged to be consistent with the official semantics.

At this point we run into a problem. In order to recover Nivat/Arnold vectors from points in a behavioural presentation, we need additional information, namely, with which processes are particular events associated? For this reason, we need to enhance the behavioural presentation model with a notion of *process structure*.

It turns out that it is always possible to impose a process structure on a behavioural presentation, providing one allows there to be an infinite number of processes. We shall also see that the existence of a process structure of a particular kind allows one to construct an occurrence net N from an asynchronous behavioural presentation B whose corresponding event structure $Es(N)$[1] is equal to $\Sigma(B)$.[2] Thus, considerations of process structure uncovers more relationships between behavioural presentations, occurrence nets and event structures.

25.2. The Model and its Vector Semantics.

It is probably simplest to begin with the treatment given by A. Arnold in [26]. Suppose A is a set of actions. A *process* on A, P, is associated with a set of behaviours $B(P) \subseteq A^*$. Since the beginning of a behaviour is also a behaviour, $B(P)$ is prefix-closed, that is

$$\downarrow B(P) = B(P) \tag{25.1}$$

If P is a process, then there we define A_P to be the smallest set A such that $B(P) \subseteq A^*$. If $\mathcal{P} = (P_i)_{i \in I}$ is an indexed family of processes, then $\mathcal{P}$ determines an indexed cover $\alpha_{\mathcal{P}}: I \to \wp(A_{\mathcal{P}})$ by $\alpha_{\mathcal{P}}(i) = A_{P_i}$, where

$$A_{\mathcal{P}} = \bigcup_{i \in I} A_{P_i}.$$

If $\alpha: I \to \wp(A)$ is any indexed cover, then define ${}_{\alpha}A$ to be the set of all functions $\underline{x}: I \to A \cup \{\Omega\}$ such that

1• $\underline{x}(i) \in \alpha(i) \cup \{\Omega\}$, each $i \in I$;
2• $\underline{x}(i) \neq \Omega$, some $i \in I$.

[1]See section 16.3.
[2]See section 5.2.

Note that ${}_{\alpha}A \subseteq M_{\alpha}$[3], so that ${}_{\alpha}A$ generates a submonoid $({}_{\alpha}A)^*$ of M_{α} which is also a partially ordered set. Indeed, if I is finite, then it is not hard to show that $({}_{\alpha}A)^* = M_{\alpha}$.

If $\mathcal{P} = (P_i)_{i \in I}$ is an indexed family of processes and $\alpha = \alpha_{\mathcal{P}}$, then define

$$B(\mathcal{P}) = \{\underline{x} \in M_{\alpha} \mid \forall i \in I : \underline{x}(i) \in B(P_i)\} \tag{25.2}$$

Intuitively, $B(\mathcal{P})$ describes the behaviours of a system of sequential processes which are reciprocally unconstrained. $\underline{x} \in B(\mathcal{P})$ describes a behaviour of the system in which process P_i has performed the sequence of actions $\underline{x}(i)$. By (25.2), each process may do whatever it wants, irrespective of what its colleagues are doing.

Synchronisation is imposed by requiring that only certain sets of process actions may take place simultaneously. Formally, we assume the existence of a set $S \subseteq {}_{\alpha}A$, where $\alpha = \alpha_{\mathcal{P}}$, of *synchronisation vectors*, and the behaviour of $\mathcal{P}$ under the constraints S is defined to be

$$B(\mathcal{P}, S) = B(\mathcal{P}) \cap S^* \tag{25.3}$$

Here, S^* denotes the submonoid of M_{α} generated by S.[4] If $\underline{x} \in B(\mathcal{P}, S)$, then after the system has performed the behaviour described by $\underline{x}$, the process P_i may perform an action $a \Leftrightarrow$

1• $\underline{x}(i).a \in B(P_i)$ - the process would be entitled to perform that action if it were on its own - and

2• Some set of processes may synchronise with P_i - there exists $\underline{s} \in S$ such that $\underline{s}(i) = a$ and for each $j \in I$, $\underline{x}(j).\underline{s}(j) \in B(P_j)$

25.2.1. EXAMPLE. Let $I = \{1,2,3\}$ and let $B(P_1) = a^*$, $B(P_2) = b^*$ and $B(P_3) = c^*$. We shall represent the vector $\underline{x} \in M_{\alpha}$ by the triple $(\underline{x}(1), \underline{x}(2), \underline{x}(3))$ Let $S = \{\underline{e}, \underline{f}, \underline{g}\}$ where

$$\underline{e} = (a, b, \Omega) \qquad \underline{f} = (\Omega, b, c) \qquad \underline{g} = (a, \Omega, c)$$

A simple induction argument shows that

$$B(\mathcal{P}, S) = \{(a^{n_a}, b^{n_b}, c^{n_c}) \mid n_a + n_b + n_c \text{ is even}\}.$$

---

[3]See 9.2.5.

[4]That is, the set of vectors which may be obtained from S by coordinatewise concatenation.

and that such a vector is a product of $(n_a+n_b-n_c)/2$ copies of $\underline{e}$, $(n_b+n_c-n_a)/2$ copies of $\underline{f}$ and $(n_c+n_a-n_b)/2$ copies of $\underline{g}$.

Let us formalise this.

25.2.2. DEFINITION. A *synchronised process system* (SPS) is pair $\Sigma=(\mathcal{P},S)$ where $\mathcal{P}=(P_i)_{i\in I}$ is an indexed family of processes and $S\subseteq {}_{\alpha}A$, where $\alpha=\alpha_{\mathcal{P}}$, such that

$$\forall \underline{s}\in S: \underline{s}\neq \underline{\Omega}_{\alpha} \wedge \left|\{i\in I \mid \underline{s}(i)\neq\Omega\}\right|<\infty \tag{25.4}$$

We should justify our inclusion of the additional constraint (25.4), which does not occur in [26]. Basically, this is due to the fact that $\underline{s}\in S$ describes a set of actions that are to be performed simultaneously. Our handling of simultaneity in chapter 22 requires that simultaneity classes be finite. We are therefore led to require that no synchronisation vector synchronises more than a finite number of processes.[5]

If $\Sigma=(\mathcal{P},S)$, then Σ determines a cover $\alpha_{\mathcal{P}}$. Define $\alpha_{\Sigma}=\alpha_{\mathcal{P}}$. In the sequel, we shall drop the subscript Σ when it is clear from the context which cover we are talking about.

The behaviour of Σ is given by the set of vectors

$$B(\Sigma)=B(\mathcal{P})\cap S^* \tag{25.5}$$

Let us show how we may use this theory to give a vector semantics to path/process programs.

Suppose that $Prog=$ **begin** $P_1\cdots P_n$ **end** is a path/process program. For each i, $1\le i\le n$, there is an initialised hybrid transition system $(H_i,q_i)=\rho[\![P_i]\!]$ which is asynchronous,[6] so that we may regard μ_i as a function from A_i to E_i. Define

$$B(P_i)=\mu_i(L(T(C(H_i))),q_i)) \tag{25.6}$$

so we may regard $Prog$ as defining an indexed family $\mathcal{P}=(P_i)_{i\in I}$ of processes in the Nivat/Arnold sense. Note that $A_{P_i}=E_i$, each i Let $(H,\underline{q})=\rho[\![Prog]\!]$ and define $J=I_{path}(Prog)$, $E_J=\bigcup_{i\in J}E_i$, $K=I_{Proc}(Prog)$ and $E_K=\bigcup_{i\in K}E_i$

For each $i\in K$, we construct vectors $\underline{e}_i:\{1,\cdots,n\}\to A\cup\{\Omega\}$, $e\in E_i$ where:

[5] It should be noted that both Nivat and Arnold are only concerned with finite sets of processes, so this restriction holds trivially for them.
[6] See 24.3.1

$$\underline{e}_i(j) = \begin{cases} e \text{ if } j = i \vee (j \in J \wedge e \in E_j) \\ \Omega \text{ otherwise} \end{cases} \tag{25.7}$$

$\underline{e}_i$ represents that execution of the event e by the process P_i (in the path/process sense); it may take place providing P_i and every path P_j which mentions e allows it to. If $e \in E_J - E_K$, that is, e belongs to some path but to no process, then we define:

$$\underline{e}_\perp(j) = \begin{cases} e \text{ if } j \in J \wedge e \in E_j \\ \Omega \text{ otherwise} \end{cases} \tag{25.8}$$

For each $i \in J$, we define

$$S_i = \{\underline{e}_i \mid e \in E_i\} \tag{25.9}$$

and we define

$$S_\perp = \{\underline{e}_\perp \mid e \in E_J - E_K\} \tag{25.10}$$

$$S = S_\perp \cup \bigcup_{i \in K} S_i \tag{25.11}$$

and putting (25.6) and (25.11) together gives us a SPS

$$\Sigma_{Prog} = (\mathcal{P}, S) \tag{25.12}$$

so that we may associate a set of vectors with $Prog$ by

$$VFS(Prog) = B(\Sigma_{Prog}) \tag{25.13}$$

This, essentially, is the vector semantics for path/process programs given in [25]; to use Arnold's terminology[7]

$$Cosy \subset Synchro.$$

As far as the semantics of [25] is concerned, Arnold's claim is certainly true. However, the semantics of [25] was not consistent with that of [33], in that it did not handle ambiguity in the same way. As we shall see, the vector semantics we have just given is consistent with that of [33] and chapter 24 only when the path/process program under consideration is unambiguous, in a sense that we shall define in section 25.6.

---

[7]See proposition 2.3 of [26].

We begin by showing that the set of vectors of an SPS may be generated using a transition system.

25.2.3. DEFINITION. Let $\Sigma = (\mathcal{P}, S)$, is a SPS, where $\mathcal{P} = (P_i)_{i \in I}$. We define $T_\Sigma = (Q_\Sigma, S_\Sigma, \rightarrow_\Sigma)$, where

1• $Q_\Sigma = B(\Sigma)$
2• $\underline{q}_1 \rightarrow_\Sigma^{\underline{s}} \underline{q}_2 \Leftrightarrow \forall i \in I : \underline{q}_2(i) = \underline{q}_1(i).\underline{s}(i)$.

T_Σ is clearly a transition system. Let $\underline{q}_\Sigma$ be defined by $\underline{q}_\Sigma(i) = \Omega$, each $i \in I$. $(T_\Sigma, \underline{q}_\Sigma)$ is thus an initialised transition system. It is clear from 25.2.3 (2) that T_Σ is unambiguous.

At this point, we need to modify our notation a little. We have used the symbol S^* to denote the submonoid of M_{α_Σ} generated by S. However, the language accepted by $(T_\Sigma, \underline{q}_\Sigma)$ will be a subset of the free monoid generated by S. (To make matters worse, we shall also need to consider trace languages defined over such free monoids). Therefore, if α is a cover, then we define

1• $\langle S \rangle_\alpha^*$ to be the submonoid of M_{α_Σ} generated by S;
2• S^* to be the free-monoid generated by S, that is to say, the set of all strings over the set S.

For example, $(a,b,\Omega).(\Omega,b,c) = (a,bb,c) = (\Omega,b,c).(a,b,\Omega)$ is in $\langle S \rangle_\alpha^*$ but not in S^*. With this new notation, (25.5) becomes $B(\Sigma) = B(\mathcal{P}) \cap \langle S \rangle_\alpha^*$

In what follows, except where explicitly stated otherwise, the elements of S will be supposed to be concatenated as in S^*.

As we have said, $L(T_\Sigma, \underline{q}_\Sigma)$ is a subset of the free monoid S^*, not the monoid $\langle S \rangle_\alpha^*$. However, the identity mapping on S extends to a unique monoid epimorphism $e : S^* \rightarrow \langle S \rangle_\alpha^*$. Thus, if $\underline{x} = \underline{s}_1 \cdots \underline{s}_n$ $\underline{s}_i \in S$, each i, then $e(\underline{x})$ is the corresponding product of the $\underline{s}_i$ in $\langle S \rangle_\alpha^*$, so that $(e(\underline{x}))(i) = \underline{s}_1(i) \cdots \underline{s}_n(i)$. This allows us to relate the initialised transition system $L(T_\Sigma, \underline{q}_\Sigma)$ to the set of vectors $B(\Sigma)$, as the following lemma shows.

25.2.4. LEMMA. If $\Sigma = (\mathcal{P}, S)$, is an SPS, then $e(L(T_\Sigma, \underline{q}_\Sigma)) = B(\Sigma)$.

PROOF. Suppose that $\mathcal{P} = (P_i)_{i \in I}$.

If $\underline{x} \in L(T_\Sigma, \underline{q}_\Sigma)$, then $\underline{x} = \underline{s}_1 \cdots \underline{s}_n$, where $\underline{q}_\Sigma \rightarrow_\Sigma^{\underline{s}_1} \cdots \rightarrow_\Sigma^{\underline{s}_n} \underline{q}_n$. and $\underline{s}_i \in S$, each i. By 25.2.3 (2) and induction, $\underline{q}_m(i) = \underline{s}_1(i) \cdots \underline{s}_m(i) = (e(\underline{s}_1 \cdots \underline{s}_m))(i)$, for each

$m \leq n$ and $i \in I$. Thus, $e(\underline{s}_1 \cdots \underline{s}_m) = \underline{q}_m \in B(\Sigma)$, for each $m \leq n$. In particular $e(\underline{x}) = \underline{q}_n \in B(\Sigma)$ and we have shown that $e(L(T_\Sigma, \underline{q}_\Sigma)) \subseteq B(\Sigma)$.

Conversely, let $\underline{x} \in B(\Sigma)$, so $\underline{x} \in B(\mathcal{P}) \cap \langle S \rangle_\alpha{}^*$, where $\alpha = \alpha_\Sigma$. Since $\underline{x} \in \langle S \rangle_\alpha{}^*$, $\underline{x} = e(\underline{s}_1 \cdots \underline{s}_n)$, for $\underline{s}_i \in S$. By induction $\underline{q}_\Sigma \rightarrow_\Sigma^{\underline{s}_1} \cdots \rightarrow_\Sigma^{\underline{s}_n} \underline{q}_n$, where for each $m \leq n$, $\underline{q}_m = e(\underline{s}_1 \cdots \underline{s}_m) \in B(\Sigma)$. Hence, $\underline{s}_1 \cdots \underline{s}_n \in L(T_\Sigma, \underline{q}_\Sigma)$ and so $\underline{x} = e(\underline{s}_1 \cdots \underline{s}_n) \in e(L(T_\Sigma, \underline{q}_\Sigma))$, Thus, $B(\Sigma) \subseteq e(L(T_\Sigma, \underline{q}_\Sigma))$.

■

25.3. Process Structure.

$L(T_\Sigma, \underline{q}_\Sigma)$ allows us to construct a behavioural interpretation of an SPS, Σ. From chapters 13 and 14, we know that it determines a sequential, possibly non-deterministic, behavioural presentation. However, that is clearly not enough: the behaviour should be capable of reflecting such parallelism consequent on the presence of several processes operating at the same time.

It is simple enough to turn T_Σ into a hybrid transition system,

$$\tilde{H}_\Sigma = (Q_\Sigma, S, \rightarrow_\Sigma, \varnothing, \tilde{\mu}_\Sigma, E_\Sigma),$$

where

$$E_\Sigma = \bigcup_{i \in I} A_{P_i} \tag{25.14}$$

$$(\tilde{\mu}_\Sigma(\underline{s}))(e) = \left|\{i \in I \mid \underline{s}(i) = e\}\right| \tag{25.15}$$

so that $\tilde{\mu}_\Sigma$ counts the number of processes that are simultaneously performing an action called e. We may now construct a behavioural presentation from $(\tilde{H}_\Sigma, \underline{q}_\Sigma)$. In general, this behavioural presentation may be shown to be purely synchronous and non-determinate.

This construction certainly reflects the idea that the vectors $\underline{s}$ represent a set of simultaneous occurrences. However, we shall take a slightly different tack.

If we use this synchronous semantics, then the claim *Cosy* $\subset$ *Synchro* will no longer hold except in a purely formal sense even in the case of unambiguous path programs, where the vector semantics coincides with the official semantics. For example, the vector (a, a, Ω) in COSY might represent a single action a executed by two paths in synchronisation, whereas in the interpretation suggested above, it would represent two distinct actions a performed simultaneously by two distinct processes.

The path/process notation was devised for the description of asynchronous systems, not synchronous ones. Perhaps this doesn't matter too much, but there is a natural way of regarding the Nivat/Arnold model as describing hybrid systems as we shall soon demonstrate. Not only does this make it rather more interesting as a model, it also 'preserves the meaning of vectors' in going from *Cosy* to *Synchro*.

Granted that we can do this, there is one further matter to deal with and that is the process structure on SPSs. As we have remarked, from a discrete behavioural presentation, we will be able to recover only part of the structure of Σ. We will know about its transition structure and simultaneity structure, but we won't know which process does what. This means that we will lose information carried by the vector semantics.

Behavioural presentations do not 'come with' a notion of process; they merely describe orderings of occurrences in time. If we wish to keep information about processes, then we need either to 'code' it into the event set E or to introduce it explicitly as structure imposed on the behavioural presentation from the outside. We take the latter approach.

First, we note that in order to have '$Cosy \subset Synchro$' at the behavioural level, we shall assume that events may be shared by certain processes. In general, several simultaneous occurrences could be labelled with the same event name, but we shall take the COSY interpretation in which a 'column vector', $\underline{a}_{\alpha}$, reflects the participation, by all 'processes' P_i such that $a \in A_{P_i}$, in a single occurrence (of an event 'a'). Thus a vector $\underline{s} \in S$ will be considered to correspond to a simultaneity class X in which each occurrence o is of a distinct event $\lambda(o)$ such that $\underline{s}(i) = \lambda(o)$, some $i \in I$.

To regain $\underline{s}$ from X, we will need to know which processes collaborated on the execution of which events. We thus need a set of process names and a function assigning occurrences to sets of processes. We would like to model sequential processes, because those are the kind that one finds in the Nivat/Arnold model. This prompts (2) of the following definition.

25.3.1. DEFINITION. Let B be a behavioural presentation. A *(sequential) process structure* on B is a pair $\underline{\Pi} = (\Pi, \beta)$, where Π is a set (of processes) and $\beta: O \to \wp(\Pi)$ is a function such that

1• $\forall o \in O: \beta(o) \neq \emptyset$;

2• $\forall o_1, o_2 \in O: \beta(o_1) \cap \beta(o_2) \neq \emptyset \wedge o_1 \neq o_2 \Rightarrow o_1 \# o_2 \vee o_1 < o_2 \vee o_2 < o_1$;

3• $\forall o \in O: |\beta(o)| < \infty$.

The intuition is that each occurrence o is performed by a specific set of processes $\beta(o)$ and that no process may perform two distinct occurrences concurrently or simultaneously. Hence the term sequential.

The events executed by process $\pi \in \Pi$ belong to the set

$E_\pi = \{e \in E \mid \exists o \in O: \pi \in \beta(o) \wedge \lambda(o) = e\}$.

We thus have an indexed cover $\alpha_{\underline{\Pi}}: \Pi \to \wp(\tilde{E})$, where $\tilde{E} = \bigcup_{\pi \in \Pi} E_\pi$, given by $\alpha_{\underline{\Pi}}(\pi) = E_\pi$, each $\pi \in \Pi$.

25.3.1 (2) entails that if $p \in P$ and $\pi \in \Pi$, then the set $p|\pi$ defined

$$p|\pi = \{o \in p \mid \pi \in \beta(o)\} = p \cap \lambda^{-1}(E_\pi) \tag{25.16}$$

is totally ordered by $\to$. If B is discrete - as from now on we shall assume that it is - then we may map $p|\pi$ to a string $\rho(p|\pi) \in E_\pi{}^*$ by defining

$$\rho(p|\pi) = \lambda(o_1) \cdots \lambda(o_n)$$

where the o_i are the distinct elements of $p|\pi$ such that $o_1 < \cdots < o_n$. This allows us to define a mapping $\rho_{\underline{\Pi}}: P \to M_\alpha$, where $\alpha = \alpha_{\underline{\Pi}}$,by

$$(\rho_{\underline{\Pi}}(p))(\pi) = \rho(p|\pi) \tag{25.17}$$

We know that a discrete behavioural presentation with process structure $(B, \underline{\Pi})$ determines a language of string vectors. Our next result shows that this language is in fact the set of behaviours of an SPS, establishing a formal connection between the two theories. First, we need a construction.

25.3.2. DEFINITION. If B is a behavioural presentation and $\underline{\Pi} = (\Pi, \beta)$ is a process structure on B, then we define $\Sigma_{(B,\underline{\Pi})} = (\mathcal{P}, S)$, where

1• $\mathcal{P} = (P_\pi)_{\pi \in \Pi}$ and $P_\pi = \{\rho(p|\pi) \mid p \in P\}$, each $\pi \in \Pi$

2• $S = \{\underline{s}_X \mid X \in O/\!\approx\}$ where

$$\underline{s}_X(\pi) = \begin{cases} \lambda(o) \text{ if } o \in X \wedge \pi \in \beta(o) \\ \Omega \text{ otherwise} \end{cases}$$

We observe that the vectors $\underline{s}_x$ are well defined by 23.3.1 (2), since if $o_1, o_2 \in X$ and $\pi \in \beta(o_1) \cap \beta(o_2)$, then as $X \in O/\!\approx$, $o_1 \approx o_2$, which is impossible if $o_1 \neq o_2$, and hence $o_1 = o_2$.

PROPOSITION. Suppose that B is a bounded, discrete behavioural presentation and $\underline{\Pi} = (\Pi, \beta)$ is a process structure on B, then $\Sigma_{(B,\underline{\Pi})}$ is an SPS and $B(\Sigma_{(B,\underline{\Pi})}) = \rho_{\underline{\Pi}}(P)$.

PROOF. Let $(\mathcal{P},S)=\Sigma=\Sigma_{(B,\Pi)}$.

Let $x\in P_\pi$, and $y\le x$ some $\pi\in\Pi$. By 25.3.2 (2), there exists $p\in P$ such that $x=\rho(p|\pi)$, that is $x=\lambda(o_1)\cdots\lambda(o_n)$, where $p|\pi=\{o_1,\cdots,o_n\}$ and $o_1<\cdots<o_n$. For some $m\le n$, $y=\lambda(o_1)\cdots\lambda(o_m)$. Let $p'=\downarrow o_m$, then by 3.3.10 (5) and 3.3.14, $p'\in P$. Now $o\in p'|\pi\Leftrightarrow o\to o_m\wedge\pi\in\beta(o)$, and as $p|\pi=\{o_1,\cdots,o_n\}$ and $o_1<\cdots<o_n$, $p'|\pi=\{o_1,\cdots,o_m\}$, and so $\rho(p'|\pi)=y$. Thus, $y\in P_\pi$ and we have shown that P_π is left closed and hence a process, by (25.1).

Next, let $X\in O/\!\approx$, then clearly $\underline{s}_X\ne\underline{\Omega}_\alpha$. Since B is bounded, $|X|<\infty$, and so by 25.3.1 (3) and (2)

$$\begin{aligned}|\{\pi\in\Pi\mid \underline{s}_X(\pi)\ne\Omega\}|&=|\{\pi\in\Pi\mid\exists o\in X:\pi\in\beta(o)\}|\\&\le|X|.\,max(\{|\beta(o)|\mid o\in X\})<\infty\end{aligned}$$

showing that (25.4) holds and that accordingly Σ is an SPS. We next show that $B(\Sigma)=\rho_{\underline{\Pi}}(P)$.

By (25.2), (25.17) and the definition of $\mathcal{P}$, $B(\Sigma)\subseteq\rho_{\underline{\Pi}}(P)\subseteq B(\mathcal{P})$. For the reverse inclusion, let $p\in P$. We prove that by induction on $|p/\!\approx|$, that $\rho_{\underline{\Pi}}(p)\in\langle S\rangle_\alpha$. Since we know that $\rho_{\underline{\Pi}}(p)\in B(\mathcal{P})$ we may conclude that $\rho_{\underline{\Pi}}(p)\in B(\Sigma)$.

Let $o\in max(p)$, $X=o/\!\approx$ and $p'=p-X$. By 7.3.3 and 3.3.1, $p'\in P$ and $p'\succ^X p$. Since $|p'/\!\approx|=|p/\!\approx|-1$, $\rho_{\underline{\Pi}}(p')\in\langle S\rangle_\alpha$, by induction. To conclude the proof, we show that $\rho_{\underline{\Pi}}(p)(\pi)=\rho_{\underline{\Pi}}(p')(\pi).\underline{s}_X(\pi)$, all $\pi\in\Pi$.

Since X is a$\approx$-class and $p|\pi$ is totally ordered by $\to$, it follows that $|X\cap p|\pi|\le 1$. If $o\in X\cap p|\pi$, then clearly $o\in max(p)$, so $\rho(p|\pi)=\rho(p'|\pi).\lambda(o)$. Otherwise, $X\cap p|\pi=\emptyset$ and consequently $\rho(p|\pi)=\rho(p'|\pi)$. The claim follows by the definition of $\underline{s}_X$.

■

We have shown that, under certain finiteness conditions, discrete behavioural presentations with process structure determine SPSs, whose vector behaviour may be derived directly from the behavioural presentations. We shall now approach the relationship from the other end, showing that, via a hybrid transition system semantics, an SPS determines a behavioural presentation with process structure.

25.4. Hybrid Transition System Semantics.

25.4.1. DEFINITION. If Σ is an SPS, we define $C_\Sigma=(Q_\Sigma,S,\to_\Sigma,\iota_\Sigma)$, where

$$\underline{s}_1 \, \iota_\Sigma \, \underline{s}_2 \Leftrightarrow \forall i \in I : \underline{s}_1(i) > \Omega \Rightarrow \underline{s}_2(i) = \Omega \qquad (25.18)$$

ι_Σ is clearly an independence relation and we know that $T(C_\Sigma) = T_\Sigma$ is unambiguous. It is straightforward to verify that if $\underline{q}_1 \to_\Sigma^{\underline{s}_1} \underline{q}_2 \to_\Sigma^{\underline{s}_2} \underline{q}_3$ and $\underline{s}_1 \, \iota_\Sigma \, \underline{s}_2$, then $\underline{q}_1 \to_\Sigma^{\underline{s}_2} \underline{q}'_2 \to_\Sigma^{\underline{s}_1} \underline{q}_3$, where

$$\underline{q}'_2(i) = \begin{cases} \underline{q}_3(i) \text{ if } \underline{s}_2(i) \neq \Omega \\ \underline{q}_1(i) \text{ otherwise} \end{cases}$$

Hence

25.4.2. LEMMA. C_Σ is an ι_Σ-asynchronous transition system.

■

Now let us extend C_Σ to a hybrid transition system. Our discussion preceding 25.3.1 leads us to define μ_Σ as follows

$$(\mu_\Sigma(\underline{s}))(e) = \begin{cases} 1 \text{ if } \exists i \in I : \underline{s}(i) = e \\ 0 \text{ otherwise} \end{cases} \qquad (25.19)$$

25.4.3. DEFINITION. If Σ is an SPS, then $H_\Sigma = (Q_\Sigma, S, \to_\Sigma, \iota_\Sigma, \mu_\Sigma, E_\Sigma)$.

25.4.4. PROPOSITION. H_Σ is a hybrid transition system satisfying the following two properties

$$\forall a \in A_\Sigma \; \forall e \in E_\Sigma : (\mu_\Sigma(a))(e) \leq 1;$$
$$\forall q \in Q_\Sigma \; \forall a, a' \in A_\Sigma : q \to_\Sigma^{a} \wedge q \to_\Sigma^{a'} \wedge \mu_\Sigma(a) = \mu_\Sigma(a') \Rightarrow a = a'.$$

PROOF. By 25.4.2, (25.19) and 25.2.3. (2).

■

From $(H_\Sigma, \underline{q}_\Sigma)$ we may construct a discrete behavioural presentation, which we shall denote by B_Σ, according to the theory of chapter 22. We shall find it useful to go over the construction in this case.

First of all, we construct $L = L(C_\Sigma, \underline{q}_\Sigma)$. This is a subset of S_ι^*, where $\iota = \iota_{\alpha_\Sigma}$, the trace language constructed from the free monoid S^*. We construct an ι-linguistic behavioural presentations $B_{L,S} = (O_L, P_L, E_L, \lambda_L)$. Recall that $O_L = Pr(L)$ so that if $o \in O_L$, then o is of the form $(u'.\underline{s})_\iota$ for $u' \in S^*$ and unique $\underline{s} \in S$ and that $\lambda_L(o) = \underline{s}$. Of course, $E_L = S$. We now wheel in the function μ_Σ which tells us how to build a simultaneity class $\varepsilon_{\mu_\Sigma}(o)$ from o. From (25.19), we see that

$$\varepsilon_{\mu_\Sigma}(o) = \{(o,e_1,1),\cdots,(o,e_k,1)\}$$

where

$$\{e_1,\cdots,e_k\} = \{e \in E_\Sigma \mid \exists i \in I : (\lambda_L(o))(i) = e\}$$

and we know from 22.2.16 (1) that $\varepsilon_{\mu_\Sigma}(o)$ is a $\approx$-class of B_Σ. Note that since $\underline{\Omega}_{\alpha_\Sigma} \notin S$ by (25.4), it follows that $\varepsilon_{\mu_\Sigma}(o) \neq \varnothing$, for all $o \in O_L$.

However, we also know from $\lambda_L(o)$ which events have been performed by which process. This prompts the following definition - to simplify the notation a little, we shall write $o*e$ for $(o,e,1)$.

25.4.5. DEFINITION. Let Σ be an SPS. Define $\underline{\Pi}_\Sigma = (\Pi_\Sigma, \beta_\Sigma)$, where

1• $\Pi_\Sigma = I$;
2• $\beta_\Sigma : O_\Sigma \to \wp(\Pi_\Sigma)$ is given by $\beta_\Sigma(o*e) = \{i \in I \mid (\lambda_L(o))(i) = e\}$.

Of course:-

25.4.6. LEMMA If Σ is an SPS, then $\underline{\Pi}_\Sigma$ is a sequential process structure on B_Σ.

PROOF. Let $o*e \in O_\Sigma$. By definition, $(\lambda_L(o))(i) = e$, some $i \in I$, and hence $\beta_\Sigma(o*e) \neq \varnothing$, giving 25.3.1 (1).

Next suppose that $o_1*e_1, o_2*e_2 \in O_\Sigma$ such that $i \in \beta_\Sigma(o_1*e_1) \cap \beta_\Sigma(o_2*e_2)$, some $i \in I$, and that $o_1*e_1 \neq o_2*e_2$. By 25.4.5 (2), $(\lambda_\Sigma(o_1))(i) = e_1$ and $(\lambda_\Sigma(o_2))(i) = e_2$, so that if $o_1 = o_2$, then $e_1 = e_2$ and hence $o_1*e_1 = o_2*e_2$, a contradiction. Thus, $o_1 \neq o_2$. By 22.2.17, $o_1*e_1 \not\approx o_2*e_2$.

Since $(\lambda_\Sigma(o_1))(i) \neq \Omega \neq (\lambda_\Sigma(o_2))(i)$, $\neg\lambda_\Sigma(o_1)\ \iota_\Sigma\ \lambda_\Sigma(o_2)$, by (25.18), and hence $\neg o_1\ co_L\ o_2$. Thus, as $o_1 \neq o_2$ and $B_{L,S}$ is asynchronous, either $o_1 \#_L o_2$, $o_1 < o_2$ or $o_2 < o_1$.

If $o_1 \# o_2$, then $o_1*e_1 \#_\Sigma o_2*e_2$, for otherwise, there exists $p \in P_L$ such that $o_1*e_1, o_2*e_2 \in \varepsilon_{\mu_\Sigma}(p)$. But then $o_1, o_2 \in p$, by 22.22, a contradiction.

If $o_1 < o_2$, then $o_1*e_1 < o_2*e_2$, for if $o_2*e_2 \in \varepsilon_{\mu_\Sigma}(p)$, then $o_2 \in p$, by 22.22, and so $o_1 \in p$, giving $o_1*e_1 \in \varepsilon_{\mu_\Sigma}(p)$. Hence, $o_1*e_1 \to_\Sigma o_2*e_2$, and since $o_1*e_1 \not\approx o_2*e_2$, $o_1*e_1 < o_2*e_2$.

Similarly, if $o_2 < o_1$, then $o_2*e_2 < o_1*e_1$, and we have established 25.3.1 (2).

Finally, let $o \in O_\Sigma$, then $o = \underline{u} * e$, some $\underline{u} \in Pr(L)$ and $\lambda_L(\underline{u}) \in S$ so, by (25.4)

$$|\beta(o)| = |\{i \in I \mid (\lambda_L(\underline{u}))(i) \neq \Omega\}| < \infty.$$

■

25.4.6 allows us to use (25.17) to define a mapping $\rho_{\Pi_\Sigma} : P_\Sigma \to M_{\alpha_\Sigma}$. Thus we have one way of getting a set of vectors from B_Σ.

However, there is another way. Suppose $p \in P_\Sigma$, then $p = \varepsilon_{\mu_\Sigma}(Pr(\underline{x}))$, for some $\underline{x} \in L(C_\Sigma, \underline{q}_\Sigma)$. We shall now show that

$$\forall u, u' \in \underline{x} : e(u) = e(u') \tag{25.20}$$

Note that $\underline{x}$ is a trace and hence a subset of S^*. Each element of $\underline{x}$ is a string of elements of S. The function e takes the string and concatenates its elements coordinatewise.

(25.20) may be proved as follows. First, we argue that

$$\forall \underline{s}_1, \underline{s}_2 \in S : \underline{s}_1 \ \iota_\Sigma \ \underline{s}_2 \Rightarrow e(\underline{s}_1 . \underline{s}_2) = e(\underline{s}_2 . \underline{s}_1)$$

which may be established by a co-ordinatewise argument, in a manner analogous to 9.2.9. We may now conclude that

$$\forall u, u' \in S^* : u \equiv_{\iota_\Sigma}^{(1)} u' \Rightarrow e(u) = e(u'),$$

using the fact that e is a monoid morphism, from which (25.20) follows as $\equiv_{\iota_\Sigma}$ is the reflexive, transitive closure of $\equiv_{\iota_\Sigma}^{(1)}$.

Therefore we may define a mapping $b_\Sigma : P_\Sigma \to \langle S \rangle_{\alpha_\Sigma}{}^*$ by

$$b_\Sigma(\varepsilon_{\mu_\Sigma}(Pr(\underline{x}_{\iota_\Sigma}))) = e(\underline{x}), \tag{25.21}$$

where $\underline{x} \in S^*$. It follows that $b_\Sigma(P_\Sigma) = e(L(T_\Sigma, \underline{q}_\Sigma))$ and we may now apply 25.2.4, to obtain the following.

25.4.7. COROLLARY. If Σ is an SPS, then $b_\Sigma(P_\Sigma) = B(\Sigma)$.

■

We have two ways of mapping points of B_Σ to vectors, that is via the functions b_Σ and ρ_{Π_Σ}. Fortunately, the two coincide. This is at the heart of the following result.

25.4.8. THEOREM. If Σ is an SPS, then B_Σ is a discrete behavioural presentation and $\underline{\Pi}_\Sigma$ is a sequential process structure on B_Σ such that

$$\rho_{\underline{\Pi}_\Sigma}(P_\Sigma) = B(\Sigma).$$

PROOF. B_Σ is a discrete behavioural presentation, by 25.4.4 and 22.2.20, and $\underline{\Pi}_\Sigma$ is a sequential process structure on B_Σ, by 25.4.6. Also, $b_\Sigma(P_\Sigma) = B(\Sigma)$, from 25.4.7. In order to complete the proof, we show that $\rho_{\underline{\Pi}_\Sigma} = b_\Sigma$.

We argue by induction on $|p/\approx|$, $p \in P_\Sigma$, that $\rho_{\underline{\Pi}_\Sigma}(p) = b_\Sigma(p)$. If $|p/\approx| = 0$, then $p \in P_\Sigma$ and hence $\rho_{\underline{\Pi}_\Sigma}(p) = \underline{\Omega}_{\alpha_\Sigma} = b_\Sigma(p)$. Suppose $|p/\approx| = n > 0$. Let $o \in max(p)$, then by 7.3.3 and 3.3.1, $p' \succ_\Sigma{}^X p$ and $p' \in P_\Sigma$, where $X = o/\approx$ and $p' = p - X$. Clearly, $|p'/\approx| = n-1$ so by induction, $\rho_{\underline{\Pi}_\Sigma}(p') = b_\Sigma(p')$.

Consider $\pi \in \Pi_\Sigma$. Of course, $(p'|\pi) = (p|\pi) - X$. If $o, o' \in (p|\pi) \cap X$, then $o \approx_\Sigma o'$ and $\pi \in \beta_\Sigma(o) \cap \beta_\Sigma(o')$, so $o = o'$, by 25.3.1 (2). Hence, either $\pi \in \beta_\Sigma(o)$, for no $o \in X$ and $p'|\pi = p|\pi$, or there exists $o \in X$ such that $(p'|\pi) = (p|\pi) - \{o\}$. Since o is clearly maximal in $p|\pi$, it easily follows that

$$\rho(p|\pi) = \rho(p'|\pi).\lambda_\Sigma(o)$$

from which we may deduce that if $\rho_{\underline{\Pi}}(p') = e(\underline{s}_1 \cdots \underline{s}_n)$, then

$$\rho_{\underline{\Pi}}(p) = e(\underline{s}_1 \cdots \underline{s}_n . \underline{s}_X) \tag{25.22}$$

where

$$\underline{s}_X = \begin{cases} \lambda_\Sigma(o) \text{ if } o \in X \wedge \pi \in \beta_\Sigma(o) \\ \Omega \text{ otherwise} \end{cases}$$

Let us consider the relationship $p' \succ_\Sigma{}^X p$ from another point of view. By construction, there exists $\underline{x}, \underline{x}' \in L(C_\Sigma, \underline{q}_\Sigma)$ and $\underline{u} \in Pr(L(C_\Sigma, \underline{q}_\Sigma))$ such that $p = \varepsilon_{\mu_\Sigma}(Pr(\underline{x}))$, $p' = \varepsilon_{\mu_\Sigma}(Pr(\underline{x}'))$ and $X = \varepsilon_{\mu_\Sigma}(\underline{u})$.

Therefore, $\varepsilon_{\mu_\Sigma}(Pr(\underline{x}')) \succ^{\varepsilon_{\mu_\Sigma}(\underline{u})} \varepsilon_{\mu_\Sigma}(Pr(\underline{x}))$ from which it easily follows that $Pr(\underline{x}') \succ^{\underline{u}_{\iota_\Sigma}} Pr(\underline{x})$. By 12.3.2, $\underline{x} = \underline{x}'.\underline{s}_{\iota_\Sigma}$, where $\underline{s} = \lambda_L(\underline{u})$.

Let $\underline{y}' \in \underline{x}'_{\iota_\Sigma}$ and let $\underline{y} = \underline{y}'.\underline{s}$, then $\underline{y} \in \underline{x}_{\iota_\Sigma}$, and by (25.21), $b_\Sigma(p) = e(\underline{y})$ and $b_\Sigma(p') = e(\underline{y}')$. We thus have

$$b_\Sigma(p) = e(\underline{s}_1 \cdots \underline{s}_n . \underline{s}) \tag{25.23}$$

where we have invoked the induction hypothesis $\rho_{\underline{\Pi}_\Sigma}(p') = b_\Sigma(p')$.

Comparing (25.22) and (25.23), it is clear that the proof will be complete if we can show that $\underline{s}_X = \underline{s}$.

If $\underline{s}(\pi) = e$, then by (25.19), $(\mu_\Sigma(\underline{s}))(e) = 1$ and $\pi \in \beta_\Sigma(\underline{u} * e)$, by 25.4.5 (2). Hence $\underline{s}_X(\pi) = \lambda_\Sigma(\underline{u} * e) = e$ and so $\underline{s}_X(\pi) = \underline{s}(\pi)$.

Finally, suppose that $\underline{s}(\pi) = \Omega$. If $\underline{s}_X(\pi) \neq \Omega$, then there exists $o \in X$, such that $\pi \in \beta_\Sigma(o)$. But then by 25.4.5 (2), $\underline{s}(\pi) = \lambda_{L(C_\Sigma, \underline{q}_\Sigma)}(\underline{u}) = \lambda_\Sigma(o) \neq \Omega$. a contradiction, so $\underline{s}_X(\pi) = \Omega = \underline{s}(\pi)$.

Hence $\underline{s}_X = \underline{s}$ and the proof is complete.

■

25.4.9. COROLLARY. If Σ is an SPS and $(B, \underline{\Pi}) = (B_\Sigma, \underline{\Pi}_\Sigma)$, then

$$B(\Sigma_{(B,\underline{\Pi})}) = B(\Sigma).$$

PROOF. By 25.3.2 and 25.4.8.

■

We conclude this section by summarising the connection between the Nivat/Arnold processes and behavioural presentations.

25.4.10. DEFINITION. If $(B_1, \underline{\Pi}_1)$ and $(B_2, \underline{\Pi}_2)$ are behavioural presentations with process structure, then an isomorphism from $(B_1, \underline{\Pi}_1)$ to $(B_2, \underline{\Pi}_2)$ is a pair (f, δ), where

1• $f : B_1 \equiv B_2$;

2• $\delta : \Pi_1 \to \Pi_2$ is a bijection satisfying

$$\forall o \in O_1 : \delta(\beta_1(o)) = \beta_2(f_O(o)).$$

Write $(f, \delta) : (B_1, \underline{\Pi}_1) \equiv (B_2, \underline{\Pi}_2)$ to indicate that (f, δ) is an isomorphism from $(B_1, \underline{\Pi}_1)$ to $(B_2, \underline{\Pi}_2)$. Write $(B_1, \underline{\Pi}_1) \equiv (B_2, \underline{\Pi}_2)$ to indicate that there exists (f, δ) such that $(f, \delta) : (B_1, \underline{\Pi}_1) \equiv (B_2, \underline{\Pi}_2)$.

25.4.11. DEFINITION. If Σ_1 and Σ_2 are SPSs, then an isomorphism from Σ_1 to Σ_2 is a pair (ϕ, δ), where

1• $\phi : B(\Sigma_1) \to B(\Sigma_2)$ is a bijection;

2• $\delta : \Pi_1 \to \Pi_2$ is a bijection satisfying

$$\forall \underline{x} \in B(\Sigma_1)\ \forall \pi \in \Pi_1 : (\phi(\underline{x}))(\delta(\pi)) = \underline{x}(\pi).$$

Write $(\phi,\delta):\Sigma_1 \equiv \Sigma_2$ to indicate that (ϕ,δ) is an isomorphism from $(B_2,\underline{\Pi}_2)$ to $(B_2,\underline{\Pi}_2)$. Write $\Sigma_1 \equiv \Sigma_2$ to indicate that there exists (ϕ,δ) such that $(\phi,\delta):\Sigma_1 \equiv \Sigma_2$.

Let **BPB** denote the class of all ≡-classes of bounded, discrete behavioural presentations with process structure. Write $[\![B,\underline{\Pi}]\!]$ from the ≡-class of $(B,\underline{\Pi})$.

Let **NA** denote the class of all ≡-classes of SPSs. Write $[\![\Sigma]\!]$ from the ≡-class of Σ.

25.4.12. THEOREM. There is a surjective mapping $Beh_{NA}:\mathbf{BPB}\to\mathbf{NA}$ such that $Beh_{NA}([\![B,\underline{\Pi}]\!]) = [\![\Sigma_{(B,\underline{\Pi})}]\!]$ for all $(B,\underline{\Pi})\in\mathbf{BPB}$

PROOF. Suppose that $(f,\delta):(B_1,\underline{\Pi}_1)\equiv(B_2,\underline{\Pi}_2)$ and let $\Sigma_i = \Sigma_{(B_i,\underline{\Pi}_i)}$, each i.

For all $\underline{x}\in B(\Sigma_1)$, define $\phi(\underline{x}):\Pi_2 \to E_1{}^*$ by $(\phi(\underline{x}))(\pi) = \underline{x}(\delta^{-1}(\pi))$. We shall show that $(\phi,\delta):\Sigma_1\equiv\Sigma_2$. By construction, $\delta:\Pi_1\to\Pi_2$ is a bijection satisfying $(\phi(\underline{x}))(\delta(\pi)) = \underline{x}(\pi)$, all $\underline{x}\in B(\Sigma_1)$ and $\pi\in\Pi_1$, so it remains to show that ϕ is a bijection $\phi:B(\Sigma_1)\to B(\Sigma_2)$.

If $p\in P_1$ and $\pi\in\Pi_1$, then

$$\begin{aligned} f_O(p|\pi) &= f_O(\{o\in p \mid \pi\in\beta_1(o)\}, \text{by definition of } p|\pi \\ &= \{f_O(o)\in f_P(p) \mid \delta(\pi)\in\delta(\beta_1(o))\}, \text{by 19.4.1 (2)} \\ &= \{o'\in f_P(p)\mid \delta(\pi)\in\beta_2(o')\}, \text{by 25.4.10} \\ &= f_P(p)|\delta(\pi). \end{aligned}$$

and since $o <_1 o' \Leftrightarrow f_O(o) <_2 f_O(o')$, it follows that $\rho(f_O(p|\pi)) = \rho(f_P(p)|\delta(\pi))$. But, $\lambda_2(f_O(o)) = \lambda_1(o)$ for all $o\in O_1$ and so $\rho(p|\pi) = \rho(f_P(p)|\delta(\pi))$. We have shown that $\phi(\rho_{\underline{\Pi}_1}(p)) = \rho_{\underline{\Pi}_2}(f_P(p))$, all $p\in P_1$, and hence $\phi(B(\Sigma_1))\subseteq B(\Sigma_2)$. By symmetry $\phi^{-1}(B(\Sigma_2))\subseteq B(\Sigma_1)$ so $\phi(B(\Sigma_1)) = B(\Sigma_2)$. Thus, ϕ is surjective. If $\phi(\underline{x}) = \phi(\underline{y})$, then $\underline{x}(\pi) = (\phi(\underline{x}))(\delta(\pi)) = (\phi(\underline{y}))(\delta(\pi)) = \underline{y}(\pi)$ and so ϕ is injective. This completes the proof that $(\phi,\delta):\Sigma_1\equiv\Sigma_2$. We have proved that $(B_1,\underline{\Pi}_1)\equiv(B_2,\underline{\Pi}_2)\Rightarrow\Sigma_{(B_1,\underline{\Pi}_1)}\equiv\Sigma_{(B_2,\underline{\Pi}_2)}$, and so Beh_{NA} is well-defined.

Finally, let $\Sigma\in\mathbf{NA}$ and let $(B,\underline{\Pi}) = (B_\Sigma,\underline{\Pi}_\Sigma)$. By 25.4.9, $B(\Sigma_{(B,\underline{\Pi})}) = B(\Sigma)$ and so $\Sigma_{(B,\underline{\Pi})}\equiv\Sigma$, where ϕ and δ are identity functions, from which we deduce that $Beh_{NA}([\![B,\underline{\Pi}]\!]) = [\![\Sigma]\!]$.

■

25.5. Comparison with the Path/Process Notation.

In this section, we examine Arnold's claim that $Cosy \subset Synchro$, which appears to suggest that the process model is less expressive than the path/process model. In one sense, this is correct, as we have pointed out already. This relationship between the semantics is misleading, however, for the simple reason that a vector semantics is not suitable in general for path/process programs and this is because they may possess ambiguity. The Nivat/Arnold process model suffers, as do all purely language based models from its inability to express this. We shall show that Arnold's claim does hold for the class of *unambiguous* path/process programs.

Let us fix a path/process program $Prog = \textbf{begin}\ P_1 \cdots P_n\ \textbf{end}$. We know that $Prog$ determines two initialised hybrid transition systems $(H,\underline{q}) = \rho[\![Prog]\!]$, and $(H_\Sigma,\underline{q}_\Sigma)$, where $\Sigma = \Sigma_{Prog}$. Let us examine the relationship between these two.

Let $I = \{1,\cdots,n\}$, $J = I_{path}(Prog)$ and $K = I_{proc}(Prog)$. If $i \in I$, then we define $(H_i,\underline{q}_i) = \rho[\![P_i]\!]$ and let $\overline{Q} = \bigcup_{i\in I} Q_i$ and $\overline{A} = \bigcup_{i\in I} A_i$.

The following sums up what we know about the structure of $(H,\underline{q})$

25.5.1. LEMMA. With the above notation

1• $\underline{q} \in Q \Leftrightarrow$ there exists $\underline{q}_J : J \to \overline{Q}$ and $\underline{q}_J : K \to \overline{Q}$ such that $\underline{q} = (\underline{q}_J, \underline{q}_K)$ and

(a) $\forall i \in J: \underline{q}_J(i) \in Q_i$

(b) $\forall j \in K: \underline{q}_J(j) \in Q_j$

2• $\underline{a} \in A \Leftrightarrow$ there exists $\underline{y}_a : J \to \overline{A} \cup \{\Omega\}$, $z_a \in (\overline{A} \times K) \cup \{\Omega\}$, $k \in K$ and $e \in E$ such that $\underline{a} = (\underline{y}_a, \underline{z}_a)$ and

(a) $\underline{z}_a = \Omega \Rightarrow \exists i \in J: \underline{y}_a(i) \neq \Omega$

(b) $\forall i \in J: (e \in E_i \Rightarrow \underline{y}_a(i) \in A_i \wedge \mu_i(\underline{y}_a(i)) = e) \wedge (e \notin E_i \Rightarrow \underline{y}_a(i) = \Omega)$

(c) $\underline{z}_a = (a,k) \Rightarrow a \in A_k \wedge \mu_k(a) = e$ and $\mu_k(a) = e$.

3• $\underline{q} \to^{(\underline{y},(a,k))} \underline{q}' \Leftrightarrow$

(a) $\forall i \in J: (\underline{y}_a(i) \neq \Omega \Rightarrow \underline{q}_J(i) \to_i^{\underline{y}(i)} \underline{q}'_J(i)) \wedge (\underline{y}_a(i) \neq \Omega \Rightarrow \underline{q}_J(i) = \underline{q}'_J(i))$

(b) $\forall k \in K: \underline{z}_a = (a,k) \Rightarrow \underline{q}_K(k) \to_j^{(a,k)} \underline{q}'_K(k)$.

(c) $\forall k \in K\ \forall a \in A_k: \underline{z}_a \neq (a,k) \Rightarrow \underline{q}_K(k) = \underline{q}'_K(k))$

4• $\underline{a}\ \iota\ \underline{b} \Leftrightarrow$

(a) $\forall i \in J: \underline{y}_a(i) > \Omega \Rightarrow \underline{y}_b(i) = \Omega$

(b) $\underline{z}_a = (a,j) \wedge \underline{z}_b = (b,k) \Rightarrow j \neq k$.

5• $\mu(\underline{a}) = e \Leftrightarrow (\exists i \in J : \mu_i(\underline{y}_a) = e) \vee ((a,k) \in \overline{A} \times K : \underline{z}_a = (a,k) \wedge \mu_k(a) = e)$.

PROOF. By 24.3.1, 24.3.3 and 24.3.5. ∎

We shall first construct an isomorphic copy of H, as follows

25.5.2. DEFINITION. With the above notation, define $\hat{H} = (\hat{Q}, \hat{A}, \hat{\rightarrow}, \hat{\iota}, E, \hat{\mu})$, where:

1• $\hat{Q} = \{\hat{\underline{q}} \mid \underline{q} \in Q\}$, where if $\underline{q} \in Q$ and $i \in I$, then

$$\hat{\underline{q}}(i) = \begin{cases} \underline{q}_J(i) \text{ if } i \in J \\ \underline{q}_K(i) \text{ if } i \in K \end{cases}$$

2• $\hat{A} = \{\hat{\underline{a}} \mid \underline{a} \in A\}$, where if $\underline{a} \in A$ and $i \in I$, then

$$\hat{\underline{a}}(i) = \begin{cases} \underline{y}_a(i) \text{ if } i \in J \\ a \text{ if } \underline{z}_a = (a,i) \\ \Omega \text{ otherwise} \end{cases}$$

3• $\hat{\underline{q}}_1 \hat{\rightarrow}^{\hat{a}} \hat{\underline{q}}_2 \Leftrightarrow \hat{\underline{q}}_1(i) \hat{\rightarrow}_i^{\hat{a}(i)} \hat{\underline{q}}_2(i)$ when $\hat{\underline{q}}_1(i) = \hat{\underline{q}}_2(i)$; otherwise, $\hat{\underline{a}}(i) \neq \Omega$;

4• $\hat{\underline{a}} \,\hat{\iota}\, \hat{\underline{b}} \Leftrightarrow$ for all $i \in I$, $\hat{\underline{a}}(i) \neq \Omega \Rightarrow \hat{\underline{b}}(i) = \Omega$

5• $\hat{\mu}(\hat{\underline{a}}) = \mu(\underline{a})$.

The relationship between $(\hat{H}, \hat{\underline{q}})$ and $(H, \underline{q})$ is summed up in the following lemma.

25.5.3. LEMMA. With the above notation

1• The functions $\underline{x} \mapsto \hat{\underline{x}}$ are bijections $Q \rightarrow \hat{Q}$ and $A \rightarrow \hat{A}$;

2• $\hat{\underline{q}}_1 \hat{\rightarrow}^{\hat{a}} \hat{\underline{q}}_2 \Leftrightarrow \underline{q}_1 \rightarrow^{\underline{a}} \underline{q}_2$;

3• $\hat{\underline{a}} \,\hat{\iota}\, \hat{\underline{b}} \Leftrightarrow \underline{a} \,\iota\, \underline{b}$;

4• $(\hat{H}, \hat{\underline{q}}) \equiv (H, \underline{q})$.

PROOF. (1), (2) and (3) may easily be checked by a comparison between 25.5.1 and 25.5.2. For (4), we define a relation $Q \, R \, \hat{Q}$ by $\underline{q} \, R \, \underline{q}' \Leftrightarrow \underline{q}' = \hat{\underline{q}}$ and a family F of functions $f_{(\underline{q},\hat{\underline{q}})}:|\underline{q}\rangle = |\hat{\underline{q}}\rangle$ given by $f_{(\underline{q},\hat{\underline{q}})}(\underline{a},\underline{q}') = (\hat{\underline{a}},\hat{\underline{q}}')$ and use (1), (2) and (3) to show that (R,F) is a strict equivalence.

■

We now set about relating the structure of $(\hat{H},\hat{\underline{q}})$ to that of $(H_\Sigma,\underline{q}_\Sigma)$ beginning with a number of observations about the relationship between components of these structure. We define a function $\phi(\underline{a}):I \to E\cup\{\Omega\}$ as follows. If $\underline{a}\in\hat{A}$, then,

$$(\phi(\underline{a}))(i) = \begin{cases} \mu_i(\underline{a}(i)) \text{ if } \underline{a}(i)\neq\Omega \\ \Omega \text{ otherwise} \end{cases} \tag{25.24}$$

25.5.4. LEMMA. ϕ is an onto function $\hat{A}\to S_\Sigma$ satisfying:

$$\phi(\hat{\underline{a}}) = \begin{cases} \underline{e}_j \text{ if } \underline{z}_a = (a,j) \\ \underline{e}_\perp \text{ otherwise} \end{cases}$$

where $e = \hat{\mu}(\hat{\underline{a}})$.

■

If $\underline{x} = \underline{a}_1\cdots\underline{a}_n$, where $\underline{a}_k\in\hat{A}$, each k, then for each $i\in I$, define $\underline{x}(i) = \underline{a}_1(i)\cdots\underline{a}_n(i)$.

25.5.5. LEMMA. There exists an onto function $\theta: L(C(\hat{H}),\hat{\underline{q}})\to B(\Sigma)$ such that if $\underline{x}\in L(C(\hat{H}),\hat{q})$, then for each $i\in I$, $\theta(\underline{x}_\iota)(i) = \mu_i(\underline{x}(i))$.

PROOF. If $\underline{x}\in\hat{A}^*$, then we define $h(\underline{x}):I\to E^*$ by $(h(\underline{x}))(i) = \mu_i(\underline{x}(i))$, each $i\in I$. Now, if $\underline{a},\underline{b}\in\hat{A}$ such that $\underline{a}\,\hat{\iota}\,\underline{b}$, then $(\underline{a}.\underline{b})(i) = (\underline{b}.\underline{a})(i)$, each $i\in I$, by 25.5.2 (4). It easily follows that if $\underline{x},\underline{y}\in\hat{A}^*$ with $\underline{x}\equiv_\iota^{(1)}\underline{y}$, then $\underline{x}(i) = \underline{y}(i)$. Since $\equiv_\iota$ is the equivalence relation generated by $\equiv_\iota^{(1)}$, it follows that $\underline{x}\equiv_\iota\underline{y}\Rightarrow\underline{x}(i) = \underline{y}(i)$. Thus, if $\underline{x},\underline{y}\in\hat{A}^*$, then $\underline{x}\equiv_\iota\underline{y}\Rightarrow h(\underline{x}) = h(\underline{y})$, so if $\underline{x}\in\hat{A}^*$ we may unambiguously define $\theta(\underline{x}_\iota) = h(\underline{x})$. We shall show that $\theta(\underline{x}_\iota)\in B(\Sigma)$ for all $\underline{x}\in L(T(C(\hat{H})),\hat{\underline{q}})$.

First, we note that if $\underline{x} = \underline{a}_1\cdots\underline{a}_k$, with $\underline{a}_1,\cdots,\underline{a}_k\in\hat{A}_{Prog}$, then

$$\begin{aligned}\mu_i(\underline{x}(i)) &= (\phi(\underline{a}_1))(i)\cdots(\phi(\underline{a}_k))(i), \text{ by (25.24)} \\ &= e(\phi(\underline{a}_1)\cdots\phi(\underline{a}_k))(i)\end{aligned}$$

and so $\theta(\underline{x}_i) = e(\phi(\underline{a}_1)\cdots\phi(\underline{a}_k))$. But $\phi(\underline{a}_1)\cdots\phi(\underline{a}_k) \in S^*$, by 25.5.4. Therefore, $\theta(\underline{x}_i) \in e(S^*) = \langle S \rangle_\alpha$, where $\alpha = \alpha_\Sigma$.

Finally, if $\underline{x} \in L(T(C(\hat{H})),\underline{\hat{q}})$, then $\underline{x}(i) \in L(T(C(H_i)),\underline{q}_i)$ for each $i \in I$, by 25.5.1 (3). Hence,

$$(\theta(\underline{x}_i))(i) = (h(\underline{x}))(i) = \mu_i(\underline{x}(i)) \in \mu_i(L(T(C(H_i)),q_i)) = B(P_i)$$

so that $\theta(\underline{x}_i) \in B(\mathcal{P})$. Therefore:

$$\theta(\underline{x}_i) \in \mathcal{P} \cap \langle S \rangle_\alpha = B(\Sigma)$$

■

We use this result to define a relation between states of $(\hat{H},\underline{\hat{q}})$ and of $(H_\Sigma,\underline{q}_\Sigma)$, recalling from 25.2.3, that the latter are the elements of $B(\Sigma)$.

25.5.6. DEFINITION. We define a relation $\hat{Q}\,\hat{R}\,Q_\Sigma$ by

$$\underline{\hat{q}}\,\hat{R}\,\underline{m} \Leftrightarrow \exists \underline{x} \in \hat{A}^*: \underline{\hat{q}} \hat{\rightarrow}^{\underline{x}} \underline{q} \wedge \theta(\underline{x}_i) = \underline{m}.$$

Let $\underline{\hat{q}'}\,\hat{R}\,\underline{m}$ and let $(\underline{a},\underline{\hat{q}''}) \in \left|\underline{\hat{q}'}\right\rangle$. By 25.5.6, there exists $\underline{x} \in \hat{A}^*$ such that $\underline{\hat{q}} \hat{\rightarrow}^{\underline{x}} \underline{\hat{q}'}$ and $\theta(\underline{x}_i) = \underline{m}$. Therefore $\underline{\hat{q}} \hat{\rightarrow}^{\underline{x}.\underline{a}} \underline{\hat{q}''}$ and, by comparison of co-ordinates, $\theta(\underline{x}_i.\underline{a}_i) = e(\theta(\underline{x}_i).\phi(\underline{a}))$. By 25.2.3, $\underline{m} = \theta(\underline{x}_i) \rightarrow_\Sigma^{\phi(\underline{a})} \theta(\underline{x}_i\underline{a}_i)$. Hence, if we define

$$f_{(\underline{\hat{q}'},\underline{m})}(\underline{a},\underline{\hat{q}''}) = (\phi(\underline{a}),\theta(\underline{x}_i.\underline{a}_i)) \tag{25.25}$$

then $f_{(\underline{\hat{q}'},\underline{m})}(\underline{a},\underline{\hat{q}''}) \in |\underline{m}\rangle$. Let $\hat{F} = \{f_{(\underline{\hat{q}},\underline{m})} \mid (\underline{\hat{q}},\underline{m}) \in \hat{R}\}$.

25.5.7. LEMMA. If all the functions $f_{(\underline{\hat{q}'},\underline{m})}$, $(\underline{\hat{q}'},\underline{m}) \in \hat{R}$ are bijective, then $(\hat{R},\hat{F})$ is a strict equivalence from $(\hat{H},\underline{\hat{q}})$ to $(H_\Sigma,\underline{q}_\Sigma)$.

PROOF. $\underline{\hat{q}} \hat{\rightarrow}^{\Omega} \underline{\hat{q}}$ and $(\theta(\Omega_i))(i) = \Omega$, each $i \in I$ so that $\theta(\Omega_i) = \underline{q}_\Sigma$, by 25.2.3, and so $\underline{\hat{q}}\,\hat{R}\,\underline{q}_\Sigma$, giving (1) of 22.2.13. Also, by (25.25)

$$f_{(\underline{\hat{q}'},\underline{m})}(\underline{a},\underline{\hat{q}''}) = (\underline{e},\underline{m'}) \Rightarrow \mu(\underline{a}) = \mu_\Sigma(\underline{e}) \wedge \underline{\hat{q}''}\,\hat{R}\,\underline{m'}$$

which gives (2) of 22.2.13. By (25.24), 25.5.1 (4) and (25.18), it is clear that if $\underline{a},\underline{b} \in \hat{A}$, then $\underline{a}\,\hat{\iota}\,\underline{b} \Rightarrow \phi(\underline{a})\,\iota_\Sigma\,\phi(\underline{b})$. Hence:

$$(f_{(\hat{\underline{q}}_0,\underline{m}_0)}(\underline{a}_1,\hat{\underline{q}}_1)=(\underline{e}_1,\underline{m}_1)\wedge f_{(\hat{\underline{q}}_1,\underline{m}_1)}(\underline{a}_2,\hat{\underline{q}}_2)=(\underline{e}_2,\underline{m}_2))\Rightarrow(\underline{a}_1\ \hat{\iota}\ \underline{a}_2\Rightarrow\underline{e}_1\ \iota_\Sigma\ \underline{e}_2)$$

giving (3) of 22.2.13. Finally, suppose that $(f_{(\hat{\underline{q}}_0,\underline{m}_0)}(\underline{a}_1,\hat{\underline{q}}_1)=(\underline{e}_1,\underline{m}_1)$, $f_{(\hat{\underline{q}}_1,\underline{m}_1)}(\underline{a}_2,\hat{\underline{q}}_2)=(\underline{e}_2,\underline{m}_2))$, $(f_{(\hat{\underline{q}}_0,\underline{m}_0)}(\underline{a}_2,\hat{\underline{q}}'_1)=(\underline{e}'_2,\underline{m}'_1)$, $f_{(\hat{\underline{q}}'_1,\underline{m}'_1)}(\underline{a}_1,\hat{\underline{q}}'_2)=(\underline{e}'_1,\underline{m}'_2))$ and $\underline{a}_1\ \hat{\iota}\ \underline{a}_2$. Then for each i, $\underline{e}_i=\phi(\underline{a}_i)=\underline{e}'_i$, by (25.25).

■

25.5.7 shows the limits of validity of the claim $Cosy\subset Synchro$, the problem being that in general the functions $f_{(\hat{\underline{q}},\underline{m})}$ are not bijective. The bugbear here is ambiguity, as we have explained We shall now show that lack of ambiguity is sufficient to guarantee that $(\hat{H},\hat{\underline{q}})\equiv(H_\Sigma,\underline{q}_\Sigma)$.

25.5.8. DEFINITION. Suppose that H is a hybrid transition system, then H is *unambiguous* $\Leftrightarrow$

$$\forall q,q_1,q_2\in Q\ \forall a_1,a_2\in A: q\rightarrow^{a_1}q_1\wedge q\rightarrow^{a_2}q_2\wedge\mu(a_1)=\mu(a_2)\Rightarrow a_1=a_2.$$

If $Prog=$ **begin** $P_1\cdots P_n$ **end** is a path/process program, then $Prog$ is unambiguous $\Leftrightarrow$ $\rho[\![P_i]\!]$ is unambiguous, each i.

25.5.9. PROPOSITION. If H_{Prog} is unambiguous, then $(H,\underline{q})\equiv(H_\Sigma,\underline{q}_\Sigma)$.

PROOF. By 25.5.3, if H is unambiguous, then $\hat{H}$ is unambiguous and if $(\hat{H},\hat{\underline{q}})\equiv(H_\Sigma,\underline{q}_\Sigma)$, then $(H,\underline{q})\equiv(H_\Sigma,\underline{q}_\Sigma)$, so in view of 25.5.8, it suffices to show that if $\hat{H}$ is unambiguous, then the functions $f_{(\hat{\underline{q}}',\underline{m})}$, $(\hat{\underline{q}}',\underline{m})\in\hat{R}$, are bijective.

Let $\hat{\underline{q}}'\hat{R}\underline{m}$ and let $\underline{x}\in\hat{A}^*$ such that $\hat{\underline{q}}\hat{\rightarrow}^{\underline{x}}\hat{\underline{q}}'$ and $\theta(\underline{x}_\iota)=\underline{m}$. If $(\underline{a}_1,\hat{\underline{q}}_1),(\underline{a}_2,\hat{\underline{q}}_2)\in\left|\hat{\underline{q}}'\right\rangle$ with $f_{(\hat{\underline{q}}',\underline{m})}(\underline{a}_1,\hat{\underline{q}}_1)=f_{(\hat{\underline{q}}',\underline{m})}(\underline{a}_2,\hat{\underline{q}}_2)$, then $\theta(\underline{a}_1)=\theta(\underline{a}_2)$ by (25.25) and $\hat{\underline{q}}'\hat{\rightarrow}^{\underline{a}_1}\hat{\underline{q}}_1$ and $\hat{\underline{q}}'\hat{\rightarrow}^{\underline{a}_2}\hat{\underline{q}}_2$. Thus, if $i\in I$, then by (25.24)

$$\underline{a}_1(i)\neq\Omega\Leftrightarrow(\theta(\underline{a}_1))(i)\neq\Omega\Leftrightarrow(\theta(\underline{a}_2))(i))\neq\Omega\Leftrightarrow\underline{a}_2(i)\neq\Omega$$

Thus if $\underline{a}_1(i)\neq\Omega$, then $\underline{a}_2(i)\neq\Omega$ so $\hat{\underline{q}}'(i)\hat{\rightarrow}^{\underline{a}_1(i)}\hat{\underline{q}}_1(i)$ and $\hat{\underline{q}}'(i)\hat{\rightarrow}^{\underline{a}_2(i)}\hat{\underline{q}}_2(i)$, by 25.5.2 (3), and since $\theta(\underline{a}_1)=\theta(\underline{a}_2)$, $\mu_i(\underline{a}_1(i))=\mu_i(\underline{a}_2(i))$, so $\underline{a}_1(i)=\underline{a}_2(i)$ as H_i is unambiguous. Also, as $T(C(H_i))$ is unambiguous, if $\underline{a}_1(i)\neq\Omega$, then $\hat{\underline{q}}_1(i)=\hat{\underline{q}}_2(i)$. On the other hand, if $\underline{a}_1(i)=\Omega$, then $\underline{a}_2(i)=\Omega$ and so $\underline{a}_1(i)=\underline{a}_2(i)$, and hence $\hat{\underline{q}}_1(i)=\hat{\underline{q}}'(i)=\hat{\underline{q}}_2(i)$, by 25.5.2 (3). Therefore, $(\underline{a}_1,\hat{\underline{q}}_1)=(\underline{a}_2,\hat{\underline{q}}_2)$ and we have shown that the functions $f_{(\hat{\underline{q}}',\underline{m})}$ are injective.

Now, suppose that $(\underline{e},\underline{m}') \in |\underline{m}\rangle$, then $\underline{m} \rightarrow_{\Sigma}^{\underline{e}} \underline{m}'$ and so by 25.2.3 (2), if $i \in I$, then $\underline{m}'(i) = \underline{m}(i).\underline{e}(i)$. By (25.6), there exists $x_i \in L(T(C(H_i)),q_i)$ such that $\underline{m}(i) = \mu_i(x_i)$ and if $\underline{e}(i) \neq \Omega$, then there exists $a_i \in A_i$ such that $\mu_i(a_i) = \underline{e}(i)$ and $x_i.a_i \in L(T(C(H_i)),q_i)$. Define $\underline{a}$ by

$$\underline{a}(i) = \begin{cases} a_i \text{ if } \underline{e}(i) \neq \Omega \\ \Omega \text{ otherwise} \end{cases}$$

and we note that $\phi(\underline{a}) = \underline{e}$.

But, we also have, $\underline{x}(i) \in L(T(C(H_i)),q_i)$ with $\underline{m}(i) = (\theta(\underline{x}))(i) = \mu_i(\underline{x}(i))$, and as H_i is unambiguous, it follows that $\underline{x}(i) = x_i$. Therefore, if $\underline{e}(i) \neq \Omega$, then $\underline{x}(i).a_i \in L(T(C(H_i)),q_i)$ and so there exists $q_i \in Q_i$ such that $\underline{\hat{q}}'(i) \rightarrow^{a_i}{}_i\, q_i$. Thus, if we define

$$\underline{\hat{q}}''(i) = \begin{cases} q_i \text{ if } \underline{e}(i) \neq \Omega \\ \underline{\hat{q}}'(i) \text{ otherwise} \end{cases}$$

then $\underline{\hat{q}}'' \in \hat{Q}$ and $\underline{\hat{q}}' \hat{\rightarrow}^{\underline{a}} \underline{\hat{q}}''$. We also observe that $\underline{m}'(i) = h(\underline{x}.\underline{a}) = (\theta(\underline{x}_i.\underline{a}_i))(i)$. Hence, $(\underline{a},\underline{\hat{q}}'') \in |\underline{\hat{q}}'\rangle$ and $f_{(\underline{\hat{q}}',\underline{m})}(\underline{a},\underline{\hat{q}}'') = (\phi(\underline{a}),\theta(\underline{x}_i.\underline{a}_i)) = (\underline{e},\underline{m}')$ and we have shown that the functions $f_{(\underline{\hat{q}}',\underline{m})}$ are surjective.

∎

25.6. Process Structure and Occurrence Nets.

In this section, we show that the existence of a special kind of process structure on a discrete asynchronous behavioural presentation allows one to embed it into an occurrence net with the same synchronic structure. As we are no longer concerned to match behavioural presentations with SPSs, we relax the definition of process structure. For the rest of this session, a process structure is an entity satisfying (1) and (2) of 25.3.1.

We begin with the construction. First, if $o_1, o_2 \in O$, then define

$$o_1 \lhd o_2 \Leftrightarrow o_1 < o_2 \wedge (\forall o \in O: o_1 \rightarrow o \rightarrow o_2 \Rightarrow o = o_1 \vee o = o_2).$$

25.6.1. DEFINITION. If $(B,\underline{\Pi})$ is a behavioural presentation with process structure, then define $N_{(B,\underline{\Pi})} = (S,P,F)$ where $F \subseteq (S \times O) \cup (O \times S)$ and

1• $S = \{(o,\pi) \mid o \in O \wedge \pi \in \beta(o)\} \cup \{(\Omega,\pi) \mid \pi \in \bigcup_{o \in min(O)} \beta(o)\}$;

2• $o\,F\,(x,\pi) \Leftrightarrow x = o$;

3• $(x,\pi)\,F\,o \Leftrightarrow (x \in O \wedge x \lhd o \wedge \pi \in \beta(o) \cap \beta(x))$
$$\vee (x = \Omega \wedge o \in min(O) \wedge \pi \in \beta(o)).$$

25.6.2. LEMMA. If $(B, \underline{\Pi})$ is a discrete behavioural presentation with process structure such that $P \neq \emptyset$, then $N_{(B,\underline{\Pi})}$ is a net.

PROOF. Let $(S,P,F) = N_{(B,\underline{\Pi})}$. We check the conditions of 16.2.1.

$S \cup P \neq \emptyset$ by hypothesis and $S \cap P = \emptyset$, giving (a).

If $o \in O$, then $\beta(o) \neq \emptyset$, by 25.3.1, and so $o\,F\,(o,\pi)$, by 25.6.1 (2), some $\pi \in \Pi$ and hence $P \subseteq dom(F)$. If $(o,\pi) \in S$, with $o \in O$, then again $o\,F\,(o,\pi)$, by 25.6.1 (2), so $(o,\pi) \in ran(F)$. If $(\Omega,\pi) \in S$, then there exists $o \in min(O)$ such that $\pi \in \beta(o)$ and $(\Omega,\pi)\,F\,o$ by 25.6.1 (2) and $(\Omega,\pi) \in dom(F)$. We have shown that $S \cup P \subseteq dom(F) \cup ran(F)$. Since the reverse inclusion holds, we have equality, giving 16.2.1 (b).

■

25.6.3. DEFINITION. If $\underline{\Pi}$ is a process structure on a behavioural presentation B, then $\underline{\Pi}$ is *complete* $\Leftrightarrow$

$$\forall o_1, o_2 \in O : o_1 \lhd o_2 \vee o_1 \# o_2 \Rightarrow \beta(o_1) \cap \beta(o_2) \neq \emptyset.$$

25.6.4. LEMMA. If $\underline{\Pi}$ is a complete process structure on a behavioural presentation B and $(S,P,F) = N_{(B,\underline{\Pi})}$, then

$$\forall o_1, o_2 \in O : o_1 \lhd o_2 \Leftrightarrow \exists s \in S : o_1\; F\; s\; F\; o_2$$

PROOF. Suppose that $o_1 \lhd o_2$, then by 25.6.3, there exists $\pi \in \beta(o_1) \cap \beta(o_2)$ and so if $s = (o_1,\pi)$, then $o_1\,F\,s\,F\,o_2$, by 25.6.1 (2) and (3). Conversely, if $o_1\,F\,s\,F\,o_2$, then $s = (o_1,\pi)$, by, some $\pi \in \beta(o_1)$, by 25.6.1 (2) and $o_1 \lhd o_2$, by 25.6.1 (3).

■

25.6.5. LEMMA. If $\underline{\Pi}$ is a complete process structure on a discrete, asynchronous behavioural presentation B and $(S,P,F) = N_{(B,\underline{\Pi})}$, then $F^* \cap (O \times O) = \rightarrow$ and F^* is a partial order.

PROOF. As in the proof of 17.4.4, it suffices to show that $F^* \cap (O \times O) = \rightarrow$, as $\rightarrow$ is a partial order, by asynchronicity of B. Suppose that $o, o' \in O$. If $o \rightarrow o'$, then either $o = o'$, in which case $o\,F^*\,o'$, or $o = o_1 \lhd \cdots \lhd o_n = o'$, some $o_1, \cdots, o_n \in O$, since B satisfies both chain conditions, and so $o\,F^*\,o'$ by 25.6.4. Conversely, if $o\,F^*\,o'$, then either $o = o'$, in which case $o \rightarrow o'$, or

$o = o_1\ F\ s_1 \cdots s_{n-1}\ F\ o_n = o'$, for some $o_1, \cdots, o_n \in O$ and $s_1, \cdots, s_{n-1} \in S$, and so $o = o_1 \lhd \cdots \lhd o_n = o'$, by 25.6.4, giving $o \to o'$.

■

25.6.6. LEMMA. If $\underline{\Pi}$ is a complete process structure on a discrete, asynchronous behavioural presentation B and $N = N_{(B,\underline{\Pi})}$, then

$$\forall o_1, o_2 \in O : o_1 \# o_2 \Leftrightarrow o_1 \#_N o_2.$$ [8]

PROOF. Suppose first that $o_1 \# o_2$, then by 25.6.3, there exists $\pi \in \beta(o_1) \cap \beta(o_2)$. Let $X_i = \downarrow o_i \cap \beta(o_i)$, each i, then by discreteness, the X_i are finite sets totally ordered by $\to$. We have two cases.

If $X_1 \cap X_2 \neq \varnothing$, then let $\bar{o} = max(X_1 \cap X_2)$ and $\hat{o}_i = min(X_i - \downarrow o_i)$, each i. As the X_i are finite $\bar{o} \lhd \hat{o}_i$ and hence $\bar{o}\ F\ (\bar{o}, \pi)\ F\ \hat{o}_i$, each i, by 25.6.1 (3).

Otherwise $X_1 \cap X_2 = \varnothing$. Let $\hat{o}_i = min(X_i)$, then $(\Omega, \pi)\ F\ \hat{o}_i$, each i, by 25.6.1 (3).

Thus, in both cases $\hat{o}_1 \#_N^{(1)} \hat{o}_2$. As $\hat{o}_i \to o_i$, $\hat{o}_i\ F^*\ o_i$, each i, by 25.6.5 and so $o_1 \#_N o_2$.

Conversely, suppose that $o_1 \#_N o_2$, then there exists $s \in S$ and $\hat{o}_1, \hat{o}_2 \in T$ such that $\hat{o}_1 \neq \hat{o}_2$ and $s\ F\ \hat{o}_i\ F^*\ o_i$, each i. By 25.6.1 (1) $s = (x, \pi)$, some $\pi \in \Pi$ and $x \in O \cup \{\Omega\}$. By 25.6.1 (3), $\pi \in \beta(\hat{o}_1) \cap \beta(\hat{o}_2)$. By 25.3.1 (2), $\hat{o}_1 < \hat{o}_2$ or $\hat{o}_2 < \hat{o}_1$ or $\hat{o}_1 \# \hat{o}_2$.

Suppose that $\hat{o}_1 < \hat{o}_2$, then $\hat{o}_2 \notin min(O)$, so $s = (\bar{o}, \pi)$, some $\bar{o} \in O$ and $\bar{o} \lhd \hat{o}_1$ and $\bar{o} \lhd \hat{o}_2$, by 25.6.1 (3). But $\bar{o} \lhd \hat{o}_1 < \hat{o}_2$, a contradiction. Hence, $\hat{o}_1 \not< \hat{o}_2$. Similarly, $\hat{o}_2 \not< \hat{o}_1$. The only other possibility is $\hat{o}_1 \# \hat{o}_2$, as required.

■

We may now show that a behavioural presentation with a complete process structure may be turned into an occurrence net.

25.6.7. PROPOSITION. If $\underline{\Pi}$ is a complete process structure on a discrete, asynchronous behavioural presentation B and $N = N_{(B,\underline{\Pi})}$, then N is an occurrence net.

PROOF. We check the conditions of 16.3.1. F^* is a partial order, by 26.6.5, giving (1). If $o_1, o_2\ F\ (o, \pi)$, then $o_1 = o = o_2$, by 25.6.1 (2), giving (2). Finally, since # is irreflexive, $\#_N$ is, by 25.6.6, giving (3)

■

[8] See section 16.3.

Occurrence nets also give rise to behavioural presentations; one throws away the S elements and retains the T elements and the ordering and conflict relation.

25.6.8. DEFINITION. Suppose that $N=(S,T;F)$ is an occurrence net and $\lambda:T\rightarrow E$, then we define $S_N=(T,F^*\cap(T\times T),\#_N)$ and $B_{(N,\lambda)}=(T,P,E,\lambda)$, where $(T,P)=\Lambda(B_N)$.

So given a complete process structure on a discrete, asynchronous behavioural presentation, we may construct an occurrence net and a labelling function on its elements, and from that net and function we may construct a labelled events structure and hence a closed behavioural presentation. The relationship between the two behavioural presentations is what one might expect, given the results of chapter 5.

25.6.9. PROPOSITION. If $\underline{\Pi}$ is a complete process structure on a discrete, asynchronous behavioural presentation B and $N=N_{(B,\underline{\Pi})}$, then $B_{(N,\lambda)}=\overline{B}$.

PROOF. By 25.6.5 and 25.6.6, $S_N=(O,\rightarrow,\#)$ and so $S_N=\Sigma(B)$, by (5.1) and hence $\Lambda(S_N))=(O,\overline{P})$, by 5.3.1.

■

25.6.10. PROPOSITION. If B is a behavioural presentation then there exists a complete process structure on B.

PROOF. Let $\mathcal{P}$ denote the set of all subsets of $X\subseteq O$ such that

$$\forall o_1,o_2\in X: o_1\rightarrow o_2 \vee o_2\rightarrow o_1 \vee o_1\# o_2 \qquad (25.26)$$

Now, $\bigcup\mathcal{P}=O$, since $\{o\}\in\mathcal{P}$, each $o\in O$. Order $\mathcal{P}$ by inclusion. If $\mathcal{X}$ is an ascending chain in $\mathcal{P}$ then $\bigcup\mathcal{X}\in\mathcal{P}$, for if $o_1,o_2\in\bigcup\mathcal{X}$, then there exists $X_1,X_2\in\mathcal{X}$ such that $o_1\in X_1$ and $o_2\in X_2$. Since $\mathcal{X}$ is an ascending chain in $\mathcal{P}$, we may assume without loss of generality that $X_1\subseteq X_2$. But now, $o_1,o_2\in X_2$ and so $o_1\rightarrow o_2$ or $o_2\rightarrow o_1$ or $o_1\# o_2$ and we have shown that $\bigcup\mathcal{X}$ satisfies (25.26).

By Zorn's lemma, $\mathcal{P}$ has maximal elements. Let Π denote the set of all maximal elements of $\mathcal{P}$ and define a function $\beta:O\rightarrow\wp(\Pi)$ by $\pi\in\beta(o)\Leftrightarrow o\in\pi$. We show that (Π,β) is a complete process structure on B.

Since $\{o\}\in\mathcal{P}$, each $o\in O$, $o\in\pi$, some $\pi\in\Pi$ and so $\beta(o)\neq\emptyset$, so 25.3.1 (1) holds. If $o_1,o_2\in O$ such that $\pi\in\beta(o_1)\cap\beta(o_2)$, then $o_1\rightarrow o_2$ or $o_2\rightarrow o_1$ or $o_1\# o_2$ by (25.26) and so if $o_1\neq o_2$, then $o_1<o_2$ or $o_2<o_1$ or $o_1\# o_2$, whence 25.3.1 (2) holds.

Finally, suppose that $o_1, o_2 \in O$ such that $o_1 \lhd o_2$ or $o_1 \# o_2$, then $\{o_1, o_2\} \subseteq \mathcal{P}$ and so $o_1, o_2 \in \pi$ some $\pi \in \Pi$ and so $\beta(o_1) \cap \beta(o_2) \neq \emptyset$.

■

Putting the pieces together, we obtain the following result.

25.6.11. THEOREM. If B is the closure of a discrete, asynchronous behavioural presentation, then there exists an occurrence net N, such that $B_{(N,\lambda)} = B$.

■

Notes.

We have presented one way of constructing a discrete behavioural presentation with sequential process structure from an SPS, Σ, but it is not the only way. For example, we could define

$$\mu'_\Sigma(\underline{s}) = |\{i \in I \mid \underline{s}(i) = e\}|$$

and check that H'_Σ obtained from H_Σ by replacing μ_Σ by μ'_Σ is a hybrid transition system. We construct a discrete behavioural presentation B'_Σ from (H'_Σ, μ'_Σ) in the usual way. Let $L = L(C(H'), q')$.

(25.30) entails that if $\underline{u} \in O_L$, then there exists a bijection between $J_{e,\underline{u}} = \{j \in I \mid \lambda_L(\underline{u}) = e\}$ and $O_{e,\underline{u}} = \{o \in \varepsilon_{\mu'_\Sigma}(\underline{u}) \mid (\lambda_\Sigma(o)) = e\}$. Since O'_Σ is the disjoint union of the sets $O_{e,\underline{u}}$, it follows that we may construct a mapping $\beta'_\Sigma : O_\Sigma \to \wp(I)$ by defining $\beta'_\Sigma(o) = J_{e,\underline{u}}$ if $o \in O_{e,\underline{u}}$. We may show that (I, β'_Σ) is a sequential process structure on B'_Σ and that the analogue to theorem 25.4.8 holds.

Chapter 26
Constructions on Hybrid Transition Systems

At the still point of the turning world. Neither flesh nor fleshless.
Neither from nor towards; at the still point, there the dance us.

T. S. ELIOT
Four Quartets

26.1. Introduction.

In the following two chapters, we shall give non-interleaving semantics to two algebraic specification languages, those of Milner [11, 3] and Hoare [2]. The semantics involves mapping certain well-defined expressions in the language in question to an initialised hybrid transition system. The semantics will be *denotational,* in the following sense.

To each non-recursive operator in each language we shall define a corresponding operator on asynchronous hybrid transition systems with initial states; this allows us to give the semantics of non-recursive expressions *compositionally*. These operators are defined in the section 6.3.

The meaning of recursive expressions will be given as the least fixed point of a related operator on initialised, asynchronous hybrid transition systems.

We consider recursive definitions of hybrid transition systems in section 26.2. This effectively involves solving a set of simultaneous equations of the form $X = f(X_1, \cdots, X_n)$, where f is a function of a particular kind. In section 26.3, we examine constructions on hybrid transition systems corresponding to certain operators in CCS and CSP and show that these may be used to define functions such as f.

26.2. Recursion.

CCS, CSP and their numerous relatives all possess facilities for the recursive definition of systems. Essentially this is by sets of simultaneous equations of the form:

$$X_j = Exp(X_1, \cdots, X_n) \tag{26.1}$$

where the X_i are variables and *Exp* is an expression with free variables $X_1, \cdots, X_n$.

In order to give a non-interleaving semantic to these languages, we need to be able to interpret expressions such as *Exp* so that given initialised hybrid transition systems $H_1^*, \cdots, H_n^*$, substitution into (26.1) gives an initialised hybrid transition system $Exp(H_1^*, \cdots, H_n^*)$. We shall then be in a position to talk about solutions of such sets of equations.

We therefore need to assume the existence of a countably infinite set $\mathcal{X}$ of *variables*. A solution to a set of equations (26.1) will consist of an assignment of values to variables, where the values are taken from the class **AHTS*** of all initialised, asynchronous hybrid transition systems. Such an assignment is therefore an element of the set **Env** (for *environment*) of all functions η from $\mathcal{X}$ to **AHTS***.

Formally, then, a recursive description is a pair $(X, \mathcal{F})$ where $X \subseteq \mathcal{X}$ and $\mathcal{F}$ is a family of functions f_x indexed by the elements of X. A solution to the description is an assignment $\eta \in \mathbf{Env}$ which is such that $\eta(x) = f_x(\eta)$, for each $x \in X$. Hence, if we define a function $\mathcal{F}: \mathbf{Env} \to \mathbf{Env}$ by

$$(\mathcal{F}(\eta))(x) = \begin{cases} (f_x(\eta))(x) \text{ if } x \in X \\ \eta(x) \text{ otherwise} \end{cases} \tag{26.2}$$

then a solution must be a *fixed point* of $\mathcal{F}$, that is $\mathcal{F}(\eta) = \eta$.

We wish to find solutions which are in some sense minimal. This involves defining an ordering on **AHTS*** and subsequently on **Env**.

26.2.1. DEFINITION. Suppose that (H_1, q_1) and (H_2, q_2) are initialised hybrid transition systems. Define $H_1 \preceq H_2 \Leftrightarrow$

1• $Q_1 \subseteq Q_2$;
2• $A_1 \subseteq A_2$;
3• $\rightarrow_1 \subseteq \rightarrow_2$;
4• $\iota_1 \subseteq \iota_2$;
5• $E_1 \subseteq E_2$;
6• $\mu_1 = \mu_2 | A_1$.

and we define $(H_1, q_1) \preceq (H_2, q_2) \Leftrightarrow H_1 \preceq H_2 \wedge q_1 = q_2$.

It is clear that $\preceq$ is a partial order on **AHTS***.

26.2.2. DEFINITION. If $\eta_1, \eta_2 \in \mathbf{Env}$, then define

$$\eta_1 \preceq \eta_2 \Leftrightarrow \forall x \in \mathcal{X}: \eta_1(x) \preceq \eta_2(x)$$

It is clear that $\preceq$ a partial order on **Env**. If $X \subseteq \mathcal{X}$, then we also define

$$\eta_1 =_X \eta_2 \Leftrightarrow \forall x \in X: \eta_1(x) = \eta_2(x) \tag{26.3}$$

A family of initialised, asynchronous hybrid transition systems, $(H_n, q_n)_{n\in\omega}$ is an *ascending chain* $\Leftrightarrow$

$$\forall m, n \in \omega: m \leq n \Rightarrow (H_m, q_m) \preceq (H_n, q_n)$$

and a family of environments, $(\eta_n)_{n\in\omega}$ is an *ascending chain* $\Leftrightarrow$

$$\forall m, n \in \omega: m \leq n \Rightarrow \eta_m \preceq \eta_n$$

It will turn out that least fixed points of functions $\mathcal{F}$, where they exist, are least upper bounds of certain ascending chains. The general construction is as follows.

26.2.3. DEFINITION. If $(H_i)_{i\in I}$ is a family of initialised, asynchronous hybrid transition systems, then define

$$\bigcup_{i\in I} H_i = (\bigcup_{i\in I} Q_i, \bigcup_{i\in I} A_i, \bigcup_{i\in I} \rightarrow_i, \bigcup_{i\in I} \iota_i, \bigcup_{i\in I} E_i, \bigcup_{i\in I} \mu_i).$$

If $(\eta_i)_{i\in I}$ is a family with $\eta_i \in$ **Env**, each $i \in I$. then define $\bigcup_{i\in I} \eta_i$, by

$$(\bigcup_{i\in I} \eta_i)(x) = \bigcup_{i\in I} (\eta_i(x)), \text{ all } x \in \mathcal{X}.$$

26.2.4. PROPOSITION. Let $(H_n, q_n)_{n\in\omega}$ be an ascending chain of initialised, asynchronous hybrid transition systems and let $H = \bigcup_{n\in\omega} H_n$ then

1• (H, q_0) is an initialised, asynchronous hybrid transition system;
2• (H, q_0) is the least upper bound of the family $(H_n, q_n)_{n\in\omega}$.

PROOF. We first observe that if $q, q', \hat{q}, \hat{q}' \in Q$ and $a, \hat{a} \in A$ such that $q \rightarrow^a q'$ and $\hat{q} \rightarrow^{a'} \hat{q}'$, then for some $m, n \in \omega$, $q \rightarrow_m^a q'$ and $\hat{q} \rightarrow_n^{a'} \hat{q}'$ and hence $q \rightarrow_k^a q'$ and $\hat{q} \rightarrow_k^{\hat{a}} \hat{q}'$, where $k = max(m, n)$.

Thus, if $q, q', q'' \in Q$ and $a \in A$ such that $q \rightarrow^a q'$ and $q \rightarrow^a q''$, then for some $k \in \omega$, $q \rightarrow_k^a q'$ and $q \rightarrow_k^a q''$, and so $q' = q''$, by unambiguity of $T(C(H_k))$, establishing the unambiguity of $T(C(H))$.

Also if $q \to^a q' \to^b q''$ with $a \iota b$, then for some $m,n \in \omega$, $q \to_m^a q' \to_m^b q''$ and $a \iota_n b$ and hence $q \to_k^a q' \to_k^b q''$ and $a \iota_k b$, where $k = max(m,n)$. Thus, there exists $\hat{q} \in Q_k$ such that $q \to_k^b \hat{q} \to_k^a q''$, and so $\hat{q} \in Q$ with $q \to^b \hat{q} \to^a q''$ showing that $C(H)$ obeys the lozenge rule.

(2) is trivial.

■

26.2.5. COROLLARY. If $(\eta_n)_{n\in\omega}$ is an ascending chain in **Env** and $\eta = \bigcup_{n\in\omega}\eta_n$ then

1• $\eta \in \mathbf{Env}$;
2• η is the least upper bound of the family $(\eta_n)_{n\in\omega}$.

■

We now describe the class of functions which will turn out to be associated with expressions (26.1). First, if (H,q) is an initialised hybrid transition system, then we define $init(H,q) = q$.

26.2.6. DEFINITION. $f{:}\,\mathbf{Env} \to \mathbf{AHTS}^*$ is *admissible* $\Leftrightarrow$

1• f is monotonic;
2• If $(\eta_n)_{n\in\omega}$ is an ascending chain in **Env**, then $f(\bigcup_{n\in\omega}\eta_n) = \bigcup_{n\in\omega} f(\eta_n)$
3• There exists $init(f)$ such that for all $\eta \in \mathbf{Env}$, $init(f(\eta)) = init(f)$.

If f is admissible, then we define $\bot_f = NIL_{init(f)}$, where in general:

$$NIL_u = ((\{u\},\varnothing,\varnothing,\varnothing,\varnothing,\varnothing),u). \tag{26.4}$$

It is clear from 26.2.1 and 26.2.6 (3) that

$$\forall\eta \in \mathbf{Env}{:}\,\bot_f \preceq f(\eta) \tag{26.5}$$

26.2.7. DEFINITION. A *recursion scheme* is a pair $(X, \mathcal{F})$ where:

1• $X \subseteq \mathcal{X}$;
2• $\mathcal{F} = \{f_x \mid x \in X\}$ is a family of admissible functions.

Given $(X, \mathcal{F})$, we may define a function $\mathcal{F}{:}\,\mathbf{Env} \to \mathbf{Env}$ by (26.2). It is clear that $\mathcal{F}$ is monotonic and preserves least upper bounds of ascending chains, since each f_x has this property. We shall use $\mathcal{F}$ to define a function $fix(X = \mathcal{F}){:}\,\mathbf{Env} \to \mathbf{Env}$ as follows.

First, if $\eta \in \mathbf{Env}$, define $\eta|X \in \mathbf{Env}$ by

$$(\eta|X)(x) = \begin{cases} \bot_{f_x} & \text{if } x \in X \\ \eta(x) & \text{otherwise} \end{cases} \tag{26.6}$$

By (26.2), (26.5) and (26.6), we see that

$$\forall \eta \in \mathbf{Env}: \eta|X \preceq \mathcal{F}(\eta|X) \tag{26.7}$$

and hence by monotonicity of $\mathcal{F}$ and induction we have an ascending chain $(\mathcal{F}^n(\eta|X))_{n\in\omega}$, where in general if $\eta \in \mathbf{Env}$, then we define

$$\begin{aligned} &\mathcal{F}^0(\eta) = \eta \\ &\forall n \in \omega: \mathcal{F}^{n+1}(\eta) = \mathcal{F}(\mathcal{F}^n(\eta)) \end{aligned} \tag{26.8}$$

We may therefore define:

$$(fix(X = \mathcal{F}))(\eta) = \bigcup_{n\in\omega} \mathcal{F}^n(\eta|X) \tag{26.9}$$

We investigate the properties of $fix(X = \mathcal{F})$. If $x \in X$, then

$$\begin{aligned} (\mathcal{F}(fix(X = \mathcal{F})(\eta)))(x) &= f_x(\bigcup_{n\in\omega} \mathcal{F}^n(\eta|X)), \text{ by (26.2) and (26.9)} \\ &= \bigcup_{n\in\omega} f_x(\mathcal{F}^n(\eta|X)), \text{ by 26.2.6 (2)} \\ &= \bigcup_{n\in\omega} (\mathcal{F}(\mathcal{F}^n(\eta|X)))(x), \text{ by (26.2)} \\ &= \bigcup_{n>0} \mathcal{F}^n(\eta|X))(x) = (fix(X = \mathcal{F})(\eta))(x), \text{ by (26.2)} \end{aligned}$$

while if $x \notin X$, then $(\mathcal{F}(fix(X = \mathcal{F})(\eta)))(x) = \eta(x) = (fix(X = \mathcal{F})(\eta))(x)$, by (26.2). Therefore

$$\mathcal{F}(fix(X = \mathcal{F})(\eta)) = fix(X = \mathcal{F})(\eta) \tag{26.10}$$

so that $fix(X = \mathcal{F})(\eta)$ is a fixed point of $\mathcal{F}$.

Suppose next that $\mathcal{F}(\eta) = \eta$. By (26.2) if $x \notin X$, then $\eta(x) = (\mathcal{F}(\eta))(x)$ and if $x \in X$, then $\bot_{f_x} \preceq f_x(\eta) = (\mathcal{F}(\eta))(x)$. Thus, by (26.2) and (26.6), $\eta|X \preceq \mathcal{F}(\eta)$ and so $\eta|X \preceq \eta$, that is $\mathcal{F}^0(\eta|X) \preceq \eta$. If $\mathcal{F}^n(\eta|X) \preceq \eta$, then by monotonicity of $\mathcal{F}$, $\mathcal{F}^{n+1}(\eta|X) \preceq \mathcal{F}(\eta) = \eta$. By induction, $\mathcal{F}^n(\eta|X) \preceq \eta$, all $n \in \omega$. Thus:

$$\mathcal{F}(\eta) = \eta \Rightarrow fix(X = \mathcal{F})(\eta) \preceq \eta \tag{26.11}$$

so that $fix(X = \mathcal{F})(\eta)$ is a minimal fixed point of $\mathcal{F}$.

26.2.8. PROPOSITION. If $(X, \mathcal{F})$ is a recursion scheme, then $fix(X = \mathcal{F})$ is monotonic and preserves least upper bounds of ascending chains.

PROOF. If $\eta_1, \eta_2 \in \mathbf{Env}$ such that $\eta_1 \preceq \eta_2$, then by (26.6), $\eta_1|X \preceq \eta_2|X$, so by monotonicity of $\mathcal{F}$, $\mathcal{F}^n(\eta_1|X) \preceq \mathcal{F}^n(\eta_2|X)$, all $n \in \omega$ and so by (26.9), $fix(X = \mathcal{F})(\eta_1) \preceq fix(X = \mathcal{F})(\eta_2)$. Thus, $fix(X = \mathcal{F})$ is monotonic.

Next, suppose that $(\eta_n)_{n\in\omega}$ is an ascending chain in $\mathbf{Env}$. Since $fix(X = \mathcal{F})$ is monotonic

$$\bigcup_{n\in\omega} fix(X = \mathcal{F})(\eta_n) \preceq fix(X = \mathcal{F})(\bigcup_{n\in\omega} \eta_n) \tag{26.12}$$

For each $n \in \omega$, define $\hat{\eta}_n = \mathcal{F}^n(\eta_n|X)$. If $m \le n$, then $\eta_m|X \preceq \eta_n|X$ and thus $\hat{\eta}_m = \mathcal{F}^m(\eta_m|X) \preceq \mathcal{F}^m(\eta_n|X) \preceq \mathcal{F}^n(\eta_n|X) = \hat{\eta}_n$, so $(\hat{\eta}_n)_{n\in\omega}$ is an ascending chain in $\mathbf{Env}$. Let $\hat{\eta} = \bigcup_{k\in\omega} \hat{\eta}_k$.

If $n \in \omega$, then $\hat{\eta}_n = \mathcal{F}^n(\eta_n|X) \preceq \bigcup_{k\in\omega} \mathcal{F}^k(\eta_n|X) = fix(X = \mathcal{F})(\eta_n)$, by (26.9), so

$$\hat{\eta} \preceq \bigcup_{k\in\omega} fix(X = \mathcal{F})(\eta_k) \tag{26.13}$$

We shall prove that $fix(X = \mathcal{F})(\bigcup_{k\in\omega} \eta_k) \preceq \hat{\eta}$. Together with (26.12), this will complete the proof.

Since $\mathcal{F}$ preserves least upper bounds, then by induction, so does $\mathcal{F}^n$, each $n \in \omega$, and so $\mathcal{F}^n(\bigcup_{k\in\omega}(\eta_k|X)) = \bigcup_{k\in\omega} \mathcal{F}^n(\eta_k|X) \preceq \bigcup_{k\ge n} \mathcal{F}^k(\eta_k|X) = \hat{\eta}$. But, by (26.6), $\bigcup_{k\in\omega}(\eta_k|X) = (\bigcup_{k\in\omega} \eta_k)|X$, and hence

$$(fix(X = \mathcal{F})(\bigcup_{k\in\omega} \eta_k) = \bigcup_{n\in\omega} \mathcal{F}^n((\bigcup_{k\in\omega} \eta_k)|X) \preceq \hat{\eta}$$

completing the proof. ■

The following corollary shows that we may use recursion schemes to define admissible functions. This will allow us to deal with nested recursive definitions.

26.2.9. COROLLARY. If $(X, \mathcal{F})$ is a recursion scheme and $x \in X$, then the function $fix_x(X = \mathcal{F})\colon \mathbf{Env} \to \mathbf{AHTS}^*$ given by

$$(fix_x(X = \mathcal{F}))(\eta) = (fix(X = \mathcal{F})(\eta))(x) \tag{26.14}$$

is admissible and $init(fix_x(X = \mathcal{F})) = init(f_x)$.

PROOF. $fix_x(X = \mathcal{F})$ is monotonic and preserves least upper bounds of ascending chains, by 26.2.2 and 26.2.3, because $fix(X = \mathcal{F})$ does, by 26.2.9. If $x \in X$, then for all $n \in \omega$, $\bot_{f_x} = (\eta|X)(x) \preceq (\mathcal{F}^n(\eta|X))(x)$ and hence

$$init(fix_x(X = \mathcal{F})) = init((\bigcup_{n\in\omega} \mathcal{F}^n(\eta|X))(x)) = init(f_x)$$

■

Recall that we are trying to use recursion schemes to define initialised hybrid transition systems. $fix_x(X = \mathcal{F})$ is not in itself an initialised hybrid transition system since the expressions corresponding to the functions in $\mathcal{F}$ may contain variables not occurring in X and hence not assigned values by the equations. $fix_x(X = \mathcal{F})$ is then a function whose image depends on the values assigned to these 'free variables' by a given environment. If, in contrast, every variable appearing in the expressions corresponding to the functions in $\mathcal{F}$ does occur in X, that is, $(X, \mathcal{F})$ is closed, then, as we shall see, $fix_x(X = \mathcal{F})$ is a constant function.

26.2.10. DEFINITION. If $f: \mathbf{Env} \to \mathbf{AHTS}^*$, then $X \subseteq \mathcal{X}$ is *essential* to $f \Leftrightarrow$

$$\forall \eta_1, \eta_2 \in \mathbf{Env}: \eta_1 =_X \eta_2 \Rightarrow f(\eta_1) = f(\eta_2).$$[1]

A recursion scheme $(X, \mathcal{F})$ is *closed* if and only if for each $x \in X$, X is essential to f_x. In other words, if η_1 and η_2 agree on X, then for every $x \in X$, $f_x(\eta_1) = f_x(\eta_2)$.

26.2.11. PROPOSITION. If $(X, \mathcal{F})$ is a closed recursion scheme then there exists $\eta_{\mathcal{F}} \in \mathbf{Env}$ such that

1• $fix(X = \mathcal{F})(\eta) = \eta_{\mathcal{F}}$, all $\eta \in \mathbf{Env}$;
2• $\eta_{\mathcal{F}}$ is the least fixed point of $\mathcal{F}$.

PROOF. (1) If $\eta_1, \eta_2 \in \mathbf{Env}$, then $\eta_1|X =_X \eta_2|X$, and as $(X, \mathcal{F})$ is closed, $\mathcal{F}(\eta_1|X) = \mathcal{F}(\eta_2|X)$, by 26.2.10 and (26.2). Therefore, for all $n > 0$, $(\mathcal{F}^n(\eta_1|X)) = (\mathcal{F}^n(\eta_2|X))$ and thus by (26.9),

$$fix(X = \mathcal{F})(\eta_1) = fix(X = \mathcal{F})(\eta_2),$$

---

[1]See (26.3)

So if we define $\eta_{\mathcal{F}} = fix_x(X = \mathcal{F})(\eta_1)$, then $\eta_{\mathcal{F}}$ satisfies (1).

(2) By (26.10) and (26.11)

■

26.3 Non-Recursive Constructions.

In the following section, we examine constructions on hybrid transition systems which in particular will allow us to construct admissible functions from admissible functions .

For the purpose of these constructions, we shall assume a non-empty set $\mathcal{A}ct$ (of actions). We require that $\mathcal{A}ct$ has the following additional properties.

1• $\mathcal{A}ct$ contains an element τ, ('tacit'), the silent or internal action.

2• There is a bijection $bar: \mathcal{A}ct - \{\tau\} \to \mathcal{A}ct - \{\tau\}$ such that if $a \in \mathcal{A}ct$ then $\bar{a} \neq a$ and $\bar{\bar{a}} = a$, where we are writing $\bar{a}$ for $bar(a)$.

26.3.1. DEFINITION. Suppose that (H,q_0) is an initialised, asynchronous hybrid transition system and $e \in \mathcal{A}ct$. We define $e \to (H,q_0) = (H',q_0')$, where

1• $q_0' = 0$ and $Q' = (Q \times \{1\}) \cup \{0\}$;

2• $A' = (A \times \{1\}) \cup \{0\}$;

3• $0 \to'^{a} q \Leftrightarrow a = 0 \wedge q = (q_0, 1)$

$(q,1) \to'^{\hat{a}} \hat{q} \Leftrightarrow \exists q' \in Q\ \exists a \in A: \hat{q} = (q',1) \wedge \hat{a} = (a,1) \wedge q \to^{a} q'$;

4• $\hat{a}\, \iota'\, \hat{b} \Leftrightarrow \exists a,b \in A: \hat{a} = (a,1) \wedge \hat{b} = (b,1) \wedge a\, \iota\, b$;

5• $E' = E \cup \{e\}$;

6• $\mu'(0) = e$ and if $a \in A$ then $\mu'(a,1) = \mu(a)$.

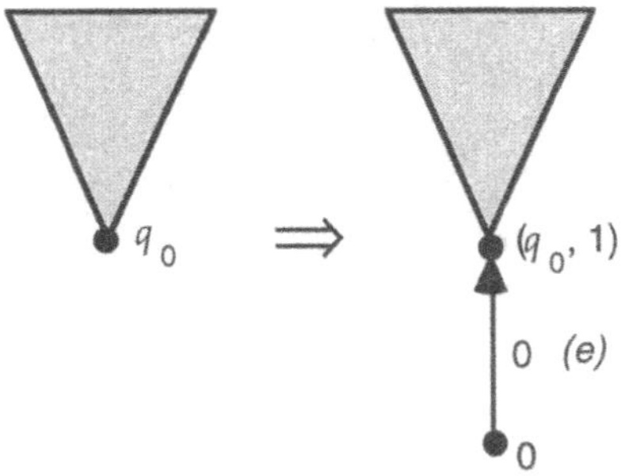

Figure 26.1.

The construction is illustrated in figure 26.1 above.

26.3.2. LEMMA. If (H,q_0) is an initialised, asynchronous hybrid transition system and $e \in \mathcal{A}ct$, then $e \to (H,q_0)$ is an asynchronous initialised hybrid transition system.

PROOF. Let $(H',q_0') = e \to (H,q_0)$. Suppose first that $\hat{q},\hat{q}_1,\hat{q}_2 \in Q'$ and $\hat{a} \in A'$ such that $\hat{q} \to'^{\hat{a}} \hat{q}_1$ and $\hat{q} \to'^{\hat{a}} \hat{q}_2$, then either $\hat{q} = 0$, in which case $\hat{q}_1 = (q_0,1) = \hat{q}_2$, or there exists $q,q_1,q_2 \in Q$ and $a \in A$ such that $\hat{q} = (q,1)$, $\hat{q}_1 = (q_1,1)$, $\hat{q}_2 = (q_2,1)$, $\hat{a} = (a,1)$, $q \to^a q_1$ and $q \to^a q_2$, in which case $q_1 = q_2$, by unambiguity of $T(C(H))$, and so $\hat{q}_1 = \hat{q}_2$. Thus $T(C(H'))$ is unambiguous.

Suppose next that $\hat{q},\hat{q}_1,\hat{q}_2 \in Q'$ and $\hat{a},\hat{b} \in A'$ such that $\hat{q} \to'^{\hat{a}} \hat{q}_1 \to'^{\hat{b}} \hat{q}_2$ and $\hat{a} \,\hat{\iota}\, \hat{b}$. By 26.3.1 (4), there exists $a,b \in A$ such that $\hat{a} = (a,1)$, $\hat{b} = (b,1)$ and $a \,\iota\, b$. By 26.3.1 (3), we cannot have $\hat{q} = q_0'$ and so there exists $q \in Q$ such that $\hat{q} = (q,1)$. Again, by 26.3.1 (3), there exists $q_1,q_2 \in Q$ such that $\hat{q}_1 = (q_1,1)$, $\hat{q}_2 = (q_2,1)$ and $q \to^a q_1 \to^b q_2$. Since $a \,\iota\, b$, there exists $q_1' \in Q$ such that $q \to^b q_1' \to^a q_2$. Thus, by 26.3.1, (1) and (3), there exists $\hat{q}_1' \in Q'$, namely $(q_1',1)$, such that $\hat{q} \to'^{\hat{b}} \hat{q}_1' \to'^{\hat{a}} \hat{q}_2$. Thus $C(H')$ satisfies the lozenge rule.

■

26.3.3. LEMMA. If $f\colon \mathbf{Env} \to \mathbf{AHTS}^*$ is monotonic and preserves least upper bounds of ascending chains and $e \in \mathcal{A}ct$, then the function $(e \to f)\colon \mathbf{Env} \to \mathbf{AHTS}^*$ given by

$$\forall \eta \in \mathbf{Env}\colon (e \to f)(\eta) = e \to (f(\eta)) \tag{26.15}$$

is admissible and $init(e \to f) = 0$.

PROOF It is immediate from 26.2.2, 26.2.3, 26.3.2 and (26.15) that since f is monotonic and preserves least upper bounds of ascending chains than so does $e \to f$. That $init(e \to f) = 0$ follows immediately from 26.3.1 (1).

■

26.3.4. DEFINITION. Let $(H_i,q_i)_{i\in I}$ be an non-empty indexed family of asynchronous initialised hybrid transition systems. Define $\sum_{i\in I}(H_i,q_i) = (H,\underline{q}_0)$ where

1• $Q = \bigcup_{i\in I}(Q_i \times \{i\}) \cup \{0\}$ and $q_0 = 0$

2• $A = \bigcup_{i\in I}(A_i \times \{i\})$

3• $E = \bigcup_{i\in I} E_i$

4• $q_0 \rightarrow'^{(a,i)} (q,j) \Leftrightarrow i = j \wedge q_i \rightarrow_i^a q$

$(q,i) \rightarrow'^{(a,j)} \hat{q} \Leftrightarrow i = j \wedge (\exists q' \in Q_i : \hat{q} = (q',i) \wedge q \rightarrow_i^a q')$;

5• $(a,i) \iota (b,j) \Leftrightarrow i = j \wedge a \iota_i b$;

6• $\mu(a,i) = \mu_i(a)$.

If $I = \varnothing$, then define $\sum_{i\in I}(H_i, q_i) = NIL_0$.

This construction is illustrated in figure 26.2.

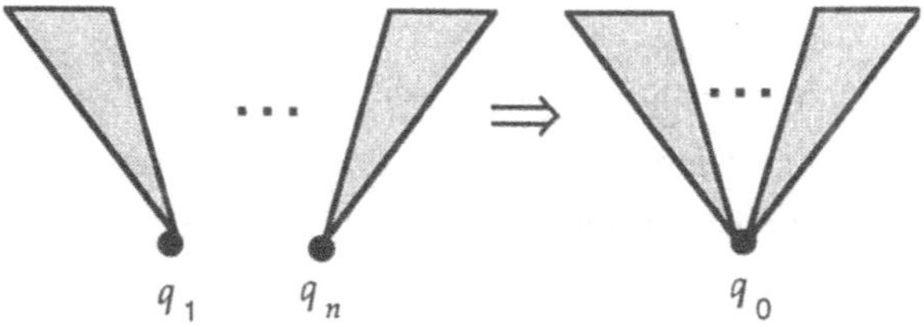

Figure 26.2.

26.3.5. LEMMA. If $(H_i, q_i)_{i\in I}$ is a non-empty indexed family of asynchronous initialised hybrid transition systems, then $\sum_{i\in I}(H_i, q_i)$ is an initialised, asynchronous hybrid transition system.

PROOF. Let $(H', \underline{q}'_0) = \sum_{i\in I}(H_i, q_i)$. Suppose first that $\hat{q}, \hat{q}_1, \hat{q}_2 \in Q'$ and $\hat{a} \in A'$ such that $\hat{q} \rightarrow'^{\hat{a}} \hat{q}_1$ and $\hat{q} \rightarrow'^{\hat{a}} \hat{q}_2$, then by 26.3.4 (3), there exists $i \in I$, $q, q_1, q_2 \in Q_i$ and $a \in A_i$ such that $\hat{q}_1 = (q_1, i)$, $\hat{q}_2 = (q_2, i)$, $\hat{a} = (a, i)$, $q \rightarrow_i^a q_1$ and $q \rightarrow_i^a q_2$, in which case $q_1 = q_2$, by unambiguity of $T(C(H_i))$, and so $\hat{q}_1 = \hat{q}_2$. Thus $T(C(H'))$ is unambiguous.

Suppose next that $\hat{q}, \hat{q}_1, \hat{q}_2 \in Q'$ and $\hat{a}, \hat{b} \in A'$ such that $\hat{q} \rightarrow'^{\hat{a}} \hat{q}_1 \rightarrow'^{\hat{b}} \hat{q}_2$ and $\hat{a}\, \hat{\iota}\, \hat{b}$. By 26.3.4 (3) and (4), there exists $i \in I$, $q, q_1, q_2 \in Q_i$ and $a, b \in A_i$ such that $\hat{a} = (a, i)$, $\hat{b} = (b, i)$, $q \rightarrow_i^a q_1 \rightarrow_i^b q_2$ and $a \iota b$, where either $q = q_i$ and $\hat{q} = \underline{q}'_0$ or $\hat{q} = (q, i)$. Since $a \iota b$, there exists $q'_1 \in Q_i$ such that $q \rightarrow_i^b q'_1 \rightarrow_i^a q_2$. Thus, by 26.3.4 (1) and (3), there exists $\hat{q}'_1 \in Q'$, namely (q'_1, i) such that $\hat{q} \rightarrow'^{\hat{b}} \hat{q}'_1 \rightarrow'^{\hat{a}} \hat{q}_2$. Thus $C(H')$ satisfies the lozenge rule.

■

26.3.6. LEMMA. If $(f_i)_{i\in I}$, is a family of admissible functions f_i: **Env** $\rightarrow$ **AHTS***, each $i \in I$, then the function $(\sum_{i\in I} f_i)$: **Env** $\rightarrow$ **AHTS*** given by

$$\forall \eta \in \mathbf{Env}: (\textstyle\sum_{i\in I} f_i)(\eta) = \sum_{i\in I}(f_i(\eta)) \tag{26.16}$$

is admissible and $init(\sum_{i\in I} f_i) = 0$.

PROOF. It is immediate from 26.2.2, 26.2.3, 26.3.5 and (26.16) that since each f_i is monotonic and preserves least upper bounds of ascending chains than so does $\sum_{i\in I} f_i$. That $init(\sum_{i\in I} f_i) = 0$ follows immediately from 26.3.4 (1).

■

The next construction is based on the parallel composition operation in CCS. It is for this construction that we need the τ symbol and the *bar* function.

26.3.7. DEFINITION. If (H_1, q_1) and (H_2, q_2) are initialised, asynchronous hybrid transition systems, such that $E_i \subseteq \mathcal{A}ct$, each i, then we define $(H_1, q_1) | (H_2, q_2) = (H, q)$, where

1• $Q = Q_1 \times Q_2$ and $q = (q_1, q_2)$;

2• $A = (A_1 \times \{\Omega\}) \cup (\{\Omega\} \times A_2) \cup A_\tau$, where

$A_\tau = \{(a_1, a_2) \in A_1 \times A_2 \mid \exists e \in \mathcal{A}ct - \{\tau\} : \mu_1(a_1) = e \wedge \mu_2(a_2) = \bar{e}\}$;

3• $(q_1', q_2') \rightarrow^{(a_1,\Omega)} (q_1'', q_2'') \Leftrightarrow q_1' \rightarrow_1^{a_1} q_1'' \wedge q_2' = q_2''$;

$(q_1', q_2') \rightarrow^{(\Omega,a_2)} (q_1'', q_2'') \Leftrightarrow q_1' = q_1'' \wedge q_2' \rightarrow_2^{a_2} q_2''$;

and if $(a_1, a_2) \in A_\tau$, then

$(q_1', q_2') \rightarrow^{(a_1,a_2)} (q_1'', q_2'') \Leftrightarrow q_1' \rightarrow_1^{a_1} q_1'' \wedge q_2' \rightarrow_2^{a_2} q_2''$;

4• $(a_1, a_2)\, \iota\, (b_1, b_2) \Leftrightarrow (a_1 \neq \Omega \neq b_1 \Rightarrow a_1\, \iota_1\, b_1) \wedge (a_2 \neq \Omega \neq b_2 \Rightarrow a_2\, \iota_2\, b_2)$;

5• $\mu(a_1, a_2) = \begin{cases} \mu_1(a_1) \text{ if } a_2 = \Omega \\ \mu_2(a_2) \text{ if } a_1 = \Omega \\ \tau \text{ otherwise} \end{cases}$

6• $E = E_1 \cup E_2 \cup \mu(A)$.

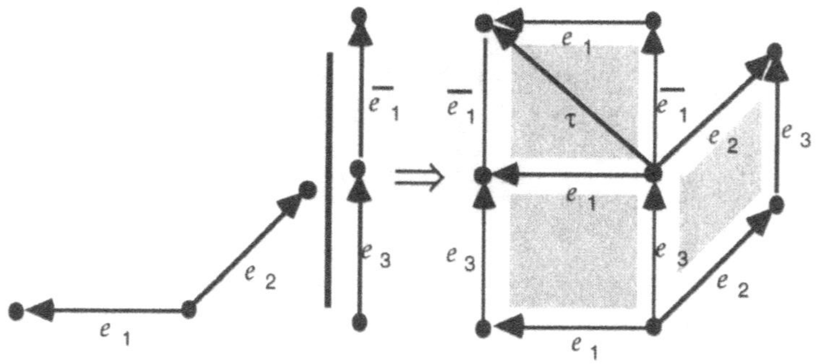

Figure 26.3.

This construction is illustrated in figure 26.3.

26.3.8. LEMMA. Suppose that (H_1,q_1) and (H_2,q_2) are initialised, asynchronous hybrid transition systems then $(H_1,q_1)|(H_2,q_2)$ is an initialised, asynchronous hybrid transition system.

PROOF. Let $(H,q)=(H_1,q_1)|(H_2,q_2)$. Suppose first that $(q_1',q_2')\rightarrow^{(a_1,a_2)}(\hat{q}_1,\hat{q}_2)$ and $(q_1',q_2')\rightarrow^{(a_1,a_2)}(\tilde{q}_1,\tilde{q}_2)$. If $a_1=\Omega$, then $\tilde{q}_1=q_1'=\hat{q}_1$ and $q_2'\rightarrow_2^{a_2}\hat{q}_2$ and $q_2'\rightarrow_2^{a_2}\tilde{q}_2$ so $\tilde{q}_2=\hat{q}_2$, as $T(C(H_2))$ is unambiguous, so $(\tilde{q}_1,\tilde{q}_2)=(\hat{q}_1,\hat{q}_2)$.

The other cases, namely $a_2=\Omega$ and $a_1\neq\Omega\neq a_2$ are similar. Hence $T(C(H))$ is unambiguous.

Next let $(q_1',q_2')\rightarrow^{(a_1,a_2)}(\hat{q}_1,\hat{q}_2)\rightarrow^{(b_1,b_2)}(\tilde{q}_1,\tilde{q}_2)$ with $(a_1,a_2)\,\iota\,(b_1,b_2)$. If $a_1\neq\Omega\neq b_1$, then by 26.3.7 (3) and (4), $q_1'\rightarrow_1^{a_1}\hat{q}_1\rightarrow^{b_1}\tilde{q}_1$ with $a_1\,\iota_1\,b_1$ and so there exists $\bar{q}_1\in Q_1$ such that $q_1'\rightarrow_1^{b_1}\bar{q}_1\rightarrow_1^{a_1}\tilde{q}_1$. Likewise, if $a_2\neq\Omega\neq b_2$, then there exists $\bar{q}_2\in Q_2$ such that $q_2'\rightarrow_2^{b_2}\bar{q}_2\rightarrow_2^{a_2}\tilde{q}_2$. Thus, if we define $\bar{q}\in Q_1\times Q_2$ by

$$\bar{q}=\begin{cases}(q_1',\bar{q}_2)\text{ if }a_1=b_1=\Omega\\(\bar{q}_1,q_2')\text{ if }a_1=b_2=\Omega\\(\bar{q}_1,\bar{q}_2)\text{ if }a_1=\Omega\text{ and }b_1\neq\Omega\neq b_2\\(q_1',\bar{q}_2)\text{ if }a_2=b_1=\Omega\\(\bar{q}_1,q_2')\text{ if }a_2=b_2=\Omega\\(\bar{q}_1,\bar{q}_2)\text{ if }a_2=\Omega\text{ and }b_1\neq\Omega\neq b_2\\(q_1',\bar{q}_2)\text{ if }a_1\neq\Omega\neq a_2\text{ and }b_1=\Omega\\(\bar{q}_1,q_2')\text{ if }a_1\neq\Omega\neq a_2\text{ and }b_2=\Omega\\(\bar{q}_1,\bar{q}_2)\text{ if }a_1\neq\Omega\neq a_2\text{ and }b_1\neq\Omega\neq b_2\end{cases}$$

then we may check that in each case, $(q_1',q_2')\rightarrow^{(b_1,b_2)}\bar{q}\rightarrow^{(a_1,a_2)}(\tilde{q}_1,\tilde{q}_2)$. For example, if $a_2=b_1=\Omega$, then $q_1'\rightarrow^{a_1}\hat{q}_1=\tilde{q}_1$ and $q_2'=\hat{q}_2\rightarrow^{b_2}\tilde{q}_2$, and so $(q_1',q_2')\rightarrow^{(\Omega,b_2)}(\bar{q}_1,\bar{q}_2)\rightarrow^{(a_1,\Omega)}(\tilde{q}_1,\tilde{q}_2)$. Thus, $C(H)$ satisfies the lozenge rule.

■

26.3.9. LEMMA. If f_1 and f_2 are admissible functions f_i: **Env** $\rightarrow$ **AHTS***, each i then the function $(f_1|f_2)$: **Env** $\rightarrow$ **AHTS*** given by

$$\forall\eta\in\textbf{Env}:(f_1|f_2)(\eta)=f_1(\eta)|f_2(\eta) \qquad (26.17)$$

is admissible and $init(f_1|f_2)=(init(f_1),init(f_2))$.

PROOF. It is immediate from 26.2.2, 26.2.3, 26.3.8 and (26.17) that since each f_i is monotonic and preserves least upper bounds of ascending chains than so does $(f_1 | f_2)$. That $init(f_1 | f_2) = (init(f_1), init(f_2))$ follows immediately from 26.3.7 (1).

■

26.3.10. DEFINITION. If (H,q) is an initialised, asynchronous hybrid transition system such that $E \subseteq \mathcal{A}ct$ and $\phi: \mathcal{A}ct \to \mathcal{A}ct$, then we define $(H,q)[\phi] = (H',q')$, where:

1• $(C(H'),q') = (C(H),q)$;
2• $E' = f(E)$;
3• $\mu' = \phi \circ \mu$.

26.2.11. LEMMA. If (H,q) is an initialised, asynchronous hybrid transition systems such that $E \subseteq \mathcal{A}ct$ and $\phi: \mathcal{A}ct \to \mathcal{A}ct$, then $(H,q)[\phi]$ is an initialised, asynchronous hybrid transition system.

■

26.3.12. LEMMA. If $\phi: \mathcal{A}ct \to \mathcal{A}ct$ and $f: \mathbf{Env} \to \mathbf{AHTS}^*$ is an admissible function such that $f(\eta)[\phi]$ is defined, each $\eta \in \mathbf{Env}$, then the function $(f[\phi]): \mathbf{Env} \to \mathbf{AHTS}^*$ given by

$$\forall \eta \in \mathbf{Env}: (f[\phi])(\eta) = f(\eta)[\phi] \tag{26.18}$$

is admissible and $init(f[\phi]) = init(f)$.

PROOF. It is immediate from 26.2.2, 26.2.3, 26.3.11 and (26.18) that since f is monotonic and preserves least upper bounds of ascending chains than so does $f[\phi]$. That $init(f[\phi]) = init(f)$ follows immediately from 26.3.10 (1).

■

26.3.13. DEFINITION. If (H,q) is an initialised, asynchronous hybrid transition system with $E, L \subseteq \mathcal{A}ct$, then define $(H,q) \setminus L = (H',q')$, where:

1• $Q' = Q$ and $q' = q$;
2• $E' = \{e \in E \mid e \neq \tau \Rightarrow \{e, \bar{e}\} \cap L = \emptyset\}$;
3• $A' = \{a \in A \mid \mu(a) \notin L\}$;
4• $\to' = \to \cap (Q \times A' \times Q)$;
5• $\iota' = \iota \cap (A' \times A')$;
6• $\mu = \mu|A'$.

Figure 26.4 illustrates the initialised hybrid transition system derived from that of figure 26.3 by applying the operator $\setminus e_1$.

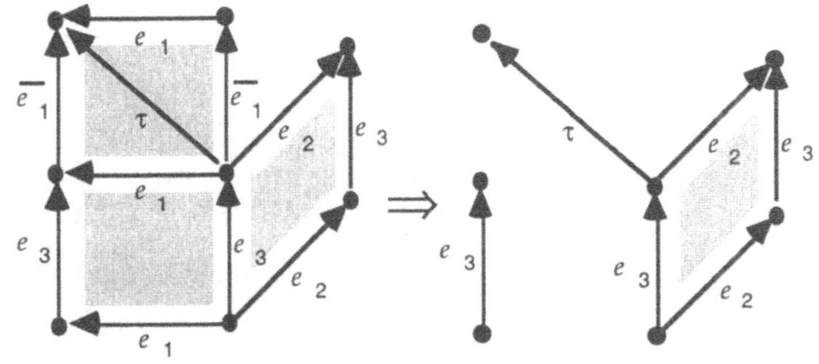

Figure 26.4.

26.3.14. LEMMA. If (H,q) is an initialised, asynchronous hybrid transition system such that $E, L, \subseteq \mathcal{A}ct$, then $(H,q)\setminus L$ is an initialised hybrid transition system.

PROOF. Let $(\hat{H},\hat{q})=(H,q)\setminus L$. Suppose first that $\hat{q},\hat{q}_1,\hat{q}_2 \in \hat{Q}$ and $\hat{a}\in A'$ such that $\hat{q} \mathrel{\hat{\to}}^{\hat{a}} \hat{q}_1$ and $\hat{q} \mathrel{\hat{\to}}^{\hat{a}} \hat{q}_2$, then $\hat{q},\hat{q}_1,\hat{q}_2 \in Q$ and $\hat{a}\in A$ with $\hat{q} \to^{\hat{a}} \hat{q}_1$ and $\hat{q} \to^{\hat{a}} \hat{q}_2$, by 26.3.13 (1), (3) and (4), so $\hat{q}_1=\hat{q}_2$, as $T(C(H))$ is unambiguous. Hence $T(C(\hat{H}))$ is unambiguous.

Next, suppose that $\hat{q},\hat{q}_1,\hat{q}_2 \in \hat{Q}$ and $\hat{a},\hat{b}\in A'$ such that $\hat{q} \mathrel{\hat{\to}}^{\hat{a}} \hat{q}_1 \mathrel{\hat{\to}}^{\hat{b}} \hat{q}_2$ and $\hat{a}\,\hat{\iota}\,\hat{b}$, then $\hat{q},\hat{q}_1,\hat{q}_2 \in Q$ and $\hat{a},\hat{b}\in A$ with $\hat{q} \to^{\hat{a}} \hat{q}_1 \to^{\hat{b}} \hat{q}_2$ and $\hat{a}\,\iota\,\hat{b}$, by 26.3.13. (1), (3), (4) and (5) Hence, there exists $q_1 \in Q$ such that $\hat{q} \to^{\hat{b}} q_1 \to^{\hat{a}} \hat{q}_2$. Since $\hat{a},\hat{b}\in A'$, $\hat{q} \mathrel{\hat{\to}}^{\hat{b}} q_1 \mathrel{\hat{\to}}^{\hat{a}} \hat{q}_2$, by 26.3.13 (4). Thus, $C(\hat{H})$ satisfies the lozenge rule. ∎

26.3.15 LEMMA. If $L \subseteq \mathcal{A}ct$ and $f\colon \mathbf{Env} \to \mathbf{AHTS}^*$ is an admissible function such that $f(\eta)\setminus L$ is defined, each $\eta \in \mathbf{Env}$, then the function $(f\setminus L)\colon \mathbf{Env} \to \mathbf{AHTS}^*$ given by

$$\forall \eta \in \mathbf{Env}\colon (f\setminus L)(\eta) = f(\eta)\setminus L \tag{26.19}$$

is admissible and $init(f\setminus L)=init(f)$.

PROOF. It is immediate from 26.2.2, 26.2.3, 26.3.13 and (26.19) that since f is monotonic and preserves least upper bounds of ascending chains than so does $(f\setminus L)$.

That $init(f\setminus L)=init(f)$ follows immediately from 26.3.13. ∎

Notes.

Of course, corresponding to each construction on initialised hybrid transition systems, there is a corresponding construction on behavioural presentations with the property that applying the construction commutes with the construction of a behavioural presentation from the initialised hybrid transition system.

For example, if B is a behavioural presentation and $e \in \mathcal{A}ct$, then we may define $e \to B = (O', P', E', \lambda')$ where

- $O' = (O \times \{0\}) \cup \{0\}$;
- $P' = \{p \times \{0\} \mid p \in P\} \cup \{0\}$;
- $E' = E \cup \{e\}$;
- $\lambda'(0) = e$ and if $o \in O$, then $\lambda'(o, 0) = \lambda(o)$

and then one may argue that if (H, q) accepts B, then $e \to (H, q)$ accepts $e \to B$. The proof involves taking an isomorphism from $B_{(H,q)}$ to B and extending it to an isomorphism from $B_{e \to (H,q)}$ to $e \to B$.

It is also not hard to believe that the constructions preserve equivalence between initialised hybrid transition systems. For example, if $(H_1, q_1) \equiv (H_2, q_2)$, then $e \to (H_1, q_1) \equiv e \to (H_2, q_2)$. The proof involves taking a strong equivalence from (H_1, q_1) to (H_2, q_2) and extending it to a strong equivalence from $e \to (H_1, q_1)$ to $e \to (H_2, q_2)$.

Thus, one could operate at the level of behavioural presentations. This is essentially the approach taken by Winskel in [19].

Chapter 27
Process Calculus (CCS)

> Apart from articulacy, we aim at an underlying theory whose basis is a small well-knit collection of ideas which justifies the manipulations of the calculus. This is as important as generality - perhaps even more important. Any theory will be superseded sooner or later; during its life, understanding it and assessing it are only possible and worthwhile if it is seen as a logical growth from rather few basic assumptions and concepts.
>
> R. MILNER [3]

27.1. Introduction.

Milner's work on parallelism theory has been expounded in two books, [11, 3]. In the first, he presents a formal theory known as CCS, standing for Calculus of Communicating Systems. The second contains an improved version of CCS, which he now prefers to call 'Process Calculus'. We shall use the term CCS for brevity; our presentation here is based on [3].

The introduction of CCS was extremely important in the history of parallelism theory, in that it inaugurated a major line of research in the area, concerned with the use of abstract algebra to describe systems. The ideas that Milner introduced in [11] have had a profound influence on the creation of new notations and, because of its elegance and simplicity, as well as its expressive power, has found acceptance among the denizens of that strange place known to academics as 'the real world', where people actually have to build real systems and sell them, and need to be sure that the things they design and build work sufficiently well to satisfy their customers.

In the process calculus, systems are described by sets of equations based on an algebra of terms in which the operators correspond to the constructions of guarding, non-deterministic choice, parallel composition, renaming and hiding. Significantly, in the process calculus there is a strong emphasis on behavioural equivalence of various kinds and on the deduction of equivalence by way of a set of equational laws. It is the ability to manipulate the expressions according to such laws that gives the notation the features of a calculus.

27.2. Syntax and Transition System Semantics.

We begin by describing the calculus. We use the notation of Milner's more recent book [3].

As in chapter 26, we assume the existence of a countably infinite set $\mathcal{X}$ of variables and a non-empty set $\mathcal{Act}$ (of actions) such that.

1• $\mathcal{Act}$ contains an element τ, ('tacit'), the silent or internal action.

2• There is a bijection $bar: \mathcal{Act}-\{\tau\} \to \mathcal{Act}-\{\tau\}$ such that if $a \in \mathcal{Act}$ then $\bar{a} \neq a$ and $\bar{\bar{a}} = a$, where we are writing $\bar{a}$ for $bar(a)$.

A *relabelling function* is a function $\phi: \mathcal{Act} \to \mathcal{Act}$ such that $\phi(\tau) = \tau$ and

$$\forall a \in \mathcal{Act}-\{\tau\}: \phi(\bar{a}) = \overline{\phi(a)}$$

27.2.1. DEFINITION. The set **Exp** of agent expressions is the smallest set containing $\mathcal{X}$ and such that if $Ex, Ex_1, Ex_2 \in \mathbf{Exp}$, $(Ex_j)_{j\in J}$ is a family of elements of **Exp** and $(x)_{j\in J}$ is a family of elements of $\mathcal{X}$ then the following expressions also belong to **Exp**.

1• $a.Ex$, where $a \in \mathcal{Act}$;

2• $\sum_{j\in J} Ex_j$;

3• $Ex_1 \mid Ex_2$;

4• $Ex \setminus L$, where $L \subseteq \mathcal{Act}-\{\tau\}$;

5• $Ex[\phi]$, where ϕ is a relabelling function;

6• $fix_i(\{x_j = Ex_j : j \in J\})$

The combinators bind in the following decreasing order of strength: guarding, summation, parallel composition, restriction and relabelling. Normal binding precedence will be overridden using parentheses, as usual.

$a.Ex$ represents a system that may perform the action a after which it performs according to the expression Ex.

$\sum_{j\in J} Ex_j$ represents a system that may non-deterministically choose to behave according to any single one of the expressions Ex_j. If $J = \emptyset$, then $\sum_{j\in J} Ex_j$ describes a system that cannot choose to behave in any way at all. Milner gives this agent a special name; **0**.

The expression $Ex_1 \mid Ex_2$ denotes a form of parallel composition. The agent $Ex_1 \mid Ex_2$ has at most three options; the agent Ex_1 may perform an action independently from Ex_2, or vice versa, or the two agents may synchronise. This may only happen if Ex_1 may perform an action a and Ex_2 may perform an action b such that $a=\bar{b}$. In this case, the resulting communication is internal to the two systems and the outcome is the silent action τ.

The expression $Ex \setminus L$ describes a system which may act according to the expression Ex, except that it may never perform any action from the set L. Since $\tau \notin L$, $Ex \setminus L$ may perform τ actions which result from synchronisation of elements belonging to L.

The expression $Ex[\phi]$ describes a system which behaves exactly like Ex except that where Ex would perform an action a, $Ex[\phi]$ performs the action $\phi(a)$.

Finally, $fix_i(\{x_j = Ex_j : j \in J\})$ introduces recursion. Here, we have a family of simultaneous recursion equations in the variables x_j and the expression denotes the value of the solution for x_i, if one exists.

In the full calculus, actions correspond to transmission and reception of data through communication ports belonging to a set A. If $a \in A$, then the action $a(v_1,\cdots,v_n)$ describes the *input* of the values $v_1,\cdots,v_n$ through an input port a and the action $\bar{a}(v_1,\cdots,v_n)$ describes the *output* of the values $v_1,\cdots,v_n$ through an output port $\bar{a}$. Data may pass from output port $\bar{a}$ to input port b only if $a=b$.

In [3], the semantics of the full calculus is defined in terms of that of the non value-passing calculus as follows. If a is an input port which will accept data from a set D, then the expression $a(x).Ex(x)$ - which is to be understood to mean 'input some value through port a and bind it to all free occurrences of x in $Ex(\,)$' - is associated with an expression $\sum_{d\in D} a_d . Ex_d$, while the expression $\bar{a}(d).Ex$ - which defines the export of the value d from port $\bar{a}$ is associated with an expression $\overline{a_d}.Ex$. In other words, the semantics of the value-passing calculus may be defined in terms of the non value-passing calculus. We shall therefore concentrate exclusively on the latter.

The behavioural semantics of this language of expressions is given in terms of initialised transition systems. The transition relation is defined using the method of Plotkin [73]. The rules for the transition relation all have the form:

$$\frac{\textit{premise}}{Ex \rightarrow^{a} Ex'}$$

with the interpretation that if *premise* is true then $Ex \rightarrow^{a} Ex'$ in the transition relation of the transition system. For example, rule 27.2.4 (1) allows us to infer that $a.\mathbf{0} \rightarrow^{a} \mathbf{0}$ - the agent $a.\mathbf{0}$ may perform the action a and then stop.

In order to define the transition rule for recursion expressions, we need to be able to describe the effect of substituting expressions for free occurrences of variables. Informally, if $\vartheta : \mathcal{X} \rightarrow \mathbf{Exp}$, then $Ex[\vartheta]$ is the expression derived from Ex by replacing each free instance of $x \in \mathcal{X}$ by $\vartheta(x)$. Of course, we first need to define what a free variable is.

27.2.2. DEFINITION. We define a map $free$: $\mathbf{Exp} \to \wp(\mathcal{X})$ as follows

1• If $x \in \mathcal{X}$ then $free(x) = \{x\}$;
2• $free(\mathbf{0}) = \varnothing$;
3• $free(a.Ex) = free(Ex)$
4• $free(\sum_{j\in J} Ex_j) = \bigcup_{j\in J} free(Ex_j)$;
5• $free(Ex_1 \mid Ex_2) = free(Ex_1) \cup free(Ex_2)$;
6• $free(Ex \setminus L) = free(Ex)$;
7• $free(Ex[\phi]) = free(Ex)$;
8• $free(fix_i(\{x_j = Ex_j : j \in J\})) = (\bigcup_{j\in J} free(Ex_j)) - \{x_j \mid j \in J\}$.

An expression Ex will be said to be *closed* if $free(Ex) = \varnothing$.

27.2.3. DEFINITION. If Ex is an expression and $\vartheta: \mathcal{X} \to \mathbf{Exp}$, then we define the expression $Ex[\vartheta]$ recursively as follows:

1• If $Ex = x$, $x \in \mathcal{X}$, then $Ex[\vartheta] = \vartheta(x)$;
2• If $Ex = \mathbf{0}$, then $Ex[\vartheta] = \mathbf{0}$;
3• If $Ex = a.Ex'$, then $Ex[\vartheta] = a.(Ex'[\vartheta])$;
4• If $Ex = \sum_{j\in J} Ex_j$, then $Ex[\vartheta] = \sum_{j\in J}(Ex_j[\vartheta])$;
5• If $Ex = Ex_1 \mid Ex_2$, then $Ex[\vartheta] = (Ex_1[\vartheta]) \mid (Ex_2[\vartheta])$;
6• If $Ex = Ex' \setminus L$, then $Ex[\vartheta] = (Ex'[\vartheta]) \setminus L$;
7• If $Ex = Ex'[\phi]$, then $Ex[\vartheta] = (Ex'[\vartheta])[\phi]$;
8• If $Ex = fix_i(\{x_j = Ex_j : j \in J\})$, then $Ex[\vartheta] = fix_i(\{x_j = Ex_j[\vartheta|J] : j \in J\})$,

where

$$(\vartheta|J)(x) = \begin{cases} x \text{ if } x = x_j, \text{some } j \in J \\ \vartheta(x) \text{ otherwise} \end{cases}$$

27.2.4. DEFINITION. Suppose that Ex is an expression, then we define $\sigma[\![Ex]\!] = ((\mathbf{Exp}, \to, \mathcal{A}ct), Ex)$, where $\to$ is defined by recursion over the structure of expressions, as follows.

1• $\dfrac{\textbf{true}}{a.Ex \to^{a} Ex}$;

2• $\dfrac{i \in J \wedge Ex_i \to^a Ex}{\sum_{j \in J} Ex_j \to^a Ex}$;

3• $\dfrac{Ex_1 \to^a Ex}{Ex_1 \mid Ex_2 \to^a Ex \mid Ex_2}$ $\dfrac{Ex_2 \to^a Ex}{Ex_1 \mid Ex_2 \to^a Ex_1 \mid Ex}$ $\dfrac{Ex_1 \to^a Ex_1' \wedge Ex_2 \to^{\bar{a}} Ex_2'}{Ex_1 \mid Ex_2 \to^\tau Ex_1' \mid Ex_2'}$

4• $\dfrac{Ex \to^a Ex' \wedge \{a, \bar{a}\} \cap L = \varnothing}{Ex \setminus L \to^a Ex' \setminus L}$;

5• $\dfrac{Ex \to^a Ex'}{Ex[\phi] \to^{\phi(a)} Ex'[\phi]}$;

6• $\dfrac{Ex_i[\vartheta_{\mathcal{E}}] \to^a Ex'}{fix_i(\mathcal{E}) \to^a Ex'}$.

where if $\mathcal{E} = \{x_j = Ex_j : j \in J\}$, then we define

$$\vartheta_{\mathcal{E}}(x) = \begin{cases} fix_j(\mathcal{E}) \text{ if } x = x_j, \text{some } j \\ x \text{ otherwise} \end{cases} \tag{27.1}$$

As an illustration of (6), if $\mathcal{E} = \{x_1 = Ex_1, x_2 = Ex_2\}$, where $Ex_1 = a.x_1 + b.x_2$ and $Ex_2 = c.x_1 + d.x_2$ then

$$Ex_1[\vartheta_{\mathcal{E}}] = a.fix_1(\mathcal{E}) + b.fix_2(\mathcal{E})$$

so $Ex_1[\vartheta_{\mathcal{E}}] \to^b fix_2(\mathcal{E})$ and hence $fix_1(\mathcal{E}) \to^b fix_2(\mathcal{E})$.

Since each closed expression is associated with a transition system, we may apply the various relations defined for transition systems to expressions. For example, we may define

$$Ex_1 \sim Ex_2 \Leftrightarrow \sigma[\![Ex_1]\!] \sim \sigma[\![Ex_2]\!] \tag{27.2}$$

Indeed, the ~ relation of section 13.3 was originally defined by Park in [51] and adapted by Milner for CCS. In particular, the interested reader may care to compare definition 13.4.1 (strong bisimulation) with definition 2, section 4.2, of [3], to see that the definition of ~ given in (27.2) corresponds exactly to that used by Milner. Observe that the condition $A_1 = A_2$ of 13.4.1 is satisfied automatically because all the transition systems arising from Milner's semantics have the same set of actions, namely $\mathcal{A}ct$.

Milner also defines ~ on expressions which may contain free variables.[1] We define $Ex_1 \sim Ex_2$ if and only if $\sigma[\![Ex_1[\vartheta]]\!] \sim \sigma[\![Ex_2[\vartheta]]\!]$ for all $\vartheta : \mathcal{X} \to \mathbf{Exp}$ such that $\vartheta(x)$ is closed, all $x \in \mathcal{X}$.

---

[1]See [3], section 4.4, definition 5.

27.3. Hybrid Transition System Semantics.

In this section we shall give a denotational semantics to CCS using the constructions described in the previous chapter. The semantics is based on a partial function

$$\rho: \mathbf{Exp} \rightarrow [\mathbf{Env} \rightarrow \mathbf{AHTS^*}]$$

which is defined recursively over the structure of expressions. The function will have the property that if $\rho[\![Ex]\!]$ is defined, then it will be admissible in the sense of 26.2.6 and that, furthermore, if Ex is closed, then $\rho[\![Ex]\!]$ will be a constant. Hence, for closed expressions, we have a mapping $\tilde{\sigma}: \mathbf{Exp} \rightarrow \mathbf{AHTS^*}$ given by

$$\tilde{\sigma}[\![Ex]\!] = H^* \Leftrightarrow \forall \eta \in \mathbf{Env}: \rho[\![Ex]\!](\eta) = H^* \tag{27.3}$$

First, if $x \in \mathcal{X}$ then the x *projection function*, $proj_x$ is given by

$$\forall \eta \in \mathbf{Env}: proj_x(\eta) = \eta(x) \tag{27.4}$$

and the nil function, nil is given by

$$\forall \eta \in \mathbf{Env}: nil(\eta) = NIL_0{}^2 \tag{27.5}$$

27.3.1. DEFINITION. We define a partial function $\rho: \mathbf{Exp} \rightarrow [\mathbf{Env} \rightarrow \mathbf{AHTS^*}]$ recursively, using the constructions of chapter 26, as follows:

1• If $x \in \mathcal{X}$ then $\rho[\![x]\!] = proj_x$;
2• $\rho[\![\mathbf{0}]\!] = nil$;
3• If $\rho[\![Ex]\!]$ is defined, then $\rho[\![a.Ex]\!] = a \rightarrow \rho[\![Ex]\!]$;
4• If $\rho[\![Ex_j]\!]$ is defined, each $j \in J$, then $\rho[\![\sum_{j\in J} Ex_j]\!] = \sum_{j\in J} \rho[\![Ex_j]\!]$;
5• If $\rho[\![Ex_j]\!]$ is defined, $j = 1, 2$, then $\rho[\![Ex_1 \mid Ex_2]\!] = \rho[\![Ex_1]\!] \mid \rho[\![Ex_2]\!]$;
6• If $\rho[\![Ex]\!]$ is defined, then $\rho[\![Ex \setminus L]\!] = \rho[\![Ex]\!] \setminus L$;
7• If $\rho[\![Ex]\!]$ is defined, then $\rho[\![Ex[\phi]]\!] = \rho[\![Ex]\!][\phi]$;
8• If $\mathcal{E} = \{x_j = Ex_j : j \in J\}$, such that for each $j \in J$, $\rho[\![Ex_j]\!]$ is defined and admissible, then

$$\rho[\![fix_i(\mathcal{E})]\!] = fix_{x_i}(X_{\mathcal{E}}, \mathcal{F}_{\mathcal{E}}).$$

[2]See (26.5)

where $X_{\mathcal{E}} = \{x_j \mid j \in J\}$ and $\mathcal{F}_{\mathcal{E}} = \{\rho[\![Ex_j]\!] : j \in J\}$.

We present the basic properties of the functions $\rho[\![Ex]\!]$.

27.3.2. PROPOSITION. If Ex is an expression such that $\rho[\![Ex]\!]$ is defined, then

1• $\rho[\![Ex]\!]$ is monotonic and preserves least upper bounds of ascending chains;

2• $free(Ex)$ is essential to $\rho[\![Ex]\!]$.[3]

PROOF. The proof is by structural induction on Ex.[4]

(1) If $Ex = x \in \mathcal{X}$, then by (1) of 27.3.1and 27.2.2, $\rho[\![Ex]\!] = proj_x$ and $free(Ex) = \{x\}$. By (27.4), $\rho[\![Ex]\!]$ is monotonic and preserves least upper bounds of ascending chains and $free(Ex)$ is essential to $\rho[\![Ex]\!]$.

(2) If $Ex = \mathbf{0}$, then by (2) of 27.3.1 and 27.2.2, $\rho[\![Ex]\!] = nil$ and $free(Ex) = \varnothing$. By (27.5), $\rho[\![Ex]\!]$ is monotonic and preserves least upper bounds of ascending chains and $free(Ex)$ is essential to $\rho[\![Ex]\!]$.

(3) If $Ex = a.\,Ex'$, then $\rho[\![Ex]\!] = a \to \rho[\![Ex']\!]$ and $free(Ex) = free(Ex')$, by (3) of 27.3.1 and 27.2.2, so $\rho[\![Ex']\!]$ is defined. By induction, $\rho[\![Ex']\!]$ is monotonic and preserves least upper bounds of ascending chains and $free(Ex')$ is essential to $\rho[\![Ex']\!]$. By 26.3.3, $\rho[\![Ex]\!]$ is monotonic and preserves least upper bounds of ascending chains. If $\eta_1, \eta_2 \in \mathbf{Env}$ such that $\eta_1 =_{free(Ex)} \eta_2$, then $\eta_1 =_{free(Ex')} \eta_2$ so $\rho[\![Ex']\!](\eta_1) = \rho[\![Ex']\!](\eta_2)$. Therefore

$$\rho[\![Ex]\!](\eta_1) = a \to \rho[\![Ex']\!](\eta_1) = a \to \rho[\![Ex']\!](\eta_2) = \rho[\![Ex]\!](\eta_2).$$

Thus, $free(Ex)$ is essential to $\rho[\![Ex]\!]$.

(4) If $Ex = \sum_{j\in J} Ex_j$ then $\rho[\![Ex]\!] = \sum_{j\in J} \rho[\![Ex_j]\!]$ and $free(Ex) = \bigcup_{j\in J} free(Ex_j)$, by (4) of 27.3.1 and 27.2.2, so for each $j \in J$, $\rho[\![Ex_j]\!]$ is defined. By induction, for each $j \in J$, $\rho[\![Ex_j]\!]$ is monotonic and preserves least upper bounds of ascending chains and $free(Ex_j)$ is essential to $\rho[\![Ex_j]\!]$. By 26.3.6 $\rho[\![Ex]\!]$ is monotonic and preserves least upper bounds of ascending chains. If $\eta_1, \eta_2 \in \mathbf{Env}$ such that $\eta_1 =_{free(Ex)} \eta_2$, then for each $j \in J$, $\eta_1 =_{free(Ex_j)} \eta_2$, and hence $\rho[\![Ex_j]\!](\eta_1) = \rho[\![Ex_j]\!](\eta_2)$. Therefore

$$\rho[\![Ex]\!](\eta_1) = \textstyle\sum_{j\in J} \rho[\![Ex_j]\!](\eta_1) = \sum_{j\in J} \rho[\![Ex_j]\!](\eta_2) = \rho[\![Ex]\!](\eta_2).$$

---

[3]See 26.2.10.

[4]We shall be using structural induction for many proofs in this chapter. An outline of this proof technique is given in appendix A.4.

Thus, $free(Ex)$ is essential to $\rho[\![Ex]\!]$.

(5) If $Ex = Ex_1 \mid Ex_2$ then by (5) of 27.3.1 and 27.2.2, $\rho[\![Ex]\!] = \rho[\![Ex_1]\!] \mid \rho[\![Ex_2]\!]$ and $free(Ex) = free(Ex_1) \cup free(Ex_2)$, so for each j, $\rho[\![Ex_j]\!]$ is defined. By induction, for each j, $\rho[\![Ex_j]\!]$ is monotonic and preserves least upper bounds of ascending chains and $free(Ex_j)$ is essential to $\rho[\![Ex_j]\!]$. By 26.3.9, $\rho[\![Ex]\!]$ is monotonic and preserves least upper bounds of ascending chains. If $\eta_1, \eta_2 \in \mathbf{Env}$ such that $\eta_1 =_{free(Ex)} \eta_2$, then for each j, $\eta_1 =_{free(Ex_j)} \eta_2$, and hence $\rho[\![Ex_j]\!](\eta_1) = \rho[\![Ex_j]\!](\eta_2)$. Therefore

$$\rho[\![Ex]\!](\eta_1) = \rho[\![Ex_1]\!](\eta_1) \mid \rho[\![Ex_2]\!](\eta_1) = \rho[\![Ex_1]\!](\eta_2) \mid \rho[\![Ex_2]\!](\eta_2) = \rho[\![Ex]\!](\eta_2)$$

and so $free(Ex)$ is essential to $\rho[\![Ex]\!]$.

(6) If $Ex = Ex' \setminus L$, then $\rho[\![Ex]\!] = \rho[\![Ex']\!] \setminus L$ and $free(Ex) = free(Ex')$, by (6) of 27.3.1 and 27.2.2, so $\rho[\![Ex']\!]$ is defined. By induction, $\rho[\![Ex']\!]$ is monotonic and preserves least upper bounds of ascending chains and $free(Ex')$ is essential to $\rho[\![Ex']\!]$. By 26.3.15, $\rho[\![Ex]\!]$ is monotonic and preserves least upper bounds of ascending chains. If $\eta_1, \eta_2 \in \mathbf{Env}$ such that $\eta_1 =_{free(Ex)} \eta_2$, then $\eta_1 =_{free(Ex')} \eta_2$ so $\rho[\![Ex']\!](\eta_1) = \rho[\![Ex']\!](\eta_2)$. Therefore

$$\rho[\![Ex]\!](\eta_1) = \rho[\![Ex']\!](\eta_1) \setminus L = \rho[\![Ex']\!](\eta_2) \setminus L = \rho[\![Ex]\!](\eta_2).$$

Thus, $free(Ex)$ is essential to $\rho[\![Ex]\!]$.

(7) If $Ex = Ex'[\phi]$, then $\rho[\![Ex]\!] = \rho[\![Ex']\!][\phi]$ and $free(Ex) = free(Ex')$, by (7) of 27.3.1 27.2.2, so $\rho[\![Ex']\!]$ is defined. By induction, $\rho[\![Ex']\!]$ is monotonic and preserves least upper bounds of ascending chains and $free(Ex')$ is essential to $\rho[\![Ex']\!]$. By 26.3.12, $\rho[\![Ex]\!]$ is monotonic and preserves least upper bounds of ascending chains. If $\eta_1, \eta_2 \in \mathbf{Env}$ such that $\eta_1 =_{free(Ex)} \eta_2$ then $\eta_1 =_{free(Ex')} \eta_2$ so $\rho[\![Ex']\!](\eta_1) = \rho[\![Ex']\!](\eta_2)$. Therefore

$$\rho[\![Ex]\!](\eta_1) = \rho[\![Ex']\!](\eta_1)[\phi] = \rho[\![Ex']\!](\eta_2)[\phi] = \rho[\![Ex]\!](\eta_2).$$

Thus, $free(Ex)$ is essential to $\rho[\![Ex]\!]$.

(8) If $Ex = fix_i(\{x_j = Ex_j : j \in J\})$, then by 27.3.1 (8) for each $j \in J$, $\rho[\![Ex_j]\!]$ is defined and admissible, so that $(X, \mathcal{F})$ is a recursion scheme, where $X = \{x_j \mid j \in J\}$ and $\mathcal{F} = \{\rho[\![Ex_j]\!] \mid j \in J\}$. Hence, $\rho[\![Ex]\!] = fix_{x_i}(X, \mathcal{F})$ is defined and is monotonic and preserves least upper bounds of ascending chains, by 26.2.9. By 27.2.2 (8), $free(Ex) = \bigcup_{j \in J} free(Ex_j) - X$. Suppose that $\eta_1, \eta_2 \in \mathbf{Env}$ such that $\eta_1 =_{free(Ex)} \eta_2$, then for each $j \in J$, $\eta_1 =_{free(Ex_j)-X} \eta_2$ so $\eta_1|X =_{free(Ex_j)} \eta_2|X$. Thus, $\rho[\![Ex_j]\!](\eta_1|X) = \rho[\![Ex_j]\!](\eta_2|X)$, each $j \in J$, so that $\mathcal{F}(\eta_1|X) = \mathcal{F}(\eta_2|X)$,

by (26.2) and the definition of $\mathcal{F}$. Therefore $fix(X, \mathcal{F})(\eta_1) = fix(X, \mathcal{F})(\eta_2)$, by (26.9). Hence

$$\rho[\![Ex]\!](\eta_1) = fix_{x_i}(X, \mathcal{F})(\eta_1) = fix_{x_i}(X, \mathcal{F})(\eta_2) = \rho[\![Ex]\!](\eta_2).$$

Thus, $free(Ex)$ is essential to $\rho[\![Ex]\!]$.

■

We note that ρ is not a total function. For example, if $x_1 \in \mathcal{X}$ then $\rho[\![x_1]\!]$ is not admissible since if fails to satisfy condition 26.2.6 (3). Consequently, $fix_1(\{x_1 = x_1\})$ is not defined. Let us investigate this problem further. First, we need a useful structural notion.

27.3.3 DEFINITION. If $Ex, \overline{Ex} \in \mathbf{Exp}$, then $\overline{Ex}$ will be said to be a *subexpresssion* of Ex if and only if $Ex = \overline{Ex}$ or one of the following holds

1• $Ex = a.Ex'$ and $\overline{Ex}$ is a subexpression of Ex';
2• $Ex = \sum_{j \in J} Ex_j$ and $\overline{Ex}$ is a subexpression of Ex_j, some $j \in J$;
3• If $Ex = Ex_1 \mid Ex_2$ and $\overline{Ex}$ is a subexpression of Ex_j, some j
4• $Ex = Ex' \setminus L$ and $\overline{Ex}$ is a subexpression of Ex';
5• $Ex = Ex'[\phi]$ and $\overline{Ex}$ is a subexpression of Ex';
6• $Ex = fix_i(\{x_j = Ex_j : j \in J\})$ and $\overline{Ex}$ is a subexpression of Ex_j, some $j \in J$;

An easy structural induction argument shows that the relation 'is a subexpression of' is a partial order on expressions.

27.3.4. LEMMA. Suppose that $Ex \in \mathbf{Exp}$ then $\rho[\![Ex]\!]$ is not defined if and only if Ex contains a subexpression of the form $fix_i(\{x_j = Ex_j : j \in J\})$ such that for some $j \in J$, $\rho[\![Ex_j]\!]$ is defined and not admissible.

PROOF. The proof is by structural induction on Ex. If $Ex = x \in \mathcal{X}$ or $Ex = \mathbf{0}$, then $\rho[\![Ex]\!]$ is defined, so the lemma holds trivially for the base cases.

Let us say that Ex has property $\mathcal{G}$ if and only if Ex contains a subexpression of the form $fix_i(\{x_j = Ex_j : j \in J\})$ such that for some $j \in J$, $\rho[\![Ex_j]\!]$ is defined and not admissible. Note that if Ex' is a subexpression of Ex and Ex' has property $\mathcal{G}$ then so has Ex.

(1) Let $Ex = a.Ex'$. If $\rho[\![Ex]\!]$ is not defined, then $\rho[\![Ex']\!]$ is not defined. By induction, Ex' has property $\mathcal{G}$ so Ex has property $\mathcal{G}$. Conversely, if Ex has property $\mathcal{G}$, then clearly, Ex' has property $\mathcal{G}$, so by induction, $\rho[\![Ex']\!]$ is not defined and so $\rho[\![Ex]\!]$ is not defined.

(2) Let $Ex = \sum_{j\in J} Ex_j$. If $\rho[\![Ex]\!]$ is not defined, then $\rho[\![Ex_j]\!]$ is not defined, some $j \in J$. By induction, Ex_j has property $\mathcal{G}$, so Ex has property $\mathcal{G}$. Conversely, if Ex has property $\mathcal{G}$, then clearly, Ex_j has property $\mathcal{G}$, some $j \in J$, so by induction $\rho[\![Ex_j]\!]$ is not defined. and so $\rho[\![Ex]\!]$ is not defined.

(3) Let $Ex = Ex_1 \mid Ex_2$. If $\rho[\![Ex]\!]$ is not defined, then $\rho[\![Ex_j]\!]$ is not defined, some j. By induction, Ex_j has property $\mathcal{G}$ so Ex has property $\mathcal{G}$. Conversely, if Ex has property $\mathcal{G}$, then clearly, Ex_j has property $\mathcal{G}$, some j, so by induction $\rho[\![Ex_j]\!]$ is not defined and so $\rho[\![Ex]\!]$ is not defined.

(4) If $Ex = Ex' \setminus L$, then the argument is analogous to (1).

(5) If $Ex = Ex'[\phi]$, then the argument is analogous to (1).

(6) Let $Ex = fix_i(\{x_j = Ex_j : j \in J\})$. If $\rho[\![Ex]\!]$ is not defined, then either $\rho[\![Ex_j]\!]$ is defined but not admissible, some $j \in J$ and so Ex has property $\mathcal{G}$, or else $\rho[\![Ex_j]\!]$ is not defined, some $j \in J$. By induction, Ex_j has property $\mathcal{G}$, so Ex has property $\mathcal{G}$. Conversely, if Ex has property $\mathcal{G}$, then either $\rho[\![Ex_j]\!]$ is defined but not admissible, some $j \in J$, or $\rho[\![Ex_j]\!]$ is not defined, some $j \in J$. In either case, $\rho[\![Ex]\!]$ is not defined.

■

We note that by 27.3.2 (1), if $\rho[\![Ex]\!]$ is defined but not admissible, then it must fail to satisfy 26.2.6 (3). We may identify a class of expressions Ex for which $\rho[\![Ex]\!]$ is defined and also admissible; they are *weakly guarded*. This condition is actually due to Milner; he shows in proposition 14, section 4.5 of [3], that the property is sufficient to guarantee uniqueness of solutions to recursion equations up to strong congruence. We shall actually show that $\rho[\![Ex]\!]$ is defined and also admissible for a larger class of expressions, which we call *faintly guarded*. The distinction between the two classes is to do with free variables, and it is not hard to show that a closed expression is weakly guarded if and only if it is faintly guarded. It should also be pointed out that Milner is mainly concerned with closed expressions.

27.3.5 DEFINITION. Let $Ex \in \mathbf{Exp}$. Then Ex is *weakly guarded* if and only if one of the following holds

1• $Ex = \mathbf{0}$;

2• $Ex = a.Ex'$ and each subexpression of Ex' of the form $fix_i(\{x_j = Ex_j : j \in J\})$ is weakly guarded;

3• $Ex = \sum_{j\in J} Ex_j$ and each Ex_j is weakly guarded;

4• $Ex = Ex_1 \mid Ex_2$ and each Ex_j is weakly guarded;

5• $Ex = Ex' \setminus L$ and Ex' is weakly guarded;

6• $Ex = Ex'[\phi]$ and Ex' is weakly guarded;

7• $Ex = fix_i(\{x_j = Ex_j : j \in J\})$ and each Ex_j is weakly guarded.

We shall say that Ex is *faintly guarded*; if and only if each subexpression of Ex of the form $fix_i(\{x_j = Ex_j : j \in J\})$ is weakly guarded. Observe that if Ex is a faintly guarded expression, then every subexpression of Ex is also faintly guarded and that if $Ex = a.Ex'$, then Ex is weakly guarded if and only if Ex' is faintly guarded. We now come to the main result in this section.

27.3.6. THEOREM. If Ex is a faintly guarded expression then

1• $\rho[\![Ex]\!]$ is defined and admissible;
2• $free(Ex)$ is essential to $\rho[\![Ex]\!]$.

PROOF. By virtue of 27.3.2 it suffices to show (1). The proof is by structural induction on Ex.

If $Ex = \mathbf{0}$, then $\rho[\![Ex]\!]$ is defined and admissible, by (27.5).

If $Ex = x \in \mathcal{X}$, then $\rho[\![Ex]\!]$ is defined and admissible, by (27.4);

If $Ex = a.Ex'$, then Ex' is faintly guarded and so for each subexpression of the form $fix_i(\{x_j = Ex_j : j \in J\})$ and for each $j \in J$, Ex_j is faintly guarded and hence $\rho[\![Ex_j]\!]$ is defined and admissible, by induction. By 27.3.4, $\rho[\![Ex']\!]$ is defined and so is monotonic and preserves least upper bounds of ascending chains, by 27.3.2 (1). By 26.3.3, $\rho[\![Ex]\!]$ is defined and admissible.

If $Ex = \sum_{j \in J} Ex_j$, then each Ex_j is faintly guarded and so each $\rho[\![Ex_j]\!]$ is defined and admissible, by induction. Thus, $\rho[\![Ex]\!]$ is defined and admissible, by 26.3.6.

The remaining non-recursive cases are similar.

Finally, if $Ex = fix_i(\{x_j = Ex_j : j \in J\})$, then each Ex_j is faintly guarded and so each $\rho[\![Ex_j]\!]$ is defined and admissible, by induction. Thus, $\rho[\![Ex]\!]$ is defined and admissible, by 26.2.9.

∎

27.3.7. COROLLARY. If Ex is a closed, faintly guarded expression then $\rho[\![Ex]\!]$ is defined and a constant function.

PROOF. $\rho[\![Ex]\!]$ is defined, by 27.3.6 (1), and $free(Ex)$ is essential to $\rho[\![Ex]\!]$, by 27.3.6 (2), so by (26.3), $\varnothing$ is essential to $\rho[\![Ex]\!]$. By 26.2.2, $\eta_1 =_\varnothing \eta_2$, for all $\eta_1, \eta_2 \in \mathbf{Env}$. Thus, $\rho[\![Ex]\!](\eta_1) = \rho[\![Ex]\!](\eta_2)$, for all $\eta_1, \eta_2 \in \mathbf{Env}$, by 26.2.10.

∎

27.4. Properties of Expressions.

In section 27.5, we show that our semantics is consistent with Milner's in the sense that given a closed, faintly guarded expression, the transition system, obtained from its hybrid transition system by removing concurrency is strongly equivalent to that given by Milner's semantics. The proof rests on certain properties of weakly and faintly guarded expressions that we collect together in this section.

First, we establish a standard property of ~.

27.4.1. PROPOSITION. ([3]) ~ is a congruence relation.

PROOF. We have already shown, following 13.4.1, that strong bisimulation is an equivalence relation. The following claims may be easily checked using 27.2.4 and 13.4.1.

Suppose, that $Ex_1 \sim Ex_2$ and let R be a strong bisimulation from $\sigma[\![Ex_1]\!]$ to $\sigma[\![Ex_2]\!]$, then

- $R \cup \{(Ex_1, Ex_2)\}$ is a strong bisimulation from $\sigma[\![e.Ex_1]\!]$ to $\sigma[\![e.Ex_2]\!]$;
- $\{(Ex_1 \backslash L, Ex_2 \backslash L) \mid (Ex_1, Ex_2) \in R\}$ is a strong bisimulation from $\sigma[\![Ex_1 \backslash L]\!]$ to $\sigma[\![Ex_2 \backslash L]\!]$;
- $\{(Ex_1[\phi], Ex_2[\phi]) \mid (Ex_1, Ex_2) \in R\}$ is a strong bisimulation from $\sigma[\![Ex_1[\phi]]\!]$ to $\sigma[\![Ex_2[\phi]]\!]$.

Suppose, that for each $j \in J$, $Ex_j \sim Ex_j'$ and that R_j is a strong bisimulation from $\sigma[\![Ex_j]\!]$ to $\sigma[\![Ex_j']\!]$. then

$$\bigcup_{j \in J} R_j \cup \{(\textstyle\sum_{j \in J} Ex_j, \sum_{j \in J} Ex_j')\}$$

is a strong bisimulation from $\sigma[\![\sum_{j \in J} Ex_j]\!]$ to $\sigma[\![\sum_{j \in J} Ex_j']\!]$

Suppose, that for each $j \in \{1,2\}$, $Ex_j \sim Ex_j'$ and that R_j is a strong bisimulation from $\sigma[\![Ex_j]\!]$ to $\sigma[\![Ex_j']\!]$, then

$$\{(\overline{Ex_1} \mid \overline{Ex_2}, \overline{Ex_1'} \mid \overline{Ex_2'}) \mid (\overline{Ex_1}, \overline{Ex_1'}) \in R_1 \wedge (\overline{Ex_2}, \overline{Ex_2'}) \in R_2\}$$

is a strong bisimulation from $\sigma[\![Ex_1 \mid Ex_2]\!]$ to $\sigma[\![Ex_1' \mid Ex_2']\!]$.

Finally, suppose that $\mathcal{E} = \{x_j = Ex_j : j \in J\}$ and that $\mathcal{E}' = \{x_j = Ex_j' : j \in J\}$ where for $j \in J$, $Ex_j \sim Ex_j'$ and that R_j is a strong bisimulation from $\sigma[\![Ex_j]\!]$ to $\sigma[\![Ex_j']\!]$. then

$$\{(fix_j(\mathcal{E}), fix_j(\mathcal{E}'))\} \cup \{(Ex[\vartheta_{\mathcal{E}}], Ex'[\vartheta_{\mathcal{E}'}]) \mid \exists j \in J : (Ex, Ex') \in R_j\}$$

is a strong bisimulation from $\sigma[\![fix_j(\mathcal{E})]\!]$ to $\sigma[\![fix_j(\mathcal{E}')]\!]$.

■

27.4.2. LEMMA. Suppose that $\mathcal{E}=\{x_j=Ex_j : j\in J\}$ with each Ex_j weakly guarded, then for every expression Ex

$$Ex[\vartheta_{\mathcal{E}}] \sim (Ex[\vartheta^1_{\mathcal{E}}])[\vartheta_{\mathcal{E}}]$$

where

$$\vartheta^1_{\mathcal{E}}(x) = \begin{cases} Ex_j & \text{if } x = x_j, \text{ some } j\in J \\ x & \text{otherwise} \end{cases}$$

PROOF. We argue by induction over the structure of Ex. The case $Ex = \mathbf{0}$ is trivial as is the case $Ex = x$, with $x = x_j$ for no $j\in J$, for in each case, $Ex[\vartheta_{\mathcal{E}}] = Ex = (Ex[\vartheta^1_{\mathcal{E}}])[\vartheta_{\mathcal{E}}]$. If $x = x_j$, some $j\in J$, then $Ex[\vartheta_{\mathcal{E}}] = fix_j(\mathcal{E})$ and $(Ex[\vartheta^1_{\mathcal{E}}])[\vartheta_{\mathcal{E}}] = Ex_j[\vartheta_{\mathcal{E}}]$ and so $Ex[\vartheta_{\mathcal{E}}] \to^e Ex' \Leftrightarrow (Ex[\vartheta^1_{\mathcal{E}}])[\vartheta_{\mathcal{E}}] \to^e Ex'$, by 27.2.4. (6). Therefore, $Ex[\vartheta_{\mathcal{E}}] \sim (Ex[\vartheta^1_{\mathcal{E}}])[\vartheta_{\mathcal{E}}]$.

If $Ex = e.Ex'$ then $Ex[\vartheta_{\mathcal{E}}] = e.(Ex[\vartheta_{\mathcal{E}}])$, by 27.2.3 (3), and

$$(Ex[\vartheta^1_{\mathcal{E}}])[\vartheta_{\mathcal{E}}] = e.((Ex'[\vartheta^1_{\mathcal{E}}])[\vartheta_{\mathcal{E}}]).$$

$Ex'[\vartheta_{\mathcal{E}}] \sim (Ex'[\vartheta^1_{\mathcal{E}}])[\vartheta_{\mathcal{E}}]$, by induction, and so $Ex[\vartheta_{\mathcal{E}}] \sim (Ex[\vartheta^1_{\mathcal{E}}])[\vartheta_{\mathcal{E}}]$, by 27.4.1.

The remaining non-recursive cases are similar.

Finally suppose that $Ex = fix_j(\overline{\mathcal{E}})$ where $\overline{\mathcal{E}} = \{x_j = \overline{Ex_j} : j\in \overline{J}\}$. By renaming variables, if necessary, we may assume that $J\cap\overline{J} = \varnothing$ so that $\vartheta_{\mathcal{E}}|\overline{J} = \vartheta_{\mathcal{E}}$ and $\vartheta^1_{\mathcal{E}}|\overline{J} = \vartheta^1_{\mathcal{E}}$. Thus, $Ex[\vartheta_{\mathcal{E}}] = fix_j(\underline{\mathcal{E}})$ and $(Ex[\vartheta^1_{\mathcal{E}}])[\vartheta_{\mathcal{E}}] = fix_j(\underline{\underline{\mathcal{E}}})$, by 27.2.3 (8), where $\underline{\mathcal{E}} = \{x_j = Ex_j[\vartheta_{\mathcal{E}}] : j\in\overline{J}\}$ and $\underline{\underline{\mathcal{E}}} = \{x_j = (Ex_j[\vartheta^1_{\mathcal{E}}])[\vartheta_{\mathcal{E}}] : j\in\overline{J}\}$. And now, $Ex_j[\vartheta_{\mathcal{E}}] \sim (Ex_j[\vartheta^1_{\mathcal{E}}])[\vartheta_{\mathcal{E}}]$, for all $j\in\overline{J}$ by induction, and so by 27.4.1, $fix_j(\underline{\mathcal{E}}) \sim fix_j(\underline{\underline{\mathcal{E}}})$, that is $Ex[\vartheta_{\mathcal{E}}] \sim (Ex[\vartheta^1_{\mathcal{E}}])[\vartheta_{\mathcal{E}}]$.

■

27.4.3. LEMMA. If $Ex_1 \in \mathbf{Exp}$ is weakly guarded, $e\in\mathcal{A}ct$ and $\vartheta : \mathcal{X} \to \mathbf{Exp}$, then

1• If $Ex_1 \to^e Ex_2$, then $Ex_1[\vartheta] \to^e Ex_2[\vartheta]$

2• If $Ex_1[\vartheta] \to^e Ex'$, then there exists Ex_2 such that $Ex_1 \to^e Ex_2$ and $Ex' = Ex_2[\vartheta]$.

PROOF. We argue by induction over the structure of Ex_1. The lemma holds trivially in the cases $Ex_1 = \mathbf{0}$ and $Ex_1 \in \mathcal{X}$

Suppose that $Ex_1 = \hat{e}.Ex_1'$, where $\hat{e} \in \mathcal{A}ct$, then by 27.2.4 (1), $Ex_1 \to^e Ex_2$ if and only if $e = \hat{e}$ and $Ex_2 = Ex_1'$. So $Ex_2[\vartheta] = Ex_1'[\vartheta]$. By 27.2.3 (3), $Ex_1[\vartheta] = \hat{e}.(Ex_1'[\vartheta])$ so by 27.2.4 (1), $Ex_1[\vartheta] \to^e Ex_2[\vartheta]$ and so (1) holds. By 27.2.4 (1), $Ex_1[\vartheta] \to^e Ex'$ if and only if $e = \hat{e}$ and $Ex' = Ex_1'[\vartheta]$, and so (2) holds with $Ex_2 = Ex_1'$.

Next, let $Ex_1 = \sum_{j \in J} Ex_j$, then for some $j \in J$, $Ex_j \to^e Ex_2$, by 27.2.4 (2), and Ex_j is weakly guarded, by 27.3.5. By induction, $Ex_j[\vartheta] \to^e Ex_2[\vartheta]$, and by 27.2.3 (3), $Ex_1[\vartheta] \to^e Ex_2[\vartheta]$, so (1) holds. If $Ex_1[\vartheta] \to^e Ex'$, then $Ex_j[\vartheta] \to^e Ex'$ for some $j \in J$, by 27.2.4 (2), and Ex_j is weakly guarded, by 27.3.5 (3), so by induction, there exists Ex_2 such that $Ex_j \to^e Ex_2$ and $Ex' = Ex_2[\vartheta]$. But $Ex_1 \to^e Ex_2$, by 27.2.4 (2).

The other non-recursive cases are similar.

Finally, suppose that $Ex_1 = fix_j(\mathcal{E})$ where $\mathcal{E} = \{x_i = Ex_i : i \in I\}$. By 27.2.4 (6), $Ex_j[\vartheta_{\mathcal{E}}] \to^e Ex_2$, and Ex_j is weakly guarded, by 27.3.5 (7). By induction, there exists Ex_2' such that $Ex_j \to^e Ex_2'$ and $Ex_2 = Ex_2'[\vartheta_{\mathcal{E}}]$. Also by induction, $Ex_j[\vartheta_{\overline{\mathcal{E}}}] \to^e Ex_2'[\vartheta_{\overline{\mathcal{E}}}]$, where $\overline{\mathcal{E}} = \{x_i = Ex_i[\vartheta|I] : i \in I\}$. By 27.2.4 (6) and 27.2.3 (8), $Ex_1[\vartheta] \to^e Ex_2'[\vartheta_{\overline{\mathcal{E}}}]$. But, $Ex_2'[\vartheta_{\overline{\mathcal{E}}}] = (Ex_2'[\vartheta_{\mathcal{E}}])[\vartheta] = Ex_2[\vartheta]$, by 27.2.3, so $Ex_1[\vartheta] \to^e Ex_2[\vartheta]$ and (1) holds. If $Ex_1[\vartheta] \to^e Ex'$, then $Ex_j[\vartheta_{\overline{\mathcal{E}}}] \to^e Ex'$, by 27.2.4 (6), and Ex_j is weakly guarded, by 27.3.5 (7) and 27.2.3 (8). By induction, there exists Ex_2' such that $Ex_j \to^e Ex_2'$ and $Ex' = Ex_2[\vartheta_{\overline{\mathcal{E}}}]$. Also by induction, $Ex_j[\vartheta_{\mathcal{E}}] \to^e Ex_2'[\vartheta_{\mathcal{E}}]$, so if we define $Ex_2 = Ex_2'[\vartheta_{\mathcal{E}}]$, then $Ex_1 \to^e Ex_2$, by 27.2.4 (6), and $Ex_2[\vartheta] = (Ex_2'[\vartheta_{\mathcal{E}}])[\vartheta] = Ex_2'[\vartheta_{\overline{\mathcal{E}}}] = Ex'$, by 27.2.3, so (2) holds.

■

27.4.4. COROLLARY. Suppose that $Ex_1 \in \mathbf{Exp}$ is faintly guarded, $e \in \mathcal{A}ct$ and $\vartheta : \mathcal{X} \to \mathbf{Exp}$, then

$$Ex_1 \to^e Ex_2 \Rightarrow Ex_1[\vartheta] \to^e Ex_2[\vartheta]$$

PROOF. The proof is by structural induction on Ex_1. The cases $Ex_1 = \mathbf{0}$ and $Ex_1 \in \mathcal{X}$ are trivial, while the arguments for the non-recursive cases are similar to those of 27.4.3. Finally, if $Ex_1 = fix_j(\mathcal{E})$, then by 27.3.5, Ex_1 must be weakly guarded and we may therefore appeal to 27.4.3 (1).

■

27.4.5. LEMMA Suppose that $\mathcal{E} = \{x_j = Ex_j : j \in J\}$ with each Ex_j weakly guarded, then if $Ex_1, Ex_2 \in \mathbf{Exp}$ and $e \in \mathcal{A}ct$ such that

1• Ex_1 is faintly guarded:
2• $Ex_1[\vartheta^1_{\mathcal{E}}] \rightarrow^e Ex_2$;
3• If $Ex_1 \rightarrow^e Ex'_2$, then $Ex'_2[\vartheta^1_{\mathcal{E}}] \neq Ex_2$,

then $Ex_1[\vartheta_{\mathcal{E}}] \rightarrow^e Ex_2[\vartheta_{\mathcal{E}}]$.

PROOF. The proof is by structural induction on Ex_1. The case $Ex_1 = \mathbf{0}$ is trivial as is the case $Ex_1 = x$ with $x = x_j$, no $j \in J$, for in each case, $Ex_1[\vartheta^1_{\mathcal{E}}] \rightarrow^e Ex_2$ is false.

If $Ex_1 = x_j$, then $Ex_1[\vartheta^1_{\mathcal{E}}] = Ex_j$ and so $Ex_j \rightarrow^e Ex_2$, so by 27.4.4, $Ex_j[\vartheta_{\mathcal{E}}] \rightarrow^e Ex_2[\vartheta_{\mathcal{E}}]$ and so $fix_j(\mathcal{E}) \rightarrow^e Ex_2[\vartheta_{\mathcal{E}}]$, by 27.2.4 (6). But by 27.2.3 (1), $fix_j(\mathcal{E}) = Ex_1[\vartheta_{\mathcal{E}}]$ Thus, $Ex_1[\vartheta_{\mathcal{E}}] \rightarrow^e Ex_2[\vartheta_{\mathcal{E}}]$.

If $Ex_1 = \hat{e}.Ex'_1$, then $Ex_1[\vartheta^1_{\mathcal{E}}] = \hat{e}.(Ex'_1[\vartheta^1_{\mathcal{E}}])$, by 27.2.3 (3), and so if $Ex_1[\vartheta^1_{\mathcal{E}}] \rightarrow^e Ex_2$, then $e = \hat{e}$ and $Ex_2 = Ex'_1[\vartheta^1_{\mathcal{E}}]$. But now, we have $Ex_1 \rightarrow^e Ex'_1$ and $Ex_2 = Ex'_1[\vartheta^1_{\mathcal{E}}]$ and so (3) fails to hold in this case.

If $Ex_1 = \sum_{j \in J} \overline{Ex_j}$, then by 27.2.4, (2) $\overline{Ex_j}[\vartheta^1_{\mathcal{E}}] \rightarrow^e Ex_2$ for some $j \in J$, and $\overline{Ex_j}$ is faintly guarded, by 27.3. If $\overline{Ex_j} \rightarrow^e Ex'_2$, then $Ex_1 \rightarrow^e Ex'_2$ and so $Ex'_2[\vartheta^1_{\mathcal{E}}] \neq Ex_2$. We may thus apply the induction hypothesis to $\overline{Ex_j}$, giving $\overline{Ex_j}[\vartheta_{\mathcal{E}}] \rightarrow^e Ex_2[\vartheta_{\mathcal{E}}]$, so by 27.2.4 (2), $\sum_{j \in J}(\overline{Ex_j}[\vartheta_{\mathcal{E}}]) \rightarrow^e Ex_2[\vartheta_{\mathcal{E}}]$, and thus, by 27.2.3 (4), $Ex_1[\vartheta_{\mathcal{E}}] \rightarrow^e Ex_2[\vartheta_{\mathcal{E}}]$

The remaining non-recursive cases are similar.

Finally, suppose that $Ex_1 = fix_i(\overline{\mathcal{E}})$, then as Ex_1 is faintly guarded, it must be weakly guarded, by 27.3.5, so by 27.4.3, if $Ex_1[\vartheta^1_{\mathcal{E}}] \rightarrow^e Ex_2$, then there exists Ex'_2 such that $Ex_1 \rightarrow^e Ex'_2$ and $Ex'_2[\vartheta^1_{\mathcal{E}}] = Ex_2$.and so (3) fails to hold in this case.

■

As an illustration, let $\mathcal{E} = (x' = e.x'', x'' = f.x')$ and $Ex_1 = x' \mid g.\mathbf{0}$, then every expressions in $\mathcal{E}$ is weakly guarded and Ex_1 is faintly guarded. Furthermore, $Ex_1[\vartheta^1_{\mathcal{E}}] = e.x'' \mid g.\mathbf{0} \rightarrow^e x'' \mid g.\mathbf{0}$ and $Ex_1 \rightarrow^e Ex'_2$ for no Ex'_2. And $Ex_1[\vartheta_{\mathcal{E}}] = fix_{x'}(\mathcal{E}) \mid g.\mathbf{0} \rightarrow^e fix_{x''}(\mathcal{E}) \mid g.\mathbf{0} = (x'' \mid g.\mathbf{0})[\vartheta_{\mathcal{E}}]$, predicted by the lemma.

27.4.6. LEMMA. Suppose that $\mathcal{E} = \{x_j = Ex_j : j \in J\}$ with each Ex_j weakly guarded, then for all $Ex_1, Ex' \in \mathbf{Exp}$ and $e \in \mathcal{A}ct$ such that

1• Ex_1 is faintly guarded;

2• $Ex_1[\vartheta_{\mathcal{E}}] \to^{e} Ex'$;

3• If $Ex_1 \to^{e} Ex_2$, then $Ex' \neq Ex_2[\vartheta_{\mathcal{E}}]$,

then there exists Ex_2 such that $Ex' = Ex_2[\vartheta_{\mathcal{E}}]$ and $Ex_1[\vartheta^1_{\mathcal{E}}] \to^{e} Ex_2$.

PROOF. The proof is by structural induction on Ex_1. The case $Ex_1 = \mathbf{0}$ is trivial as is the case $Ex_1 = x$, with $x = x_j$ for no $j \in J$, for in each case, $Ex_1[\vartheta_{\mathcal{E}}] \to^{e} Ex'$ is false.

If $Ex_1 = x_j$, some $j \in J$, then $Ex_1[\vartheta_{\mathcal{E}}] = fix_j(\mathcal{E})$ and so $fix_j(\mathcal{E}) \to Ex'$ and so by 27.2.4 (6) and 27.2.3 (8), $Ex_j[\vartheta_{\mathcal{E}}] \to^{e} Ex'$. As the Ex_j are weakly guarded, there exists Ex_2 such that $Ex' = Ex_2[\vartheta_{\mathcal{E}}]$ and $Ex_j \to^{e} Ex_2$, by 27.4.3 (2). But, $Ex_j = Ex_1[\vartheta^1_{\mathcal{E}}]$, and so $Ex_1[\vartheta^1_{\mathcal{E}}] \to^{e} Ex_2$.

If $Ex_1 = \hat{e}.Ex_1'$, then $Ex_1[\vartheta_{\mathcal{E}}] = \hat{e}.(Ex_1'[\vartheta_{\mathcal{E}}])$, by 27.2.3 (3), so if $Ex_1[\vartheta_{\mathcal{E}}] \to^{e} Ex'$, then $e = \hat{e}$ and $Ex' = Ex_1'[\vartheta_{\mathcal{E}}]$. But, now we have $Ex_1 \to^{e} Ex_2$ and $Ex' = Ex_2[\vartheta_{\mathcal{E}}]$, where $Ex' = Ex_2[\vartheta_{\mathcal{E}}]$, so (3) fails to hold.

If $Ex_1 = \sum_{j \in J} \overline{Ex_j}$, then by 27.2.3 (3) and 27.2.4 (2), $\overline{Ex_j}[\vartheta_{\mathcal{E}}] \to^{e} Ex'$, some $j \in J$. If $\overline{Ex_j} \to^{e} Ex_2$, then $Ex_1 \to^{e} Ex_2$ and so $Ex' \neq Ex_2[\vartheta_{\mathcal{E}}]$. Thus, we may apply the induction hypothesis to $\overline{Ex_j}$ so there exists Ex_2 such that $Ex' = Ex_2[\vartheta_{\mathcal{E}}]$ and $\overline{Ex_j}[\vartheta^1_{\mathcal{E}}] \to^{e} Ex_2$ and if $\overline{Ex_j} \to^{e} Ex''$, then $Ex''[\vartheta^1_{\mathcal{E}}] \neq Ex_2$. By 27.2.4, (2) $\sum_{j \in J}(\overline{Ex_j}[\vartheta^1_{\mathcal{E}}]) \to^{e} Ex_2$, and hence by 27.2.3 (4), $Ex_1[\vartheta^1_{\mathcal{E}}] \to^{e} Ex_2$.

The remaining non-recursive cases are similar.

Finally, suppose that $Ex_1 = fix_i(\overline{\mathcal{E}})$, then as Ex_1 is faintly guarded, it must also be weakly guarded and so by 27.4.3, if $Ex_1[\vartheta_{\mathcal{E}}] \to^{e} Ex'$ then there exists Ex_2 such that $Ex_1 \to^{e} Ex_2$ and $Ex' = Ex_2[\vartheta_{\mathcal{E}}]$, so (3) fails to hold.

■

To illustrate this result, we may use the example following 27.4.5. which, as we have seen, satisfies all the hypotheses of the lemma. We also have $Ex_1[\vartheta_{\mathcal{E}}] = fix_{x'}(\mathcal{E}) \mid g.\mathbf{0} \to^{e} Ex$, where $Ex = fix_{x''}(\mathcal{E}) \mid g.\mathbf{0}$. And if we define $Ex_2 = x'' \mid g.\mathbf{0}$, then $Ex = (x'' \mid g.\mathbf{0})[\vartheta_{\mathcal{E}}]$ and $Ex_1[\vartheta^1_{\mathcal{E}}] = e.x'' \mid g.\mathbf{0} \to^{e} Ex_2$.

Finally, we show that the property of being faintly guarded is hereditary.

27.4.7. LEMMA. If $Ex_1 \to^{e} Ex_2$ and Ex_1 is faintly guarded, then Ex_2 is faintly guarded.

PROOF. The proof is by structural induction on Ex_1. The cases $Ex_1 = \mathbf{0}$ and $Ex_1 \in \mathcal{X}$ are trivial.

If $Ex_1 = \hat{e}.Ex'$, then $Ex' = Ex_2$, by 27.2.4 (1) and so Ex_2 is faintly guarded by 27.3.5.

If $Ex_1 = \sum_{j \in J} Ex_j$, then $Ex_j \to^e Ex_2$, some j. By 27.3.5, Ex_j is faintly guarded, so by induction, Ex_2 is faintly guarded.

If $Ex_1 = Ex_1' \mid Ex_1''$, then $Ex_2 = Ex_2' \mid Ex_2''$, where either $Ex_1' \to^e Ex_2'$ and $Ex_1'' = Ex_2''$, or $Ex_1' = Ex_2'$, and $Ex_1'' \to^e Ex_2''$, or $Ex_1' \to^{e'} Ex_2'$ and $Ex_1'' \to^{\bar{e}'} Ex_2''$, some e'. By 27.3.5, Ex_1' and Ex_1'' are faintly guarded. By induction, Ex_2' and Ex_2'' are faintly guarded. By 27.3.5, Ex_2 is faintly guarded.

The other non-recursive cases are similar.

Finally, if $Ex_1 = fix_i(\mathcal{E})$, where $\mathcal{E} = \{x_i = Ex_i : i \in J\}$ then $Ex_j[\vartheta_{\mathcal{E}}] \to^e Ex_2$. As the Ex_j are weakly guarded, by 27.3.5, by 27.4.3 (2), there exists Ex_2' such that $Ex_2 = Ex_2'[\vartheta_{\mathcal{E}}]$ and $Ex_j \to^e Ex_2'$. By induction, Ex_2' is faintly guarded and so by 27.3.5, Ex_2 is faintly guarded.

■

27.5. Comparison of the Semantics.

We conclude this chapter by demonstrating that our hybrid transition systems semantics is consistent with the of Milner. As in chapter 25, we do this by showing that if our semantics associates an expression with an initialised hybrid transition system H^* and Milner's semantics associates that expression with an initialised transition system T^*, then throwing away the concurrency structure of H^* produces something equivalent to T^*.

Recall from 23.5.1 that if H is an asynchronous hybrid transition system, then its interleaving is the transition system $T_H = (Q, E, \to_H)$, where $q_1 \to_H^e q_2$ if and only if there exists $a \in A$ such that $\mu(a) = e$ and $q_1 \to^a q_2$. If (H, q) is an asynchronous, initialised hybrid transition system, we also define $T_{(H,q)} = (T_H, q)$.

If follows that if Ex is a closed, faintly guarded expression then there are two initialised transition systems that we may construct from it, namely $\sigma[\![Ex]\!]$, given by 27.2.4, and $T_{\tilde{\sigma}[\![Ex]\!]}$, where $\tilde{\sigma}[\![Ex]\!]$ is given by (27.3). In this section, we show that these two semantics are equivalent, in the sense that

$$T_{\tilde{\sigma}[\![Ex]\!]} \sim \sigma[\![Ex]\!] \tag{27.6}$$

To simplify the notation, we shall define $((Q_{Ex}, E_{Ex}, \to_{Ex}), q_{Ex}) = T_{\tilde{\sigma}[\![Ex]\!]}$, wherever Ex is a closed, weakly guarded expression. The strong bisimulations we construct from $T_{\tilde{\sigma}[\![Ex]\!]}$ to $\sigma[\![Ex]\!]$ will actually be functions. Let us establish some terminology.

27.5.1. DEFINITION. Suppose that (H,q) is an asynchronous, initialised hybrid transition system, then a *bisimulation function* is a function $\chi: Q \to \mathbf{Exp}$ such that if $a \in A$, $e \in \mathcal{A}ct$ and $q_1, q_2 \in R(T(C(H)),q)$,[5] then

1• $q_1 \to^a q_2 \Rightarrow \chi(q_1) \to^{\mu(a)} \chi(q_2)$;
2• $\chi(q_1) \to^e Ex_2 \Rightarrow \exists q_2 \in Q\, \exists a \in \mu^{-1}(e): q_1 \to^a q_2 \wedge \chi(q_2) \sim Ex_2$.

We write $\chi:(H,q) \sim Ex$ to indicate that there is a bisimulation function $\chi: Q \to \mathbf{Exp}$ with $\chi(q) = Ex$.

27.5.2. LEMMA Suppose that (H,q) is an asynchronous, initialised hybrid transition system and that Ex is a closed, faintly guarded expression, then $\chi:(H,q) \sim Ex \Rightarrow T_{\bar{\sigma}[[Ex]]} \sim \sigma[[Ex]]$.

PROOF. By 27.5.1 and 13.4.1 and 27.4.1, the relation R defined by $q' \mathrel{R} Ex' \Leftrightarrow \chi(q) \sim Ex'$ is a strong bisimulation.

■

We now set about the proof that if Ex is a closed, faintly guarded expression, then there is a bisimulation function $\chi:(H,q) \sim Ex$. The argument will be rather complicated as we will need to be able to deal with the general case in which Ex may contain free variables and recursion subexpressions. To cope with such cases, we need the following generalisation of definition 27.5.1.

27.5.3. DEFINITION. If $\eta \in \mathbf{Env}$ then a *bisimulation family* on η is a family $\underline{\underline{\chi}} = (\chi_x)_{x \in X}$ of functions such that for each $x \in X$, there exists $Ex_x \in \mathbf{Exp}$ such that $\chi_x : \eta(x) \sim Ex_x$. We note that by 27.5.1, $Ex_x = \chi_x(q_{\eta(x)})$, where we recall that $q_{\eta(x)}$ is the initial state of $\eta(x)$, and so Ex_x is unique.

If $\underline{\underline{\chi}} = (\chi_x)_{x \in X}$ is a bisimulation family, then we define $\underline{\underline{\chi}}_x = \chi_x$.

Our argument will be that if $\eta \in \mathbf{Env}$ and $\underline{\underline{\chi}}$ is a bisimulation family on η, and if Ex is faintly guarded, then from these we may construct a bisimulation function on $\rho[[Ex]](\eta)$. The construction may be made in a systematic manner using the following concept:

27.5.4. DEFINITION. If Ex is an expression such that $\rho[[Ex]]$ is defined, then a *bisimulation transform* for Ex is a function X such that if $\eta \in \mathbf{Env}$ and

---

[5]See 13.2.3.

$\underline{\underline{\chi}} = (\chi_x)_{x\in X}$ is a bisimulation family on η, then $X(\underline{\underline{\chi}})$ is a function from the set of states of $\rho[\![Ex]\!](\eta)$ to **Exp** satisfying

$$X(\underline{\underline{\chi}}): \rho[\![Ex]\!](\eta) \sim Ex[\underline{\underline{\chi}} * \eta]$$

where $\underline{\underline{\chi}} * \eta : \mathcal{X} \to \mathbf{Exp}$ is given by $(\underline{\underline{\chi}} * \eta)(x) = \underline{\underline{\chi}}_x (q_{\eta(x)})$.

27.5.4 deals with the following situation. If $\eta \in \mathbf{Env}$ and $\underline{\underline{\chi}} = (\chi_x)_{x\in X}$ is a bisimulation family on η, then for each $x \in \mathcal{X}$ there exists a unique expression Ex_x such that $\chi_x : \eta(x) \sim Ex_x$ We may now make the substitution Ex_x for x in Ex and obtain an expression $Ex[\underline{\underline{\chi}} * \eta]$. We may also apply $\rho[\![Ex]\!]$ to η, giving an initialised, asynchronous hybrid transition system $\rho[\![Ex]\!](\eta)$. A bisimulation transform for Ex may then be applied to $\underline{\underline{\chi}}$ to give a bisimulation function from $\rho[\![Ex]\!](\eta)$ to $Ex[\underline{\underline{\chi}} * \eta]$.

Recall that the semantics of recursion expressions is given in terms of ascending chains of elements of **Env**. Each of these will turn out to the image under the semantic mapping of an expression which is a 'finite approximation' to the recursion expression and its bisimulation transform is then constructed from the bisimulation transforms of the approximations in a manner to be explained. The process requires that the ascending chain is regular in the sense of the following definitions.

27.5.5. DEFINITION. Suppose that $\underline{\underline{H}}^* = (H_n, q_n)_{n=0}^N$ is a finite or infinite ascending chain of asynchronous, initialised hybrid transition systems, $\tilde{\underline{\underline{\chi}}} = (\chi_n)_{n=0}^N$ is a family of bisimulation functions $\chi_n : Q_n \to \mathbf{Exp}$ and $\vartheta : \mathcal{X} \to \mathbf{Exp}$, then we shall say that $\underline{\underline{H}}^*$ is *regular* with respect to $\underline{\underline{\chi}}$ and $\vartheta \Leftrightarrow$ for all n:

1• If $q \in R(T(C(H_n)), q_n)$ then for all $m > n$, $|q\rangle_m = |q\rangle_{n+1}$;

2• If $q \in R(T(C(H_n)), q_n)$, then $\chi_{n+1}(q) = \chi_n(q)[\vartheta]$;

3• If $q \to_{n+1}^a q'$, then $q' \in R(T(C(H_n)), q_n) \Rightarrow q \to_n^a q'$.

Regularity means that in passing from H_n to H_{n+1}, only the outermost part of the structure of the transition system is effected. Thus, once we know the transitions from a given state of H_n, we know the transitions from this state for all H_m with $m > n$, and hence in the least upper bound of the H_n. Regularity also ensures, via 27.5.5 (2), that the action of χ_{n+1} on H_{n+1} is actually determined by the action of χ_n on H_n.

27.5.6. DEFINITION. Suppose that $\underline{\underline{\eta}} = (\eta_n)_{n=0}^{N}$ is a finite or infinite ascending chain of elements of **Env**, $\tilde{\underline{\underline{\chi}}} = (\underline{\underline{\chi}}_n)_{n=0}^{N}$ is a family such that $\underline{\underline{\chi}}_n$ is a bisimulation family of η_n, each n, and $\vartheta : \mathcal{X} \to \mathbf{Exp}$ and then we shall say that $\underline{\underline{\eta}}$ is *regular* with respect to $\underline{\underline{\chi}}$ and $\vartheta \Leftrightarrow$ for all $x \in \mathcal{X}$ one of the following holds

1• $\eta_n(x) = \eta_0(x)$, all n

2• $(\eta_n(x))_{n=0}^{N}$ is regular with respect to $\underline{\underline{\chi}}$ and ϑ.

The claim that if $\rho[\![Ex]\!]$ is defined then Ex has a bisimulation transform is proved by induction over the structure of Ex. The induction hypothesis needs to be more complex than this claim, however, for reasons which will hopefully become clear later; we shall actually prove that if $\rho[\![Ex]\!]$ is defined then Ex is normal, in the sense of the following definition.

27.5.7. DEFINITION. We shall say that an expression Ex is *normal* $\Leftrightarrow$

1• $\rho[\![Ex]\!]$ is defined and Ex has a bisimulation transform;

2• If $\underline{\underline{\eta}} = (\eta_n)_{n=0}^{N}$ is regular with respect to $\tilde{\underline{\underline{\chi}}} = (\underline{\underline{\chi}}_n)_{n=0}^{N}$ and $\vartheta : \mathcal{X} \to \mathbf{Exp}$, then for every bisimulation transform X, of Ex, $(\rho[\![Ex]\!](\eta_n))_{n=0}^{N}$ is regular with respect to $(\mathrm{X}(\underline{\underline{\chi}}_n))_{n=0}^{N}$ and ϑ.

Our aim in the rest of this section is to show that every faintly guarded expression is normal. We know already from 27.3.6 that $\rho[\![Ex]\!]$ is defined for faintly guarded expressions Ex, so that the task is to show that every faintly guarded expression has an associated bisimulation transform and satisfies 27.5.7 (2). The proof is, inevitably, by structural induction over Ex; we begin with the non-recursive constructions.

27.5.8. LEMMA. If $Ex \in \mathbf{Exp}$ and $e \in \mathcal{Act}$, then $e.Ex$ is normal if Ex is normal. In particular, if X is a bisimulation transform for Ex, then $e \to \mathrm{X}$ is a bisimulation transform for $e.Ex$ where if $\eta \in \mathbf{Env}$ and $\underline{\underline{\chi}}$ is a bisimulation family on η, then

$$((e \to \mathrm{X})(\underline{\underline{\chi}}))(\hat{q}) = \begin{cases} e.Ex[\underline{\underline{\chi}} * \eta] \text{ if } \hat{q} = 0 \\ (\mathrm{X}(\underline{\underline{\chi}}))(q) \text{ if } \hat{q} = (q,1) \end{cases} \tag{27.7}$$

PROOF. Let $(H, q_0) = \rho[\![Ex]\!](\eta)$, $(H', q') = \rho[\![e.Ex]\!](\eta)$, $\chi = \mathrm{X}(\underline{\underline{\chi}})$ and $\chi' = (e \to \mathrm{X})(\underline{\underline{\chi}})$. We note that by 27.3.1 (3), $(H', q') = e \to (H, q_0)$.

Suppose that $\hat{q}_1 \rightarrow'^{\hat{a}} \hat{q}_2$, then by 26.3.1 either $\hat{q}_1 = 0$, $\hat{q}_2 = (q_0, 1)$, $\hat{a} = 0$ and $\mu'(\hat{a}) = e$, or there exists $q_1, q_2 \in Q$ and $a \in A$ such that $\hat{q}_1 = (q_1, 1)$, $\hat{q}_2 = (q_2, 1)$, $\hat{a} = (a, 1)$, $\mu'(\hat{a}) = \mu(a)$ and $q_1 \rightarrow^{a} q_2$.

In the first case $\chi'(\hat{q}_1) = e.Ex[\underline{\underline{\chi}} * \eta]$ and $\chi'(\hat{q}_2) = \chi(q_0)$, by (27.7), while $\chi(q_0) = Ex[\underline{\underline{\chi}} * \eta]$, as $\chi{:}(H, q_0) \sim Ex[\underline{\underline{\chi}} * \eta]$, by 27.5.4, so by 27.2.4 (1), $\chi'(\hat{q}_1) \rightarrow^{\mu'(\hat{a})} \chi'(\hat{q}_2)$.

In the second case, using the fact that $\mu'(\hat{a}) = \mu(a)$, $\chi(q_1) \rightarrow^{\mu(a)} \chi(q_2)$, by 27.5.1 (1), as $q_1 \rightarrow^{a} q_2$ and hence $\chi'(\hat{q}_1) = \chi(q_1) \rightarrow^{\mu'(\hat{a})} \chi(q_2) = \chi'(\hat{q}_2)$. Thus 27.5.1 (1) holds for χ'.

Next, suppose that $\chi'(\hat{q}_1) \rightarrow^{\hat{e}} Ex_2$, then by 26.3.1 either $\hat{q}_1 = 0$, or there exists $q_1 \in Q$ such that $\hat{q}_1 = (q_1, 1)$.

In the first case, $\chi'(\hat{q}_1) = e.Ex[\underline{\underline{\chi}} * \eta]$, by (27.7), so by 27.2.4 (1), $Ex_2 = Ex[\underline{\underline{\chi}} * \eta]$ and $\hat{e} = e$. But, $\chi(q_0) = Ex[\underline{\underline{\chi}} * \eta]$, since $\chi{:}(H, q_0) \sim Ex[\underline{\underline{\chi}} * \eta]$, so $\chi'(q_0, 1) = Ex[\underline{\underline{\chi}} * \eta]$, and by 26.3.1 $0 \rightarrow'^{0} (q_0, 1)$ with $\mu'(0) = e$.

In the second case, $(\mathrm{X}(\underline{\underline{\chi}}))(q_1) \rightarrow^{\hat{e}} Ex_2$, so there exists $q_2 \in Q$ and $a \in \mu^{-1}(\hat{e})$ such that $q_1 \rightarrow^{a} q_2$ and $(\mathrm{X}(\underline{\underline{\chi}}))(q_2) = Ex_2$. Hence, by 26.3.1, there exists, $(q_1, 1) \rightarrow'^{(a,1)} (q_2, 1)$ with $(a, 1) \in \mu'^{-1}(\hat{e})$ and by (27.7), $\chi'(q_2, 1) \sim Ex_2$. Thus, in both cases, 27.5.1 (2) holds for χ'.

We have shown that $e \rightarrow \mathrm{X}$ is a bisimulation transform for $e.Ex$; it remains to show that 27.5.7 (2) holds.

Let $(H'_n, q'_n) = \rho[\![Ex]\!](\eta^n_{\mathcal{F}})$ and $(H_n, q_n) = \rho[\![e.Ex]\!](\eta^n_{\mathcal{F}})$, so that by 27.3.1, $(H_n, q_n) = e \rightarrow (H'_n, q'_n)$.

If $q \in Q_n$ and $m > n$, then by 26.3.1 either $q = 0$ or $q = (q', 1)$, with $q' \in Q'_n$. In the first case, $|q\rangle_m = \{((q_m, 1), 0)\} = \{((q_{n+1}, 1), 0)\} = |q\rangle_{n+1}$ and in the second case , by 26.3.1 and induction,

$$\begin{aligned} |q\rangle_m &= \{((q'', 1), (a, 1)) \mid (q'', a) \in |q'\rangle_m\} \\ &= \{((q'', 1), (a, 1)) \mid (q'', a) \in |q'\rangle_{n+1}\} = |q\rangle_{n+1} \end{aligned}$$

hence, 27.5.5 (1) holds.

Next, suppose that X' is a bisimulation transform for $e.Ex$ and define $(\mathrm{X}(\underline{\underline{\chi}}))(q') = (\mathrm{X}'(\underline{\underline{\chi}}))(q', 1)$. It is easy to show that X is a bisimulation transform for Ex and that $\mathrm{X}' = e \rightarrow \mathrm{X}$.

If $q \in Q_{n+1}$, then again by 26.3.1, either $q = 0$ or $q = (q', 1)$, with $q' \in Q'_n$. In the first case, we have

$$(\mathrm{X}'(\underline{\underline{\chi}}_{n+1}))(q) = e.(\mathrm{X}(\underline{\underline{\chi}}_{n+1}))(q), \text{by (27.7)}$$

$$= e.((X(\underline{\underline{\chi}}_n))(q))[\vartheta]), \text{ by hypothesis}$$
$$= (e.(X(\underline{\underline{\chi}}))(q))[\vartheta], \text{ by 27.2.3 (3)}$$
$$= (X'(\underline{\underline{\chi}}_n))(q)[\vartheta]$$

and in the second case, we have

$$(X'(\underline{\underline{\chi}}_{n+1}))(q) = (X(\underline{\underline{\chi}}_{n+1}))(q'), \text{ by (27.7)}$$
$$= (X(\underline{\underline{\chi}}_n))(q')[\vartheta], \text{ by hypothesis}$$
$$= (X'(\underline{\underline{\chi}}_n))(q)[\vartheta]$$

and hence, 27.5.5 (2) holds.

Suppose that $q_1 \to_{n+1}^{a} q_2$ and that $q_2 \in Q_n$, then we cannot have $q_2 = 0$ and so $q_2 = (q_2', 1)$, $q_2' \in Q_n'$. Again, we have two cases. In the first, $q_1 = 0$ and so $q_2' = a = 0$, and so $q_1 \to_n^{a} q_2$, using 26.3.1. In the second, $q_1 = (q_1', 1)$ and $a = (a', 1)$ with $q_1' \to_{n+1}^{a'} q_2'$, and since $q_2' \in Q_n'$, $q_1' \to_n^{a'} q_2'$, by induction, and so $q_1 \to_n^{a} q_2$. Hence, 27.5.5 (3) holds.

■

We have similar results for the other non-recursive constructs, with analogous proofs, which we accordingly omit.

27.5.9. LEMMA. Suppose that $(Ex_j)_{j\in J}$ is a family of normal expressions, then $\sum_{j\in J} Ex_j$ is normal. In particular, if X_j is a bisimulation transform for Ex_j, each j, then $\sum_{j\in J} X_j$ is a bisimulation transform for $\sum_{j\in J} Ex_j$, where if $\eta \in \mathbf{Env}$ and $\underline{\underline{\chi}}$ is a bisimulation family on η, then

$$(\textstyle\sum_{j\in J} X_j(\underline{\underline{\chi}}))(\hat{q}) = \begin{cases} \sum_{j\in J} Ex_j[\underline{\underline{\chi}} * \eta] \text{ if } \hat{q} = 0 \\ (X_j(\underline{\underline{\chi}}))(q) \text{ if } \hat{q} = (q, j) \end{cases} \tag{27.8}$$

■

27.5.10. LEMMA. Suppose that Ex_1 and Ex_2 are normal expressions, then $Ex_1 \mid Ex_2$ is normal. In particular, if X_j is a bisimulation transform for Ex_j, each j, then $X_1 \mid X_2$ is a bisimulation transform for $Ex_1 \mid Ex_2$, where if $\eta \in \mathbf{Env}$ and $\underline{\underline{\chi}}$ is a bisimulation family on η, then

$$((X_1 \mid X_2)(\underline{\underline{\chi}}))(q_1', q_2') = (X_1(\underline{\underline{\chi}}))(q_1') \mid (X_2(\underline{\underline{\chi}}))(q_2') \tag{27.9}$$

■

27.5.11. LEMMA. Suppose that Ex is a normal expression, then $Ex \setminus L$ is normal. In particular, if X is a bisimulation transform for Ex, then $\mathrm{X} \setminus L$ is a bisimulation transform for $Ex \setminus L$, where if $\eta \in \mathbf{Env}$ and $\underline{\underline{\chi}}$ is a bisimulation family on η, then

$$((\mathrm{X} \setminus L)(\underline{\underline{\chi}}))(q) = (\mathrm{X}(\underline{\underline{\chi}}))(q) \setminus L \tag{27.10}$$

■

27.5.12. LEMMA. Suppose that Ex is a normal expression, then $Ex[\phi]$ is normal. In particular, if X is a bisimulation transform for Ex, then $\mathrm{X}[\phi]$ is a bisimulation transform for $Ex[\phi]$, where if $\eta \in \mathbf{Env}$ and $\underline{\underline{\chi}}$ is a bisimulation family on η, then

$$((\mathrm{X}[\phi])(\underline{\underline{\chi}}))(q) = (\mathrm{X}(\underline{\underline{\chi}}))(q)[\phi] \tag{27.11}$$

■

27.5.13. LEMMA. If $\eta \in \mathbf{Env}$ and $\underline{\underline{\chi}}$ is a bisimulation family on η, then

1• **0** is normal with bisimulation transform X given by $(\mathrm{X}(\underline{\underline{\chi}}))(0) = \mathbf{0}$;

2• $x \in \mathcal{X}$ is normal with bisimulation transform X given by $\mathrm{X}(\underline{\underline{\chi}}) = \underline{\underline{\chi}}_x$.

■

Finally, we deal with recursion. Let $\mathcal{E} = \{x_j = Ex_j : j \in J\}$, where for each j, Ex_j is normal and weakly guarded with bisimulation transform X_j and suppose that $\eta \in \mathbf{Env}$ and that $\underline{\underline{\chi}}$ is a bisimulation family on η. Define

$$\eta^0_{\mathcal{E}}(x) = \begin{cases} \perp_{\rho[\![Ex_j]\!]} \text{ if } x = x_j, \text{ some } j \\ \eta(x) \text{ otherwise} \end{cases} \tag{27.12}$$

$$\eta^{n+1}_{\mathcal{E}}(x) = \begin{cases} \rho[\![Ex_j]\!](\eta^n_{\mathcal{E}}) \text{ if } x = x_j, \text{ some } j \\ \eta(x) \text{ otherwise} \end{cases} \tag{27.13}$$

We note that, by 27.3.1 (8) and (26.8), for each i

$$\rho[\![fix_i(\mathcal{E})]\!](\eta) = \bigcup_{n \in \omega} \eta^n_{\mathcal{E}}(x_i) \tag{27.14}$$

Define a sequence of families of functions $(\underline{\underline{\chi}}^n_{\mathcal{E}})_{n \in \omega}$ on $\mathcal{X}$, as follows:

$$(\underline{\underline{\chi}}^0_{\mathcal{E}})_x = \begin{cases} \chi^0_x \text{ if } x = x_j \text{ some } j \\ \underline{\underline{\chi}}_x \text{ otherwise} \end{cases} \tag{27.15}$$

$$(\underline{\underline{\chi}}^{n+1}_{\mathcal{E}})_x = \begin{cases} X_j(\underline{\underline{\chi}}^{n}_{\mathcal{E}}) \text{ if } x = x_j \text{ some } j \\ \underline{\underline{\chi}}_x \text{ otherwise} \end{cases} \tag{27.16}$$

where χ^0_x is the constant function $\chi^0_x(q) = x$.

It is clear that $\underline{\underline{\chi}}^0_{\mathcal{E}}$ is a bisimulation family on $\eta^0_{\mathcal{E}}$. Suppose by induction that $\underline{\underline{\chi}}^n_{\mathcal{E}}$ is a bisimulation family on $\eta^n_{\mathcal{E}}$. Since each Ex_j is normal, by 27.5.7, (27.16) and 27.5.4,

$$(\underline{\underline{\chi}}^{n+1}_{\mathcal{E}})_{x_j} : \rho[\![Ex_j]\!](\eta^n_{\mathcal{E}}) \sim Ex_j[\underline{\underline{\chi}}^n_{\mathcal{E}} * \eta^n_{\mathcal{E}}] \tag{27.17}$$

But by (27.13), $(\rho[\![Ex_j]\!](\eta^n_{\mathcal{E}}))(x_j) = \eta^{n+1}_{\mathcal{E}}(x_j)$ so, $\underline{\underline{\chi}}^{n+1}_{\mathcal{E}}$ is a bisimulation family on $\eta^{n+1}_{\mathcal{E}}$. By induction, $\underline{\underline{\chi}}^n_{\mathcal{E}}$ is a bisimulation family on $\eta^n_{\mathcal{E}}$ for all $n \in \omega$.

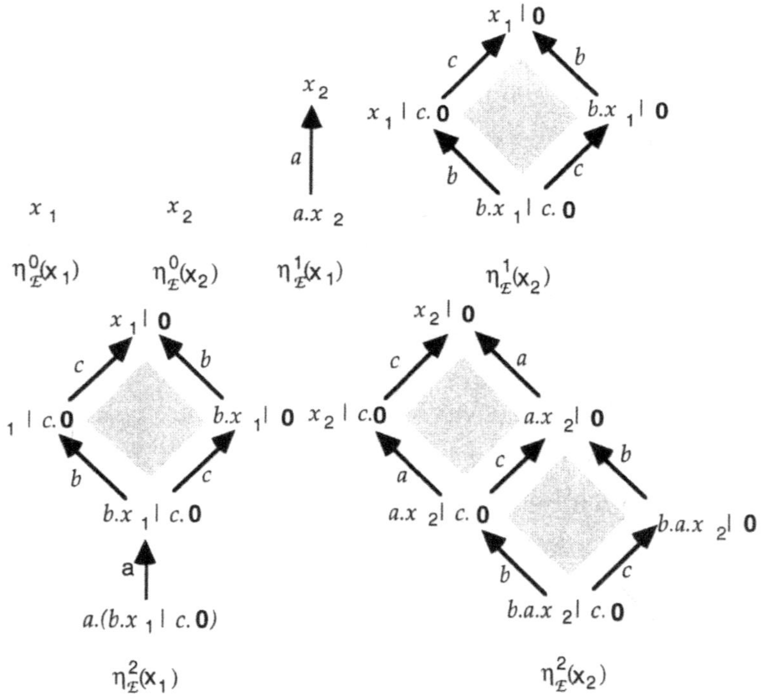

Figure 27.1.

Define

$$(X_{fix_j(\mathcal{E})}(\underline{\underline{\chi}}))(q) = ((\underline{\underline{\chi}}^{index^j_{\mathcal{E}}(q)}_{\mathcal{E}})_{x_j}(q))[\vartheta_{\mathcal{E},\underline{\chi},\eta}] \tag{27.18}$$

where $index^j_{\mathcal{E}}(q)$ is the smallest integer n such that q is a state of $\eta^n_{\mathcal{E}}(x_j)$ and

$$\vartheta_{\mathcal{E},\underline{\underline{\chi}},\eta}(x) = \begin{cases} fix_i(\mathcal{E})[\underline{\underline{\chi}}*\eta] \text{ if } x = x_i \\ x \text{ otherwise} \end{cases} \tag{27.19}$$

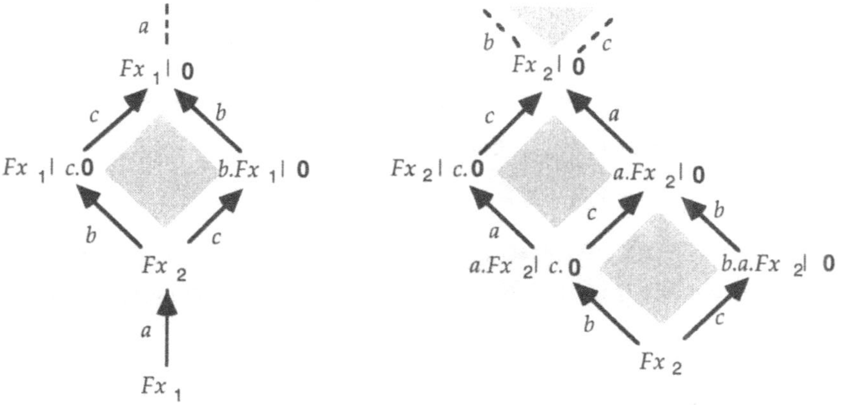

Figure 27.2.

This construction is illustrated in figure 27.1, which shows $\eta^i_{\mathcal{E}}(x_1)$ and $\eta^i_{\mathcal{E}}(x_2)$ for $i = 0,1,2$, and figure 27.2, which shows part of the transition systems $\rho[\![Fx_i]\!]$, where $Fx_i = fix_i(\mathcal{E})$ and $\mathcal{E} = \{x_1 = a.x_2, x_2 = b.x_1 \mid c.\mathbf{0}\}$.

In both diagrams, the expression on each node is its image under the appropriate bisimulation function.

Let us now set about establishing that under the stated conditions, the functions $X_{fix_j(\mathcal{E})}$ are indeed bisimulation transforms for the expressions $fix_j(\mathcal{E})$. First, we establish an important property of the sequence $(\eta^n_{\mathcal{E}})_{n\in\omega}$.

27.5.14. LEMMA. $(\eta^n_{\mathcal{E}})_{n\in\omega}$ is regular with respect to $(\underline{\underline{\chi}}^n_{\mathcal{E}})_{n\in\omega}$ and $\vartheta^1_{\mathcal{E}}$.

PROOF. We argue by induction on N that $(\eta^n_{\mathcal{E}})^N_{n=0}$ is regular with respect to $(\underline{\underline{\chi}}^n_{\mathcal{E}})^N_{n=0}$ and $\vartheta^1_{\mathcal{E}}$ for all $N \geq 1$.

Since $\eta^n_{\mathcal{E}}(x) = \eta^0_{\mathcal{E}}(x)$, when $x = x_j$ for no $j \in J$, by (27.12) and (27.13), it suffices to show that for each $j \in J$, $(\eta^n_{\mathcal{E}}(x_j))^N_{n=0}$ is regular with respect to $(\underline{\underline{\chi}}^n_{\mathcal{E}})^N_{n=0}$ and $\vartheta^1_{\mathcal{E}}$.

Bearing in mind that $\eta^0_{\mathcal{E}}(x_j) = \bot_{\rho[\![Ex_j]\!]}$, inspection of 27.5.5 shows that to settle the case $N = 1$ it suffices to show that $(\underline{\underline{\chi}}^1_{\mathcal{E}})_{x_j}(q_j) = (\underline{\underline{\chi}}^0_{\mathcal{E}})_{x_j}(q_j)[\vartheta^1_{\mathcal{E}}]$, where q_j is the initial state of all of the $\eta^n_{\mathcal{E}}(x_j)$. But by (27.16) and 27.5.4, since $\underline{\underline{\chi}}^0_{\mathcal{E}} * \eta^0_{\mathcal{E}}(x_j) = q_j$,

$$(\underline{\underline{\chi}}_{\mathcal{E}}^{1})_{x_j}(q_j)=X_j(\underline{\underline{\chi}}_{\mathcal{E}}^{0})(q_j)=Ex_j[\underline{\underline{\chi}}_{\mathcal{E}}^{0}*\eta_{\mathcal{E}}^{0}]=Ex_j=(\underline{\underline{\chi}}_{\mathcal{E}}^{0})_{x_j}(q_j)[\vartheta_{\mathcal{E}}^{1}]$$

For the induction step, we note that since each Ex_j is normal, then by 27.2.7 (2), the sequences $((\rho[\![Ex_j]\!](\eta_{\mathcal{E}}^{n}))(x_j))_{n=0}^{N}$ are regular with respect to $(X_j(\underline{\underline{\chi}}_{\mathcal{E}}^{n}))_{n=0}^{N}$ and $\vartheta_{\mathcal{E}}^{1}$, that is by (27.13) and (27.16), $(\eta_{\mathcal{E}}^{n}(x_j))_{n=1}^{N+1}$ is regular with respect to $(\underline{\underline{\chi}}_{\mathcal{E}}^{n})_{n=1}^{N+1}$ and $\vartheta_{\mathcal{E}}^{1}$. Since the pairs of sequences $(\eta_{\mathcal{E}}^{n}(x_j))_{n=0}^{N}$ and $(\eta_{\mathcal{E}}^{n}(x_j))_{n=1}^{N+1}$, and $(\underline{\underline{\chi}}_{\mathcal{E}}^{n})_{n=0}^{N}$ and $(\underline{\underline{\chi}}_{\mathcal{E}}^{n})_{n=1}^{N+1}$ overlap, it follows that $(\eta_{\mathcal{E}}^{n}(x_j))_{n=0}^{N+1}$ is regular with respect to $(\underline{\underline{\chi}}_{\mathcal{E}}^{n})_{n=0}^{N+1}$ and $\vartheta_{\mathcal{E}}^{1}$.

■

27.5.15. PROPOSITION If $\mathcal{E}=\{x_j=Ex_j:j\in J\}$, where for each $j\in J$, Ex_j is normal and weakly guarded, then $X_{fix_j(\mathcal{E})}$ defined in (27.18) is a bisimulation transform for $fix_j(\mathcal{E})$.

PROOF. By 27.5.1, 27.5.3 and 27.5.4, we it suffices to show that if $\chi_j^n=(\underline{\underline{\chi}}_{\mathcal{E}}^{n})_j$, $\chi_j=X_{fix_j(\mathcal{E})}(\underline{\underline{\chi}})$, $(H_{j,n},q_{j,n})=\eta_{\mathcal{E}}^{n}(x_j)$ and $(H_j,q_j)=\rho[\![fix_j(\mathcal{E})]\!](\eta)$, then:

$$\chi_j(q_j)=fix_j(\mathcal{E})[\underline{\underline{\chi}}*\eta] \tag{27.20}$$

$$\forall q,q'\in r_{q_j}(Q_j):q\to_j^{a}q'\Rightarrow\chi_j(q)\to^{\mu_j(a)}\chi_j(q') \tag{27.21}$$

$$\forall q\in r_{q_j}(Q_j):\chi_j(q)\to^{e}Ex'\Rightarrow\exists a\in\mu_j^{-1}(e)\,\exists q'\in\chi_j^{-1}(Ex'):q\to_j^{a}q' \tag{27.22}$$

(27.20) holds because by definition, $index_{\mathcal{E}}^{j}(q_j)=0$ and $\chi_j^{0}(q_j)=x_j$, so by (27.18), $\chi_j(q_j)=x_j[\vartheta_{\mathcal{E},\underline{\underline{\chi}},\eta}]=fix_j(\mathcal{E})[\underline{\underline{\chi}}*\eta]$.

Next, suppose that $q\to_j^{a}q'$, where, $q,q'\in r_{q_j}(Q_j)$ and let n be minimal such that $q\to_{j,n}^{a}q'$. Since χ_j^n is a bisimulation function on $(H_{j,n},q_{j,n})$, $\chi_j^n(q)$ and $\chi_j^n(q')$, are both faintly guarded, by 27.4.7, and $\chi_j^n(q)\to^{\mu_{j,n}(a)}\chi_j^n(q')$, and hence, $\chi_j^n(q)\to^{\mu_j(a)}\chi_j^n(q')$, By 27.5.5 (3), we cannot have $q'\in Q_{j,n-1}$, and so $index_{\mathcal{E}}^{j}(q')=n$. By 27.5.5 (1), we cannot have $q\in Q_{j,n-2}$, and so either $index_{\mathcal{E}}^{j}(q)=n$ or $index_{\mathcal{E}}^{j}(q)=n-1$. By 27.4.4, $\chi_j^n(q)[\vartheta_{\mathcal{E},\underline{\underline{\chi}},\eta}]\to^{\mu_j(a)}\chi_j^n(q')[\vartheta_{\mathcal{E},\underline{\underline{\chi}},\eta}]$, so in the first case $\chi_j(q)\to^{\mu_j(a)}\chi_j(q')$, by (27.18), as $index_{\mathcal{E}}^{j}(q)=index_{\mathcal{E}}^{j}(q')=n$. In the second case, $\chi_j^{n-1}(q)[\vartheta_{\mathcal{E}}^{1}]\to^{\mu_j(a)}\chi_j^n(q')$, by 27.5.5 (2), and so by 27.4.5, $\chi_j^{n-1}(q)[\vartheta_{\mathcal{E},\underline{\underline{\chi}},\eta}]\to^{\mu_j(a)}\chi_j^n(q')[\vartheta_{\mathcal{E},\underline{\underline{\chi}},\eta}]$, that is $\chi_j(q)\to^{\mu_j(a)}\chi_j(q')$ by (27.18). We have shown that (27.21) holds.

Finally, suppose that $\chi_j(q)\to^{e}Ex'$, then $\chi_j^n(q)[\vartheta_{\mathcal{E},\underline{\underline{\chi}},\eta}]\to^{e}Ex'$ by (27.18), where $index_{\mathcal{E}}^{j}(q)=n$, We have two cases:

Case 1. There exists Ex_2 such that $\chi_j^n(q) \to^e Ex_2$ and $Ex_2[\vartheta_{\mathcal{E},\underline{\chi},\eta}] = Ex'$;

Case 2. For all Ex_2, if $\chi_j^n(q) \to^e Ex_2$, then $Ex_2[\vartheta_{\mathcal{E},\underline{\chi},\eta}] \neq Ex'$

In case 1, there exists $q' \in Q_{j,n}$ and $a \in (\mu_{j,n})^{-1}(e)$ such that $q \to_{j,n}^a q'$ and $\chi_j^n(q') \sim Ex_2$. We note that $a \in (\mu_j)^{-1}(e)$ and $q \to_j^a q'$. We cannot have $q' \in Q_{j,n-1}$, as this would entail that $q \in Q_{j,n-2}$, because the sequence $(\eta_{\mathcal{E}}^n)_{n\in\omega}$ is regular with respect to $(\underline{\chi}^n_{\mathcal{E}})_{n\in\omega}$ and $\vartheta_{\mathcal{E}}^1$, contradicting $index_{\mathcal{E}}^j(q) = n$. Hence, $index_{\mathcal{E}}^j(q') = n$ and so $\chi_j(q') = \chi_j^n(q')[\vartheta_{\mathcal{E},\underline{\chi},\eta}] \sim Ex_2[\vartheta_{\mathcal{E},\underline{\chi},\eta}] = Ex'$.

In case 2, by 27.5.6 (2) there exists Ex_2 such that $\chi_j^n(q)[\vartheta_{\mathcal{E}}^1] \to^e Ex_2$ and $Ex_2[\vartheta_{\mathcal{E},\underline{\chi},\eta}] = Ex'$. Since $(\eta_{\mathcal{E}}^n)_{n\in\omega}$ is regular with respect to $(\underline{\chi}^n_{\mathcal{E}})_{n\in\omega}$ and $\vartheta_{\mathcal{E}}^1$ by 27.5.14, $\chi_j^n(q)[\vartheta_{\mathcal{E}}^1] = \chi_j^{n+1}(q)$, so there exists $q' \in Q_{j,n+1}$ and $a \in (\mu_{j,n+1})^{-1}(e)$ such that $a \in (\mu_j)^{-1}(e)$ and $\chi_j^{n+1}(q') \sim Ex_2$. Again, we note that $a \in (\mu_j)^{-1}(e)$ and $q \to_j^a q'$.

We cannot have $q \in Q_{j,n-2}$, by 27.5.5 (1), so either $index_{\mathcal{E}}^j(q') = n+1$ or $index_{\mathcal{E}}^j(q') = n$. In the first case, $\chi_j(q') = \chi_j^{n+1}(q')[\vartheta_{\mathcal{E},\underline{\chi},\eta}] \sim Ex_2[\vartheta_{\mathcal{E},\underline{\chi},\eta}] = Ex'$. In the second case, since $(\eta_{\mathcal{E}}^n)_{n\in\omega}$ is regular with respect to $(\underline{\chi}^n_{\mathcal{E}})_{n\in\omega}$ and $\vartheta_{\mathcal{E}}^1$, by 27.5.14, $q \to_{j,n}^a q'$ and so $\chi_j^n(q) \to^e \chi_j^n(q')$ and thus

$$\begin{aligned}
\chi_j(q') &= \chi_j^n(q')[\vartheta_{\mathcal{E},\underline{\chi},\eta}] \sim (\chi_j^n(q')[\vartheta_{\mathcal{E}}^1])[\vartheta_{\mathcal{E},\underline{\chi},\eta}] \\
&= \chi_j^{n+1}(q')[\vartheta_{\mathcal{E},\underline{\chi},\eta}] \\
&= Ex_2[\vartheta_{\mathcal{E},\underline{\chi},\eta}] = Ex'
\end{aligned}$$

This completes the proof of (27.22) and therefore of the proposition. ∎

27.5.16. COROLLARY. If $\mathcal{E} = \{x_j = Ex_j : j \in J\}$, where for each j, Ex_j is weakly guarded and contains no recursion subexpressions, then $X_{fix_j(\mathcal{E})}$ defined in (27.18) is a bisimulation transform for $fix_j(\mathcal{E})$.

PROOF. By structural induction and lemmas 27.5.8 to 27.5.13, each Ex_j is normal. We now appeal to 27.5.15. ∎

The main theorem of this section say that every faintly guarded expression Ex has a bisimulation transform. The proof will be by induction on the number of recursion subexpressions within Ex. The idea behind the induction step is that we may replace a recursion subexpression Ex', which itself has no recursion subexpressions, and hence a bisimulation transform, by a variable, giving an expression Ex'' to which the induction hypothesis applies and which therefore also has a bisimulation transform. The two transforms are then pieced together to give one for Ex. The justification for this is contained in 27.5.17. First, we need some notation. If $x_0 \in \mathcal{X}$, $\eta \in \mathbf{Env}$, $\underline{\underline{\chi}}$ is a bisimulation family, H^* is an initialised, asynchronous hybrid transition system and X is a bisimulation transform, then we define:

$$Ex[x_0 \setminus Ex'] = Ex[\vartheta]$$

$$\text{where } \vartheta(x) = \begin{cases} Ex' \text{ if } x = x_0 \\ x \text{ otherwise} \end{cases}$$

so $Ex[x_0 \setminus Ex']$ is the expression obtained from Ex by replacing each free occurrence of x_0 by Ex'. We also define:

$$(\eta[x_0 \setminus H^*])(x) = \begin{cases} H^* \text{ if } x = x_0 \\ \eta(x) \text{ otherwise} \end{cases}$$

so that $\eta[x_0 \setminus H^*]$ is the function which differs from η only in that it maps x_0 to H^*. Finally, we define

$$(\underline{\underline{\chi}}[x_0 \setminus \mathrm{X}])_x = \begin{cases} \mathrm{X}(\underline{\underline{\chi}})_x \text{ if } x = x_0 \\ \underline{\underline{\chi}}_x \text{ otherwise} \end{cases}$$

so that the family of functions $\underline{\underline{\chi}}[x_0 \setminus \mathrm{X}]$ differs from $\underline{\underline{\chi}}$ only in that its x_0 member is $\mathrm{X}(\underline{\underline{\chi}})_{x_0}$.

27.5.17. PROPOSITION. If Ex' and Ex'' are expressions possessing bisimulation transforms X' and X'', respectively, then $Ex = Ex''[x_0 \setminus Ex']$ has a bisimulation transform X, where if $\underline{\underline{\chi}}$ is a bisimulation family then

$$\mathrm{X}(\underline{\underline{\chi}}) = \mathrm{X}''(\underline{\underline{\chi}}[x_0 \setminus \mathrm{X}'(\underline{\underline{\chi}})]).$$

PROOF. Suppose that $\eta \in \mathbf{Env}$ and $\underline{\underline{\chi}}$ is a bisimulation family on η, so that for each $x \in \mathcal{X}$, there exists $Ex_x \in \mathbf{Exp}$ such that $\underline{\underline{\chi}}_x : \eta(x) \sim Ex_x$, by 27.5.3. Let $\underline{\underline{\chi}}' = \underline{\underline{\chi}}[x_0 \setminus \mathrm{X}'(\underline{\underline{\chi}})]$ and $\eta' = \eta[x_0 \setminus \rho[\![Ex']\!](\eta)]$. If $x \neq x_0$ then $\eta'(x) = \eta(x)$ and $\underline{\underline{\chi}}'_x = \underline{\underline{\chi}}_x$ and so $\underline{\underline{\chi}}'_x : \eta'(x) \sim Ex_x$. Otherwise, $\mathrm{X}(\underline{\underline{\chi}}) : \rho[\![Ex']\!](\eta) \sim Ex_{x_0}[\underline{\underline{\chi}} * \eta]$, by

27.5.4. But, $\underline{\underline{X}}(\underline{\underline{\chi}}) = \underline{\underline{\chi}}'_{x_0}$ and $\rho[\![Ex']\!](\eta) = \eta(x_0)$, so $\underline{\underline{\chi}}' : \eta(x_0) \sim Ex_{x_0}[\underline{\underline{\chi}} * \eta]$ and we have shown that $\underline{\underline{\chi}}'$ is a bisimulation family on η'.

By 27.5.4, $\underline{\underline{X}}''(\underline{\underline{\chi}}') : \rho[\![Ex'']\!](\eta') \sim Ex''[\underline{\underline{\chi}}' * \eta']$. But, $\underline{\underline{X}}(\underline{\underline{\chi}}) = \underline{\underline{X}}''(\underline{\underline{\chi}}')$ and so in order to complete the proof, we need to show that $\rho[\![Ex'']\!](\eta') = \rho[\![Ex]\!](\eta')$ and $Ex'[\underline{\underline{\chi}}' * \eta'] = Ex[\underline{\underline{\chi}} * \eta]$, that is, we must show that whenever $\eta \in \mathbf{Env}$ and $\underline{\underline{\chi}}$ is a bisimulation family on η, then

$$\rho[\![Ex'']\!](\eta[x_0 \backslash \rho[\![Ex']\!](\eta)) = \rho[\![Ex''[x_0 \backslash Ex']]\!](\eta) \tag{27.23}$$
$$Ex''[\underline{\underline{\chi}}[x_0 \backslash \underline{\underline{X}}'] * \eta[x_0 \backslash \rho[\![Ex']\!](\eta)]] \sim (Ex''[x_0 \backslash Ex'])[\underline{\underline{\chi}} * \eta] \tag{27.24}$$

We prove (27.23) by induction over the structure of Ex''.

The cases $Ex'' = \mathbf{0}$ and $Ex'' = x \neq x_0$ are trivial. If $Ex'' = x_0$, then

$$\rho[\![Ex'']\!](\eta') = \eta'(x_0) = \rho[\![Ex']\!](\eta) = \rho[\![Ex''[x_0 \backslash Ex']]\!](\eta).$$

If $Ex'' = e.Ex'''$, then:

$$\begin{aligned}\rho[\![Ex'']\!](\eta') &= \rho[\![e.Ex''']\!](\eta[x_0 \backslash \rho[\![Ex']\!](\eta)]) \\ &= e \to (\rho[\![Ex''']\!](\eta[x_0 \backslash \rho[\![Ex']\!](\eta)])), \text{ by 27.3.1 (3)} \\ &= e \to (\rho[\![Ex'''[x_0 \backslash Ex']]\!](\eta)), \text{ by induction} \\ &= \rho[\![e.Ex'''[x_0 \backslash Ex']]\!](\eta) = \rho[\![Ex''[x_0 \backslash Ex']]\!](\eta)\end{aligned}$$

where the penultimate equality follows by 27.2.3 (3) and 27.3.1 (3)

The other non-recursive cases are similar.

Finally, suppose that $Ex'' = fix_j(\mathcal{E})$ where $\mathcal{E} = \{x_j = Ex_j : j \in J\}$ and let $X = \{x_j \mid j \in J\}$. Note that $\rho[\![fix_j(\mathcal{E})]\!]$ is defined, as Ex'' has a bisimulation transform. By induction, for each $j \in J$:

$$\rho[\![Ex_j]\!](\eta[x_0 \backslash H^*]) = \rho[\![x_j[x_0 \backslash Ex']]\!](\eta) \tag{27.25}$$

where $H^* = \rho[\![Ex']\!](\eta)$. By renaming the elements of X if necessary, we may assume that $x_0 \notin X$. Thus, $\rho[\![Ex''[x_0 \backslash [Ex']]\!](\eta) = \bigcup_{n \in \omega} \eta_n(x_j)$, where $\eta_0 = \eta | X$ and

$$\eta_{n+1}(x) = \begin{cases} \rho[\![Ex_j[x_0 \backslash Ex']]\!](\eta_n) \text{ if } x = x_j \in X \\ \eta_n(x) \text{ otherwise} \end{cases}$$

and $\rho[\![Ex'']\!](\eta[x_0 \backslash H^*]) = \bigcup_{n \in \omega} \eta'_n(x_j)$, where $\eta'_0 = (\eta[x_0 \backslash H^*]) | X$ and

$$\eta'_{n+1}(x) = \begin{cases} \rho[\![Ex_j]\!](\eta'_n) \text{ if } x = x_j, x_j \in X \\ \eta'_n(x) \text{ otherwise} \end{cases}$$

We shall show by induction that $\eta'_n = \eta_n[x_0 \backslash H^*]$ for all $n \in \omega$, and hence that $\eta'_n(x_j) = \eta_n(x_j)$, as $x_j \neq x_0$, from which (27.23) follows.

If $x_j \in X$, then $x_j \neq x_0$ and if $x_j \notin X$, then $\eta_0(x) = \eta_0(x) = \eta'_0(x)$ when $x \neq x_0$, and $\eta'_0(x_0) = \eta[x_0 \backslash H^*](x_0) = H^* = \eta_0[x_0 \backslash H^*](x_0)$. Thus, the claim holds when $n = 0$.

For the induction step we note that if $x_j \in X$, then

$$\begin{aligned} \eta'_{n+1}(x_j) &= \rho[\![Ex_j]\!](\eta'_n) \\ &= \rho[\![Ex_j]\!](\eta_n[x_0 \backslash H^*]) = \rho[\![Ex_j[x_0 \backslash Ex']]\!](\eta_n) \\ &= \eta_{n+1}(x_j) = \eta_{n+1}(x_j)[x_0 \backslash H^*] \end{aligned}$$

while if $x \notin X$ and $x \neq x_0$, then

$$\eta'_{n+1}(x) = \eta'_n(x) = (\eta_n[x_0 \backslash H^*])(x) = \eta_n(x) = \eta_{n+1}(x) = (\eta_{n+1}[x_0 \backslash H^*])(x)$$

while

$$\eta'_{n+1}(x_0) = \eta'_n(x_0) = (\eta_n[x_0 \backslash H^*])(x_0) = H^* = (\eta_{n+1}[x_0 \backslash H^*])(x_0)$$

completing the proof of the induction step and of (27.23).

Let us now turn to the proof of (27.24). We observe that

$$(\underline{\underline{\chi}}[x_0 \backslash \mathrm{X}'] * \eta[x_0 \backslash H^*])(x) = \begin{cases} \mathrm{X}'(\underline{\underline{\chi}})_{x_0}(q_{H^*}) \text{ if } x = x_0 \\ \underline{\underline{\chi}}_x(q_{\eta(x)}) \text{ otherwise} \end{cases}$$

But X' is a bisimulation transform for Ex' and so $\mathrm{X}'(\underline{\underline{\chi}})_{x_0}(q_{H^*}) \sim Ex'[\underline{\underline{\chi}} * \eta]$. It follows that:

$$\begin{aligned} Ex''(\underline{\underline{\chi}}[x_0 \backslash \mathrm{X}'] * \eta[x_0 \backslash H^*]) &= (Ex''[[x_0 \backslash Ex'[\underline{\underline{\chi}} * \eta]]])[\underline{\underline{\chi}} * \eta] \\ &= (Ex''[x_0 \backslash Ex'])[\underline{\underline{\chi}} * \eta]. \end{aligned}$$

■

At last, we come to the main results of this section.

27.5.18. THEOREM. Every faintly guarded expression Ex has a bisimulation transform.

PROOF. We argue by induction on the number of recursion subexpressions of Ex. The base case follows by structural induction, using lemmas 27.5.8 to 27.5.13.

For the induction step, we may clearly find a recursion subexpression Ex' which itself contains no recursion subexpressions. Let Ex'' be obtained from Ex by replacing each instance of Ex' by x_0, where x_0 is a variable which is not free in Ex. But now, $Ex = Ex''[x_0 \setminus Ex']$, and we may apply 27.5.17.

■

27.5.18 allows us finally to establish our claim, namely that when Ex has a well defined meaning in the interleaving semantics, then it also has a well-defined meaning in the non-interleaving semantics and that the former may be recovered from the latter, up to strong equivalence, by throwing away all information about concurrency.

27.5.19. THEOREM. If Ex is faintly guarded and closed, then $T_{\bar{\sigma}[\![Ex]\!]} \sim \sigma[\![Ex]\!]$.

PROOF. By 27.5.18 and 27.5.2.

■

Notes.

Process Algebraists may be surprised, perhaps outraged, to find that we have not addressed one of the most important aspects of process calculus, namely that it is a calculus, having notions of equivalence and equational laws for reasoning about equivalences between expressions. Of course, existing equivalences would be inappropriate for a non-interleaving view of processes. For example, the equivalence

$$a.\mathbf{0} \mid b.\mathbf{0} \approx a.b.\mathbf{0} + b.a.\mathbf{0}$$

would not hold in a non-interleaving context. It seems fairly clear, however, that laws of strong equivalence dealing with combinators other than $\mid$ should hold.

We have concentrated on equivalences relating to isomorphism. Process Algebraists tend to be more interested in relations which express the fact that the behaviours of the two systems in question are indistinguishable. Thus, in CCS, there are two further relations, namely observational equivalence and observational congruence. It is this author's belief that extending these notations to handle concurrent systems is a non-trivial task owing to the interaction between concurrency, non-determinacy and τ actions, and he has made no attempt here to do so.

In an earlier version of this chapter, we took an alternative approach to dealing with nested recursion which avoided all the paraphernalia of bisimulation transforms and which we illustrate below. The system described by the expression

$$fix_x(x = a.fix_y(y = b.x)) \tag{27.26}$$

could also be defined using a pair of simultaneous equations

$$x = a.y \qquad y = b.x$$

which could also be derived from the expression

$$fix_x(x = a.y, y = b.x) \tag{27.27}$$

without nested recursion. In other words, every recursion expression may be transformed into one without nested recursion. Of course, one still is under the obligation to prove that the resulting expression is strongly equivalent to the one with which one started. In the end, we decided to go the whole hog.

Chapter 28
Theoretical CSP

> When concurrency confronted the computing community about a quarter of a century ago, it caused an endless confusion, partly by the technically very different circumstances in which it emerged, partly by the accident of history that it introduced non-determinism at the same time. The disentanglement of that confusion required the hard work of a mature and devoted scientist who, with luck, would clarify the situation.
>
> E. W. DIJKSTRA
> Foreword to [2]

28.1. Introduction.

In this chapter we discuss the process algebra, presented by Hoare in chapter three of [2], which is sometimes known as 'Theoretical CSP'. CSP stands for Communicating Sequential Processes [74], a design for a programming language of which a refined version is published in [2]. Theoretical CSP is an abstract version of this language without value passing. Hoare presents a semantics to this language in terms of sets of failures [75] and divergences [76]. Failures are pairs (x,X) where x is a sequence of actions that the system may perform and X is a set of actions that the system may refuse to perform after some execution of the sequence x. A divergence is a sequence of actions that can leave the system in an undefined or chaotic state.

We shall construct an alternative semantics involving hybrid transition systems, using some of the operators of chapter 25. We shall do this for a sublanguage of TCSP; we have already spent a lot of time on one process algebra.

We then show that Hoare's semantics is in some sense an interleaved version of our non-interleaving semantics for this sublanguage.

To do this, we show that there is a construction which takes transition systems and delivers sets of failures and divergences and that the sets given in Hoare's semantics coincide with those obtained by the construction applied to an interleaved version of the appropriate hybrid transition system.

We conclude by showing how the non-interleaving semantics may be extended to cope with the whole language.

28.2. TCSP and its Failure/Divergence Semantics.

As for CCS, we assume that we have denumerable sets $\mathcal{A}ct$ of actions and $\mathcal{X}$ of process variables.

28.2.1. DEFINITION. The set **TCSP** of TCSP expressions is defined to be the smallest set containing $\mathcal{X}$ such that if $P, P_1, P_2 \in \mathbf{TCSP}$, $a \in \mathcal{A}ct$, $\varnothing \neq A \subseteq \mathcal{A}ct$ and $x \in \mathcal{X}$, then the following expressions also belong to **TCSP**

1• $Stop_A$;
2• $Chaos_A$
3• $a \to P$.
4• $(P_1 \| P_2)$;
5• $(P_1 ||| P_2)$;
6• $(P_1 \, [] \, P_2)$;
7• $(P_1 \sqcap P_2)$;
8• $P \backslash A$;
9• $(\mu x{:}A.P)$.

As in the case of CCS, we shall say that an instance of a variable $x \in \mathcal{X}$ in an expression P is bound if it occurs in a sub-expression of the form $(\mu x{:}A.P)$, otherwise this instance is free. For example, in the expression

$$(x \| (\mu x{:}\{a\}.(a \to x)))$$

the first instance of x is free; the others, bound. An expression will be said to be closed $\Leftrightarrow$ if contains no free instances of any variable.

We define a TCSP *program* to be a closed TCSP expression.

The semantics presented in [2] maps TCSP into a set of what Hoare calls processes and which we shall call CSP processes to distinguish this usage from the other many and inconsistent uses of this word. As we indicated in the introduction to this chapter, a process is made up of failures and divergences. Let us present the definition and then explain the intuition behind it.

28.2.2. DEFINITION. A CSP Process (henceforth abbreviated to 'process') is a triple $P = (A, F, D)$, where

- $A \subseteq \mathcal{A}ct$ is a finite and non-empty set of actions;
- $F \subseteq A^* \times \wp(A)$ is the set of *failures* of P;
- $D \subseteq A^*$ is the set of *divergences* of P:

satisfying:-

1• $(\Omega, \varnothing) \in F$;

2• $(x, X) \in F \wedge Y \subseteq X \Rightarrow (x, Y) \in F$;

3• $(x, X) \in F \wedge y \leq x \Rightarrow (y, \varnothing) \in F$

4• $(x, X) \in F \wedge a \in A \Rightarrow (x.a, \varnothing) \in F \vee (x \cup \{a\}, X) \in F$;

5• $D \subseteq \{x \in A^* \mid \exists X \subseteq A : (x, X) \in F\}$;

6• $x \in D \wedge x \leq y \Rightarrow y \in D$;

7• $x \in D \wedge X \subseteq A \Rightarrow (x, X) \in F$.

We denote by **HProc** the set of all CSP processes.

If $(x, X) \in F$, then we interpret this to mean that the program may perform the sequence of actions x and that there is some execution of the sequence x following which the program will refuse to execute any action from the set X. Thus, if the program's environment can for some reason insist that the program must perform some action from that set, then the program will fail. If we look at the axioms in order, we see that F must be non-empty (from (1)) and that if it is possible to refuse any element of X, then the program must certainly refuse any element of $Y \subseteq X$ (from (2)).

If x is a possible execution, then any prefix y of it must also be a possible execution and hence $(y, Y) \in F$, some $Y \subseteq A$. By (2), we certainly have $(y, \varnothing) \in F$. This explains (3).

(4) says that if (x, X) is a failure of P, then either a may execute after x, in which case $(x.a, X) \in F$, or not, in which case we may conclude that P may refuse $X \cup \{a\}$ after x.

D is a set of executions of P which lead P to a state in which it behaves in the same way as $CHAOS_A$. To quote [2], $CHAOS_A$ 'may behave like any process whatsoever. It is the most non-deterministic of all processes, the least predictable, the least controllable, and in short the worst.' It is in some sense the totally undefined process and in that respect provides the set of processes with actions A with a bottom element. This is useful because the set becomes a complete partial order, which means that one may take a least-fixed-point approach in giving the semantics of recursion By its very nature it is operationally meaningless.

(6) says that once a process becomes chaotic, it must remain so.

We may tidy up this definition a little by introducing a relation $\preceq$ on $A^* \times \wp(A)$. Define

$$(x, X) \preceq (y, Y) \Leftrightarrow (x = y \wedge X \subseteq Y) \vee (x \leq y \wedge X = \varnothing) \tag{28.1}$$

As usual, a set F will be said to be left-closed in $A^* \times \wp(A)$ if

$$(y,Y) \in F \wedge (x,X) \preceq (y,Y) \Rightarrow (x,X) \in F.$$

28.2.3. LEMMA. (A,F,D) is a CSP process $\Leftrightarrow$ A is a finite, non-empty set and

1• F is non-empty and left-closed in $A^* \times \wp(A)$;
2• $D \times \wp(A) \subseteq F$ and $D.A \subseteq D$;
3• $(x,X) \in F \wedge a \in A \Rightarrow (x.a,\emptyset) \in F \vee (x \cup \{a\}, X) \in F$.

PROOF. (1) is equivalent to 28.2.2 (1), (2) and (3); (2) is equivalent to 28.2.2 (5), (6) and (7) and (3) is identical to 28.2.2 (4).

■

The semantics of TCSP programs is given by a *partial* function σ: **TCSP** $\rightarrow$ **HProc** and is compositional in the sense that σ is defined by recursion on the structure of terms in TCSP.

First, we note that $x \in \mathcal{X}$ is not closed and hence not a TCSP program.

28.2.4. DEFINITION. If A is a finite, non-empty set, then

1• $\sigma[\![Stop_A]\!] = (A, \{\Omega\} \times \wp(A), \emptyset)$;
2• $\sigma[\![Chaos_A]\!] = (A, A^* \times \wp(A), A^*)$;

$Stop_A$ is a process which has an alphabet of actions but which can do nothing, not even diverge. We have touched on the nature of $Chaos_A$ already. That $\sigma[\![Stop_A]\!]$ and $\sigma[\![Chaos_A]\!]$ are processes when A is finite and non-empty is an easy consequence of 28.2.3.

Next we consider $a \rightarrow P$, which is to be interpreted intuitively as a program that executes the action a, after which it behaves like P. It corresponds to the guarding constructor of chapters 26 and 27.

On this and other constructions, Hoare places various constraints. For example, he requires that if $\sigma[\![P]\!] = (A,F,D)$, then $a \in A$ ([2], D4, page 131). The constraints are required in order that the constructions on processes which correspond to the constructors on expressions do produce processes; expressions satisfying them have well-defined semantics. Thus, although we define the constructions without constraints, the propositions that assert that the result of a construction is a process will need the appropriate constraint as a hypothesis.

We do not require the constraints when we give our non-interleaving semantics, but of course when we examine the relationship between the two semantics, we will have to restrict ourselves to expressions satisfying the constraints.

28.2.5. DEFINITION. If $(A,F,D) \in \mathbf{HProc}$ and $a \in \mathcal{A}ct$, then we define $a \rightarrow (A,F,D) = (A \cup \{a\}, F', a.D)$, where

$$F' = (\{\Omega\} \times \wp(A - \{a\})) \cup \{(a.x, X) \mid (x, X) \in F\}.$$

28.2.6. LEMMA. If $(A,F,D) \in \mathbf{HProc}$ and $a \in A$ then $a \rightarrow (A,F,D) \in \mathbf{HProc}$.

PROOF. Let $(A,F,D) \in \mathbf{HProc}$. and define $(A,F',a.D) = a \rightarrow (A,F,D)$. By 28.2.3 (1), F is non-empty and left-closed in $A^* \times \wp(A)$. Thus, F' is non-empty. Next, suppose that $(y,Y) \in F'$ and $(x,X) \preceq (y,Y)$. If $y = \Omega$, then $x = \Omega$ and $X \subseteq Y \subseteq A - \{a\}$, so $(x,X) \in F'$. Otherwise, $y = a.y'$, with $(y',Y) \in F$, and so either $x = \Omega$, in which case $X = \emptyset$, by (28.1) and certainly $(\Omega, \emptyset) \in F'$, or $x = a.x'$, some $x' \in A^*$. it easily follows, using (28.1), that $(x',X) \preceq (y',Y)$ and so $(x',X) \in F$, as F is left-closed, and hence $(x,X) = (a.x',X) \in F'$. Thus, F' is left-closed and we have shown that $(A,F',a.D)$ satisfies 28.2.3 (1).

Since $D \times \wp(A) \subseteq F$ and $D.A \subseteq D$, by 28.2.3 (2), it easily follows that $a.D \times \wp(A) \subseteq F'$ and $a.D.A \subseteq a.D$, so $(A,F',a.D)$ satisfies 28.2.3 (3).

Finally, suppose that $(x,X) \in F'$ and $b \in A$ such that $(x.b, \emptyset) \notin F'$. If $x = \Omega$, then $X \subseteq A - \{a\}$ and $b \neq a$ and so $(x, X \cup \{b\}) \in F'$. Otherwise $x = a.x'$, some $x' \in A^*$ and $(x',X) \in F$. We cannot have $(x'.b, X) \in F$, for then it would follow that $(x.b,X) = (a.x'.b,X) \in F'$ and so $(x', X \cup \{b\}) \in F$, by 28.2.3 (3), and hence $(x, X \cup \{b\}) = (a.x', X \cup \{b\}) \in F'$. Thus, $(A,F',a.D)$ satisfies 28.2.3 (3).

■

$(P_1 \parallel P_2)$ describes a system composed of subsystems P_1 and P_2 running in parallel and synchronising on shared actions. Recall that if A and B are sets $x, y \in B^*$ and $a \in B$, then the function $\pi_A : B^* \rightarrow A^*$ is given by

$$\pi_A(\Omega) = \Omega$$
$$\pi_A(x.y) = \pi_A(x).\pi_A(y)$$
$$\pi_A(a) = \begin{cases} a \text{ if } a \in A \\ \Omega \text{ otherwise} \end{cases}$$

28.2.7. DEFINITION. If $(A_1,F_1,D_1),(A_2,F_2,D_2) \in \mathbf{HProc}$, then we define $(A_1,F_1,D_1) \parallel (A_2,F_2,D_2) = (A,F,D)$, where

1• $A = A_1 \cup A_2$;

2• $D = \{y \in A^* \mid \exists x \leq y : (\pi_{A_1}(x) \in D_1 \wedge \pi_{A_2}(x) \in dom(F_2))$
$\vee (\pi_{A_2}(x) \in D_2 \wedge \pi_{A_1}(x) \in dom(F_1)\};$

3• $F=\{(x,X\cup Y)\mid(\pi_{A_1}(x),X)\in F_1\wedge(\pi_{A_2}(x),Y)\in F_2\}\cup(D\times\wp(A))$,

where if $F\subseteq A^*\times\wp(A)$, then

$$dom(F)=\{x\in A^*\mid\exists X\subseteq A:(x,X)\in F\}.$$

The elements of $dom(F)$ are known as *traces*.[1]

28.2.8. LEMMA. Suppose that $(A_1,F_1,D_1),(A_2,F_2,D_2)\in\mathbf{HProc}$, then $(A_1,F_1,D_1)\,||\,(A_2,F_2,D_2)\in\mathbf{HProc}$.

PROOF. Let $(A,F,D)=(A_1,F_1,D_1)\,||\,(A_2,F_2,D_2)$. F is non-empty since it contains the pair $(\Omega,\emptyset)$ which certainly satisfies 28.2.7 (3). If $y\in D$, then $x\le y$ for x with the properties given in (2), and so if $a\in A$, then $x\le y.a$ and so $y.a\in D$. We also note that $D\times\wp(A)\subseteq F$ by (3). Hence 28.2.3 (2) holds.

For the remainder of the proof, we abbreviate $\pi_{A_i}(x)$ to x_i.

Suppose that $(y,Y)\in F$ and $(x,X)\preceq(y,Y)$. If $x\in D$, then $(x,X)\in F$, so assume $x\notin D$. Thus, if $x=y$ then $y\notin D$ and so $X\subseteq Y$ and there exists $Y_1,Y_2\subseteq A$, such that $(y_i,Y_i)\in F_i$, each i, and $Y_1\cup Y_2=Y$. But now, for each i, $(x_i,X\cap Y_i)\preceq(y_i,Y_i)$, so $(x_i,X\cap Y_i)\in F_i$, by left-closure of F_i, and consequently $(x,X)=(x,(X\cap Y_1)\cup(X\cap Y_2))\in F$. Otherwise, $x<y$ and $X=\emptyset$, So $x_i\le y_i$, where $y_i\in dom(F_i)$, each i, and thus by 28.2.2, (3) $(x_i,\emptyset)\in F_i$, each i so $(x,X)=(x,\emptyset)\in F$. We have shown that F is left-closed and so 28.2.3 (1) holds.

Finally, suppose that $(x,X)\in F$ and $a\in A$ such that $(x.a,\emptyset)\notin F$, then $x\notin D$ and so there exists $X_1,X_2\subseteq A$, such that $(x_i,X_i)\in F_i$, each i and $X_1\cup X_2=X$. Without loss of generality, we may assume that $a\in A_1$ and $(x_1.a,\emptyset)\notin F_1$, so $(x_1,X_1\cup\{a\})\in F_1$ and so $(x,X_1\cup\{a\}\cup X_2)=(x,X\cup\{a\})\in F$. We have shown that 28.2.3 (3) holds.

■

$(P_1\,|||\,P_2)$ is a system composed of processes P_1 and P_2 running in parallel without synchronising at all. To define the semantics of , we need to define the set of interleavings of two strings. Suppose $a,b\in A$ and $x,y\in A^*$. Define:

$$x\,|||\,y=\begin{cases}\{x\}\text{ if } y=\Omega\\ \{y\}\text{ if } y=\Omega\\ a.(x'\,|||\,y)\cup b.(x\,|||\,y')\text{ if } x=a.x'\text{ and } y=b.y'\end{cases}$$

---

[1] Unfortunately.

For example $ab \mathbin{|||} cb = \{abcb, acbb, cabb, cbab\}$.

The following properties of the $|||$ operator may easily be established from the definition and induction on $|x|$.

$$x \in y \mathbin{|||} z \Rightarrow x.a \in (y.a) \mathbin{|||} z \tag{28.2}$$
$$x \in y \mathbin{|||} z \wedge x' \leq x \Rightarrow \exists y' \leq y\ \exists z' \leq z : x' \in y' \mathbin{|||} z' \tag{28.3}$$

As in the case of guarding, there is a constraint on the programs that may be combined in this way. ([2], page 119 and **D7**, page 131) requires that if $\sigma[\![P_1]\!] = (A_1, F_1, D_1)$ and $\sigma[\![P_2]\!] = (A_2, F_2, D_2)$ then $A_1 = A_2$.

28.2.9. DEFINITION. Suppose that $(A_1, F_1, D_1), (A_2, F_2, D_2) \in$ **HProc**, then $(A_1, F_1, D_1) \mathbin{|||} (A_2, F_2, D_2) = (A, F, D)$, where

1• $A = A_1 \cup A_2$;

2• $D = \{x \in A^* \mid \exists y, z \in A^* : x \in y \mathbin{|||} z \wedge ((y \in D_1 \wedge z \in dom(F_2)) \vee (z \in D_2 \wedge y \in dom(F_1))\}$;

3• $F = \{(x, X) \in A^* \mid \exists y, z \in A^* : x \in y \mathbin{|||} z \wedge (y, X) \in F_1 \wedge (z, X) \in F_2\} \cup (D \times \wp(A))$.

28.2.10. LEMMA. Suppose that $(A_1, F_1, D_1), (A_2, F_2, D_2) \in$ **HProc** and $A_1 = A_2$ then $(A_1, F_1, D_1) \mathbin{|||} (A_2, F_2, D_2) \in$ **HProc**.

PROOF. Let $(A, F, D) = (A_1, F_1, D_1) \mathbin{|||} (A_2, F_2, D_2)$. F is non-empty since it contains the pair $(\Omega, \varnothing)$ which certainly satisfies 28.2.9 (3). If $x \in D$, then without loss of generality, there exists $y \in D_1$ and $z \in dom(F_2)$ such that $x \in y \mathbin{|||} z$. If $a \in A$, then $y.a \in D_1$ and $x.a \in (y.a) \mathbin{|||} z$, by (28.2), so $x.a \in D$. We also note that $D \times \wp(A) \subseteq F$ by (3). Hence 28.2.3 (2) holds.

Suppose that $(y, Y) \in F$ and $(x, X) \preceq (y, Y)$. If $x \in D$, then $(x, X) \in F$, so assume $x \notin D$. Thus, if $x = y$ then $X \subseteq Y$ and $y \notin D$ and so there exists $z_1, z_2 \in A^*$, such that $(z_i, Y) \in F_i$, each i, and $y \in z_1 \mathbin{|||} z_2$. But now, for each i, $(z_i, X) \preceq (z_i, Y)$, so $(z_i, X) \in F_i$, by left-closure of F_i, and $x \in z_1 \mathbin{|||} z_2$, so $(x, X) \in F$. Otherwise, $x < y$ and $X = \varnothing$, and so there exists $z_i \in dom(F_i)$, each i such that $y \in z_1 \mathbin{|||} z_2$. By (28.3), there exists $x_i \leq z_i$, each i, such that $x \in x_1 \mathbin{|||} x_2$. By 29.2.2 (3), $(x_i, \varnothing) \in F_i$, each i, so $(x, X) = (x, \varnothing) \in F$. We have shown that F is left-closed and so 28.2.3 (1) holds.

Finally, suppose that $(x, X) \in F$ and $a \in A$ such that $(x.a, \varnothing) \notin F$, then $x \notin D$ and so there exists $z_1, z_2 \in A^*$, such that $(z_i, X) \in F_i$, each i, and $x \in z_1 \mathbin{|||} z_2$. Hence for each i, $a \in A_i$ and $(z_i.a, \varnothing) \notin F_i$, so $(x_i, X_i \cup \{a\}) \in F_i$. Therefore, $(x, X \cup \{a\}) \in F$. We have shown that 28.2.3 (3) holds.

■

$[]$ is a choice operator. $P_1 [] P_2$ is a program that can elect to behave either like P_1 or like P_2. As in the case of guarding and the $|||$ operator, there is a constraint on the programs that may be combined in this way. ([2], page 106 and **D 7**, page 131) requires that if $\sigma[\![P_1]\!] = (A_1, F_1, D_1)$ and $\sigma[\![P_2]\!] = (A_2, F_2, D_2)$ then $A_1 = A_2$.

28.2.11. DEFINITION. Suppose that $(A_1, F_1, D_1), (A_2, F_2, D_2) \in \mathbf{HProc}$, then $(A_1, F_1, D_1) [] (A_2, F_2, D_2) = (A, F, D)$, where

1• $A = A_1 \cup A_2$;
2• $D = D_1 \cup D_2$;
3• $F = (F_1 \cap F_2) \cup \{(x, X) \in F_1 \cup F_2 \mid x > \Omega\} \cup (D \times \wp(A))$.

28.2.12. LEMMA. Suppose that $(A_1, F_1, D_1), (A_2, F_2, D_2) \in \mathbf{HProc}$ and $A_1 = A_2$ then $(A_1, F_1, D_1) [] (A_2, F_2, D_2) \in \mathbf{HProc}$.

PROOF. Let $(A, F, D) = (A_1, F_1, D_1) [] (A_2, F_2, D_2)$. $D \times \wp(A) \subseteq F$, by 28.2.11 (3), and $D.A = (D_1 \cup D_2).A = D_1.A \cup D_2.A \subseteq D_1 \cup D_2 = D$, by 28.2.11 (2). Hence 28.2.3 (2) holds.

F is non-empty, since it contains $(\Omega, \varnothing) \in F_1 \cap F_2$. Suppose that $(y, Y) \in F$ and $(x, X) \preceq (y, Y)$. If $(y, Y) \in F_1 \cap F_2$, then $(x, X) \in F_i$, each i, by left-closure of F_i, and so $(x, X) \in F_1 \cap F_2 \subseteq F$. If $(y, Y) \in F_1 \cup F_2$ with $y > \Omega$, then $(y, Y) \in F_i$, some i, and so $(x, X) \in F_i$ and either $x > \Omega$, which gives $(x, X) \in F_1 \cup F_2$ with $x > \Omega$, that is $(x, X) \in F$, or $x = \Omega$, in which case $x < y$ and so $X = \varnothing$, and $(x, X) = (\Omega, \varnothing) \in F$. If $y \in D$, then either $x = y$, in which case $(x, X) \in F$, or $x < y$ and so $X = \varnothing$. But $y \in D_i$, some i, by 28.2.11 (2), and so $(x, X) \in F_i$, by 28.2.2 (3). In the latter case, either $x > \Omega$, and so $(x, X) \in F_1 \cup F_2$ with $x > \Omega$, that is $(x, X) \in F$, or $x = \Omega$, and $(x, X) \in F$, by 28.2.2 (1). We have shown that F is left-closed and so 28.2.3 (1) holds.

Finally, suppose that $(x, X) \in F$ and $a \in A$ such that $(x.a, \varnothing) \notin F$, then $x \notin D$ and hence either $(x, X) \in F_1 \cap F_2$ or $(x, X) \in F_1 \cup F_2$ with $x > \Omega$. If $(x.a, \varnothing) \in F_i$, either i, then $(x.a, X) \in F_1 \cup F_2$ with $x.a > \Omega$, a contradiction. Therefore, $(x.a, \varnothing) \notin F_i$, each i. Consequently if $(x, X) \in F_1 \cap F_2$, then $(x, X \cup \{a\}) \in F_1 \cap F_2 = F$. Otherwise, $(x, X) \in F_1 \cup F_2$ with $x > \Omega$, and so $(x, X) \in F_i$ and $(x.a, \varnothing) \notin F_i$, whence $(x, X \cup \{a\}) \in F_i$ with $x.a > \Omega$, that is $(x, X \cup \{a\}) \in F$. We have shown that 28.2.3 (3) holds.

■

$\sqcap$ is also a choice operator. $P_1 \sqcap P_2$ is a program that can elect to behave either like P_1 or P_2 without the environment being capable of influencing the choice. As in the case of guarding and the $|||$ operator, there is a constraint on the programs that may be combined in this way. ([2], page 102) requires that if $\sigma[\![P_1]\!]=(A_1,F_1,D_1)$ and $\sigma[\![P_2]\!]=(A_2,F_2,D_2)$ then $A_1=A_2$.

28.2.13. DEFINITION. Suppose that $(A_1,F_1,D_1),(A_2,F_2,D_2)\in \mathbf{HProc}$, then $(A_1,F_1,D_1)\sqcap(A_2,F_2,D_2)=(A,F,D)$, where

1• $A=A_1\cup A_2$;
2• $D=D_1\cup D_2$;
3• $F=F_1\cup F_2$.

28.2.14. LEMMA. Suppose that $(A_1,F_1,D_1),(A_2,F_2,D_2)\in \mathbf{HProc}$ and $A_1=A_2$ then $(A_1,F_1,D_1)\sqcap(A_2,F_2,D_2)\in \mathbf{HProc}$.

PROOF. Let $(A,F,D)=(A_1,F_1,D_1)\sqcap(A_2,F_2,D_2)$.

$D.A=(D_1\cup D_2).A=D_1.A\cup D_2.A\subseteq D_1\cup D_2=D$, by 28.2.13 (2) and $D\times\wp(A)=(D_1\times\wp(A_1))\cup(D_2\times\wp(A_2))\subseteq F_1\cup F_2=F$ by 28.2.13 (3). Hence 28.2.3 (2) holds.

F is non-empty, since the F_i are. If $(y,Y)\in F$ and $(x,X)\preceq(y,Y)$, then $(y,Y)\in F_i$, some i, and so $(x,X)\in F_i$ and so $(x,X)\in F_1\cup F_2=F$. We have shown that 28.2.3 (1) holds.

Finally, suppose that $(x,X)\in F$ and $a\in A$ such that $(x.a,\varnothing)\notin F$, then $(x,X)\in F_i$ and $(x.a,\varnothing)\notin F_i$, some i, whence $(x,X\cup\{a\})\in F_i$, that is $(x,X\cup\{a\})\in F$. We have shown that 28.2.3 (3) holds .

■

$(P\setminus C)$ is the program that which behaves like P except that all actions belonging to the set C are hidden. Hoare does not seem to impose any constraints on its use, in contrast to the cases of the non-deterministic choice operators or the ||| operator. A constraint is needed, nontheless, for in view of 28.2.15 (1) and 28.2.2, $(A,F,D)\setminus C$ with $A-C=\varnothing$ would fail to be a process.

28.2.15 DEFINITION. Suppose that $(A,F,D)\in \mathbf{HProc}$ and $C\subseteq \mathcal{A}ct$ then $(A,F,D)\setminus C=(A',F',D')$, where

1• $A'=A-C$;
2• $D'=\{x\in A'^*\mid \exists y\in A^*:\pi_{A'}(y)\le x$
$\qquad\wedge(y\in D\vee\forall n\in\omega\,\exists u\in C^n:(y.u,\varnothing)\in F)\}$;
3• $F'=\{(x',X)\in A'^*\times\wp(A')\mid\exists(x,X)\in F:$
$\qquad x'=\pi_{A'}(x)\wedge X=X'\cup C\}\cup(D'\times\wp(A'))$.

28.2.16. LEMMA. If $(A,F,D) \in \mathbf{HProc}$ and $C \subseteq \mathcal{Act}$ such that $A - C \neq \varnothing$, then $(A,F,D) \setminus C \in \mathbf{HProc}$.

PROOF. Let $(A',F',D') = (A,F,D) \setminus C$. We note that if $x \in D'$, then $y \leq x$ for y with the properties given in 28.2.15 (2), and so if $a \in A'$, then $y \leq x.a$ and so $x.a \in D'$. We also note that $D' \times \wp(A') \subseteq F'$ by 28.2.15 (3). Hence 28.2.3 (2) holds.

For the next part of the proof, we establish the following general property.

$$\forall x \in dom(F): \pi_{A'}(x) \notin D' \Rightarrow (\pi_{A'}(x), \varnothing) \in F' \tag{28.4}$$

By 28.2.15 (2), since $\pi_{A'}(x) \notin D'$, there exists $n \in \omega$ such that $dom(F) \cap x.C^n \neq \varnothing$ and $dom(F) \cap x.C^{n+1} = \varnothing$. Let $x.u \in dom(F) \cap C^n$. If $(x.u, C) \notin F$, then there exists $c \in C$ and $X \subseteq C - \{c\}$ such that $(x.u, X) \in F$ and $(x.u, X \cup \{c\}) \notin F$. But then, $(x.u.c, \varnothing) \in F$, by 28.2.2 (4), contradicting $dom(F) \cap x.C^{n+1} = \varnothing$. Thus, $(x.u, C) \in F$, so $(\pi_{A'}(x), \varnothing) = (\pi_{A'}(x.u), \varnothing) \in F'$. As

$$\forall x \in dom(F): \pi_{A'}(x) \in D' \Rightarrow (\pi_{A'}(x), \varnothing) \in F',$$

by 28.2.15 (3), it follows that

$$\forall x \in dom(F): (\pi_{A'}(x), \varnothing) \in F' \tag{28.5}$$

In particular, $(\Omega, \varnothing) \in F'$ and F' is non-empty. Next, let $(y', Y') \in F'$ and $(x', X') \preceq (y', Y')$. If $x' \in D'$, then certainly $(x', X') \in F'$, so suppose $x' \notin D'$.

Suppose first that $x' < y'$, then $X' = \varnothing$. If $y' \notin D'$, then there exists $z \in dom(F)$ such that $y' = \pi_{A'}(z)$. Otherwise, $y' \in D'$ and there exists $x' \leq z' < y'$ and $a \in A'$ such that $z'.a \leq y'$, $z' \notin D'$ and $z'.a \in D'$. By 28.2.15 (2), there exists $z \in dom(F)$ such that $z'.a = \pi_{A'}(z)$. In each case $x' \leq \pi_{A'}(z)$, so there exists $x \leq z$ such that $x' = \pi_{A'}(x)$. Since $z \in dom(F)$, $x \in dom(F)$ and so $(x', X') = (\pi_{A'}(x), \varnothing) \in F'$, by (28.5).

Otherwise, $x' = y'$ and $X' \subseteq Y'$. Since $x' \notin D'$, there exists $(y, Y) \in F$ such that $y' = \pi_{A'}(y)$ and $Y = Y' \cup C$. If $X = X' \cup C$, then $X \subseteq Y$ and so by 28.2.2 (2), $(y, X) \preceq (y, Y)$. Hence, $(y, X) \in F$ and since $x' = \pi_{A'}(y)$ and $X = X' \cup C$, it follows that $(x', X') \in F'$. We have shown that 28.2.3 (1) holds.

Finally, suppose that $(x', X') \in F'$ and $a \in A'$ such that $(x'.a, \varnothing) \notin F'$, then $x' \notin D'$, so there exists $(x, X) \in F$ such that $x' = \pi_{A'}(x)$ and $X = X' \cup C$. If $(x.a, \varnothing) \in F$, then $(x'.a, \varnothing) = (\pi_{A'}(x.a), \varnothing) \in F'$, by (28.5), a contradiction, and so $(x.a, \varnothing) \notin F$ and hence $(x, X \cup \{a\}) \in F$. But now we have $x' = \pi_{A'}(x)$ and $X \cup \{a\} = (X' \cup \{a\}) \cup C$, so $(x', X' \cup \{a\}) \in F'$. We have shown that 28.2.3 (3) holds.

■

Finally, we come to expressions of the form $(\mu x{:}A.P)$, which introduces recursion into the language. $(\mu x{:}A.P)$ describes a process which behaves like P until it 'hits' the variable x, at which point it behaves like P until it 'hits' the variable x, at which point ...

Technically, $\sigma[\![(\mu x{:}A.P)]\!]$ is defined as a fixed point of a certain monotonic function on **HProc** derived from P. Of course, to make sense of such a statement, we need to make **HProc** into some kind of partial order. Define

$$(A,F,D)\leq(A',F',D')\Leftrightarrow A'\subseteq A\wedge F'\subseteq F\wedge D'\subseteq D \tag{28.6}$$

It is clear that this relation makes **HProc** into a partial order with $\sigma[\![Chaos_{Act}]\!]$ as bottom element. It is not hard to show that it is a complete partial order. The least upper bound of a directed set $\mathbf{X}\subseteq\mathbf{HProc}$, denoted by $\cap\mathbf{X}$, is (A,F,D) where

$$A=\bigcap_{(A,F,D)\in\mathbf{X}}A \qquad F=\bigcap_{(A,F,D)\in\mathbf{X}}F \qquad D=\bigcap_{(A,F,D)\in\mathbf{X}}D.$$

28.2.17. LEMMA. If $(A,F_n,D_n)_{n\in\omega}$ is monotonic increasing sequence of elements of **HProc** then $\bigcap_{n\in\omega}(A,F_n,D_n)\in\mathbf{HProc}$.

PROOF. Let $(A,F,D)=\bigcap_{n\in\omega}(A,F_n,D_n)$.

Since $(\Omega,\varnothing)\in F_n$, each $n\in\omega$, by 28.2.2(1), $(\Omega,\varnothing)\in F$ and so F is non-empty. If $(y,Y)\in F$ and $(x,X)\preceq(y,Y)$, then for each $n\in\omega$, $(y,Y)\in F_n$ and hence $(x,X)\in F_n$, so that $(x,X)\in F$. Thus, 28.2.3 (1) holds.

We also note that

$$D.A=(\bigcap_{n\in\omega}D_n).A=\bigcap_{n\in\omega}(D_n.A)\subseteq\bigcap_{n\in\omega}D_n=D$$
$$D\times\wp(A)=(\bigcap_{n\in\omega}D_n)\times\wp(A)=\bigcap_{n\in\omega}(D_n\times\wp(A))\subseteq\bigcap_{n\in\omega}F_n=F$$

so 28.2.3 (2) holds.

Finally, suppose that $(x,X)\in F$ and $a\in A$ such that $(x.a,\varnothing)\notin F$, then for some $N\in\omega$, $(x.a,\varnothing)\notin F_N$ and as $(A,F_n,D_n)_{n\in\omega}$ is monotonic increasing, $(x.a,\varnothing)\notin F_n$, all $n\geq N$, But, as $(x,X)\in F$, $(x,X)\in F_n$, all $n\in\omega$, and so $(x,X\cup\{a\})\in F_n$, all $n\geq N$. Thus, again using monotonicity of $(A,F_n,D_n)_{n\in\omega}$

$$(x,X\cup\{a\})\in\bigcap_{n\geq N}F_n=\bigcap_{n\in\omega}F_n=F$$

so 28.2.3 (3) holds.

∎

We may now give the CSP semantics for programs. First, we need some notation. If $(\mu x{:}A.P)$ is a program, then it is closed, and so the only free variable of P is x. Hence, if $\tilde{P}$ is a program, then the expression $P[x\backslash\tilde{P}]$ formed from P by substituting each free instance of x by $\tilde{P}$ is also closed. We may therefore define a sequence of programs $P^{(n)}$ by

$$P^{(0)} = Chaos_A$$
$$P^{(n+1)} = P[x\backslash P^{(n)}]$$

For the purpose of the next definition, we define $\alpha(P)=A$ if $\sigma[\![P]\!]=(A,F,D)$.

28.2.18 DEFINITION. We define a partial function $\sigma{:}\mathbf{TCSP}\to\mathbf{HProc}$ as follows:

1• If $a\to P$ is a program such that $\sigma[\![P]\!]$ is defined and $a\in\alpha(P)$, then $\sigma[\![a\to P]\!]=a\to\sigma[\![P]\!]$;
2• If $P_1 \| P_2$ is a program such that $\sigma[\![P_1]\!]$ and $\sigma[\![P_2]\!]$ are defined, then $\sigma[\![P_1 \| P_2]\!]=\sigma[\![P_1]\!]\|\sigma[\![P_2]\!]$;
3• If $P_1 ||| P_2$ is a program such that $\sigma[\![P_1]\!]$ and $\sigma[\![P_2]\!]$ are defined and $\alpha(P_1)=\alpha(P_2)$, then $\sigma[\![P_1 ||| P_2]\!]=\sigma[\![P_1]\!]|||\sigma[\![P_2]\!]$;
4• If $P_1 \,[]\, P_2$ is a program such that $\sigma[\![P_1]\!]$ and $\sigma[\![P_2]\!]$ are defined and $\alpha(P_1)=\alpha(P_2)$, then $\sigma[\![P_1 \,[]\, P_2]\!]=\sigma[\![P_1]\!]\,[]\,\sigma[\![P_2]\!]$;
5• If $P_1 \sqcap P_2]\!]$ is a program such that $\sigma[\![P_1]\!]$ and $\sigma[\![P_2]\!]$ are defined and $\alpha(P_1)=\alpha(P_2)$, then $\sigma[\![P_1 \sqcap P_2]\!]=\sigma[\![P_1]\!]\sqcap\sigma[\![P_2]\!]$;
6• If $P\backslash C$ is a program such that $\sigma[\![P]\!]$ is defined and $\alpha(P)-C\neq\varnothing$ then $\sigma[\![P\backslash C]\!]=\sigma[\![P]\!]\backslash C$
7• If $\mu x{:}A.P$ is a program such that the sequence $(\sigma[\![P^{(n)}]\!])_{n\in\omega}$ is defined and monotonic, then $\sigma[\![\mu x{:}A.P]\!]=\bigcap_{n\in\omega}\sigma[\![P^{(n)}]\!]$.

We are now almost ready to prove that the TCSP semantics is defined for all programs. The proof is by (guess what!) structural induction, for which the following lemma provides the final ingredient of the induction step. In the statement of the lemma, the substitution $P[\vartheta]$ is defined is a manner analogous to that of chapter 27

28.2.19. LEMMA. Suppose that P is an expression and $\vartheta_1,\vartheta_2{:}\mathcal{X}\to\mathbf{TCSP}$ such that for each $x\in\mathcal{X}$

1• $\sigma[\![\vartheta_i(x)]\!]\in\mathbf{HProc}$, each i;

2• $\sigma[\![\vartheta_1(x)]\!] \leq \sigma[\![\vartheta_2(x)]\!]$,

then $\sigma[\![P[\vartheta_i]]\!] \in \mathbf{HProc}$, each i, and $\sigma[\![P[\vartheta_1]]\!] \leq \sigma[\![P[\vartheta_2]]\!]$.

PROOF. We argue by induction over the structure of P. The base cases, $P = Stop_A$, $P = Chaos_A$ and $P \in \mathcal{X}$ are trivial. We consider one non-recursive case of the induction step; the others are similar.

Let $P = a \to P'$. By induction, $\sigma[\![P'[\vartheta_i]]\!] \in \mathbf{HProc}$, each i, and $\sigma[\![P'[\vartheta_1]]\!] \leq \sigma[\![P'[\vartheta_2]]\!]$. $\sigma[\![P[\vartheta_i]]\!] = \sigma[\![a \to P'[\vartheta_i]]\!] \in \mathbf{HProc}$, by 28.2.6, while it is a trivial consequence of 28.2.5 and (28.6) that if $\sigma[\![P'[\vartheta_1]\!] \leq \sigma[\![P'[\vartheta_2]]\!]$, then $\sigma[\![a \to P'[\vartheta_1]]\!] \leq \sigma[\![a \to P'[\vartheta_2]]\!]$.

Finally, let $P = (\mu y{:}A.\overline{P})$, then $P[\vartheta_i] = (\mu y{:}A.\overline{P}[\hat{\vartheta}_i])$, where

$$\hat{\vartheta}_i(x) = \begin{cases} \vartheta_i(x) \text{ if } x \neq y \\ y \text{ otherwise} \end{cases}$$

and $\sigma[\![P[\vartheta_i]]\!] \in \mathbf{HProc}$ providing that the sequence $(\sigma[\![\overline{P}_i^n]\!])_{n\in\omega}$ is defined and monotonic increasing, where $\overline{P}_i^0 = Chaos_A$ and $\overline{P}_i^{n+1} = (\overline{P}[\hat{\vartheta}_i])[y \setminus \overline{P}_i^n]$. But it is easy to see that $\overline{P}_i^{n+1} = \overline{P}[\tilde{\vartheta}_i^n]$, $n \in \omega$, where $\tilde{\vartheta}_i^0 = \tilde{\vartheta}_i$ and

$$\tilde{\vartheta}_i^{n+1}(x) = \begin{cases} \vartheta_i(x) \text{ if } x \neq y \\ \overline{P}[\tilde{\vartheta}_i^n] \text{ otherwise} \end{cases}$$

so by structural and mathematical induction, for each i, $\sigma[\![\tilde{\vartheta}_i^n(x)]\!] \in \mathbf{HProc}$ and $\sigma[\![\vartheta_1^n(x)]\!] \leq \sigma[\![\vartheta_2^n(x)]\!]$, all $n > 0$. Thus, by induction, for each i, $\sigma[\![\overline{P}_i^n]\!] \in \mathbf{HProc}$ and $\sigma[\![\overline{P}_i^n]\!] \leq \sigma[\![\overline{P}_i^{n+1}]\!]$ and $\sigma[\![\overline{P}_1^n]\!] \leq \sigma[\![\overline{P}_2^n]\!]$. Thus, $(\sigma[\![\overline{P}_i^n]\!])_{n\in\omega}$ is defined and monotonic increasing and so $\sigma[\![P[\vartheta_i]]\!] \in \mathbf{HProc}$, each i, And since $\sigma[\![\overline{P}_1^n]\!] \leq \sigma[\![\overline{P}_2^n]\!]$, each $n \in \omega$. $\sigma[\![P[\vartheta_1]]\!] \leq \sigma[\![P[\vartheta_2]]\!]$.

∎

Finally, we have:

28.2.20. THEOREM. If P is a program such that $\sigma[\![P]\!]$ is defined, then $\sigma[\![P]\!] \in \mathbf{HProc}$.

PROOF. By induction over the structure of P. The only base cases to consider are $P = Stop_A$ and $P = Chaos_A$, for which we know the claim of the theorem to hold, as the case $P \in \mathcal{X}$ cannot arise.

The non-recursive parts of the induction step are covered in 28.2.6, 28.2.8, 28.2.10, 28.2.12, 28.2.14 and 28.2.16.

Suppose finally, that $P = (\mu y{:}A.\overline{P})$. We inductively define a sequence of functions $\vartheta_n : \mathcal{X} \to \mathbf{TCSP}$ by $\vartheta_0(x) = Chaos_A$ and $\vartheta_{n+1}(x) = \overline{P}[\vartheta_n]$, all $x \in \mathcal{X}$. The functions ϑ_n are constant, say $P_n = \vartheta_n(x)$, all $x \in \mathcal{X}$, and if the sequence $(\sigma[\![P_n]\!])_{n\in\omega}$ is defined and monotonic increasing, then $\sigma[\![P]\!] = \bigcap_{n\in\omega} \sigma[\![P_n]\!]$.

But, $\sigma[\![P_0]\!]$ is certainly defined. Suppose by induction that $\sigma[\![P_n]\!]$ is defined for each $n \le N$ and that $\sigma[\![\vartheta_n(x)]\!] \le \sigma[\![\vartheta_{n+1}(x)]\!]$ each $n < N$, then for all $x \in \mathcal{X}$, $\sigma[\![\vartheta_{N-1}(x)]\!], \sigma[\![\vartheta_N(x)]\!] \in \mathbf{HProc}$ and $\sigma[\![\vartheta_{N-1}(x)]\!] \le \sigma[\![\vartheta_N(x)]\!]$, Thus by 28.2.19

$$\sigma[\![P_{N+1}]\!] = \sigma[\![\overline{P}[\vartheta_N]]\!] \in \mathbf{HProc}$$
$$\sigma[\![P_N]\!] = \sigma[\![\overline{P}[\vartheta_{N-1}]]\!] \le \sigma[\![\overline{P}[\vartheta_N]]\!] = \sigma[\![P_{N+1}]\!].$$

By induction, $(\sigma[\![P_n]\!])_{n\in\omega}$ is defined and monotonic increasing and so $\sigma[\![P]\!] \in \mathbf{HProc}$ by 28.2.18.

■

28.2.20 allows us to interpret each closed expression as a process. We may also interpret arbitrary expressions, but as functions. Let us denote the set of programs by **Prog**, then we may interpret an expression P as a function

$$\tilde{\sigma}[\![P]\!] : [\mathcal{X} \to \mathbf{Prog}] \to \mathbf{HProc}$$

defined by $\tilde{\sigma}[\![P]\!](\vartheta) = \sigma[\![P[\vartheta]]\!]$. We note that $P[\vartheta]$ is a program and that consequently $\tilde{\sigma}[\![P]\!]$ is well-defined, by 28.2.20.

We conclude this section by showing that we may extend the process semantics to the whole of TCSP; recall that Hoare imposes restrictions on the application of combinators related to the alphabet $\alpha(P)$ of P. The following construction takes a process and increases its alphabet.

28.2.21. DEFINITION. If P is a program such that $\sigma[\![P]\!]$ is defined and $\alpha(P) \subseteq B$, then we define $ext_B(P) = (B, F', D')$, where

1• $D' = D.B^*$;
2• $F' = \{(x, X) \in B^* \times \wp(B) \mid (x, X \cap \alpha(P)) \in F\} \cup (D' \times \wp(B))$.

28.2.22. PROPOSITION. If P is a program such that $\sigma[\![P]\!]$ is defined and $\alpha(P) \subseteq B$, then $ext_B(P) \in \mathbf{HProc}$.

PROOF. Let $(B, F', D') = ext_B(P)$, then clearly, F' is non-empty, while 28.2.3 (2) follows immediately from 28.2.21.

Suppose that $(y, Y) \in F'$ and $(x, X) \preceq (y, Y)$. If $x \in D'$, then $(x, X) \in F'$ by 28.2.21 (2), so suppose that $x \notin D'$. We shall prove that there exists

$(z,Z)\in F'-(D'\times\wp(B))$ such that $(x,X)\preceq(z,Z)$. This is certainly true if $y\notin D'$, for then we may simply take $(z,Z)=(y,Y)$, so suppose that $y\in D'$, then $y\neq x$ and so by (28.1), $x<y$ and $X=\varnothing$. Since $x\notin D'$, there exists $x\leq z<y$ such that $z\notin D'$, but $z.b\in D'$, some $b\in B$ with $z.b\leq y$. By 27.2.21 (1), $z.b=u.w$ with $u\in D$ and $w\in B^*$. If $u\leq z$, then by 27.2.21 (1), $z\in D'$, a contradiction, so $z.b=u$. Thus, $(z.b,\varnothing)\in F$ and as $(z,\varnothing)\preceq(z.b,\varnothing)$, by (28.1), $(z,\varnothing)\in F$, and since $x\leq z$, $(x,X)=(x,\varnothing)\preceq(z,\varnothing)$.

Now, if $(y,Y)\in F'$ and $(x,X)\preceq(y,Y)$, then by the foregoing, we may assume that $y\notin D'$, and hence $(y,Y\cap\alpha(P))\in F$. Since $(x,X)\preceq(y,Y)$, it follows that $(x,X\cap\alpha(P))\preceq(y,Y\cap\alpha(P))$ and so $(x,X\cap\alpha(P))\in F$, giving $(x,X)\in F'$. We have shown that 28.2.3 (1) holds.

Finally, suppose that $(x,X)\in F'$ but that $x.b\notin dom(F')$, some $b\in B$, then $x\notin D'$ and so $(x,X\cap\alpha(P))\in F$. If $x.b\in dom(F)$, then $(x.b,\varnothing)\in F$ and so $(x.b,\varnothing\cap\alpha(P))\in F$, contradicting $x.b\notin dom(F')$. Therefore by 28.2.3 (3), $(x,(X\cap\alpha(P))\cup\{b\})\in F$. In particular $b\in\alpha(P)$, so $(x,(X\cup\{b\})\cap\alpha(P))\in F$ and hence $(x,X\cup\{b\})\in F'$. We have shown that 28.2.3 (3 holds.

■

We may now extend the semantics of 28.2.18, by replacing rules (1), (3), (4) and (5) by

1'• $\sigma[\![a\rightarrow P]\!]=a\rightarrow ext_B(P)$, where $B=\alpha(P)\cup\{a\}$;

3'• $\sigma[\![P_1 \mid\mid\mid P_2]\!]=ext_B(P)_1 \mid\mid\mid ext_B(P_2)$, where $B=\alpha(P)_1\cup\alpha(P_2)$;

4'• $\sigma[\![P_1 \,\square\, P_2]\!]=ext_B(P)_1 \,\square\, ext_B(P_2)$, where $B=\alpha(P)_1\cup\alpha(P_2)$;

5'• $\sigma[\![P_1 \sqcap P_2]\!]=ext_B(P)_1 \sqcap ext_B(P_2)$, where $B=\alpha(P)_1\cup\alpha(P_2)$.

28.3. Hybrid Transition System Semantics.

We shall now present a non-interleaving semantics for a subset of CSP, which we shall call FCSP, involving hybrid transition systems. To each FCSP program P, we shall associate an asynchronous, initialised hybrid transition system $\rho[\![P]\!]$.

FCSP is the subset of CSP generated using rules (1), (2), (3), (4), (5), (7) and (8) only. We extend the semantics to cover the whole language in section 28.5. Our reasons for restricting ourselves to FCSP at this point is that we only wish to compare semantics for the sublanguage. To do so for the full language would involve a great deal of work of the kind seen at the end of chapter 27.

Before we present our semantics, we need to make an observation about the kind about initialised hybrid transition systems with which we shall be dealing. This observation involves the hiding operator $\backslash C$, which makes all actions belonging to the set C invisible. We shall use the symbol τ, which we suppose not to belong to A, to denote invisible actions. The set E of events of the hybrid transition systems under consideration in this chapter may contain τ.

These transition systems may also be shown to be asynchronous hybrids. Recall that this means that $(\mu(a))(e)=1$ for some $e \in E$ and $(\mu(a))(e')=0$ for $e \neq e'$. As usual, we shall abuse terminology by writing '$\mu(a)=e$' in such a case. μ extends to a monoid morphism $\mu^*: A^* \to E^*$ in the usual way. If $x \in A^*$, then we define $\mu^{-\tau}(x)=\pi_{E-\{\tau\}}(\mu^*(x))$. In other words $\mu^{-\tau}(x)$ is the sequence of visible events corresponding to the sequence of transitions x. We also define $E^{-\tau}=E-\{\tau\}$.

The definition of ρ will be compositional, so we need operations on hybrid transition systems corresponding to the operators in FCSP. Some of these have been defined already.

First, we turn to the $\|$ operator. This will be handled almost in the same way as in chapter 24 except that we need to take τ actions into account. The reason is that τ actions do not synchronise with any other and hence only involve the behaviour of one process. We shall give the definition for families of asynchronous hybrid transition systems. For finite sets of transition systems, we use an infix notation, as we shall also for the $|||$, and $\sqcap$ operators.

Our definition uses 24.3.2. We first modify each of the components so that each of its τ elements are renamed to something which cannot synchronise with anything in any of the other transition systems. Then we carry out the composition, then we reverse the renaming. As renaming and composing delivers initialised, asynchronous hybrid transition systems from initialised, asynchronous hybrid transition systems, it follows that our new composition operator does the same. Define

$$(H,q)[a\backslash b]=(H,q)[f]$$

where

$$f(e)=\begin{cases} e \text{ if } e \neq a \\ b \text{ otherwise} \end{cases}$$

Suppose that $(H_i,q_i)_{i\in I}$ is a family of initialised, asynchronous hybrid transition systems and for each $i \in I$, let a_i be an element which belongs to no A_j and which are such that $a_i \neq a_j$ when $i \neq j$. It is not hard to show that the initialised, asynchronous hybrid transition system defined by

$$\|^{\tau}_{i\in I}(H_i,q_i)=(\|_{i\in I}(H_i,q_i)[\tau\backslash a_i])[f] \qquad (28.7)$$

where

$$f(e)=\begin{cases}\tau \text{ if } e=a_i\text{, some } i\in I\\ e \text{ otherwise}\end{cases}$$

does not depend on the choice of a_i. We take (28.7) as the definition of the composition operator.

Note that 28.3.1 reduces to 24.3.1 in the case of hybrid transition systems without τ actions.

28.3.1. COROLLARY. Let $(H_i,q_i)_{i\in I}$ be a family of initialised, asynchronous hybrid transition systems Then $\|^{\tau}_{i\in I}(H_i,q_i)$ is an initialised, asynchronous hybrid transition system.

■

Next, we turn to the hiding operator.

28.3.2. DEFINITION. If (H,q) is an asynchronous, initialised hybrid transition system and $C\subseteq E^{-\tau}$, then we define $(H,q)\backslash\backslash C=(H,q)[f]$, where

$$f(e)=\begin{cases}e \text{ if } e\notin C\\ \tau \text{ otherwise}\end{cases}$$

28.3.3. COROLLARY. If (H,q) is an asynchronous, initialised hybrid transition system and $C\subseteq E^{-\tau}$, then $(H,q)\backslash\backslash C$ is an asynchronous initialised hybrid transition system.

■

28.3.4 DEFINITION. We define a function ρ: **FCSP** $\to$ **AHTS*** as follows.

1• $\rho[\![Stop_A]\!]=((\{0\},\varnothing,\varnothing,\varnothing,A,\varnothing),0);$
2• $\rho[\![Chaos_A]\!]=((\{0\},\{0\},\{(0,0,0)\},\varnothing,A,\{(0,\tau)\}),0)$
3• $\rho[\![a\to P]\!]=a\to\rho[\![P]\!];$
4• $\rho[\![P_1 \,\|\, P_2]\!]=\rho[\![P_1]\!]\,\|^{\tau}\,\rho[\![P_2]\!];$
5• $\rho[\![P_1 \,|||\, P_2]\!]=\rho[\![P_1]\!]\,|||\,\rho[\![P_2]\!]$
6• $\rho[\![P_1 \sqcap P_2]\!]=\rho[\![P_1]\!]+\rho[\![P_2]\!];$
7• $\rho[\![P\backslash C]\!]=\rho[\![P\backslash\backslash C]\!].$

28.3.5. PROPOSITION If P is a FCSP program, then $\rho[\![P]\!]$ is an initialised, asynchronous hybrid transition system.

PROOF. By structural induction on P. It is clear that $\rho[\![Stop_A]\!]$ and $\rho[\![Chaos_A]\!]$ are initialised, asynchronous hybrid transition systems. The induction step involving $\rho[\![a \to P]\!]$ is covered by 26.3.2; that involving $\rho[\![P_1 \| P_2]\!]$, by 28.3.1; that involving $\rho[\![P_1 ||| P_2]\!]$, by 24.3.5; that involving $\rho[\![P_1 \sqcap P_2]\!]$, by 26.3.5 and that involving $\rho[\![P \setminus C]\!]$ by 28.3.3.

■

28.4. Comparison of the Semantics.

We show that, at least in the case of FCSP programs, our semantics agrees with that of [2]. We do so by constructing a function γ from initialised, asynchronous hybrid transition systems to CSP processes such that for each program P,

$$\gamma(\rho[\![P]\!]) = \sigma[\![P]\!] \tag{28.8}$$

From (28.8), it follows that our hybrid transition system semantics provides at least as much information as does Hoare's. Indeed, Hoare's semantics is included in ours, at least for the sublanguage. The inclusion is strict in the sense that we consider aspects of structure which Hoare ignores, such as the presence of concurrency.

The construction makes use of the following notations. Suppose that H is a hybrid transition systems, then $q \in Q$ is a *divergent state* of H providing it belongs to $div(H)$, where

$$div(H) = \{q \in Q \mid \forall n \in \omega\ \exists u \in A^n : \mu^{-\tau}(u) = \Omega\} \tag{28.9}$$

A state is divergent if, from it, the system is uncontrollable; it may forever ignore the environment by making silent moves. A state $q \in Q$ is *stable* providing it belongs to $stbl(H)$ where

$$stbl(H) = \{q \in Q \mid \forall a \in A : q \to^a \Rightarrow \mu(a) \neq \tau\} \tag{28.10}$$

28.4.1. DEFINITION. If (H,q) is an initialised, asynchronous hybrid transition system, then we define $\gamma(H,q) = (E^{-\tau}, F, D)$, where $F = F' \cup (D \times \wp(E^{-\tau}))$ and

1• $D = \{x \in E^{-\tau}* \mid \exists u \in A^* \exists q' \in div(H) : q \to^u q' \wedge \mu^{-\tau}(u) \leq x\}$;

2• $F' = \{(x, X) \mid \exists u \in A^* \exists q' \in stbl(H) :$
$$q \to^u q' \wedge x = \mu^{-\tau}(u) \wedge X \subseteq E^{-\tau} - nxt(H, q')\}$$

and

$$nxt(H,q') = \{\mu(a) \mid \exists q'' \in Q:(a,q'') \in [q'\rangle \wedge \mu(a) \neq \tau\}.$$

Of course, we need to show that $\gamma(H,q)$ is a CSP process. First, we collect together some useful observations.

28.4.2. LEMMA. Let (H,q) be an initialised, asynchronous hybrid transition system and let $\gamma(H,q) = (E^{-\tau}, F, D)$. If $Tr(H,q) = \mu^{-\tau}(L(T(C(H)),q))$ then

1• $Tr(H,q)$ is left-closed in $E^{-\tau}{}^*$;
2• If $x \in Tr(H,q) - D$, then there exists $u \in A^*$ and $q' \in stbl(H)$ such that $q \rightarrow^u q'$ and $\mu^{-\tau}(u) = x$;
3• If $x \in Tr(H,q)$, then $(x,\varnothing) \in F$;
4• If $x \in D$ and $y \leq x$ then either $y \in D$ or $y \in Tr(H,q)$.

PROOF. (1) Trivial.
(2) By definition, we may find $u \in L(T(C(H)),q)$ such that $\mu^{-\tau}(u) = x$. We argue by contradiction. Suppose that whenever $q \rightarrow^w q''$ and $\mu^{-\tau}(w) = x$ then $q'' \notin stbl(H)$. We shall show that in that case, whenever $q \rightarrow^w q''$ and $\mu^{-\tau}(w) = x$ then $q'' \in div(H)$ from which it follows that $x \in D$, by 28.4.1 (1). The induction hypothesis, that if $n \in \omega$ there exists $v \in L(T(C(H)),q'')$ such that $|v| = n$ and $\mu^{-\tau}(v) = \Omega$, is clearly true when $n = 0$ and if it holds for n with $v \in L(T(C(H)),q'')$, let $q'' \rightarrow^v \tilde{q}$ so that $q \rightarrow^{w.v} \tilde{q}$ and $\mu^{-\tau}(w.v) = x$. BY our assumption, $\tilde{q} \notin stbl(H)$. Hence, there exists $a \in A$ such that $\tilde{q} \rightarrow^a$ and $\mu(a) = \tau$. And now, $v.a \in L(T(C(H)),q'')$ with $|v.a| = n+1$ and $\mu^{-\tau}(v.a) = \Omega$, completing the induction step.
(3) If $x \in D$, then $(x,\varnothing) \in F$ by 28.4.1 (1). Otherwise, by (2) of the present lemma, we may find $q' \in stbl(H)$ and $u \in A^*$ such that $q \rightarrow^u q'$ and $\mu^{-\tau}(u) = x$. Since $\varnothing \subseteq E^{-\tau} - nxt(H,q')$, the assertion follows in this case, also by 28.4.1 (2).
(4) By 28.4.1 (1), if $x \in D$, then there exists $q' \in div(H)$ and $u \in A^*$ such that $q \rightarrow^u q'$ and $\mu^{-\tau}(u) \leq x$. If $\mu^{-\tau}(u) \leq y$ then $y \in D$. Otherwise $y \leq \mu^{-\tau}(u)$. But $\mu^{-\tau}(u) \in Tr(H,q)$ and so $y \in Tr(H,q)$ by (1).

■

28.4.3. PROPOSITION. If (H,q) is an initialised, asynchronous hybrid transition system, then $\gamma(H,q) \in$ **HProc**.

PROOF. Let $\gamma(H,q) = (E^{-\tau}, F, D)$. Since $\Omega \in Tr(H,q)$, it follows from 28.4.2 (3) that $(\Omega,\varnothing) \in F$ and so $F \neq \varnothing$. Let us show that F is left-closed. Suppose that

$(y,Y) \in F$ and $(x,X) \preceq (y,Y)$. By 28.4.1, either $y \in D$ or $y \in Tr(H,q)$. There are two cases corresponding to the two clauses of (28.1).

In the first case, $x < y$ and $X = \emptyset$. If $y \in D$ then by 28.4.2 (4), either $x \in D$ in which case $(x,\emptyset) \in F$ by 28.4.1 (1) or $x \in Tr(H,q)$ in which case $(x,\emptyset) \in F$, by 28.4.2 (3). If $y \in Tr(H,q)$, then $(x,\emptyset) \in F$, by 28.4.2 (1) and (3).

In the second case, $x = y$ and $X \subseteq Y$. If $y \in D$ then $(x,X) \in F$ by 28.4.1 (1). Otherwise $y \in Tr(H,q)$ so there exists $u \in A^*$ and $q' \in stbl(H)$ such that $q \to^u q'$, $\mu^{-\tau}(u) = y$ and $Y \subseteq E^{-\tau} - nxt(H,q')$. As $X \subseteq Y$, $X \subseteq E^{-\tau} - nxt(H,q')$ and so $(x,X) = (y,X) \in F$ by 28.4.1 (2).

That $D.E^{-\tau} \subseteq D$ and $D \times \wp(E^{-\tau}) \subseteq F$ follows immediately from 28.4.1.

Finally, suppose that $(x,X) \in F$ and $e \in E^{-\tau}$ such that $(x.e,\emptyset) \notin F$. Hence, $x \in D$ and so there exists $u \in A^*$ and $q' \in stbl(H)$ such that $q \to^u q'$, $\mu^{-\tau}(u) = x$ and $X \subseteq E^{-\tau} - nxt(H,q')$. If $e \in nxt(H,q')$ then $x.e \in Tr(H,q)$ and consequently $(x.e,\emptyset) \in F$ by 28.4.2 (3), a contradiction. So, $e \notin nxt(H,q')$ which means that $X \cup \{e\} \subseteq E^{-\tau} - nxt(H,q')$, and now $(x, X \cup \{e\}) \in F$, by 28.4.1 (2).

■

Let us now proceed with a proof of (28.8). As we said, this is by structural induction. The induction step is the subject of the following string of lemmas.

28.4.4. LEMMA. If (H,q) is an initialised, asynchronous hybrid transition system and $e \in \mathcal{A}ct$, then

$$\gamma(e \to (H,q)) = e \to \gamma(H,q).$$

PROOF. Let

$$\gamma(H,q) = (E,F,D)$$
$$\gamma(e \to (H,q)) = (E',F',D')$$
$$(E'',F'',D'') = e \to (E,F,D).$$

If $a \in A$, then define $\hat{a} = (a,1)$ and if $u = a_1 \cdots a_n$, $a_1, \cdots, a_n \in A$, then define $\hat{u} = (a_1,1) \cdots (a_n,1)$.

By 26.3.1 (5) and 28.2.6, $E'' = E \cup \{e\} = E'$.

By 26.3.1, (28.9) and (28.10)

$$div(H') = \{(q,1) \mid q \in div(H)\}$$
$$stbl(H') = \{(q,1) \mid q \in stbl(H)\} \cup \{0\}$$
$$\forall q' \in Q: nxt(H,q') = nxt(H',(q',1))$$

Now, by 28.2.5 and 28.4.1, $y \in D'' \Leftrightarrow$ there exists $x \in D$ such that $y = a.x$ $\Leftrightarrow$ $y = a.x$ and there exists $u \in A^*$ and $q' \in div(H)$ such that $q \to^u q'$ and $\mu^{-\tau}(u) \leq x$ $\Leftrightarrow$ there exists $u \in A^*$ and $q' \in div(H)$ such that $0 \to'^0 (q,1) \to'^{\hat{u}} (q',1)$ and $\mu'^{-\tau}(\hat{u}) \leq x \Leftrightarrow$ there exists $u' \in A^*$ (namely $0.\hat{u}$) and $q'' \in div(H')$ (namely $(q',1)$) such that $0 \to'^{\hat{u}} q''$ and $\mu'^{-\tau}(\hat{u}) \leq a.x \Leftrightarrow$ $y \in D'$ and we have shown that $D' = D''$.

By 26.3.1 (3) and 28.4.1, $(\Omega, X) \in F'' \Leftrightarrow X \subseteq E'' - \{e\} \Leftrightarrow X \subseteq E' - \{e\} \Leftrightarrow$ $(\Omega, X) \in F'$.

$(y, X) \in F''$ with $y \neq \Omega$ and $y \notin D'' \Leftrightarrow$ there exists $(x, X) \in F$ such that $y = a.x \Leftrightarrow y = a.x$ and there exists $u \in A^*$ and $q' \in stbl(H)$ such that $q \to^u q'$, $\mu^{-\tau}(u) = x$ and $X \subseteq E^{-\tau} - nxt(H, q') \Leftrightarrow$ there exists $u \in A^*$ and $q' \in stbl(H)$ such that $0 \to'^0 (q,1) \to'^{\hat{u}} (q',1)$, $\mu^{-\tau}(u) = x$ and $X \subseteq E^{-\tau} - nxt(H, q') \Leftrightarrow$ there exists $u' \in A^*$ (namely $0.\hat{u}$) and $q'' \in stbl(H')$ (namely $(q',1)$) such that $0 \to'^{\hat{u}} q''$, $\mu'^{-\tau}(\hat{u}) = a.x$ and $X \subseteq E'^{-\tau} - nxt(H', q'') \Leftrightarrow (y, X) \in F'$ with $y \neq \Omega$ and $y \notin D'$ and we have shown that $F' = F''$.

■

Now we turn to the || operator. Let (H_1, q_1) (H_2, q_2) be asynchronous, initialised hybrid transition systems and let $(H, q) = (H_1, q_1) \,||^\tau\, (H_2, q_2)$.

From 24.3.2 and (28.7), we may deduce that

$$Q = Q_1 \times Q_2 \tag{28.11}$$

and that A is the set of all $(a_1, a_2) \in (A_1 \cup \{\Omega\}) \times (A_2 \cup \{\Omega\})$ such that

$$\begin{aligned} & a_1 \neq \Omega \neq a_2 \wedge \mu_1(a_1) = \mu_2(a_2) \neq \tau \\ & a_1 = \Omega \neq a_2 \wedge \mu_2(a_2) \in E_2 \\ & a_1 \neq \Omega = a_2 \wedge \mu_1(a_1) \in E_1 \end{aligned} \tag{28.12}$$

and that if $(q_1', q_2'), (q_1'', q_2'') \in Q$, then

$$(q_1', q_2') \to^{(a_1, a_2)} (q_1'', q_2'') \Leftrightarrow q_1' \to_1^{a_1} q_1'' \wedge q_2' \to^{a_2} q_2'' \tag{28.13}$$

where $q_1 \to^\Omega q_2 \Leftrightarrow q_1 = q_2$

and that if $(a_1, a_2), (b_1, b_2) \in (A_1 \cup \{\Omega\}) \times (A_2 \cup \{\Omega\})$ then

$$(a_1, a_2)\,\iota\,(b_1, b_2) \Leftrightarrow (a_1 \neq \Omega \neq b_1 \Rightarrow a_1\, \iota_1\, b_1) \wedge (a_2 \neq \Omega \neq b_2 \Rightarrow a_2\, \iota_2\, b_2) \tag{28.14}$$

and that

$$E = \mu(A) \cup E_1 \cup E_2 \tag{28.15}$$

and finally that

$$\mu(a_1, a_2) = \begin{cases} \mu_1(a_1) \text{ if } a_1 \neq \Omega \\ \mu_2(a_2) \text{ otherwise} \end{cases} \tag{28.16}$$

First, we need some notation. If $u = a_1 \cdots a_n$ with $a_1, \cdots, a_n \in A$, then define $u(i) = a_1(i) \cdots, a_n(i)$. By (28.12), $u(i) \in A_i^*$.

If $u = a_1 \cdots a_n$ with $a_1, \cdots, a_n \in A_1$, then we shall write (u, Ω) for the string $(a_1, \Omega) \cdots (a_n, \Omega)$. (Ω, u) for $u \in A_2^*$, is defined analogously.

28.4.5. LEMMA. If $(u_1, u_2) \in A_1^* \times A_2^*$, then

$$(\exists u \in A^* \ \forall i \in \{1,2\}: u(i) = u_i) \Leftrightarrow (\exists x \in (E^{-\tau})^* \ \forall i \in \{1,2\}: \pi_{E_i^{-\tau}}(x) = \mu_i^{-\tau}(u_i)).$$

PROOF. Suppose first that $u = a_1 \cdots a_n$, with $a_1, \cdots, a_n \in A$, such that $u(i) = u_i$, each i. Let $x = \mu^{-\tau}(u)$. We show by induction on $|u|$ that $\pi_{E_i^{-\tau}}(x) = \mu_i^{-\tau}(u_i)$, each i. If $|u| = 0$, that is $u = \Omega$, then the result is clear. If $|u| > 0$, then by induction $\pi_{E_i^{-\tau}}(a_1 \cdots a_{n-1}) = \mu_i^{-\tau}(a_1(i) \cdots a_{n-1}(i))$, each i, and so it suffices to show that $\pi_{E_i^{-\tau}}(a_n) = \mu_i^{-\tau}(a_n(i))$, each i. But this is an immediate consequence of (28.12) and (28.16).

Next suppose there exists $x \in (E^{-\tau})^*$ such that $\pi_{E_1^{-\tau}}(x) = \mu_1^{-\tau}(u_1)$ and $\pi_{E_2^{-\tau}}(x) = \mu_2^{-\tau}(u_2)$.

We argue by induction on $|u_1| + |u_2|$. If $|u_1| + |u_2| = 0$, then take $u = \Omega$. For the induction step, we shall show that for each i

$$\exists a \in A: \mu^{-\tau}(a) \leq x \wedge a(i) \leq u_i \tag{28.17}$$

Once we have proved (28.17), then we may argue as follows. Since both $\pi_{E_i^{-\tau}}$ and $\mu_i^{-\tau}$ are monoid morphisms, by construction, it follows from (28.16) and (28.17) that $\pi_{E_i^{-\tau}}(x / \mu^{-\tau}(a)) = \mu_i^{-\tau}(u_i / a(i))$, each i, and since $a(i) \neq \Omega$ for some i, by (28.12), $|u_1| + |u_2| < |u_1 / a(1)| + |u_2 / a(2)|$. By induction, there exists $u' \in A^*$ such that $u'(i) = u_i / a(i)$, each i. It is now clear that $u = a.u'$ satisfies the requirements of the lemma. It remains, therefore, to prove (28.17).

We have two cases:

1• $\exists i \in \{1,2\} \ \exists a_i \in A_i: a_i \leq u_i \wedge \mu_i(a_i) = \tau;$

2• $\exists a_1 \in A_1 \ \exists a_2 \in A_2: a_1 \leq u_1 \wedge a_2 \leq u_2 \wedge \mu_1(a_1) \neq \tau \neq \mu_2(a_2)$

In case (1), suppose without loss of generality that $a_1 \in A_1$ and $a_1 \leq u_1$ with $\mu_1(a_1) = \tau$. By (28.12), $a = (a_1, \Omega) \in A$. Clearly for each i, $a(i) \leq u_i$, and since $\mu(a) = \tau$, $\mu^{-\tau}(a) = \Omega \leq x$. Thus (28.17) holds in case (1).

In case (2), we cannot have $x = \Omega$ and so suppose $e \in E^{-\tau}$ with $e \leq x$. Since $\pi_{E_1^{-\tau}}(x) = \mu_1^{-\tau}(u_1)$ and $\pi_{E_2^{-\tau}}(x) = \mu_2^{-\tau}(u_2)$, if $e \in E_i^{-\tau}$ and $a_i \leq u_i$, then $\mu_i(a_i) = e$. (28.12) allows us to define $a = (a_1, a_2)$ and it is easy to check that a satisfies (28.17).

■

LEMMA 28.4.6. With the above notation

1• $\forall q, q' \in Q\ \forall u \in A^*: q \rightarrow^u q' \Leftrightarrow q(1) \rightarrow^{u(1)} q'(1) \wedge q(2) \rightarrow^{u(2)} q'(2);$
2• $div(H) = (div(H_1) \times Q_2) \cup (Q_1 \times div(H_2));$
3• $stbl(H) = stbl(H_1) \times stbl(H_2).$

PROOF. (1) follows from (28.13) and induction on $|u|$.

For (2), suppose that $q \in div(H_1) \times Q_2$, then by (28.3) for each $n \in \omega$ there exists $u_n \in A_1^n$ such that $q(1) \rightarrow_1^{u_n}$ and $\mu_1^{-\tau}(u_n) = \Omega$, and so $\mu_1(u_n) = \tau^n$. If $u_n = a_1 \ldots a_n$, with $a_1, \ldots, a_n \in A_1$, then $(u_n, \Omega) = (a_1, \Omega) \ldots (a_n, \Omega) \in A^*$, by (28.12), and by (28.13), $q \rightarrow^{(u_n, \Omega)}$. By (28.16), $\mu(u_n, \Omega) = \tau^n$, that is $\mu^{-\tau}(u_n, \Omega) = \Omega$. Since n is arbitrary, it follows by (28.3) that $q \in div(H)$. We have proved that $div(H_1) \times Q_2 \subseteq div(H)$ By symmetry, $Q_1 \times div(H_2) \subseteq div(H)$. Thus, $(div(H_1) \times Q_2) \cup (Q_1 \times div(H_2)) \subseteq div(H)$.

Conversely, let $q \in div(H)$ We shall show that if $q(2) \notin div(H_2)$, then $q(1) \in div(H_1)$. By symmetry, $q(1) \notin div(H_1) \Rightarrow q(2) \in div(H_2)$. This will establish the converse inclusion.

If $q(2) \notin div(H_2)$, then there is a number n_0 such that if $q(2) \rightarrow_2^u$ and $|u| > n_0$, then $\mu_2^{-\tau}(u) \neq \Omega$. Hence, if $q \rightarrow^u$ and $|u| > n + n_0$ with $\mu^{-\tau}(u) = \Omega$, then $|\pi_{A_2}(u)| \leq n_0$, as $\mu_2^{-\tau}(\pi_{A_2}(u)) = \pi_{E_2^{-\tau}}(\mu^{-\tau}(u)) = \Omega$, by (28.16). Thus $|\pi_{A_1}(u)| > n$ and $\mu_1^{-\tau}(\pi_{A_1}(u)) = \Omega$ Consequently, for any n, there exists $u' \in A_1^*$ such that $q(1) \rightarrow_1^{u'}$, $|u'| \geq n$ and $\mu_1^{-\tau}(u') = \Omega$. Hence, $q(1) \in div(H_1)$.

Finally, we show (3). Suppose $q(1) \rightarrow_1^a q'$ with $\mu_1(a) = \tau$. By (28.12), $(a, \Omega) \in A$ and by (28.13), $q \rightarrow^{(a,\Omega)} (q', q(2))$ and $\mu(a, \Omega) = \tau$. Hence $q(1) \notin stbl(H_1) \Rightarrow q \notin stbl(H)$. By symmetry, $q(2) \notin stbl(H_2) \Rightarrow q \notin stbl(H)$. Hence $stbl(H) \subseteq stbl(H_1) \times stbl(H_2)$. Conversely, if $q \notin stbl(H)$ then $q \rightarrow^a q'$ with $\mu(a) = \tau$. By (28.12) and (28.16), either $a(1) = \Omega$, $a(2) \neq \Omega$ and $\mu_2(a(2)) = \tau$ or $a(2) = \Omega$, $a(1) \neq \Omega$ and $\mu_1(a(1)) = \tau$. Without loss of generality, suppose the latter, then $q(1) \to_1^{a(1)} q'$ and $\mu_1(a(1)) = \tau$. Therefore, $q \notin stbl(H) \Rightarrow q(1) \notin stbl(H1) \vee q \notin stbl(H2)$.

Thus $stbl(H_1) \times stbl(H_2) \subseteq stbl(H)$, completing the proof.

■

28.4.7. PROPOSITION. Suppose (H_1, q_1) and (H_2, q_2) are initialised, asynchronous hybrid transition systems. Then

$$\gamma(H_1, q_1) \| \gamma(H_2, q_2) = \gamma((H_1, q_1) \| (H_2, q_2))$$

PROOF. Let

$$\begin{aligned}
&(H, q_0) = (H_1, q_1) \| (H_2, q_2) \\
&\gamma(H_1, q_1) = (E_1^{-\tau}, F_1, D_1) \\
&\gamma(H_2, q_2) = (E_2^{-\tau}, F_2, D_2) \\
&\gamma(H_1, q_1) \| \gamma(H_2, q_2) = (E'^{-\tau}, F', D') \\
&\gamma(H, q_0) = (E^{-\tau}, F, D).
\end{aligned}$$

We show that $(E^{-\tau}, F, D) = (E'^{-\tau}, F', D')$.

By 28.2.7 (1) and (28.15), $E^{-\tau} = E_1^{-\tau} \cup E_2^{-\tau} = E'^{-\tau}$.

Let $y \in D$, then by 28.4.1 (1) there exists $u \in A^*$ and $q' \in div(H)$ such that $q_0 \to^u q'$ and $\mu^{-\tau}(u) \le y$. Let $x = \mu^{-\tau}(u)$. By 28.4.6 (1), $q_0(i) \to_i^{u(i)} q'(i)$, each i. By 28.4.2 (3), $\mu^{-\tau}(u(i)) \in dom(F_i)$, each i. Also, by 28.4.6 (2), $q'(1) \in div(H_1)$ or $q_2' \in div(H_2)$ and hence $\mu_1^{-\tau}(u(1)) \in D_1$ or $\mu_2^{-\tau}(u(2)) \in D_2$. As in the proof of 28.4.5, $\pi_{E_i}^{-\tau}(x) = \mu_i^{-\tau}(u(i))$. Therefore, $x \le y$ and either $\pi_{E_1}^{-\tau}(x) \in dom(F_1)$ and $\pi_{E_2}^{-\tau}(x) \in D_2$ or $\pi_{E_1}^{-\tau}(x) \in D_1$ and $\pi_{E_2}^{-\tau}(x) \in dom(F_2)$. In either case, $y \in D'$ by 28.2.7 (2). Hence, $D \subseteq D'$.

Now, suppose that $y \in D'$. By 28.2.7 (2) without loss of generality, we may assume that there exists $x \le y$ such that $\pi_{E_1}^{-\tau}(x) \in D_1$ and $\pi_{E_2}^{-\tau}(x) \in dom(F_2)$. Thus, by 28.4.1 (1) we may find $u_1 \in A_1^*$ and $q_1' \in div(H_1)$ such that $q_1 \to_1^{u_1} q_1'$ and $\mu_1^{-\tau}(u_1) \le \pi_{E_1}^{-\tau}(x)$. By 28.4.1, we have three cases:

(a) $\pi_{E_2}^{-\tau}(x) = \Omega$;

(b) $\exists u_2 \in A_2^* \exists q_2' \in div(H_2): q_2 \to_2^{u_2} q_2' \wedge \mu_2^{-\tau}(u_2) \le \pi_{E_2}^{-\tau}(x)$;

(c) $\exists u_2 \in A_2^* \exists q_2' \in stbl(H_2): q_2 \to_2^{u_2} q_2' \wedge \mu_2^{-\tau}(u_2) = \pi_{E_2}^{-\tau}(x)$.

In case (a), suppose that $u = a_1 \cdots a_n$. Since $\mu_1^{-\tau}(u_1) \le \pi_{E_1}^{-\tau}(x)$ and $\pi_{E_2}^{-\tau}(x) = \Omega$ it follows that if $\mu_1(a_i) \ne \Omega$ then $\mu_1(a_i) \notin E_2^{-\tau}$ and so by (28.12), $(a_i, \Omega) \in A$, each i, and by 28.4.6 (1), $q_0 \to^{(u,\Omega)} (q_1', q_0(2))$. By (28.16),

$\mu^{-\tau}(u)=\mu_1^{-\tau}(u_1)\leq\pi_{E_1}^{-\tau}(x)=x$ (since $\pi_{E_2}^{-\tau}(x)=\Omega\leq y$). By 28.4.6 (2), $(q_1',q_0(2))\in div(H)$ and now $y\in D$, by 28.4.1 (1).

In case (b) let $x'\leq x$ such that $\mu_2^{-\tau}(u_2)=\pi_{E_2}^{-\tau}(x')$. Let $u_1'\leq u_1$ such that $\mu_1^{-\tau}(u_1')=\pi_{E_1}^{-\tau}(x')$ and let $q_1''\in Q_1$ such that $q_1\rightarrow_1^{u_1'}q_1''$. By 28.4.5, there exists $u\in A^*$ such that $u(1)=u_1'$ and $u(2)=u_2$. An examination of the proof of that lemma shows that we may choose u so that $\mu^{-\tau}(u)=x'$. By 28.4.6 (1), $q_0\rightarrow^u(q_1'',q_2')$. Since $(q_1'',q_2')\in div(H)$, by 28.4.6 (2), and $\mu^{-\tau}(u)=x'\leq x\leq y$, it follows from 28.4.1 (1) that $y\in D$.

In case (c), we may also use 28.4.5 and 28.4.6 to find $u\in A^*$ such that $q_0\rightarrow^u(q_1',q_2')$ and $\mu^{-\tau}(u)=x$, with $(q_1',q_2')\in div(H)$. Since $\mu^{-\tau}(u)=x\leq y$, it follows from 28.4.1 (1) that $y\in D$. We have proved that $D'\subseteq D$ and conclude that $D'=D$.

Next, let $(x,X)\in F$. If $x\in D$, then $x\in D'$ so that $(x,X)\in F'$, so suppose that $x\notin D$. Then there exists $q\in stbl(H)$ and $u\in A^*$ such that $q_0\rightarrow^u q$ and $\mu^{-\tau}(u)=x$. By 28.4.5 and 28.4.6, for each i, $q_i\rightarrow_i^{u(i)}q(i)\in stbl(H_i)$ with $\pi_{E_i}^{-\tau}(x)=\pi_{E_i}^{-\tau}(\mu^{-\tau}(u))=\mu_i^{-\tau}(u(i))$ and so $(\pi_{E_i}^{-\tau}(\mu^{-\tau}(u)),X_i)\in F_i$ for any $X_i\subseteq E_i^{-\tau}-nxt(H_i,q(i))$. Define $X_i=X\cap(E_i^{-\tau}-nxt(H_i,q(i)))$, each i. so that $X_i\subseteq E_i^{-\tau}-nxt(H_i,q(i))$, each i. By 28.2.7 (3), in order to show that $(x,X)\in F'$, it suffices to show that $X=X_1\cup X_2$. This will follow by a simple set theoretic argument, the definition of the X_i and the following

$$E^{-\tau}-nxt(H,q)=(E_1^{-\tau}-nxt(H_1,q(1)))\cup(E_2^{-\tau}-nxt(H_2,q(1))) \qquad (28.18)$$

Suppose that $e\in E^{-\tau}-nxt(H,q)$, but that $e\notin E_1^{-\tau}-nxt(H_1,q(1))$ We shall show that $e\in E_2^{-\tau}-nxt(H_2,q(2))$. This will show that the left hand side of (28.18) is contained in the right hand side.

If $e\notin E_1^{-\tau}$, then $e\in E_2^{-\tau}$, since $E^{-\tau}=E_1^{-\tau}\cup E_2^{-\tau}$. If $e\in nxt(H_2,q(2))$, then there exists $a\in A_2$ such that $q(2)\rightarrow_2^a$ and $\mu_2(a)=e$. Since $e\notin E_1^{-\tau}$, it follows from (28.12) and (28.13) that $(\Omega,a)\in A$ and that $q\rightarrow^{(\Omega,a)}$ and, $\mu^{-\tau}(a,\Omega)=e$ contradicting $e\in E^{-\tau}-nxt(H,q)$. So if $e\notin E_1^{-\tau}$ then $e\in E_2^{-\tau}-nxt(H_2,q(2))$. The other possibility is that $e\in nxt(H_1,q(1))$. In this case, $q(1)\rightarrow_1^a$, with $\mu_1(a)=e$. If $e\notin E_2^{-\tau}$, then as before $(\Omega,a)\in A$ and $q\rightarrow^{(\Omega,a)}$ with $\mu^{-\tau}(a,\Omega)=e$, contradicting $e\in E^{-\tau}-nxt(H,q)$. If $e\in nxt(H_2,q(2))$, then $q(2)\rightarrow_2^{a'}$ with $\mu_2(a')=e$ and then $(a,a')\in A$ with $q\rightarrow^{(a,a')}$ with $\mu^{-\tau}(a,a')=e$, again contradicting $e\in E^{-\tau}-nxt(H,q)$. Hence, again, $e\in E^{-\tau}-nxt(H,q)$.

The converse inclusion is a simple consequence of (28.15) and (28.13).

Finally, let us show that $F'\subseteq F$. Let $(x,X)\in F'$. Again, if $x\in D'$, then $x\in D$ so that $(x,X)\in F$, so suppose that $x\notin D'$. By 28.2.7, there exist

$X_i \subseteq E_i^{-\tau} - nxt(H_i, q(i))$ such that $X = X_1 \cup X_2$ and $(\pi_{E_i}^{-\tau}(x), X_i) \in F_i$. We also note that $\pi_{E_i}^{-\tau}(x) \notin D_i$ each i, because of 28.2.7 (2). Hence, $q_i \rightarrow^{u_i} q_i'$ with $q_i' \in stbl(H_i)$, $\mu_i^{-\tau}(u_i) = \pi_{E_i}^{-\tau}(x)$, $X_i \subseteq E_i^{-\tau} - nxt(H_i, q_i')$ and $X = X_1 \cup X_2$. $X \subseteq E^{-\tau} - nxt(H,(q_1', q_2'))$, by (28.18) and so if we may find $u \in A^*$ such that $\mu^{-\tau}(u) = x$ and $u(i) = u_i$, each i, then by 28.4.6 and (28.18), $q_0 \rightarrow^{u} (q_1', q_2')$ with $(q_1', q_2') \in stbl(H)$, $\mu^{-\tau}(u) = x$ and $X \subseteq E^{-\tau} - nxt(H,(q_1', q_2'))$, so that $(x, X) \in F$. Thus, to conclude the proof, we must construct this u. But its existence is guaranteed by 28.4.5.

■

Next, we turn to the ||| operator. Again, we need some notation. Let $(H, q_0) = (H_1, q_1) ||| (H_2, q_2)$. If $u \in A^*$ and $u \neq \Omega$, then $u = (a_1, i_1) \ldots (a_n, i_n)$, where $a_1, \ldots, a_n \in A$ and $i_1, \ldots, i_n \in \{1,2\}$. If $\{j_1, \ldots, j_k\} = \{l \mid i_l = i\}$ with $j_1 < \ldots < j_k$ then let $u(i) = a_{j_1} \ldots a_{j_k}$. Intuitively, $\mu^{-\tau}(u)$ is an interleaving of the sequences $\mu_i^{-\tau}(u(i))$. Define $\Omega(i) = \Omega$.

28.4.8. LEMMA. With the above notation, if $u_i \in A_i^*$ and $x \in (E^{-\tau})^*$, then

$$x \in \mu_1^{-\tau}(u_1) ||| \mu_2^{-\tau}(u_2) \Leftrightarrow \exists u \in A^* : \mu^{-\tau}(u) = x \wedge u(1) = u_1 \wedge u(2) = u_2$$

PROOF. First, let us suppose that $x \in (E^{-\tau})^*$ such that $x \in \mu_1^{-\tau}(u_1) ||| \mu_2^{-\tau}(u_2)$. We argue by induction on $n = |u_1| + |u_2|$ that there exists $u \in A^*$ such that $\mu^{-\tau}(u) = x$ and $u(i) = u_i$, each i.

If $n = 0$, then we may take $u = \Omega$.

Suppose that $n > 0$. We have two cases.

In the first case, we have $a \in A_i$ with $a \leq u_i$ and $\mu_i(a) = \tau$ some i. Suppose, without loss of generality that $i = 1$. Let $u_1' = u_1 / a$ and $u_2' = u_2$. Clearly, $\mu_1^{-\tau}(u_1') = \mu_1^{-\tau}(u_1)$ and so $x \in \mu_1^{-\tau}(u_1') ||| \mu_2^{-\tau}(u_2')$. Since $|u_1'| + |u_2'| < n$, we may apply the induction hypothesis and conclude that there exists $u' \in A^*$ such that $\mu^{-\tau}(u') = x$ and $u'(i) = u_i'$, each i. Let $u = (a, 1).u'$.

In the second case, for each i, if $a_i \in A_i$ and $a_i \leq u_i$ then $\mu_i(a_i) \neq \Omega$. If $e \in E^{-\tau}$, with $e \leq x$, then from the definition of ||| on sequences, we must have $e \leq \mu_i^{-\tau}(u_i)$ for some i. Suppose, without loss of generality that $i = 1$, then $\mu_1^{-\tau}(a_1) = e$ As above, we may find $u' \in A^*$ such that $\mu^{-\tau}(u') = x / e$, $u'(1) = u_1 / a_1$ and $u'(2) = u_2$. Take $u = (a_1, 1).u'$.

Now we prove the converse. Suppose that $u \in A^*$ such that $\mu^{-\tau}(u) = x$ and $u(i) = u_i$, each i. If $u = \Omega$, then $u(i) = \Omega$, each i, and certainly $\Omega \in \Omega ||| \Omega$. If $u > \Omega$, then $(a, i) \leq u$, some $a \in A_i$. Suppose, without loss of generality, that $i=1$. Let $u'=u/(a,1)$, $u_1'=u_1/a$ and $u_2'=u_2$. By induction, $\mu^{-\tau}(u')\in\mu_1^{-\tau}(u_1')\,|||\,\mu_2^{-\tau}(u_2')$. Hence,

$$\mu^{-\tau}(a,1).\mu^{-\tau}(u')\in\mu_1^{-\tau}(a).(\mu_1^{-\tau}(u_1')\,|||\,\mu_2^{-\tau}(u_2')) \tag{28.19}$$

But, by definition of $|||$ on sequences,

$$\mu_1^{-\tau}(a).(\mu_1^{-\tau}(u_1')\,|||\,\mu_2^{-\tau}(u_2'))\subseteq\mu_1^{-\tau}(u_1)\,|||\,\mu_2^{-\tau}(u_2). \tag{28.20}$$

Let $x=(a,1).u'$, then $\mu^{-\tau}(x)=\mu^{-\tau}(a,1).\mu^{-\tau}t(u')$, since $\mu^{-\tau}$ is a monoid morphism. From (28.19) and (28.20), we see that $x\in\mu_1^{-\tau}(u_1)\,|||\,\mu_2^{-\tau}(u_2)$, as required

■

We also have the following result which is analogous to 28.4.6.

28.4.9. LEMMA. With the above notation,

1• $q\rightarrow^{u} q'\Leftrightarrow q(1)\rightarrow_1^{u(1)} q'(1)\wedge q(2)\rightarrow_2^{u(2)} q'(2);$
2• $div(H)=(div(H_1)\times Q_2)\cup(Q_1\times div(H_2));$
3• $stbl(H)=stbl(H_1)\times stbl(H_2).$

PROOF. The proof of this lemma is entirely analogous to that of 28.4.6.

■

28.4.10 PROPOSITION. If (H_1,q_1) and (H_2,q_2) are asynchronous initialised hybrid transition systems such that $E_1^{-\tau}=E_2^{-\tau}$, then

$$\gamma(H_1,q_1)\,|||\,\gamma(H_2,q_2)=\gamma((H_1,q_1)\,|||\,(H_2,q_2))$$

PROOF. Let

$$(H,q_0)=(H_1,q_1)\,|||\,(H_2,q_2)$$
$$\gamma(H_1,q_1)=(E_1^{-\tau},F_1,D_1)$$
$$\gamma(H_2,q_2)=(E_2^{-\tau},F_2,D_2)$$
$$\gamma(H_1,q_1)\,|||\,\gamma(H_2,q_2)=(E'^{-\tau},F',D')$$
$$\gamma(H,q_0)=(E^{-\tau},F,D).$$

We show that $(E^{-\tau},F,D)=(E'^{-\tau},F',D')$.

Certainly, $E'^{-\tau}=E_1^{-\tau}\cup E_2^{-\tau}=E^{-\tau}$ by 24.2.4 (2) and 28.2.9 (1).

Let $y\in D$, then there exists $u\in A^*$ and $q\in div(H)$ such that $q_0\rightarrow^{u} q$ and $\mu^{-\tau}(u)\le y$. By 28.4.9 (1), $q_i\rightarrow_i^{u(i)} q(i)$ and by 28.4.2 (3), $\mu_i^{-\tau}(u(i))\in dom(F_i)$, each i. By 28.4.9 (2), either $q(1) \in div(H_1)$ or $q(2) \in div(H_2)$. Without loss of generality, assume the former, so that $\mu_1^{-\tau}(u(1)) \in D_1$ by 28.4.1 (1). But $\mu^{-\tau}(u) \le y$, and so $\mu^{-\tau}(u) \in \mu_1^{-\tau}(u(1)) ||| \mu_2^{-\tau}(u(2))$ with $\mu_1^{-\tau}(u(1)) \in D_1$ and $\mu_2^{-\tau}(u(2)) \in dom(F_2)$, by 28.4.8. By 28.2.9 (2), $y \in D'$.

Conversely, let $y \in D'$ then there exists $x \le y$ with $x \in x_1 ||| x_2$ for $x_i \in D_i$ some i, and $x_i \in dom(F_i)$, each i. As in the proof of 28.4.7, we may find $u_i \in A_i^*$ and $q_i' \in Q_i$ such that $q_i \to_i^{u_i} q_i'$, $\mu_i^{-\tau}(u_i) \le x_i$, each i, and $q_i' \in div(H_i)$ for at least one i. By definition of $|||$ on sequences,

$$z_1 \le x_1 \wedge z_2 \le x_2 \wedge x \in x_1 ||| x_2 \Rightarrow \exists z \le x : z \in z_1 ||| z_2$$

and so there exists $z \le x$ with $z \in \mu_1^{-\tau}(u_1) ||| \mu_2^{-\tau}(u_2)$. By 28.4.8, there exists $u \in A^*$ such that $\mu^{-\tau}(u) = z$ and $u(i) = u_i$, each i. By 28.4.9 (1), $q_0 \to^u (q_1', q_2')$ Since $q_i' \in div(H_i)$ for at least one i, it follows that $(q_1', q_2') \in div(H)$, by 28.4.9 (2). Hence, $z \in D$ and since $z \le x \le y$, it follows that $y \in D$. We have shown that $D = D'$.

Next, suppose that $(x, X) \in F$. As usual, we only consider $x \notin D$, so that there exists $u \in A^*$ and $q \in stbl(H)$ such that $q_0 \to^u q$ with $\mu^{-\tau}(u) = x$ and $X \subseteq E^{-\tau} - nxt(H, q)$. Let $x_i = \mu_i^{-\tau}(u(i))$, then $q_i \to_i^{u(i)} q(i) \in stbl(H_i)$, by 28.4.9 (1) and (3). Furthermore, $x \in x_1 ||| x_2$, by 28.4.8. Referring to 28.2.9 (3), we see that in order to complete the proof that $(x, X) \in F'$ it suffices to show that $X \subseteq E_i^{-\tau} - nxt(H_i, q(i))$, each i. But it is clear from 28.3.3 (4) that $nxt(H, q) = nxt(H_1, q(1)) \cup nxt(H_2, q(2))$ and so from $X \subseteq E^{-\tau} - nxt(H, q)$ we may deduce that $X \subseteq E^{-\tau} - nxt(H_i, q(i))$, each i. Hence $(x, X) \in F'$.

Conversely, suppose that $(x, X) \in F'$. Again we may assume that $x \notin D'$. By 28.2.9 (3), there exists $x_i \in (E_i^{-\tau})^*$ such that $x \in x_1 ||| x_2$ and $(x_i, X_i) \in F_i$, each i.

We cannot have $x_i \in D_i$, since by 28.2.9 (2) it would follow that $x \in D'$, contradicting our assumption. Thus, for each i. there exists $q_i' \in stbl(H_i)$ and $u_i \in A_i^*$ such that $\mu_i^{-\tau}(u_i) = x_i$, $q_i \to_i^{u_i} q_i'$ and $X \subseteq E_i^{-\tau} - (H_i, q_i')$ By 28.4.8, there exists $u \in A^*$ such that $u(i) = u_i$ and $\mu^{-\tau}(u) = x$. Hence, by 28.4.9 (1) and (3), $q_0 \to^u (q_1', q_2') \in stbl(H)$ and so we may conclude that $(x, X) \in F$ if we can show that $X \subseteq E^{-\tau} - nxt(H', (q_1', q_2'))$. But, $X \subseteq E_i^{-\tau} - nxt(H_i, q_i')$, each i, and so

$$\begin{aligned} X &\subseteq (E^{-\tau} - nxt(H_1, q_1')) \cap (E^{-\tau} - nxt(H_2, q_2')) \\ &= E^{-\tau} - (nxt(H_1, q_1') \cup nxt(H_2, q_2')) = E^{-\tau} - nxt(H', (q_1', q_2')) \end{aligned}$$

and the proof is complete.

■

Next, we turn to the internal choice operator. Again, we need some notation. If $a_1, \cdots, a_n \in A_i$ and $v = a_1 \cdots a_n$, then define $(v,i) = (a_1,i) \cdots (a_n,i)$.

28.4.11. PROPOSITION. If (H_1,q_1) and (H_2,q_2) are asynchronous initialised hybrid transition systems such that $E_1^{-\tau} = E_2^{-\tau}$. Then

$$\gamma(H_1,q_1) \sqcap \gamma(H_2,q_2) = \gamma((H_1,q_1)+(H_2,q_2)).$$

PROOF. Let

$$\begin{aligned}
&(H,q_0) = (H_1,q_1) \sqcap (H_2,q_2)\\
&\gamma(H_1,q_1) = (E_1^{-\tau}, F_1, D_1)\\
&\gamma(H_2,q_2) = (E_2^{-\tau}, F_2, D_2)\\
&\gamma(H_1,q_1) \sqcap \gamma(H_2,q_2) = (E'^{-\tau}, F', D')\\
&\gamma(H,q_0) = (E^{-\tau}, F, D).
\end{aligned}$$

We show that $(E^{-\tau},F,D) = (E'^{-\tau},F',D')$. We first observe that by 26.3.4 (3) and 28.2.13 (1), $E^{-\tau} = E_1^{-\tau} \cup E_2^{-\tau} = E'^{-\tau}$.

By 28.3.10, if $u \in A^*$ and $u > \Omega$ then

$$q_0 \rightarrow^u q \Leftrightarrow \exists i \in \{1,2\}\, \exists v \in A_i^*\, \exists q_i' \in Q_i : u = (v,i) \wedge q_i \rightarrow_i^v q \wedge q = (q_i',i)$$

and if $q' \neq q_0$, then

$$\begin{aligned}
q' \rightarrow^u q'' \Leftrightarrow \exists i \in \{1,2\}\, \exists v \in A_i^*\, \exists q_i', q_i'' \in Q_i :&\\
u = (v,i) \wedge q_i' \rightarrow_i^v q_i'' \wedge q' = (q_i',i) \;\; q'' = (q_i'',i)&
\end{aligned}$$

and if $v \in L(T(C(H_i)),q_i)$ and $q' \in Q_i$ then

$$\mu^{-\tau}(v,i) = \mu_i^{-\tau}(v) \text{ and } nxt(H,(q',i)) = nxt(H_i,q')$$

from which we may easily deduce that

$$stbl(H) = (stbl(H_1) \times \{1\}) \cup (stbl(H_2) \times \{2\}) \tag{28.21}$$

if $q_1 \notin div(H_1) \wedge q_2 \notin div(H_2)$, then

$$div(H) = (div(H_1) \times \{1\}) \cup (div(H_2) \times \{2\}) \tag{28.22}$$

and if $q_1 \in div(H_1) \vee q_2 \in div(H_2)$, then

$$div(H) = (div(H_1) \times \{1\}) \cup (div(H_2) \times \{2\}) \cup \{q_0\} \tag{28.23}$$

Given (28.23), if either $q_1 \in div(H_1)$ or $q_2 \in div(H_2)$ then

$$\gamma(H, q_0) = \gamma(H_1, q_1) \sqcap \gamma(H_2, q_2) = \sigma[\![Chaos_{E^{-\tau}}]\!] = \gamma((H_1, q_1) + (H_2, q_2))$$

Assume, therefore, that $q_1 \notin div(H_1)$ and $q_2 \notin div(H_2)$, then

$$\begin{aligned} x \in D &\Leftrightarrow \exists u \in A^* \, \exists q \in div(H): q_0 \rightarrow^u q \wedge \mu^{-\tau}(u) \leq x \\ &\Leftrightarrow \exists i \in \{1,2\} \, \exists v \in A_i^* \, \exists q \in div(H_i): q_i \rightarrow_i^u q \wedge \mu_i^{-\tau}(v) \leq x \\ &\Leftrightarrow \exists i \in \{1,2\}: x \in D_i \end{aligned}$$

and therefore $D = D_1 \cup D_2 = D'$, (by 28.2.13(2)). We also have for $x \notin D$,

$$\begin{aligned} (x, X) \in F &\Leftrightarrow \exists u \in A^* \, \exists q \in stbl(H): \\ &\qquad q_0 \rightarrow^u q \wedge \mu^{-\tau}(u) = x \wedge X \subseteq E^{-\tau} - nxt(H, q) \\ &\Leftrightarrow \exists i \in \{1,2\} \, \exists u \in A_i^* \, \exists q \in Q_i: \\ &\qquad q_i \rightarrow_i^u q \wedge \mu_i^{-\tau}(v) = x \wedge X \subseteq E_i^{-\tau} - nxt(H_i, q) \\ &\Leftrightarrow \exists i \in \{1,2\}: (x, X) \in F_i \\ &\Leftrightarrow (x, X) \in F_1 \cup F_2 \end{aligned}$$

By 28.2.13 (3), $F = F_1 \cup F_2 = F'$.

■

Next, we turn to the hiding operator.

28.4.12. PROPOSITION. Let (H, q) is an initialised, asynchronous hybrid transition system and let $C \subset E^{-\tau}$. Then

$$\gamma(H, q) \backslash\backslash C = \gamma((H, q) \backslash C).$$

PROOF. Let $(H', q') = (H, q) \backslash C$, $\gamma(H, q) = (E^{-\tau}, F, D)$, $\gamma(H', q') = (E'^{-\tau}, F', D')$ and $\gamma(H, q) \backslash\backslash C = (E''^{-\tau}, F'', D'')$. We show that $(E'^{-\tau}, F', D') = (E''^{-\tau}, F'', D'')$.

First, by 28.2.15 (1) and 28.3.2, $E'^{-\tau} = E^{-\tau} - C = E''^{-\tau}$.

Next, by 28.3.2,

$$div(H') = div(H) \cup \{q \in Q \mid \forall n \in \omega \, \exists u \in C^n : q \rightarrow^u\}$$

Hence:

$$x \in D'' \Leftrightarrow \exists y \in (E^{-\tau})^*: \pi_{E''}^{-\tau}(y) \leq x \wedge (y \in D \vee \forall n \in \omega \, \exists u \in C^n : y.u \in dom(F))$$

$$\Leftrightarrow \exists u \in A^* \exists q_1 \in Q: q \rightarrow^u q_1 \wedge \mu'^{-\tau}(u) \leq x$$
$$\wedge (q_1 \in div(H) \vee \forall n \in \omega\, \exists v \in C^n : q_1 \rightarrow^v)$$
$$\Leftrightarrow \exists u \in A^* \exists q_1 \in Q: q \rightarrow^u q_1 \wedge q_1 \in div(H') \wedge \mu'^{-\tau}(u) \leq x \;\Leftrightarrow x \in D'$$

so $D' = D''$.

Next, by 28.3.2,

$$stbl(H') = \{\tilde{q} \in stbl(H) \mid \forall a \in A: \tilde{q} \rightarrow^a \Rightarrow \mu^{-\tau}(a) \notin C\}$$

It follows from this that if $\tilde{q} \in stbl(H')$, then $nxt(H',\tilde{q}) = nxt(H,\tilde{q})$ and $X \subseteq E'^{-\tau} - nxt(H',\tilde{q}) \Leftrightarrow X \cup C \subseteq E^{-\tau} - nxt(H,\tilde{q})$. Hence if $x \notin D' = D''$ then

$$(x,X) \in F'' \Leftrightarrow \exists y \in (E^{-\tau})^*: \pi_{E''}{}^{-\tau}(y) = x \wedge (y, X \cup C) \in F$$
$$\Leftrightarrow \exists u \in A^* \exists \tilde{q} \in stbl(H): q \rightarrow^u \tilde{q} \wedge \pi_{E''}{}^{-\tau}(\mu^{-\tau}(u)) = x$$
$$\wedge X \cup C \subseteq E^{-\tau} - nxt(H,\tilde{q})$$
$$\Leftrightarrow \exists u \in A^* \exists \tilde{q} \in stbl(H'): q \rightarrow^u \tilde{q} \wedge \mu'^{-\tau}(u) = x \wedge X \cup C \subseteq E^{-\tau} - nxt(H,\tilde{q})$$
$$\Leftrightarrow (x,X) \in F'.$$

so $F = F'$ and the proof is complete.

■

Finally, we deal with the two constant processes.

28.4.13. LEMMA. If $\varnothing \subset E \subseteq \mathcal{Act}$ then

1• $\gamma(\rho[\![Stop_E]\!]) = \sigma[\![Stop_E]\!]$;
2• $\gamma(\rho[\![Chaos_E]\!]) = \sigma[\![Chaos_E]\!]$

PROOF. (1) Let $(H',q') = \rho[\![Stop_E]\!]$ and $(E',F',D') = \gamma(H',q')$. By 28.3.4 (1), $E' = E$, $Q = stbl(H') = \{0\}$ and $q' = 0$. Thus, $div(H') = \varnothing$ and so by 28.4.1 (1), $D' = \varnothing$. Also by 28.3.4 (1), $q' \rightarrow'^u q'' \Leftrightarrow u = \Omega \wedge q'' = q'$ and therefore by 28.4.1 (2), $F' = \{(\Omega, X) \mid X \subseteq E - nxt(H',q')\}$. Hence, by 28.2.4 (1)

$$\gamma(\rho[\![Stop_E]\!]) = (E, \{\Omega\} \times \wp(E), \varnothing) = \sigma[\![Stop_E]\!].$$

(2) Let $(H',q') = \rho[\![Chaos_E]\!]$ and $(E',F',D') = \gamma(H',q')$. By 28.3.4 (1), $E' = E$, $Q = div(H') = \{0\}$ and $q' = 0$. By 28.4.1 (1), $D' = \wp(E)$. Hence, by 28.2.4 (1),

$$\gamma(\rho[\![Chaos_E]\!]) = (E, E^* \times \wp(E), E^*) = \sigma[\![Chaos_E]\!].$$

■

28.4.14. THEOREM. Suppose that P is a FCSP program, then such that $\sigma[\![P]\!]$ is defined , then $\rho[\![P]\!]$ is an initialised, asynchronous hybrid transition system and $\gamma(\rho[\![P]\!]) = \sigma[\![P]\!]$.

PROOF. The proof is by structural induction on P. using 28.4.4, 28.4.7, 28.4.10, 28.4.11, 28.4.12 and 28.4.13.

■

28.5. Extending the Hybrid Transition System Semantics.

In this section, we indicate how the hybrid transition system semantics may be extended to deal with the other choice operator and recursion.

28.5.1. DEFINITION. If one of the following are the case

1• If $q_1 \in div(H_1)$ or $q_2 \in div(H_2)$;
2• If $q_1 \in stbl(H_1)$ and $q_2 \in stbl(H_2)$

then we define

$$(H_1, q_1) [\!] (H_2, q_2) = (H_1, q_1) + (H_2, q_2)$$

The other case is more complicated. Recall that if $q \in Q$ then

$$|q\rangle = \{(a, q') \in A \times Q \mid q \to^a q'\}$$

28.5.2. DEFINITION. If neither (1) nor (2) of 28.5.1 hold, then we define

$$(H_1, q_1) [\!] (H_2, q_2) = \sum_{(a,q) \in |q_1\rangle_1} (\mu_1(a) \to H_1/q) + \sum_{(b,q') \in |q_2\rangle_2} (\mu_2(b) \to H_2/q')$$

where

1• H_1/q is (H_1, q) if $\mu_1(a) \neq \tau$ and $(H_1, q) [\!] (H_1, q_2)$, otherwise;
2• H_2/q' is (H_2, q') if $\mu_2(b) \neq \tau$ and $(H_1, q_1) [\!] (H_2, q')$, otherwise.

In actually constructing $(H_1, q_1) [\!] (H_2, q_2)$, we may omit those parts of the transition system not reachable from the initial state, as this obviously gives rise to the same behavioural presentation. With this proviso, the construction is illustrated in figure 28.1 below.

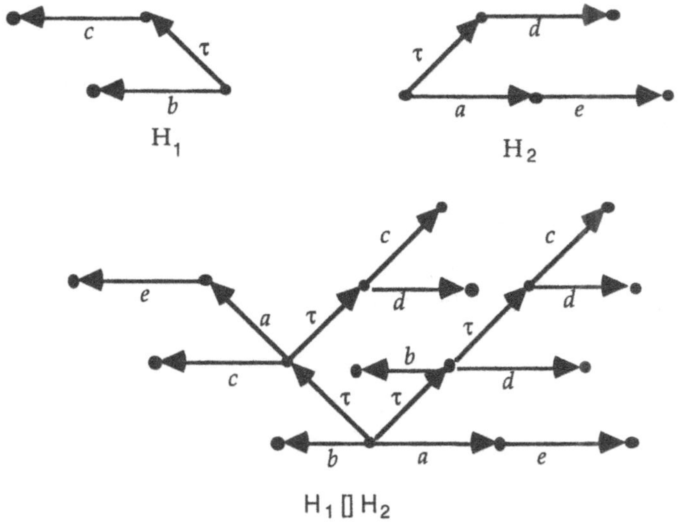

Figure 28.1.

28.5.3. LEMMA. If (H_1,q_1) and (H_2,q_2) are asynchronous initialised hybrid transition systems, then $(H_1,q_1)\square(H_2,q_2)$ is an asynchronous initialised hybrid transition system.

PROOF. By structural induction, 26.3.2 and 26.3.5.

■

Finally, we come to recursion. This will be handled much in the same way as in chapter 26, except that we need to modify the ordering on hybrid transition systems to conform to that defined for processes. The modification is to cope with divergences.

28.5.4. DEFINITION. Let (H,q) be an initialised hybrid transition system. We define $stbl(H,q)=(\tilde{Q},A,\tilde{\rightarrow},\iota,E,\mu)$ where $\tilde{\rightarrow}=\rightarrow\cap((\tilde{Q}-div(H))\times A\times\tilde{Q})$ and $\tilde{Q}$ is the set of all $\tilde{q}\in Q$ such that

- •1 There exists $x\in A^*$ such that $q\rightarrow^x\tilde{q}$;
- •2 For all $u<x$, $q\rightarrow^u q'\Rightarrow q'\notin div(H)$.

In what follows, we shall write $(\tilde{H},q)$ for $(stbl(H,q),q)$ with the usual convention regarding subscripts.

Informally, $stbl(H,q)$ is obtained from that part of H reachable from the initial state by pruning away all arcs which lead from divergent states and consequently, those parts of H which are no longer reachable from the initial state.

Let $\tilde{q} \in \tilde{Q}$ and let $q \rightarrow^{x} \tilde{q}$ such that for all $u < x$, $q \rightarrow^{u} q' \Rightarrow q' \notin div(H)$. We shall say that $\tilde{q} \in \tilde{Q}$ is a fringe state if there exists $a \in A$ such that $\tilde{q} \rightarrow^{a}$ but $\tilde{q} \tilde{\nrightarrow}^{a}$. In such a case $\tilde{q} \rightarrow^{a} q'$, some $q' \in Q$. If $q' \in \tilde{Q}$, then $(\tilde{q}, a, q') \in (\tilde{Q} \times A \times \tilde{Q}) - \tilde{\rightarrow}$, so $\tilde{q} \in div(H)$. If $q' \notin \tilde{Q}$, then $q \rightarrow^{x.a} q'$ and so there exists $u < x.a$ such that $q \rightarrow^{u} q''$ and $q'' \in div(H)$. We cannot have $u < x$, and so $u = x$ and by unambiguity, $\tilde{q} = q''$. Hence, all fringe states are divergent states.

28.5.5. DEFINITION. If (H_1, q_1) and (H_2, q_2) are initialised hybrid transition systems then we define $(H_1, q_1) \leq_H (H_2, q_2) \Leftrightarrow$

1• $E_2^{-\tau} \subseteq E_1^{-\tau}$;

2• $(\tilde{H}_1, q_1) \preceq (\tilde{H}_2, q_2)$;

3• $div(H_2) \cap \tilde{Q}_1 \subseteq div(H_1)$;

4• If $\tilde{q} \in \tilde{Q}_1$, $\tilde{q} \tilde{\rightarrow}_2^{a}$ and $\tilde{q} \tilde{\nrightarrow}_1^{a}$, then $\tilde{q} \in div(H_1)$.

28.5.6. LEMMA. $\leq_H$ is a pre-order.

PROOF. $\leq_H$ is clearly reflexive. We prove that $\leq_H$ is transitive. Suppose that $(H_1, q_1) \leq_H (H_2, q_2)$ and $(H_2, q_2) \leq_H (H_3, q_3)$. Then $E_3^{-\tau} \subseteq E_2^{-\tau} \subseteq E_1^{-\tau}$, giving 28.5.5 (1), and $(\tilde{H}_1, q_1) \preceq (\tilde{H}_2, q_2) \preceq (\tilde{H}_3, q_3)$, giving 28.5.5 (2), in view of the transitivity of $\preceq$. Next since $(H_1, q_1) \leq_H (H_2, q_2)$, $\tilde{Q}_1 \subseteq \tilde{Q}_2$ and so

$$div(H_3) \cap \tilde{Q}_1 = (div(H_3) \cap \tilde{Q}_2) \cap \tilde{Q}_1$$
$$\subseteq div(H_2) \cap \tilde{Q}_1 = div(H_1).$$

Finally, suppose that $\tilde{q} \in \tilde{Q}_1$, $\tilde{q} \tilde{\rightarrow}_3^{a}$ and $\tilde{q} \tilde{\nrightarrow}_1^{a}$. If $\tilde{q} \tilde{\rightarrow}_2^{a}$, then $\tilde{q} \in div(H_1)$, because $(H_1, q_1) \leq_H (H_2, q_2)$. Otherwise, $\tilde{q} \tilde{\nrightarrow}_2^{a}$, and so $\tilde{q} \in div(H_2)$, because $(H_2, q_2) \leq_H (H_3, q_3)$, and hence $\tilde{q} \in div(H_2) \cap \tilde{Q}_1 \subseteq div(H_1)$. Therefore, (4) holds.

■

The next lemma shows that this ordering is consistent with that on processes.

28.5.7. LEMMA. If (H_1,q_1) and (H_2,q_2) are initialised hybrid transition systems then $(H_1,q_1)\leq_H(H_2,q_2)\Rightarrow\gamma(H_1,q_1)\leq\gamma(H_2,q_2)$.

PROOF. For each i, let $\gamma(H_i,q_i)=(E_i^{-\tau},F_i,D_i)$.

Let $(H_1,q_1)\leq_H(H_2,q_2)$, then $(\tilde{H}_1,q_1)\preceq(\tilde{H}_2,q_2)$ and so $q_1=q_2=q$

Suppose that $x\in D_2$, then there exists $u\in A_2{}^*$ and $q'\in div(H_2)$ such that $q\rightarrow_2{}^u q'$ and $\mu_2{}^{-\tau}(u)\leq x$. Let u be minimal in the sense that if $v<u$ and $q\rightarrow_2{}^v q''$, then $q''\notin div(H_2)$. Then $q'\in\tilde{Q}_2$. Let $v\leq u$ be maximal with respect to $q\tilde{\rightarrow}_1{}^v q''$, some $q''\in\tilde{Q}_1$. If $v=u$, then $q\tilde{\rightarrow}_2{}^u q''$, as $(\tilde{H}_1,q_1)\preceq(\tilde{H}_2,q_2)$ and so $q'=q''$ by unambiguity. Thus, $q''\in div(H_2)\cap\tilde{Q}_1\subseteq div(H_1)$, by 28.5.5 (3). Otherwise, $q''\tilde{\rightarrow}_2{}^a$ but $q''\tilde{\nrightarrow}_1{}^a$, where $v.a\leq u$, and so $q''\in div(H_1)$, by 28.5.5 (4). We have shown that $q\tilde{\rightarrow}_1{}^v q''$ with $q''\in div(H_1)$ and

$$\mu_1{}^{-\tau}(v)=\mu_2{}^{-\tau}(v)\leq\mu_2{}^{-\tau}(u)\leq x$$

so $x\in D_1$ and we have shown that $D_2\subseteq D_1$.

Next, let $(x,X)\in F_2$. If $x\in D_1$, then as $X\subseteq E_2{}^{-\tau}\subseteq E_1{}^{-\tau}$, $(x,X)\in F_1$, so suppose that $x\notin D_1$. Then $x\notin D_2$, so there exists $u\in A_2{}^*$ and $q'\in stbl(H_2)$ such that $q\rightarrow_2{}^u q'$, $\mu_2{}^{-\tau}(u)=x$ and $X\subseteq E_2{}^{-\tau}-nxt(H_2,q')$. If $q'\notin\tilde{Q}_2$, then for some $v<u$, $q\rightarrow_2{}^v q''$ with $q''\in div(H_2)$. But then $\mu_2{}^{-\tau}(v)\leq\mu_2{}^{-\tau}(u)\leq x$, and so $x\in D_2$, a contradiction. So, $q'\in\tilde{Q}_2$. Let $v\leq u$ be maximal with respect to $q\tilde{\rightarrow}_1{}^v q''$, some $q''\in\tilde{Q}_1$. If $v<u$, then as before $q''\in div(H_1)$ and we have the contradiction $x\in D_1$, so $q\tilde{\rightarrow}_1{}^u q''$, and consequently $q\tilde{\rightarrow}_2{}^u q'$ as $(\tilde{H}_1,q_1)\preceq(\tilde{H}_2,q_2)$. By unambiguity, $q'=q''$. Since $(\tilde{H}_1,q_1)\preceq(\tilde{H}_2,q_2)$, $\mu_1{}^{-\tau}(u)=\mu_2{}^{-\tau}(u)=x$ and since $q'\in stbl(H_2)$, $nxt(H_1,q')=nxt(H_2,q')$ and since $q'\in stbl(H_1)$, by 28.5.5 (4), $X\subseteq E_2{}^{-\tau}-nxt(H_2,q')\subseteq E_1{}^{-\tau}-nxt(H_1,q')$. Hence, $(x,X)\in F_1$.448 449

■

Since $(H_1,q_1)\leq_H(H_2,q_2)$ implies $(\tilde{H}_1,q_1)\preceq(\tilde{H}_2,q_2)$, we may apply the theory of section 26.2. In particular if f: **Env** $\rightarrow$ **AHTS*** is H-admissible, in the sense that

1• $(\forall x\in X:\eta_1(x)\leq_H\eta_2(x))\Rightarrow f(\eta_1)\leq_H f(\eta_2)$

2• there exists $init_H(f)$ such that $init_H(f(\eta))=init_H(f)$,

then we may define $fix(\tilde{x}=f)(\eta)=\bigcup_{n\in\omega}stbl(\eta_n(x))$, where

$$\eta_0(x) = \begin{cases} \bot_{init_H(f)} \text{ if } x = \tilde{x} \\ \eta(x) \text{ otherwise} \end{cases}$$

$$\eta_{n+1}(x) = \begin{cases} f(\eta_n) \text{ if } x = \tilde{x} \\ \eta(x) \text{ otherwise} \end{cases}$$

28.5.8 DEFINITION. We define a function $\rho: \mathbf{TCSP} \to [\mathbf{Env} \to \mathbf{AHTS}^*]$ as follows. Suppose that $\eta \in \mathbf{Env}$, then

1• $\rho[\![Stop_A]\!](\eta) = ((\{0\}, \varnothing, \varnothing, \varnothing, A, \varnothing), 0);$
2• $\rho[\![Chaos_A]\!](\eta) = ((\{0\}, \{0\}, \{(0,0,0)\}, \varnothing, A \cup \{\tau\}, \{(0,\tau)\}, 0);$
3• $\rho[\![a \to P]\!](\eta) = a \to \rho[\![P]\!](\eta);$
4• $\rho[\![P_1 \,||\, P_2]\!](\eta) = \rho[\![P_1]\!](\eta) \,||^{\tau}\, \rho[\![P_2]\!](\eta);$
5• $\rho[\![P_1 \,|||\, P_2]\!](\eta) = \rho[\![P_1]\!](\eta) \,|||\, \rho[\![P_2]\!](\eta)\ ;$
6• $\rho[\![P_1 \,\square\, P_2]\!](\eta) = \rho[\![P_1]\!](\eta) \,\square\, \rho[\![P_2]\!](\eta);$
7• $\rho[\![P_1 \sqcap P_2]\!](\eta) = \rho[\![P_1]\!](\eta) + \rho[\![P_2]\!](\eta);$
8• $\rho[\![P \backslash C]\!](\eta) = \rho[\![P]\!](\eta) \backslash\backslash C;$
9• $\rho[\![\mu x{:}A.\,P]\!](\eta) = \begin{cases} fix(x = \rho[\![P]\!])(\eta) \text{ if } \rho[\![P]\!] \text{ is H-admissible} \\ \rho[\![Chaos_A]\!](\eta) \text{ otherwise} \end{cases}$

28.5.9. PROPOSITION. If $P \in \mathbf{TCSP}$ and $\eta \in \mathbf{Env}$, then $\rho[\![P]\!](\eta) \in \mathbf{AHTS}^*$. Furthermore, if $P \in \mathbf{Prog}$, then $\rho[\![P]\!]$ is a constant function.

PROOF. $\rho[\![P]\!](\eta) \in \mathbf{AHTS}^*$ by structural induction on P. The argument that if $P \in \mathbf{Prog}$, then $\rho[\![P]\!]$ is a constant function is entirely analogous to that in section 27.3.

■

Notes.

There is one important difference between the functions ρ of this and the preceding chapter. The latter is only a partial function; the former is total. The totality of the TCSP ρ is possible, however, only because the semantic domain of TCSP contains the (almost) totally undefined processes $\rho[\![Chaos_A]\!](\eta)$ (almost, because each such process has at least an 'alphabet').

That the CCS semantic function is partial is due to the fact that some expressions are not faintly guarded, and this in its turn follows whenever a recursion sub-expression is not completely defined. It is not hard to see (another structural induction argument here) that in the case of TCSP, when a recursion sub-expression is not completely defined, then the resulting process will have a non-empty set of divergences, and (yet another structural induction argument here) that a recursion subexpression is not completely defined when it is in some sense not faintly guarded. We thus have:

- In CCS, Ex faintly guarded means that $\rho[\![Ex]\!](\eta)$ is defined;
- In TCSP, P faintly guarded means that $\rho[\![P]\!](\eta)$ is divergence-free.

Chapter 29
Conclusions: Theories of Parallelism

What, finally, is the practical application of all this?

KINGSLEY AMIS
Lucky Jim

In the preface, we spoke of theories of parallelism; such theories consist of a syntactic element, generating a class of objects, together with a semantic element, a means of associating a meaning with each object. We have now met a variety of these.

Our survey is, naturally, incomplete. However, our aim has not been to compile an encyclopaedia of current theories of parallelism, rather to illustrate the variety among notations and semantic models that have been proposed over the past twenty years or so.

Taking stock, there seems to be three broad classes under consideration.

- Notations: This is the syntactic part of a theory. For example, there are the graphical/set theoretic notations of computation graphs, Petri nets and Keller parallel computation graphs and the textual notations of path/process programs, Process Calculus and CSP.

- Automata: This is part of the semantics of a theory. In some cases such as Keller parallel computation graphs and Process Calculus, automata lie in the theory's semantic domain. In all but one case,[1] we have used our own non-sequential automata to give an alternative semantics to the theory, thereby putting them all into a single context.

- Behaviour: In some cases, the semantics of the theory has been couched in terms of objects representing behaviour. For example, occurrence nets in the case of C/E systems, sets of sequences of non-empty sets of events in the case of computation graphs and vectors in the case of Nivat/Arnold

[1]That of Karp and Miller Computation Graphs.

processes. We have also considered models such as behavioural presentations, event structures and vector and trace languages.

The relationships between these may be expressed pictorially as follows.

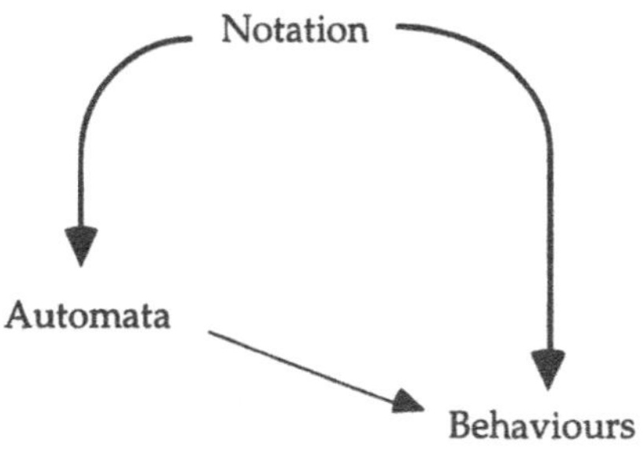

Figure 29.1.

Our concern is with the two latter classes. As for the automata, we have thoroughly investigated the relationships between them in section 22.3. As far as behavioural models are concerned, we have the situation pictured in figure 29.2.

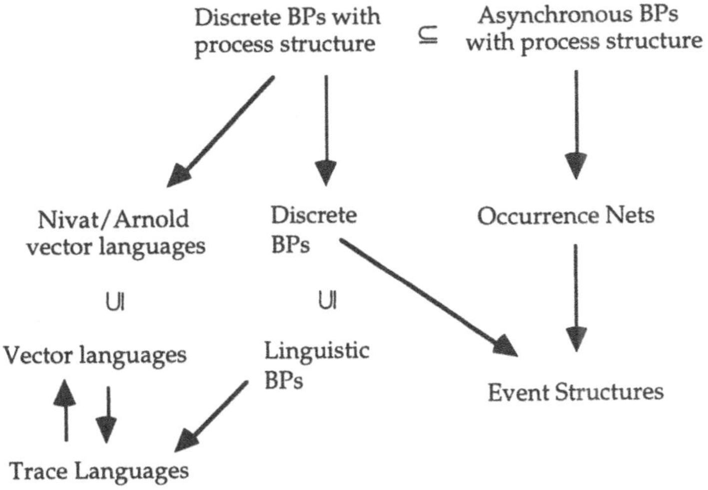

Figure 29.2

As far as the theories considered in this book are concerned, the model consisting of discrete behavioural presentations with process structure serves as a semantic domain for the lot. Thus, we are led to consider a model based on tuples (O,P,E,D,λ,δ), where (O,P,E,λ) is a (discrete) behavioural presentation, D is a set of participants or agents and $\delta{:}O \to \wp(D)$ with the interpretation that if $\pi \in \delta(o)$, then π participates in the occurrence o.

Actually, such a structure is really only a (discrete) behavioural presentation in disguise, for it may be represented by a structure $(O,P,\tilde{E},\tilde{\lambda})$ where $\tilde{E}=E\times\wp(D)$ and $\tilde{\lambda}:O\rightarrow\tilde{E}$ is given by $\tilde{\lambda}(o)=(\lambda(o),\delta(o))$. One point of making this identification is that the theory of discrete behavioural presentations becomes immediately applicable to these structures.

Indeed, there is nothing to stop the elements of the sets O and E being as complex as one wishes. For example, if one wishes to represent the fact that occurrences have duration, then one may construct O of triples (x,t_1,t_2) where $t_1,t_2\in\mathbb{R}$ and $t_1<t_2$, representing the fact that this occurrence began at local time t_1 and ended at local time t_2. Imposing on this conditions such as

$$(x,t_1,t_2)<(y,s_1,s_2)\Rightarrow x\neq y\wedge t_2\leq s_1$$

delivers something like Murphy's interval event structure model [77]. It is a mistake to believe that models such as behavioural presentations or event structure can only deal with 'atomic actions'.

Our investigations suggest the possibility of a unification at the semantic level, as pictured in figure 29.3.

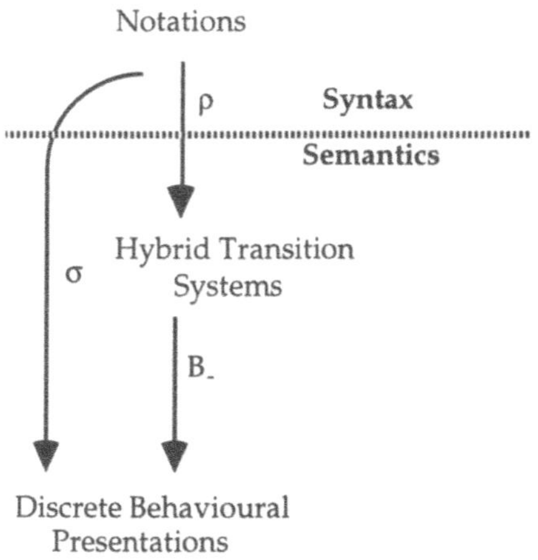

Figure 29.3

An immediate consequence of this unification has been the ability to compare the expressive power of notations, that is, to deduce relationships between structures at the syntactic level. For example, we have seen that the class of loosely coupled systems embeds into the class of C/E systems and the class of path/process programs embeds into the class of CSP expressions. The results of chapter 21 may also be used to show that every finite C/E system may be described by a path program in which no path contains any repeated event names.

Another possible advantage of a unification at the semantic level is that attention is thereby concentrated on one brand of mathematical theory. In this case, the mathematics is based on vector languages and it is this author's (hardly impartial!) view that this is the 'right kind' of mathematics, and that based on interleaving representations of behaviour is the 'wrong kind' of mathematics. For example, a vector representation of certain kinds of determinate system has a characteristically simple structure and this structure is lost when one interleaves the behaviour.

The point of introducing these theories, however, has been to support the design of concurrency systems; the semantics should allow one to define behavioural properties and to provide the basis for techniques for deducing the existence of such properties from the structure of the syntactic entity.

We have seen that behavioural properties may be described in terms of first order logic and behavioural presentations (which is why we have not needed to discuss temporal logic). We have also seen that in the cases we have considered, it is not hard to deduce some property of $B_{(H,q)}$ from a corresponding property of (H,q). The hard part is deducing properties of a hybrid transition system $\rho[\![Spec]\!]$ from a specification $Spec$?

The specific nature of this problem depends on the notation and its particular semantic function. Perhaps, given a fixed semantic domain, we might try designing a notation with it particularly in mind - first decide what it is that you're trying to reason about and then design a language in which you can talk and reason about it conveniently.

Appendix

Review of Mathematical Prerequisites

For completeness, we give here definitions and elementary properties of mathematical objects of which we make frequent use. We assume a knowledge of basic Set Theory and Logic.

A.1. Semigroups.

A *semigroup* is a pair $(S,*)$, where S is a set and $*$ is an associative binary operation, that is a function $*: S\times S \to S$ such that, if we write $s_1 * s_2$ for $*(s_1, s_2)$, then

$$\forall s_1, s_2, s_3 \in S: s_1 * (s_2 * s_3) = (s_1 * s_2) * s_3 \tag{A.1}$$

If $(S,*)$ is a semigroup then a *subsemigroup* of $(S,*)$ is set $X \subseteq S$ satisfying

$$\forall x_1, x_2 \in X \Rightarrow x_1 * x_2 \in X$$

If $(S,*)$ is a semigroup and $X \subseteq S$, then the subsemigroup of $(S,*)$ generated by X is the intersection of all subsemigroups of $(S,*)$ which contain X.

If $(S,*)$ is a semigroup then an *identity* for $(S,*)$ is an element $e \in S$ satisfying

$$\forall s \in S: e * s = s = s * e \tag{A.2}$$

It is easy to show, using (A.1), that if $(S,*)$ has an identity, then that identity is unique.

A *monoid* is a semigroup with identity.

If $(M,*)$ is a monoid then a *submonoid* of $(M,*)$ is subsemigroup of $(M,*)$ containing the identity.

If $(M,*)$ is a semigroup and $X \subseteq M$, then the submonoid of $(M,*)$ generated by X is the intersection of all submonoids of $(M,*)$ which contain X.

A semigroup $(S,*)$ is *left-cancellable* if and only if

$$\forall s_1, s_2, s \in S: s * s_1 = s * s_2 \Rightarrow s_1 = s_2 \tag{A.3}$$

and is *right-cancellable* if and only if

$$\forall s_1, s_2, s \in S: s_1 * s = s_2 * s \Rightarrow s_1 = s_2 \tag{A.4}$$

If $(S,*)$ and $(T,\bullet)$ are semigroups, then a *homomorphism* from $(S,*)$ to $(T,\bullet)$ is a function $\phi: S \to T$ satisfying

$$\forall s_1, s_2 \in S: \phi(s_1 * s_2) = \phi(s_1) \bullet \phi(s_2) \tag{A.5}$$

An onto homomorphism is called an *epimorphism*, a 1-1 homomorphism is called a *monomorphism* and a bijective homomorphism is called an *isomorphism*. If there is an isomorphism from $(S,*)$ to $(T,\bullet)$, then $(S,*)$ and $(T,\bullet)$ are said to be *isomorphic*.

If $(S,*)$ is a semigroup, then a *congruence* on $(S,*)$ is an equivalence relation $\equiv$ on S satisfying

$$\forall s_1, s_2, s \in S: s_1 \equiv s_2 \Rightarrow s * s_1 \equiv s * s_2 \wedge s_1 * s \equiv s_2 * s \tag{A.6}$$

If $\equiv$ is a congruence on $(S,*)$, then we shall write $s/\equiv$ for the equivalence class of $s \in S$ and $S/\equiv$ for the set of all such equivalence classes. As a consequence of (A.6), the following unambiguously defines a binary relation on $S/\equiv$ which turns it into a semigroup.

$$\forall s_1, s_2 \in S: (s_1/\equiv) * (s_2/\equiv) = (s_1 * s_2)/\equiv \tag{A.7}$$

If $(S,*)$ has an identity e, then $e/\equiv$ is an identity for $(S/\equiv,*)$.

If $(S,*)$ and $(T,\bullet)$ are semigroups and $\phi: S \to T$ is a homomorphism from $(S,*)$ to $(T,\bullet)$, then the relation

$$ker(\phi) = \{(s_1, s_2) \in S_1 \times S_2 \mid \phi(s_1) = \phi(s_2)\} \tag{A.8}$$

called the *kernel* of ϕ, may be shown to be a congruence on $(S,*)$, and it may be shown that there is a unique monomorphism $\tilde{\phi}$ from $(S/Ker(\phi),*)$ to $(T,\bullet)$ which satisfies $\tilde{\phi}(s/Ker(\phi)) = \phi(s)$, all $s \in S$. If ϕ, is an epimorphism, then $\tilde{\phi}$ is an isomorphism.

If $\equiv$ is a congruence on $(S,*)$, then there is an epimorphism $\phi_{\equiv}$ from $(S,*)$ to $(S/\equiv,*)$ given by $\phi_{\equiv}(s) = s/\approx$, and $Ker(\phi_{\equiv}) = \equiv$.

A.2 Ordered Sets.

A *pre-ordered set* is a pair $(X,\rightarrow)$, where X is a set and $\rightarrow$ is a pre-order, that is, a reflexive, transitive relation, that is

$$\forall x \in X: x \rightarrow x \tag{A.9}$$
$$\forall x,y,z \in X: x \rightarrow y \wedge y \rightarrow z \Rightarrow x \rightarrow z \tag{A.10}$$

If in addition, $\rightarrow$ is antisymmetric, that is

$$\forall x,y \in X: x \rightarrow y \wedge y \rightarrow x \Rightarrow x = y \tag{A.11}$$

then it is a *partial order* and $(X,\rightarrow)$ is a *partially ordered set* (sometimes abbreviated to *poset*). We usually use $\leq$ for the order relation of a poset. If we also have

$$\forall x,y \in X: x \leq y \vee y \leq x \tag{A.12}$$

then $\leq$ is a total order and $(X,\leq)$ is a *totally ordered set*.

If $(X,\rightarrow)$ is a pre-ordered set, then the relation $\approx$ defined by

$$o_1 \approx o_2 \Leftrightarrow o_1 \rightarrow o_2 \wedge o_2 \rightarrow o_1 \tag{A.13}$$

is an equivalence relation and that there is a partial order $\leq$ on the set $X/\approx$ of $\approx$-equivalence classes of X satisfying

$$U_1 \leq U_2 \Leftrightarrow \forall o_1 \in U_1\ \forall o_2 \in U_2 : o_1 \rightarrow o_2 \tag{A.14}$$

If $(X,\leq)$ is a poset then the *least upper bound* of a set $Y \subseteq X$, denoted by $\bigsqcup Y$ is the (necessarily unique) element $y \in X$ such that

$$\forall x \in X: y \leq x \Leftrightarrow Y \leq x \tag{A.15}$$

and the *greatest lower bound* of a Y, denoted by $\bigsqcap Y$, is the (necessarily unique) element $y \in X$ such that

$$\forall x \in X: x \leq y \Leftrightarrow x \leq Y \tag{A.16}$$

where

$$Y \le x \Leftrightarrow (\forall y \in Y : y \le x) \tag{A.17}$$
$$x \le Y \Leftrightarrow (\forall y \in Y : x \le y) \tag{A.18}$$

If $x, y \in X$, then we write $x \sqcup y$ for $\sqcup\{x,y\}$ and $x \sqcap y$ for $\sqcap\{x,y\}$. If $x \sqcup y$ and $x \sqcap y$ are defined for each $x, y \in X$, then $(X, \le)$ is a *lattice*. If $\sqcap Y$, and $\sqcup Y$, are defined for every $Y \subseteq X$, then $(X, \le)$ is a *complete lattice.*

It may be shown that $(X, \le)$ is a lattice if and only if $(X, \sqcap)$ and $(X, \sqcup)$ are both semigroups in which $\sqcap$ distributes over $\sqcup$ and vice versa.

If $(X, \le)$ and $(Y, \le')$ are partially ordered sets, then a *morphism* from $(X, \le)$ to $(Y, \le')$ is a function $f : X \to Y$ satisfying

$$\forall x_1, x_2 \in X : x_1 \le x_2 \Rightarrow f(x_1) \le f(x_2) \tag{A.19}$$

An *isomorphism* from $(X, \le)$ to $(Y, \le')$ is a bijective morphism whose inverse is a morphism from $(Y, \le')$ to $(X, \le)$. If there exists an isomorphism from $(X, \le)$ to $(Y, \le')$, then we say that $(X, \le)$ and $(Y, \le')$ are *isomorphic.*

We make the general observation that if $(X, \le)$ and $(Y, \le')$ are isomorphic, then they have identical order theoretic properties.

For example, if $U \subseteq X$ and $\sqcup U$ is defined, then $\sqcup f(U)$ is defined. Furthermore, $\sqcup f(U) = f(\sqcup U)$.

A.3. Strings.

If X is a set, then the set of strings (also known as sequences) over X, denoted by X^* is the set of all functions $f : n \to X$, $n \in \omega$, where we are regarding $n > 0$ as the set $\{0, \cdots, n-1\}$ and 0 as $\varnothing$. We define the *length* of a string f, denoted by $|f|$, to be its domain. For example, if $|f| = 0$, then f is the unique function $f : \varnothing \to X$, known as the empty string. We shall denote the empty string by Ω.

We define $X^+ = X^* - \{\Omega\}$.

$f \in X^+$ is represented on the page as a sequence $f(0) \cdots f(|f|-1)$.

We may turn X^* into a semigroup by defining for each $f, g \in X^*$ a function $f * g : |f| + |g| \to X$ by

$$(f * g)(n) = \begin{cases} f(n) \text{ if } n < |f| \\ g(n - |f|) \text{ otherwise} \end{cases} \tag{A.20}$$

and by (A.20), Ω is an identity of $(X^*, *)$, which is therefore a monoid.

From (A.20), we see that $f*g$ would be represented on the page as a sequence $f(0)\cdots f(|f|-1).g(0)\cdots g(|f|+|g|-1)$.

It may be shown that $(X^*,*)$ is right and left cancellable.

By definition, $|f*g|=|f|+|g|$ and so the function which maps $f\in X^+$ to $|f|$ is a monoid homomorphism from $(X^*,*)$ to $(\omega,+)$.

We may also impose a partial order on X^*, by defining

$$f\le g \Leftrightarrow \exists h\in X^*: f*h=g \tag{A.21}$$

Equivalently:

$$f\le g \Leftrightarrow |f|\le|g| \wedge f=g\|f| \tag{A.22}$$

If $f\le g$, then we shall say that f is a *prefix* of g.

Ω is the *bottom element* of $(X^*,\le)$ in the sense that

$$\forall f\in X^*: \Omega\le f \tag{A.23}$$

Since $(X^*,*)$ is left cancellable, if $f\le g$, then the string h of (A.21) is unique. We shall denote it by g/f.

If $f\in X^+$, then the set $\downarrow f=\{g\in X^*\ g\le f\}$ of prefixes of f may be shown to be totally ordered by $\le$. As a consequence,

$$\exists g\in X^*: f_1,f_2\le g \Rightarrow f_1\le f_2 \vee f_2\le f_1 \tag{A.24}$$

A.4. Structural Induction.

In chapters 27 and 28, we shall be making frequent use of a technique known as *structural induction,* which is used to prove properties of sets of expressions. It is assumed that we have

- A non-empty set **C** of *constants;*
- A non-empty set **F** of *function symbols;*
- A function $arity: \mathrm{F}\to\omega-\{0\}$

The sets **C** and **F** and the function *arity* constitute a *signature,* Σ.

Let $\hat{\Sigma}$ denote the set consisting of $\mathbf{C}\cup\mathbf{F}$ together with commas and parentheses. A subset C of $\hat{\Sigma}^*$ is *closed* if and only if

- $\mathbf{C}\subseteq C$

- If $f \in \mathbf{F}$ with $arity(f) = n$ and $e_1, \cdots, e_n \in C$, then $f(e_1, \cdots, e_n) \in C$

Since $\hat{\Sigma}^*$ is closed, the set of closed sets is non-empty and hence we may form its intersection, $expr(\Sigma)$, the set of all expressions generated by Σ. $expr(\Sigma)$ is non-empty since it contains $\mathbf{C}$.

Now suppose that $\mathcal{P}$ is a property of $\hat{\Sigma}^*$ such that

S1• Every element of $\mathbf{C}$ satisfies $\mathcal{P}$;
S2• If $e_1, \cdots, e_n \in \Sigma^*$ all satisfy $\mathcal{P}$ and $f \in \mathbf{F}$ with $arity(f) = n$, then $f(e_1, \cdots, e_n)$ satisfies $\mathcal{P}$.

Then the set of all elements of $\hat{\Sigma}^*$ which satisfy $\mathcal{P}$ is clearly closed and must therefore contain the intersection of all closed sets, namely $expr(\Sigma)$. Thus, we have the

PRINCIPLE OF STRUCTURAL INDUCTION. If $\mathcal{P}$ is a property of $\hat{\Sigma}^*$ satisfying S1 and S2, then every element of $expr(\Sigma)$ satisfies $\mathcal{P}$.

We illustrate the use of this principle with an example. If $E_1, E_2 \in expr(\Sigma)$, then we define $E_1 \preceq E_2$ if and only if one of the following holds:

1• $E_1 = E_2$;
2• $\exists f \in \mathbf{F}\, \exists E_1', \cdots, E_n' \in expr(\Sigma)\, \exists i\colon arity(f) = n \wedge E_2 = f(E_1', \cdots, E_n') \wedge E_1 \preceq E_i'$.

If $E_1 \preceq E_2$, then E_1 is a *subexpression* of E_2.

A.1. PROPOSITION. $\preceq$ is transitive.

PROOF. We shall say that E_3 has property $\mathcal{P}$ if and only if whenever $E_1 \preceq E_2$ and $E_2 \preceq E_3$, then $E_1 \preceq E_3$.

We observe that if $E_2 = E_3$, then $E_1 \preceq E_3$ because $E_1 \preceq E_2$, and in particular, if $E_3 \in \mathbf{C}$, then (2) cannot hold and so $E_1 \preceq E_2$. Hence very element of $\mathbf{C}$ satisfies $\mathcal{P}$.

Next, suppose that $E_3 = f(E_1', \cdots, E_n')$, where $E_1', \cdots, E_n' \in expr(\Sigma)$ and $f \in \mathbf{F}$ with $arity(f) = n$, and each E_i' satisfies $\mathcal{P}$. If $E_2 = E_3$, then $E_1 \preceq E_3$, so suppose that $E_2 \preceq E_i'$, some i, then $E_1 \preceq E_2$, $E_2 \preceq E'$ and E_i' satisfies $\mathcal{P}$, and so $E_1 \preceq E'$, and so by (2), $E_1 \preceq E_3$. Thus E_3 has property $\mathcal{P}$.

■

References

[1] Pnueli, A.: Applications of Temporal Logic to the Specification and Verification of Reactive Systems: a Survey of Current Trends, in Current Trends in Concurrency, ed. de Bakker, E J. W., de Roever, W. P. and Rozenberg, G., Lecture Notes in Computer Science, vol. 224. pp. 510-584, Springer Verlag, 1986.

[2] Hoare, C. A. R.: *Communicating Sequential Processes*, Prentice-Hall International Series in Computer Science,1985.

[3] Milner, A. R. J. G.: *Communication and Concurrency*, Prentice Hall, 1989

[4] Parnas, D. L.: Software Aspects of Strategic Defence Systems, Technical Report DCS-47-IR, Department of Computer Science, University of Victoria, July 1985.

[5] Dijkstra, E. W.: *Selected Writings on Computing: A Personal Perspective*, Springer Verlag, 1982.

[6] Turing, A. M.: Checking a Large Routine, Mathematical Laboratory, Cambridge University, 1950.

[7] Petri, C.A.: Kommunikation mit Automaten, Ph.D. Thesis, Rheinisch-Westfalisches Institut fur Instrumentelle Mathematik an der Universitat Bonn, 1962

[8] Petri, Fundamentals of a Theory of Asynchronous Information Flow, Proc. IFIP Conference, Popplewell, 1962.

[9] Karp, R. M., Miller, R.E.: Properties of a Model of Parallel Computations: Determinacy, Termination, Queuing, SIAM J. Appl. Math., vol. 14 (6), 1966.

[10] Holt, A. W. et al.: Information System Theory Project: Final Report, RADC-TR-68-305, NTIS AD 676972, Applied Data Research Inc., Princeton, NJ, 1968

[11] Milner, A. R. J. G.: *A Calculus of Communicating Systems,* Lecture Notes in Computer Science, vol. 92, Springer Verlag, 1980

[12] Zimmerman, H.: OSI Reference Model - The ISO Model of Archetecture for Open Systems Interconnection, IEEE Transactions on Communications, vol. COM-28(4), pp. 425-432, 1980

[13] ISO/TC 97/SC 21/WG-FDT/SG-C: LOTOS, a Temporal Ordering Specification, December 1986

[14] ISO/TC 97/SC 21/WG-FDT/SG-C: Estelle, a Formal Description Technique Based on an Extended State Transition Model, December 1986

[15] Lauer, P. E., Campbell, R. H.: Formal Semantics for a Class of High Level Primitives for Coordinating Concurrent Processes, Acta Informatica, vol. 5, pp. 247-322, 1975

[16] Mazurkiewicz, A.: Concurrent Program Schema and their Interpretations, Procedings Aarhus Workshop on Verification of Parallel Programs, 1977.

[17] Einstein, A. *The Meaning of Relativity,* Princeton University Press, 1921.

[18] Petri. C. A.: Non-Sequential Processes, Interner Bericht ISF-77-05, Gesellschaft fur Mathematik und Datenverarbeitung,1977

[19] G. Winskel: Event Structure Semantics for CCS and Related Languages, Computer Science Department, University of Aarhus, Tech. Rep. DAIMIPB-159, April, 1983

[20] Neilsen, M., Plotkin, G. and Winskel, G.: Petri Nets, Event Structures and Domains,Proceedings, Symposium on the Semantics of Concurrent Computation, Lecture Notes in Computer Science, vol. 70, pp. 266-283, Springer Verlag, 1979

[21] Keller, R. M.: Formal Verification of Parallel Programs, Journal of CACM, vol. 19(7), 1975

[22] Starke, P.: Processes in Petri Nets, Elektronische Informationsverarbeitung und Kybernetik, vol. 17(8-9), pp. 389-416, 1981

[23] Genrich, H. J., Lautenbach, K. and Thiagarajan, P. S.: Elements of General Net Theory, Proceedings, Advanced Course on General Net Theory of Processes and Systems,Lecture Notes in Computer Science, vol. 84, pp. 21-164, Springer Verlag, 1979

[24] Winkowski, J.: Algebras of Arrays - a Tool to Deal with Concurrency, Tech. Rep. 287, Institute of Computer Science, Polish Academy of Sciences, 1977

[25] Shields, M. W.: Adequate Path Expressions, Proceedings, Symposium on the Semantics of Concurrent Computation, Lecture Notes in Computer Science, vol. 70, pp. 249-265, Springer Verlag, 1979

[26] Arnold, A.: Synchronized Behaviours of Processes and Rational Relations, Acta Informatica, vol. 17, pp. 21-29, 1982

[27] Milner, A. R. J. G.: Calculi for Synchrony and Asynchrony,Tech. Rep. CSR-104-82, Computer Science Department, University of Edinburgh, 1982

[28] Genrich, H. J.: Sychronisationsgraphen, Acta Informatica, vol. 2, pp. 143-161, 1973

[29] Best, E.: A Theorem on the Characteristics of Non-Sequential Processes, Annales Societatis Mathematicae Polonae: Fundamenta Informaticae, vol. III.1, pp. 77-94, 1980

[30] Winskel. G.: Events in Computation, Ph.D. Thesis, Computer Science Department, University of Edinburgh, October 1980

[31] Davey, B. A. and Priestley, H. A.: *Introduction to Lattices and Order*, Cambridge University Press, 1990

[32] Janicki, R.: A Characterisation of Concurrency-like Relations, Proceedings, Symposium on the Semantics of Concurrent Computation, Lecture Notes in Computer Science, vol. 70, pp. 109-122, Springer Verlag

[33] Lauer, P. E., Shields, M.W., Torrigiani, P.: COSY: A Specification Language based on Path Expressions, Acta Informatica, vol. 12, pp. 109-158, 1979

[34] Janicki, R.: A Formal Semantics for Concurrent Systems with a Priority Relation. Tech. Rep. R 84-16,Institut for Elektroniske Systemer, University of Aalborg, 1984

[35] Janicki, R.: Transforming Sequential Systems into Concurrent Systems, Theoretical Computer Science,vol. 36, pp. 27-58, 1985

[36] Janicki, R., Lauer, P. E, Kouney, M. and Devillers, R.: Concurrency and Maximally Concurrent Evolution of Nonsequential Systems, Theoretical Computer Science, vol. 43, pp. 213-238, 1986

[37] Nivat, M.: Sur la synchronisation des processus, Revue Technique Thomson-CSF, vol. 11, pp. 889-919, 1979

[38] Shields, M. W.: Non-sequential Behaviours I, Technical Report CRS-119-82, Computer Science Department, University of Edinburgh, June, 1982

[39] Shields, M. W.: Deterministic Asynchronous Automata, Automata on Infinite Words, Lecture Notes in Computer Science, vol. 192, pp. 89-98, ed. M. Nivat and D. Perrin, Springer Verlag,1984

[40] Reid, F. X.: *Categories for the Unemployed Mathematician*, Utopia Press, 1984

[41] Reid, F. X.: Collected Laundry Lists, Utopia Press, 1991.

[42] Shields, M. W.: Concurrent Machines, Computer Journal, vol. 28, pp. 449-465, 1985

[43] Bednarcztk, M. A.: Categories of Asynchronous Systems, Ph.D. Thesis, University of Sussex, October 1987

[44] Aarlbersberg, I. J. and Rozenberg, G.: Theory of Traces, Technical Report 86-16, Department of Computer Science, University of Leiden, September 1986

[45] Aarlbersberg, I. J.: Studies in Trace Theory, Ph.D. Thesis, University of Leiden, November 1987

[46] Mazurkiewicz, A.: Multitraces: an Alternative Model of Non-Sequential Behaviour, 1990

[47] Shields, M. W.: Multitraces, Hypertraces and Partial Order Semantics, Formal Aspects of Computing, vol. 4, pp. 649-672, 1992.

[48] Lauer, P. E., Shields, M.W., Best, E.: Formal Theory of the Basic COSY Notation, Tech. Rep. 143, Computing Laboratory, University of Newcastle-upon-Tyne, November 1979.

[49] Kwong, Y. S.: On Reduction of Asynchronous Systems, Theoretical Computer Science, vol. 5, pp. 25-50, 1977

[50] Kwong, Y. S.: On the Absence of Livelocks in Parellel Programs, Proceedings, Symposium on the Semantics of Concurrent Computation, Lecture Notes in Computer Science, vol. 70, pp. 172-190, Springer Verlag

[51] Sifakis, J.: A Unified Approach for Studying the Properties of Transition Systems, Theoretical Computer Science, vol. 18, pp. 1-32, 1982

[52] Kwiatkowska, M. Z.: Fairness for Non-Interleaving Concurrency, Ph.D. Thesis, Computer Science Department, University of Leicester, 1989

[53] Park, D. R. M.: Concurrency and Automata on Infinite Sequences, Lecture Notes in Computer Science, vol. 104, Springer Verlag, 1980

[54] Shields, M. W.: Deterministic, Asynchronous Automata: Formal Models in Programming, ed. Neuhold, E. J. and Chroust, G., pp. 317-246, North Holland, 1985

[55] Handbook on Logic for Computer Science, Vol IV.

[56] Sassone, V.: Transition Systems with Independence vs. Asynchronous Transition Systems, Proceedings, POMIV'96 Workshop on Partial Order Methods in Verification, AMS, 1997

[57] Peterson, J. L.: *Petri Net Theory and the Modelling of Systems*, Prentice-Hall, 1981

[58] Bramms, J. W.: *Reseaux de Petri: Theorie et Practique*, Masson, 1983

[59] Genrich, H. J., Lautenbach, A. K.: The Analysis of Distributed Systems by Means of Predicate/Transition Nets, Proceedings, Symposium on the Semantics of Concurrent Computation, Lecture Notes in Computer Science, Springer Verlag, vol. 70, 1979

[60] Genrich, H. J.: Predicate/Transition Nets, Petri Nets: Central Models and their Properties, eds. Brauer, E. W., Reisig, W. and Rozenberg, G., Lecture Notes in Computer Science, vol. 254, pp. 207-247, Springer Verlag, 1986

[61] Jensen, K: Coloured Petri Nets, Petri Nets: Central Models and their Properties, eds. Brauer, E W., Reisig, W. and Rozenberg, G.: Lecture Notes in Computer Science, vol. 254, pp. 248-299, Springer Verlag, 1986

[62] Rozenberg, G.: The Behaviour of Elementary Net Systems, Petri Nets: Central Models and their Properties, eds. Brauer, E W., Reisig, W. and Rozenberg, G.: Lecture Notes in Computer Science, vol. 254, pp. 60-94, Springer Verlag, 1986

[63] Maggiolo-Schettini, A., Wedde, H., Winkowski, J.: Modelling a Solution for a Control Problem in Distributed Systems by Restriction, Proceedings, Symposium on the Semantics of Concurrent Computation, Lecture Notes in Computer Science, vol. 70, pp. 226-248, Springer Verlag, 1979

[64] Lautenbach, K., Wedde, H.: Generating Control Mechanisms by Restrictions, Proceedings, MFCS Symposium, Lecture Notes in Computer Science, vol. 45, 1976

[65] Hoare, C. A. R.: An Axiomatic Basis for Computer Programming, CACM, vol. 12, 1969

[66] de Roever, W. P.: The Quest for Compositionality, Formal Methods in Programming, ed. E. J. Neuhold and G. Chroust, North-Holland, pp. 181-205, 1985

[67] Mac Lane, S.: *Categories for the Working Mathematician*, Springer Verlag, 1971

[68] Shields, M. W.: Behavioural Presentations, Lecture Notes in Computer Science, vol. 354, Procedings REX Workshop on Linear Time, Branching Time and Partial Order in Logics and Models for Concurrency, Springer Verlag, 1989

[69] U. Goltz, U., Reisig W. : The Non-Sequential Behaviour of Petri Nets, Information and Control, vol. 57, 1983.

[70] Janicki, R., Lauer, P. E.: Specification and Analysis of Concurrent Systems: the COSY Approach, EATCS Monographs in Theoretical Computer Science, Springer Verlag, 1992.

[71] Campbell, R. H., Habermann, A. N.: The Specification of Process Synchronisation by Path Expressions, Lecture Notes in Computer Science, ed. G. Goos and J. Hartmanis, vol. 16, pp. 89-102. Springer Verlag, 1974.

[72] Hoare, C. A. R.: Monitors: An Operating System Structuring Concept, Comm. ACM, 549-557, vol. 17 (10), 1974

[73] Plotkin, G. D.: A Structural Approach to Operational Semantics, report DAIMI-FN-19, Department of Computer Science, University of Aarhus, 1961

[74] Hoare, C. A. R.: Communicating Sequential Processes, CACM, vol. 21, 1978.

[75] Brookes, S. D., Hoare, C. A. R., Roscoe, A. W.: A Theory of Communication Sequential Processes, JACM, vol. 31, pp. 560-599, 1984.

[76] Brookes, S. D., Roscoe, A. W.: An Improved Failures Model for Communication Sequential Processes, Proceedings NSF-SERC Seminar on Concurrency, Lecture Notes in Computer Science, Springer Verlag, 1985.

[77] Murphy, D. V. J.: Time, Causality and Concurrency, Ph.D. Thesis, University of Surrey, December 1989.

Index

acceptance 175, 185, 285
action sequence 8, 155
adjoint 238
admissible function 371
ambiguity 153, 156
Arnold, A. 25, 342
arrows 231
ascending chain 370
ascending chain condition 28
asynchrony 1, 9, 24, 66, 313
bags 283
Bednarcztk, M. 1120, 232
behavioural presentation 12, 17
    bagged 283
    bounded 385
    closed 14, 49
    discrete 12, 26, 67
    hybrid 24
    linguistic 141, 148
    plain 86
    replete 29
    sequential 24
    unlabelled 45
    with process structure 343
bisimulation 172
    function 399
    family 398
    transform 398
bottom element 290
Best, E. 37, 1133
Campbell, R. H. 7, 320
cases
    of a condition/event system 194
    of a loosely coupled system 226
category 231
CCS 5, 6, 320, 383
closure 50
codomain 231
coherence 51, 57
coincidence free 57
compatibility 51, 56
complete prime 40
compositionality 368
computation graph 5, 76
concession 194, 311
concurrency 7, 12, 23
concurrent 1, 8, 10, 12
condition/event systems 7, 193
configuration 57, 307
conflict 195
conflict relation 45
conflict-free 46
confusion 170, 190
consistent completeness 38, 105
contact-free 204
COSY 321
CSP 320, 415
    process 416
    traces 419
denotational semantics 368
derivation 30
derivation sequence 8, 155
descending chain condition 28
determinacy 13, 24, 68, 122, 149
Dijkstra, E. W. 415
discrete 12, 30
divergences 417
divergent state 432
dolomite 4
domain 231
embedding 267
    isomorphisms 267
enabled transition 194
environment 369
equivalence of categories 246
ESTELLE 5
event 16, 194
event structure 14, 44, 57
event vector 98
extensionality principle 273
failures 417
faintly guarded expression 393
$\approx$-finitary 32
firing sequences 190
fixed point 369
flow notation 3337
flow relation 191
formal methods 4

forward stability 69, 149
full event structure 57
functor 237
Goltz, I. 315
Habermann, N. 320
Hoare, C. A. R. 25, 320, 415
Holt, A. W. 4
hypertrace 121
identity
  functor 246
  morphism 232
independence relation 21, 101
indexed cover 97
inverse 232,247
ISO 5
isomorphism 14, 138, 146, 232
  of nets 196
Karp, R. M. 4, 7, 25, 75
K-density 37
Keller, R. M. 25, 172, 304
kernel 256
Kwiatkowska, M,. Z 303
Kwong, Y. S. 170
λ-reduced 138
label isomorphism 141
Lauer, P. E. 7, 321
left-closure 34, 67, 117
Levi's lemma 120
loosely coupled system 228
  with a transition relation 229
LOTOS 5
lozenge property 73, 184
marking 307
maximal configuration 225
Mazurkiewicz, A. 9, 121
Miller, R. E. 4, 7, 25, 75
Milner, R. 5, 14, 25, 161, 172, 383
monoid 457
monitor 320
morphism 231, 232, 234
multitrace 1121
Murphy, D. 454
mutual exclusion 13, 21, 225
natural equivalence 246
natural transformation 246
Neilsen, M. 25
net 5, 190
  acyclic 195
  labelled 323
  marked-graph 195
  marked, labelled 323
  occurrence 197
  place/transition193
  predicate/transition 193
  process 195
Net theory 7, 23, 190, 305
Nivat, M. 25, 342
non-determinacy 13, 21, 22
occurrence 17
Open Systems Interconnection 5
operational semantics 7
pairwise compatibility 51
parallel composite 119, 269
parallel computation 2
parallel decomposition 268, 272
  singular 269
parallel programs 306
parallelism 1
Park, D. 172, 387
Parnas, D. 3
path expression 5, 320, 322
Petri, C. A. 4, 23, 37, 190
place complementation 272
Plotkin, G. 25, 385
Pnueli, A. 2
points 12, 17
possible worlds 12, 17
prefix closure 119
pre-order 12, 17, 459
prime 11, 40
prime algebraicity 40, 113
process
  in CSP 416
  in an event structure 46
  of a net 195
  of the path/process notation 324
product 258
projection function 258, 388
Q-congruence 252
reactive 2
realisation by nets 268
recursion scheme 378
reduction morphism 251
Reid, F. X. 38, 56, 66, 470
Reisig, W. 319
relabelling function 384
safe firing rule 201
SDI 3
S-element 191
semantics theorems 14
semi-commutative monoid 9
semigroup 451
sequential interleaving 142
Sifakis, J. 170
simple net 194
simultaneity 1, 12, 23
stable state 432
Starke , P. 42
state machine net 270
states 154
step 30
strict equivalence 178, 187,290
string language
  sorted 118
  linear 123
strong bisimulation 164, 177
strongly determinate 161
strong equivalence 14, 165, 177
strongly forward stable 122, 150
structural induction 461
subexpresssion 391, 462
synchronisation vectors 344

synchronised process system 344
synchrony 14, 24
system 1
TCSP 6, 416
T-element 191
time ordering 21
token mathematics 190
trace 9
trace language 11, 38, 93
transformational 2
transition 191
transition system 7, 8, 154
    asynchronous 7, 11, 38, 184
    hybrid 7, 228
    initialised transition 154
    singly labelled 268
unambiguity 8, 161, 362
unseparated 139
vector language 93
    finite state 126
    left-closed 117
    simple 126
weakly guarded expression 392
Winkowski, J. 25, 42
Winskel, G. 25, 56
Zeno's Paradoxes 26

GPSR Compliance

The European Union's (EU) General Product Safety Regulation (GPSR) is a set of rules that requires consumer products to be safe and our obligations to ensure this.

If you have any concerns about our products, you can contact us on ProductSafety@springernature.com

In case Publisher is established outside the EU, the EU authorized representative is:

Springer Nature Customer Service Center GmbH
Europaplatz 3
69115 Heidelberg, Germany

Batch number: 10844669

Printed by Printforce, the Netherlands